Rigor and Reproducibility in Genetics and Genomics

Translational and Applied Genomics

Rigor and Reproducibility in Genetics and Genomics
Peer-reviewed, Published, Cited

Edited by

Series Editor

George P. Patrinos
Professor, Department of Pharmacy, University of Patras, School of Health Sciences, Patras, Greece; United Arab Emirates University, College of Medicine and Health Sciences, Department of Pathology, Al-Ain, UAE; United Arab Emirates University, Zayed Center of Health Sciences, Al-Ain, UAE; Erasmus University Medical Center, School of Medicine and Health Sciences, Department of Pathology – Bioinformatics Unit, Rotterdam, The Netherlands

Series Volume Editors

Douglas F. Dluzen
Visiting Professor of Biology, Morgan State University, Baltimore, MD, United States; Office of Graduate Biomedical Education, Johns Hopkins University School of Medicine, Baltimore, MD, United States

Monika H.M. Schmidt
Genetics and Genome Biology, SickKids Research Institute, Toronto, ON, Canada

Academic Press is an imprint of Elsevier
125 London Wall, London EC2Y 5AS, United Kingdom
525 B Street, Suite 1650, San Diego, CA 92101, United States
50 Hampshire Street, 5th Floor, Cambridge, MA 02139, United States
The Boulevard, Langford Lane, Kidlington, Oxford OX5 1GB, United Kingdom

Copyright © 2024 Elsevier Inc. All rights reserved.

No part of this publication may be reproduced or transmitted in any form or by any means, electronic or mechanical, including photocopying, recording, or any information storage and retrieval system, without permission in writing from the publisher. Details on how to seek permission, further information about the Publisher's permissions policies and our arrangements with organizations such as the Copyright Clearance Center and the Copyright Licensing Agency, can be found at our website: www.elsevier.com/permissions.

This book and the individual contributions contained in it are protected under copyright by the Publisher (other than as may be noted herein).

Notices

Knowledge and best practice in this field are constantly changing. As new research and experience broaden our understanding, changes in research methods, professional practices, or medical treatment may become necessary.

Practitioners and researchers must always rely on their own experience and knowledge in evaluating and using any information, methods, compounds, or experiments described herein. In using such information or methods they should be mindful of their own safety and the safety of others, including parties for whom they have a professional responsibility.

To the fullest extent of the law, neither the Publisher nor the authors, contributors, or editors, assume any liability for any injury and/or damage to persons or property as a matter of products liability, negligence or otherwise, or from any use or operation of any methods, products, instructions, or ideas contained in the material herein.

ISBN 978-0-12-817218-6

For information on all Academic Press publications
visit our website at https://www.elsevier.com/books-and-journals

Publisher: Stacy Masucci
Acquisitions Editor: Peter B. Linsley
Editorial Project Manager: Susan E. Ikeda
Production Project Manager: Jayadivya Saiprasad
Cover Designer: Mark Rogers

Typeset by STRAIVE, India

Contents

Contributors xv
Preface xix

SECTION 1 Introduction

CHAPTER 1 Rigor and reproducibility in genetic research and the effects on scientific reporting and public discourse 3
Monika H.M. Schmidt and Douglas F. Dluzen

Introduction 3
What is the rigor and reproducibility crisis? 5
    The issue of waning public trust in scientific research 7
What are the contributing factors to the reproducibility crisis? 9
The societal importance of open science 12
Conclusions 15
References 17

CHAPTER 2 Unveiling the hidden curriculum: Developing rigor and reproducibility values through teaching and mentorship 23
Marina E. Tourlakis

Introduction 23
In the classroom 24
    In practice 27
In advising 34
In the research lab 35
    In practice 37
Conclusion and future work 42
References 43

SECTION 2 Genotyping

CHAPTER 3 Genome-wide association studies (GWAS): What are they, when to use them? 51
Fan Wang

Introduction and history 51
Study design and statistical methods 52
    Frequentist vs Bayesian modeling 52
    GWAS for case and control association testing 54
    GWAS for quantitative traits 56

Quality control for GWAS 58
    Technical QC..59
    Genetic QC...59
Concluding note 60
Using R and PLINK for GWAS 61
    R primer 61
    Plink primer..61
    Plink data formats 64
    Genome-wide example...67
References..76

CHAPTER 4   GWAS in the learning environment 83
Amy L. Stark

Introduction...83
Theoretical background for designing and interpreting GWAS..............83
    What is GWAS?..83
    What is a GWAS stated simply? 84
    GWAS evaluation 85
Populations included in the GWAS...86
    Population diversity 86
Population size 87
Number of tests in a GWAS 87
The trait in question: The phenotype 89
Summary 90
Reference 90

CHAPTER 5   Polygenic risk scores and comparative genomics: Best practices and statistical considerations 91
Sally I-Chun Kuo and Fazil Aliev

Genome-wide association studies 91
The polygenic scoring approach 92
Preparing data for polygenic score calculation 93
Software programs for PRS calculation...97
Sample script for calculating PRS using PRS-CS 99
Challenges associated with creating predictive PRS and interpreting
    PRS results 99
Conclusion 101
Appendix A 101
    Example script for creating polygenic scores 101
    Using single R script to prepare and send all commands to the
        system to create PRS-CS score 104
References..108

CHAPTER 6   Sequence analysis and genomics in the classroom **115**
Rebecca C. Burgess, Rivka Glaser, and Kimberly Pause Tucker

Crowd-sourced undergraduate research 115
    The Genomics Education Partnership 116
    SEA-PHAGES 119
Standalone DNA sequencing activities 120
    16S rRNA bacterial genotyping/identification activity 120
    Building and interpreting molecular phylogenies 123
    Bitter tasting ability (PTC) genotyping activity 127
Other resources for DNA sequencing and analysis activities 132
    National Center for Case Study Teaching in Science 132
    Association for Biology Laboratory Education 132
    Genetics Society of America 132
    HHMI biointeractive 133
    QUBES 133
    GCAT-SEEK 133
    Bio-rad cloning and sequencing explorer series 133
References 134

CHAPTER 7   Classroom to career: Implementation considerations for engaging students with meaningful DNA sequencing learning opportunities **137**
Charles Wray

Introduction 137
DNA sequencing at the high school level 138
    Essential content and learning goals 138
Technology considerations for high schools 141
DNA sequencing at the undergraduate level 142
    Engaging and exciting students 144
    Promoting research careers 145
    Skills and career training 147
Future considerations 149
References 149

SECTION 3   **Next-generation sequencing & gene expression**

CHAPTER 8   Review of gene expression using microarray and RNA-seq **159**
Ana B. Villaseñor-Altamirano, Yalbi Itzel Balderas-Martínez, and Alejandra Medina-Rivera

Introduction 159
High-throughput techniques to assess gene expression 161

      Microarrays—What are they, and how are they conducted?............................161
      Sequencing..163
      How does one overcome read length challenges?..165
  Applications 166
      Gene expression profiling 166
      Splicing detection...168
      Expression quantitative trait loci assessment 170
      Single-cell RNA-seq 172
  Public databases 173
      Databases 173
      Reproducibility across studies 176
  Key point reproducibility, replicability, robustness, generalizability....................176
      Batch effect 177
      Metaanalysis..177
  Conclusion and remarks..178
  Acknowledgments...178
  References..179

CHAPTER 9   **Guidelines and important considerations for 'omics-level studies** **189**
Francesca Luca and Athma A. Pai

Overview of the RNA-sequencing experimental protocol 190
Study design considerations—Definitions..192
      RNA quality 192
      Confounders 193
      Replicates 193
      Sequencing depth 195
      Read length and type..196
Analysis of RNA-seq data..197
      Quality control 197
      Mapping of reads 197
      Quantifying gene and isoform expression..199
      Quantifying absolute mRNA abundance..200
      Comparing gene expression between groups...201
      Studying the genetic determinants of gene expression 202
      Analysis of alternative splicing events...205
Summary 206
References..207

CHAPTER 10 Rigor and reproducibility of RNA sequencing analyses 211
Dominik Buschmann, Tom Driedonks, Yiyao Huang, Juan Pablo Tosar, Andrey Turchinovich, and Kenneth W. Witwer

Rigor and reproducibility of RNA sequencing analyses 211
    Introduction to RNA sequencing 211
    Extracellular RNA: A case study for the challenges of low-input RNA-Seq 212
Experimental design and preanalytical variables 213
    The measurement process 213
    Sample preservation: What is the original state? 214
    Contamination introduced before or during library preparation 215
    Quality control 216
    Experimental design and validation 217
The impact of library preparation methods 218
    RNA and cDNA fragmentation 218
    Reverse transcription 220
    Adapter attachment 222
    Library amplification and unique molecular identifiers 223
    Impact of size selection on RNA fragment distribution and detected RNA biotypes 223
    Data analysis 225
    Preprocessing 225
    Read mapping 226
    Overlapping annotations and database quality 227
    Read quantification 228
    Data quality control 228
    Normalization and differential expression analysis 229
    Functional analysis of RNA-Seq data 230
Conclusions 230
References 231

CHAPTER 11 Validation of gene expression by quantitative PCR 247
Arundhati Das, Debojyoti Das, and Amaresh C. Panda

Introduction 247
Materials 248
    RNA extraction 248
    Assessment of RNA 248
    cDNA synthesis and qPCR 248
Method 249
    Total RNA isolation 249

Analysis of RNA quantity, purity, and integrity	250
Reverse transcription of RNA	250
Designing RNA-specific primers	252
Validation of expression by quantitative (q)PCR	254
Technical notes	255
Acknowledgments	256
Conflict of interest	256
References	256

SECTION 4 Epigenetic analyses

CHAPTER 12 Best practices for epigenome-wide DNA modification data collection and analysis 261

Joseph Kochmanski and Alison I. Bernstein

Introduction	261
DNA modifications	261
DNA modifications in health and disease	262
Methods for detection of DNA modifications	264
Bisulfite conversion methods	264
Alternatives to BS-based methods	265
Challenges to reproducibility in DNA modification EWAS research	265
Biology	265
Methodology	267
Statistics	268
Experimental planning & reporting: Methods, code, and data	272
Conclusion	274
References	274

CHAPTER 13 Best practices for the ATAC-seq assay and its data analysis 285

Haibo Liu, Rui Li, Kai Hu, Jianhong Ou, Magnolia Pak, Michael R. Green, and Lihua Julie Zhu

An overview of ATAC-seq	285
Generating high-quality ATAC-seq data	287
Experimental design	287
Nuclei preparation and quality control	288
Tagmentation, PCR amplification, and quality control	289
Sequencing	290
Analyzing ATAC-seq data: From stringent data quality control to comprehensive data mining	291
Raw read quality control and preprocessing	291
Alignment, postalignment processing, and quality control	291
Peak calling	293

        Differential peak analysis 296
        Annotation and functional analysis 297
        Visualization 297
        Nucleosome positioning 298
        TF occupancy inference 299
        Motif mapping 303
        Differential TF binding activity analysis 303
        Reconstruction of gene regulatory network 304
    Summary 305
    Acknowledgments 309
    References 309

CHAPTER 14   Best practices for ChIP-seq and its data analysis 319
Huayun Hou, Matthew Hudson, and Minggao Liang

    Introduction 319
    Crucial considerations for a rigorous ChIP-seq experiment 320
        To fix or not to fix? 320
        Nuclear isolation for ChIP-seq 321
        Fragmenting chromatin for ChIP-seq 321
        Single-cell and low-input ChIP-seq 322
        Appropriate controls in ChIP-seq 322
        Antibody incubation, chromatin washing, and elution 323
    Analysis of ChIP-eluted chromatin 323
        Library preparation 323
        Sequencing a ChIP-seq experiment 325
        Sequencing read preprocessing and alignment 326
        Peak calling 326
        Quality control (QC) 327
    Visualization of ChIP-seq data 328
    Peak annotation and functional enrichment analysis 329
    Differential binding analysis 331
    Motif analysis 332
    Key take-away 333
    Conclusion 334
    References 335

CHAPTER 15   A practical guide for essential analyses of Hi-C data 343
Yu Liu and Erica M. Hildebrand

    A brief summary of Hi-C and an example analysis pipeline 343
    Hi-C data processing and quality control 345
    Visualization of Hi-C data 346
    Scaling plots and chromosome folding 348

  Compartment analysis 349
  Insulation and TAD boundaries 351
  Dot calling and CTCF-CTCF loops 352
  Integration of Hi-C data with ChIP-seq and RNA-seq data 354
  Data use 357
  Computational resources 357
  Acknowledgments 357
  References 357

CHAPTER 16 Epigenetics in the classroom 363
Khadijah Makky

  Epigenetics in undergraduate biology curriculum: Why, when,
   where, and how 363
    How is the chapter organized? 364
    Where is it appropriate to introduce epigenetics in the biology curriculum? 364
    How to design your epigenetics unit 366
  Approach to teaching epigenetics using high-impact practices 368
  Learning outcome 1: Building the epigenetics foundation knowledge 368
    Regulation of gene expression and epigenetics 369
  Epigenetic marks and chromatin conformation 370
    How does the chromatin conformation change? 370
    Epigenetic marks 370
  Learning outcome 2: Application using basic knowledge to critically
   understand many epigenetic phenomena 373
    Genomic imprinting—Using compare/contrast 373
    X-inactivation—Using concept maps 373
    Summarizing information—Using case studies 375
  Learning outcome 3: Integration-connecting the knowledge from this
   unit to the realms of life 377
    Epigenetics and cancer 377
    Culminating case study 377
    Case presentation in the classroom 378
    Epigenetics and human behavior: Nature versus nurture 384
  Conclusion and future considerations 391
  References 391

SECTION 5 Gene editing technologies

CHAPTER 17 Genome editing technologies 397
Dana Vera Foss and Alexis Leigh Norris

  Introduction 397
  Zinc finger nucleases 399
    Background 399

Applications	400
Future directions	402
Transcriptional activator-like effector nucleases	402
What they are	402
Applications	404
Future directions	405
CRISPR-Cas systems	405
Background	405
Applications	406
Future directions	407
Delivery of genome editing systems	407
Background	407
Methods	408
Future directions	410
Gene drive systems	411
What they are	411
How they are used	412
Future directions	412
Unintended genomic alterations	413
Background	413
Methods	414
Future directions	416
Conclusion	416
References	417

CHAPTER 18 Genetic modification of mice using CRISPR-Cas9: Best practices and practical concepts explained 425

Vishnu Hosur, Benjamin E. Low, and Michael V. Wiles

Genetically engineered mouse models of human disease	425
International Knockout Mouse Consortium (IKMC)	426
Cancer	426
Alzheimer's disease (AD)	427
COVID-19	427
Generating mouse models using CRISPR-Cas9	427
Methodology	428
Genetic diversity in mice	430
Reagent delivery to the mouse zygote	431
Guide design	431
Key considerations for generating KO alleles	434
Verification of KO alleles	436
Verification of dropout alleles	436
Key considerations for generating small knock-in alleles	438
Verification of small knock-in alleles	438

Key considerations for generating large knock-in alleles 440
    Verification of large knock-in alleles 440
Practical methods 441
Mosaicism 442
Unintended consequences 443
    Off-targeting events can occur but are circumventable...................................443
    Unintended on-target effects are likely but preventable..................................443
Conclusions and future perspective 444
    Cas9 variants 444
    Delivery of CRISPR-Cas9 components 444
    Base editing..445
    Prime editing 445
Acknowledgments..446
References..446

CHAPTER 19 CRISPR classroom activities and case studies 453
TyAnna L. Lovato and Richard M. Cripps

Importance of course-based undergraduate research experiences 453
Importance of CRISPR as a teaching tool 453
Approaches to teaching CRISPR 454
    Setting the stage 454
    Strategies for short time frames in bacteria...455
    16week teaching strategies...456
    General considerations for developing CRISPR in the classroom...................456
Classroom activities at the University of New Mexico...457
    Designing and cloning CRISPR targets in Drosophila 457
    Recent innovations at the University of New Mexico......................................460
Considerations of rigor and reproducibility in CRISPR classes 462
Conclusions..464
Appendix: Detailed methods used to create guide RNA plasmids for use in
    Drosophila..464
    Exercise 1: Annealing oligonucleotides to generate short dsDNA inserts........464
    Exercise 2: Phosphorylating your dsDNA, and ligating into pBFv-U6.2.........465
    Exercise 3: Transformation of ligation products into *E. coli* 465
    Exercise 4: Picking colonies, minipreps, and sequencing................................466
    Exercise 5: Sequencing clean-up 467
    Exercise 6: Analysis of sequences and transformation of successful clones........468
Acknowledgments..470
References..470

Index 473

Contributors

Fazil Aliev
Department of Psychiatry, Robert Wood Johnson Medical School, Rutgers Behavioral and Health Sciences, Piscataway, NJ, United States

Yalbi Itzel Balderas-Martínez
Instituto Nacional de Enfermedades Respiratorias Ismael Cosio Villegas, Laboratorio de Biología Computacional, Mexico City, Mexico

Alison I. Bernstein
Department of Pharmacology and Toxicology; Environmental and Occupational Health Science Institute, Rutgers University, Piscataway, NJ, United States

Rebecca C. Burgess
Stevenson University, Owings Mills, MD, United States

Dominik Buschmann
Department of Molecular and Comparative Pathobiology, Johns Hopkins University School of Medicine, Baltimore, MD, United States

Richard M. Cripps
Department of Biology, San Diego State University, San Diego, CA, United States

Arundhati Das
Institute of Life Sciences, Bhubaneswar, Odisha, India

Debojyoti Das
Institute of Life Sciences, Bhubaneswar, Odisha, India

Douglas F. Dluzen
Office of Graduate Biomedical Education, Johns Hopkins University School of Medicine, Baltimore, MD, United States

Tom Driedonks
Department of Molecular and Comparative Pathobiology, Johns Hopkins University School of Medicine, Baltimore, MD, United States

Dana Vera Foss
Wilson Lab, University of California Berkeley, Berkeley, CA, United States

Rivka Glaser
Stevenson University, Owings Mills, MD, United States

Michael R. Green
Department of Molecular, Cell and Cancer Biology, University of Massachusetts Chan Medical School, Worcester, MA, United States

Erica M. Hildebrand
Department of Systems Biology, University of Massachusetts Chan Medical School, Worcester, MA, United States

Vishnu Hosur
The Jackson Laboratory, Bar Harbor, ME, United States

Huayun Hou
Genetics and Genome Biology, SickKids Research Institute, Toronto, ON, Canada

Kai Hu
Department of Molecular, Cell and Cancer Biology, University of Massachusetts Chan Medical School, Worcester, MA, United States

Yiyao Huang
Department of Molecular and Comparative Pathobiology, Johns Hopkins University School of Medicine, Baltimore, MD, United States; Department of Laboratory Medicine, Nanfang Hospital, Southern Medical University, Guangzhou, Guangdong, China

Matthew Hudson
Genetics and Genome Biology, SickKids Research Institute, Toronto, ON, Canada

Joseph Kochmanski
Rancho BioSciences, San Diego, CA; Department of Translational Neuroscience, Michigan State University, East Lansing, MI, United States

Sally I-Chun Kuo
Department of Psychiatry, Robert Wood Johnson Medical School, Rutgers Behavioral and Health Sciences, Piscataway, NJ, United States

Rui Li
Department of Molecular, Cell and Cancer Biology, University of Massachusetts Chan Medical School, Worcester, MA, United States

Minggao Liang
Genetics and Genome Biology, SickKids Research Institute, Toronto, ON, Canada

Haibo Liu
Department of Molecular, Cell and Cancer Biology, University of Massachusetts Chan Medical School, Worcester, MA, United States

Yu Liu
Department of Systems Biology, University of Massachusetts Chan Medical School, Worcester, MA, United States

TyAnna L. Lovato
Department of Biology, University of New Mexico, Albuquerque, NM, United States

Benjamin E. Low
The Jackson Laboratory, Bar Harbor, ME, United States

Francesca Luca
Center for Molecular Medicine and Genetics; Department of Obstetrics and Gynecology, Wayne State University, Detroit, MI, United States; Department of Biology, University of Rome "Tor Vergata", Rome, Italy

Khadijah Makky
Department of Biomedical Sciences, Marquette University, Milwaukee, WI, United States

Alejandra Medina-Rivera
Laboratorio Internacional de Investigación sobre el Genoma Humano, UNAM, Querétaro, Mexico

Alexis Leigh Norris
Food and Drug Administration, Bioinformatician, Center for Veterinary Medicine, Rockville, MD, United States

Jianhong Ou
Department of Cell Biology, Duke University School of Medicine, Duke Regeneration Center, Duke University, Durham, NC, United States

Athma A. Pai
RNA Therapeutics Institute, University of Massachusetts Chan Medical School, Worcester, MA, United States

Magnolia Pak
Department of Molecular, Cell and Cancer Biology, University of Massachusetts Chan Medical School, Worcester, MA, United States

Amaresh C. Panda
Institute of Life Sciences, Bhubaneswar, Odisha, India

Monika H.M. Schmidt
Genetics and Genome Biology, SickKids Research Institute, Toronto, ON, Canada

Amy L. Stark
University of Notre Dame, Notre Dame, IN, United States

Juan Pablo Tosar
Nuclear Research Center, School of Science, Universidad de la República; Functional Genomics Laboratory, Institut Pasteur de Montevideo, Montevideo, Uruguay

Marina E. Tourlakis
Biology Department, University of the Fraser Valley, Abbotsford, BC, Canada

Kimberly Pause Tucker
Stevenson University, Owings Mills, MD, United States

Andrey Turchinovich
Division of Cancer Genome Research, German Cancer Research Center (DKFZ) and German Cancer Consortium (DKTK); Heidelberg Biolabs GmbH, Heidelberg, Germany

Ana B. Villaseñor-Altamirano
Laboratorio Internacional de Investigación sobre el Genoma Humano, UNAM, Querétaro, Mexico

Fan Wang
University of Chicago, Center for Translational Data Science, Chicago, IL, United States

Michael V. Wiles
The Jackson Laboratory, Bar Harbor, ME, United States

Kenneth W. Witwer
Department of Molecular and Comparative Pathobiology, Johns Hopkins University School of Medicine, Baltimore, MD, United States

Charles Wray
The Jackson Laboratory, Genomic Education, Bar Harbor, ME, United States

Lihua Julie Zhu
Department of Molecular, Cell and Cancer Biology; Department of Molecular Medicine, Program in Bioinformatics and Integrative Biology, University of Massachusetts Chan Medical School, Worcester, MA, United States

Preface

The growth of scientific knowledge is rarely linear. Historically, the pace of discoveries was sufficiently gradual to permit revision of proposed theories in a timely manner or, at least, without massive investment of resources. In recent years, the landscape of how genetic and genomic research is conducted has rapidly changed with the advent of the age of computing. In silico research and computational experiments complementing traditional at-the-bench research now represent a significant portion of newly published research, and it is published at an astonishing pace. Moreover, new computational methods and their related bench techniques are continuously under development, discussed at conferences, and, increasingly, promoted on preprint servers for public consumption.

Preprint servers in and of themselves present a new challenge to the field of biomedical science: although these servers increase accessibility of scientific research, particularly for the public, they also provide opportunity for nonpeer-reviewed content to be widely disseminated, irrespective of the quality or reproducibility of data or methods presented. Issues relating to incomplete or incorrect reporting of such findings by news outlets and other nonexpert media personalities are merely one consideration of the importance of rigorous, reproducible methods and reporting standards. For genomics researchers, rigorous methods and detailed documentation pertaining to computational tools are absolutely crucial at all times: during critical evaluation of preprint publications by fellow scientists, during peer review, and long into the future, should another researcher choose to adopt the same computational method or tool in their work.

The rapid pace of new developments in genetics and genomics comes with an additional caveat: It makes educational textbooks, like this one, seemingly out-of-date by publication. Yet, providing cutting-edge methods is not the goal of this book; this book is concerned with providing guidelines and principles for conducting reproducible, high-quality genomic research. It is neither a reference manual nor an encyclopedia of methods, as the staggering number of computational tools and in silico techniques querying ever more complex ideas cannot be captured within the physical constraints of a book, or even an anthology of books!

This (e-)book seeks to provide one of the first compilations of genomic techniques with a focus on addressing the reproducibility crisis currently faced by biomedical research. Admittedly, the mountain to climb in this regard is enormous and will require coordinated efforts from granting bodies, publishers, and researchers themselves. Nonetheless, it begins—as with all systemic changes in a society—with educating the newest members of the genetics and genomics research community: trainees, early career investigators, and lecturers teaching this material. This is our intended audience, and the contents of each chapter will reflect this angle.

Rigor and Reproducibility in Genetics and Genomics is chiefly concerned with laying a foundation of basic "dry lab" methodologies and providing thoughtful examples of how to pivot to new approaches while still upholding rigorous scientific practice to produce reproducible outcomes. This book originated as an Invited Session at the 2017 American Society of Human Genetics Annual Meeting in Orlando, Florida. We attempted to include as many topics as we felt this book could reasonably discuss, and selected methods and computational research areas that are rapidly growing or already widely adopted. Our authorship is reflective of the diversity and global nature of genetic and genomic researchers, a key principle we kept in mind during the recruitment phase for this book.

We assume that most readers have a basic understanding of genetics and genomics, but have nonetheless attempted to include one or more review chapters in each section (see Chapters 3, 8, 12, and 17) providing a brief overview of the techniques to be discussed in subsequent chapters. Where possible, we have included teaching resource chapters written by expert undergraduate educators (Chapters 2, 4, 6, 7, 16, and 19). The intervening chapters provide relevant examples and protocols for some of the most au courant approaches in genetic and genomic research. These chapters also highlight the merits and drawbacks to any particular methodology or computational tool, as well as key considerations when developing a research pipeline using the technique under examination. This book will put readers on solid footing when looking to apply the discussed genomic techniques to their work.

The greatest thanks and acknowledgments are owed to each of the chapter authors: for their time, patience, and expert contributions. The COVID-19 pandemic extended the project timeline on the development of this book in unimaginable ways. The first year (or more) of the pandemic paused facets of research and complicated everyone's personal lives, yet our authors pushed through—this speaks volumes about the importance they placed on the written contents between these covers. Many of these chapters were coauthored by doctoral trainees or postdoctoral researchers, who are often at the leading edge of research and developing improved research methods. This book was written by them with you, the reader, at the forefront.

We would also like to thank the editing team at Elsevier, in particular Peter Linsley, who recognized the importance of this topic and approached us with this opportunity to educate. As well, our senior editorial project managers, Susan Ikeda and Kristi Anderson, who worked tirelessly to keep this project moving toward completion. In particular, a special thank you to Susan for her patient understanding and warm encouragement as we faced various editing hurdles.

Finally, a huge thanks to our families, who were considerate in their time and patience as we worked on this book at all hours of the day (and night). We have each navigated the wonderful arrival of two children apiece, further motivating our desire to set up young trainees with a new resource that can serve as a guide during their research careers, establishing a brighter future for biomedical research.

We hope you find this book knowledge-dense and resource-intensive in a directly applicable sense, and wish you the best in your genetic and genomic research journey!

Douglas F. Dluzen
Monika H.M. Schmidt

SECTION 1

Introduction

CHAPTER 1

Rigor and reproducibility in genetic research and the effects on scientific reporting and public discourse

Monika H.M. Schmidt[a] and Douglas F. Dluzen[b]

[a]*Genetics and Genome Biology, SickKids Research Institute, Toronto, ON, Canada,* [b]*Office of Graduate Biomedical Education, Johns Hopkins University School of Medicine, Baltimore, MD, United States*

Introduction

The scientific method has been practiced by humankind throughout our evolution, as we engaged in *trial and error* and worked toward finding better ways to survive and thrive. At its most basic, the scientific method requires the observer to integrate known information about a situation and process through influencing factors as the observer puts into motion a plan to obtain a desired outcome. Whether or not one is scientifically trained, everyone has practiced this form of logical thinking at some point in their lives. For example, imagine coming home after a long day at work, sitting down in a favorite chair or couch, and turning on the television, but the television does not turn on. You hit the power button on the remote again—nothing. Disbelief and frustration might begin. You must now work through the different factors inhibiting your relaxation and enjoyment.

> Anything blocking the signal path between remote and television? No? Check.
> Power to the television? Yes. Check.
> Power to the living space? Yes. Check. (And likely integrated into consideration already).
> Batteries in the remote dead? Swap and replace—then retest. Bingo!

The scientific method is a problem-solving tool designed to give us a certain degree of confidence when we finally obtain a result, whether it was predicted or not. The conclusions drawn from even the simplest of experiments are only as strong as the weakest point in the underlying approach to generating, collecting, and analyzing the data from that approach or experimental design. The same is true in genetic and genomic research.

Strong experimental designs accounting for confounding variables are needed to untangle the complex factors that may influence the outcomes of any given genetic or genomic study. This is especially true for analyses that incorporate information from large population data sets. This chapter and those that follow in this textbook resource examine some of the ways in which we can structure the most widely used experimental approaches in genetic and genomic research to increase the confidence, replicability, and applicability of the results.

Historically speaking, scientific "proof" used to require that one could demonstrate a scientific phenomenon in front of other scientists. There would be documentation of experiments with written word, and illustrations came later to allow readers to imagine being in the room, observing the experiment, and thereby accelerating the pace of dissemination of scientific research and its outcomes. While the general public may have often been invited to these discussions, debates, and lectures, they were not usually involved in the interpretation and advancement of the work. That has changed in the last few decades as news media, patient advocates, and those interested in the societal impact of publicly or privately funded science have become a necessary and essential component of the discussion of scientific advancement. This is especially true when we consider how the knowledge generated in the laboratory or clinic is *applied* in daily life.

The scientific discourse and review that validate new research results are tiered:

1. The first tier—the choice of methodology and the approaches taken by the authors of a given work and their collaborators.
2. The second tier—the review by the research community (grant review panels, conferences, and journal manuscript peer reviewers).
3. The third tier—feedback from the wider research community once a manuscript is submitted to a preprint service and/or formally accepted for publication in a peer-reviewed journal.
4. Fourth tier—delivery of research findings to the broader public where they may interact with the data, interpreting the applicability of the results to public policy or healthcare practices, or even providing the foundation to answer subsequent questions unearthed in the original study.

Breakdowns anywhere within or between these tiers have historically contributed to the publication of results that may have been misinterpreted, overly conflated, falsified, or fabricated, and have allowed methodologies inappropriately chosen to give a false sense of confidence with a study's results. *Research in many areas has gently shifted from a culture of "show me" to "trust me"—a defining reason for the need to ensure reproducibility of scientific works.*

In the field of genetics and genomics, advancing technology and statistical methods can be so diverse and complex that it is difficult to describe them even to a technical audience. Peer reviewers and journal editors are required to review enormous volumes of submissions and to have a wide breadth of expertise, without having sufficient information (or time) to do their jobs thoroughly, thereby inadvertently permitting problematic research to slip through the peer review processes. The myriad reasons underlying this problem relate to funding challenges and a *publish or perish* attitude that underlies much of biomedical research—but some of these systemic issues are beyond the scope of this book.

There are numerous other concerns in the scientific community that can contribute to published research that is not methodologically sound or able to be reproduced by other laboratories. In the past two decades, the subfield of meta-research has emerged, in which statisticians, researchers, and clinicians have examined the nature of the scientific method itself within biomedical and genetic research in order to identify key factors that influence the reliability and replicability of peer-reviewed science [1,2]. Meta-research has identified a possible rigor and reproducibility crisis in peer review and publishing processes as more and more manuscripts are published containing science that cannot be replicated and/or using inappropriate approaches for the given context. Further, due to the aforementioned *publish or perish* culture that is particularly prevalent in competitive research environments, combined with digitally rendered data figures, the publication of difficult-to-detect but completely falsified data has had a marked uptick. A collaborative effort by researchers to identify and report such falsifications is necessary—an excellent example of image forensics is the work of Dr. Elisabeth Bik (Twitter: @ MicrobiomDigest) [3], discussed in further detail here.

This chapter is dedicated to introducing the historical context of this potential crisis (which some argue is also an *opportunity* for change), identifying systemic factors that may have contributed to the lack of replication within scientific studies and reproducibility by other groups, and suggestions for geneticists on key steps to improve upon communicating with the public on these issues.

> **Key point: reproducibility versus replicability**
>
> The terms reproducibility and replicability are used in this chapter and throughout this book. The difference between these terms is subtle, so much so that these terms are often used interchangeably—albeit incorrectly. Toward fostering rigorous attention to all details in scientific research, including language, we suggest that the definitions as outlined by the National Academy of Sciences in their 2019 book Reproducibility and Replicability in Science [4] be adopted across scientific communities. Thus:
>
> **Reproducibility** is the ability to consistently obtain the <u>same</u> results using identical input data or reagents examined via the same experimental conditions and analyses.
>
> **Replicability** is the ability to obtain consistent (but not necessarily identical) results when using different input data or reagents with the goal of answering the same scientific question.
>
> If this seems confusing, consider an analogy involving baking a chocolate chip cookie: A reproducible batch of cookies will use identical ingredients (the same flour, same butter, same chocolate chips, same sugar, same water) and identical apparati (the same oven with the same cookie sheet) and identical baking conditions (same bake time and temperature). Assuming the recipe instructions are clear and detailed (no "add a thimbleful of baking powder") and that the ingredients are pure (the flour should not have any contaminants in it), the baker will likely be able to consistently produce the same delicious batch of chocolate chip cookies. A replicable batch of cookies will strive to consistently achieve delicious golden-on-the-outside and gooey-in-the-centre chocolate chip cookies, but may use ingredients produced by different companies, apparati with slight differences (air bake sheets versus plain aluminum bake sheets, for example) and may even follow slightly different instructions. Presumably though, with the same question of achieving the aforementioned cookie, a replicable chocolate chip cookie (not an oatmeal cookie) will be achieved.

What is the rigor and reproducibility crisis?

Rigorous and reproducible research practices are the bedrock of scientific advancement. One of the more thorough and recent reexaminations of the scientific method began in 2005 with an essay written by Dr. John Ioannidis. Dr. Ioannidis made a claim with far-reaching implications: that much of the published research findings were false [5]. He discussed that most studies were too small, underpowered, and/or included biases in study design, implementation, data collection and/or analysis, interpretation, and reporting. Ioannidis argued, "most research questions are addressed by many teams, and it is misleading to emphasize the statistically significant findings of any single team. What matters is the totality of the evidence. Diminishing bias through enhanced research standards and curtailing of prejudices may also help."

Ioannidis' work, and that of others, initiated a much-needed conversation identifying the qualities of a successful research study. Most scientific disciplines have now re-examined standard research protocols and practices, and found varying degrees of replication of prior studies. For example, the <u>Reproducibility Project: Cancer Biology</u> replicated 50 experiments from 23 high-impact cancer-related research papers [6]. The study investigators replicated less than half of the experiments that provided positive results, but nearly 80% of the experiments that exhibited null results. As well, for those studies replicated, the effect sizes were smaller than initially reported.

In 2015, *Nature* conducted a survey of over 1500 researchers on issues related to reproducibility. In the fields of biology and medicine, over 50% of researchers failed to replicate their own experiments, and at least 60% reported failing to reproduce the work of someone else. Two-thirds of those surveyed also reported establishing procedures in the laboratory to support reproducible work [7]. While some of these numbers seem quite high, this report may also highlight an aspect of the very nature of the scientific method, in which correction within research subfields is a necessary component of validating essential results.

Scientific discourse concerning research results is a natural component of the scientific method. A recent analysis of "disagreement" within four million scientific research articles found that 0.41% of papers published in the broad category of "biomedical and health sciences" references disagreements with prior published work [8]. This disagreement with prior literature was categorized as either "paper-level disagreement" or "community-level disagreement" and included a definition of disagreement that encompassed discussion of controversy, dissonance, explicit disagreement with prior work, or lack of consensus with prior work or works [8]. These and other data naturally lead to a discussion of whether this is acceptable "noise" within the scientific community or not. Hypotheses and theorems that may be supported by evidence can always be toppled by new, stronger data or ideas. Providing new evidence that questions prior ideas is an imperative role the research community plays in monitoring its own advancement.

Alternative approaches have been taken to address the reported reproducibility crisis. Retraction Watch began as a citizen science website in 2010 to document and track retractions of research papers or other scholarly work in research. Between the beginning of 2012 and the end of September 2022, over 1200 research articles related to the keyword "genetics" had been retracted due to concerns or errors with the data. Similar results occur when searching the same time period for papers related to "cancer" or "oncology". Dr. Elisabeth Bik has made a second career out of identifying fraudulent research via her Science Integrity Digest, highlighting manipulated figure images on her social media accounts [9]. In 2019, she led a study examining 960 research papers published in *Molecular and Cellular Biology* between 2009 and 2016 and found that 6% had inappropriately duplicated figure images [10,p. 20]. This was a follow-up to an earlier study of over 20,000 papers published within 40 journals between 1995 and 2014. She and her colleagues found that almost 4% of these papers had problems with one or more figures and that at least half of these, 2% of all the papers, had evidence of visual manipulation [11].

In the field of genetics and genomics, structural problems contribute to a lack of rigorous research practice. Historically, nearly 96% of all participants in all genome-wide association studies (GWAS) are of European ancestry, with a paltry 3% of Asian ancestry being the next most represented ancestral population [12]. Lack of ancestral representation in GWAS and related genomic analyses limits the ability to identify physiologically- or disease-relevant variation in the human genome—the true variation the human genome is not being accurately captured. How can geneticists infer the genetic contributors to disease processes if the complexity of variation that contributes to the said diseases is largely ignored? Presently, the shocking lack of representation in data sets limits the ability to extrapolate our understanding of genetic contributors to disease to populations outside of Western European ancestries.

Initiatives such as the National Institutes of Health's (NIH) *All of Us* research program has been developed to increase the diversity of biomedical research studies [13] and promote new opportunities to expand our knowledge about genomic diversity. The H3Africa (Human Hereditary and Health in Africa) Initiative is a leading consortium of researchers and laboratories in Africa to further address the disparity in our knowledge about variation in the human genome [14]. While these essential databases and others like it catch up on the collection of diverse biospecimens, detailed health history, and necessary representative sample sizes, geneticists have based most of the field's knowledge of fundamental diseases processes on the Western European genome.

Numerous statistical approaches exist for inferring associative and causal DNA variants related to disease development, environmental response, and other physiological pathways. These approaches include polygenic risk scoring (PRS), Mendelian randomization, estimates of heritability, genome-wide copy number variant (CNV) analysis, identifying variation in allele variation to estimate human migration, and others [15,16]; however, the past decade has seen GWAS dominate this realm of "big data" statistical genomic research. Most of these analyses are built on the foundation of databases such as the UK Biobank, which have >90% European ancestry in their sampled populations [17]. The 1000 Genomes Project Consortium is more diverse, with samples from 26 different ethnic populations; however, there are on average only ~100 samples per population in the database [18–20]. The small sample size per ethnic population means that most studies will be severely underpowered, limiting the ability to detect novel variants and smaller, but still physiologically relevant, effect sizes.

There are thus a number of additional factors contributing to the rigor and reproducibility crisis in biomedical research, with specific concerns for genetic and genomic researchers. These factors include funding challenges and an unhealthy culture around publishing results, structural challenges in genetic research and diverse sample collection/patient recruitment (and ethical compensation), and a lack of rigorous reporting and data sharing standards. These factors and more are detailed in "What are the contributing factors to the reproducibility crisis?" section and discussed at length.

This textbook endeavors to identify and address technical and methodological issues in genomic research that negatively impact reproducibility of data, and rigorous research practices. Additionally, corollary factors that impact rigor and reproducibility in research are discussed, including: improving genetic education at the secondary and post-secondary levels as well as in graduate training; communication in collaboration and study design; methodology and data sharing; and general transparency and open science practices. These considerations together strengthen the methodology of a research study and increase the confidence and replicability of results [2].

The issue of waning public trust in scientific research

Unfortunately, the era of social media and sensationalized headlines, combined with financial interests by competing groups, including "Big Natural" (a term coined by Dr. Jen Gunter, a self-proclaimed fighter for evidence-based women's health), leads to disagreement within and beyond the scientific community. The scientific process is naturally self-correcting. As evidence accumulates and results are replicated (or not), every bit of incorrect, non-rigorously conducted or reported research that makes its way to the public prior to being identified as such contributes to the confusion and misinformation campaigns that fuel the media's economic engine (including social media influencers), sowing distrust among the public. The time and space to conduct science and verify results has thus shrunk considerably and demands that researchers adhere to the highest standards of rigorous research and reporting (see case studies in Box 1.1 and Box 1.2 for more).

Box 1.1 The SARS-CoV-2 Pandemic

The Coronavirus Disease 2019 (COVID-19) pandemic put the scientific method under immense public scrutiny, changing perceptions globally of what can be accomplished when researchers are provided adequate funding resources, minimal bureaucratic hurdles, and practice Open Science. Unfortunately, the push to publish COVID-19 related information also meant that a small percentage of these published papers (72 papers, or 0.03%, at the time of this writing) were later found to be inaccurate [21]; two of these retractions came from high-profile peer-reviewed journals (*The Lancet* and *New England Journal of Medicine*). For members of the public who understand this to be part of the self-monitoring and self-correcting aspect of the scientific method, changing information based on new data strengthens their belief in the biomedical research machine. In contrast, for those who already feel alienated or lack familiarity with the scientific method or the wider biomedical establishment, changing discourse can breed discomfort and fear. The ongoing societal discourse between researchers promoting their work, the non-scientific public, and advancement of misinformation campaigns has both helped and hindered the global understanding of the scientific method at large, and severe acute respiratory syndrome coronavirus 2 (SARS-CoV-2).

The genomic sequence of SARS-CoV-2 had been identified and published by the end of January 2020, just months into the early stages of the pandemic [22,23], paving the way for a deeper understanding of the nature of the virus and the development and testing of multiple COVID-19 vaccines a few months later. Within 10 months of the first publicly-confirmed case of COVID-19, there were over 125,000 scientific articles published in the scientific literature, of which 30,000 were on preprint services such as the bioRxiv and medRxiv [24]. It was an incredible burst of scientific focus, discovery, and examination.

The public understanding of COVID-19 early in the pandemic was shaped by these preprint servers. Social media and news reporting of preprint COVID-19 findings escalated quickly during the spring of 2020 [25,26] and public understanding and misinformation was influenced by where the public accessed COVID-19-related information [27]. Additionally, journalistic reporting and public misunderstanding about the differences between "preprint" manuscripts and "peer-reviewed" articles fueled misinformation about both COVID-19 itself, and the scientific need to use preprints for rapid sharing of new results, while still waiting for the formal peer review process to be conducted [28,29].

For example, early preprint manuscripts in bioRxiv suggested that the COVID-19 spike protein had genetic sequence similarities with several human immunodeficiency virus (HIV) proteins, which were unlikely to have evolved naturally, suggesting that SARS-CoV-2 might have been engineered [30]. The paper was quickly retracted given the numerous issues with the sequencing approach, the data produced, and its analysis. Nonetheless, conspiracy theorists, and individuals who stood to gain financially from dissemination of misinformation/conspiracies, continued to use preprint articles like this one to promote COVID-19 misinformation and generate public distrust around COVID-19 research, and the medical establishment at large.

This highlights a delicate balance between public engagement with open-source, preprint scientific research and the time it takes for researchers to validate, correct, and review new scientific literature. Further discussion has been called for regarding use of the term "preprint" in news reports on PDFs uploaded to preprint servers so that it is clearer to the non-scientific community that peer review and validation of the results are still required [25]. Mainstream news media seems to be generally cognizant of this important difference and journalists are improving with their adherence to highlight that a preprint article is a non-peer-reviewed PDF published online. Given the accessibility to and rapid promotion of preprint manuscripts, peer-reviewed validation of research within the genetics and genomics community will ultimately have to catch up to insulate against misinformation.

Within the genetics research community, safeguards have been used to validate sequences from SARS-CoV-2 samples and must continue to be used efficiently. The National Center for Biotechnology Information (NCBI) began using the Viral Annotation DefineR (VADR) system to analyze SARS-CoV-2 systems to ensure sequence quality [31]. As well, the NIH hosts an open-access data dashboard to support COVID-19 researchers, including access to the COVID-19 Genome Sequence Dataset to submit sequencing information to the Short Read Archive hosted by NCBI, or the GISAID database supported by Freunde von GISAID e.V. and other partners. These repositories are instrumental in helping the scientific community validate sequencing findings and results, identify novel SARS-CoV-2 variants, as well outline what must be identified in related preprint manuscripts so as to inform journalists and others reporting the results of a particular study.

> **Box 1.2 The Advancement of CRISPR**
>
> Aside from polymerase chain reaction (PCR), nothing has ushered in a tsunami of new genomic and molecular biology research more than the development of Clustered Regularly Interspaced Short Palindromic Repeats (CRISPR) gene editing [32]. CRISPR has already revolutionized approaches to therapy in the clinic to treat sickle cell anemia and β-Thalasemmia [33] and cancer [34], establish new crops [35], and pave our way of understanding our own development [36,37]. There have also been significant advances in the methodologies of using CRISPR in the laboratory and clinic, including the expansion into several different types of CRISPR-associated (Cas) proteins, the ability to make precise single-base edits, and editing of RNA transcripts [38].
>
> Given the fundamental nature and power of CRISPR approaches, there has been considerable debate within the scientific and public communities on how best to use these potentially generation-altering genomic tools. There are no definitive answers on how best to juggle the moral and ethical implications of CRISPR alterations with the goal of improving human and agricultural health and well-being. This is further complicated by using CRISPR to study embryonic development [39] or editing the human germline.
>
> In 2018, researcher He Jiankui announced the birth of the first human babies born with germline genetic modifications using CRISPR [40]. This news sent the world into shock given that the procedures for the use of CRISPR for heritable transmission in humans had hardly been formalized, or even agreed upon in the international community. The three babies born in China exemplifies one of the major ethical debates in the public related to genomic research. What began just over a decade ago in bacterium has now influenced the lives of children born without a say in the procedure performed on them. Regardless of where geneticists fall on the spectrum of the acceptable use of CRISPR gene editing, answers must be found on a number of issues, including:
> - Who should govern the use of CRISPR in the research environment?
> - Who has a say in what types of cells are used and what types of experiments are performed using CRISPR?
> - How do we navigate off-target DNA edits and management of resources to validate new approaches? [41,42]
> - What roles do researchers and the media play in communicating novel findings?
> - What role should CRISPR-regulated gene drives play in shaping or modifying the environment? [43]
>
> As more and more research becomes accessible, more and more the non-scientific public will need to be educated on this issue to ensure productive public discourse when answering these and other questions.

What are the contributing factors to the reproducibility crisis?

The scientific method naturally leads scientists to engage in criticism of one another's work—ideally constructively, although this is not always the case. This self-monitoring dynamic is intended to strengthen our foundational knowledge and percolate interest in new research avenues. Despite informal feedback from colleagues and the formalized peer review process, hundreds of peer-reviewed publications, many on topics related to genetics and genomics, are retracted each year. How is it that so many studies "miss the mark," whether intentionally or unintentionally? Why are inaccuracies or flat-out falsehoods missed? What are the intrinsic factors that influence replication and reproduction of research data—particularly positive experimental results?

From the inception of a research question through to publication and subsequent critiquing by members of the field, each tier along the way contributes to whether a study produces the highest quality research in the most reproducible manner or not. The first of these tiers is to ask the correct question—leading or biased questions will inherently give rise to biased conclusions. Next, it is necessary to conduct a thorough review of the literature, to know what has been done and found previously, where the gaps in our knowledge may exist, and whether any of these previous studies have drawn conclusions that incongruous with their results, or in the context of the field. Methodology and study design are critical—selecting the appropriate samples, controls, techniques, and tests will strengthen the quality of the data produced and the conclusions that can be drawn.

Data analysis is the next significant point where many research studies stumble, establishing a significant finding where one might not exist due to the use of inappropriate statistical tests. Interpretations of these analyses can be challenging, and at times over-interpretation despite weak evidence leads authors to propose causality where it does not clearly exist. Accurate and transparent reporting of methods and all results (not just the positive results), free sharing of code and data sets used in computational work, and publication of raw (unprocessed) research data (whether through a publisher's data repository, supplementary results, GitHub, or via a privately hosted website) is a basic tenet of rigorous, open science. Finally, we come to peer review and publication—where, in theory, oversights or flat-out mistakes in the aforementioned stages should be caught, revised, and re-submitted for review. Unfortunately, given the complexity of much genetic and genomic research, and the time pressures faced by researchers, peer review is not the *silver bullet* to solving the rigor and reproducibility crisis. The most crucial of these stages and factors affecting reproducibility are expanded upon below.

Numerous methodological factors contribute to the validity of a research study. Munafo et al. reviews that factors such as publication bias, failure to control for bias, low statistical power, poor quality control, and *P*-hacking can all contribute to undermining the validity of research studies and inhibiting other laboratories' ability to reproduce work [2]. It is also becoming increasingly important for geneticists to have at least some foundational understanding of biostatistics and statistical science. Appropriate tests must be chosen, given a specific context, for correction of false positives [44], variant imputation [45,46], population structure and confounding variables [47], or even within pipelines to account and control for internal technical errors caused by the sequencing platform [48]. There can also be important considerations when combining different data sets and admixture of samples [49], or even deciding upon an appropriate threshold for significance [50].

As mentioned earlier in this chapter, the lack of diverse representation in most GWAS and/or study populations can also impair efforts to replicate findings. Homogenous cohorts fail to capture functional variants in the human genome that are important for physiological processes or disease progression. Downstream, this homogeneity creates problems when building new protocols or platform technologies for sequencing and variant calling of new samples, as it utilizes assumptions or known variants identified only in a single population. This is especially relevant when using polygenic risk scores (PRS) to assess and predict predisposition to different conditions (discussed further in Chapter 5). Given a majority of PRS calculations were performed using underlying variant data from individuals of European ancestry, PRS in individuals from other backgrounds are less accurate and useful in the clinic [51–53].

A corollary contributing factor to the reproducibility crisis, supplementary to the lab bench itself, is the culture of career advancement within academic research, highlighted by the proverbial "publish or perish" narrative. This narrative and reality in academic science pressures early career investigators to show their research productivity by publishing multiple papers as a means of establishing job security. While there are many other components to the tenure package in academia, the ability to show productivity from grant funding and the ability to deliver research results is the primary consideration for tenure review committees. While it seems superficially sensible that promotion should be tied to scholarship, particularly the ability to conduct and publish impacting research, there is a disconnect between this requirement for job security and the culture of how research is reported in the literature.

The primary example of this bias in published literature is the fact that there exists a systemic reporting bias that emphasizes positive results in peer-reviewed literature and disfavors the reporting of negative results, even among biomedical and clinical research trials [54,55]. In turn, this influences the approaches that investigators (particularly early-career investigators) take to validate their research, knowing their livelihoods and those in their labs are dependent upon showing successful outcomes in their work. This disconnect can be perpetuated by review, promotion, and tenure (RPT) committees dependent on the institutional metrics used to define the scholarly success of faculty members under consideration for promotion.

Inappropriate measures of scholarship, such as impact factor (IF) or rewarding quantity over quality (which can lead to a lack of reproducibility) can also inappropriately incentivize biomedical researchers to publish work that reinforces job protection and less-than-excellent scholarship [56–58]. Responsibility for training the next generation of researchers also falls heavily on principal investigators. Genetic researchers at all levels, and particularly research associates and principal investigators, can help develop strong scholarly habits in trainees via the demonstration and reinforcement of responsible, rigorous research conduct. One should encourage open and honest communication regarding reporting preliminary findings and during meetings with collaborators. Further, setting and upholding laboratory policies for recording thorough and accurate lab notes, and reporting research misconduct when it occurs, provide valuable tools and lessons to graduate trainees. The latter requires mandatory and extensive training regarding responsible conduct of research and also requires that trainees are provided institutional and field-specific resources to access when needed [59].

There should also be articulated institutional-specific policies for early-career investigators to follow when questions related to research integrity arise that can be professionally explored without necessarily being automatically punitive. These internal review policies of institutions may also play a role in the repercussions for researchers who falsify or fabricate data.

Across US and global institutions, the policies for investigating cases of fabricated or falsified data vary widely. Best practices for reviewing these cases that are more widely adopted may help reduce the frequency of retractions in the scientific literature [60,61]. An analysis of 1316 papers published from US institutions across multiple scientific disciplines found that the competitive environment of the authors' institution biased against reporting negative research results [62]. This and other work has spurred discussion on how best to remedy the bias that influences reliable result reporting.

Some journals have taken a new approach to emphasize the methodology of the science as opposed to the results or findings. *Cell Press*, a peer-reviewed journal within the Elsevier portfolio, launched *STAR Protocols* in 2016 to identify reproducible protocols in the life sciences that were accessible and validated [63]. STAR stands for Structured Transparent Accessible Reproducible, and the journal articles are reviewed by core facility and technologically experienced research scientists. The Center for Open Science initiated the use of **Registered Reports** to re-emphasize peer review on the methodology of the study as opposed to the final results of the analysis.

In a Registered Report, researchers submit their idea and study design for an initial round of peer review, in which reviewers weigh the integrity and strength of the research idea and methodology. If the report passes this round, the paper is conditionally accepted, regardless of the results of the study, pending adherence to the reviewed protocol [64,65]. Select journals will accept and publish genetic studies that are pre-registered reports as part of their publishing model, include *Scientific Reports*, *PLOS ONE*, *PLOS Biology*, *BMC Biology*, and *BMC Medicine*.

eLife recently adopted a new peer review protocol that requires all reviewed articles to first be published as a preprint. Next, the reviewed article is automatically published by the journal regardless of the peer review process. This new form of acceptance also includes the views of the reviewing experts, those who have discussed the work on the preprint forums, and the author's reply (if necessary). This radical change has removed the accept/revise/reject model of formal peer review [66] and already sparked considerable and healthy debate within the scientific community.

Given the complex nature of some genomic analysis, additional resources will be needed to help trainees and early-career investigators develop the necessary intuition and skillset to ask appropriate questions that challenge the integrity of a given methodology, whether with their own work or another's. These questions should become second nature for newly trained researchers; perhaps as ingrained into graduate training as is the emphasis on identifying a research question, developing a testable hypothesis, or designing and analyzing a more inclusive (diverse) cohort. If there is more openness up front on how to develop the best methodological approach to a particular experiment or question, or how to best review it, there will be fewer concerns about the results if they are not able to be replicated elsewhere.

The societal importance of open science

> When Jonas Salk was asked who owned the patent to his new polio vaccine, he famously replied, "Well, the people, I would say. There is no patent. Could you patent the sun?"

In all the years since 1955, Dr. Jonas Salk's idea that his and his team's science be available solely for the betterment of humanity is still a high bar to achieve given the current systemic infrastructure of research, publishing, patenting, and health care. With the advent of modern technologies that reduce cost and time, the ideal of "open science" has inspired the creation of large, public, and free databases that have promoted research and considerable secondary research worldwide.

Unfortunately, given the enormous influence of profit-driven privatization of medical care and insurance, particularly in the United States, and elsewhere in the world, there are many economic factors that prevent the latest breakthroughs from establishing themselves for free or with widespread usage in the public domain. One need to look no further than the patent disputes between MIT's Broad Institute and the University of California, Berkeley (alongside Dr. Emmanuelle Charpentier) regarding ownership of CRISPR gene-editing technology—a legal drama that continues to unfold. Each institute is keenly aware of the economic boon from owning control of CRISPR and the downstream licensing of this approach, and this is just within the United States. The issue becomes even more complex when looking at patent ownership of CRISPR technologies in the European Union and elsewhere.

Dramatic steps have been taken toward the democratization of science and unrestricted access of research results and large data sets. A prime example of this is the UK Biobank, an open-access database with greater than half a million genomes (with phenotypic data), to which any qualified scientist on the planet can apply for ethical approval access. The UK Biobank is a not-for-profit organization, supported by various levels of UK government and charitable foundations. Although not a perfect resource—the database lacks samples of ethnic diversity (as discussed above)—it continues to add new genomes regularly and provided a wealth of information to mine for large-scale genomic studies. An unusual example of the democratization of biology comes in the form of 3D printing technologies, which are increasingly allowing researchers to design tools or modify those that they have already, eliminating the high costs of biotech sales and increasing specificity tailored to their needs. In addressing public access to published-behind-a-paywall articles, all research that is federally-funded by the United States government will be required to be immediately available and open access upon publication by 2026 [67]. Steps like these ensure that all researchers, as well as the general public, have access to essential data and analysis as quickly as possible.

The field of genomic research has seen an exponential growth in the amount of data generated and made available to researchers and the public. Open science and data sharing agreements have become increasingly important in managing this data. One of the key challenges is balancing the need for data sharing with protecting patient privacy. The 1000 Genomes Project Consortium [18] and the National Cancer Institute's Genomic Data Commons [68] are two examples of successful data sharing initiatives.

The Genomic Data Commons integrates clinical data from individual studies by harmonizing inputs on sample collection, the alignment of sequencing data to a common reference genome, and standardizing protocols on variant calling, and other metrics. There are also controlled and restricted data sets within this public database (and others) that are curated in accordance with the informed consent documents or other guidelines delineated when participants are recruited into participating studies. This identifiable data may be embargoed or behind a secure wall such that only those who apply to access this data are granted permission to use it. While not entirely open access, these restrictions reflect necessary precautions needed for patient privacy.

Data availability is also determined by the country hosting the database. In the United States, there are numerous federal and state laws that regulate the collection, usage, and disclosure of genomic data. For example, The Genetic Information Nondiscrimination Act (GINA) prohibits employers, health insurance companies, and others from using genetic information to discriminate against individuals. The European Union has adopted the General Data Protection Regulation (GDPR) which protects the privacy of personal data, including genomic data [69]. The GDPR requires that individuals must provide informed consent for the collection and use of their data, and it gives individuals the right to access, rectify, and erase their data. The GDPR also requires that organizations implement appropriate technical and organizational measures to protect personal data. The law prohibits processing this data in such a way that could even indirectly reveal sensitive information about an individual.

In China, the Cybersecurity Law, Data Security Law, and Personal Information Protection Law (PIPL) have been implemented to govern how personal identifiable information (both biological and digital) are collected, protected, and stored in China. These regulations also delineate that consent for this information to be collected must be freely given and informed and that it can be withdrawn.

While these laws have made it challenging for geneticists and researchers to access and use genetic data [70], they are essential to protect personal information in a rapidly changing research environment. Additional guidance has been needed for open access of genetic data beyond these laws. For example, in the United States, there has been historically many cases of data mismanagement and lack of consent when it comes to the collection and use of samples from indigenous communities, and other racial and ethnic populations historically underrepresented in genomic studies. New guidelines that focus on trust, accountability, and equity must be implemented to ensure protection of this information and safeguard against sample misuse, along with including the input of the participants in the study who are providing the samples [71]. Data consortiums must also be sensitive to our changing understanding of the intersection of race, ethnicity, and ancestry, especially when samples are being collated together from different genomic databases [72,73].

These and other guidelines should always be continually revisited to ensure equitable access and protection of genomic information. Ideally, open science ensures that researchers and bioethicists always have the opportunity to shore up problems in research pipelines, the process of study participant recruitment, consent, and engagement, and in reporting analysis outcomes.

The non-scientific public must also continue to have a stronger voice in how this data is used and discussed. Social media platforms such as Twitter, Facebook, and Mastodon allow researchers to engage directly with the public and the media. In the first months of the COVID-19 pandemic, hundreds of thousands of tweets on Twitter discussed a variety of topics related to the information from and perception of the Centers for Disease Control and Prevention (CDC) regarding COVID-19. The most discussed topics included the credibility of the CDC and the CDC guidelines related to COVID-19 exposure and response [74].

This rapid fire promotion of the latest in scientific discovery is a boost to equitable access to research results and informed policy but can also promote mistrust in the process of science and aid in the spread of misinformation or false information [75,76]. Twitter bots and other malware can spread misinformation or sow the appearance of disagreements within a scientific field when there is large consensus, as what has happened concerning the discussion focused on the safety and efficacy of vaccinations [77].

Genetic and genomic studies are not immune to these trends. When news of He Jinkaiu's experiment using CRISPR and the birth of the first CRISPR-edited humans, Twitter, Chinese social media platform Weibo, and other social media platforms explored with discourse related to the ethical controversy and societal implications of its use [78,79] (see Box 1.2). These conversations appear to be linked with the news cycle in that conversations can be tied with when news breaks related to a specific event or key development in genetics research [78,80].

Additional consequences of genetic and genomic information being so easily accessible have extended far beyond the halls of academia and industry. Direct-to-consumer (DTC) DNA testing has grown in the last decade and contributed to mainstream discussion of genetic variation, ancestry, and susceptibility to disease. However, not all of the perceived health information related to some of these products are discussed by trained professionals, which opens the public discourse up for the spread of misinformation or basing healthcare decisions based on non-clinical test results [81–83].

Participants of DTC DNA testing are also concerned about opaque privacy protection related to their DNA testing results [84]. DTC testing has influenced family dynamics and relationships when ancestry results return, often without much support from the company providing the service [85]. There are also questions concerning who can give permission to have their DNA tested. This is a particularly complex issue when that individual does not know or authorize the test or is deceased [86]. The results of these analyses can have profound consequences and the impacts on society are still not completely understood.

Arguably one of the most controversial cases of DNA privacy in DTC testing is the use of genetic test results by law enforcement. In 2018, news broke in the United States that the famous Golden State Killer, a serial killer who committed murders in the 1970s, had been identified by police by using the public genealogy website GEDMatch [87]. Law enforcement officials had uploaded DNA from a crime scene and identified a relative of the killer in GEDMatch, ultimately arresting a retired police officer who had committed those terrible crimes. As an additional consequence, the case immediately brought up questions related to the ethical use of DTC testing, including data privacy, public safety, DNA ownership, and other complicated bioethical questions. These questions are further confounded when weighing personal privacy and protection versus public safety, including ensuring criminals are found. Since 2018, GEDMatch and other genealogy databases have helped solve hundreds of cold cases and crimes.

Negotiating these and other complicated bioethics of genetic research is not formally part of the training of many geneticists in the US and around the world. The NIH has mandated that institutions receiving NIH funding implement RCR training for grant awardees and trainees [88]. RCR training can highlight many different issues including navigating trainee power dynamics, responsible data collection and reporting, conflict of interest, the peer review process, and even "the scientist as a responsible member of society, contemporary ethical issues in biomedical research, and the environmental and societal impacts of scientific research" [88].

However, institutions are generally free to implement RCR training as they see fit, and there is little uniformity across the US or within the international community. There should be incentives at the institutional or national level in graduate training and with early-career faculty development that stresses the importance of a societal-conscious biomedical researcher. Given genetic technology and discovery has become a part of everyday conversation, additional training is needed to help researchers navigate how to discuss their work with a broad and diverse community. Bioethics, rigorous methods, and RCR need to become more integrated into undergraduate and graduate training such that researchers are prepared for these conversations either among themselves, with lawmakers or other members of society, or even within their or social networks of family and friends.

Conclusions

Given the rapid pace of genetic research, it is likely that exciting new advancements in our understanding of the genome will continue to emerge, along with bold interventions in clinical practice. These developments may have unforeseen ramifications, making it critical for geneticists, clinicians, trainees at all levels, patients, and the public to have a voice in how we apply and expand our knowledge. The emerging use of artificial intelligence, like ChatGPT and other AI-driven programs, are rapidly gaining traction in numerous software platforms. These AI programs are in their infancy—the "training" stage—but this is a critical time for AI as the data sets used for training will inform the biases inherent to these platforms. The implications are enormous and wide-reaching in all fields in the context of scientific writing. For example, the ability for an AI to produce scientific literature that *sounds* correct but in fact misconstrues the facts or simply is incorrect leads to an enormous black box about regulating the use of AI in preparation of manuscripts and other publications. Just as this book was preparing to go to press, ChatGPT and other AIs took the internet by storm, so much so that Italy temporarily banned ChatGPT [89] and publishers were forced to quickly respond with guidance to authors on the matter. Elsevier Group (the publisher of this book) issued guidelines in March 2023 stating:

> Where authors use generative AI and AI-assisted technologies in the writing process, these technologies should only be used to improve readability and language of the work....Authors should disclose in their manuscript the use of AI and AI-assisted technologies and a statement will appear in the published work. [66].

In spirit of these guidelines, the authors of this manuscript can reveal that the fourth paragraph of the prior section in this chapter ("The field of genomic research has seen an exponential growth....)" was imported into ChatGPT to improve readability, and as an example of the power of AI to write scientific works that are indistinguishable from human-authored work.

To better facilitate public discourse of genetic research, it is imperative that the scientific literature reflect the highest level of rigorous methodology. As evidenced by the daily updates of our knowledge of SARS-CoV-2 during the COVID-19 pandemic (see Box 1.1) and the growing clinical use of CRISPR gene-editing (Box 1.2), researchers must ensure that the information they bring forth is meritorious and reproducible, with a responsibility to both scientific and broader communities. The era of open science offers a unique opportunity for collaboration and encourages researchers to work together to define best practices in order to improve the transparency and accessibility of research outcomes.

Reproducible research does not necessarily mean that the results of any given experiment or project will always be correct. Rather, it endeavors to foster the careful consideration required such that the underlying hypothesis, approach of testing the said hypothesis, and the data collected and analyzed are meaningfully interpretable. Geneticists and researchers should approach their work such that it can grow with the changing knowledge of the community at large and that others can go back to ensure our bedrock principles and knowledge are sturdy.

The scientific research enterprise is flawed in that it is limited in part by our preceding knowledge of the world, and in part by the naïve mistakes of the untrained or ignorance of those willing to take short cuts. The convalescence of these aspects can lead to incorrect scientific conclusions, which are at times inappropriately disseminated via the use of preprint servers and AI-supported technologies before researchers are able to discuss and self-correct the science. While there are many ways to tackle these issues to ensure progress in our work and for the betterment of the society, they can be summarized into three strategic goals that the genetic research community should always strive for:

1. The genetics research community should always work to improve the general public's understanding of the scientific process so that open science and public discourse are less reactionary or misinformed.
2. The genetics research community should continue to establish reproducible research practices to strengthen the research findings and make them more representative of the diverse global population.
3. The genetics research community should promote the development of strong science communication skills within the next generation of the research and clinical workforce.

This chapter has outlined a few of the individual and collective actions that can be taken to achieve these aforementioned goals. Institutional and departmental commitment to these or similar ideals will also solidify the genetic research infrastructure as a whole and reinforce the need to continue to execute strong research practices. The subsequent chapters in this textbook are meant to provide a deeper knowledge into reproducible research practices using a variety of widely used approaches in genetics and genomics, from PCR to CRISPR. Additionally, this textbook also provides guidance on how faculty, mentors, or others in instructional positions can infuse and promote rigorous practices into their work and curriculums so that future research trainees achieve the highest standard of reproducible research.

By instilling these practices at all levels of the scientific enterprise, we can continue to push our knowledge of genetics in new and meaningful directions, helping researchers achieve the goal of their studies being peer reviewed, published, and cited!

References

[1] J.P.A. Ioannidis, D. Fanelli, D.D. Dunne, S.N. Goodman, Meta-research: evaluation and improvement of research methods and practices, PLoS Biol. 13 (2015) e1002264, https://doi.org/10.1371/journal.pbio.1002264.

[2] M.R. Munafò, B.A. Nosek, D.V.M. Bishop, K.S. Button, C.D. Chambers, N. Percie du Sert, U. Simonsohn, E.-J. Wagenmakers, J.J. Ware, J.P.A. Ioannidis, A manifesto for reproducible science, Nat. Hum. Behav. 1 (2017) 1–9, https://doi.org/10.1038/s41562-016-0021.

[3] H. Shen, Meet this super-spotter of duplicated images in science papers, Nature 581 (2020) 132–136, https://doi.org/10.1038/d41586-020-01363-z.

[4] National Academies of Sciences, Engineering, and Medicine, Understanding reproducibility and replicability, in: Reproducibility and Replicability in Science, The National Academies Press, Washington, DC, 2019, https://doi.org/10.17226/25303.

[5] J.P.A. Ioannidis, Why most published research findings are false, PLoS Med. 2 (2005) e124, https://doi.org/10.1371/journal.pmed.0020124.

[6] Science, C. for O, Reproducibility Project: Cancer Biology, 2023 (WWW Document) https://www.cos.io/rpcb. (Accessed 27 February 2023).

[7] M. Baker, 1,500 scientists lift the lid on reproducibility, Nature 533 (2016) 452–454, https://doi.org/10.1038/533452a.

[8] W.S. Lamers, K. Boyack, V. Larivière, C.R. Sugimoto, N.J. van Eck, L. Waltman, D. Murray, Investigating disagreement in the scientific literature, elife 10 (2021) e72737, https://doi.org/10.7554/eLife.72737.

[9] Science Integrity Digest, Science Integrity Digest, Sci. Integr. Dig, 2022 (WWW Document) https://scienceintegritydigest.com/. (Accessed 27 February 2023).

[10] E.M. Bik, F.C. Fang, A.L. Kullas, R.J. Davis, A. Casadevall, Analysis and correction of inappropriate image duplication: the molecular and cellular biology experience, Mol. Cell. Biol. 38 (2018) e00309–e00318, https://doi.org/10.1128/MCB.00309-18.

[11] E.M. Bik, A. Casadevall, F.C. Fang, The prevalence of inappropriate image duplication in biomedical research publications, MBio 7 (2016) e00809–e00816, https://doi.org/10.1128/mBio.00809-16.

[12] M.C. Mills, C. Rahal, The GWAS diversity monitor tracks diversity by disease in real time, Nat. Genet. 52 (2020) 242–243, https://doi.org/10.1038/s41588-020-0580-y.

[13] All of Us Research Program Overview, Us Res. Program NIH, 2020 (WWW Document) https://allofus.nih.gov/about/program-overview. (Accessed 27 February 2023).

[14] About – H3Africa, 2023. URL https://h3africa.org/index.php/about/. (Accessed 27 February 2023).

[15] N. Chatterjee, J. Shi, M. García-Closas, Developing and evaluating polygenic risk prediction models for stratified disease prevention, Nat. Rev. Genet. 17 (2016) 392–406, https://doi.org/10.1038/nrg.2016.27.

[16] P.M. Visscher, N.R. Wray, Q. Zhang, P. Sklar, M.I. McCarthy, M.A. Brown, J. Yang, 10 years of GWAS discovery: biology, function, and translation, Am. J. Hum. Genet. 101 (2017) 5–22, https://doi.org/10.1016/j.ajhg.2017.06.005.

[17] T.A. Manolio, Using the data we have: improving diversity in genomic research, Am. J. Hum. Genet. 105 (2019) 233–236, https://doi.org/10.1016/j.ajhg.2019.07.008.

[18] 1000 Genomes Project Consortium, A. Auton, L.D. Brooks, R.M. Durbin, E.P. Garrison, H.M. Kang, J.O. Korbel, J.L. Marchini, S. McCarthy, G.A. McVean, G.R. Abecasis, A global reference for human genetic variation, Nature 526 (2015) 68–74, https://doi.org/10.1038/nature15393.

[19] S. Fairley, E. Lowy-Gallego, E. Perry, P. Flicek, The international genome sample resource (IGSR) collection of open human genomic variation resources, Nucleic Acids Res. 48 (2020) D941–D947, https://doi.org/10.1093/nar/gkz836.

[20] P.H. Sudmant, T. Rausch, E.J. Gardner, R.E. Handsaker, A. Abyzov, J. Huddleston, Y. Zhang, K. Ye, G. Jun, M.H.-Y. Fritz, M.K. Konkel, A. Malhotra, A.M. Stütz, X. Shi, F.P. Casale, J. Chen, F. Hormozdiari,

G. Dayama, K. Chen, M. Malig, M.J.P. Chaisson, K. Walter, S. Meiers, S. Kashin, E. Garrison, A. Auton, H.Y.K. Lam, X.J. Mu, C. Alkan, D. Antaki, T. Bae, E. Cerveira, P. Chines, Z. Chong, L. Clarke, E. Dal, L. Ding, S. Emery, X. Fan, M. Gujral, F. Kahveci, J.M. Kidd, Y. Kong, E.-W. Lameijer, S. McCarthy, P. Flicek, R.A. Gibbs, G. Marth, C.E. Mason, A. Menelaou, D.M. Muzny, B.J. Nelson, A. Noor, N.F. Parrish, M. Pendleton, A. Quitadamo, B. Raeder, E.E. Schadt, M. Romanovitch, A. Schlattl, R. Sebra, A.A. Shabalin, A. Untergasser, J.A. Walker, M. Wang, F. Yu, C. Zhang, J. Zhang, X. Zheng-Bradley, W. Zhou, T. Zichner, J. Sebat, M.A. Batzer, S.A. McCarroll, 1000 Genomes Project Consortium, R.E. Mills, M.B. Gerstein, A. Bashir, O. Stegle, S.E. Devine, C. Lee, E.E. Eichler, J.O. Korbel, An integrated map of structural variation in 2,504 human genomes, Nature 526 (2015) 75–81, https://doi.org/10.1038/nature15394.

[21] Retracted Coronavirus (COVID-19) Papers, 2023. https://retractionwatch.com/retracted-coronavirus-covid-19-papers/.

[22] R. Lu, X. Zhao, J. Li, P. Niu, B. Yang, H. Wu, W. Wang, H. Song, B. Huang, N. Zhu, Y. Bi, X. Ma, F. Zhan, L. Wang, T. Hu, H. Zhou, Z. Hu, W. Zhou, L. Zhao, J. Chen, Y. Meng, J. Wang, Y. Lin, J. Yuan, Z. Xie, J. Ma, W.J. Liu, D. Wang, W. Xu, E.C. Holmes, G.F. Gao, G. Wu, W. Chen, W. Shi, W. Tan, Genomic characterisation and epidemiology of 2019 novel coronavirus: implications for virus origins and receptor binding, Lancet 395 (2020) 565–574, https://doi.org/10.1016/S0140-6736(20)30251-8.

[23] F. Wu, S. Zhao, B. Yu, Y.-M. Chen, W. Wang, Z.-G. Song, Y. Hu, Z.-W. Tao, J.-H. Tian, Y.-Y. Pei, M.-L. Yuan, Y.-L. Zhang, F.-H. Dai, Y. Liu, Q.-M. Wang, J.-J. Zheng, L. Xu, E.C. Holmes, Y.-Z. Zhang, A new coronavirus associated with human respiratory disease in China, Nature 579 (2020) 265–269, https://doi.org/10.1038/s41586-020-2008-3.

[24] N. Fraser, L. Brierley, G. Dey, J.K. Polka, M. Pálfy, F. Nanni, J.A. Coates, The evolving role of preprints in the dissemination of COVID-19 research and their impact on the science communication landscape, PLoS Biol. 19 (2021) e3000959, https://doi.org/10.1371/journal.pbio.3000959.

[25] R. Ravinetto, C. Caillet, M.H. Zaman, J.A. Singh, P.J. Guerin, A. Ahmad, C.E. Durán, A. Jesani, A. Palmero, L. Merson, P.W. Horby, E. Bottieau, T. Hoffmann, P.N. Newton, Preprints in times of COVID19: the time is ripe for agreeing on terminology and good practices, BMC Med. Ethics 22 (2021) 106, https://doi.org/10.1186/s12910-021-00667-7.

[26] S.L. Taneja, M. Passi, S. Bhattacharya, S.A. Schueler, S. Gurram, C. Koh, Social media and research publication activity during early stages of the COVID-19 pandemic: longitudinal trend analysis, J. Med. Internet Res. 23 (2021) e26956, https://doi.org/10.2196/26956.

[27] D. De Coninck, T. Frissen, K. Matthijs, L. d'Haenens, G. Lits, O. Champagne-Poirier, M.-E. Carignan, M.D. David, N. Pignard-Cheynel, S. Salerno, M. Généreux, Beliefs in conspiracy theories and misinformation about COVID-19: comparative perspectives on the role of anxiety, depression and exposure to and Trust in Information Sources, Front. Psychol. 12 (2021) 646394, https://doi.org/10.3389/fpsyg.2021.646394.

[28] L. Brierley, Lessons from the influx of preprints during the early COVID-19 pandemic, Lancet Planet. Health 5 (2021) e115–e117, https://doi.org/10.1016/S2542-5196(21)00011-5.

[29] M.S. Majumder, K.D. Mandl, Early in the epidemic: impact of preprints on global discourse about COVID-19 transmissibility, Lancet Glob. Health 8 (2020) e627–e630, https://doi.org/10.1016/S2214-109X(20)30113-3.

[30] P. Pradhan, A.K. Pandey, A. Mishra, P. Gupta, P.K. Tripathi, M.B. Menon, J. Gomes, P. Vivekanandan, B. Kundu, Uncanny similarity of unique inserts in the 2019-nCoV spike protein to HIV-1 gp120 and Gag, bioRxiv (2020) 927871, https://doi.org/10.1101/2020.01.30.927871.

[31] A.A. Schäffer, E.L. Hatcher, L. Yankie, L. Shonkwiler, J.R. Brister, I. Karsch-Mizrachi, E.P. Nawrocki, VADR: validation and annotation of virus sequence submissions to GenBank, BMC Bioinformatics 21 (2020) 211, https://doi.org/10.1186/s12859-020-3537-3.

[32] M. Jinek, K. Chylinski, I. Fonfara, M. Hauer, J.A. Doudna, E. Charpentier, A programmable dual-RNA-guided DNA endonuclease in adaptive bacterial immunity, Science 337 (2012) 816–821, https://doi.org/10.1126/science.1225829.

[33] H. Frangoul, D. Altshuler, M.D. Cappellini, Y.-S. Chen, J. Domm, B.K. Eustace, J. Foell, J. de la Fuente, S. Grupp, R. Handgretinger, T.W. Ho, A. Kattamis, A. Kernytsky, J. Lekstrom-Himes, A.M. Li, F. Locatelli, M.Y. Mapara, M. de Montalembert, D. Rondelli, A. Sharma, S. Sheth, S. Soni, M.H. Steinberg, D. Wall, A. Yen, S. Corbacioglu, CRISPR-Cas9 gene editing for sickle cell disease and β-thalassemia, N. Engl. J. Med. 384 (2021) 252–260, https://doi.org/10.1056/NEJMoa2031054.

[34] A. Dimitri, F. Herbst, J.A. Fraietta, Engineering the next-generation of CAR T-cells with CRISPR-Cas9 gene editing, Mol. Cancer 21 (2022) 78, https://doi.org/10.1186/s12943-022-01559-z.

[35] K. Chen, Y. Wang, R. Zhang, H. Zhang, C. Gao, CRISPR/Cas genome editing and precision plant breeding in agriculture, Annu. Rev. Plant Biol. 70 (2019) 667–697, https://doi.org/10.1146/annurev-arplant-050718-100049.

[36] B. Artegiani, D. Hendriks, J. Beumer, R. Kok, X. Zheng, I. Joore, S. Chuva de Sousa Lopes, J. van Zon, S. Tans, H. Clevers, Fast and efficient generation of knock-in human organoids using homology-independent CRISPR-Cas9 precision genome editing, Nat. Cell Biol. 22 (2020) 321–331, https://doi.org/10.1038/s41556-020-0472-5.

[37] A. Martinez-Silgado, F.A. Yousef Yengej, J. Puschhof, V. Geurts, C. Boot, M.H. Geurts, M.B. Rookmaaker, M.C. Verhaar, J. Beumer, H. Clevers, Differentiation and CRISPR-Cas9-mediated genetic engineering of human intestinal organoids, STAR Protoc. 3 (2022) 101639, https://doi.org/10.1016/j.xpro.2022.101639.

[38] J. Tao, D.E. Bauer, R. Chiarle, Assessing and advancing the safety of CRISPR-Cas tools: from DNA to RNA editing, Nat. Commun. 14 (2023) 212, https://doi.org/10.1038/s41467-023-35886-6.

[39] M. Morrison, S. de Saille, CRISPR in context: towards a socially responsible debate on embryo editing, Palgrave Commun. 5 (2019) 1–9, https://doi.org/10.1057/s41599-019-0319-5.

[40] New Scientist, CRISPR Babies: What's Next for the Gene-Edited Children from Trial in China?, New Scientist, 2022 (WWW Document) https://www.newscientist.com/article/mg25533930-700-whats-next-for-the-gene-edited-children-from-crispr-trial-in-china/. (Accessed 27 February 2023).

[41] P.J. Chen, D.R. Liu, Prime editing for precise and highly versatile genome manipulation, Nat. Rev. Genet. 24 (2023) 161–177, https://doi.org/10.1038/s41576-022-00541-1.

[42] B. Wienert, M.K. Cromer, CRISPR nuclease off-target activity and mitigation strategies, Front. Genome Ed. 4 (2022) 1050507, https://doi.org/10.3389/fgeed.2022.1050507.

[43] W.T. Garrood, N. Kranjc, K. Petri, D.Y. Kim, J.A. Guo, A.M. Hammond, I. Morianou, V. Pattanayak, J.K. Joung, A. Crisanti, A. Simoni, Analysis of off-target effects in CRISPR-based gene drives in the human malaria mosquito, Proc. Natl. Acad. Sci. U. S. A. 118 (2021) e2004838117, https://doi.org/10.1073/pnas.2004838117.

[44] K. Korthauer, P.K. Kimes, C. Duvallet, A. Reyes, A. Subramanian, M. Teng, C. Shukla, E.J. Alm, S.C. Hicks, A practical guide to methods controlling false discoveries in computational biology, Genome Biol. 20 (2019) 118, https://doi.org/10.1186/s13059-019-1716-1.

[45] Q. Sun, W. Liu, J.D. Rosen, L. Huang, R.G. Pace, H. Dang, P.J. Gallins, E.E. Blue, H. Ling, H. Corvol, L.J. Strug, M.J. Bamshad, R.L. Gibson, E.W. Pugh, S.M. Blackman, G.R. Cutting, W.K. O'Neal, Y.-H. Zhou, F.A. Wright, M.R. Knowles, J. Wen, Y. Li, Cystic Fibrosis Genome Project, Leveraging TOPMed imputation server and constructing a cohort-specific imputation reference panel to enhance genotype imputation among cystic fibrosis patients, HGG Adv. 3 (2022) 100090, https://doi.org/10.1016/j.xhgg.2022.100090.

[46] Q. Sun, Y. Yang, J.D. Rosen, M.-Z. Jiang, J. Chen, W. Liu, J. Wen, L.M. Raffield, R.G. Pace, Y.-H. Zhou, F.A. Wright, S.M. Blackman, M.J. Bamshad, R.L. Gibson, G.R. Cutting, M.R. Knowles, D.R. Schrider, C. Fuchsberger, Y. Li, MagicalRsq: machine-learning-based genotype imputation quality calibration, Am. J. Hum. Genet. 109 (2022) 1986–1997, https://doi.org/10.1016/j.ajhg.2022.09.009.

[47] J.H. Sul, L.S. Martin, E. Eskin, Population structure in genetic studies: confounding factors and mixed models, PLoS Genet. 14 (2018) e1007309, https://doi.org/10.1371/journal.pgen.1007309.

[48] Y. Yin, C. Butler, Q. Zhang, Challenges in the application of NGS in the clinical laboratory, Hum. Immunol. 82 (2021) 812–819, https://doi.org/10.1016/j.humimm.2021.03.011.

[49] K.E. Grinde, L.A. Brown, A.P. Reiner, T.A. Thornton, S.R. Browning, Genome-wide significance thresholds for admixture mapping studies, Am. J. Hum. Genet. 104 (2019) 454–465, https://doi.org/10.1016/j.ajhg.2019.01.008.
[50] P.C. Sham, S.M. Purcell, Statistical power and significance testing in large-scale genetic studies, Nat. Rev. Genet. 15 (2014) 335–346, https://doi.org/10.1038/nrg3706.
[51] L. Duncan, H. Shen, B. Gelaye, J. Meijsen, K. Ressler, M. Feldman, R. Peterson, B. Domingue, Analysis of polygenic risk score usage and performance in diverse human populations, Nat. Commun. 10 (2019) 3328, https://doi.org/10.1038/s41467-019-11112-0.
[52] A.R. Martin, M. Kanai, Y. Kamatani, Y. Okada, B.M. Neale, M.J. Daly, Clinical use of current polygenic risk scores may exacerbate health disparities, Nat. Genet. 51 (2019) 584–591, https://doi.org/10.1038/s41588-019-0379-x.
[53] Y. Wang, K. Tsuo, M. Kanai, B.M. Neale, A.R. Martin, Challenges and opportunities for developing more generalizable polygenic risk scores, Annu. Rev. Biomed. Data Sci. 5 (2022) 293–320, https://doi.org/10.1146/annurev-biodatasci-111721-074830.
[54] M. Mitra-Majumdar, A.S. Kesselheim, Reporting bias in clinical trials: progress toward transparency and next steps, PLoS Med. 19 (2022) e1003894, https://doi.org/10.1371/journal.pmed.1003894.
[55] E.H. Turner, A. Cipriani, T.A. Furukawa, G. Salanti, Y.A. de Vries, Selective publication of antidepressant trials and its influence on apparent efficacy: updated comparisons and meta-analyses of newer versus older trials, PLoS Med. 19 (2022) e1003886, https://doi.org/10.1371/journal.pmed.1003886.
[56] J.P.A. Ioannidis, S. Greenland, M.A. Hlatky, M.J. Khoury, M.R. Macleod, D. Moher, K.F. Schulz, R. Tibshirani, Increasing value and reducing waste in research design, conduct, and analysis, Lancet 383 (2014) 166–175, https://doi.org/10.1016/S0140-6736(13)62227-8.
[57] E.C. McKiernan, L.A. Schimanski, C. Muñoz Nieves, L. Matthias, M.T. Niles, J.P. Alperin, Use of the journal impact factor in academic review, promotion, and tenure evaluations, elife 8 (2019) e47338, https://doi.org/10.7554/eLife.47338.
[58] D.B. Rice, H. Raffoul, J.P.A. Ioannidis, D. Moher, Academic criteria for promotion and tenure in biomedical sciences faculties: cross sectional analysis of international sample of universities, BMJ 369 (2020) m2081, https://doi.org/10.1136/bmj.m2081.
[59] A.L. Antes, L.B. Maggi, How to conduct responsible research: a guide for graduate students, Curr. Protoc. 1 (2021) e87, https://doi.org/10.1002/cpz1.87.
[60] H. Bauchner, P.B. Fontanarosa, A. Flanagin, J. Thornton, Scientific misconduct and medical journals, JAMA 320 (2018) 1985–1987, https://doi.org/10.1001/jama.2018.14350.
[61] C.K. Gunsalus, A.R. Marcus, I. Oransky, Institutional research misconduct reports need more credibility, JAMA 319 (2018) 1315–1316, https://doi.org/10.1001/jama.2018.0358.
[62] D. Fanelli, Do pressures to publish increase scientists' Bias? An empirical support from US states data, PLoS One 5 (2010) e10271, https://doi.org/10.1371/journal.pone.0010271.
[63] E. Marcus, A STAR is born, Cell 166 (2016) 1059–1060, https://doi.org/10.1016/j.cell.2016.08.021.
[64] C.D. Chambers, E. Feredoes, S.D. Muthukumaraswamy, P.J. Etchells, C.D. Chambers, E. Feredoes, S.D. Muthukumaraswamy, P.J. Etchells, Instead of "playing the game" it is time to change the rules: registered reports at *AIMS Neuroscience* and beyond, AIMS Neurosci. 1 (2014) 4–17, https://doi.org/10.3934/Neuroscience.2014.1.4.
[65] B.A. Nosek, D. Lakens, Registered reports: a method to increase the credibility of published results, Soc. Psychol. 45 (2014) 137, https://doi.org/10.1027/1864-9335/a000192.
[66] Elsevier, Publishing Ethics for Editors, 2023 (WWW Document) https://www.elsevier.com/about/policies/publishing-ethics. (Accessed 21 March 2023).
[67] White House, OSTP Issues Guidance to Make Federally Funded Research Freely Available Without Delay | OSTP, White House, 2022 (WWW Document) https://www.whitehouse.gov/ostp/news-updates/2022/08/25/ostp-issues-guidance-to-make-federally-funded-research-freely-available-without-delay/. (Accessed 21 March 2023).

[68] Z. Zhang, K. Hernandez, J. Savage, S. Li, D. Miller, S. Agrawal, F. Ortuno, L.M. Staudt, A. Heath, R.L. Grossman, Uniform genomic data analysis in the NCI genomic data commons, Nat. Commun. 12 (2021) 1226, https://doi.org/10.1038/s41467-021-21254-9.

[69] B.M. Knoppers, A. Bernier, S. Bowers, E. Kirby, Open data in the era of the GDPR: lessons from the human cell atlas, Annu. Rev. Genomics Hum. Genet. 24 (2023) null, https://doi.org/10.1146/annurev-genom-101322-113255.

[70] E.W. Clayton, A.M. Tritell, A.M. Thorogood, Avoiding liability and other legal land mines in the evolving genomics landscape, Annu. Rev. Genomics Hum. Genet. 24 (2023) null, https://doi.org/10.1146/annurev-genom-100722-021725.

[71] M. Hudson, N.A. Garrison, R. Sterling, N.R. Caron, K. Fox, J. Yracheta, J. Anderson, P. Wilcox, L. Arbour, A. Brown, M. Taualii, T. Kukutai, R. Haring, B. Te Aika, G.S. Baynam, P.K. Dearden, D. Chagné, R.S. Malhi, I. Garba, N. Tiffin, D. Bolnick, M. Stott, A.K. Rolleston, L.L. Ballantyne, R. Lovett, D. David-Chavez, A. Martinez, A. Sporle, M. Walter, J. Reading, S.R. Carroll, Rights, interests and expectations: indigenous perspectives on unrestricted access to genomic data, Nat. Rev. Genet. 21 (2020) 377–384, https://doi.org/10.1038/s41576-020-0228-x.

[72] J.K. Wagner, J.-H. Yu, D. Fullwiley, C. Moore, J.F. Wilson, M.J. Bamshad, C.D. Royal, Guidelines for genetic ancestry inference created through roundtable discussions, Hum. Genet. Genomics Adv. 4 (2023), https://doi.org/10.1016/j.xhgg.2023.100178.

[73] A.T. Khan, S.M. Gogarten, C.P. McHugh, A.M. Stilp, T. Sofer, M.L. Bowers, Q. Wong, L.A. Cupples, B. Hidalgo, A.D. Johnson, M.-L.N. McDonald, S.T. McGarvey, M.R.G. Taylor, S.M. Fullerton, M.P. Conomos, S.C. Nelson, Recommendations on the use and reporting of race, ethnicity, and ancestry in genetic research: experiences from the NHLBI TOPMed program, Cell Genomics 2 (2022) 100155, https://doi.org/10.1016/j.xgen.2022.100155.

[74] J.C. Lyu, G.K. Luli, Understanding the public discussion about the Centers for Disease Control and Prevention during the COVID-19 pandemic using twitter data: text mining analysis study, J. Med. Internet Res. 23 (2021) e25108, https://doi.org/10.2196/25108.

[75] W.-Y.S. Chou, A. Oh, W.M.P. Klein, Addressing health-related misinformation on social media, JAMA 320 (2018) 2417–2418, https://doi.org/10.1001/jama.2018.16865.

[76] S. Vosoughi, D. Roy, S. Aral, The spread of true and false news online, Science 359 (2018) 1146–1151, https://doi.org/10.1126/science.aap9559.

[77] D.A. Broniatowski, A.M. Jamison, S. Qi, L. AlKulaib, T. Chen, A. Benton, S.C. Quinn, M. Dredze, Weaponized health communication: twitter bots and Russian trolls amplify the vaccine debate, Am. J. Public Health 108 (2018) 1378–1384, https://doi.org/10.2105/AJPH.2018.304567.

[78] J. Ji, M. Robbins, J.D. Featherstone, C. Calabrese, G.A. Barnett, Comparison of public discussions of gene editing on social media between the United States and China, PLoS One 17 (2022) e0267406, https://doi.org/10.1371/journal.pone.0267406.

[79] C. Ni, Z. Wan, C. Yan, Y. Liu, E.W. Clayton, B. Malin, Z. Yin, The public perception of the #gene EditedBabies event across multiple social media platforms: observational study, J. Med. Internet Res. 24 (2022) e31687, https://doi.org/10.2196/31687.

[80] C.G. Allen, B. Andersen, M.J. Khoury, M.C. Roberts, Current social media conversations about genetics and genomics in health: a twitter-based analysis, Public Health Genomics 21 (2018) 93–99, https://doi.org/10.1159/000494381.

[81] C.H. Basch, G.C. Hillyer, L. Samuel, E. Datuowei, B. Cohn, Direct-to-consumer genetic testing in the news: a descriptive analysis, J. Community Genet. 14 (2023) 63–69, https://doi.org/10.1007/s12687-022-00613-z.

[82] J.S. Roberts, M.C. Gornick, D.A. Carere, W.R. Uhlmann, M.T. Ruffin, R.C. Green, Direct-to-consumer genetic testing: user motivations, decision making, and perceived utility of results, Public Health Genomics 20 (2017) 36–45, https://doi.org/10.1159/000455006.

[83] M. Smith, S. Miller, A principled approach to cross-sector genomic data access, Bioethics 35 (2021) 779–786, https://doi.org/10.1111/bioe.12919.

[84] G.L. Ruhl, J.W. Hazel, E.W. Clayton, B.A. Malin, Public attitudes toward direct to consumer genetic testing, in: AMIA Annu. Symp. Proc. AMIA Symp. 2019, 2019, pp. 774–783.
[85] S. Zhang, When a DNA Test Reveals Your Daughter Is Not Your Biological Child, The Atlantic, 2018 (WWW Document) https://www.theatlantic.com/science/archive/2018/10/dna-test-divorce/571684/. (Accessed 10 March 2023).
[86] S. Zhang, Is DNA Left on Envelopes Fair Game for Testing?, The Atlantic, 2019 (WWW Document) https://www.theatlantic.com/science/archive/2019/03/dna-tests-for-envelopes-have-a-price/583636/. (Accessed 10 March 2023).
[87] S. Zhang, How a Tiny Website Became the Police's Go-To Genealogy Database, The Atlantic, 2018 (WWW Document) https://www.theatlantic.com/science/archive/2018/06/gedmatch-police-genealogy-database/561695/. (Accessed 10 March 2023).
[88] NOT-OD-22-055: FY 2022, Updated Guidance: Requirement for Instruction in the Responsible Conduct of Research, 2022 (WWW Document) https://grants.nih.gov/grants/guide/notice-files/NOT-OD-22-055.html. (Accessed 13 March 2023).
[89] AI Application ChatGPT Temporarily Banned in Italy over Data Collection Concerns, 2023. https://www.cbc.ca/news/world/italy-openai-chatgpt-ban-1.6797963.

CHAPTER 2

Unveiling the hidden curriculum: Developing rigor and reproducibility values through teaching and mentorship

Marina E. Tourlakis
Biology Department, University of the Fraser Valley, Abbotsford, BC, Canada

Introduction

Dr. John Ioannides' 2005 article arguing that much of the published science record was fit for the trash heap [1] set off a major cascade of introspection, finger pointing, counter arguments, and calls for reform [2] (see also Chapter 1). The question of how to maintain the integrity of the scientific enterprise is not a new one [3], nor have we lacked guidelines on how to conduct ourselves as scientists [4–6]. Nevertheless, Ioannides' study and several that followed (in addition to a variety of other alarm bells rung in academia [2,7–11]) demonstrate that revisiting how we train scientists is warranted.

Many institutions include graduate courses and seminars on academic integrity and the responsible conduct of research (RCR) in line with regulations from government and funding bodies [6,12–14]. However, programs are inconsistent and (ironically) lack rigor in their structure, communication, and application [13], and there is little evidence of their efficacy [12,15]. They typically consist of singular offerings despite calls to integrate RCR training across the curriculum [5,16]. Standalone offerings do not support long-term learning gains and can signal low importance [17,18]. Most importantly, this training is typically aimed at postgraduates, despite the increasing involvement of undergraduates in research [17]. RCR training should feature early and regularly in a scientists' training.

It is possible that training scientists to be rigorous and responsible practitioners is taken for granted, part of the "hidden curriculum"—if you made it to the PI position, you must be doing it right. However, the reproducibility crisis tells us otherwise. To address concerns regarding a lack of rigor in how research in genetics and genomics is conducted and communicated, we must make visible the "hidden curriculum" by providing overt and explicit training in RCR [19] in the classroom and the research lab. We are all responsible for supporting this culture change [13,20] (Fig. 2.1).

This chapter builds upon the problems and potential solutions discussed in Chapter 1, focusing primarily on the responsibilities of supervisors and trainees in genetics and genomics classrooms and research labs. These venues are where socialization of trainees into the scientific enterprise occurs. Expectations are set, behaviors developed, and reward (and punishment) systems established. Topics, activities, and strategies are presented for cultivating a culture of rigor. The perspective presented in this chapter is informed by a review of the literature from the past 20 years, focusing primarily on the last decade, on the

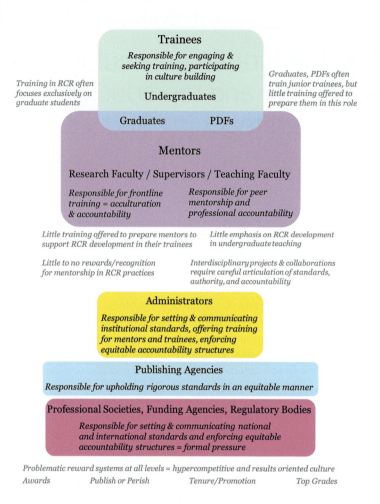

FIG. 2.1

Responsibility for RCR culture. All members of the academic and research community are responsible for supporting RCR training through appropriate engagement and upholding accountability. *PDFs,* post-doctoral fellows.

subjects of mentoring undergraduates and postgraduates (multiple disciplines), undergraduate teaching, learning and research experiences (primarily in bio/med disciplines), and responsible research practices in the sciences, as well as the author's own experiences as a trainee, educator, and mentor.

In the classroom

Science education has been subjected to considerable scrutiny in the past three decades. Prominent scientific societies have advocated for the relevance of science education for every member of society and called for a radical rethink of how we teach sciences [21–23]. Biology, and, by extension, genetics education has enjoyed a virtual explosion of research in teaching and learning in recent years [23] (Box 2.1).

Box 2.1 Professional development resources for teaching and learning biology

Teacher Training
- HHMI Summer Institutes on Scientific Teaching
- IRACDA (Institutional Research and Academic Career Development Awards program)
- CIRTL (Center for the Integration of Research, Teaching and Learning)
- MERIT (McMaster Education Research, Innovation and Theory program)
- PALM (promoting active learning and mentoring)

Communities of practice:
Communities of practice or faculty learning communities are an established mechanism for communal professional development [24]. Create a community of practice at your home institution or in your region to discuss, collaborate, and commiserate on your efforts to develop rigor and reproducibility values and competencies in the classroom.

For evaluating your teaching practice:
- The Classroom Observation Protocol for Undergraduate STEM
- Generalized Observation and Reflection Platform

Online teaching resources:
- Learn.Genetics and Teach.Genetics
- CourseSource.org
- Cell Collective
- HHMI Biointeractive
- Genomics Education Partnership
- Science Education
- National Center for Case Study Teaching in Science
- AAAS Vision & Change in Undergraduate Biology Education

Academic integrity focused organizations:
- Quality Assurance Agency (for Higher Education) Academic Integrity Advisory Group (UK)
- International Center for Academic Integrity

Examples of biology (or science) education journals: *CBE-Life Sciences Education, Frontiers in Education, Journal of Microbiology and Biology Education, American Biology Teacher, Science Education, Advances in Physiology Education, Cell Biology Education, International Journal of STEM Education*

Specific articles on teaching genetics:
- Redfield, "Why do we have to learn this stuff?" – A new genetics for 21st century students. Teach the Genetics of Our Time. *Plos Biology* 2012
- Radick, Teach students the biology of their time. *Nature News* 2016
- Plunkett-Rondeau et al., Training future physicians in the era of genomic medicine: trends in undergraduate medical genetics education. *Genetics in Medicine* 2015
- Series of Q&As with genetics educators in Volume 34 (Issues 1–5) of *Trends in Genetics* 2018
- Todd and Romine, The Learning Loss Effect in Genetics: What Ideas Do Students Retain or Lose after Instruction? *CBE-Life Sciences Education* 2018
- Hales, Signaling Inclusivity in Undergraduate Biology Courses through Deliberate Framing of Genetics Topics Relevant to Gender Identity, Disability and Race. *CBE-Life Sciences Education* 2020

Biology education societies and conferences: Society for the Advancement of Biology Education Research, The Western Conference on Science Education, Royal Canadian Institute for Science

Despite calls for change, undergraduate biology classrooms in many institutions persist in following the traditional structure of lecture and cookbook labs (students reproduce classic experiments with known outcomes). This focus on content mastery and memorization does not paint an accurate picture of science [25] nor adequately prepare learners for the rigors of research. By focusing on facts, we overlook (and even discourage) developing the rigorous thinking habits we expect from scientists: how did we arrive at this fact? how was confidence in this fact established? what did we think before this, and are there other theories? and why do we even care to know this? This training in scientific thinking continues to primarily occur at the postgraduate level.

In particular, biology education research supports the implementation of scientific teaching pedagogies [23,26] and authentic research experiences (AREs) at the undergraduate level [27–29]. Scientific teaching asks the educator to iteratively evaluate and revise lessons toward establishing an evidence-based teaching practice [23,26]. It takes a bottom-up approach that considers students' prior knowledge and tailors the approach to support students' progress toward achieving learning outcomes [28,30] (Fig. 2.2).

If we imbue our genetics and genomics undergraduate classrooms with scientific thinking and offer students opportunities to engage in research, we develop the habits of mind and professionalism that we expect from trainees and future scientists [23,28,31,32]. Integrating activities throughout the curriculum with explicit framing as "research integrity" or "professional conduct in science" can foster a community of learning that better reflects the scientific enterprise earlier in the training process. Moreover, providing a curricular framework can allow us to codify, regulate, and evaluate instruction in RCR (Fig. 2.2). Meanwhile, the collaborative learning environments, clear communication of expectations, and overt instruction on scholarly norms that student-centered teaching practices afford dovetail

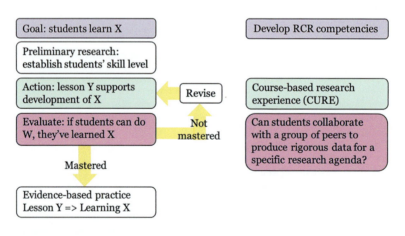

FIG. 2.2

Evidence-based pedagogies to support RCR training. Scientific teaching works toward establishing an evidence-based practice, in other words, a rigorous and reproducible practice. Similarly, authentic research experiences, particularly CUREs provide students with opportunities to wrestle with the realities of open-ended inquiry in collaboration with peers in a relatively low-stakes context. CUREs thus provide experiential learning opportunities not only to develop RCR competencies, but they contextualize these skill sets rather than making them a tired list of rules one is meant to know and follow. Besides being effective in terms of learning gains, these evidence-based pedagogical approaches allow instructors to model in their teaching practice the rigor they expect of their trainees.

with recommendations from research on academic dishonesty and contract cheating [13,19,33] as well as research on improving recruitment and retention of diverse learners to our field [34–36].

In practice

How then, does one apply best practices to integrate RCR instruction in genetics and genomics teaching contexts?

No single course can develop the complete set of skills and knowledge base required to cultivate a value for rigor and RCR competencies. Instead, all courses should include explicit RCR learning outcomes (Box 2.2).

Box 2.2 Applying a scientific teaching approach to your course (re)design

If we wish to apply the scientific teaching approach of backwards design to our curriculum (re)design, we need to contemplate a series of questions (adapted from [22,26]):

1. What are the learning goals of the lesson/course/curriculum?
2. What knowledge and skills are relevant to the subject area?
3. What should students know and be able to do at the end of the unit/course?

For example, we want to develop trainees who:
- *understand the importance of careful and honest data collection and reporting*
- *can explain sources of false positives and false negatives in an RNA-seq experiment*
- *can explain the concept of falsifiability and apply it to a specific experiment*, etc.

1. What evidence would suffice to indicate learning, i.e., what are the learning outcomes of the lesson/course/curriculum? (should be discrete, concrete, and measurable)
2. What do proficiency and mastery in the subject area at this level (intro course vs senior seminar) in the curriculum look like?
3. What evidence would I accept that a student has achieved proficiency or mastery across the relevant content and skills identified in (1)?
4. What evidence would convince my colleagues?

For example, at the end of this course, students will be able to:
- Introductory level:
- *Explain and provide examples of the scientific method*
- *Draft basic scientific reports that pass a replicability rubric: provide a thorough accounting of their research methods (e.g., what search terms and filters were used with what databases to conduct literature review) and their experimental methods (if applicable) and what papers and data were included and which were not.*
- *Distinguish between error and misconduct in research and provide specific examples to support their argument.*
- *Define and distinguish between conflicts of interest, commitment and trust using specific examples to support their argument.*
- *Define responsible research practices and explain their importance to the scientific literature.*
- Advanced level:
- *Critique published literature using a replicability and reproducibility rubric: how did the authors address bias, did the authors consider power and effect size, did the authors provide a thorough accounting of their methods to facilitate replicability, do the authors discuss what data was included and what was not, does the study include any enhanced research standards, e.g. was it pre-registered, do the authors provide open access to their data collections?*

Continued

> **Box 2.2 Applying a scientific teaching approach to your course (re)design—cont'd**
> - *Discuss epistemologies in science including the use of proof, the use of statistics, and the scientific method using historical and contemporary contexts.*
> - *Design experiments and defend their design using a rigor framework: demonstrate how they are addressing bias, how they've considered power and effect size, how they are ensuring consistency in technique, how they will report their findings.*
>
> 1. What learning activities will support students to meet goals?
> 2. What experiential learning supports developing understanding and competency in content and skills identified in (1)?
> 3. What topics should be covered to support knowledge and skill development?
> 4. What skills should be overtly taught and/or scaffolded?
> 5. What types of disciplinary practice should be overtly taught and/or scaffolded?
> 6. What feedback will support progress toward learning goals and when should it be given?
>
> *For example, students will engage in an inquiry based research project to measure allele frequency for a heritable and easily scorable trait in* Drosophila *over multiple generations.*
>
> For an excellent example of designing RCR learning goals, outcomes, activities and assessments, see Diaz-Martinez et al. [17].
>
> For specific recommendations on how to implement an RCR conceptual framework to your course design, see Clements et al. [12].

1. Make explicit the expectations of your classroom, the expectations scientists have of scientists, society has of scientists, and why this is so.
2. Support students with frequent feedback and mentorship to meet goals you set for them.
3. Evaluate whether your teaching activities are having the desired outcome.

The stakes: Responsible research practices, ethics, and societal context

The first generation learner in their first science classroom may be hearing the words "rigor" and "reproducibility" for the first time. Explicit discussion of the stakes contributes to students' abilities to understand the societal context of the scientific enterprise [22] (Box 2.3).

To introduce RCR in the classroom, you can include papers in your reading list that specifically address RCR [1,5,37,38]. Structure a discussion on why reproducibility can be challenging. Make clear that researchers need to work hard to be rigorous; a well-intentioned researcher can still produce sloppy science if they lack technical knowledge, do not thoroughly consider sources of bias, lack strong communication skills or a strong knowledge of statistics. Note how culture contributes, and how we learn from mistakes made.

> **Box 2.3 Specific Lesson Idea: Incidental findings and the duty to report**
> **Learn Goals:** *illustrate a relatable example of how excellent record keeping impacts real people.*
> **Student Learning Outcomes:** *students will be able to (1) explain why risk assessment of variants changes over time using specific examples, (2) defend a position on long-term data storage policy in context of reporting incidental findings.*
> **Suggested plan**: *assign policy statements on incidental findings and use case studies to illustrate considerations and challenges associated with reporting incidental findings over short and long-time frames.*

The stakes can be articulated at 3 levels:

1. **Personal** sloppy work will not serve the individual scientist well in the long run and models bad behaviors to the next generation of scientists.
2. **Local** science is a cumulative endeavor; it is critical that the foundation on which others build their work is sound.
3. **Societal** society depends on sound science: it provides evidence for decision making at the individual to government levels and fosters wonder in the world around us; trust in the scientific enterprise is needed to support these roles (from *On Being a Scientist*, pp. 2–3 [5]).

Importantly, make explicit the alignment of these stakes with behaviors and skills you value in your course assessments including good citation practices, academic integrity, strong communications skills, ethics, etc.

The nature of science

Many undergraduate students view science as an enterprise based on absolute answers, conducted by objective scientists. When we teach science as a series of memorizable facts and following protocols toward known outcomes (i.e., cookbook labs) we feed into this falsehood. Instead, show students that revision and correction are the norm and that resiliency in the face of repeated failure are fundamental traits of the successful scientist [31,39].

In the genetics classroom, provide a more accurate representation of science by featuring dogmatic revolutions (retroviruses, internal ribosome entry sites, etc.), calling attention to how theories evolved based on new evidence, rigorous testing, and skepticism. In the teaching lab, opt for open-ended, inquiry-based lab activities (see *Experiential Learning and CUREs* below).

Teaching the true nature of science will better prepare students for the rigors of research and foster responsible consumption of scientific information in the media [40,41] (Box 2.4).

Science communication, publishing and peer review

Developing a value for RCR practices is incomplete without a thorough examination of communication practices in science. Explicit instruction in best practices for sharing scientific findings and how sharing of scientific findings is the primary mechanism for correction, progress, and reward in science [2,43] are necessary components of a genetics curriculum.

POP OUT (Quote): "Instances in which scientists detect and address flaws in work constitute evidence of success, not failure, because they demonstrate the underlying protective mechanisms of science at work." Alberts et al. [43].

Box 2.4 Specific Lesson Idea: Multiple hypotheses

Learning goals: *introduce science philosophy and develop a value for rigorous hypothesis testing.*
Student learning outcome: *students will be able to connect evidence with claims in context of hypothesis testing.*
Suggested plan: *assign Platt's Strong Inference essay [42] as a first day reading for a lab course. Assign questions in prelabs referring to this text and prompting students to generate multiple hypotheses and what evidence they'd accept as refuting each.*

Providing explicit instruction in how to engage with scientific literature is a must in any introductory science course. Foster critical reading skills by devoting lecture time to a tour of a scientific paper, explaining the author list (and decision-making thereof), the different components of the paper and their ordering, citation of figures and tables in the text, the acknowledgement and disclosure statements, and the reference list. We do a disservice to trainees if we assume they enter the postsecondary classroom with this knowledge.

Similarly, provide multiple opportunities for students to develop and practice science communication (written, visual, quantitative, oral). Explain to students that producing coherent data presentations and drafting thorough methods sections in their lab reports develop skills scientists need to support their claims. Provide students with a target audience and specific genre-exemplars (research article, perspective, review) to guide their rhetoric [44]. And use draft revision cycles to emulate peer review and foster a value for iterative improvement post feedback (Box 2.5).

Peer review also opens the door to learning about the sociocultural contexts of errors in the literature and academic misconduct. Engage students with case studies [45] that discuss famous errors or retractions, or have students research the history, motivation, and effectiveness of preprint servers [43] (Box 2.6).

A note on narrative: motivated by the rise of anti-science rhetoric (e.g., climate-change denial, anti-vaxxers), and an impression that scientists are poor communicators, many scientists are encouraged to employ narrative tools to communicate their work to the broader public. This could form the basis of a vigorous discussion with students: debating the merit of using narrative as a science communication tool. Possible sources to inform such an activity:

- Dahlstrom and Scheufele's (Escaping) the paradox of scientific storytelling [46]
- Dahlstrom's Using narratives and storytelling to communicate science with non-expert audiences [47]
- NPR Scicommers
- ASAP Science
- Science Sam

Box 2.5 Specific Lesson Idea: Critical reading of literature

Learning goals: *orient students in the reading and critique of literature in molecular genetics AND practice writing in this genre.*

Student learning outcome: *students will be able to name the purpose of the structural components of a research article in the discipline of molecular genetics, and emulate this in their own writing.*

Scaffolded activities:
(1) *students work in small groups to interrogate the rhetorical structure of a research article and articulate the purpose of each section*
(2) *students emulate the style in their own writing of a scientific report based on a provided context and data set,*
(3) *students engage in a focused peer review and reflect on how their writing style was impacted by their rhetorical analysis and peer review feedback.*
(4) *instructor-facilitated debrief with the class to highlight key takeaways regarding the genre's rhetorical structure, the utility of that structure for a) communicating rigorously and b) facilitating replication, and any process takeaways for students (this activity helped me recognize, challenged me to, introduced me to, etc.).*

> **Box 2.6 Preprint servers as educational tools**
>
> The topic of peer review can be used in the classroom to develop an understanding of the process of science while critically engaging with literature. In preparation for a class on peer review, scientific method or other aspects of the process of science, set students a homework task to visit the website of a preprint server (e.g., bioRxiv, medRxiv, arXiv) and their corresponding Wikipedia pages. Ask them to prepare brief answers to the questions,
> - What problem(s) do pre-print servers aim to address and how?
> - Based on your preliminary investigations, is there evidence that they are achieving their aim(s)?
>
> Have students bring their answers to class to contribute to small group discussions. For a larger class, have students submit one question the activity raised on the subject of rigor in science to a question bank prior to class which you can then draw from to inform your large group discussion. To follow the class discussion, have students submit a brief reflection answering the prompt: Reflecting on your ideas of how science proceeds, did pre-print servers confirm or challenge your prior notions?
>
> *For a more advanced course, consider the following essay prompt: The essay* On Being a Scientist [5] *discusses the importance of not releasing results prior to them being peer-reviewed. The recent surge of pre-print servers appears to subvert this cultural expectation. Further, during the COVID-19 pandemic, it became commonplace to read the phrase "These findings have not been peer-reviewed" in media reporting science for public and expert audiences. How do traditional and contemporary peer review structures support or hinder the progress of science during an urgent public health crisis?*

Quantitative reasoning and analysis

Quantitative reasoning (QR) is a foundational skill for success in genetics and genomics [22]. Increasingly, there are calls to integrate QR and mathematics into the biology classroom rather than relying on prerequisites [22].

As genetics and genomics instructors, focus on the most common types of quantitative analyses in our field (e.g., William Noble provides a great primer on multiple-test correction) and how these connect with your course's goals. Should students memorize an equation? Should students be able to explain in words (an important component of QR) what a test compares and what information it yields about the data, or the claims the data is being used to test? Perhaps, they should be able to select an appropriate statistical test and explain their reasoning.

Connect lessons to RCR by using real data examples to show how confidence is built and how it might be abused (see here for a great reading). Helping students understand the meaning behind equations and statistics can help build a healthy skepticism of data and data presentations, which is key to understanding issues of reproducibility.

It is crucial to establish the prior knowledge and capabilities of your students and, where possible, tailor your lessons or tutorial support to meet them appropriately. Provide many opportunities for practice and multiple examples. To help make expectations clear, use a QR Rubric (see here for an example).

Beyond teaching quantitative analysis, strive to engage students in QR in every class: ask them to explain a graph to the student sitting next to them, or what an odds ratio means, or *what exactly a frequency of 0.27 means* (Box 2.7).

Sloppy science and academic misconduct

Consistent with the call to be overt and explicit in RCR training, we must engage with missteps in the scientific record. Several examples of published and retracted papers in the literature afford the

> **Box 2.7 Specific Lesson Idea: Why do GWAS studies require a replication study?**
> **Learning goal:** *illustrate the problem of false associations in genome-wide research and introduce methods for multiple test correction.*
> **Student learning outcome:** *students will be able to explain the quantitative reasoning for replication populations in GWAS studies using probability and false discovery concepts.*
> **Suggested plan**: *assign a GWAS study and work with students to answer the following: what is the purpose of a replication population? what, if any, criteria differ between the original and replication population, and why? what is false discovery and why does it happen in GWAS studies? What are the stakes? How is an FDR established and what are permissible levels and why?*

instructor contexts for teaching biological concepts while also making explicit **expectations for RCR**, **mechanisms for correction**, and **consequences for academic misconduct**.

Specific cases can be used to illustrate lapses in rigor and to discuss peer review processes, mentorship, retraction procedures, and the impact of academic misconduct on trainees (refer to [43,48,49]) (Table 2.1; Box 2.8).

Experiential learning and CUREs

Experiential learning (EL) is a crucial component of RCR training. Virtually, all undergraduate biology courses include some EL component, most typically in a laboratory setting. Labs offer an invaluable mechanism for demonstrating the fallibility of experimental sciences, the importance of reference measures, consistent technique, variables, sources of error, the concept of falsifiability, and multiple hypotheses, and the importance of rigorous reporting (Box 2.9).

As noted earlier, cookbook labs continue to feature prominently in biology courses [27,34]. Cookbook labs have been derided for their rigid structure and confirmatory approaches which contribute to a notion that science is about collecting evidence to fit your hypothesis [21,22]. This mainstay of the science curriculum can instill in students a sense of failure when the evidence they collect does not line up with what the lab manual tells them [26]. What is more, it does not afford opportunities to foster creativity and flexibility, necessary attributes of scientists [30]. Therefore, biology educators are mobilizing to increase the prevalence of inquiry-based labs in biology curricula [27] (Box 2.10).

CUREs are rich with RCR training opportunities [52–54]:

- Students struggle productively and receive an authentic experience of how science progresses through discovery, iteration, collaboration [31,32,55]

Table 2.1 Student accessible readings on reproducibility.

D. Butler, Biomedical researchers lax in checking for imposter cell lines. *Nature News* 12 October 2015

Journals unite for reproducibility. *Nature Editorial* 5 November 2014

L.P. Freedman, et al. Reproducibility: changing the policies and culture of cell line authentication. *Nature Methods* 12(6) June 2015 493–497

Challenges in irreproducible research *Nature Special* collection of commentary and articles

D.B. Allison et al., Reproducibility: A tragedy of errors. *Nature Comment*. 3 February 2016

Kimmelman et al., Distinguishing between Exploratory and Confirmatory Preclinical Research Will Improve Translation. *PLOS Biology Perspective* 20 May 2014

> **Box 2.8 Case-based learning**
>
> Case studies are a powerful tool in education as they situate abstract concepts in specific contexts through storytelling, discussion, and role play. Case studies can be used to make RCR issues come alive for trainees. They can be as simple as reviewing a specific study in one lecture period with guided questions to foster concept mastery and discussion, or, they can be multi-day forays into an issue. Generally, they work best if tackled within small groups. The National Centre for Case Study Teaching in Science is an excellent resource for a variety of cases, many of which can be pitched to different levels and adapted to one or multiple lecture periods. NB: to maximize relatability, make sure to include diverse identities in the Case Study.
>
> Examples of famous retractions that can be investigated in a case study format:
> 1998: Wakefield and a link between MMR and autism.
> 2010: The #arseniclife affair (see ref. [45] for a case study on this topic).
> 2014: Obakata stem cell scandal and tragedy.
> 2020: AlShebli et al., mentoring study.

- Students wrestle with reproducibility by embedding iteration in their design [56]
- Students develop a sense of ownership and identity as a scientist due to a perception of contributing to bonafide research projects [56]
- Students develop an appreciation for collaboration due to the multi-site aspect of many CUREs [27]
- Students develop RCR practices due to the skills required for collaboration, e.g., careful and detailed record keeping, frequent discussion of process and progress, and peer review

> **Box 2.9 Cultivating a culture of rigor in the teaching lab**
> - Promote patience and attention to detail as key qualities of the scientist: tell students that science requires a cool head. If they find themselves too excited in an inspired or anxious way, they should take a 5-min break to cool down.
> - Cultivate strong observational skills: train students to use their eyes, ears, touch and nose (not taste!), and not just rely on measurements.
>
> *Case in point: students in an introductory biology courses were exposing beet root samples to various solvents to investigate the chemical makeup of membranes. They measured betacyanin leakage into the solvent as a proxy for membrane integrity using a spectrophotometer. A group of students had the misfortune of using a faulty instrument. In their report, the group noted their measurements did not make sense and they pronounced the experiment a failure. However, photographs they had taken during the exercise revealed betacyanin leakage did indeed correlate with solvent concentration. This prompted me to make explicit to students that they reconcile their qualitative and quantitative observations in their reports. One of the many times I have realized that what might seem obvious to me is not necessarily so to the novice scientist.*
>
> - Prioritize rigorous practices in your assessments: **drop the long form lab report** and have students utilize evidence-based writing-to-learn practices to develop understanding and scientific writing skills [44]. For example, have students focus solely on drafting data presentations, using genre exemplars from the literature, iterative attempts, and feedback, to master data analysis and data presentations. See this book's DataVerse for a sample short form lab report (https://dataverse.scholarsportal.info/dataverse/RandRinGenomics).
> - Provide overt and explicit instruction on record keeping (i.e., lab notebooks), including formative and summative assessment thereof.
>
> Importantly, we need to recognize that the bulk of teaching in the undergraduate lab is conducted by trainees (i.e., teaching assistants). Teaching faculty must commit to training TAs on how to support RCR practices [50,51].

> **Box 2.10 A quick guide to converting a cookbook lab into a basic inquiry-based lab**
>
> 1. Review your established protocol and determine which components are flexible (and safe) for students to modify
> 2. Articulate a specific learning outcome for the lab and a specific objective "In this lab, we will investigate how X impacts Y"
> 3. Make a list of available reagents and instruments, including safety precautions and constraints (e.g., give students a range of substrate amount to use in a reaction, or a range of time to expose samples to a stimulus).
> 4. Task students (in small groups) to complete a prelab that they will discuss with their TA/instructor prior to commencing their work. This discussion is a key step to correct for concept errors while engaging students on the subject of experimental design, e.g., no, you cannot reasonably achieve this during our lab period, or yes, that could be a source of false positives. Prelab questions:
> 5. What is your research question and hypothesis, including rationale thereof?
> 6. What is your strategy for collecting sufficient data to build confidence in your result?
> 7. What controls will you include in your experiment? What could give you a false positive or false negative?
> 8. What is your proposed protocol? Specify reagent amounts and incubation times, as well as how many biological replicates and technical replicates you'll employ.
>
> This strategy works best if you have static groups and 3 or more labs of increasing complexity for students to practice designing experiments and iteratively improve.
>
> *Example:* convert a PCR-based lab by allowing students to determine primer, template and Mg^{2+} amounts and/or template sources. Importantly, can included design aspects in the prelab that you then keep firm in the actual lab (e.g., annealing and extension time could be part of the design but for logistical reasons be constrained to one protocol during the lab)

Further, evidence is building that CUREs play an important role in recruitment and retention of diverse learners to STEM fields [54] and provide a more equitable structure for developing tacit knowledge in how to do research [27,29], particularly as students from historically underrepresented groups are less likely to land a position in a research lab, or even seek one [27,28,35]. CUREs offer an economical avenue to provide virtually all undergraduate students with an opportunity to develop research acumen [31]. For more on the benefits of CUREs and how they can support genetics and genomics teaching, see Chapter 7.

It is important to note that labs and CUREs typically operate as group projects. Group work is an important venue for developing RCR behaviors. The skills developed during group work can be leveraged during future difficult conversations with lab members or supervisors. For excellent resources on navigating group work dynamics, see here.

In advising

Undervalued or at best newly valued, academic advising is an increasingly prominent aspect of the undergraduate experience [57], especially in honors programs [58]. Teaching faculty engage in some form of advising during office hours, informally after class, and sometimes by e-mail. Many programs assign all undergrads to a faculty advisor, but advising remains a component of the faculty members portfolio that has little structure, training, oversight or recognition, despite being increasingly associated with student success [57]. These often close relationships with faculty members provide a venue where soft skill development, culture, values, and habits of mind are formed [57]. Setting goals, reflecting on if/how those goals were met, and ultimately connecting evidence (study habits, performance) with outcomes, we develop habits in learners that bear fruit in the classroom, lab, and beyond. By making the implicit explicit in discussing our own training and work experience, we establish values and expectations early. Further, these interactions help trainees develop interpersonal skills with faculty that will prepare them for positive and negative interactions in their later training and careers. In essence, advising is a form of mentoring and should be approached with an intention to foster specific outcomes, in our case, a value for rigor (Fig. 2.3).

In the research lab

The research lab and its immediate environment (department, university, research institute) is where the stated causes of the reproducibility crisis manifest: poor experimental design [1], communication breakdown [59,60], problematic reward systems [61,62], dismal career prospects, and academic misconduct [13]. Moreover, these causes often converge on issues of mentorship; mentoring networks

Set structured expectations

Advisors are expected to:
- Establish benchmarks for goal meeting,
- Socialize advisees by talking about their career trajectory and/or daily activities,
- Help advisee make meaning of their grades and academic performance, encouraging strong but sustainable study habits,
- Assess goal progress regularly and encourage revision as needed, and reflection on why things went well, or why they didn't
- Model honest, collaborative and respectful dialogue

Advisees are expected to:
- Set an agenda for all meetings,
- Establish professional goals
  short term small goals, e.g., term paper
  medium term larger goals, e.g. graduating
  long term large goals, e.g. career
- Reflect on courses, internships, etc. with regular frequency to identify strengths, growth areas and connect to goals.

Start an advising record

✓ Draft a two-column table:
  column A = *mentoring activity*
  column B = *mentoring goal*
✓ Populate each column with a list of activities and goals (include realized and aspirational items)
✓ Make connections between activities and goals (use a different colour to draw connecting lines)
✓ Keep the table handy and revisit it daily, weekly or monthly for a set period

FIG. 2.3

Simple activities to cultivate habits of mind consistent with rigorous practice in the advising relationship. Interested in improving or professionalizing advising at your institution? A longstanding undervaluing of advising and mentorship might be attributed to the difficulty in "measuring" the work and the impact of the work. Creating structures and accountability can formalize the work and make it measurable. This exercise has the added bonus of providing participants with material to include in their next grant, promotion documents, or job application. This exercise can also be adapted to use with trainees, either asking them to reflect on mentorship they provide to junior members or asking them to reflect on what support they expect and need from their advisor/mentor and what activities they notice.

play a major role in socializing the next generation of scientists [20]. As such, mentoring features prominently in recommendations for improving the quality of research [63] and has become a major focus of higher education literature and university administrations.

Who is a mentor? Mentors are typically more senior to their mentee and serve as a source of knowledge, inspiration, emotional support, and networking. Mentorship goes beyond co-authorship or collaboration or who happens to be your assigned supervisor. A mentor is someone who has tacit knowledge of a system and *uses this knowledge to help a novice navigate that system.* In academia, many individuals serve as mentors and individuals at all levels of career should seek multiple mentors to support different components of their professional identity (Box 2.11).

In addition to mentorship from senior staff/supervisors, peer mentoring plays a critical role [7,65], particularly in retention of persons from historically underrepresented groups who are less likely to have a supervisor in tune with their needs [64]. In the trainee context, mentorship of junior trainees (undergrads) often falls to senior *trainees* (postgrads) [66]. Interestingly, the triad structure of faculty-postgrad-undergrad, provided there is direct and frequent interaction between faculty and undergrad, outperforms the dyad structure of faculty-undergrad when reporting gains in scientific thinking [7]. The same study suggested undergrads mentored solely by postgrads may develop technical skills but not enjoy the same exposure to the broader context of research and lifestyle of a scientist and thus not relate or be exposed to the realities of a career in science [7]. On the other hand, if trained exclusively be a faculty mentor, undergrads may ask fewer questions and generally be more cautious in their interactions, less willing to demonstrate vulnerability or ignorance for fear of jeopardizing their standing with their supervisor [7].

It follows that we need to support faculty supervisors to be the mentors they want to be, and part of this equation is for faculty to foster peer mentoring networks in a manner that includes their oversight and direct involvement [65]. In short, training the mentors, be they faculty, senior staff, or senior trainees, is important [64].

Box 2.11 Developing a mentorship network

Social networks increase the resources available to individuals which in turn help them gain institutional knowledge and confidence in navigating new cultures [64]. For example, a trainee may form a mentoring relationship with their supervisor to support their professional identity as a scientific researcher, but seek out a different mentor to aid them in navigating academia as a Person of Color. Trainees and junior faculty should seek out a network of mentors to support the different facets of their professional identity.

For trainees and junior faculty: seek out mentors who embody your values and inspire you. Remember your official supervisor should not be the only person you seek mentorship from - you can and *should* seek mentorship from individuals at varying levels of seniority, positions, and locales.

It is important to have at least one mentor who you see yourself in. Think about what your professional goals are and seek a mentor who mirrors them back and has ostensibly had a similar personal journey getting there [20].

The journals *Nature* and *Science* have mentoring resources on their career pages that can help you identify the type of mentoring you require and how to get it, including advice on how to man.

For mentors: seek to support your mentees according to your position and lived experience. Be honest about where and how you can help. Set clear boundaries so that mentees have reasonable expectations for the type of support they will receive, and can decide accordingly how to expand their mentorship network.

In practice
Keeping things simple, one can focus on two approaches for mentoring RCR practices:

Habits of mind
The responsible researcher reflects on and interrogates data, but also their own actions and motives. Cultivate a reflective research practice: by continuously asking questions about your groups' process, reviewing the reproducibility literature with your research group, and participating in quality assurance programs, you will be forced to reflect on your own policies and procedures and as a result, perhaps identify spotty practices or weak areas that might leave you vulnerable to an integrity breach [67].

Protocols
Foster rigorous organizational practices. The responsible researcher develops practices to make their research easily replicable. Easy is important here: the more one separates their findings and communication thereof from the process, the more difficult it will be for an independent researcher to replicate and build upon the work and hence establish the reproducibility of the key claims [68].

As in the classroom, take a scientific approach to your mentoring:

1. determine goals,
2. determine what evidence will suffice to demonstrate goals are met,
3. provide training that will support development,
4. assess through observation and feedback from your team and colleagues, and
5. revise as needed.

Importantly, do not let the tenor of RCR training be punitive or judgmental, cultivating a culture of rigor benefits everyone—the habits of mind and protocols will likely result in more efficient work, more rigorous work, and more efficient communication with your team members and collaborators [67,68]. Remember to be explicit, foster honesty, provide training/support, and be kind [69]. The remainder of this section outlines key components of a successful ongoing mentoring program.

Communication of expectations
It is the responsibility of all members of a research group to build a culture of rigor. However, given their authority and power, mentors set the tone and provide leadership, training opportunities, and clear communication of expectations for RCR practices.

A new junior member of a lab is tasked with a lot of 'meaning making'. They need to assimilate information about the specifics of the biology, the technicalities of experiments, the context of their project goals, and how that all fits within the team, the field, and the cultural norms of the team [69]. Mentors can guide meaning making through their actions and words. As a mentor, be sure to include explicit instruction on the expectations of team members and lab standards (Box 2.12).

Be a predictable mentor. Make expectations clear early on [69] (Box 2.13).

Lab meetings and journal clubs
Do not rely on the trainee to ask all the right questions, or for you to have remembered to cover everything in an initial training session. This is particularly relevant for trainees with prior research experience or with interdisciplinary projects where the specific expertise may lie outside the immediate

Box 2.12 Training to emphasize RCR

Explicit training on how to collect data, make measurements, record your procedures and report your data is critical. Take the time to enumerate and communicate to new members of the lab (regardless of their training level) the practices that you do not even think about, having committed to muscle-memory many years ago, and conduct regular peer-peer (or senior peer/mentor-peer) audits of lab notebooks [67].

An excellent way to train that provides the learner with hands-on training in techniques *and* first-hand experience with replication is to task trainees with replicating a key result published by your group [69].

Ground rules:
- safety: workplace hazards, including harassment
- ethics: conflict of interest, commitment or trust, scientists' responsibility to society, human and animal research ethics
- maintaining equipment in good working order

Data collection:
- how *and when* are lab notes recorded? (offering a template or exemplar is important, see here for an excellent resource on lab notebook practices)
- how many technical and biological replicates are needed to build confidence in a result?
- how are new reagents such as antibodies tested for quality assurance?
- how are biological samples such as RNA tested for quality assurance?
- what statistical tests are typical and what error rates/confidence scores acceptable?
- what imaging software do you use?
- what settings (if any) are permissible to manipulate (and to what degree) when acquiring images, and how is this recorded and reported?
- how are mistakes/contaminated reagents/etc. reported?

Data communication:

Engage in ongoing discussions about science communication practices. Take time to highlight rhetoric used in presentations during lab meeting and papers during journal clubs. Debate concerns about "clean stories" in publishing eroding the accuracy of the record [63]. When it comes time to draft a manuscript with a trainee, discuss your process, ask them about their comfort level with writing and data presentation drafting to identify what they will need from you in terms of additional communications training support. Make clear your expectations for writing and timelines, and how you will approach revision cycles.

Collaborations:

Take care to share your data collection and communication standards at the outset of collaborations in a meeting with collaborators, ideally with trainees from all groups in attendance, and follow up in writing. Mentor your trainees in respectful dialogue to navigate disagreements and empower them to speak with you if they have concerns by creating spaces for one-on-one discussions and signaling your confidence in their judgment through actions and words.

Box 2.13 Incorporating RCR into the interview process

Potential supervisor:
- Outline training that new members of the research group should complete prior to engaging fully in lab activities
- Share a "Welcome letter" with all new lab members (regardless of level or position) that outlines expectations, responsibilities of all members of the lab, including the faculty member [70]

Aspiring trainee:
- Ask what type of training will occur (if this is not offered)
- Ask questions about rigor, accountability, and integrity, e.g., at what stage in the research does the supervisor typically seeks to publish a study, how authorship is determined, and who is involved in writing manuscripts.

research group. Lab meetings and journal clubs are great mechanisms for identifying and correcting problematic behaviors.

Lab meetings Here is where team members demonstrate their rigor in record keeping, analysis, communication, and respectful dialogue. It is recommended that lab meetings include presentations that provide the full context (what, why, how) for the data being shared. This serves to develop presentation skills of trainees while also prompting team members (including the supervisor) to recall the key questions and hypotheses at play. Taking it one step further, make it a habit to ask at these meetings "what else might explain what you saw?" to force everyone to always consider other explanations, including but also beyond human or instrument error.

Journal clubs Here is where team members demonstrate their critical thinking and communication skills. Beyond critiquing the rigor and replicability of featured publications keep an eye out for publications on reproducibility or other aspects of research integrity to include featured articles. It is important to not hide these issues away, and instead to discuss them openly, so trainees see your values and your expectations, and have an opportunity to ask questions in a safe space.

Trainee engagement
Trainees in turn must engage meaningfully with training, including asking questions, providing feedback, and likewise modeling appropriate behaviors for junior members (see also [69]) (Box 2.14).

Accountability structures
Surveys of trainees have demonstrated dissatisfaction with current RCR programs in terms of training along with perceptions of inconsistent consequences based on position [71]. To address this within the research space, supervisors can provide documentation in the form of contracts that specify roles and responsibilities of lab members (including those in leadership), non-negotiable practices, resources for support, and consequences for misconduct. Many institutions have such contracts as part of their graduate programs; however, generating a personalized contract that is unique to your circumstances helps communicate your specific values and expectations as the team leader and can include all members of the lab, not just graduate students.

Include in your document portfolio **accepted workflows for research design, file management, data sharing, protocol documentation, reagent and biologicals storage**, etc. Importantly, **set revision dates to ensure that documentation reflects feedback from members** and stays current with

---

Box 2.14 What can trainees do to contribute to RCR practices?
- Review the literature on philosophy of science to refamiliarize oneself with the foundational tenets of science.
- Make time for developing RCR skills (e.g., attend workshops on data management, data presentation, scientific integrity, reproducibility, etc.) and employ tools that can assess methods and reporting quality (e.g., SciScore).
- Model rigorous behaviors: practice excellent record keeping, be a meeting wizard, conduct data analyses promptly, consult with experts (your supervisor, collaborators, statisticians), consider multiple hypotheses, be a critic of your own work and that of your team… but a constructive critic that attacks data and ideas, not people.
- If your institute or department does not offer workshops on RCR practices, ask for them or design one yourself (in consultation with experts)—looks good on your CV too!
- Seek out opportunities to assess rigor (e.g., engage meaningfully in lab meetings, journal clubs)

best practices. Similarly, ensure team members have access to and are aware of your institution's research integrity policy and relevant regulatory bodies in your region.

Finally, create accountability by developing goals *in consultation* with your trainees. Commit to a mutually agreed upon framework and timeline for evaluating benchmarks, including mechanisms for review and revision.

Go further

- *Post a signed contract that stipulates your rights and responsibilities as the team leader in a prominently visible spot in the lab space to demonstrate that cultivating a rigorous research and learning space requires everyone, including the "boss", act with integrity.*
- *Some institutes include a graduate student oath and ceremony that pledges integrity.*
- *Does your institution have a whistleblower policy that applies to research integrity breaches? If not, advocate for one* [16,48].

Administrators, including **deans and department or program chairs**, have an important role to play in supporting RCR by creating institutional structures to foster dialogue and accountability (see Ref [13] for core elements of effective integrity policies). Administrators should delineate when and how formalized discussion of RCR should occur in training. This should include language about RCR in syllabi for any courses or programs that involve research (e.g., honors thesis programs) and directions to faculty on RCR components that must be included at formal checkpoints in the training process. Structures might include (adapted from X):

regular informal drop-in hours for trainees to discuss RCR issues with a program administrator

- a checklist (technical and biological replicates, controls, sample size, etc.) for reporting on RCR practices at each committee meeting
- a diverse group of faculty and/or senior trainees with specific training on conflict management and RCR practices who serve as advisors to trainees
- structured time in each committee meeting for the committee, *sans* supervisor, to have a frank discussion with the trainee

Conflicts of interest or commitment

Create a space that welcomes difficult conversations about conflicts of interest and commitment. We must also consider implicit and explicit biases reflective of individuals' lived experiences that play out in the collection, interpretation and reporting of data, and actively work to recognize, correct, and whenever possible, mitigate the impact of such biases on our work. Make a habit of acknowledging biases you may have to ensure these do not creep off your radar, and to encourage self-checking in your team. And ensure team members have access to and understand your institution's COI policy.

Feedback and reflection

Foster an open and honest conversation through feedback and reflection on your team's research practices. For example, provide specific and targeted feedback to trainees on their progress toward benchmarks and schedule monthly blue-sky lab meetings to discuss research directions. Use the latter as a means to revisit landmark research papers in your field, in reproducibility, or philosophy of science, encouraging everyone to engage in vigorous critique of the scientific process, including your own. Demonstrate your commitment to evidence-based mentoring practices and your openness to correction by incorporating feedback from these sessions into your practice.

What to do if you are worried about integrity in your lab

If you are worried about integrity practices in your lab or research unit, your first move should be to find and review the institutional research integrity policy and identify the appropriate reporting structure. You should also start a written record of dated observations and collect any correspondence relevant to the concern. Human resources departments and ombuds offices can share resources for having difficult conversations. If you are the supervisor, it is likely that you should first discuss the matter with your trainee or staff member. If you are a trainee, it is likely that you will need to identify a trusted mentor to assist you in moving forward. Start familiarizing yourself with policy at your institution and in your region/country [48].

Mentoring for equity, diversity and inclusion (EDI)

It is worth noting the argument that increased regulation and oversight to support RCR can be make-work and inhibit progress and academic freedom [6]. But a lack of policy tends to favor individuals in privileged circumstances and disadvantage those in vulnerable positions. A particular area that needs attention and research is to determine how to support RCR competencies in diverse learners through mentorship. For example, there are unique considerations for supporting international English as a second language (EAL) trainees who tend to be underserved by current support systems for RCR training [71]. Further, there is a growing body of literature focused on how mentoring practices impact retention of diverse individuals to STEM fields [20,28,35]. It is the responsibility of the mentor to engage with the literature on the experiences of historically underrepresented groups in STEM fields [20,35]. Moreover, it is imperative that institutions and their leadership incorporate best practices into their policy frameworks for mentorship and integrity training along with affording professional development opportunities for their research faculty to enact these changes. Part of modeling RCR behaviors includes educating oneself on EDI issues that might impact your mentoring and the research outcomes of your trainees so you can offer appropriately tailored support.

Support for mentorship in RCR practices

Several resources are available, aimed at various levels of engagement, to improve mentoring relationships, and develop RCR practices (see Box 2.15).

A key challenge is that the realities of the contemporary scientific enterprise may preclude the (otherwise) simplest intervention to improve rigor and reproducibility in research: slow down [2,72]. Fast science and pushes from government, funding agencies, and institution executives to grow the graduate student population strain effective mentoring. Effective mentorship requires meaningful interactions between trainees and faculty [65]. Unfortunately, the reality is that supervisors have myriad obligations that can sideline mentoring of junior members. It is the responsibility of the faculty member to recognize when they cannot fulfill this obligation [6] and the responsibility of government, funding agencies, and institutions to support responsible mentorship practices.

> **Box 2.15 Mentoring resources for mentors and mentees**
> - The "Welcome letter"
> - Dr. Tracey Bretag: internationally recognized expert on evidence-based academic and research integrity programming. University of South Australia Business School, Adelaide, Australia
> - National Research Mentoring Network
> - STEMM mentoring
> - UCSF Office of Career and Professional Development
> - Science Careers
>   - The Individual Development Plan platform is an important tool for all postgraduates. Mentors can also use as a framework to support RCR. Trainees can use this as a conversation starter about the mentoring they need, or to broach difficult conversations on RCR.
>   - Articles provide advice for scientists at many stages of career and de-stigmatize difficult topics, e.g., Navigating conflict with your supervisor (2020)
> - Nature's Mentoring Collection
>   - Notable posts include "Why you need an agenda for meetings with your principal investigator" (2018) and "Great mentoring is key for the next generation of scientists" (2019)
> - Gandrud, C. Reproducible Research with R and RStudio (2020)
>   - a handbook for establishing R and RStudio workflows
> - Responsible Science: Ensuring the Integrity of the Research Process Volume I NSF 1992 (especially Chapters 2, 6, and 7)
>   - Includes definitions, historical context, and recommendations for teaching, mentoring, institutional policies to cultivate rigorous research practices and advocates for a national (US) body responsible for fostering, monitoring and assessing research integrity.
> - NAS, On Being a Scientist: A Guide to Responsible Conduct in Research: Third Edition
>   - Can serve as a primer for new lab members. Includes case studies that can serve as topics in journal clubs, lab meetings, workshops. Or, simply post to the cork board in the lunchroom to spark some introspection or discussion across research groups!
> - Committee on Science, Engineering, and Public Policy, Philip A. Griffiths, Chair, *Adviser, Teacher, Role Model, Friend: On Being a Mentor to Students in Science and Engineering*, National Academy Press, 1997
>   - A companion piece to On Being a Scientist, this collection outlines responsibilities and recommendations for mentors.
> - M. Baker How quality control could save your science. *Nature News Feature* 27 January 2016

Conclusion and future work

It is an exciting time to be a biology educator, especially in the fields of genetics and genomics. There are an unprecedented number of resources and mentoring networks to support rigorous teaching practices, and no shortage of inspiring research stories to contextualize the knowledge foundations of our field. It is perhaps an overwhelming time as well, as precarious labor conditions and increased scrutiny raise the standards of the profession.

Similarly, today's research faculty are overwhelmed with a number of obligations besides their commitment to research. Research faculty are teachers, mentors, business and people managers, and get little to no training in any of these responsibilities [20]. The robust literature advocating for reforms to how we teach and how we mentor is daunting when squared with these additional responsibilities.

However, there are incremental changes one can make, and evidence-based practices (learning communities) that can assist in implementing changes to teaching and mentoring practices.

The iterative improvement and building of ideas that we take for granted as foundational to scientific inquiry should likewise be reflected in how we train. Conditions change: what was once status quo may no longer be appropriate, for any number of reasons. Do not simply use your training experience as the benchmark, reflect on your experience as a trainee, as a mentor, and improve upon it. Better yet, consult the literature and get involved. The education research community has yet to systematically tackle issues of RCR training and no doubt could benefit from the participation of researchers at multiple institutes. Work needs to be done to establish whether or not the approaches reviewed here and elsewhere yield desired outcomes [20].

A vital component of the equation is that the considerable work to build and maintain evidence-based practices in teaching, learning, and mentoring is valued through recognition and compensation. Despite a growing body of literature supporting the importance of undergraduate research experiences and the importance of direct interactions between faculty and undergraduates, this mentoring work continues to go unrewarded and unrecognized in many institutions [66]. Similarly, despite wide acknowledgement that science's myopic view of rewarding research output in the form of publications [2,16,72], change is slow. If we are serious about maintaining the integrity of our research enterprises, we need seismic shifts in policy to value the work needed to revitalize research culture. People in positions of power (administration, professional societies, and funding agencies, Fig. 2.1) need to put words into action and actively promote and value (with promotions and remuneration) the considerable work being done to prepare the next generation of scientists that does not directly contribute to publication record.

References

[1] J.P.A. Ioannidis, Why most published research findings are false, PLoS Med. 2 (2005) e124.

[2] B. Alberts, M.W. Kirschner, S. Tilghman, H. Varmus, Rescuing US biomedical research from its systemic flaws, Proc. Natl. Acad. Sci. U. S. A. 111 (2014) 5773–5777, https://doi.org/10.1073/pnas.1404402111.

[3] K. Popper, Conjectures and Refutations: The Growth of Scientific Knowledge, Routledge, London, 1963.

[4] K.E. Bettridge, A.L. Cook, R.C. Ziegelstein, P.J. Espenshade, A Scientist's oath, Mol. Cell 71 (2018) 879–881, https://doi.org/10.1016/j.molcel.2018.08.026.

[5] National Academy of Sciences (US), National Academy of Engineering (US) and Institute of Medicine (US) Committee on Science, Engineering, and Public Policy, On Being a Scientist: A Guide to Responsible Conduct in Research, third ed., National Academies Press (US), Washington, DC, 2009, https://doi.org/10.17226/12192.

[6] National Academy of Sciences (US), National Academy of Engineering (US) and Institute of Medicine (US) Panel on Scientific Responsibility and the Conduct of Research, Responsible Science: Ensuring the Integrity of the Research Process, National Academies Press (US), Washington, DC, 1993, https://doi.org/10.17226/1864.

[7] M.L. Aikens, S. Sadselia, K. Watkins, M. Evans, L.T. Eby, E.L. Dolan, A social capital perspective on the mentoring of undergraduate life science researchers: an empirical study of undergraduate-postgraduate-faculty triads, CBE Life Sci. Educ. 15 (2016), https://doi.org/10.1187/cbe.15-10-0208.

[8] J. Graves, E.D. Jarvis, An Open Letter: Scientists and Racial Justice, The Scientist, 2020. https://www.the-scientist.com/editorial/an-open-letter-scientists-and-racial-justice-67648.

[9] C.S. Hayter, M.A. Parker, Factors that influence the transition of university postdocs to non-academic scientific careers: an exploratory study, Res. Policy 48 (2019) 556–570, https://doi.org/10.1016/j.respol.2018.09.009.

[10] National Academies of Sciences, Engineering, and Medicine, Policy and Global Affairs, Committee on Women in Science, Engineering, and Medicine, Committee on the Impacts of Sexual Harassment in Academia, Sexual Harassment of Women: Climate, Culture, and Consequences in Academic Sciences, Engineering, and Medicine, National Academies Press (US), Washington, DC, 2018.
[11] R. Van Noorden, Some hard numbers on science's leadership problems, Nature 557 (2018) 294–296, https://doi.org/10.1038/d41586-018-05143-8.
[12] J.D. Clements, N.D. Connell, C. Dirks, M. El-Faham, A. Hay, E. Heitman, J.H. Stith, E.C. Bond, R.R. Colwell, L. Anestidou, J.L. Husbands, J.B. Labov, Engaging actively with issues in the responsible conduct of science: lessons from international efforts are relevant for undergraduate education in the United States, CBE Life Sci. Educ. 12 (2013) 596–603, https://doi.org/10.1187/cbe.13-09-0184.
[13] S. Mahmud, T. Bretag, Fostering integrity in postgraduate research: an evidence-based policy and support framework, Account. Res. 21 (2014) 122–137, https://doi.org/10.1080/08989621.2014.847668.
[14] N.H. Steneck, R.E. Bulger, The history, purpose, and future of instruction in the responsible conduct of research, Acad. Med. 82 (2007) 829–834.
[15] P. Satalkar, D. Shaw, How do researchers acquire and develop notions of research integrity? A qualitative study among biomedical researchers in Switzerland, BMC Med. Ethics 20 (2019) 72, https://doi.org/10.1186/s12910-019-0410-x.
[16] D.S. Kornfeld, Perspective: research misconduct: the search for a remedy, Acad. Med. 87 (2012) 877–882, https://doi.org/10.1097/ACM.0b013e318257ee6a.
[17] L.A. Diaz-Martinez, G.R. Fisher, D. Esparza, J.M. Bhatt, C.E. D'Arcy, J. Apodaca, S. Brownell, L. Corwin, W.B. Davis, K.W. Floyd, P.J. Killion, J. Madden, P. Marsteller, T. Mayfield-Meyer, K.K. McDonald, M. Rosenberg, M.A. Yarborough, J.T. Olimpo, Recommendations for effective integration of ethics and responsible conduct of research (E/RCR) education into course-based undergraduate research experiences: a meeting report, CBE Life Sci. Educ. 18 (2019) mr2, https://doi.org/10.1187/cbe.18-10-0203.
[18] T. Phillips, F. Nestor, G. Beach, E. Heitman, America COMPETES at 5 years: an analysis of research-intensive universities' RCR training plans, Sci. Eng. Ethics 24 (2018) 227–249, https://doi.org/10.1007/s11948-017-9883-5.
[19] T. Bretag, Challenges in addressing plagiarism in education, PLoS Med. 10 (2013) e1001574, https://doi.org/10.1371/journal.pmed.1001574.
[20] L. Clement, K.N. Leung, J.B. Lewis, N.M. Saul, The supervisory role of life science research faculty: the missing link to diversifying the academic workforce? J. Microbiol. Biol. Educ. 21 (2020), https://doi.org/10.1128/jmbe.v21i1.1911.
[21] AAAS, The Liberal Art of Science: Agenda for Action, AAAS, Washington, DC, 1990.
[22] AAAS, Vision and Change: A Call to Action, AAAS, Washington, DC, 2010. https://live-visionandchange.pantheonsite.io/wp-content/uploads/2013/11/aaas-VISchange-web1113.pdf. (Accessed 25 June 2020).
[23] J. Handelsman, D. Ebert-May, R. Beichner, P. Bruns, A. Chang, R. DeHaan, J. Gentile, S. Lauffer, J. Stewart, S.M. Tilghman, W.B. Wood, Education. Scientific teaching, Science 304 (2004) 521–522.
[24] L.C. Parker, A.M. Gleichsner, O.A. Adedokun, J. Forney, Targeting change: assessing a faculty learning community focused on increasing statistics content in life science curricula, Biochem. Mol. Biol. Educ. 44 (2016) 517–525, https://doi.org/10.1002/bmb.20974.
[25] C.I. Petersen, P. Baepler, A. Beitz, P. Ching, K.S. Gorman, C.L. Neudauer, W. Rozaitis, J.D. Walker, D. Wingert, The tyranny of content: "content coverage" as a barrier to evidence-based teaching approaches and ways to overcome it, CBE Life Sci. Educ. 19 (2020) ar17, https://doi.org/10.1187/cbe.19-04-0079.
[26] J. Handelsman, S. Miller, C. Pfund, Scientific Teaching, W. H. Freeman and Company, New York, NY, 2007.
[27] S.C.R. Elgin, G. Bangera, S.M. Decatur, E.L. Dolan, L. Guertin, W.C. Newstetter, E.F. San Juan, M.A. Smith, G.C. Weaver, S.R. Wessler, K.A. Brenner, J.B. Labov, Insights from a convocation: integrating discovery-based research into the undergraduate curriculum, CBE Life Sci. Educ. 15 (2016), https://doi.org/10.1187/cbe.16-03-0118.

[28] M.C. Linn, E. Palmer, A. Baranger, E. Gerard, E. Stone, Education. Undergraduate research experiences: impacts and opportunities, Science 347 (2015) 1261757, https://doi.org/10.1126/science.1261757.

[29] C.A. Wei, T. Woodin, Undergraduate research experiences in biology: alternatives to the apprenticeship model, CBE Life Sci. Educ. 10 (2011) 123–131, https://doi.org/10.1187/cbe.11-03-0028.

[30] D.B. Luckie, J.R. Aubry, B.J. Marengo, A.M. Rivkin, L.A. Foos, J.J. Maleszewski, Less teaching, more learning: 10-yr study supports increasing student learning through less coverage and more inquiry, Adv. Physiol. Educ. 36 (2012) 325–335, https://doi.org/10.1152/advan.00017.2012.

[31] D. Lopatto, A.G. Rosenwald, J.R. DiAngelo, A.T. Hark, M. Skerritt, M. Wawersik, A.K. Allen, C. Alvarez, S. Anderson, C. Arrigo, A. Arsham, D. Barnard, C. Bazinet, J.E.J. Bedard, I. Bose, J.M. Braverman, M.G. Burg, R.C. Burgess, P. Croonquist, C. Du, S. Dubowsky, H. Eisler, M.A. Escobar, M. Foulk, E. Furbee, T. Giarla, R.L. Glaser, A.L. Goodman, Y. Gosser, A. Haberman, C. Hauser, S. Hays, C.E. Howell, J. Jemc, M.L. Johnson, C.J. Jones, L. Kadlec, J.D. Kagey, K.L. Keller, J. Kennell, S.C.S. Key, A.J. Kleinschmit, M. Kleinschmit, N.P. Kokan, O.R. Kopp, M.M. Laakso, J. Leatherman, L.J. Long, M. Manier, J.C. Martinez-Cruzado, L.F. Matos, A.J. McClellan, G. McNeil, E. Merkhofer, V. Mingo, H. Mistry, E. Mitchell, N.T. Mortimer, D. Mukhopadhyay, J.L. Myka, A. Nagengast, P. Overvoorde, D. Paetkau, L. Paliulis, S. Parrish, M.L. Preuss, J.V. Price, N.A. Pullen, C. Reinke, D. Revie, S. Robic, J.A. Roecklein-Canfield, M.R. Rubin, T. Sadikot, J.S. Sanford, M. Santisteban, K. Saville, S. Schroeder, C.D. Shaffer, K.A. Sharif, D.E. Sklensky, C. Small, M. Smith, S. Smith, R. Spokony, A. Sreenivasan, J. Stamm, R. Sterne-Marr, K.C. Teeter, J. Thackeray, J.S. Thompson, S.T. Peters, M. Van Stry, N. Velazquez-Ulloa, C. Wolfe, J. Youngblom, B. Yowler, L. Zhou, J. Brennan, J. Buhler, W. Leung, L.K. Reed, S.C.R. Elgin, Facilitating growth through frustration: using genomics research in a course-based undergraduate research experience, J. Microbiol. Biol. Educ. 21 (2020), https://doi.org/10.1128/jmbe.v21i1.2005.

[32] C.D. Shaffer, C. Alvarez, C. Bailey, D. Barnard, S. Bhalla, C. Chandrasekaran, V. Chandrasekaran, H.-M. Chung, D.R. Dorer, C. Du, T.T. Eckdahl, J.L. Poet, D. Frohlich, A.L. Goodman, Y. Gosser, C. Hauser, L.L.M. Hoopes, D. Johnson, C.J. Jones, M. Kaehler, N. Kokan, O.R. Kopp, G.A. Kuleck, G. McNeil, R. Moss, J.L. Myka, A. Nagengast, R. Morris, P.J. Overvoorde, E. Shoop, S. Parrish, K. Reed, E.G. Regisford, D. Revie, A.G. Rosenwald, K. Saville, S. Schroeder, M. Shaw, G. Skuse, C. Smith, M. Smith, E.P. Spana, M. Spratt, J. Stamm, J.S. Thompson, M. Wawersik, B.A. Wilson, J. Youngblom, W. Leung, J. Buhler, E.R. Mardis, D. Lopatto, S.C.R. Elgin, The genomics education partnership: successful integration of research into laboratory classes at a diverse group of undergraduate institutions, CBE Life Sci. Educ. 9 (2010) 55–69, https://doi.org/10.1187/09-11-0087.

[33] T. Bretag, Contract cheating will erode trust in science, Nature 574 (2019) 599, https://doi.org/10.1038/d41586-019-03265-1.

[34] G.L. Connell, D.A. Donovan, T.G. Chambers, Increasing the use of student-centered pedagogies from moderate to high improves student learning and attitudes about biology, CBE Life Sci. Educ. 15 (2016) ar3, https://doi.org/10.1187/cbe.15-03-0062.

[35] S. Hurtado, N.L. Cabrera, M.H. Lin, L. Arellano, L.L. Espinosa, Diversifying science: underrepresented student experiences in structured research programs, Res. High. Educ. 50 (2009) 189–214.

[36] E.J. Theobald, M.J. Hill, E. Tran, S. Agrawal, E.N. Arroyo, S. Behling, N. Chambwe, D.L. Cintrón, J.D. Cooper, G. Dunster, J.A. Grummer, K. Hennessey, J. Hsiao, N. Iranon, L. Jones, H. Jordt, M. Keller, M.E. Lacey, C.E. Littlefield, A. Lowe, S. Newman, V. Okolo, S. Olroyd, B.R. Peecook, S.B. Pickett, D.L. Slager, I.W. Caviedes-Solis, K.E. Stanchak, V. Sundaravardan, C. Valdebenito, C.R. Williams, K. Zinsli, S. Freeman, Active learning narrows achievement gaps for underrepresented students in undergraduate science, technology, engineering, and math, Proc. Natl. Acad. Sci. U. S. A. 117 (2020) 6476–6483, https://doi.org/10.1073/pnas.1916903117.

[37] K.J. Gilbert, R.L. Andrew, D.G. Bock, M.T. Franklin, N.C. Kane, J.-S. Moore, B.T. Moyers, S. Renaut, D.J. Rennison, T. Veen, T.H. Vines, Recommendations for utilizing and reporting population genetic analyses: the reproducibility of genetic clustering using the program STRUCTURE, Mol. Ecol. 21 (2012) 4925–4930, https://doi.org/10.1111/j.1365-294X.2012.05754.x.

[38] S. Senn, Statistical pitfalls of personalized medicince, Nature 563 (7733) (2018) 619–621, https://doi.org/10.1038/d41586-018-07535-2.
[39] X. Lin-Siegler, J.N. Ahn, J. Chen, F.-F.A. Fang, M. Luna-Lucero, Even Einstein struggled: effects of learning about great scientists' struggles on high school students' motivation to learn science, J. Educ. Psychol. 108 (2016) 314–328, https://doi.org/10.1037/edu0000092.
[40] S.M. Hiebert, The strong-inference protocol: not just for grant proposals, Adv. Physiol. Educ. 31 (2007) 93–96.
[41] L. Laplane, P. Mantovani, R. Adolphs, H. Chang, A. Mantovani, M. McFall-Ngai, C. Rovelli, E. Sober, T. Pradeu, Opinion: why science needs philosophy, Proc. Natl. Acad. Sci. U. S. A. 116 (2019) 3948, https://doi.org/10.1073/pnas.1900357116.
[42] J.R. Platt, Strong inference: certain systematic methods of scientific thinking may produce much more rapid progress than others, Science 146 (1964) 347–353.
[43] B. Alberts, R.J. Cicerone, S.E. Fienberg, A. Kamb, M. McNutt, R.M. Nerem, R. Schekman, R. Shiffrin, V. Stodden, S. Suresh, M.T. Zuber, B.K. Pope, K.H. Jamieson, SCIENTIFIC INTEGRITY. Self-correction in science at work, Science 348 (2015) 1420–1422, https://doi.org/10.1126/science.aab3847.
[44] C. Moskovitz, D. Kellogg, Science education. Inquiry-based writing in the laboratory course, Science 332 (2011) 919–920, https://doi.org/10.1126/science.1200353.
[45] A. Prud'homme-Généreux, Aliens on Earth? The #arseniclife affair, 2014.
[46] M.F. Dahlstrom, D.A. Scheufele, (Escaping) the paradox of scientific storytelling, PLoS Biol. 16 (2018) e2006720, https://doi.org/10.1371/journal.pbio.2006720.
[47] M.F. Dahlstrom, Using narratives and storytelling to communicate science with nonexpert audiences, Proc. Natl. Acad. Sci. U. S. A. 111 (Suppl 4) (2014) 13614–13620, https://doi.org/10.1073/pnas.1320645111.
[48] K. Zimmer, When Your Supervisor Is Accused of Research Misconduct, The Scientist, 2020. https://www.the-scientist.com/careers/when-your-supervisor-is-accused-of-research-misconduct-67581.
[49] J.W. Tsai, F. Muindi, Towards sustaining a culture of mental health and wellness for trainees in the biosciences, Nat. Biotechnol. 34 (2016) 353–355, https://doi.org/10.1038/nbt.3490.
[50] J.W. Reid, G.E. Gardner, Navigating tensions of research and teaching: biology graduate students' perceptions of the research-teaching Nexus within ecological contexts, CBE Life Sci. Educ. 19 (2020) ar25, https://doi.org/10.1187/cbe.19-11-0218.
[51] E.E. Schussler, Q. Read, G. Marbach-Ad, K. Miller, M. Ferzli, Preparing biology graduate teaching assistants for their roles as instructors: an assessment of institutional approaches, CBE Life Sci. Educ. 14 (2015), https://doi.org/10.1187/cbe.14-11-0196.
[52] J. Lave, E. Wenger, Situated Learning: Legitimate Peripheral Participation, Cambridge University Press, Cambridge, 1991, https://doi.org/10.1017/CBO9780511815355.
[53] National Academies of Sciences, Engineering, and Medicine, Integrating Discovery-Based Research into the Undergraduate Curriculum: Report of a Convocation, National Academies Press, Washington, DC, 2015.
[54] S.E. Rodenbusch, P.R. Hernandez, S.L. Simmons, E.L. Dolan, Early engagement in course-based research increases graduation rates and completion of science, engineering, and mathematics degrees, CBE Life Sci. Educ. 15 (2016), https://doi.org/10.1187/cbe.16-03-0117.
[55] L.A. Corwin, C. Runyon, A. Robinson, E.L. Dolan, The laboratory course assessment survey: a tool to measure three dimensions of research-course design, CBE Life Sci. Educ. 14 (2015) ar37, https://doi.org/10.1187/cbe.15-03-0073.
[56] L.A. Corwin, C.R. Runyon, E. Ghanem, M. Sandy, G. Clark, G.C. Palmer, S. Reichler, S.E. Rodenbusch, E.L. Dolan, Effects of discovery, iteration, and collaboration in laboratory courses on undergraduates' research career intentions fully mediated by student ownership, CBE Life Sci. Educ. 17 (2018) ar20, https://doi.org/10.1187/cbe.17-07-0141.
[57] T.L. Strayhorn, Reframing academic advising for student success: from advisor to cultural navigator, NACADA J. 35 (2015) 56–63, https://doi.org/10.12930/NACADA-14-199.

[58] K.D. Huggett, Advising in undergraduate honors programs: a learner-centered approach, NACADA J. 24 (2004) 75–87, https://doi.org/10.12930/0271-9517-24.1-2.75.

[59] R. Giner-Sorolla, Science or art? How aesthetic standards grease the way through the publication bottleneck but undermine science, Perspect. Psychol. Sci. 7 (2012) 562–571, https://doi.org/10.1177/1745691612457576.

[60] Y. Katz, Against storytelling of scientific results, Nat. Methods 10 (2013) 1045, https://doi.org/10.1038/nmeth.2699.

[61] R. Heesen, Why the reward structure of science makes reproducibility problems inevitable, J. Philos. 115 (2018) 661–674, https://doi.org/10.5840/jphil20181151239.

[62] F. Romero, Novelty versus replicability: virtues and vices in the reward system of science, Philos. Sci. 84 (2017) 1031–1043.

[63] A. Casadevall, L.M. Ellis, E.W. Davies, M. McFall-Ngai, F.C. Fang, A framework for improving the quality of research in the biological sciences, MBio 7 (2016), https://doi.org/10.1128/mBio.01256-16.

[64] V. Lewis, C.A. Martina, M.P. McDermott, L. Chaudron, P.M. Trief, J.G. LaGuardia, D. Sharp, S.R. Goodman, G.D. Morse, R.M. Ryan, Mentoring interventions for underrepresented scholars in biomedical and behavioral sciences: effects on quality of mentoring interactions and discussions, CBE Life Sci. Educ. 16 (2017), https://doi.org/10.1187/cbe.16-07-0215.

[65] M. Joshi, M.L. Aikens, E.L. Dolan, Direct ties to a faculty mentor related to positive outcomes for undergraduate researchers, Bioscience 69 (2019) 389–397, https://doi.org/10.1093/biosci/biz039.

[66] E.L. Dolan, D. Johnson, The undergraduate-postgraduate-faculty triad: unique functions and tensions associated with undergraduate research experiences at research universities, CBE-Life Sci. Educ. 9 (2010) 543–553, https://doi.org/10.1187/cbe.10-03-0052.

[67] M. Baker, How quality control could save your science, Nature 529 (2016) 456–458, https://doi.org/10.1038/529456a.

[68] C. Gandrud, Reproducible Research with R and RStudio, second ed., Chapman and Hall/CRC, 2015.

[69] P.S. Lukeman, A guide to mentoring undergraduates in the lab, Nat. Nanotechnol. 8 (2013) 784–786, https://doi.org/10.1038/nnano.2013.237.

[70] L.M. Bennett, R. Maraia, H. Gadlin, The "welcome letter": a useful tool for laboratories and teams, J. Transl. Med. Epidemiol. 2 (2014) 1035.

[71] S. Mahmud, T. Bretag, Integrity in postgraduate research: the student voice, Sci. Eng. Ethics 21 (2015) 1657–1672, https://doi.org/10.1007/s11948-014-9616-y.

[72] R. Benedictus, F. Miedema, M.W.J. Ferguson, Fewer numbers, better science, Nature 538 (2016) 453–455, https://doi.org/10.1038/538453a.

SECTION 2

Genotyping

CHAPTER 3

Genome-wide association studies (GWAS): What are they, when to use them?

Fan Wang

University of Chicago, Center for Translational Data Science, Chicago, IL, United States

Introduction and history

Genome-wide association studies (GWAS) are one approach to discovering candidate genetic variants associated with complex phenotypes. Using GWAS, a wide variety of traits in the field of genetic medicine, including human attributes, common diseases, lifestyle-associated diseases, and therapeutic drug responses, have been extensively studied with a view to providing potential genetic targets to elucidate their mechanisms. GWAS have led to an increased understanding of the physiological mechanisms behind human characteristics and have also succeeded in identification of genetic susceptibilities to many diseases [1].

Along with the development of new sequencing technologies, the first GWAS in the world was led by Dr. Tanaka Toshihiro in 2002 [2]. Tanaka's group identified functional single nucleotide polymorphisms (SNPs) in the lymphotoxin-α (LTA) gene as risk factors for myocardial infarction (commonly known as a heart attack).

Although the core architectural knowledge holds steady, the very first GWAS was relatively primitive compared to present-day GWAS research due to the limited number of subjects and SNP coverage of the genome. Nonetheless, this frontier study laid the foundation for current GWAS in terms of (a) using a pooled control method for a significant increase in statistical power and (b) linkage disequilibrium mapping for identifying the most associated SNPs.

The Human Genome Project (HGP) was completed in April 2003 [3,4]. The International Human Genome Sequencing Consortium reported a sequence, which reads approximately 99% of the gene-containing part of the human sequence with high accuracy and nearly complete coverage [3]. Along with the completion of the sequencing of the human genome, the International HapMap Project was launched in October 2002 to create a public resource of common patterns of human genetic variations, providing insights into genetic studies of clinical outcomes and disease risks [5]. As opposed to the HGP providing informative knowledge on the invariant sequence across individuals, the HapMap Project developed a haplotype map of the human genome to describe the common variation in human DNA sequence [3].

The HapMap resource contains one million genetic variations from four representative populations: European Caucasian, African, Chinese, and Japanese. Using the HapMap database, researchers have elucidated the generality of recombination hotspots, which results in substantial correlations of SNPs with their neighbors [6]. Inspired by these early findings, genomic variation studies became more efficient with the association information examined by HapMap genotype data. Moreover, the further selection of candidate SNPs was facilitated by utilizing the linkage disequilibrium information identified through HapMap data, eliminating the necessity of sequencing every single base pair of the genome.

With HapMap available, GWAS became easier to perform, sped along by the growth of genotyping systems. More and more researchers have been collecting DNA from patients, genotyping them by commercial arrays, and performing statistical analysis. To facilitate GWAS for the genetic community, a growing number of projects and groups have been sharing control cases, which makes the collection of control cases unnecessary, except for diseases with high prevalence (>10%). Prominent databases used widely in GWAS include the NCBI's Database of Genotypes and Phenotypes (dbGaP) [7], the Broad Institute's Genome Aggregation Database (gnomAD) [8], and the UK Biobank [9].

As a result, an enormous number of susceptibility genes and loci have been identified by GWAS for many common diseases, including but not limited to diabetes [10,11], obesity [12,13], hypertension [14], coronary artery disease [15], inflammatory bowel disease [16], and osteoporosis [17]. Through functional analysis of disease-associated variants and genes, we can potentially reveal the mechanism of the disease and develop novel treatments.

Study design and statistical methods

GWAS is intended to understand the relationship between SNP genotypes and observed traits or phenotypes. To this end, it is necessary to robustly detect genomic effects with sufficient statistical power. Early GWAS publications characterized a number of spurious associations for complex diseases [18–20] due to the lack of controls for population structure, low power of detection, and significant statistical evidence prior to the association hypotheses.

The key to robust detection is to have strong evidence; however, the question is: what measure of evidence does one choose? To answer this question, it is crucial to understand the fundamentals of two major statistical methods, (1) frequentist and (2) Bayesian measures of evidence. It is also important to characterize the phenotype of interest. To better understand the key concepts of GWAS design, the underlying statistical basis is required. We strongly recommend that readers learn some basic statistical knowledge, including forming and testing a hypothesis, understanding levels of significance, the null hypothesis, Pearson's Chi-square test, F-test, Student's t-test, and Z-test.

The Handbook of Biological Statistics by McDonald [21] provides clear clarification and examples of the most commonly used statistical tests in the GWAS analysis.

Frequentist vs Bayesian modeling

The long-debated question between frequentist (classical approach) and Bayesian approach has been much discussed, with particular interest in GWAS.

Frequentist modeling is based on frequency distributions of the hypothetical replications and uses a P-value to make inference. Unlike Bayesian statistics, which relies on prior probability distributions, frequentists care whether the test statistic is more extreme than the observed probability under random experiments of the null hypothesis H_0.

A "significant" P-value is usually defined as smaller than an arbitrary threshold. The advantage of the frequentist is that the P-value can be easily calculated. Another major advantage is that the test statistics including t, F, z, χ^2, etc., are well-standardized by P-value. However, the strength of statistical evidence can vary across different sample sizes and initial setups, which is hard to standardize by P-value. Another misleading question on P-value is how to interpret it a genetics association, which was demonstrated by Ball [22].

Unlike frequentists, Bayesian statistics is based on probability theory. Bayesian theorem has been widely accepted by the GWAS community. To better understand Bayesian theorem, we need to introduce two important terminologies: a prior probability (commonly known as prior odds or prior) and posterior probability (commonly known as posterior odds or posterior). Prior probability is the evidence to weight in the effect of a case with a given knowledge. Posterior probability is the final predicted value calculated from the basic likelihood function in combination with the prior probability distribution. The major element of Bayesian theorem is to specify a prior probability distribution that summarizes the knowledge about the hypothesis and then to calculate the posterior probability using the observed data.

Let's look at an example of ice cream sales and weather to better understand how the Bayesian theorem allows us to incorporate prior probability and calculate posterior probability. We might be interested to know the probability of selling ice cream on any given day given the weather type. This is a typical Bayesian problem. Event A represents the ice cream sale; and event B represents the weather type. And the posterior probability, which is the result of this Bayesian question, could be written mathematically as: $P(A=\text{ice cream sale}|B=\text{weather type})$.

In this example, the prior probability could be the probability of selling ice cream before knowing anything about the weather type, which could be written mathematically as $P(A=\text{ice cream sale})$. The marginal likelihood would be the conditional probability of observing the weather type under an ice cream sale, which could be written mathematically as $P(B=\text{weather type}|A=\text{ice cream sale})$. The probability of the data in this case is the probability of the weather type, which could be written mathematically as $P(B=\text{weather type})$.

Based on Bayesian theorem's equation $P(A|B) = \dfrac{P(B|A) \times P(A)}{P(B)}$, we can compute the posterior probability of selling ice cream given the weather type.

If we talk about all the statistical terminology in a genetics context, we are given the genotype data for thousands of samples, and we are interested in learning, which locus out of the millions of total loci are the causal variants (posterior probabilities of association), given the biological information for each SNP (prior distribution), and a linear regression or logistic regression (likelihood of data). The commonly used biological information to specify the prior probability includes expression quantitative trait loci (eQTL) and DNAseI hypersensitive sites annotation. Let's take eQTL analysis as an example.

The aim of eQTL mapping is to identify genetic variants (specifically SNPs) that explain variations in gene expression. The likelihood can be a linear regression of a quantitative or binary trait on the SNPs, which could be calculated from a linear regression model $y = \beta x + \varepsilon$. Stranger et al. [23] proposed this straightforward linear regression model with additive genetic effects from SNPs. And since then, this idea has been commonly used to map eQTL in the field of GWAS. In this linear regression equation, y indicates a gene expression trait; x indicates the predictor variables, i.e., SNPs, with the additive genetic effects; β indicates the regression coefficients for the SNPs, which encode the effect of predictor variables. This is where priors are incorporated: the variance of β reflects the assumption that dominant and recessive alleles have different contributions to the phenotype (i.e., gene expression); ε indicates the residual error term.

In summary, frequentist method doesn't require explicit prior probability but the key limitation is the strong dependency on conventional *P*-value. Comparing to the frequentist approach, the Bayesian method is able to test subtle effects with low prior odds and huge sample sizes, since it can benefit from the elicitation of biologically informative priors [24,25] and improve the statistical power in some scenarios [26].

GWAS for case and control association testing

Association mapping is now routinely being used to identify loci that are involved with complex traits [27], including type 1 [28,29], type 2 diabetes [10,18,30–33], prostate cancer [34–37], and asthma [38]. Among these complex traits, there are two major categories of phenotypes: binary case/control (also called categorical) and quantitative. Case/control or categorical phenotypes include disease status and ordinal categorical data collected from surveys, questionnaires, and medical tests. Technological advances have made it feasible to perform case/control association studies on a genome-wide basis with hundreds of thousands of markers in a single study. We consider testing a genetic marker for association with a disease in thousands of unrelated subjects. Case and control association methods essentially test for independence between trait and allele/genotype.

The case/control traits are typically analyzed by the contingency table method. Contingency table methods test and measure the deviation under the null hypothesis of independence that there is no association between the phenotype and allele types. The genotype-based contingency table for GWAS is well-suited for categorical traits. And the most popular contingency table method is the Chi-square (χ^2) method (see Table 3.1).

The calculation of the expected factor is straightforward by comparing the proportions of each SNP allele in the cases and the controls. For example, the expected number of cases with allele A under the null hypothesis is: Proportion$_{case}$ × Proportio n$_{allele\ A}$ × Total alleles

where Proportion$_{case}$ = Total cases/Total alleles

Similarly, we could calculated the expected number of cases with allele G, controls with allele A and controls with allele G. To examine your calculation, the total number of allele A, allele G, cases, and controls could be compared with the observed table. The results should be consistent (see Table 3.2).

The Chi-square statistic is described in the equation: $\chi^2 = \sum_{i=1}^{n} \frac{(O_i - E_i)^2}{E_i}$

Where the O_i is observed frequency for *i*th outcome; the E_i is expected frequency for *i*th outcome; the *n* is total number of outcomes.

Table 3.1 Case control Chi-square example (expected data).

Allele type	Allele A	Allele G	Total
Case	45	953	998
Control	45	955	1000
Total	90	1908	1998

In this example, we suppose that there are two alleles, A and G for a given locus. To get the table of expected allele counts, we first sum rows and columns from the observed table. Then for each cell determined by a particular row and column, we multiply the row total by the column total and divide by the overall total.

Table 3.2 Case control Chi-square example (observed data).

Allele type	Allele A	Allele G	Total
Case	22	976	998
Control	68	932	1000
Total	90	1908	1998

Observed or actual values should be the values you get from data directly. We only need to sum rows and columns. No extra calculation is needed.

The *P*-value of Chi-square test indicates the probability of the observed data under the null hypothesis. Low *P*-value (*P*-value $<\alpha$, where α is a user-defined value) means the observed data are unlikely under the null hypothesis. Thus, we reject the null hypothesis and declare the significant association between the SNP alleles and the phenotype of interest. In general cases, $\alpha = 0.01$ or 0.05 is used as threshold.

A more preferable statistical method for case/control phenotype is logistic regression. Logistic regression is a Bayesian statistical-based method used for binary phenotypes instead of continuous traits, and is an extension of linear regression. Let's walk through how logistic regression works for case/control traits in GWAS.

We assume Y_i is the phenotype for sample i, and $Y_i=0$ stands for control phenotype, and $Y_i=1$ stands for case phenotype.

We assume X_i is the genotype of sample i at a specific locus. If we count the dosage of allele a, we will get $X_i=0$, which stands for genotype AA, and $X_i=1$, which stands for genotype Aa, and $X_i=2$, which stands for genotype aa.

The basic Bayesian logistic regression model is to assume the expected probability of phenotype ($y=1$) given a genotype (x), where the probability $\theta(x) = Pr\{y=1|x\}$, and then define the estimated log odds ratio (OR).

The goal of logistic regression is to estimate the odds ratio, and examine whether β_1 significantly deviated from 0. The odds ratio is basically the ratio of odds that a given phenotype with allele a and the same phenotype with allele A. To help us better interpret the odds ratio, let's look at some examples of the odds ratio.

For example, the odds of phenotype with genotype AA is $\dfrac{Pr\{y=1|x=0\}}{Pr\{y=0|x=0\}} = e^{\beta_0}$

For example, the odds of phenotype with genotype Aa is $\dfrac{Pr\{y=1|x=1\}}{Pr\{y=0|x=1\}} = e^{\beta_0+\beta_1}$

Based on the above two odds, we can derive the odds ratio of genotype Aa $OR_{Aa} = \dfrac{\text{Odds of phenotype in a sample with Aa genotype}}{\text{Odds of phenotype in a sample with AA genotype}} = \dfrac{Pr\{y=1|x=1\}/Pr\{y=0|x=1\}}{Pr\{y=1|x=0\}/Pr\{y=0|x=0\}} = e^{\beta_1}$

The odds ratio is a good measure of the genotype effect size. For example, $OR_{Aa}=1$ implies there is no association between genotype and phenotype; $OR_{Aa}>1$ implies that the genotype Aa is positively associated with the phenotype; $OR_{Aa}<1$ implies that the genotype Aa is negatively associated with the phenotype.

Comparing to the Chi-square test, logistic regression is usually the preferred approach in the field of GWAS, because logistic regression provides better adjustment for potential confounders such as clinical covariates, genotyping batches, genotypes at other locus, and demographic factors. By adding extra terms to the logistic model, the effect of these covariates could be adjusted:

$$f(x) = \log \frac{\theta(x)}{1-\theta(x)} = \beta_0 + \beta_1 x_1 + \beta_2 x_2 + \beta_3 x_3 + \cdots + \beta_n x_n.$$

GWAS for quantitative traits

Another common set of traits are quantitative traits. When we talk about a quantitative or complex trait, we mean that a combination of multiple genes and environmental factors exerts a small influence to determine a trait. Usually, there is no apparent Mendelian basis or mechanism that can cause or explain the variation of the trait, but together with genes and environmental influences, they combine to define an individual outcome. Most common diseases or phenotypes work this way, including cholesterol levels, diabetes, obesity, blood pressure, etc.

The most common association tests used for quantitative traits are generalized linear models (GLMs). Linear regression models are a simple way to examine the linear relationship between a dependent variable and an independent variable. In GWAS, a linear regression model can be fitted for each locus using the number of minor alleles and phenotype. The assumption of linear regression is that the phenotype depends additively on the number of minor alleles in the genotype. The linear regression equation is defined as $y = x\beta_1 + \beta_0 + \varepsilon$, where,

> y is the continuous-valued trait.
> x is the SNP genotype (allele dosage) at a particular locus.
> β_1 is the regression coefficient of the linear regression model, which represents how strong the association is between the SNP genotype and phenotype. In Fig. 3.1, the regression coefficient β_1 can be visualized as the slope of the regression line, and it stands for the effect of each allele copy on the phenotype value.
> β_0 is the Y-intercept that is a constant value. β_0 help anchor the regression line in the right position. At special circumstances that both x (SNP genotype dosage) and ε (noise) are 0, the value of β_0 will be the phenotypic score (y). If neither criterion was satisfied, then β_0 doesn't have meaningful interpretation.
> ε is the noise term (also called error term) of the linear regression model. We put the part of phenotype than can't be explained by $x\beta_1 + \beta_0$ as ε. In GWAS, the noise could be environmental factors that determine a person's risk or probability of developing a phenotype.

In the linear regression equation, y and x are observed values in a GWAS dataset; β_1, β_0 are parameters to be estimated using y and x. The null hypothesis is that there is no linear relationship between a genotype and a phenotype, which means β_1, β_0 parameters are 0 and errors ε are uncorrelated in the regression equation. We need to test whether the null hypothesis is rejected or not. Another assumption of linear regression is that the sampling distribution of t-statistic follows a t-distribution. If the P-values are smaller than the threshold (i.e., 0.01, 0.001, etc.), we conclude that the association between a genotype and a phenotype is statistically significant. And a larger β_1 indicates a bigger influence of a genotype.

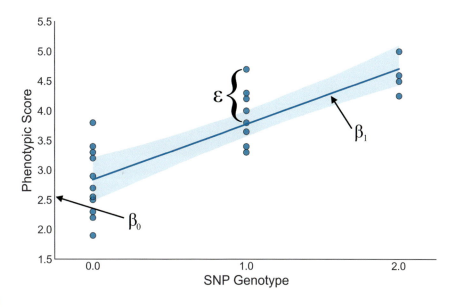

FIG. 3.1

Linear regression model for GWAS. Using linear regression method, phenotype is modeled as a linear function of a specific SNP genotype, regression coefficient, Y-intercept, and noise.

All the above clarification considers one single SNP at a time, which is called the univariate regression. However, most common diseases are believed to be caused by a set of SNPs that work together to present an additive effect. In addition, in GWAS analyses, we typically have millions of SNPs (working in an additive manner) in a single linear regression model to find out the set of SNPs influencing the phenotype (e.g., disease susceptibility) the most. Likewise, in the univariate regression model (single genotype at a time), we could consider all J SNPs at one time to build a multivariate regression model, which is defined as: $y = \sum_{j=1}^{J}(x_j \beta_{1j} + \beta_0 + \varepsilon)$.

As described in Fig. 3.2, the multivariate regression model evaluates the joint effect of J SNPs across N individuals

To evaluate how suitable the linear regression (both univariate and multivariate) is for the GWAS data, we can refer to R^2, the proportion of genetic variance that can be explained by the linear regression model. To calculate R^2, we would need to compute the residual sum of square (RSS) using the equation: $RSS = \sum_{j=1}^{J}(y_{j-\hat{\beta}_0-x_j\hat{\beta}_0})^2$. Then we could compute the R^2 using the equation: $R^2 = 1 - \frac{RSS}{j-1}/\widehat{Var}(y)$. The R^2 values range from 0 to 1. And a larger R^2 indicates more genetic variance is explained by the regression model. However, R^2 is not the only criterion to assess how suitable the model is for your GWAS data. Another straightforward way is to plot the genotype dosages versus phenotypic scores as a scatter plot and a regression line (e.g., Fig. 3.1). From this scatter plot, we will easily spot if any

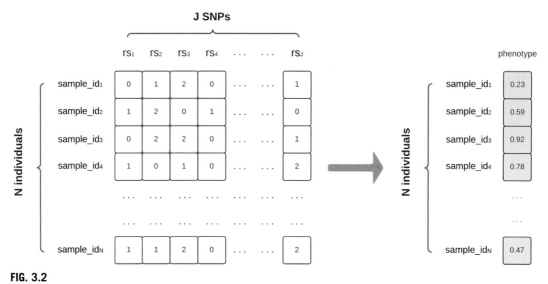

FIG. 3.2

Multivariate regression. Using multivariate regression method, phenotype is modeled as a joint effect of J SNP genotypes, regression coefficients, Y-intercepts, and noises.

From E.J. Topol, S.S. Murray, K.A. Frazer, The genomics gold rush, JAMA 298 (2007) 218–221.

deviations from the additive effect assumption exist. Besides that, we can visualize the residual terms (ε) between difference genotype dosage groups in the scatter plot to speculate whether there is any interaction term between genotype and environmental factor.

Quantile normalization (QN) is a crucial statistical technique for preprocessing the quantitative phenotype in GWAS. Quantile normalization transforms the phenotypic data to follow a normal distribution by diminishing the systematic bias and undesirable technical outliers. Quantile normalization is also required when you have multiple GWAS data cohorts, for which you may want to harmonize the phenotypic data by making their distributions identical or similar.

Quality control for GWAS

GWAS aims to discover the true biological relationships underlying a phenotype of interest and genetic variations; however, the success of a GWAS has largely depended on the overall quality of the input data. Errors such as improper study design and experimental errors during genotype calling, will potentially lead to spurious associations and false-positive associations. This is discussed further in Chapter 6. This section will clarify how to apply rigorous quality control (QC) procedures to your data prior to conducting GWAS.

Most of the QC procedures could be performed using PLINK [39], an open-source, cross-platform software for QC and analysis of GWAS data. PLINK is used for implementing a GWAS on a pharmacogenomics (PGx) dataset later in this section, and command options to conduct the following QC steps using PLINK will also be listed. See the PLINK tutorial of operations page for more details at https://zzz.bwh.harvard.edu/plink/index.shtml.

Technical QC

Technical QC involves the check of individual and SNP missingness. First, one must exclude SNPs that are missing in a large proportion (self-defined threshold) of the total samples. The threshold could be relaxed, i.e., st, i.e., 2%, depending on the level of missingness to filter out SNPs. The genotype missingness can be checked with the "--geno" option in PLINK.

After the SNP filtering is performed, the second part of technical QC is individual filtering. Individuals are excluded who have high rates of SNP missingness. In this procedure, individuals with low genotype calls are removed. Individual missingness can be checked with the "--mind" option in PLINK.

Genetic QC

Genetic QC examines your genotype data together with the samples' clinical information such as pedigree, gender, race, to detect if there are any genotyping errors assuming the clinical information is correct.

Mendelian's law of segregation QC

The famous Mendel's Law of Segregation proves that pair of alleles segregate during gamete formation and randomly unite at fertilization. This principle could also be applied in GWAS. If the GWAS samples are trio samples (related samples, or parent-offspring duo), it is possible to check the Mendelian inconsistency for all the genotypes, which indicate the genotyping errors, if the provided pedigree information is correct. The Mendelian inconsistency can be checked by the "--mendel" option in PLINK.

Hardy-Weinberg equilibrium QC

The Hardy-Weinberg equilibrium (HWE) states the probabilities for dominant homozygous (AA), heterozygous (Aa), and recessive homozygous (aa), given the minor allele (a) frequency of a biallelic locus. The Hardy-Weinberg equilibrium is defined as $p^2 + 2pq + q^2 = 1$ where p is the major allele (A) frequency; q is the dominant homozygous (AA) frequency; $2pq$ is the recessive homozygous (aa) frequency. Since $p + q = 1$, we can also write the Hardy-Weinberg equilibrium as $(1-q)^2 + 2q(1-q) + q^2 = 1$. The assumptions underlying HWE is that p and q are stable from generation to generation in a large and randomly mating population. If a genotype deviates from HWE, it is a common indicator of genotyping issue, or evolutionary selection. The HWE can be checked with the "--hwe" option in PLINK.

Relatedness QC

If the samples in your GWAS data are not unrelated and have self-reported relationship information available, you may be able to further check for the potential DNA sample mix-ups. Using the "--genome" option in PLINK, we can quickly estimate the pairwise kinship coefficient for each pair of samples in your GWAS dataset. By comparing the kinship coefficients against the self-reported relationships, we could identify the potential relatedness errors. Kinship coefficient reflects the relatedness of paired samples: $\frac{1}{2}$ (monozygotic twins), $\frac{1}{4}$ (parent-offspring pairs or full sibling pairs), $\frac{1}{8}$ (2nd-degree relative pairs, such as half-sibs, avuncular pairs, and grandparent-grandchild pairs), $\frac{1}{16}$ (3rd-degree relative pairs, such as first cousins), and 0 (unrelated pairs).

Another handy software package to evaluate the correspondence between self-reported relationship and those inferred from the genotype is Peddy [40]. Other than relatedness, Peddy can also check the self-reported gender and ancestry information. The inputs for Peddy are VCF and PED fi, which is bit different from PLINK. One of the advantages of Peddy compared with PLINK is Peddy can generate an interactive web page to visualize the genotype-derived clinical information against the self-reported information.

Gender discrepancy QC
Next, one should seek to verify if a gender discrepancy exists between genotype-derived information and the information recorded in the GWAS dataset. Gender discordance is based on the fact that males have a single copy of the X and Y chromosomes. Therefore, in theory, there shouldn't be any heterozygotes except the pseudo-autosomal region of the Y chromosome for males. We can predict a sample's gender based on the heterozygosity/homozygosity rate of X chromosome. The homozygosity estimate should be more than 0.8 for a male sample; and less than 0.2 for a female sample. The gender discrepancy QC can be checked by the "--check-sex" option in PLINK. Samples show significant gender discrepancy will be labeled as "PROBLEM" in the generated *.sexcheck file.

Population structure QC
The population structure (or stratification) check is a must when the samples of your GWAS data comprise individuals with different ancestries, since genetic differences among ancestries could lead to potential spurious associations (ancestrally informative genotype) rather than true biological associations. The strategy to characterize population structure is through the computation of identity by descent (IBD) for each pair of samples. If a DNA segment is an IBD segment in two or more individuals, it means the segment is inherited from a common ancestor of these individuals. PLINK implements IBD estimation using multidimensional scaling (MDS) approach. Please note that the IBD estimation limits to autosomal chromosomes only. In general, individuals sharing one allele IBD at every locus are parent-offspring relationships; and individuals sharing zero, one, and two alleles IBD at 25%, 50%, and 25% of the genome respectively are sibling relationships. In PLINK, the IBD metrics can be generated by "--genome" option. Then PLINK will conduct linkage clustering of individuals based on the estimated IBD. This step can be done using "--cluster" option in PLINK. Finally, by referring to the clustering results, we can reduce the effect of population structure through the removal of individuals with a divergent ancestry.

In summary, the collection of QC methods mentioned above allows us to measure the similarity between pedigree estimates of ethnicity and genetic estimates of ethnicity. However, pedigree information could be quite different from what is estimated using an algorithm like Peddy. Because self-reported pedigree information show only the locations of sample's known ancestors, whereas in genetic ethnicity estimation, we estimate the unknown amount of genetic variants inherited from all of a sample's ancestors. Genetic estimates of ethnicity also go back thousands of years, beyond the end of a pedigree paper trail. Nevertheless, the agreement between a pedigree and our genetic ethnicity estimate helps us to have a general idea on the overall quality of the GWAS data.

Concluding note
There is no doubt that GWAS has been a milestone in the field of human genetics. Millions of genetic loci have surfaced over the past decade with previously unknown associations with a good deal of complex traits and diseases, which definitely broadens the biological knowledge on the mechanisms

and related pathways behind the scenes. With the development of next generation sequencing, the new discoveries by GWAS are expected to surge with larger sample size and more comprehensive sequence readouts.

In this chapter, we briefly review the history, study design, and analytical methods that underlie GWAS approach. Next, we are going to perform our own GWAS analysis systematically. The following workshop is designed to introduce the readers' typical scenarios that they may come across in their own GWAS. We will introduce several commonly used software and tools, which are especially useful for high-dimensional GWAS data. Then we will use these tools to investigate the relations between different traits and SNPs in the genome-wide scans. Finally, we will introduce the tools to either visualize the GWAS results in a genome-wide view or a specific region graphically. Now let's get right into it.

Using R and PLINK for GWAS

Next, readers may work with a set of pharmacogenomics (PGx) data files containing thousands of SNPs from chromosome 19 genotyped in 175 patients with clinical drug response as a phenotype. The goal of our pharmacogenomics GWAS is to identify the genetic regions/locus linked to the drug response. Please note that typical GWAS data would contain variants across the entire genome; however, here the focus will be upon one chromosome only to make the workshop more tractable.

R primer

R (http://www.r-project.org/) is a free software environment for statistical computing and graphics. It compiles and runs on a wide variety of platforms, including Windows and MacOS. To download R, go to http://www.r-project.org/ and click download R. Click on a CRAN mirror location close to you (e.g., http://cran.wustl.edu/).

For MacOS X:
Click on "Download R for (Mac) OS X" and then "R-4.0.5.pkg" and follow the on-screen instructions to install. After installation R will be located in your Applications folder.
For Windows:
Click on "Download R for Windows" and then base and then Download R 4.0.5 for Windows and follow the onscreen instructions. Make note of where R gets installed on your machine.

Once installed, open R. It should look something like in Fig. 3.3 (this screen capture is from MacOS X, the Windows version will look slightly different). Try entering the following at the command line to verify R is working.

Your plot should look like what is presented in Fig. 3.4.

Plink primer

Plink [39] is a free, open-source whole genome association analysis toolset, designed to perform a range of basic, large-scale analyses in a computationally efficient manner. If you want to learn more about the analysis options available beyond what we cover in this tutorial, extensive documentation is available at: https://www.cog-genomics.org/plink/1.9/general_usage. To download Plink, please go to https://www.cog-genomics.org/plink2 and download the appropriate version for your computer platform. It should appear as in Fig. 3.5.

FIG. 3.3

R console. Screen capture of the R console.

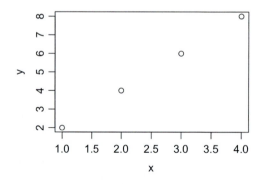

FIG. 3.4

R output. If the commands have worked as expected, this dot plot is the output one should expect.

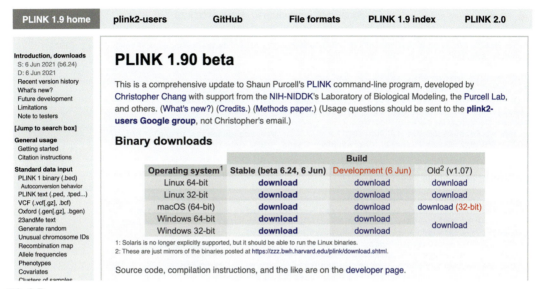

FIG. 3.5

Plink download page. Plink website has a download page for different operating systems.

For MacOS X

Click the stable version for macOS (64-bit) and unzip the downloaded file. Move the file called *plink* to your *GWAS_workshop* folder. Plink is a command line program (clicking on an icon with the mouse will get you nowhere). Open a Terminal window (if you have not used Terminal before, search for "Terminal" in a Find window to see where it located, double click to open). Try typing these commands in your Terminal window:

The first command to change the active directory to the *GWAS_workshop* folder:

    cd Desktop/GWAS_workshop/

The second command to run plink:

    ./plink

We have not directed plink to any input files, which is why we get the warning message, but this screen verifies plink is working on your computer (as in Fig. 3.6).

For windows

Click the stable version for Windows system and unzip the downloaded file. Move the file called *plink.exe* to your GWAS_workshop folder. Plink is a command line program (clicking on an icon with the mouse will get you nowhere). Open DOS windows by selecting "Command Prompt" from the start menu, or searching for "cmd" in the "Run" option of the start menu. Try typing these commands in your DOS window:

The first command to change the active directory to the *GWAS_workshop* folder:

    cd Desktop/GWAS_workshop/

The second command to run plink:

    ./plink

```
GWAS_workshop — -bash — 97×28
Last login: Sun Jun 27 22:18:47 on ttys001
4096 SHA256:dn9GzvzjHgdyHzmmIWPl7DucTFLJJ3MUfZRVzkxlUvs fanwang@MacBook-Pro-2.local (RSA)
ssh key checked

The default interactive shell is now zsh.
To update your account to use zsh, please run `chsh -s /bin/zsh`.
For more details, please visit https://support.apple.com/kb/HT208050.
(base) fanwang@macbook-pro-2:~$ cd Desktop/GWAS_workshop/
(base) fanwang@macbook-pro-2:~/Desktop/GWAS_workshop$ ./plink
PLINK v1.90b6.24 64-bit (6 Jun 2021)          www.cog-genomics.org/plink/1.9/
(C) 2005-2021 Shaun Purcell, Christopher Chang   GNU General Public License v3

  plink <input flag(s)...> [command flag(s)...] [other flag(s)...]
  plink --help [flag name(s)...]

Commands include --make-bed, --recode, --flip-scan, --merge-list,
--write-snplist, --list-duplicate-vars, --freqx, --missing, --test-mishap,
--hardy, --mendel, --ibc, --impute-sex, --indep-pairphase, --r2, --show-tags,
--blocks, --distance, --genome, --homozyg, --make-rel, --make-grm-gz,
--rel-cutoff, --cluster, --pca, --neighbour, --ibs-test, --regress-distance,
--model, --bd, --gxe, --logistic, --dosage, --lasso, --test-missing,
--make-perm-pheno, --tdt, --qfam, --annotate, --clump, --gene-report,
--meta-analysis, --epistasis, --fast-epistasis, and --score.

"plink --help | more" describes all functions (warning: long).
(base) fanwang@macbook-pro-2:~/Desktop/GWAS_workshop$ |
```

FIG. 3.6

Check if plink works properly. The plink version and simple usage instruction will appear in the R console, if the plink is installed properly.

We have not directed plink to any input files, which is why we get the warning message, but this screen verifies plink is working on your computer.

Plink data formats

PED files

To see an example PED file, open *PGx_class_data_chr19_region.ped* with any text editor. PED files end with *.ped* and are white-space (space or tab) delimited. The first six columns are required:

Family ID
Individual ID
Paternal ID
Maternal ID
Sex (1=male; 2=female; other=unknown)
phenotype

In our example of PED file, the individuals are unrelated, so the family IDs and individual IDs are the same and the paternal IDs are zeros. The phenotype in the example file is a quantitative trait and missing values are coded as -9 in PED files. The next 16 columns starting with column 7 are the genotypes for 8 SNPs. Each SNP has two alleles specified (A, T, C, or G) and 0 is the missing genotype character. Either both alleles should be missing (i.e., 0) or neither.

MAP files

To see the corresponding example MAP file, open *PGx_class_data_chr19_region.map* with any text editor. MAP files end with *.map* and are white-space (space or tab) delimited. By default, each line of the MAP file describes a single marker and must contain exactly four columns:

Chromosome (1–22, X, Y, MT or 0=unknown)
SNP identifier or rs ID
Genetic distance (position in morgans or centimorgans)
Base-pair coordinate (bp units)

In our example MAP file, 8 SNPs from chromosome 19 are listed in the order their genotypes appear in the PED file. For our purposes, genetic distance measurements are not necessary, so everything in the third column is set to 0.

BED (binary PED) files

BED file (.bed) is a binary PED file, which is more efficient than PED file in terms of disk space and processing time. The only sacrifice is that we have to store the pedigree and phenotype data in separate FAM file (.fam) and store the allele names information in an extended MAP file (.bim). Otherwise this information will be lost in the BED file. The set of .bed, .bim, and .fam files could be created using the function --make-bed in plink.

FAM files

Accompanying a BED file, a FAM (.fam) file describes a single person by row with the following six fields:

Family ID (FID)
Individual ID (IID)
Paternal ID
Maternal ID
Sex (1=male; 2=female; 0=unknown)
Phenotype value (1=control, 2=case, -9/0=missing data)

The FAM files are still plain text files: these can be viewed with a standard text editor. Please note that .fam file has no header line.

BIM files

Accompanying a BED file, BIM (.bim) file contains extended variant (SNP) information. Each row describes a variant with the following six fields:

Chromosome (1–22, X, Y, MT or 0=unknown)
SNP identifier or rs ID
Genetic distance (position in morgans or centimorgans)
Base-pair coordinate (bp units)
Allele 1 (clear bits in .bed; usually minor allele)
Allele 2 (set bits in .bed; usually major allele)

The BIM files are still plain text files: These can be viewed with a standard text editor. Please note that. bim file has no header line.

Test run

Make sure the executable plink file and the *PGx_class_data_chr19_region.ped* and *PGx_class_data_chr19_region.map* are in the same directory (folder) on your computer. Open a terminal or DOS window and change to the directory with your data. Run the following command:

```
./plink --file PGx_class_data_chr19_region --linear --out test
```

(*./plink* on MacOS X, *plink* on Windows, everything else is the same. The "./" simply tells the Terminal window to search the current directory for the plink executable).

The screen output should look like Fig. 3.7

The results of the linear regression for each SNP are in the file *test.assoc.linear*, which should look like what is seen in Fig. 3.8.

Each column means:

CHR: Chromosome
SNP: SNP identifier
BP: Physical position (base pair)
A1: Tested allele (minor allele by default)
TEST: Code for the test, default ADD meaning the additive effects of allele dosage
NMISS: Number of nonmissing individuals included in analysis
BETA: Regression coefficient (slope)
STAT: Coefficient *t*-statistic
P: *P*-value for *t*-statistic

Learn more about the significant SNPs by looking them up in the UCSC Genome Browser at http://genome.ucsc.edu/cgi-bin/hgGateway.

From the test run, you will also see a message indicating that the log file will be saved in test.log. It basically logs the above plink command you just ran.

```
[(base) fanwang@macbook-pro-2:~/Desktop/GWAS_workshop$ ./plink --file PGx_class_data_chr19_region --linear --out test
PLINK v1.90b6.24 64-bit (6 Jun 2021)           www.cog-genomics.org/plink/1.9/
(C) 2005-2021 Shaun Purcell, Christopher Chang    GNU General Public License v3
Logging to test.log.
Options in effect:
  --file PGx_class_data_chr19_region
  --linear
  --out test

16384 MB RAM detected; reserving 8192 MB for main workspace.
.ped scan complete (for binary autoconversion).
Performing single-pass .bed write (8 variants, 176 people).
--file: test-temporary.bed + test-temporary.bim + test-temporary.fam written.
8 variants loaded from .bim file.
176 people (96 males, 80 females) loaded from .fam.
175 phenotype values loaded from .fam.
Using 1 thread (no multithreaded calculations invoked).
Before main variant filters, 176 founders and 0 nonfounders present.
Calculating allele frequencies... done.
Total genotyping rate is 0.78267.
8 variants and 176 people pass filters and QC.
Phenotype data is quantitative.
Writing linear model association results to test.assoc.linear ... done.
```

FIG. 3.7

Plink output. Once plink command finished running, it tells you the results are written to an output.

```
[(base) fanwang@macbook-pro-2:~/Desktop/GWAS_workshop$ head test.assoc.linear
 CHR         SNP         BP   A1     TEST    NMISS       BETA       STAT          P
  19   rs10416706   20349698   T      ADD      166   -0.007796    -0.131      0.896
  19   rs12972967   20358400   T      ADD       89    -0.5186     -5.965   5.143e-08
  19   rs10413538   20370690   T      ADD       86    -0.6752     -7.292   1.555e-10
  19    rs7257475   20372113   T      ADD       88    -0.6708     -7.36    1.023e-10
  19     rs918442   20385941   A      ADD      166     0.0137      0.2417     0.8093
  19   rs10407670   20386099   C      ADD      169    -0.1444     -1.444      0.1506
  19   rs10414884   20386247   C      ADD      166    -0.1708     -1.897      0.05956
  19    rs2701312   20386516   G      ADD      164     0.01164     0.2017     0.8404
```

FIG. 3.8

Take a quick look at the output file test.assoc.linear. Print the top 9 rows of the output file test.assoc.linear.

Genome-wide example

Step 1. Examine phenotype data

Next, we will go through an example of a GWAS. The data come from the Yoruba population from Ibadan, Nigeria. Lymphoblastoid cell lines (LCLs) derived from individuals of this population have been treated with increasing concentration of Drug Y to determine the concentration at which 50% growth inhibition (IC_{50}) occurs for each cell line. A total of 256,896 SNPs have been genotyped in this population. Before we run the GWAS, we will examine the distribution of the phenotype (IC_{50}) in R.

To load the phenotype data, open R or Rstudio and change to the directory that contains the file *DrugY_phenotypes_for_R.txt* (MacOS X: Click "Misc," then under "Change Working Directory," select the folder that contains the files; Windows: Click on "File," then under "Change dir," select the folder that contains the files). Missing values are coded as NA. To load the data into R, run the following command (hint: if you copy and paste, do not include the initial >):

> data <- read.table("DrugY_phenotypes_for_R.txt", header=T)

To view the first few lines of the phenotype data:

```
> data[1:3,]
  FID  IID drugY_IC50
1 1001 1001   5.594256
2 1002 1002   8.525633
3 1003 1003  12.736739
```

To view the distribution of the phenotype:

> hist(data$drugY_IC50)

This command makes a histogram as seen in Fig. 3.9.

The histogram look skewed to the right (nonnormal); therefore, we can try a log_2 transformation to make the phenotype more normal using:

> log2_ic50 <- log2(data$drugY_IC50)
> hist(log2_ic50)

As seen in Fig. 3.10.

The log_2-transformed phenotype distribution looks better. We can verify the log_2-transformed phenotype is consistent with normality by running a Shapiro-Wilk test. The Shapiro-Wilk test tests the null hypothesis that a sample came from a normally distributed population. *P*-values <0.05 indicate a

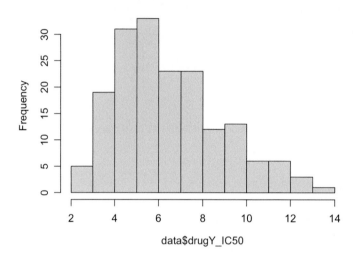

FIG. 3.9

Phenotype distribution. Use hist function in R to visualize the distribution of drug Y IC50.

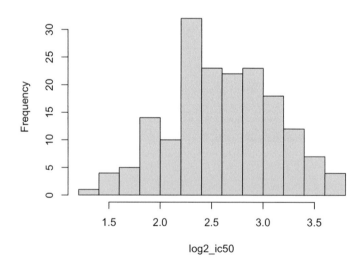

FIG. 3.10

Log transformed phenotype distribution. Use hist function in R to visualize the distribution of log-transformed drug Y IC50.

sample is likely not normally distributed. Compare the results of the Shapiro-Wilk test for the untransformed and log$_2$-transformed phenotypes:

```
> shapiro.test(data$drugY_IC50)
```

Shapiro-Wilk normality test

```
data: data$drugY_IC50
W = 0.9515, p-value = 1.033e-05
> shapiro.test(log2_ic50)
```

Shapiro-Wilk normality test

```
data: log2_ic50
W = 0.99074, p-value = 0.318
```

Step 2. Genome-wide analysis

Now that we are comfortable using the log$_2$ (*IC*50) as our phenotype, we are ready to run the GWAS in plink. To save space and time, we will use a binary PED file (ends with *.bed*). Here is a list of the files we will use:

PGx_class_data.bed (binary file with genotype information)
PGx_class_data.fam (first six columns of a .ped file)
PGx_class_data.bim (extended MAP file: two extra columns for allele names)

The *.fam* and *.bim* files are still plain text files: These can be viewed with a standard text editor. Do not try to view the *.bed* file: it is a compressed file and you'll only see lots of strange characters on the screen. The reformatting between .bed/.fam/.bim and .ped/.map is easy by using the command "plink --bfile file_name --recode --out file_name".

Rather than including the phenotype in the .fam file, this time we have a separate phenotype file called *DrugY_phenotypes_for_plink.txt*. A separate phenotype file allows multiple phenotypes to be tested for association with the same genotype data. Missing values are coded as -9. The first few lines of the phenotype file are:

```
FID IID drugY_IC50 log2_drugY_IC50
1001 1001 5.594255712. 2.4839462
1002 1002 8.52563251. 3.09180687
1003 1003 12.7367394 3.67092409
```

The first two columns must contain the family ID (FID) and individual ID (IID). The following columns contain as many phenotypes as wanted.

Run the following command:

```
./plink --bfile PGx_class_data --maf 0.05 --linear --adjust --pheno DrugY_phenotypes_for_plink.txt --pheno-name log2_drugY_IC50 --out PGx_class_data.log2_drugY_IC50
```

Here is an explanation of each option used:

--adjust: tells plink to adjust the *P*-values for each SNP using various multiple-testing correlation methods (see below)
--bfile: tells plink your data files that begin with *PGx_class_data* are in binary format

--linear: tells plink to run a linear additive association test for each SNP
--maf: tells plink to filter out any SNPs with a minor allele frequency < 0.05
--out: tells plink the text to begin each output file (*PGx_class_data.log2_drugY_IC50* in this case)
--pheno: tells plink the phenotype is located in a separate phenotype file (*DrugY_phenotypes_for_plink.txt* in this case)
--pheno-name: tells plink the column heading of the phenotype to use in the phenotype file (*log2_drugY_IC50* in this case)

This run should take a minute or two depending on your computer, and will produce three output files:

PGx_class_data.log2_drugY_IC50.log
PGx_class_data.log2_drugY_IC50.assoc.linear
PGx_class_data.log2_drugY_IC50.assoc.linear.adjusted

The .log file contains the same information that is output to your screen during the run. The *.assoc.linear* file is the same as the test run above, just this time it contains data for 254,424 SNPs. Here are the first 10 lines:

```
CHR SNP BP A1 TEST NMISS BETA STAT P
1 rs9699599 558185 G ADD 85 0.2769 1.898 0.06115
1 rs12138618 740098 A ADD 86 -0.07265 -0.3671 0.7145
1 rs3131969 744045 G ADD 164 0.06787 1.135 0.258
1 rs6672353 767376 A ADD 85 0.04227 0.2873 0.7746
1 rs13302982 851671 G ADD 166 0.08407 1.414 0.1591
1 rs3121567 933331 T ADD 166 -0.07329 -1.341 0.1818
1 rs6667248 1007079 G ADD 166 -0.05232 -0.8134 0.4172
1 rs3766191 1007450 T ADD 166 0.009789 0.1489 0.8818
1 rs6700376 1079693 G ADD 173 0.08495 1.462 0.1455
```

Notice the output is sorted by chromosome and base pair.

The .assoc.linear.adjusted file, on the other hand, is sorted by the most significant results. Here are the first 10 lines:

```
CHR SNP UNADJ GC BONF HOLM SIDAK_SS SIDAK_SD FDR_BH FDR_BY
19 rs7257475 1.023e-10 5.146e-10 2.603e-05 2.603e-05 2.603e-05 2.603e-05 1.978e-05 0.0002576
19 rs10413538 1.555e-10 7.571e-10 3.955e-05 3.955e-05 3.955e-05 3.955e-05 1.978e-05 0.0002576
19 rs12972967 5.143e-08 1.764e-07 0.01309 0.01309 0.013 0.013 0.004362 0.05681
2 rs10170982 1.344e-07 4.707e-07 0.0342 0.0342 0.03362 0.03362 0.008549 0.1113
2 rs10186803 2.097e-07 7.067e-07 0.05335 0.05335 0.05195 0.05195 0.01048 0.1365
2 rs17025871 2.472e-07 8.262e-07 0.06289 0.06289 0.06096 0.06095 0.01048 0.1365
2 rs12622974 3.609e-07 1.171e-06 0.09183 0.09182 0.08774 0.08773 0.01312 0.1708
2 rs13394005 4.294e-07 1.374e-06 0.1093 0.1093 0.1035 0.1035 0.01366 0.1779
21 rs2826383 8.907e-07 2.68e-06 0.2266 0.2266 0.2028 0.2028 0.02518 0.3279
```

The columns corresponding to:

CHR: Chromosome number
SNP: SNP identifier
UNADJ: Unadjusted *P*-value
GC: Genomic-control corrected *P*-values
BONF: Bonferroni single-step adjusted *P*-values
HOLM: Holm [41] step-down adjusted *P*-values
SIDAK_SS: Sidak single-step adjusted *P*-values
SIDAK_SD: Sidak single-down adjusted *P*-values
FDR_BH: Benjamini and Hochberg [42] step-up FDR control
FDR_BY: Benjamini and Yekutieli [43] step-up FDR control

Step 3. Plot results
Manhattan and Q-Q plots

Manhattan and quantile-quantile (QQ) plots are commonly used to summarize the genome-wide association findings. Using the output from the .assoc.linear file and an R script, we can easily make Manhattan and Q-Q plots of our data. Open R and change to the directory that contains your .assoc.linear file and the file called qqman.r.

Read your data and the script than makes the Manhattan and Q-Q plots into R:

```
> data <- read.table("PGx_class_data.log2_drugY_IC50.assoc.linear", header = T)
> source("qqman.r")
> tiff("manplot.tiff", res = 300, width = 7, height = 4, units = 'in')
> manhattan(data)
> dev.off()
```

The generated Manhattan plot will be located in your GWAS_workshop folder and should like this in Fig. 3.11.

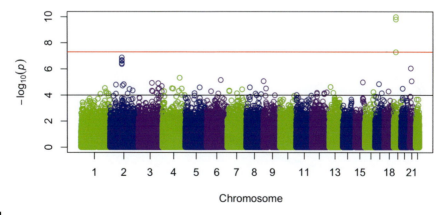

FIG. 3.11

Manhattan plot. Manhattan plot represent the −log10 of *P* values in genomic order by chromosome and position on the chromosome (*x*-axis).

To make a Q-Q plot called qqplot.tiff:

```
> tiff("qqplot.tiff", res = 300, width = 5, height = 4, units = 'in')
> qq(data$P)
> dev.off()
```

The generated Q-Q plot will be located in your GWAS_workshop folder and should like this in Fig. 3.12.

Manhattan plot interpretation

A Manhattan plot represents the association's statistical significance as $-\log_{10}$ (P-value) in the y-axis against chromosome coordinates in the x-axis. Each dot corresponds to one variant, and the higher a variant ($-\log 10$ of the association P-value), the more significantly associated it is with the corresponding SNP with the phenotype your GWAS focused on (Drug Y IC_{50} in our example). You may also specify the significant threshold (P-value=0.0001 and P-value=$-5e^{-8}$ in our case) as horizontal lines in a Manhattan plot. As you can tell, the Manhattan plot is a convenient way to display millions of genetic variants across all the chromosomes in a single plot. You can easily spot regions of the genome that above the specific significance thresholds that you defined. In our example, we can take the region in chromosome 19 for further exploration.

Using the small R script *qqman.r* in the GWAS_workshop folder, we can generate a basic Manhattan plot. However, due to the lack of further annotation information with gene names, recombination rates, linkage disequilibrium (LD) measures, etc., we only have a rough idea that there may be some high-impact consequence variants in chromosome 19. In order to overcome the limitation to annotate the Manhattan plots, researchers have developed several softwares including but not limited to LocusZoom [44], Manhattan++ [45], GWAMA [46], to add more annotations like gene names, allele frequencies, recombination rates, linkage disequilibrium, variant consequence, and so on. The use of LocusZoom will be discussed in detail in a subsequent section.

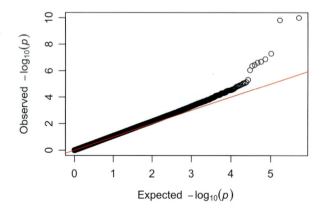

FIG. 3.12

Q-Q plot. It is common to use Q-Q plot to find the type of distribution for a random variable.

Q-Q plot interpretation

The Q-Q (quantile-quantile) plot is used to assess the number and the magnitude of observed associations between SNPs and disease or trait under a study, compared to the association statistics expected under the null hypothesis of no association. Observed $-\log_{10}$ (P-values) are ranked in order from smallest to largest on the y-axis, and plotted against the distribution that would be expected under the null hypothesis of no association on the x-axis. Deviations from the identity line ($y=x$) suggest either that the assumed distribution is incorrect or that the sample contains values arising in some other manner, as by a true association. In GWAS, it is often assumed that only a few SNP associations will be significant and thus large deviations from the identity line are thought to be due to the population structure or nonnormality of the phenotype tested. After attempts to correct for such confounders have been made, some residual inflation of the test statistics could remain. One method to correct for this is genomic control [47]. The genomic control inflation factor (λ) is computed as the median of all genome-wide observed test statistics divided by the expected median of the test statistic under the null hypothesis of no association (making the assumption that the number of true associations is very small compared to the millions of tests performed). The GWAS results are then corrected for residual inflation of the test statistic by dividing the observed test statistic at each SNP by the λ. Genomic control corrected P-values can be found in the .assoc.linear.adjusted plink output file.

LocusZoom plot

LocusZoom [44] is a web-based tool to facilitate viewing the regional association results from genome-wide association scans or candidate gene studies. Using the .assoc.linear file from the plink output, we can examine the genomic regions surrounding our top SNPs. As an example, we will look at the region surrounding one of our top hits rs7257475 in the following steps:

1. Go to the LocusZoom website http://locuszoom.org/.
2. Click "Single Plot (Your Data—Original LocusZoom)".
3. Enter the path to your PGx_class_data.log2_drugY_IC50.assoc.linear file.
4. Enter the following information into the boxes:
5. Click "Plot Data".
6. LocusZoom will generate a file called YRI.rs7257475.400 kb.pdf. It should look like this, as in Figs. 3.13–3.15.

The LocusZoom main panel shows the drugY IC50-associated P-values on the $-\log_{10}$ scale on the y-axis on the left side, and the chromosomal position along the x-axis in the lower panel. In this case, we specify the region to display via an index SNP (rs7257475) and a window size (a region of 400 kb on either side of it). From the LocusZoom lower panel, we can easily visualize the genomic region near the rs7257475 locus, which is *ZNF826* gene, including the location and orientation of the gene. The y-axis on the right side shows the local estimates of recombination rates (in centimorgan per megabase) from HapMap Phase II.

The blue line (recombination rate) gives us an overview of how robust the recombination patterns are according to the population-scale variation catalog database such as HapMap (CEU, YRI, and JPT+CHB) [5] and 1000 Genomes [48] projects. Genetic recombination is a crucial biological process that contributes to the genetic diversity by recombining alleles and segregating chromosomes [49]. Genetic recombination over generations help avoid deleterious chromosomal abnormalities such as fetal aneuploidy [49].

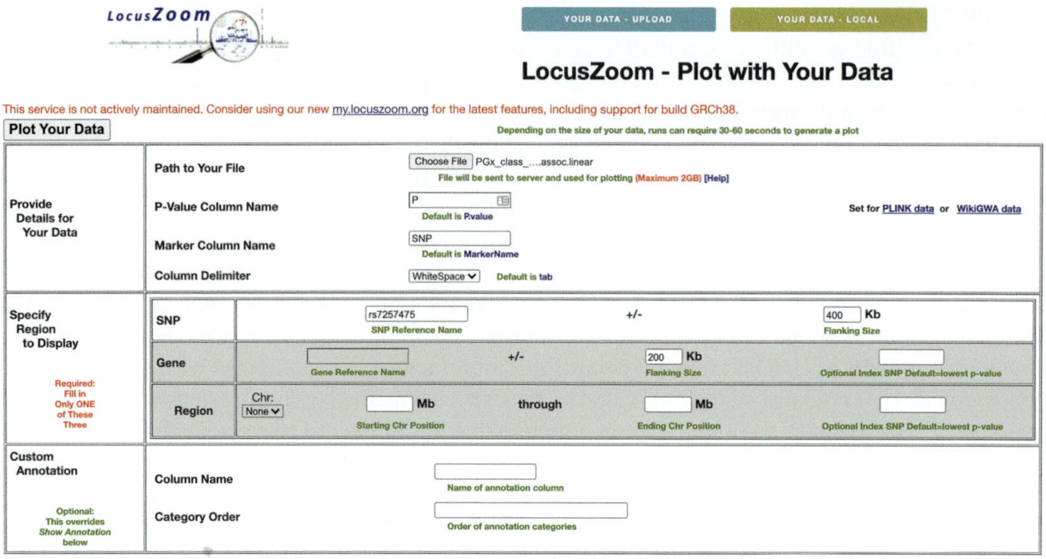

74 Chapter 3 Genome-wide association studies

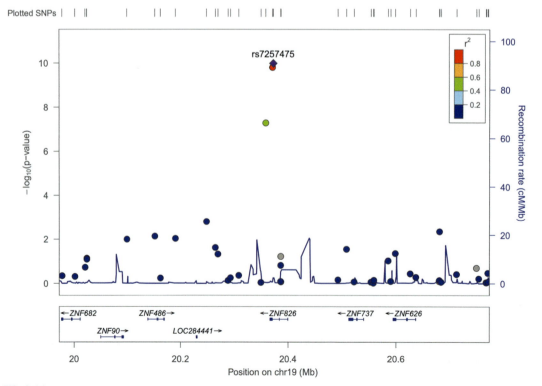

FIG. 3.13

LocusZoom initiation page. Define data sources and region to display in the LocusZoom initiation page.

FIG. 3.14

LocusZoom output. A plot of the rs7257475 locus is created.

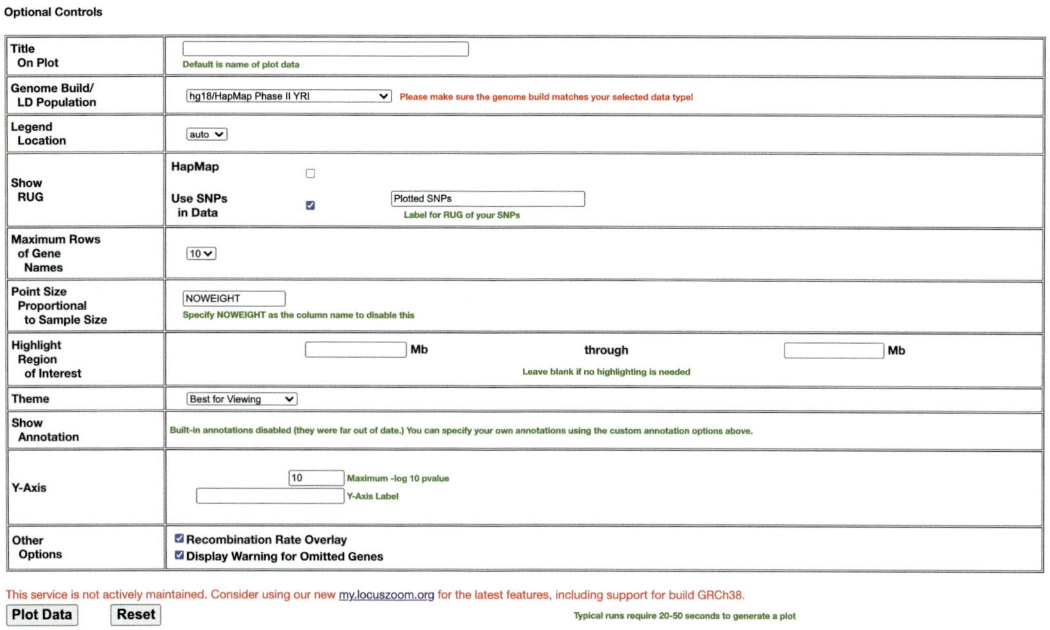

FIG. 3.15

LocusZoom initiation page (continued). Define more customized output layout in the LocusZoom initiation page.

Boxplot of top SNP

It is useful to view the phenotype-genotype relationship of your top hit SNPs. We can use R to make a boxplot of our drug phenotype against variant rs7257475 phenotype, but first, we must use plink to pull out the genotypes of rs7257475 in a format appropriate for R.

Run the following command in your plink directory:

```
./plink --bfile PGx_class_data --snp rs7257475 --recodeA --out rs7257475_genotypes
```

This will make the file called rs7257475_genotypes.raw. The --recodeA command tells plink to create a file with SNP genotypes recoded in terms of additive components. The first few lines are:

```
FID IID PAT MAT SEX PHENOTYPE rs7257475_T
1001 1001 0 0 1 -9 0
1002 1002 0 0 1 -9 0
1003 1003 0 0 2 -9 0
1004 1004 0 0 1 -9 0
```

The default calculation for the additive recoding in plink is to count the number of minor alleles per individual. The name of the last column reflects the SNP name (e.g., rs7257475) with the name of the minor allele appended (e.g., T). In our case, the column name "rs7257475_T" indicates that T is the minor allele. Homozygotes for the major allele are coded as 0, heterozygotes as 1, homozygotes for the minor allele as 2, and missing data as NA in .raw files.

Chapter 3 Genome-wide association studies

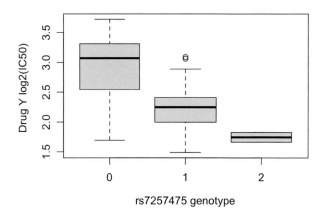

FIG. 3.16

Genotype boxplot. Boxplot of phenotype by genotype of rs7257475 represents that the more copies of minor allele T, the higher sensitivity to drug Y.

Open R and change to the directory that contains your .raw file and the file called DrugY_phenotypes_for_R.txt (MacOS X: Click "Misc", then under "Change Working Directory", select the folder that contains the files; Windows: Click on "File", then under "Change dir", select the folder that contains the files). Run the following commands:

```
> gts <- read.table("rs7257475_genotypes.raw", header = T)
> pts <- read.table("DrugY_phenotypes_for_R.txt", header = T)
> boxplot(log2(pts$drugY_IC50) ~ gts$rs7257475_T, ylab = "Drug Y log2(IC50)",
xlab = "rs7257475 genotype")
```

This should generate a boxplot like this, as in Fig. 3.16.

From the genotype boxplot, we can see that increasing copies of the minor allele T at rs7257475 locus are associated with increased sensitivity to drug Y. This plot can be saved by clicking on "File" > "Save As" in R, or "Export" > "Save as image" in Rstudio.

References

[1] E.J. Topol, S.S. Murray, K.A. Frazer, The genomics gold rush, JAMA 298 (2007) 218–221.
[2] K. Ozaki, Y. Ohnishi, A. Iida, A. Sekine, R. Yamada, T. Tsunoda, H. Sato, H. Sato, M. Hori, Y. Nakamura, T. Tanaka, Functional SNPs in the lymphotoxin-alpha gene that are associated with susceptibility to myocardial infarction, Nat. Genet. 32 (2002) 650–654.
[3] International Human Genome Sequencing Consortium, Finishing the euchromatic sequence of the human genome, Nature 431 (2004) 931–945.
[4] J. Schmutz, J. Wheeler, J. Grimwood, M. Dickson, J. Yang, C. Caoile, E. Bajorek, S. Black, Y.M. Chan, M. Denys, J. Escobar, D. Flowers, D. Fotopulos, C. Garcia, M. Gomez, E. Gonzales, L. Haydu, F. Lopez, L. Ramirez, J. Retterer, A. Rodriguez, S. Rogers, A. Salazar, M. Tsai, R.M. Myers, Quality assessment of the human genome sequence, Nature 429 (2004) 365–368.
[5] International HapMap Consortium, The international HapMap project, Nature 426 (2003) 789–796.

References

[6] International HapMap Consortium, A haplotype map of the human genome, Nature 437 (2005) 1299–1320.
[7] M.D. Mailman, M. Feolo, Y. Jin, M. Kimura, K. Tryka, R. Bagoutdinov, L. Hao, A. Kiang, J. Paschall, L. Phan, N. Popova, S. Pretel, L. Ziyabari, M. Lee, Y. Shao, Z.Y. Wang, K. Sirotkin, M. Ward, M. Kholodov, K. Zbicz, J. Beck, M. Kimelman, S. Shevelev, D. Preuss, E. Yaschenko, A. Graeff, J. Ostell, S.T. Sherry, The NCBI dbGaP database of genotypes and phenotypes, Nat. Genet. 39 (2007) 1181–1186.
[8] M. Lek, K.J. Karczewski, E.V. Minikel, K.E. Samocha, E. Banks, T. Fennell, A.H. O'Donnell-Luria, J.S. Ware, A.J. Hill, B.B. Cummings, T. Tukiainen, D.P. Birnbaum, J.A. Kosmicki, L.E. Duncan, K. Estrada, F. Zhao, J. Zou, E. Pierce-Hoffman, J. Berghout, D.N. Cooper, N. Deflaux, M. DePristo, R. Do, J. Flannick, M. Fromer, L. Gauthier, J. Goldstein, N. Gupta, D. Howrigan, A. Kiezun, M.I. Kurki, A.L. Moonshine, P. Natarajan, L. Orozco, G.M. Peloso, R. Poplin, M.A. Rivas, V. Ruano-Rubio, S.A. Rose, D.M. Ruderfer, K. Shakir, P.D. Stenson, C. Stevens, B.P. Thomas, G. Tiao, M.T. Tusie-Luna, B. Weisburd, H.-H. Won, D. Yu, D.M. Altshuler, D. Ardissino, M. Boehnke, J. Danesh, S. Donnelly, R. Elosua, J.C. Florez, S.B. Gabriel, G. Getz, S.J. Glatt, C.M. Hultman, S. Kathiresan, M. Laakso, S. McCarroll, M.I. McCarthy, D. McGovern, R. McPherson, B.M. Neale, A. Palotie, S.M. Purcell, D. Saleheen, J.M. Scharf, P. Sklar, P.F. Sullivan, J. Tuomilehto, M.T. Tsuang, H.C. Watkins, J.G. Wilson, M.J. Daly, D.G. MacArthur, Analysis of protein-coding genetic variation in 60,706 humans, Nature 536 (2016) 285–291, https://doi.org/10.1038/nature19057.
[9] C. Sudlow, J. Gallacher, N. Allen, V. Beral, P. Burton, J. Danesh, P. Downey, P. Elliott, J. Green, M. Landray, B. Liu, P. Matthews, G. Ong, J. Pell, A. Silman, A. Young, T. Sprosen, T. Peakman, R. Collins, UK biobank: an open access resource for identifying the causes of a wide range of complex diseases of middle and old age, PLoS Med. 12 (2015) e1001779, https://doi.org/10.1371/journal.pmed.1001779.
[10] J. Dupuis, C. Langenberg, I. Prokopenko, R. Saxena, N. Soranzo, A.U. Jackson, E. Wheeler, N.L. Glazer, N. Bouatia-Naji, A.L. Gloyn, C.M. Lindgren, R. Mägi, A.P. Morris, J. Randall, T. Johnson, P. Elliott, D. Rybin, G. Thorleifsson, V. Steinthorsdottir, P. Henneman, H. Grallert, A. Dehghan, J.J. Hottenga, C.S. Franklin, P. Navarro, K. Song, A. Goel, J.R.B. Perry, J.M. Egan, T. Lajunen, N. Grarup, T. Sparsø, A. Doney, B.F. Voight, H.M. Stringham, M. Li, S. Kanoni, P. Shrader, C. Cavalcanti-Proença, M. Kumari, L. Qi, N.J. Timpson, C. Gieger, C. Zabena, G. Rocheleau, E. Ingelsson, P. An, J. O'Connell, J. Luan, A. Elliott, S.A. McCarroll, F. Payne, R.M. Roccasecca, F. Pattou, P. Sethupathy, K. Ardlie, Y. Ariyurek, B. Balkau, P. Barter, J.P. Beilby, Y. Ben-Shlomo, R. Benediktsson, A.J. Bennett, S. Bergmann, M. Bochud, E. Boerwinkle, A. Bonnefond, L.L. Bonnycastle, K. Borch-Johnsen, Y. Böttcher, E. Brunner, S.J. Bumpstead, G. Charpentier, Y.-D.I. Chen, P. Chines, R. Clarke, L.J.M. Coin, M.N. Cooper, M. Cornelis, G. Crawford, L. Crisponi, I.N.M. Day, E.J.C. de Geus, J. Delplanque, C. Dina, M.R. Erdos, A.C. Fedson, A. Fischer-Rosinsky, N.G. Forouhi, C.S. Fox, R. Frants, M.G. Franzosi, P. Galan, M.O. Goodarzi, J. Graessler, C.J. Groves, S. Grundy, R. Gwilliam, U. Gyllensten, S. Hadjadj, G. Hallmans, N. Hammond, X. Han, A.-L. Hartikainen, N. Hassanali, C. Hayward, S.C. Heath, S. Hercberg, C. Herder, A.A. Hicks, D.R. Hillman, A.D. Hingorani, A. Hofman, J. Hui, J. Hung, B. Isomaa, P.R.V. Johnson, T. Jørgensen, A. Jula, M. Kaakinen, J. Kaprio, Y.A. Kesaniemi, M. Kivimaki, B. Knight, S. Koskinen, P. Kovacs, K.O. Kyvik, G.M. Lathrop, D.A. Lawlor, O. Le Bacquer, C. Lecoeur, Y. Li, V. Lyssenko, R. Mahley, M. Mangino, A.K. Manning, M.T. Martínez-Larrad, J.B. McAteer, L.J. McCulloch, R. McPherson, C. Meisinger, D. Melzer, D. Meyre, B.D. Mitchell, M.A. Morken, S. Mukherjee, S. Naitza, N. Narisu, M.J. Neville, B.A. Oostra, M. Orrù, R. Pakyz, C.N.A. Palmer, G. Paolisso, C. Pattaro, D. Pearson, J.F. Peden, N.L. Pedersen, M. Perola, A.F.H. Pfeiffer, I. Pichler, O. Polasek, D. Posthuma, S.C. Potter, A. Pouta, M.A. Province, B.M. Psaty, W. Rathmann, N.W. Rayner, K. Rice, S. Ripatti, F. Rivadeneira, M. Roden, O. Rolandsson, A. Sandbaek, M. Sandhu, S. Sanna, A.A. Sayer, P. Scheet, L.J. Scott, U. Seedorf, S.J. Sharp, B. Shields, G. Sigurethsson, E.J.G. Sijbrands, A. Silveira, L. Simpson, A. Singleton, N.L. Smith, U. Sovio, A. Swift, H. Syddall, A.-C. Syvänen, T. Tanaka, B. Thorand, J. Tichet, A. Tönjes, T. Tuomi, A.G. Uitterlinden, K.W. van Dijk, M. van Hoek, D. Varma, S. Visvikis-Siest, V. Vitart, N. Vogelzangs, G. Waeber, P.J. Wagner, A. Walley, G.B. Walters, K.L. Ward, H. Watkins, M.N. Weedon, S.H. Wild, G. Willemsen, J.C.M. Witteman, J.W.G. Yarnell, E. Zeggini, D. Zelenika, B. Zethelius, G. Zhai, J.H. Zhao, M.C. Zillikens, I.B. Borecki, R.J.F. Loos, P. Meneton, P.K.E. Magnusson, D.M. Nathan, G.H. Williams, A.T. Hattersley, K. Silander, V.

Salomaa, G.D. Smith, S.R. Bornstein, P. Schwarz, J. Spranger, F. Karpe, A.R. Shuldiner, C. Cooper, G.V. Dedoussis, M. Serrano-Ríos, A.D. Morris, L. Lind, L.J. Palmer, F.B. Hu, P.W. Franks, S. Ebrahim, M. Marmot, W.H.L. Kao, J.S. Pankow, M.J. Sampson, J. Kuusisto, M. Laakso, T. Hansen, O. Pedersen, P.P. Pramstaller, H.E. Wichmann, T. Illig, I. Rudan, A.F. Wright, M. Stumvoll, H. Campbell, J.F. Wilson, R.N. Bergman, T.A. Buchanan, F.S. Collins, K.L. Mohlke, J. Tuomilehto, T.T. Valle, D. Altshuler, J.I. Rotter, D.S. Siscovick, B.W.J.H. Penninx, D.I. Boomsma, P. Deloukas, T.D. Spector, T.M. Frayling, L. Ferrucci, A. Kong, U. Thorsteinsdottir, K. Stefansson, C.M. van Duijn, Y.S. Aulchenko, A. Cao, A. Scuteri, D. Schlessinger, M. Uda, A. Ruokonen, M.-R. Jarvelin, D.M. Waterworth, P. Vollenweider, L. Peltonen, V. Mooser, G.R. Abecasis, N.J. Wareham, R. Sladek, P. Froguel, R.M. Watanabe, J.B. Meigs, L. Groop, M. Boehnke, M.I. McCarthy, J.C. Florez, I. Barroso, New genetic loci implicated in fasting glucose homeostasis and their impact on type 2 diabetes risk, Nat. Genet. 42 (2010) 105–116, https://doi.org/10.1038/ng.520.

[11] A. Xue, Y. Wu, Z. Zhu, F. Zhang, K.E. Kemper, Z. Zheng, L. Yengo, L.R. Lloyd-Jones, J. Sidorenko, Y. Wu, A.F. McRae, P.M. Visscher, J. Zeng, J. Yang, Genome-wide association analyses identify 143 risk variants and putative regulatory mechanisms for type 2 diabetes, Nat. Commun. 9 (2018) 2941, https://doi.org/10.1038/s41467-018-04951-w.

[12] R.J.F. Loos, G.S.H. Yeo, The genetics of obesity: from discovery to biology, Nat. Rev. Genet. (2021), https://doi.org/10.1038/s41576-021-00414-z.

[13] T.M. Frayling, N.J. Timpson, M.N. Weedon, E. Zeggini, R.M. Freathy, C.M. Lindgren, J.R.B. Perry, K.S. Elliott, H. Lango, N.W. Rayner, B. Shields, L.W. Harries, J.C. Barrett, S. Ellard, C.J. Groves, B. Knight, A.-M. Patch, A.R. Ness, S. Ebrahim, D.A. Lawlor, S.M. Ring, Y. Ben-Shlomo, M.-R. Jarvelin, U. Sovio, A.J. Bennett, D. Melzer, L. Ferrucci, R.J.F. Loos, I. Barroso, N.J. Wareham, F. Karpe, K.R. Owen, L.R. Cardon, M. Walker, G.A. Hitman, C.N.A. Palmer, A.S.F. Doney, A.D. Morris, G.D. Smith, A.T. Hattersley, M.I. McCarthy, A common variant in the FTO gene is associated with body mass index and predisposes to childhood and adult obesity, Science 316 (2007) 889–894.

[14] D. Levy, G.B. Ehret, K. Rice, G.C. Verwoert, L.J. Launer, A. Dehghan, N.L. Glazer, A.C. Morrison, A.D. Johnson, T. Aspelund, Y. Aulchenko, T. Lumley, A. Köttgen, R.S. Vasan, F. Rivadeneira, G. Eiriksdottir, X. Guo, D.E. Arking, G.F. Mitchell, F.U.S. Mattace-Raso, A.V. Smith, K. Taylor, R.B. Scharpf, S.-J. Hwang, E.J.G. Sijbrands, J. Bis, T.B. Harris, S.K. Ganesh, C.J. O'Donnell, A. Hofman, J.I. Rotter, J. Coresh, E.J. Benjamin, A.G. Uitterlinden, G. Heiss, C.S. Fox, J.C.M. Witteman, E. Boerwinkle, T.J. Wang, V. Gudnason, M.G. Larson, A. Chakravarti, B.M. Psaty, C.M. van Duijn, Genome-wide association study of blood pressure and hypertension, Nat. Genet. 41 (2009) 677–687, https://doi.org/10.1038/ng.384.

[15] C.M. Lusk, G. Dyson, A.G. Clark, C.M. Ballantyne, R. Frikke-Schmidt, A. Tybjærg-Hansen, E. Boerwinkle, C.F. Sing, Validated context-dependent associations of coronary heart disease risk with genotype variation in the chromosome 9p21 region: the Atherosclerosis Risk in Communities study, Hum. Genet. 133 (2014) 1105–1116, https://doi.org/10.1007/s00439-014-1451-3.

[16] K.M. de Lange, L. Moutsianas, J.C. Lee, C.A. Lamb, Y. Luo, N.A. Kennedy, L. Jostins, D.L. Rice, J. Gutierrez-Achury, S.-G. Ji, G. Heap, E.R. Nimmo, C. Edwards, P. Henderson, C. Mowat, J. Sanderson, J. Satsangi, A. Simmons, D.C. Wilson, M. Tremelling, A. Hart, C.G. Mathew, W.G. Newman, M. Parkes, C.W. Lees, H. Uhlig, C. Hawkey, N.J. Prescott, T. Ahmad, J.C. Mansfield, C.A. Anderson, J.C. Barrett, Genome-wide association study implicates immune activation of multiple integrin genes in inflammatory bowel disease, Nat. Genet. 49 (2017) 256–261, https://doi.org/10.1038/ng.3760.

[17] X. Zhu, W. Bai, H. Zheng, Twelve years of GWAS discoveries for osteoporosis and related traits: advances, challenges and applications, Bone Res. 9 (2021) 23, https://doi.org/10.1038/s41413-021-00143-3.

[18] D. Altshuler, J.N. Hirschhorn, M. Klannemark, C.M. Lindgren, M.C. Vohl, J. Nemesh, C.R. Lane, S.F. Schaffner, S. Bolk, C. Brewer, T. Tuomi, D. Gaudet, T.J. Hudson, M. Daly, L. Groop, E.S. Lander, The common PPARgamma Pro12Ala polymorphism is associated with decreased risk of type 2 diabetes, Nat. Genet. 26 (2000) 76–80.

[19] J.D. Terwilliger, K.M. Weiss, Linkage disequilibrium mapping of complex disease: fantasy or reality? Curr. Opin. Biotechnol. 9 (1998) 578–594.
[20] T. Emahazion, L. Feuk, M. Jobs, S.L. Sawyer, D. Fredman, D.S. Clair, J.A. Prince, A.J. Brookes, SNP association studies in Alzheimer's disease highlight problems for complex disease analysis, Trends Genet. 17 (2001) 407–413.
[21] J.H. McDonald, Handbook of Biological Statistics, third ed., Sparky House Publishing, Baltimore, Maryland, 2014. http://www.biostathandbook.com.
[22] R.D. Ball, Bayesian methods for quantitative trait loci mapping based on model selection: approximate analysis using the Bayesian information criterion, Genetics 159 (2001) 1351–1364.
[23] B.E. Stranger, A.C. Nica, M.S. Forrest, A. Dimas, C.P. Bird, C. Beazley, C.E. Ingle, M. Dunning, P. Flicek, D. Koller, S. Montgomery, S. Tavaré, P. Deloukas, E.T. Dermitzakis, Population genomics of human gene expression, Nat. Genet. 39 (2007) 1217–1224.
[24] A. Das, M. Morley, C.S. Moravec, W.H.W. Tang, H. Hakonarson, K.B. Margulies, T.P. Cappola, S. Jensen, S. Hannenhalli, Bayesian integration of genetics and epigenetics detects causal regulatory SNPs underlying expression variability, Nat. Commun. 6 (2015) 8555, https://doi.org/10.1038/ncomms9555.
[25] S.-I. Lee, A.M. Dudley, D. Drubin, P.A. Silver, N.J. Krogan, D. Pe'er, D. Koller, Learning a prior on regulatory potential from eQTL data, PLoS Genet. 5 (2009) e1000358, https://doi.org/10.1371/journal.pgen.1000358.
[26] D.J. Balding, A tutorial on statistical methods for population association studies, Nat. Rev. Genet. 7 (2006) 781–791.
[27] The Wellcome Trust Case Control Consortium, Genome-wide association study of 14,000 cases of seven common diseases and 3,000 shared controls, Nature 447 (2007) 661–678.
[28] H. Hakonarson, S.F.A. Grant, J.P. Bradfield, L. Marchand, C.E. Kim, J.T. Glessner, R. Grabs, T. Casalunovo, S.P. Taback, E.C. Frackelton, M.L. Lawson, L.J. Robinson, R. Skraban, Y. Lu, R.M. Chiavacci, C.A. Stanley, S.E. Kirsch, E.F. Rappaport, J.S. Orange, D.S. Monos, M. Devoto, H.-Q. Qu, C. Polychronakos, A genome-wide association study identifies KIAA0350 as a type 1 diabetes gene, Nature 448 (2007) 591–594.
[29] J.A. Todd, N.M. Walker, J.D. Cooper, D.J. Smyth, K. Downes, V. Plagnol, R. Bailey, S. Nejentsev, S.F. Field, F. Payne, C.E. Lowe, J.S. Szeszko, J.P. Hafler, L. Zeitels, J.H.M. Yang, A. Vella, S. Nutland, H.E. Stevens, H. Schuilenburg, G. Coleman, M. Maisuria, W. Meadows, L.J. Smink, B. Healy, O.S. Burren, A.A.C. Lam, N.R. Ovington, J. Allen, E. Adlem, H.-T. Leung, C. Wallace, J.M.M. Howson, C. Guja, C. Ionescu-Tîrgovişte, M.J. Simmonds, J.M. Heward, S.C.L. Gough, D.B. Dunger, L.S. Wicker, D.G. Clayton, Robust associations of four new chromosome regions from genome-wide analyses of type 1 diabetes, Nat. Genet. 39 (2007) 857–864.
[30] L.J. Scott, K.L. Mohlke, L.L. Bonnycastle, C.J. Willer, Y. Li, W.L. Duren, M.R. Erdos, H.M. Stringham, P.S. Chines, A.U. Jackson, L. Prokunina-Olsson, C.-J. Ding, A.J. Swift, N. Narisu, T. Hu, R. Pruim, R. Xiao, X.-Y. Li, K.N. Conneely, N.L. Riebow, A.G. Sprau, M. Tong, P.P. White, K.N. Hetrick, M.W. Barnhart, C.W. Bark, J.L. Goldstein, L. Watkins, F. Xiang, J. Saramies, T.A. Buchanan, R.M. Watanabe, T.T. Valle, L. Kinnunen, G.R. Abecasis, E.W. Pugh, K.F. Doheny, R.N. Bergman, J. Tuomilehto, F.S. Collins, M. Boehnke, A genome-wide association study of type 2 diabetes in Finns detects multiple susceptibility variants, Science 316 (2007) 1341–1345.
[31] R. Sladek, G. Rocheleau, J. Rung, C. Dina, L. Shen, D. Serre, P. Boutin, D. Vincent, A. Belisle, S. Hadjadj, B. Balkau, B. Heude, G. Charpentier, T.J. Hudson, A. Montpetit, A.V. Pshezhetsky, M. Prentki, B.I. Posner, D.J. Balding, D. Meyre, C. Polychronakos, P. Froguel, A genome-wide association study identifies novel risk loci for type 2 diabetes, Nature 445 (2007) 881–885.
[32] V. Steinthorsdottir, G. Thorleifsson, I. Reynisdottir, R. Benediktsson, T. Jonsdottir, G.B. Walters, U. Styrkarsdottir, S. Gretarsdottir, V. Emilsson, S. Ghosh, A. Baker, S. Snorradottir, H. Bjarnason, M.C.Y. Ng, T. Hansen, Y. Bagger, R.L. Wilensky, M.P. Reilly, A. Adeyemo, Y. Chen, J. Zhou, V. Gudnason, G. Chen, H. Huang, K. Lashley, A. Doumatey, W.-Y. So, R.C.Y. Ma, G. Andersen, K. Borch-Johnsen, T. Jorgensen, J.V. van Vliet-Ostaptchouk, M.H. Hofker, C. Wijmenga, C. Christiansen, D.J. Rader, C. Rotimi, M. Gurney,

J.C.N. Chan, O. Pedersen, G. Sigurdsson, J.R. Gulcher, U. Thorsteinsdottir, A. Kong, K. Stefansson, A variant in CDKAL1 influences insulin response and risk of type 2 diabetes, Nat. Genet. 39 (2007) 770–775.

[33] E. Zeggini, M.N. Weedon, C.M. Lindgren, T.M. Frayling, K.S. Elliott, H. Lango, N.J. Timpson, J.R.B. Perry, N.W. Rayner, R.M. Freathy, J.C. Barrett, B. Shields, A.P. Morris, S. Ellard, C.J. Groves, L.W. Harries, J.L. Marchini, K.R. Owen, B. Knight, L.R. Cardon, M. Walker, G.A. Hitman, A.D. Morris, A.S.F. Doney, M.I. McCarthy, A.T. Hattersley, Replication of genome-wide association signals in UK samples reveals risk loci for type 2 diabetes, Science 316 (2007) 1336–1341.

[34] J. Gudmundsson, P. Sulem, A. Manolescu, L.T. Amundadottir, D. Gudbjartsson, A. Helgason, T. Rafnar, J.T. Bergthorsson, B.A. Agnarsson, A. Baker, A. Sigurdsson, K.R. Benediktsdottir, M. Jakobsdottir, J. Xu, T. Blondal, J. Kostic, J. Sun, S. Ghosh, S.N. Stacey, M. Mouy, J. Saemundsdottir, V.M. Backman, K. Kristjansson, A. Tres, A.W. Partin, M.T. Albers-Akkers, J. Godino-Ivan Marcos, P.C. Walsh, D.W. Swinkels, S. Navarrete, S.D. Isaacs, K.K. Aben, T. Graif, J. Cashy, M. Ruiz-Echarri, K.E. Wiley, B.K. Suarez, J.A. Witjes, M. Frigge, C. Ober, E. Jonsson, G.V. Einarsson, J.I. Mayordomo, L.A. Kiemeney, W.B. Isaacs, W.J. Catalona, R.B. Barkardottir, J.R. Gulcher, U. Thorsteinsdottir, A. Kong, K. Stefansson, Genome-wide association study identifies a second prostate cancer susceptibility variant at 8q24, Nat. Genet. 39 (2007) 631–637.

[35] J. Gudmundsson, P. Sulem, V. Steinthorsdottir, J.T. Bergthorsson, G. Thorleifsson, A. Manolescu, T. Rafnar, D. Gudbjartsson, B.A. Agnarsson, A. Baker, A. Sigurdsson, K.R. Benediktsdottir, M. Jakobsdottir, T. Blondal, S.N. Stacey, A. Helgason, S. Gunnarsdottir, A. Olafsdottir, K.T. Kristinsson, B. Birgisdottir, S. Ghosh, S. Thorlacius, D. Magnusdottir, G. Stefansdottir, K. Kristjansson, Y. Bagger, R.L. Wilensky, M.P. Reilly, A.D. Morris, C.H. Kimber, A. Adeyemo, Y. Chen, J. Zhou, W.-Y. So, P.C.Y. Tong, M.C.Y. Ng, T. Hansen, G. Andersen, K. Borch-Johnsen, T. Jorgensen, A. Tres, F. Fuertes, M. Ruiz-Echarri, L. Asin, B. Saez, E. van Boven, S. Klaver, D.W. Swinkels, K.K. Aben, T. Graif, J. Cashy, B.K. Suarez, O. van Vierssen Trip, M.L. Frigge, C. Ober, M.H. Hofker, C. Wijmenga, C. Christiansen, D.J. Rader, C.N.A. Palmer, C. Rotimi, J.C.N. Chan, O. Pedersen, G. Sigurdsson, R. Benediktsson, E. Jonsson, G.V. Einarsson, J.I. Mayordomo, W.J. Catalona, L.A. Kiemeney, R.B. Barkardottir, J.R. Gulcher, U. Thorsteinsdottir, A. Kong, K. Stefansson, Two variants on chromosome 17 confer prostate cancer risk, and the one in TCF2 protects against type 2 diabetes, Nat. Genet. 39 (2007) 977–983.

[36] G. Thomas, K.B. Jacobs, M. Yeager, P. Kraft, S. Wacholder, N. Orr, K. Yu, N. Chatterjee, R. Welch, A. Hutchinson, A. Crenshaw, G. Cancel-Tassin, B.J. Staats, Z. Wang, J. Gonzalez-Bosquet, J. Fang, X. Deng, S.I. Berndt, E.E. Calle, H.S. Feigelson, M.J. Thun, C. Rodriguez, D. Albanes, J. Virtamo, S. Weinstein, F.R. Schumacher, E. Giovannucci, W.C. Willett, O. Cussenot, A. Valeri, G.L. Andriole, E.D. Crawford, M. Tucker, D.S. Gerhard, J.F. Fraumeni, R. Hoover, R.B. Hayes, D.J. Hunter, S.J. Chanock, Multiple loci identified in a genome-wide association study of prostate cancer, Nat. Genet. 40 (2008) 310–315, https://doi.org/10.1038/ng.91.

[37] M. Yeager, N. Orr, R.B. Hayes, K.B. Jacobs, P. Kraft, S. Wacholder, M.J. Minichiello, P. Fearnhead, K. Yu, N. Chatterjee, Z. Wang, R. Welch, B.J. Staats, E.E. Calle, H.S. Feigelson, M.J. Thun, C. Rodriguez, D. Albanes, J. Virtamo, S. Weinstein, F.R. Schumacher, E. Giovannucci, W.C. Willett, G. Cancel-Tassin, O. Cussenot, A. Valeri, G.L. Andriole, E.P. Gelmann, M. Tucker, D.S. Gerhard, J.F. Fraumeni, R. Hoover, D.J. Hunter, S.J. Chanock, G. Thomas, Genome-wide association study of prostate cancer identifies a second risk locus at 8q24, Nat. Genet. 39 (2007) 645–649.

[38] M.F. Moffatt, M. Kabesch, L. Liang, A.L. Dixon, D. Strachan, S. Heath, M. Depner, A. von Berg, A. Bufe, E. Rietschel, A. Heinzmann, B. Simma, T. Frischer, S.A.G. Willis-Owen, K.C.C. Wong, T. Illig, C. Vogelberg, S.K. Weiland, E. von Mutius, G.R. Abecasis, M. Farrall, I.G. Gut, G.M. Lathrop, W.O.C. Cookson, Genetic variants regulating ORMDL3 expression contribute to the risk of childhood asthma, Nature 448 (2007) 470–473.

[39] C.C. Chang, C.C. Chow, L.C. Tellier, S. Vattikuti, S.M. Purcell, J.J. Lee, Second-generation PLINK: rising to the challenge of larger and richer datasets, GigaScience 4 (2015) 7, https://doi.org/10.1186/s13742-015-0047-8.
[40] B.S. Pedersen, A.R. Quinlan, Who's who? Detecting and resolving sample anomalies in human DNA sequencing studies with Peddy, Am. J. Hum. Genet. 100 (2017) 406–413, https://doi.org/10.1016/j.ajhg.2017.01.017.
[41] S. Holm, A simple sequentially rejective multiple test procedure, Scand, J. Stat. 6 (2) (1979) 65–70.
[42] Y. Benjamini, Y. Hochberg, Controlling the false discovery rate: a practical and powerful approach to multiple testing, J. R. Stat. Soc., B: Stat. Methodol. 57 (1) (1995) 289–300.
[43] Y. Benjamini, D. Yekutieli, The control of the false discovery rate in multiple testing under dependency, The Annals of Statistics 29 (4) (2001) 1165–1188.
[44] R.J. Pruim, R.P. Welch, S. Sanna, T.M. Teslovich, P.S. Chines, T.P. Gliedt, M. Boehnke, G.R. Abecasis, C.J. Willer, LocusZoom: regional visualization of genome-wide association scan results, Bioinformatics 26 (2010) 2336–2337, https://doi.org/10.1093/bioinformatics/btq419.
[45] C. Grace, M. Farrall, H. Watkins, A. Goel, Manhattan++: displaying genome-wide association summary statistics with multiple annotation layers, BMC Bioinformatics 20 (2019) 610, https://doi.org/10.1186/s12859-019-3201-y.
[46] R. Mägi, A.P. Morris, GWAMA: software for genome-wide association meta-analysis, BMC Bioinformatics 11 (2010) 288, https://doi.org/10.1186/1471-2105-11-288.
[47] B. Devlin, K. Roeder, Genomic control for association studies, Biometrics 55 (1999) 997–1004.
[48] A. Auton, L.D. Brooks, R.M. Durbin, E.P. Garrison, H.M. Kang, J.O. Korbel, J.L. Marchini, S. McCarthy, G.A. McVean, G.R. Abecasis, A global reference for human genetic variation, Nature 526 (2015) 68–74, https://doi.org/10.1038/nature15393.
[49] T. Hassold, H. Hall, P. Hunt, The origin of human aneuploidy: where we have been, where we are going, Hum. Mol. Genet. (16 Spec number 2) (2007). R203-8.

CHAPTER 4

GWAS in the learning environment

Amy L. Stark
University of Notre Dame, Notre Dame, IN, United States

Introduction

Bringing genome-wide association studies (GWAS) into a classroom presents a "good news" and "bad news" situation. The good news is that the concepts behind GWAS are both straightforward and familiar to most students; however, the bad news is that bringing the understanding of these concepts to the complexities of GWAS in the research environment represents a significant educational challenge. Students are likely to be familiar with genetic variation, and the concept of association is quite easy to grasp. Nonetheless, the reality is that there are accomplished scientists who struggle with an appropriate understanding of GWAS. This chapter aims to empower your students and early trainees to become skilled assessors of GWAS.

Prior to the beginning of teaching about GWAS, it is important to establish that all students possess the same essential knowledge base. In the context of GWAS, this means ensuring that the class is intimately familiar with the ideas related to genetic variation in human beings, specifically minor alleles, allele frequencies, and how genetic variation ranges across populations. Although it is likely that these concepts were covered in basic biology or genetics courses, a review to solidify understanding prior to bringing GWAS into the learning environment will provide for an enriched learning experience for all students (see Box 4.1).

Previous chapters in this book have focused on the conceptual background and statistical methods that are relevant in designing and executing a GWAS. This chapter focuses on how to introduce GWAS to your classroom, how to interpret GWAS research, and creating students (particularly undergraduates) who, while they may not have the computational skills to perform a GWAS, are well-prepared to discuss and understand research done in this area.

Theoretical background for designing and interpreting GWAS
What is GWAS?

After completing your review of genetic variation, it is time to transition to GWAS methodology. The computational skills required for designing and executing a GWAS are complex and will not be the focus of this chapter. Instead, we focus on introducing association studies more broadly, for example, by comparing two sets of variables and determining if there is a relationship between them. These examples can

> **Box 4.1 Teaching hint.**
> Create an assignment prior to beginning GWAS lessons. The 1000 Genomes Web Browser (operated within Ensemble http://useast.ensembl.org/Homo_sapiens) allows students to query genetic variants and review what alleles are. By providing a list of genetic variants to look up, you can demonstrate allele frequencies vary across populations. Some example variants to include could be rs699, with sharp differences observed between European populations and all other populations; rs1535, with differences observed between the African population and all other populations; rs10201985, with no variation seen in East Asian populations but nearly 50/50 in African populations; and rs171984 that is polymorphic in all populations at roughly the same frequency.

> **Box 4.2 Teaching tip.**
> In a classroom, grades are something that is always on everyone's mind. In introducing association studies, using grades as a variable and comparing with different nongenetic variables tends to get everyone on the same page. Does a grade on an assignment associate with the color car the students drive? Their hair color? How much sleep do they get? Where do they sit in class? Attendance? A discussion on these different "association studies" can help ensure all are on the same page. Specifically, while an association between the color car they drive (or hair color) is unlikely to reflect a meaningful association, even if one is found, how many hours of sleep and where a student sits in class may reflect something meaningful (or NOT!). In the discussion, ask what would happen to the association, if all science classes were included in the study. Would the associations established in the current class disappear or be confirmed?
> Optional Activity: For a previously graded assignment, collect some of the data above through a survey or iClicker and plot the association with grades to visualize the association within the class.

be simple and entertaining, to generate intrigue in the topic, and then they should start moving closer to bringing in the complexities of genome-wide studies. Box 4.2 walks through an example.

This class discussion in association studies will touch upon many of the concepts needed to understand and interpret GWAS studies. This discussion can therefore be referenced as further concepts are covered.

What is a GWAS stated simply?

At its core, a GWAS looks at genetic variation and evaluates each variant to see if it demonstrates an association with an outcome or phenotype. These variants can be medical (i.e., blood pressure, disease status) or not (i.e., height, weight, eye color). Scientists have even performed GWAS studies for personality traits and preferences such as political preferences and religion practices. If you have a large group of people's genomes and an outcome, you can perform a GWAS. Whether the results are scientifically rigorous and should be trusted is another question that we will discuss how to assess.

In order for the scientific methods to be considered rigorous, there needs to be two important factors established:

1. A genetic contribution to the trait to be studied.
2. A distribution of that trait in the population studied.

For example, if a trait has no documented or hypothesized genetic contribution (i.e., earning a grade in a class), then a GWAS would be unlikely to identify meaningful variants. Typically, the genetic contribution to a trait or phenotype has been established through some type of heritability study. Sometimes,

> **Box 4.3 Try it yourself.**
> Which of these traits are likely to have genetic variants that could reasonably alter phenotype?
> If we are evaluating the phenotypes below, which of the following traits likely has genetic variants that could reasonably be hypothesized to contribute?
> - Hair color? Texture?
> - Height?
> - Hours slept at night?
> - Arm strength as measured by how many pull-ups completed in 60 s?

a heritability study is difficult to execute for a number of reasons (such as difficulty in quantifying a trait precisely across large numbers of people), and a genetic contribution is hypothesized due to the similarity of the trait with other traits or observations of the trait running in families. While having an estimate of heritability is helpful, it is not required. Ultimately, the first step when evaluating a GWAS is to ask oneself: *Is it reasonable to hypothesize that the trait studied has genetic variants which could contribute to the phenotype being evaluated?* (Box 4.3).

It is also important that the trait is distributed throughout the population studied. Statistically speaking, the trait should be *normally distributed*, but there are many computational transformations that can allow for a distribution to achieve a distribution that is close to normal. From a noncomputational perspective, it is important that the trait is well-distributed, meaning that there are not extreme outliers, the majority of the population studied has a different phenotype, and all the "extreme" phenotypes are not focused in the same way. Critically, if the trait variation is driven by individuals with extreme versions of the trait, it is unlikely that a GWAS is an appropriate tool.

GWAS evaluation

Once it is established whether GWAS is an appropriate analysis, ideally because the trait has some genetic basis and is well-distributed within the population, this brings in the next level of analysis on the design of the GWAS study (Box 4.4).

> **Box 4.4 In class teaching activity.**
> **Correlation vs. Causation**
> One of the largest criticisms about GWAS lies in its name. Even the most optimally designed study is simply identifying associations that are correlative. There is no way any GWAS study can identify causation. Bringing genetics into the classroom with this concept can be challenging. First, present a number of scenarios with true correlative "causes" that do not seem realistic, for example, all car accidents in the country for the month of December are caused by red vehicles. Discuss each to determine whether the correlation is likely to be causative. For example, is there something about people who select red vehicles that also make them poor December drivers? Are men or women more likely to drive red cars? Discuss potential hidden variables that are included in this example. Finally, convert the example to genetics. If Gene X is associated with a trait of interest, the first step is to analyze what is known about gene X and determine if a relationship with the trait can be hypothesized. Depending on the size of the class, you may provide a number of different genes with a short summary of what is known about them, and then have each student or small group of students present why each association is more likely to correlative or causative.

Continued

> **Box 4.4 In class teaching activity—cont'd**
>
> *Questions to discuss*
> *Does a more plausible explanation determine if an association is correlative or causative?*
> No. It can help design functional follow-up experiments, but it is impossible to determine if an association is correlative or causative based on the hypothesis of how the association could be caused.
> *How can causation be proven?*
> Causation is almost impossible to **prove**. Causation can be suggested through follow-up experiments with demonstrated functional effects, but a new discovery can always change the perspective.
> *Does a lower P-value suggest causation over correlation?*
> No! The strength of the statistics suggestion an association cannot be used as evidence to suggest causation over correlation. Causation must be demonstrated through the functional experiments.

Populations included in the GWAS

The population included in a GWAS introduces many potential issues; this is an example of practical and theoretical concerns merging together. From a theoretical perspective, an ideal GWAS study would contain an equal number of participants from all human populations, and statistically control for the ancestral differences of each participant. In reality, that is nearly impossible to execute for a number of reasons, not the least of which is the challenge a researcher may have in establishing contacts across the world.

Population diversity

Most researchers focus on a research population that is easily accessible for logistical reasons. The majority of GWAS research is conducted in North America and Europe, meaning that populations of European ancestry are over-represented in these studies. Well-intentioned studies that include a small number of participants of different ancestral backgrounds must ensure that the GWAS analysis adequately addresses the different ancestries. If not, the results of the GWAS may simply identify genetic variants that are responsible for ancestral differences, not the trait of interest. For an example of that mistake, read about the chopstick acuity GWAS (Hamer and Sirota, 2000).

When considering the challenges of controlling for ancestry correctly, one might suggest that the solution would be to select a homogeneous population to conduct a GWAS. While this is commonly done, it deviates from the ideal of GWAS, which is the representative of all populations. Genetic variants identified in a single population may or may not be relevant in other populations. Furthermore, genetic variants not present in the population studied would not be interrogated in the GWAS and would thus be missed entirely. Neither of these scenarios is ideal after investing in a GWAS study and should be carefully scrutinized in published GWAS. If a GWAS was conducted in a single population, careful consideration should be given as to whether the results would be applicable to other populations. How does one address this concern? One avenue is to evaluate results in an additional population. A lack of replicability could suggest that the original GWAS is not broadly applicable to other populations, or that other design flaws were present in the original GWAS.

In the human population, it is well-established that African populations contain the greatest amount of genetic diversity. In order to query the most complete set of genetic variants, African populations would be ideal. However, there remain logistical and technical challenges to this type of study. Many researchers lack connections with collaborators who work with African populations. More pressing, however, is that most genotyping chips do not capture the full variation present in

> **Box 4.5 Discussion box.**
> Ethically, what are some of the implications for people with African ancestry being under-represented in GWAS studies? On genotyping chips, what are some of the longer-term implications?

African populations. Because genotyping chips were designed using populations of European ancestry, variants not present in that population would not be captured on said chips. Genotype imputation is one tool to overcome this hurdle, but it requires additional computational expertise and skill to execute (see example class discussion questions in Box 4.5).

Population size

In GWAS, there is never a large enough population to study. An ideal GWAS would include the genomes of every individual worldwide. From that statement, it is easy to see that a GWAS of 100 individuals is dramatically under-powered. Even 1000 individuals is a severely restricted population to study when considering the size of the human population. Due to the expense and logistics of conducting a GWAS, it is not practical for single researchers or research groups to execute studies that include tens or hundreds or thousands of individuals.

As such, many GWAS are conducted within large consortia or collaborations between several research groups. Thinking back to the original discussion about class grades and associations, what would have been the outcome, if more students were added to the study? It is inevitable that the results will change. The same is true with GWAS. Adding additional subjects will alter the results.

There are a few considerations when increasing population size for GWAS. For example, for some traits, there might be a limit on the number of people who fit the defined criteria to be recruited for a study. Alternately, studies that do have a large number of subjects recruited may not have the quality control to ensure that the phenotype is comparably defined between all subjects. This latter point is particularly relevant when accessing subject information in multiple clinical databases defined by different phenotyping standards. By adding additional subjects, researchers take a step closer to the ideal population size, but very often this can bring additional interpretation challenges, if the trait is not consistently defined or evaluated.

Number of tests in a GWAS

When a GWAS is executed, generally, 2–10 million genetic variants are evaluated for correlation with the trait of interest. This introduces a large multiple-testing problem, which can be a challenge for non-statistically familiar scientists to fully appreciate. Consider the following:

Each genetic variant generates a statistic to demonstrate the likelihood that it correlates with the trait of interest, most commonly, a *P*-value.

When a *P*-value is small enough, the researcher may choose to graph the genotypes at the variant against the trait of interest to evaluate the signal further.

Common mistake of interpretation: Seeing a genetic variant map with a trait of interest looks exciting! However, that excitement neglects the **totality** of a GWAS: If one only looked at *one* variant, then a low *P*-value and an exciting correlative graph would be a reasonable approach.

Consider: With a GWAS, the analysis was conducted millions of times. The definition of a P-value is the probability of a result rejecting the null hypothesis. Therefore, a P-value of 0.001 gives a 1 in 1000 probability of rejecting the null hypothesis.

Normally, that would be a fantastic P-value! However, with a GWAS of 2 million variants, the rejection of the null hypothesis would occur 2000 times by chance.

If the GWAS results in 1000 variants meeting the $P=0.001$ threshold, that number represents fewer hits than what would be expected by chance.

The results of the GWAS would therefore be deemed *not* particularly exciting.

This exercise can be done at different thresholds, but it is important to know what the P-value threshold used was and what the expectation would be of the number of variants that would be associated by chance rather than by biology. **The confounding reality is that some of the associations may be real and reflect legitimate biology, but through current statistical models, it is impossible to determine which are from legitimate hits and which are not**. A researcher could design additional replication cohorts or functional (validation) experiments in an attempt to identify such variants; however, there is a lower probability of success, dissuading many from doing so.

There is more than one statistical approach for how to address the many individual tests contained in a GWAS. The most appropriate one for a study can vary based on a number of factors including the trait, research goals, and computational expertise. However, a GWAS that does not address the multiple tests conducted as part of the analysis is significantly flawed. See the class activity in Box 4.6 to discuss these concepts with trainees.

Box 4.6 In class teaching activity.

Sample size

The size and makeup of the population participating in GWAS is one of the most complex parts of fully understanding GWAS. The best way to really help students grasp this concept is by having them experience it. Create a fictional population of decent size, perhaps 50–100 individuals, and provide limited genetic information and phenotypic information. Allow the students to identify genetic variants that associate with the trait of interest. Double the population size and repeat the exercise. Continue until the research population size is over 50,000. In order to most effectively demonstrate the effect of the larger sample size, it would be ideal in the fictional population, if the first 2 or 3 analyses identify the same genetic variant or variants. The next 2 or 3 analyses would then identify different genetic variants with the original signal dissipating.

Questions to discuss

Why did the same genetic variants come up after the first two analyses?

Different genetic variants will demonstrate association with different populations, even if a subset of the population was included in a previous analysis.

Why did a different genetic variant come up after the final analysis? What happened to the first genetic variant?

With the increasing size of the population analyzed, the analysis is unique from all preceding analysis. The initial variant that demonstrated association did not show any association at the end.

Which analysis do you believe?

While each analysis was completed accurately with the population provided, the analysis with the largest population is the analysis to rely on.

Why?

The larger the population, the less likely an outlier will be driving or affecting the association. The larger the population, the more representative of the general population it should be.

How does this relate to GWAS in the research realm?

In the research realm, an early association finding can disappear when a larger study comes out. Sometimes, researchers have dedicated follow-up experiments on the initial finding and have demonstrated a relevant pathway to suggest the association was accurate. However, generally, the research community relies on the results of the largest study to reach a consensus about the genetic variants relevant for a trait.

The trait in question: The phenotype

When conducting a GWAS on a large number of people, it is **imperative that the phenotype is consistently measured and summarized across all the participants**. For some traits, like height or disease status, this is a straightforward process; however, for other traits, it can be more tricky. For example, defining and assaying an adverse drug side effect can be more challenging than measuring height. Additionally, subjective interpretation can come into play; one researcher might quantify a person's reported side effect as severe, while another researcher might say that same person's side effect is within the normal range. Introducing noise at the phenotype/trait level makes interpreting GWAS results almost impossible. Genetic variation that might have a real effect would be masked by those individuals with an inconsistently measured phenotype, and similarly, genetic variation that may have no true effect could produce a signal for the same reason.

Finally, one of the greatest challenges in designing and conducting a GWAS is identifying the correct individuals to include in the study. To be clear, this challenge does not reflect the difficulties in obtaining a representative ancestral or diverse population, or accurately controlling for population ancestry. Rather, this challenge reflects the need to have the correct underlying genetic hypothesis for the phenotype to be studied.

In order to understand this challenge more fully, consider a nongenetic example in Box 4.7. If researchers wanted to study the cause of exhaustion, immediate challenges would be readily apparent; exhaustion has many diverse and distinct causes ranging from behavior choices (staying up all night to study or party) to medical causes (a virus or disease) to the unexplainable (idiopathic).

Thus, grouping people merely by phenotype (those who are exhausted), but paying no attention to their underlying cause (the "genotype" in this example), would inevitably produce flawed results. Simply put, **different causes of the same trait will not yield insight into each other**. Additionally, if the goal is to research individuals with idiopathic exhaustion, including individuals with known causes (behavioral choices, illness) would only confound the study's conclusions.

The example of exhaustion is easily relatable and understandable as to why different underlying causes should not be grouped together. The same principle applies in GWAS; however the underlying causes are much less clear. **If the phenotype or trait being studied in the GWAS has known and distinct genetic etiologies, the study must stratify based on that information**. Unfortunately, if little is known about the genetic etiology, it is impossible to design a study that stratifies different genetic causes. One of the primary reasons to conduct a GWAS is to learn about the genetic etiology, so it is quite rare to have that information prior to conducting the study. If a trait or phenotype is multigenic, meaning many genes contribute, GWAS is well-positioned to identify the many genetic variants that contribute. However, if there are multiple distinct causes, GWAS may wash the signal out for each of them, resulting in no significant findings.

For example, let's consider the trait of high blood pressure. If 50% of the people in the GWAS have high blood pressure caused by genetic variation within genes A, B, C, D, and E and the other 50% of the

---

Box 4.7 Teaching box.
In this example, what is the phenotype?
  Exhaustion.
  What are the underlying hypotheses for this phenotype?
  Behavioral choices, illness, or unknown/idiopathic causes. Use for class discussion.

people in the GWAS have high blood pressure caused by genetic variation within genes Q, R, S, T, U, and V, the GWAS is unlikely to find any significant results, despite a real genetic signal existing. If the genome-wide analysis were performed on each half of the population the signal for the respective genes would likely appear. There is, however, no way to know the underlying genetic causes prior to performing a GWAS analysis. This is a simplistic example to reflect the complexities of human biology.

The GWAS studies are regularly encouraged to be as large as possible to gain the power to detect smaller gene effects. However, as this section has demonstrated, the larger the population, the greater the chances are to confound the results through improperly controlled ancestral differences, imprecise phenotype annotation (phenotypic noise), or distinct and diverse genetic etiologies. While a researcher can attempt to use best practices to avoid the first two, the final challenge is very difficult to control. Sometimes distinct genetic causes result in distinct phenotypic differences. Different genetic mutations in the cystic fibrosis gene have been associated with different symptoms, disease manifestation, and progression. This principle can also be applied in complex traits, though often is quite subtle.

Summary

Despite these challenges, GWAS remains a powerful tool to identify genetic variants contributing to complex phenotypes. In order to fully assess GWAS analyses and studies, it is important to remember these challenges. While conceptually understanding these challenges is important, being able to effectively teach them is a secondary challenge. As well, while the computational skills to execute a GWAS require more expertise than most scientists have readily available, this chapter should provide the preparation to discuss GWAS design and results with any scientist and trainee. The technique and power behind a GWAS provide an opportunity to gain interesting and novel insights into human biology. A well-designed and performed GWAS study can be the key in developing new hypotheses and experiments to gain greater understanding in human biology. Understanding both the strengths and cautions of GWAS enables discussions across all fields of science.

Reference

D. Hamer, L. Sirota, Beware the chopsticks gene, Mol. Psychiatry 5 (2000) 11–13. https://doi.org/10.1038/sj.mp.4000662.

CHAPTER 5

Polygenic risk scores and comparative genomics: Best practices and statistical considerations

Sally I-Chun Kuo and Fazil Aliev

Department of Psychiatry, Robert Wood Johnson Medical School, Rutgers Behavioral and Health Sciences, Piscataway, NJ, United States

Gene identification has advanced rapidly over the last decades, with the primary goal of identifying genetic variants associated with complex behaviors/traits and disease outcomes. A genome-wide association study (GWAS) is a modern molecular approach in genetics to identify specific genetic variants associated with a particular phenotype. The GWAS method is a hypothesis-free approach that involves systematically testing for associations between single nucleotide polymorphisms (SNPs) across the genome and an outcome/phenotype. A polygenic scoring approach emerges as a useful method to use results from GWASs to create an aggregated measure of genetic risk—namely polygenic risk score (PRS; also referred to as genome-wide polygenic score [GPS] or genetic risk score [GRS])—in independent samples.

This chapter is structured as a primer on understanding the polygenic scoring approach and creating polygenic risk scores. We first provide an overview of the polygenic scoring methods and discuss what polygenic scores represent. Second, we outline the general process for creating polygenic risk scores, including key steps, a summary of different software for calculating risk scores, resources, and sample scripts. Finally, we discuss important considerations when conducting polygenic analyses and challenges associated with creating and interpreting polygenic risk scores.

Genome-wide association studies

Twin and family studies indicate that virtually all complex behaviors/traits and multifactorial outcomes are under genetic influences [1]. Genome-wide association study (GWAS) is a popular gene-identification design in the past decade which involves testing genetic variants across the genome to identify genotype-phenotype associations [2]. GWAS is an approach used in genetics research to associate specific genetic variants with a phenotype (e.g., diseases, traits, behavioral outcomes) and scan the genomes from many different people and examine these associations.

Once such genetic markers are identified, they can be used to understand how genes contribute to the disease and develop better prevention and treatment strategies. GWASs have identified genetic variants associated with complex phenotypes/behaviors, including substance use behaviors and related problems [3,4], psychiatric disorders [5–7], coronary artery disease [8], blood pressure [9], and educational attainment [10]. Genetics studies also reveal that complex behaviors usually have a polygenic architecture, which means that genetic influences include the effects of many common genetic variants of small effect sizes [11]. As genome-wide association studies are identifying reliable, robust genetic variants, their results can also be used to create genetic risk scores.

The polygenic scoring approach

Genome-wide association study (GWAS) is an approach that scans markers across the genome to identify genetic variants (single nucleotide polymorphisms [SNPs]) associated with a phenotype. The results from genome-wide identification studies can be used to create polygenic risk score (PRS), alternatively called genome-wide polygenic score (GPS). PRS indexes an individual's overall genetic predisposition for a phenotype based on their measured genotype. Polygenic scores are calculated by summing the number of effect alleles present weighted by their effect size as estimated in a specific discovery GWAS, resulting in a single continuous score for each individual that represents the overall genetic liability for a phenotype [12].

Fig. 5.1 provides an overview of the polygenic scoring approach. An important consideration of calculating polygenic scores is that the discovery GWAS sample and the target sample must be independent, meaning that there is no sample overlap between the discovery and the target samples. No overlap in the samples is necessary to ensure independent predictions. PRS is calculated by summing the number of risk alleles that an individual carries weighted by the parameter estimates (e.g., betas, odds ratios) identified in a discovery GWAS. As a hypothetical example illustrated in Fig. 5.1, consider three SNPs with the following alleles from the discovery GWAS: SNP 1 risk allele is T, SNP 2 risk allele is G, and SNP 3 risk allele is G. In the target independent sample, Person 1 carries one risk allele (T) at SNP 1, zero risk allele at SNP 2, and two risk alleles (G) at SNP 3. Person 1's PRS would

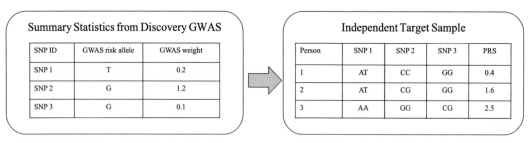

FIG. 5.1

A hypothetical example for calculating polygenic risk scores. This example uses GWAS summary statistics from a discovery sample *(left)* to calculate polygenic risk scores (PRS) in an independent sample (no sample overlap with discovery GWAS). Discovery GWAS provides the risk allele and effect sizes (weights; e.g., beta) to calculate the weighted sum of risk alleles that an individual in the target sample carries. For example, Person 1 carries one risk allele (T) at SNP 1, zero risk allele at SNP 2, and two risk alleles (G) at SNP 3. Thus, Person 1's PRS would be equivalent to $0.2*1+1.2*0+0.1*2=0.4$. This process occurs across all SNPs included for calculating a specific PRS for each person in the independent target sample.

be equivalent to 0.2*1+1.2*0+0.1*2=0.4. Person 2 carries one risk allele (T) at SNP 1, one risk allele (G) at SNP 2, and two copies of risk allele (G) at SNP 3. Person 2's PRS would be equivalent to 0.2*1+1.2*1+0.1*2=1.6. This process occurs across all SNPs included for calculating a specific PRS for each person in the independent target sample. An individual's polygenic score is calculated as the sum of the number of their risk alleles weighted by the effect estimate reported in the GWAS summary statistics. Accordingly, polygenic scores represent an aggregated genetic liability to a trait at the individual level based on a particular discovery GWAS summary statistics.

Polygenic risk scores provide a flexible way of bringing results from large-scale gene identifications (i.e., GWAS) into deeply phenotyped samples or longitudinal data to characterize genetic risk across development, and in conjunction with the environment. This allows us to study a number of interesting questions, such as:

- Does genetic risk result in different phenotypes at different developmental stages [13,14]?
- Is genetic risk associated with intermediary endophenotypes [15,16]?
- What environmental factors alter the association between genetic risk and phenotypes [17–19]?

In addition, PRS can be incorporated in Mendelian Randomization studies to infer causal relationships (e.g., [20] and in virtual parent designs to disentangle causal effects of offspring genetics from parent genetics [21]. Accordingly, PRS is a useful approach to examine a range of research questions and test nuanced hypotheses.

Preparing data for polygenic score calculation

Once summary statistics from the discovery GWAS results and data from independent samples (with individual level genotypic and phenotypic data) are identified and obtained, the next stage of the PRS calculation and analysis process requires processing and quality control of the data. We outline steps and considerations below.

Quality control. The validity of the PRS and related analyses depends in part on the quality of the discovery GWAS and independent sample data. It is important to ensure that the data collection and genotyping undergo quality control (QC) procedures based on the standard guidelines. Once the initial genotyping QC is complete, additional QC should be implemented to ensure that high-quality QC is done on the genome-wide data in independent samples. Some of the standard QC parameters include checking minor allele frequency (MAF), Hardy-Weinberg Equilibrium (HWE), pedigree (if applicable), and sex chromosomes. PLINK [22,23] is a useful software for the QC procedures.

Palindromic SNPs. Fig. 5.2 provides a schematic representation of a palindromic SNP vs. non-palindromic SNP. SNPs with A/T or G/C alleles are called palindromic SNP because their alleles are represented by the same pair of letters on the first and second strands. This introduces ambiguity in accurately identifying the effect (risk) allele in the discovery GWAS results. Therefore, removing palindromic SNPs is needed before PRS calculation. In other words, we call SNPs with A/T and G/C alleles palindromic because we cannot count the number of risk alleles accurately without extra information about the strands in both the discovery and target samples. Thus, palindromic SNPs need to be removed before counting alleles and summing risk scores. From 12 different kinds of SNP letter pairs (GC, CG, GT, TG, GA, AG, CT, TC, CA, AC, TA, AT), only four (GC, CG, TA, AT) are palindromic, making approximately one-third of all SNPs as palindromic. While most genotyping arrays try to limit

94   Chapter 5 Polygenic risk scores and comparative genomics

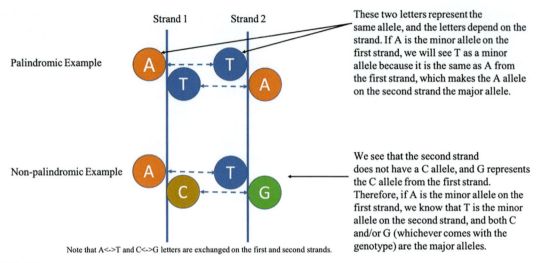

FIG. 5.2

A schematic representation of a palindromic SNP vs. nonpalindromic SNP. SNPs whose alleles correspond to nucleotides that pair with each other in a double-stranded DNA molecule are called palindromic (i.e., SNPs with A/T or G/C alleles). Palindromic SNPs need to be removed before polygenic scoring because they can create ambiguity into the identity of the effect (risk) allele in the GWAS.

the number of palindromic SNPs, all arrays and imputed genotype files, nonetheless, have a sizable number of palindromic SNPs. Removing palindromic SNPs is a critical step in preparation of risk score calculation (see the online DataVerse for these resources: https://dataverse.scholarsportal.info/dataverse/RandRinGenomics).

Strand flipping. SNPs that have mismatched alleles in the discovery and target data that are non-palindromic are relatively easy to handle. They can be addressed by strand flipping the alleles to their complementary alleles. As an example, let's look at an SNP with A and C alleles and assume that the first strand of the target sample has letters AA, then the second strand has TT alleles because A and T alleles are exchangeable. Now imagine that the discovery sample has C as a risk allele based on the first strand. For the TT genotype in the target sample, we see that it does not match the AC allele pair. This means that TT comes from the opposite strand which has T and G alleles. Accordingly, we flip TT alleles from the opposite strand to AA alleles from the matching strand and find the number of C alleles in TT genotype as zero. This allows us to determine if the target sample was genotyped using the same or opposite strands, and we flip the strand to the opposite matching strand in the target sample if needed. Replacing T and G alleles to opposite A and C alleles can be done by the --flip [file name] option in Plink [22,23] software. This option uses SNP lists from the mentioned file and swaps alleles A<->T and G<->C, which correspond to the opposite strand alleles. Similarly, if the target data's phenotype is TG, then after flipping it changes to AC and C allele and count becomes one. Thus, mismatching SNPs that are resolvable can be handled by strand flipping.

Multiallelic SNPs. Some SNPs can have more than two allele pairs. For example, the same SNP can have A and T alleles in some individuals and A and C letters in other individuals. There is not a perfect way to use this kind of SNPs in PRS scoring if discovery and target samples have different pairs.

Accordingly, the best practice is to exclude multiallelic SNPs from the polygenic scoring. However, sometimes discovery or target samples have both versions of the SNP—for example, an SNP is listed twice, but with different allele pairs such as rs12345:A:C with A/C alleles and rs12345:A:G with A/G alleles. In this case, we can keep one of the SNPs with matching alleles in the discovery and target samples. There are also SNPs with multiple allele letters like rs12345:T:TTC. These SNPs correspond to indels (short polymorphisms that correspond to the addition or removal of a small number of bases in a DNA sequence), and we also exclude SNPs with multiple letters from PRS calculation.

SNP pruning. In general, PRS calculation is based on summing the number of risk alleles that an individual carries weighted by effect sizes. Because of recombination process and length of recombination, SNPs located in close distance in the genome are correlated (referred to as linkage disequilibrium or LD). Adding effects of highly correlated SNPs could inflate risk scores. Therefore, PRS is generally calculated with a subset of independent SNPs. These SNPs can be selected using various methods. One pruning method involves including SNPs below a certain LD threshold. SNPs can be selected to prune using LD correlations of an external panel, such as 1000G [24,25]. Using an external panel allows pruning to be consistent because it does not rely on the LD correlations of the small target data.

It is important to note that external panel SNP correlations depend on the ancestry of individuals. The same pair of SNPs can have different correlations in populations of different ancestral background. It is also possible to have different minor allele frequencies among samples of individuals from different ancestries. External panels have subsamples with different ancestries. 1000 Genomes Project has more than 30 initial ancestries which are then grouped into five superpopulations. There are also special panels developed for a single ancestry. Nowadays, we see an increasing effort to create more panels representing most of the worldwide ancestry groups. It is recommended to use panels or subsamples with a matching ancestry with the target sample (see section **PRS in diverse ancestral populations**). For admixed samples, the proper method would be to use an external sample with admixed ancestries.

The use of Plink for pruning has two steps: (1) creating a list of correlated SNPs to be dropped and (2) excluding those SNPs from the target sample (or discovery GWAS). Plink can create a list of correlated SNPs independently with the option --indep-pairwise 50 5 0.2, where 50 is the window size, 5 is step size to move forward, and r^2 threshold. One of the correlated SNPs from each pair of SNPs in the window go to the <output-name>.prune.in file with the list of independent SNPs. One preferred way is to use clumping instead of indep-pairwise with the option like:

--clump-kb 500 --clump-r2 0.2 --clump <p-value file> --clump-p1 0.5 --clump-p2 0.5

The *P*-value file contains SNP and p-value list (two columns) from the discovery sample. Using this method, Plink prioritizes keeping SNPs that are highly significant. If two SNPs are correlated more than r^2, the option "indep-pairwise" prefers to keep the SNP with higher MAF. The clump method keeps the more significant SNP, allowing the most significant SNPs to be used in PRS scoring. After clumping, Plink creates the file with the list of "independent" SNPs based on --clump-r2 threshold.

In the "pruning and thresholding" method, we then use

--bfile <target-sample> --score <file-with-SNP_and_Betas> no-sum --q-score-range <file with p-threshlolds> <clumped_output_file> 3 5 header

where the clumped output file is the file from the –clump-kb command given above.

Matching discovery and target genotype files. In most cases, genotype files come in the Plink format. The most popular file format is with the following three different extensions: .bed, .bim, and .fam files. The .bed file contains genotypic information in a binary format. .fam file contains pedigree information, including sex and phenotype. .bim file contains chromosome, SNP, position information, and two other columns with SNP alleles.

PRS is calculated using weights and allele counts from the same SNPs between the discovery and the target genotype files. Depending on the genotyping platforms used for discovery and target samples, the map versions of SNP positions can be different. Some SNPs can also have synonyms, which creates additional complexities when trying to match SNPs between the discovery and the target genotype files. Human genome reference assemblies define SNP position information. The most up-to-date version of assembly is GRCh38/hg38, which was released in 2013 (https://www.ncbi.nlm.nih.gov/assembly/GCF_000001405.26/). It is important to verify that the discovery and target samples use the same assembly before any PRS calculation. GRCh38 is the recommended platform to be used to match SNP positions. If the discovery and target data used different assemblies, there are uplifting software programs and protocols (https://genome.ucsc.edu/cgi-bin/hgLiftOver) that could change assembly from one to another. In the plink standard (.bed, .bim, .fam) file formats, uplifting affects only the .bim file.

After SNP positions are matched, the next step is allele matching. Removing palindromic SNPs is critical in most of the cases (see above; also detailed in Appendix A). However, when the discovery and target samples were genotyped using the same genotyping array, removing is unnecessary because there are no mismatching SNPs. Among nonpalindromic SNPs, there are two possibilities: (a) discovery and target alleles match or (b) alleles come from the opposite strand and are flipped. In the first case, we keep the alleles unchanged and in the second case we flip the alleles in one of the discovery or target samples. For this purpose, we can use Plink's --flip option (described above) in target samples because, in general, discovery samples come without genotypic information.

Plink switches A and T alleles and G and C alleles in the provided SNP list files, which is equivalent to strand switching. Creating the list of SNPs to be flipped can be done using different techniques. R software can be used for this purpose. R can be used to match the discovery sample's SNP name, allele 1 columns and target sample's SNP name, and allele 1 and allele 2 information. Because SNPs are nonpalindromic, the discovery sample's allele 1 must match with one of allele 1 or allele 2 from the target sample. All SNPs with nonmatching alleles are listed in the SNP list file for flipping. R commands for these steps are summarized in Box 5.1.

It is becoming increasingly popular to use meta-analysis of GWASs results as the discovery GWAS for calculating PRS. METAL [26] is a commonly used computational tool for meta analyzing GWAS results. METAL lists alleles using lower case letters. Plink's –scoremethod allele matching is case sensitive, and it cannot match between lower- and upper-case letters. As a practical note, when working with meta-analyzed discovery GWAS summary statistics provided by the METAL program, switching alleles to capital letters would be important. This can be done efficiently using commands in the Unix/Linux environment with one line of command to capitalize all alleles letters:

sed -i -e 's/g/G/g; s/c/C/g; s/t/T/g; s/a/A/g' <your discovery GWAS file name>

This "sed" command capitalizes all "g", "c", "t" and "a" letters in the discovery GWAS results file. Each of "s/../../g" groups in the command replaces the phrase between the first two slashes to the phrase between the second and third slashes. Since most of the GWAS result files do not have any other text involving the four allele letters, using capitalization of all letters in the file works in almost all cases.

> **Box 5.1 Sample R script for selecting SNPs to flip and match SNPs between the discovery and target samples.**
>
> ```
> ## Selection list of SNPs to flip
> ## parameters, change 5 parameter lines below accordingly
> target_file = "target_file_full_name_with_path_no_extension"
> discovery_file = "discovery_GWAS_file_full_name_with_path"
> dis_SNP = "discovery_file_SNP_column_name" ## i.e. SNP, Marker or ID
> dis_A1 = "discovery_file_allele_1_column_name" ## i.e Allele1, Ref or A1
> flip_file="SNP_list_file_full_name_for_flip"## output for Plink --flip
> ## parameters finished here, do not change below
> ## read SNP, allele 1 and allele 2 columns of target sample
> target=read.table(paste0(target_file,"bim"),header=F)[,c(2,5,6)]
> names(target)=c("SNP","A1","A2")}## assign names to columns
> ## read SNP and allele 1 columns of discovery sample
> discovery=read.table(discovery_file,header=T)[,c(dis_SNP,dis_A1)]
> names(discovery)=c("SNP","Ref")}## assign names to columns
> ## merge files by SNP column (Note: drop duplicated SNPs before merging)
> combined=merge(x=target,y=discovery,by="SNP",all=F)
> ## select non-matching SNPs
> final = combined[combined$Ref != combined$A1 & combined$Ref != combined$A2,]
> ## print SNP column from final table to flip file
> write.table(final$SNP, file = flip_file, row.names=F, col.names=T, quote=F)
> ```

Note that if the header of a file has words like "Beta," then it would become "BeTA" after the capitalization command. One can then add ";s/BeTA/Beta/g" to the end of the command to correct the change back to "Beta":

sed -i -e 's/g/G/g; s/c/C/g; s/t/T/g; s/a/A/g; s/BeTA/Beta/g' <discov. GWAS file>

and add more "s/../../g" groups, if needed.

Software programs for PRS calculation

There are many software programs to compute polygenic risk scores. Table 5.1 provides a summary of various common programs along with a brief description.

Although there are different PRS software programs, they generally follow the same principle to construct polygenic scores using a subset of independent SNPs (due to linkage disequilibrium) with sufficient minor alleles. Plink's pruning and thresholding method [23] first prunes SNPs based on an external panel, thresholds discovery GWAS *p*-values, then adds GWAS weights multiplied by the number of risk allele thorough all remaining SNPs. This score assumes that the contribution of SNPs in the overall score is additive by SNPs and the number of alleles of each SNP. Plink P+T creates one score per each of *p*-value threshold values. PRS calculation is typically done across a series of discovery GWAS p-value thresholds at 0.0001, 0.001, 0.01, 0.05 0.1, 0.2, 0.3, 0.4, and 0.5. PRSice [28] can process and calculate PRS across a broad range of p-value thresholds and identify the optimal threshold (best-fit PRS).

Table 5.1 Summary of software programs for computing polygenic risk scores.

Software (Reference)	Description
Plink pruning and thresholding (P+T) [22,23]	Plink pruning and thresholding method generates polygenic scores using a linear combination of simple linear regression effect size estimates and allele counts at single-nucleotide polymorphisms (SNPs) that are selected via marker pruning (LD threshold) coupled with GWAS p-value thresholding.
PRSice ("precise") – PRSice-2 [27,28]	PRSice (pronounce "precise") calculates, applies, evaluates, and plots the results of PRS and can identify most predictive PRS, incorporate covariates PRS at a large number of thresholds ("high resolution") to provide the best-fit PRS, as well as provide results calculated at broad P-value thresholds. It can thin SNPs according to linkage disequilibrium and P-value or use all SNPs, handles genotyped, and imputed data, can calculate and incorporate ancestry-informative variables, and can apply PRS across multiple traits in a single run.
LDpred [29]	LDpred is a Python-based software package that adjusts GWAS summary statistics for the effects of linkage disequilibrium (LD).
Multi-ethnic PRS [30]	Creates Plink P+T PRS to each discovery GWAS separately, and linearly combines the resulting PRS. This method improved the prediction in admixed populations. When discovery GWAS is not available in diverse population, a 10-fold method is used in the target sample to run GWAS in 9/10th of the sample and create PRS in 1/10th of the sample.
LassoSum [31]	LassoSum method is based on penalized regression (an elastic net) approach.
Bigsnpr [32]	Bigsnpr is an R package for analysis of large-scale data which creates risk scores based on statistical learning.
SBayesR [33]	SbayesR uses Bayesian posterior inference through the combination of a likelihood that connects the multiple regression coefficients with summary statistics from GWAS and a finite mixture of normal distributions prior on the marker effects.
JAMPred [34]	JAMPred uses a two-step approach to construct genome-wide PRS from meta-GWAS summary statistics, first by adjusting local LD, followed by adjusting long-range LD.
PRS-CS [35] PRS-CSx [36]	PRS-CS is a Python-based command line tool that infers posterior SNP effect sizes under continuous shrinkage (CS) prior to using GWAS summary statistics and an external LD reference panel. PRS-CSx is an extension of PRS-CS to accommodate multiancestry platform.
Truncated Lasso Penalty (TLP) [37]	TLP extends the LassoSum framework to the Truncated Lasso Penalty and the elastic net.

A multiethnic PRS approach is used to create PRS in the target sample when there is no matching ancestry discovery sample available. This approach [30] combines GWAS results in one ancestry (e.g., European) with training 10-fold GWAS data from the target (e.g., non-European) population (see **PRS in diverse ancestral populations**).

Some PRS software programs (Ldpred, LassoSum, Bigsnpr, SbayesR, JAMPred, PRS-CS or PRS-CSx, Truncated Lasso Penalty) adjust the original discovery GWAS weights with methods such as machine learning, different penalized regression, and external panels to create new weights for SNPs in consideration to be used in the PRS summing, as in the case of Plink P+T method. For example, LDpred [29] uses a Bayesian approach to recompute the SNP effect sizes while accounting for LD from an external panel. PRS-CS [35] employs a Bayesian regression and continuous shrinkage method to correct for linkage disequilibrium.

Empirical data and simulation studies have compared the performance of polygenic scores calculated using different approaches and software programs. Data generally show mixed results, with fluctuating, but small, differences [38,39]. Accounting for LD between SNPs generally improves PRS prediction accuracy.

Sample script for calculating PRS using PRS-CS

We provide a sample script for calculating PRS using the PRS-CS [35] method with annotation notes in Appendix A (see this book's DataVerse[a]). In the resource document, we first explain how to implement different QC processes and prepare files for computing risk scores. The R script is an example of a PRS calculation-automated pipeline set up that users can compute PRS by entering simple parameters such as the discovery GWAS results file name/path, target sample's genotype file name/path, and preferences for software. This script is designed to have the following workflow:

i. Conduct initial file cleaning
ii. Remove palindromic SNPs
iii. Remove duplicate SNPs
iv. Generate all necessary files for the PRS-CS software program
v. Run PRS-CS
vi. Combine chromosome-based new weights to be used in Plink
vii. Run Plink to create final polygenic risk scores

Running time of this R script depends on the discovery GWAS summary statistics and target data file sizes. Because multitasking options are different for different systems and server environments, we use a single-run option here, which can take a few hours to create risk scores for one phenotype for all individuals in the target sample depending on the target sample size. Notably, by adding --chrom [chromosome number] into the Python PRS-CS command and running each chromosome separately, it is possible to create scores faster and more efficiently. The last set of steps of the script is to combine chromosome-based results and to run Plink --score using the combined chromosome results from the PRS-CS steps.

Challenges associated with creating predictive PRS and interpreting PRS results

Incorporating polygenic risk scores in research is a useful tool for understanding and characterizing genetic liability. However, it is important to note limitations, challenges, and caveats.

Explaining small portions of the variance. At present, even with discovery GWAS in large cohorts of individuals, PRS continues to explain only small portions of the variability in independent

---

[a]Or access the appendix directly via this link: https://dataverse.scholarsportal.info/dataset.xhtml?persistentId=doi:10.5683/SP2/QK5JOG.

samples for many complex outcomes of interest. This cautions against diagnostic use in clinical settings in the present form. However, the predictive power of PRS is expected to continue to increase as the discovery sample sizes continue to grow [40]. For example, PRS from a recent GWAS of approximately one million individuals explained ~2.5% of the variance in alcohol consumption [4]. PRS from a recent multivariate GWAS of externalizing problems of approximately 1.5 million individuals explained ~10% of the variance in externalizing in independent samples [41].

Meaning of PRS. Polygenic risk scores, by design, represent a weighted linear combination of risk-associated alleles. PRS assumes an additive effect of individual SNPs, and it does not take into account complex higher-order interactive relationships between genetic risk variants. Additionally, PRS do not imply causation and reveal the underlying mechanisms linking genetics and complex traits/outcomes. Further, PRS on their own are not biologically meaningful, and they do not explicitly provide insights about biological functions of the variants included in the risk score calculation.

Missing heritability. Current polygenic scores only capture a small fraction of the total known heritability of a trait/behavior [42]. This "missing heritability" issue suggests that polygenic scores do not fully capture all genetic influences.

Quality and characteristics of the discovery GWAS. The performance of PRS is only as good as the results of the discovery GWAS that used to calculate them [43]. Small discovery sample size results in nonreliable GWAS results, and this would result in PRS using biased estimates from the discovery GWAS. A common way to increase the discovery GWAS sample size is by pooling samples and studies and meta-analyze across different GWAS analyses. However, differences in samples (e.g., sample recruitment criteria, demographics like sex and age, socioeconomic background, and geographical region) and phenotypes (continuous or categorical) used in meta analyses also can result in nonreliable meta-analysis results and further contribute to the performance and portability of polygenic scores [44,45].

PRS accuracy is also a function of recent human demographic history, such that a greater proportion of phenotypic variance is explainable in target populations that are genetically more similar to the population of the discovery GWAS [46]. Thus, the performance of PRS depends in part on how much discovery and target samples are genetically and phenotypically close. For instance, GWAS results in a population-based sample that may not translate well to samples with different ascertainment strategies. Alcohol-dependence polygenic scores calculated using GWAS results from a population-based sample predicted alcohol problems in another independent population-based sample, but not in high-risk clinically ascertained samples [45]. This underscores the need to examine characteristics of discovery GWAS samples and have diverse types of discovery samples in order to enhance generalizability of PRS results.

PRS in diverse ancestral populations. As noted above, the performance of PRS depends in part on the similarities in the discovery GWAS sample and the independent target sample. One key determinant of the performance of PRS is to have matched ancestry between the discovery GWAS sample and the independent sample; PRS derived from ancestry background that differs from the target sample perform poorly with attenuated variance explained [47]. However, to date, discovery GWAS results for almost all phenotypes are mostly available in samples of European ancestry individuals [48,49]. This limits the conclusions and interpretability of PRS within samples of individuals of European ancestry because genetic associations vary across populations. Lack of diversity in GWASs is problematic because using GWAS results in individuals of mostly European ancestry to create PRS in diverse target samples can lead to inflated or deflated PRS due to differences in genotypes, minor allele frequencies, and LD patterns across populations of diverse ancestry.

Statistical methods are under rapid development to better account for patterns of LD and allele frequency to improve PRS performance in diverse samples. When discovery GWAS weights are not available in samples of diverse ancestry, or discovery sample sizes are relatively small, creating multiethnic PRS [30] could improve the predictive power of PRS. A multiethnic PRS approach involves combining results of a large discovery GWAS sample of European ancestry individuals with GWAS results in target sample using a 10-fold GWAS strategy. The 10-fold approach uses each 1/10th of the target sample as the "target" sample and conducts a "discovery" GWAS in the rest of the 9/10th of the sample and uses weights to calculate PRS in the 1/10th of the sample. Standardized scores are combined across folds to create the final multiethnic PRS. This method has shown to increase accuracy of PRS performance for diabetes phenotypes in diverse populations [30].

The recently developed PRS-CSx [36] approach is a new method to integrate GWAS summary statistics results from multiple ancestral populations to improve effect size estimation by combining summary statistics from multiple populations and leveraging LD pattern and diversity across discovery samples. While statistical methods are likely to help address the challenges associated with creating PRS in diverse populations, consorted efforts to increase the diversity of participants in genomic research are scientifically and morally imperative in order to prevent further contribution to health disparities <2019#62> [50].

Conclusion

In summary, polygenic risk scores provide a flexible way of bringing results from large-scale gene-identification GWASs into deeply phenotyped samples for modeling disease risk prediction and understanding the etiology of psychopathology or disease, as well as for addressing questions about how genetic risk unfolds over time and/or how it interfaces with the environments. As GWAS sample size increases and diversifies, polygenic risk scores will continue to play a critical role in biomedical and social science research and will lay the foundation for informing personalized and precision medicine. Methods for calculating more predictive PRS are under rapid development. Concerted efforts to have adequate representation of diversity in genetic research are also underway. The efficacy of the use of PRS necessitates careful considerations when creating risk scores as well as appropriate interpretation of PRS and recognition of polygenic risk scores' strengths and limitations.

Appendix A
Example script for creating polygenic scores

Fazil Aliev (fazil.aliev@rutgers.edu) and Sally Kuo (sally.kuo@rutgers.edu)
https://github.com/fazilaliev/Script_for_PRS
Files and software needed:

1. Discovery sample GWAS is preferably on the same ancestry individuals as target sample individuals for phenotype of interest. This file must have Chromosome, Marker/SNP name, position, effect and alternative alleles, GWAS beta or OR, *p*-value. Extra information like MAF, imputation quality also might be helpful for extra filtering and QC. Let's assume that GWAS file name is `discovery_example.txt`. We will use `[discovery_path]/discovery_example.txt` to refer this file. We use brackets ([]) to refer discovery GWAS file path on the server. In most cases

this file contains one line per marker after removing possible lines per marker - per covariate and one header line. Here is an example of top lines of `discovery_example.txt file`:

CHR	SNP	BP	A1	A2	INFO	Beta	SE	P
8	rs62513865	101592213	t	c	0.949	−1.006	0.0271	0.8086
8	rs79643588	106973048	a	g	0.997	1.0178	0.0244	0.4606
8	rs17396518	108690829	t	g	0.987	0.9612	0.0143	0.0046
8	rs983166	108681675	a	c	0.998	0.9799	0.0139	0.1452
8	rs35107696	109712249	a	at	0.999	1.0130	0.0165	0.4302
8	rs37704624	105176418	t	ttc	1	1.0085	0.0157	0.5867
8	rs7014597	104152280	c	g	0.993	1.0192	0.0182	0.2934
8	rs3134156	100479917	t	c	0.998	0.9824	0.0190	0.3526
8	rs6980591	103144592	a	c	0.997	1.0449	0.0167	0.0083
8	rs72670434	108166508	a	t	0.985	1.0126	0.0147	0.3898

2. Target sample genotype information. This information can be in any of uncompressed or compressed formats that Plink [22,23] accepts. Assume that Plink standard format .bed (.bim, .fam) is used and file names are: `[target_path]/target_example.bed, [target_path]/target_example.bim` and `[target_path]/target_example.fam` which is referred in plink as `--bfile [target_path]/target_example`.
For simplicity we also assume that markers of discovery and target files have the same allele codes and same map/positions. When maps are different, any uplifting algorithm can match discovery and target marker maps.
3. Plink (https://zzz.bwh.harvard.edu/plink/plink2.shtml) software installed (Any of Plink 1 or Plink 2 versions can work but Plink 2 is much faster).
4. PRS-CS [35] software installed (https://github.com/getian107/PRScs) (python needs to be installed to use PRS-CS, but for P+T scores, PRS-CS and python are not needed).

We explain here only commands based on the unix/linux environment, but commands for other software and programming language environment are also similar. Chromosome and position columns will not be used in analysis because chromosome number and marker positions will not be the same (originally or after possible uplifting) between the discovery and target samples. Accordingly, only marker name (SNP column above) will be used to match SNPs. In the example above allele names have small letters. Target sample usually uses capital letters. To match SNPs we need to capitalize allele letters. The command to capitalize possible allele letters is:

```
sed -i -e 's/g/G/g; s/c/C/g; s/t/T/g; s/a/A/g' [discovery_path]/discovery_example.txt
```

Results will be written to the same file. Allele letters in the file will be capitalized after running this command.

As we see in the above example rs7014597 (c, g alleles) and rs72670434 (a, t alleles), SNPs are palindromic and should be removed. Allele columns are located in columns 4 and 5. All palindromic SNPs (i.e., having allele pairs AT, TA, CG, GC) must be removed (with the exceptions when the strand in genotyping/imputation is the same across the discovery and target samples. Also SNPs rs35107696 and rs377046245 have alleles having more than one letters and must be removed. We will keep only

SNPs with allele pairs "AC", "AG", "CA", "CT", "GA", "GT", "TC" and "TG", which will ensure that all palindromic SNPs have been removed and only biallelic SNPs are included. You might need to remove all markers with no *p*-value. Usually, in a GWAS results file we see period (.) or NA when p-value or Beta/OR values are missing. *P*-value column number is 9 and $9>0 restriction in the following command will filter out all lines with no positive p-value.

Let's write both filters to one single command:

```
gawk -F " " '($4$5=="AC" || $4$5=="AG" || $4$5=="CA" || $4$5=="CT" || $4$5=="GA" || $4$5=="GT" || $4$5=="TC" || $4$5=="TG") && $9>0 || $1=="CHR" {print $0}' [discovery_path]/discovery_example.txt > [work_path]/discovery_cleaned.txt
```

Note that depending on your unix system you might need to replace `gawk` to `awk`. Gawk is a powerful version of `awk`, some unix systems might only have `awk`.

Note that in the above command $4$5 pairs corresponds to the first and second allele columns (4 and 5) which must be changed accordingly when the allele columns in the discovery GWAS file are different. Additional filtering (by, for example, MAF, Hardy-Weinberg (H-W), INFO) can be performed at this step, if necessary. Results of the command will be written to `[work_path]/discovery_cleaned.txt` file and look like:

CHR	SNP	BP	A1	A2	INFO	Beta	SE	P
8	rs62513865	101592213	T	C	0.949	−1.006	0.0271	0.8086
8	rs79643588	106973048	A	G	0.997	1.0178	0.0244	0.4606
8	rs17396518	108690829	T	G	0.987	0.9612	0.0143	0.0046
8	rs983166	108681675	A	C	0.998	0.9799	0.0138	0.1452
8	rs3134156	100479917	T	C	0.998	0.9824	0.0190	0.3526
8	rs6980591	103144592	A	C	0.997	1.0449	0.0167	0.0083

...

We can run some filtering based on INFO and missingness. If there is MAF and H-W information available in discovery sample, we can also add MAF and H-W filters.

```
gawk -F " " '$9!="NA" && $6>0.8 {print $2 $4 $5 $7 $9}' [discovery_path]/discovery_cleaned.txt > output_file.txt
```

$9 and $6 in this command refer to the P-value column (column #9) and the INFO column (column #6). $9 != "NA" means to drop lines with p-value ="NA".

{print $2 $4 $5 $7 $9} prints *SNP, A1, A2, BETA* and *P* columns, where column numbers are followed by the $ sign.

PRS-CS [35] creating method has two steps. First, we create new weights based on discovery GWAS and external panel information provided by PRS-CS software. Second we apply Plink *–score* method to new weights in target sample and create scores.

Need to make sure PRS-CS and python are installed properly.

```
python [PRS-CS directory]/PRScs.py --ref_dir=[ref files directory]/[ancestry reference file] --bim_prefix=[target_path]/target_example --sst_file=[work_path]/discovery_cleaned.txt --chrom=[chromosomome no] --n_gwas=[gwas sample size] --out_dir=[output file name with PRS-CS new weights with full path] --seed=123
```

Here [ancestry reference file] is the file with the ancestry as the discovery ancestry if discovery sample has one ancestry. Corresponding ancestry files are provided by the software. File names start with `ldblk_1kg_` followed by ancestry. PRS-CS software provides an output file with the new weights for SNPs for each chromosome. Combined file with all chromosome outputs with new weights is used with Plink -- *score* method.

Chromosome files can be combined with

```
cat [output file name with PRS-CS new weights with full path]* > [combined PRS-CS weights file]
```

Plink command to create final score looks like:

```
[Plink full path name]/plink --bfile [target_path]/target_example --allow-no-sex --out [final output PRS-CS file full name] --score [combined PRS-CS weights file] 2 4 6 no-sum
```

The last Plink command will create a final PRS-CS file [final output PRS-CS file full name]. profile which can be used for further prediction analyses.

Using single R script to prepare and send all commands to the system to create PRS-CS score

It is possible to use R to send all above mentioned commands to the unix system. R code here creates PRS-CS without any other commands. By copying, pasting and running the script below you can create PRS-CS for any discovery GWAS file and target genotype file. Discovery GWAS file and three target sample genotype files (Plink format) must be located in the "data" directory. All necessary parameters are listed in the beginning of the script. The rest of the script works using parameters entered on the top part.

```
### This R script runs basic steps of PRS-CS creating on Linux/Unix systems
### Requires installed Plink, PRS-CS, R and Python
### (Most of systems have R and Python, please check first and install if needed)
### All chromosomes run at once and it is slow.
### Add --chrom=[chr_no] option to run by chromosome for fast runs

### To run the script:
### 1) Copy this file to working directory (~/PRS-CS_test),
### 2) Copy discovery GWAS file and targer genotype files to data_dir
###    (~/PRS-CS_test/data)
### 3) Make sure that Plink and PRS-CS (R and Python) are installed on the system
### 4) Correct all parameters below before "### End of parameters" line and save
### 5) Use system command   :   cd ~/PRS-CS_test
###             and then type :    [R_ path_if_needed/]R --no-save < prs_cs_tesr.R


### Enter all parameters here:

### Working directory on root (put "/" at the end)
wd="~/PRS-CS_test/"
### Data directory: put discovery and target sample files here
data_dir=paste0(wd,"data/")
```

```
### Directory, where PRS-CS is located (has "PRScs-master" directory inside)
prs_cs_dir=paste0("[put_PRS-CS_directory_here]","/")
phi=0                                  ## PRS-CS threshold phi parameter phi=0 uses
## auto, pho>0 uses phi as threshold
plink_exe=paste0("put_plink_directory_here","plink")   ## Put full path for plink exe
file

### Discovery GWAS file parameters:
discovery_gwas="discovery_gwas.txt"    ## Discovery GWAS file name in working directory
n_discovery_gwas=10000                 ## discovery GWAS sample size
gwas_delimiter=" "                     ## gwas delimiter, use "," if csv
ancestry="EUR"                         ## discovery GWAS sample ancesty
or_is_used=0                           ## or_is_used=0 if discovery GWAS has BETA
                                       ## or_is_used=1 is for OR(Odds Ratio)
snp_column_no=1                        ## SNP/marker (rs) column number
ref_allele_column_no=4                 ## Reference (A1) column number
alt_allele_column_no=5                 ## Alternative (A2) column number
beta_column_no=7                       ## Beta (OR if binary) cloumn number
p_value_column_no=9                    ## p-value column number

### Target sample genotype file parameters:
target_gwas = "target_file_name"       ## Target sample Plink format genotype file
                                       ## name w/o extension (.bed, .bim, .fam)

### End of parameters
############# Do not change below, all necessary parameters are above ########

### Setting working directory and creating data and temp directories
setwd(wd)
if (!file.exists(data_dir)) dir.create(data_dir,recursive=T)
temp_dir=paste0(wd,"temp/")            ## Tepm directory to write temp files
if (!file.exists(temp_dir)) dir.create(temp_dir,recursive=T)

### Selecting necessary lines from discovery GWAS file
Sys.setlocale('LC_ALL','C'); ltr="\""; lsr="'"; uuu="$"; uu1="-"

### Depending on Linux version you might need to replace "gawk" below with "awk".
### Check your system's gawk/awk commands
### Keep only palindromic SNPs
comfaz=paste0(uuu,snp_column_no,",",uuu,ref_allele_column_no,",",
  uuu,alt_allele_column_no,",",uuu,beta_column_no,",",uuu,p_value_column_no)
a1a2=paste0(uuu,ref_allele_column_no,uuu,alt_allele_column_no)   ## This is A1A2
column ($ref$alt)
command1=paste0("gawk ",uu1,"F ",ltr,gwas_delimiter,ltr,
  " '",uuu,p_value_column_no,">0 && (",
  a1a2,"==",ltr,"AC",ltr," || ",a1a2,"==",ltr,"AG",ltr," || ",
```

```
  a1a2,"==",ltr,"CA",ltr," || ",a1a2,"==",ltr,"CT",ltr," || ",
  a1a2,"==",ltr,"GA",ltr," || ",a1a2,"==",ltr,"GT",ltr," || ",
  a1a2,"==",ltr,"TC",ltr," || ",a1a2,"==",ltr,"TG",ltr," || ",
  a1a2,"==",ltr,"ac",ltr," || ",a1a2,"==",ltr,"ag",ltr," || ",
  a1a2,"==",ltr,"ca",ltr," || ",a1a2,"==",ltr,"ct",ltr," || ",
  a1a2,"==",ltr,"ga",ltr," || ",a1a2,"==",ltr,"gt",ltr," || ",
  a1a2,"==",ltr,"tc",ltr," || ",a1a2,"==",ltr,"tg",ltr,
  ") {print ",comfaz,"}' ", data_dir, discovery_gwas," > ",
  temp_dir, "temp1.txt")
system(command1)

### Replace small allele letters with big letters
command2=paste0("sed -i -e 's/g/G/g; s/c/C/g; s/t/T/g; s/a/A/g' ",
  temp_dir, "temp1.txt")
system(command2)

### Read non-palindromic file with reduced columns and assign names to columns
merged0 = read.table(paste0(temp_dir, "temp1.txt"), header=F)

names(merged0) = c("SNP", "A1", "A2", ifelse(or_is_used,"OR","BETA"), "P")

### Remove duplicate snps
merged1=merged0[!duplicated(merged0[,"SNP"]),]

### Now select only common snps b/w discovery and target
command3=paste0("awk '{print $2}' ",data_dir, target_gwas, ".bim > ",
  temp_dir, "tempbim.txt")    ## Copy rs column of bim file
system(command3)
inputbim=read.table(paste0(temp_dir, "tempbim.txt"), header=F)   ## Read rs column
names(inputbim)[1]="SNP"                                          ## Assign a name

### Merge and keep only common SNP names
merged=merge(merged1, inputbim, by="SNP", all.x=F, all.y=F)

### Create score file
write.table(merged, file=paste0(temp_dir,"score_file.txt"), row.names=F,
  col.names=T, quote=F)

## Writing common SNPs
write.table(merged[,"SNP"],file=paste0(temp_dir, "common.txt"), row.names=F,
  col.names=F, quote=F)

### Using plink to extract subset of common discovery and target SNPs
command4=paste0(plink_exe," --bfile ", data_dir, target_gwas,
  " --allow-extra-chr --keep-allele-order --allow-no-sex --extract ",
  temp_dir, "common.txt --make-bed --out ", temp_dir, "newtarget")
system(command4)
```

```
### Resolving strand issue between discovery and target samples
### (Plink's --flip option)
### Read new/common target genotype file SNP, A1 and A2 columns
inputbim2=read.table(paste0(temp_dir, "newtarget.bim"), header=F)
## Assign names A to A1 and B to A2
names(inputbim2)=c("cc", "SNP", "p1", "p2", "A", "B")
tempmg=merge(merged[,c("SNP", "A1", "A2")], inputbim2[,c("SNP", "A", "B")],
  by="SNP", all.x=F, all.y=F)
tempmerge2=tempmg[as.character(tempmg$A1) !=
  as.character(tempmg$A) && as.character(tempmg$A1) != as.character(tempmg$B),]
targetname=""
if (dim(tempmerge2)[1]>0)  ## Check if there are flipped SNPs and print the list
{ write.table(tempmerge2[,c(sn2)], file=paste0(temp_dir, "flip.txt"),
    row.names=F, col.names=F, quote=F)
  command5=paste0(plink_exe, " --bfile ",
    temp_dir, "newtarget --allow-extra-chr --allow-no-sex --flip ",
    temp_dir, "flip.txt --make-bed --out ",temp_dir,"newtargett")
  targetname="t"##for target bed file name
  system(command5)
}

### Creating python PRS-CS command
prs_cs_command=paste0("python ",prs_cs_dir,"PRScs-master/PRScs.py ",
  " --ref_dir=", prs_cs_dir, "ldblk_1kg_", ifelse(ancestry=="AFR", "afr",
  ifelse(ancestry=="EAS", "eas", "eur")),
  " --bim_prefix=", temp_dir, "newtarget", targetname,
  " --sst_file=", temp_dir, "score_file.txt ",
  " --n_gwas=", n_discovery_gwas, ifelse(phi>0, paste0(" --phi=",phi," "), " "),
  " --out_dir=", temp_dir, "prs_cs_out --seed=123459921")
### --ref_dir=path_to_ref --bim_prefix=path_to_bim --sst_file=path_to_sumstats
system(prs_cs_command)                    ## Runs PRS-CS and creates new weights

### Results of PRS-CS are splitted by chromosome, combine into one file
command6=paste0("cat ",temp_dir,"prs_cs_out*phiauto_chr*.txt >",
  temp_dir,"prs_estimates.txt")
system(command6)

### Using Plink with new weights to create final PRS-CS score file
plink_command=paste0(plink_exe," --bfile ", temp_dir, "newtarget", targetname,
  " --allow-no-sex --out ", wd, "New_PRS-CS_score --score ", temp_dir,
  "prs_estimates.txt 2 4 6 no-sum")
system(plink_command)

### Scores will be written to [working directory]/New_PRS-CS_score.profile
```

References

[1] E. Turkheimer, Three laws of behavior genetics and what they mean, Curr. Dir. Psychol. Sci. 9 (5) (2000) 160–164, https://doi.org/10.1111/1467-8721.00084.

[2] P.M. Visscher, N.R. Wray, Q. Zhang, P. Sklar, M.I. McCarthy, M.A. Brown, J. Yang, 10 years of GWAS discovery: biology, function, and translation, Am. J. Hum. Genet. 101 (1) (2017) 5–22, https://doi.org/10.1016/j.ajhg.2017.06.005.

[3] H.R. Kranzler, H. Zhou, R.L. Kember, R. Vickers Smith, A.C. Justice, S. Damrauer, P.S. Tsao, D. Klarin, A. Baras, J. Reid, J. Overton, D.J. Rader, Z. Cheng, J.P. Tate, W.C. Becker, J. Concato, K. Xu, R. Polimanti, H. Zhao, J. Gelernter, Genome-wide association study of alcohol consumption and use disorder in 274,424 individuals from multiple populations, Nat. Commun. 10 (1) (2019) 1499, https://doi.org/10.1038/s41467-019-09480-8.

[4] M. Liu, Y. Jiang, R. Wedow, Y. Li, D.M. Brazel, F. Chen, G. Datta, J. Davila-Velderrain, D. McGuire, C. Tian, X. Zhan, M. Agee, B. Alipanahi, A. Auton, R.K. Bell, K. Bryc, S.L. Elson, P. Fontanillas, N.A. Furlotte, D.A. Hinds, B.S. Hromatka, K.E. Huber, A. Kleinman, N.K. Litterman, M.H. McIntyre, J.L. Mountain, C.A.M. Northover, J.F. Sathirapongsasuti, O.V. Sazonova, J.F. Shelton, S. Shringarpure, C. Tian, J.Y. Tung, V. Vacic, C.H. Wilson, S.J. Pitts, A. Mitchell, A.H. Skogholt, B.S. Winsvold, B. Sivertsen, E. Stordal, G. Morken, H. Kallestad, I. Heuch, J.-A. Zwart, K.K. Fjukstad, L.M. Pedersen, M.E. Gabrielsen, M.B. Johnsen, M. Skrove, M.S. Indredavik, O.K. Drange, O. Bjerkeset, S. Børte, S.Ø. Stensland, H. Choquet, A.R. Docherty, J.D. Faul, J.R. Foerster, L.G. Fritsche, M.E. Gabrielsen, S.D. Gordon, J. Haessler, J.-J. Hottenga, H. Huang, S.-K. Jang, P.R. Jansen, Y. Ling, R. Mägi, N. Matoba, G. McMahon, A. Mulas, V. Orrù, T. Palviainen, A. Pandit, G.W. Reginsson, A.H. Skogholt, J.A. Smith, A.E. Taylor, C. Turman, G. Willemsen, H. Young, K.A. Young, G.J.M. Zajac, W. Zhao, W. Zhou, G. Bjornsdottir, J.D. Boardman, M. Boehnke, D.I. Boomsma, C. Chen, F. Cucca, G.E. Davies, C.B. Eaton, M.A. Ehringer, T. Esko, E. Fiorillo, N.A. Gillespie, D.F. Gudbjartsson, T. Haller, K.M. Harris, A.C. Heath, J.K. Hewitt, I.B. Hickie, J.E. Hokanson, C.J. Hopfer, D.J. Hunter, W.G. Iacono, E.O. Johnson, Y. Kamatani, S.L.R. Kardia, M.C. Keller, M. Kellis, C. Kooperberg, P. Kraft, K.S. Krauter, M. Laakso, P.A. Lind, A. Loukola, S.M. Lutz, P.A.F. Madden, N.G. Martin, M. McGue, M.B. McQueen, S.E. Medland, A. Metspalu, K.L. Mohlke, J.B. Nielsen, Y. Okada, U. Peters, T.J.C. Polderman, D. Posthuma, A.P. Reiner, J.P. Rice, E. Rimm, R.J. Rose, V. Runarsdottir, M.C. Stallings, A. Stančáková, H. Stefansson, K.K. Thai, H.A. Tindle, T. Tyrfingsson, T.L. Wall, D.R. Weir, C. Weisner, J.B. Whitfield, B.S. Winsvold, J. Yin, L. Zuccolo, L.J. Bierut, K. Hveem, J.J. Lee, M.R. Munafò, N.L. Saccone, C.J. Willer, M.C. Cornelis, S.P. David, D.A. Hinds, E. Jorgenson, J. Kaprio, J.A. Stitzel, K. Stefansson, T.E. Thorgeirsson, G. Abecasis, D.J. Liu, S. Vrieze, 23andMe Research Team, & Hunt All-In Psychiatry, Association studies of up to 1.2 million individuals yield new insights into the genetic etiology of tobacco and alcohol use, Nat. Genet. 51 (2) (2019) 237–244, https://doi.org/10.1038/s41588-018-0307-5.

[5] D. Demontis, R.K. Walters, J. Martin, M. Mattheisen, T.D. Als, E. Agerbo, G. Baldursson, R. Belliveau, J. Bybjerg-Grauholm, M. Bækvad-Hansen, F. Cerrato, K. Chambert, C. Churchhouse, A. Dumont, N. Eriksson, M. Gandal, J.I. Goldstein, K.L. Grasby, J. Grove, O.O. Gudmundsson, C.S. Hansen, M.E. Hauberg, M.V. Hollegaard, D.P. Howrigan, H. Huang, J.B. Maller, A.R. Martin, N.G. Martin, J. Moran, J. Pallesen, D.S. Palmer, C.B. Pedersen, M.G. Pedersen, T. Poterba, J.B. Poulsen, S. Ripke, E.B. Robinson, F.K. Satterstrom, H. Stefansson, C. Stevens, P. Turley, G.B. Walters, H. Won, M.J. Wright, ADHD Working Group of the Psychiatric Genomics Consortium (PGC), Early Lifecourse & Genetic Epidemiology (EAGLE) Consortium, 23 and Me Research Team, O.A. Andreassen, P. Asherson, C.L. Burton, D.I. Boomsma, B. Cormand, S. Dalsgaard, B. Franke, J. Gelernter, D. Geschwind, H. Hakonarson, J. Haavik, H.R. Kranzler, J. Kuntsi, K. Langley, K.-P. Lesch, C. Middeldorp, A. Reif, L.A. Rohde, P. Roussos, R. Schachar, P. Sklar, E.J.S. Sonuga-Barke, P.F. Sullivan, A. Thapar, J.Y. Tung, I.D. Waldman, S.E. Medland, K. Stefansson, M. Nordentoft, D.M. Hougaard, T. Werge, O. Mors, P.B. Mortensen, M.J. Daly, S.V. Faraone, A.D. Børglum, B.M. Neale, Discovery of the first genome-wide significant risk loci for attention deficit/hyperactivity disorder, Nat. Genet. 51 (1) (2019) 63–75, https://doi.org/10.1038/s41588-018-0269-7.

[6] D.M. Howard, M.J. Adams, T.-K. Clarke, J.D. Hafferty, J. Gibson, M. Shirali, J.R.I. Coleman, S.P. Hagenaars, J. Ward, E.M. Wigmore, C. Alloza, X. Shen, M.C. Barbu, E.Y. Xu, H.C. Whalley, R.E. Marioni, D.J. Porteous, G. Davies, I.J. Deary, G. Hemani, K. Berger, H. Teismann, R. Rawal, V. Arolt, B.T. Baune, U. Dannlowski, K. Domschke, C. Tian, D.A. Hinds, M. Trzaskowski, E.M. Byrne, S. Ripke, D.J. Smith, P.F. Sullivan, N.R. Wray, G. Breen, C.M. Lewis, A.M. McIntosh, 23andMe Research Team, & Major Depressive Disorder Working Group of the Psychiatric Genomics Consortium, Genome-wide meta-analysis of depression identifies 102 independent variants and highlights the importance of the prefrontal brain regions, Nat. Neurosci. 22 (3) (2019) 343–352, https://doi.org/10.1038/s41593-018-0326-7.

[7] Schizophrenia Working Group of the Psychiatric Genomics Consortium, Biological insights from 108 schizophrenia-associated genetic loci, Nature 511 (7510) (2014) 421–427, https://doi.org/10.1038/nature13595.

[8] P. van der Harst, N. Verweij, Identification of 64 novel genetic loci provides an expanded view on the genetic architecture of coronary artery disease, Circ. Res. 122 (3) (2018) 433–443, https://doi.org/10.1161/CIRCRESAHA.117.312086.

[9] T.J. Hoffmann, G.B. Ehret, P. Nandakumar, D. Ranatunga, C. Schaefer, P.-Y. Kwok, C. Iribarren, A. Chakravarti, N. Risch, Genome-wide association analyses using electronic health records identify new loci influencing blood pressure variation, Nat. Genet. 49 (1) (2017) 54–64, https://doi.org/10.1038/ng.3715.

[10] J.J. Lee, R. Wedow, A. Okbay, E. Kong, O. Maghzian, M. Zacher, T.A. Nguyen-Viet, P. Bowers, J. Sidorenko, R. Karlsson Linnér, M.A. Fontana, T. Kundu, C. Lee, H. Li, R. Li, R. Royer, P.N. Timshel, R.K. Walters, E.A. Willoughby, L. Yengo, M. Agee, B. Alipanahi, A. Auton, R.K. Bell, K. Bryc, S.L. Elson, P. Fontanillas, D.A. Hinds, J.C. McCreight, K.E. Huber, N.K. Litterman, M.H. McIntyre, J.L. Mountain, E.S. Noblin, C.A.M. Northover, S.J. Pitts, J.F. Sathirapongsasuti, O.V. Sazonova, J.F. Shelton, S. Shringarpure, C. Tian, V. Vacic, C.H. Wilson, A. Okbay, J.P. Beauchamp, M.A. Fontana, J.J. Lee, T.H. Pers, C.A. Rietveld, P. Turley, G.-B. Chen, V. Emilsson, S.F.W. Meddens, S. Oskarsson, J.K. Pickrell, K. Thom, P. Timshel, R.D. Vlaming, A. Abdellaoui, T.S. Ahluwalia, J. Bacelis, C. Baumbach, G. Bjornsdottir, J.H. Brandsma, M.P. Concas, J. Derringer, N.A. Furlotte, T.E. Galesloot, G. Girotto, R. Gupta, L.M. Hall, S.E. Harris, E. Hofer, M. Horikoshi, J.E. Huffman, K. Kaasik, I.P. Kalafati, R. Karlsson, A. Kong, J. Lahti, S.J. van der Lee, C.D. Leeuw, P.A. Lind, K.-O. Lindgren, T. Liu, M. Mangino, J. Marten, E. Mihailov, M.B. Miller, P.J. van der Most, C. Oldmeadow, A. Payton, N. Pervjakova, W.J. Peyrot, Y. Qian, O. Raitakari, R. Rueedi, E. Salvi, B. Schmidt, K.E. Schraut, J. Shi, A.V. Smith, R.A. Poot, B. St Pourcain, A. Teumer, G. Thorleifsson, N. Verweij, D. Vuckovic, J. Wellmann, H.-J. Westra, J. Yang, W. Zhao, Z. Zhu, B.Z. Alizadeh, N. Amin, A. Bakshi, S.E. Baumeister, G. Biino, K. Bønnelykke, P.A. Boyle, H. Campbell, F.P. Cappuccio, G. Davies, J.-E. De Neve, P. Deloukas, I. Demuth, J. Ding, P. Eibich, L. Eisele, N. Eklund, D.M. Evans, J.D. Faul, M.F. Feitosa, A.J. Forstner, I. Gandin, B. Gunnarsson, B.V. Halldórsson, T.B. Harris, A.C. Heath, L.J. Hocking, E.G. Holliday, G. Homuth, M.A. Horan, J.-J. Hottenga, P.L. de Jager, P.K. Joshi, A. Jugessur, M.A. Kaakinen, M. Kähönen, S. Kanoni, L. Keltigangas-Järvinen, L.A.L.M. Kiemeney, I. Kolcic, S. Koskinen, A.T. Kraja, M. Kroh, Z. Kutalik, A. Latvala, L.J. Launer, M.P. Lebreton, D.F. Levinson, P. Lichtenstein, P. Lichtner, D.C.M. Liewald, A. Loukola, L.L.C. Study, P.A. Madden, R. Mägi, T. Mäki-Opas, R.E. Marioni, P. Marques-Vidal, G.A. Meddens, G. McMahon, C. Meisinger, T. Meitinger, Y. Milaneschi, L. Milani, G.W. Montgomery, R. Myhre, C.P. Nelson, D.R. Nyholt, W.E.R. Ollier, A. Palotie, L. Paternoster, N.L. Pedersen, K.E. Petrovic, D.J. Porteous, K. Räikkönen, S.M. Ring, A. Robino, O. Rostapshova, I. Rudan, A. Rustichini, V. Salomaa, A.R. Sanders, A.-P. Sarin, H. Schmidt, R.J. Scott, B.H. Smith, J.A. Smith, J.A. Staessen, E. Steinhagen-Thiessen, K. Strauch, A. Terracciano, M.D. Tobin, S. Ulivi, S. Vaccargiu, L. Quaye, F.J.A. van Rooij, C. Venturini, A.A.E. Vinkhuyzen, U. Völker, H. Völzke, J.M. Vonk, D. Vozzi, J. Waage, E.B. Ware, G. Willemsen, J.R. Attia, D.A. Bennett, K. Berger, L. Bertram, H. Bisgaard, D.I. Boomsma, I.B. Borecki, U. Bültmann, C.F. Chabris, F. Cucca, D. Cusi, I.J. Deary, G.V. Dedoussis, C.M. van Duijn, J.G. Eriksson, B. Franke, L. Franke, P. Gasparini, P.V. Gejman, C. Gieger, H.-J. Grabe, J. Gratten, P.J.F. Groenen, V. Gudnason, P. van der Harst, C. Hayward, D.A. Hinds, W. Hoffmann, E. Hyppönen, W.G. Iacono, B. Jacobsson, M.-R. Järvelin, K.-H. Jöckel, J. Kaprio, S.L.R. Kardia,

T. Lehtimäki, S.F. Lehrer, P.K.E. Magnusson, N.G. Martin, M. McGue, A. Metspalu, N. Pendleton, B.W.J.H. Penninx, M. Perola, N. Pirastu, M. Pirastu, O. Polasek, D. Posthuma, C. Power, M.A. Province, N.J. Samani, D. Schlessinger, R. Schmidt, T.I.A. Sørensen, T.D. Spector, K. Stefansson, U. Thorsteinsdottir, A.R. Thurik, N.J. Timpson, H. Tiemeier, J.Y. Tung, A.G. Uitterlinden, V. Vitart, P. Vollenweider, D.R. Weir, J.F. Wilson, A.F. Wright, D.C. Conley, R.F. Krueger, G.D. Smith, A. Hofman, D.I. Laibson, S.E. Medland, M.N. Meyer, J. Yang, M. Johannesson, P.M. Visscher, T. Esko, P.D. Koellinger, D. Cesarini, 23andMe Research Team, Cogent, & Social Science Genetic Association Consortium, Gene discovery and polygenic prediction from a genome-wide association study of educational attainment in 1.1 million individuals, Nat. Genet. 50 (8) (2018) 1112–1121, https://doi.org/10.1038/s41588-018-0147-3.

[11] R. Plomin, Blueprint: How DNA Makes Us Who We Are, MIT Press, 2019.

[12] S.M. Purcell, N.R. Wray, J.L. Stone, P.M. Visscher, M.C. O'Donovan, P.F. Sullivan, P. Sklar, Common polygenic variation contributes to risk of schizophrenia and bipolar disorder, Nature 460 (7256) (2009) 748–752, https://doi.org/10.1038/nature08185.

[13] P.R. Jansen, T.J.C. Polderman, K. Bolhuis, J. van der Ende, V.W.V. Jaddoe, F.C. Verhulst, T. White, D. Posthuma, H. Tiemeier, Polygenic scores for schizophrenia and educational attainment are associated with behavioural problems in early childhood in the general population, J. Child Psychol. Psychiatry 59 (1) (2018) 39–47, https://doi.org/10.1111/jcpp.12759.

[14] S. Mistry, V. Escott-Price, A.D. Florio, D.J. Smith, S. Zammit, Genetic risk for bipolar disorder and psychopathology from childhood to early adulthood, J. Affect. Disord. 246 (2019) 633–639, https://doi.org/10.1016/j.jad.2018.12.091.

[15] J. Harper, M. Liu, S.M. Malone, M. McGue, W.G. Iacono, S.I. Vrieze, Using multivariate endophenotypes to identify psychophysiological mechanisms associated with polygenic scores for substance use, schizophrenia, and education attainment, Psychol. Med. (2021) 1–11, https://doi.org/10.1017/S0033291721000763.

[16] S. Ranlund, S. Calafato, J.H. Thygesen, K. Lin, W. Cahn, B. Crespo-Facorro, S.M.C. de Zwarte, Á. Díez, M. Di Forti, GROUP, C. Iyegbe, A. Jablensky, R. Jones, M.-H. Hall, R. Kahn, L. Kalaydjieva, E. Kravariti, C. McDonald, A.M. McIntosh, A. McQuillin, PEIC, M. Picchioni, D.P. Prata, D. Rujescu, K. Schulze, M. Shaikh, T. Toulopoulou, N. van Haren, J. van Os, E. Vassos, M. Walshe, WTCCC2, C. Lewis, R.M. Murray, J. Powell, E. Bramon, A polygenic risk score analysis of psychosis endophenotypes across brain functional, structural, and cognitive domains, Am. J. Med. Genet. B Neuropsychiatr. Genet. 177 (1) (2018) 21–34, https://doi.org/10.1002/ajmg.b.32581.

[17] P.B. Barr, S.I.-C. Kuo, F. Aliev, A. Latvala, R. Viken, R.J. Rose, J. Kaprio, J.E. Salvatore, D.M. Dick, Polygenic risk for alcohol misuse is moderated by romantic partnerships, Addiction 114 (10) (2019) 1753–1762, https://doi.org/10.1111/add.14712.

[18] S.I.C. Kuo, J.E. Salvatore, F. Aliev, T. Ha, T.J. Dishion, D.M. Dick, The family check-up intervention moderates polygenic influences on long-term alcohol outcomes: results from a randomized intervention trial, Prev. Sci. 20 (7) (2019) 975–985, https://doi.org/10.1007/s11121-019-01024-2.

[19] J.L. Meyers, J.E. Salvatore, F. Aliev, E.C. Johnson, V.V. McCutcheon, J. Su, S.I.C. Kuo, D. Lai, L. Wetherill, J.C. Wang, G. Chan, V. Hesselbrock, T. Foroud, K.K. Bucholz, H.J. Edenberg, D.M. Dick, B. Porjesz, A. Agrawal, Psychosocial moderation of polygenic risk for cannabis involvement: the role of trauma exposure and frequency of religious service attendance, Transl. Psychiatry 9 (1) (2019) 269, https://doi.org/10.1038/s41398-019-0598-z.

[20] D.B. Rosoff, T.-K. Clarke, M.J. Adams, A.M. McIntosh, G. Davey Smith, J. Jung, F.W. Lohoff, Educational attainment impacts drinking behaviors and risk for alcohol dependence: results from a two-sample Mendelian randomization study with ~780,000 participants, Mol. Psychiatry 26 (4) (2021) 1119–1132, https://doi.org/10.1038/s41380-019-0535-9.

[21] T.C. Bates, B.S. Maher, S.E. Medland, K. McAloney, M.J. Wright, N.K. Hansell, K.S. Kendler, N.G. Martin, N.A. Gillespie, The nature of nurture: using a virtual-parent design to test parenting effects on children's educational attainment in genotyped families, Twin Res. Hum. Genet. 21 (2) (2018) 73–83, https://doi.org/10.1017/thg.2018.11.

[22] C.C. Chang, C.C. Chow, L.C. Tellier, S. Vattikuti, S.M. Purcell, J.J. Lee, Second-generation PLINK: rising to the challenge of larger and richer datasets, Gigascience 4 (1) (2015), https://doi.org/10.1186/s13742-015-0047-8.

[23] S. Purcell, B. Neale, K. Todd-Brown, L. Thomas, M.A.R. Ferreira, D. Bender, J. Maller, P. Sklar, P.I.W. de Bakker, M.J. Daly, P.C. Sham, PLINK: A tool set for whole-genome association and population-based linkage analyses, Am. J. Hum. Genet. 81 (3) (2007) 559–575, https://doi.org/10.1086/519795.

[24] P.H. Sudmant, T. Rausch, E.J. Gardner, R.E. Handsaker, A. Abyzov, J. Huddleston, Y. Zhang, K. Ye, G. Jun, M. Hsi-Yang Fritz, M.K. Konkel, A. Malhotra, A.M. Stütz, X. Shi, F. Paolo Casale, J. Chen, F. Hormozdiari, G. Dayama, K. Chen, M. Malig, M.J.P. Chaisson, K. Walter, S. Meiers, S. Kashin, E. Garrison, A. Auton, H.Y.K. Lam, X. Jasmine Mu, C. Alkan, D. Antaki, T. Bae, E. Cerveira, P. Chines, Z. Chong, L. Clarke, E. Dal, L. Ding, S. Emery, X. Fan, M. Gujral, F. Kahveci, J.M. Kidd, Y. Kong, E.-W. Lameijer, S. McCarthy, P. Flicek, R.A. Gibbs, G. Marth, C.E. Mason, A. Menelaou, D.M. Muzny, B.J. Nelson, A. Noor, N.F. Parrish, M. Pendleton, A. Quitadamo, B. Raeder, E.E. Schadt, M. Romanovitch, A. Schlattl, R. Sebra, A.A. Shabalin, A. Untergasser, J.A. Walker, M. Wang, F. Yu, C. Zhang, J. Zhang, X. Zheng-Bradley, W. Zhou, T. Zichner, J. Sebat, M.A. Batzer, S.A. McCarroll, R.E. Mills, M.B. Gerstein, A. Bashir, O. Stegle, S.E. Devine, C. Lee, E.E. Eichler, J.O. Korbel, The 1000 Genomes Project Consortium, An integrated map of structural variation in 2,504 human genomes, Nature 526 (7571) (2015) 75–81, https://doi.org/10.1038/nature15394.

[25] The 1000 Genomes Project Consortium, A global reference for human genetic variation, Nature 526 (7571) (2015) 68–74, https://doi.org/10.1038/nature15393.

[26] C.J. Willer, Y. Li, G.R. Abecasis, METAL: fast and efficient meta-analysis of genomewide association scans, Bioinformatics 26 (17) (2010) 2190–2191, https://doi.org/10.1093/bioinformatics/btq340.

[27] S.W. Choi, P.F. O'Reilly, PRSice-2: Polygenic Risk Score software for biobank-scale data, Gigascience 8 (7) (2019), https://doi.org/10.1093/gigascience/giz082.

[28] J. Euesden, C.M. Lewis, P.F. O'Reilly, PRSice: Polygenic Risk Score software, Bioinformatics 31 (9) (2014) 1466–1468, https://doi.org/10.1093/bioinformatics/btu848.

[29] B.J. Vilhjálmsson, J. Yang, H.K. Finucane, A. Gusev, S. Lindström, S. Ripke, G. Genovese, P.-R. Loh, G. Bhatia, R. Do, T. Hayeck, H.-H. Won, S. Ripke, B.M. Neale, A. Corvin, J.T.R. Walters, K.-H. Farh, P.A. Holmans, P. Lee, B. Bulik-Sullivan, D.A. Collier, H. Huang, T.H. Pers, I. Agartz, E. Agerbo, M. Albus, M. Alexander, F. Amin, S.A. Bacanu, M. Begemann, R.A. Belliveau, J. Bene, S.E. Bergen, E. Bevilacqua, T.B. Bigdeli, D.W. Black, R. Bruggeman, N.G. Buccola, R.L. Buckner, W. Byerley, W. Cahn, G. Cai, D. Campion, R.M. Cantor, V.J. Carr, N. Carrera, S.V. Catts, K.D. Chambert, R.C.K. Chan, R.Y.L. Chen, E.Y.H. Chen, W. Cheng, E.F.C. Cheung, S.A. Chong, C.R. Cloninger, D. Cohen, N. Cohen, P. Cormican, N. Craddock, J.J. Crowley, D. Curtis, M. Davidson, K.L. Davis, F. Degenhardt, J. Del Favero, L.E. DeLisi, D. Demontis, D. Dikeos, T. Dinan, S. Djurovic, G. Donohoe, E. Drapeau, J. Duan, F. Dudbridge, N. Durmishi, P. Eichhammer, J. Eriksson, V. Escott-Price, L. Essioux, A.H. Fanous, M.S. Farrell, J. Frank, L. Franke, R. Freedman, N.B. Freimer, M. Friedl, J.I. Friedman, M. Fromer, G. Genovese, L. Georgieva, E.S. Gershon, I. Giegling, P. Giusti-Rodrguez, S. Godard, J.I. Goldstein, V. Golimbet, S. Gopal, J. Gratten, J. Grove, L. de Haan, C. Hammer, M.L. Hamshere, M. Hansen, T. Hansen, V. Haroutunian, A.M. Hartmann, F.A. Henskens, S. Herms, J.N. Hirschhorn, P. Hoffmann, A. Hofman, M.V. Hollegaard, D.M. Hougaard, M. Ikeda, I. Joa, A. Julia, R.S. Kahn, L. Kalaydjieva, S. Karachanak-Yankova, J. Karjalainen, D. Kavanagh, M.C. Keller, B.J. Kelly, J.L. Kennedy, A. Khrunin, Y. Kim, J. Klovins, J.A. Knowles, B. Konte, V. Kucinskas, Z.A. Kucinskiene, H. Kuzelova-Ptackova, A.K. Kahler, C. Laurent, J.L.C. Keong, S.H. Lee, S.E. Legge, B. Lerer, M. Li, T. Li, K.-Y. Liang, J. Lieberman, S. Limborska, C.M. Loughland, J. Lubinski, J. Lnnqvist, M. Macek, P.K.E. Magnusson, B.S. Maher, W. Maier, J. Mallet, S. Marsal, M. Mattheisen, M. Mattingsdal, R.W. McCarley, C. McDonald, A.M. McIntosh, S. Meier, C.J. Meijer, B. Melegh, I. Melle, R.I. Mesholam-Gately, A. Metspalu, P.T. Michie, L. Milani, V. Milanova, Y. Mokrab, D.W. Morris, O. Mors, P.B. Mortensen, K.C. Murphy, R.M. Murray, I. Myin-Germeys, B. Mller-Myhsok, M. Nelis, I. Nenadic, D.A. Nertney, G. Nestadt, K.K. Nicodemus, L. Nikitina-Zake, L. Nisenbaum, A. Nordin, E. O'Callaghan, C. O'Dushlaine, F.A. O'Neill, S.-Y. Oh, A. Olincy, L. Olsen, J. Van Os, C.

Pantelis, G.N. Papadimitriou, S. Papiol, E. Parkhomenko, M.T. Pato, T. Paunio, M. Pejovic-Milovancevic, D.O. Perkins, O. Pietilinen, J. Pimm, A.J. Pocklington, J. Powell, A. Price, A.E. Pulver, S.M. Purcell, D. Quested, H.B. Rasmussen, A. Reichenberg, M.A. Reimers, A.L. Richards, J.L. Roffman, P. Roussos, D.M. Ruderfer, V. Salomaa, A.R. Sanders, U. Schall, C.R. Schubert, T.G. Schulze, S.G. Schwab, E.M. Scolnick, R.J. Scott, L.J. Seidman, J. Shi, E. Sigurdsson, T. Silagadze, J.M. Silverman, K. Sim, P. Slominsky, J.W. Smoller, H.-C. So, C.C.A. Spencer, E.A. Stahl, H. Stefansson, S. Steinberg, E. Stogmann, R.E. Straub, E. Strengman, J. Strohmaier, T.S. Stroup, M. Subramaniam, J. Suvisaari, D.M. Svrakic, J.P. Szatkiewicz, E. Sderman, S. Thirumalai, D. Toncheva, P.A. Tooney, S. Tosato, J. Veijola, J. Waddington, D. Walsh, D. Wang, Q. Wang, B.T. Webb, M. Weiser, D.B. Wildenauer, N.M. Williams, S. Williams, S.H. Witt, A.R. Wolen, E.H.M. Wong, B.K. Wormley, J.Q. Wu, H.S. Xi, C.C. Zai, X. Zheng, F. Zimprich, N.R. Wray, K. Stefansson, P.M. Visscher, R. Adolfsson, O.A. Andreassen, D.H.R. Blackwood, E. Bramon, J.D. Buxbaum, A.D. Børglum, S. Cichon, A. Darvasi, E. Domenici, H. Ehrenreich, T. Esko, P.V. Gejman, M. Gill, H. Gurling, C.M. Hultman, N. Iwata, A.V. Jablensky, E.G. Jonsson, K.S. Kendler, G. Kirov, J. Knight, T. Lencz, D.F. Levinson, Q.S. Li, J. Liu, A.K. Malhotra, S.A. McCarroll, A. McQuillin, J.L. Moran, P.B. Mortensen, B.J. Mowry, M.M. Nthen, R.A. Ophoff, M.J. Owen, A. Palotie, C.N. Pato, T.L. Petryshen, D. Posthuma, M. Rietschel, B.P. Riley, D. Rujescu, P.C. Sham, P. Sklar, D. St. Clair, D.R. Weinberger, J.R. Wendland, T. Werge, M.J. Daly, P.F. Sullivan, M.C. O'Donovan, P. Kraft, D.J. Hunter, M. Adank, H. Ahsan, K. Aittomäki, L. Baglietto, S. Berndt, C. Blomquist, F. Canzian, J. Chang-Claude, S.J. Chanock, L. Crisponi, K. Czene, N. Dahmen, I.D.S. Silva, D. Easton, A.H. Eliassen, J. Figueroa, O. Fletcher, M. Garcia-Closas, M.M. Gaudet, L. Gibson, C.A. Haiman, P. Hall, A. Hazra, R. Hein, B.E. Henderson, A. Hofman, J.L. Hopper, A. Irwanto, M. Johansson, R. Kaaks, M.G. Kibriya, P. Lichtner, S. Lindström, J. Liu, E. Lund, E. Makalic, A. Meindl, H. Meijers-Heijboer, B. Müller-Myhsok, T.A. Muranen, H. Nevanlinna, P.H. Peeters, J. Peto, R.L. Prentice, N. Rahman, M.J. Sánchez, D.F. Schmidt, R.K. Schmutzler, M.C. Southey, R. Tamimi, R. Travis, C. Turnbull, A.G. Uitterlinden, R.B. van der Luijt, Q. Waisfisz, Z. Wang, A.S. Whittemore, R. Yang, W. Zheng, S. Kathiresan, M. Pato, C. Pato, R. Tamimi, E. Stahl, N. Zaitlen, B. Pasaniuc, G. Belbin, E.E. Kenny, M.H. Schierup, P. De Jager, N.A. Patsopoulos, S. McCarroll, M. Daly, S. Purcell, D. Chasman, B. Neale, M. Goddard, P.M. Visscher, P. Kraft, N. Patterson, A.L. Price, Modeling linkage disequilibrium increases accuracy of polygenic risk scores, Am. J. Hum. Genet. 97 (4) (2015) 576–592, https://doi.org/10.1016/j.ajhg.2015.09.001.

[30] C. Márquez-Luna, P.-R. Loh, South Asian Type 2 Diabetes Consortium, The SIGMA Type 2 Diabetes Consortium, A.L. Price, Multiethnic polygenic risk scores improve risk prediction in diverse populations, Genet. Epidemiol. 41 (8) (2017) 811–823, https://doi.org/10.1002/gepi.22083.

[31] T.S.H. Mak, R.M. Porsch, S.W. Choi, X. Zhou, P.C. Sham, Polygenic scores via penalized regression on summary statistics, Genet. Epidemiol. 41 (6) (2017) 469–480, https://doi.org/10.1002/gepi.22050.

[32] F. Privé, H. Aschard, A. Ziyatdinov, M.G.B. Blum, Efficient analysis of large-scale genome-wide data with two R packages: bigstatsr and bigsnpr, Bioinformatics 34 (16) (2018) 2781–2787, https://doi.org/10.1093/bioinformatics/bty185.

[33] L.R. Lloyd-Jones, J. Zeng, J. Sidorenko, L. Yengo, G. Moser, K.E. Kemper, H. Wang, Z. Zheng, R. Magi, T. Esko, A. Metspalu, N.R. Wray, M.E. Goddard, J. Yang, P.M. Visscher, Improved polygenic prediction by Bayesian multiple regression on summary statistics, Nat. Commun. 10 (1) (2019) 5086, https://doi.org/10.1038/s41467-019-12653-0.

[34] P.J. Newcombe, C.P. Nelson, N.J. Samani, F. Dudbridge, A flexible and parallelizable approach to genome-wide polygenic risk scores, Genet. Epidemiol. 43 (7) (2019) 730–741, https://doi.org/10.1002/gepi.22245.

[35] T. Ge, C.-Y. Chen, Y. Ni, Y.-C.A. Feng, J.W. Smoller, Polygenic prediction via Bayesian regression and continuous shrinkage priors, Nat. Commun. 10 (1) (2019) 1776, https://doi.org/10.1038/s41467-019-09718-5.

[36] Y. Ruan, Y.-C. Anne Feng, C.-Y. Chen, M. Lam, A. Sawa, A.R. Martin, S. Qin, H. Huang, T. Ge, Improving polygenic prediction in ancestrally diverse populations, medRxiv (2021) 20248738, https://doi.org/10.1101/2020.12.27.20248738.

[37] J. Pattee, W. Pan, Penalized regression and model selection methods for polygenic scores on summary statistics, PLoS Comput. Biol. 16 (10) (2020) e1008271, https://doi.org/10.1371/journal.pcbi.1008271.

[38] G. Ni, J. Zeng, J.A. Revez, Y. Wang, Z. Zheng, T. Ge, R. Restuadi, J. Kiewa, D.R. Nyholt, J.R.I. Coleman, J.W. Smoller, J. Yang, P.M. Visscher, N.R. Wray, A comparison of ten polygenic score methods for psychiatric disorders applied across multiple cohorts, Biol. Psychiatry (2021), https://doi.org/10.1016/j.biopsych.2021.04.018.

[39] O. Pain, K.P. Glanville, S.P. Hagenaars, S. Selzam, A.E. Fürtjes, H.A. Gaspar, J.R.I. Coleman, K. Rimfeld, G. Breen, R. Plomin, L. Folkersen, C.M. Lewis, Evaluation of polygenic prediction methodology within a reference-standardized framework, PLoS Genet. 17 (5) (2021) e1009021, https://doi.org/10.1371/journal.pgen.1009021.

[40] F. Dudbridge, Power and predictive accuracy of polygenic risk scores, PLoS Genet. 9 (3) (2013) e1003348, https://doi.org/10.1371/journal.pgen.1003348.

[41] R. Karlsson Linnér, T.T. Mallard, P.B. Barr, S. Sanchez-Roige, J.W. Madole, M.N. Driver, H.E. Poore, R. de Vlaming, A.D. Grotzinger, J.J. Tielbeek, E.C. Johnson, M. Liu, S.B. Rosenthal, T. Ideker, H. Zhou, R.L. Kember, J.A. Pasman, K.J.H. Verweij, D.J. Liu, S. Vrieze, COGA Collaborators, H.R. Kranzler, J. Gelernter, K.M. Harris, E.M. Tucker-Drob, I.D. Waldman, A.A. Palmer, K.P. Harden, P.D. Koellinger, D.M. Dick, Multivariate genomic analysis of 1.5 million people identifies genes related to addiction, antisocial behavior, and health, Nat. Neurosci. 24 (10) (2021) 1367–1376.

[42] L.J. Matthews, E. Turkheimer, Across the great divide: pluralism and the hunt for missing heritability, Synthese 198 (3) (2021) 2297–2311, https://doi.org/10.1007/s11229-019-02205-w.

[43] R. de Vlaming, A. Okbay, C.A. Rietveld, M. Johannesson, P.K.E. Magnusson, A.G. Uitterlinden, F.J.A. van Rooij, A. Hofman, P.J.F. Groenen, A.R. Thurik, P.D. Koellinger, Meta-GWAS accuracy and power (MetaGAP) calculator shows that hiding heritability is partially due to imperfect genetic correlations across studies, PLoS Genet. 13 (1) (2017) e1006495, https://doi.org/10.1371/journal.pgen.1006495.

[44] H. Mostafavi, A. Harpak, I. Agarwal, D. Conley, J.K. Pritchard, M. Przeworski, Variable prediction accuracy of polygenic scores within an ancestry group, elife 9 (2020) e48376, https://doi.org/10.7554/eLife.48376.

[45] J.E. Savage, J.E. Salvatore, F. Aliev, A.C. Edwards, M. Hickman, K.S. Kendler, J. Macleod, A. Latvala, A. Loukola, J. Kaprio, R.J. Rose, G. Chan, V. Hesselbrock, B.T. Webb, A. Adkins, T.B. Bigdeli, B.P. Riley, D.M. Dick, Polygenic risk score prediction of alcohol dependence symptoms across population-based and clinically ascertained samples, Alcohol. Clin. Exp. Res. 42 (3) (2018) 520–530, https://doi.org/10.1111/acer.13589.

[46] R.E. Peterson, K. Kuchenbaecker, R.K. Walters, C.-Y. Chen, A.B. Popejoy, S. Periyasamy, M. Lam, C. Iyegbe, R.J. Strawbridge, L. Brick, C.E. Carey, A.R. Martin, J.L. Meyers, J. Su, J. Chen, A.C. Edwards, A. Kalungi, N. Koen, L. Majara, E. Schwarz, J.W. Smoller, E.A. Stahl, P.F. Sullivan, E. Vassos, B. Mowry, M.L. Prieto, A. Cuellar-Barboza, T.B. Bigdeli, H.J. Edenberg, H. Huang, L.E. Duncan, Genome-wide association studies in ancestrally diverse populations: opportunities, methods, pitfalls, and recommendations, Cell 179 (3) (2019) 589–603, https://doi.org/10.1016/j.cell.2019.08.051.

[47] A.R. Martin, C.R. Gignoux, R.K. Walters, G.L. Wojcik, B.M. Neale, S. Gravel, M.J. Daly, C.D. Bustamante, E.E. Kenny, Human demographic history impacts genetic risk prediction across diverse populations, Am. J. Hum. Genet. 100 (4) (2017) 635–649, https://doi.org/10.1016/j.ajhg.2017.03.004.

[48] A.B. Popejoy, S.M. Fullerton, Genomics is failing on diversity, Nat. News 538 (7624) (2016) 161.

[49] G.L. Wojcik, M. Graff, K.K. Nishimura, R. Tao, J. Haessler, C.R. Gignoux, H.M. Highland, Y.M. Patel, E.P. Sorokin, C.L. Avery, G.M. Belbin, S.A. Bien, I. Cheng, S. Cullina, C.J. Hodonsky, Y. Hu, L.M. Huckins, J. Jeff, A.E. Justice, J.M. Kocarnik, U. Lim, B.M. Lin, Y. Lu, S.C. Nelson, S.-S.L. Park, H. Poisner, M.H. Preuss, M.A. Richard, C. Schurmann, V.W. Setiawan, A. Sockell, K. Vahi, M. Verbanck, A. Vishnu, R.W. Walker, K.L. Young, N. Zubair, V. Acuña-Alonso, J.L. Ambite, K.C. Barnes, E. Boerwinkle, E.P. Bottinger, C.D. Bustamante, C. Caberto, S. Canizales-Quinteros, M.P. Conomos, E. Deelman, R. Do, K. Doheny, L. Fernández-Rhodes, M. Fornage, B. Hailu, G. Heiss, B.M. Henn, L.A. Hindorff, R.D. Jackson, C.A. Laurie, C.C. Laurie, Y. Li, D.-Y. Lin, A. Moreno-Estrada, G. Nadkarni, P.J. Norman, L.C. Pooler, A.P. Reiner, J. Romm, C. Sabatti, K. Sandoval, X. Sheng, E.A. Stahl, D.O. Stram, T.A. Thornton, C.L. Wassel, L.R. Wilkens, C.A. Winkler, S. Yoneyama, S. Buyske, C.A. Haiman, C. Kooperberg, L. Le Marchand, R.J.F. Loos, T.C. Matise, K.E. North, U. Peters, E.E. Kenny, C.S. Carlson, Genetic analyses of diverse populations improves discovery for complex traits, Nature 570 (7762) (2019) 514–518, https://doi.org/10.1038/s41586-019-1310-4.

[50] A.R. Martin, M. Kanai, Y. Kamatani, Y. Okada, B.M. Neale, M.J. Daly, Clinical use of current polygenic risk scores may exacerbate health disparities, Nat. Genet. 51 (4) (2019) 584–591.

CHAPTER 6

Sequence analysis and genomics in the classroom

Rebecca C. Burgess, Rivka Glaser, and Kimberly Pause Tucker
Stevenson University, Owings Mills, MD, United States

Genomic activities are a valuable addition to an undergraduate biology course at any level. Even if students do not engage in sophisticated experimentation, introducing the concepts of the genome and using real-sequence data in the course can deepen students' understanding of how genomic information functions in the cell, how it evolves over time, and how it can be used by scientists to answer research questions.

Here, we intend to provide some examples of how genomic data can be used in an undergraduate classroom—from semester-long collaborative research projects to short 1-hour activities to multilab projects. These resources are not exhaustive, but rather meant to provide a starting point for instructors looking to incorporate genomics in their classroom at any level. In PART I, we discuss two examples of crowd-sourced, semester-long undergraduate research projects, where students perform small parallel projects from large-scale research projects and contribute their work to a publishable study. These projects require instructor training, but engaging in these studies comes with profound support in the form of resources, troubleshooting, and curriculum development from their extensive "community of practice." The standalone resources on the other hand (PART II) require little instructor expertise in genomics/bioinformatics and encompass lab and nonlab genotyping activities that can be used for a variety of different audiences from high school through postsecondary studies. In PART III, we have provided a list of other online resources and repositories, as of Summer 2020, for introducing genomics into the classroom.

Crowd-sourced undergraduate research

Course-based undergraduate research experiences (CUREs) are methods by which faculty can bring authentic research experiences to their undergraduates, either by designing a CURE based on their own research or by participating in ongoing research projects. Not only do CUREs introduce students to the nature of scientific experimentation, but evidence also shows that the length of time spent participating in CUREs correlates with increased positive impact on student learning and engagement in science [1]. Discussed in the following section are two such CURES based on well-established collaborative research projects involving hundreds of universities across the nation: the Genomics Education Partnership and the SEA-PHAGES program.

The Genomics Education Partnership

The Genomics Education Partnership (GEP) (https://www.gep.wustl.edu/) is a national and international consortium of faculty engaged in comparative genomics course-based undergraduate research. A significant number of the GEP members are faculty at primarily undergraduate institutions (PUIs). GEP members collaborate with scientific partners at Washington University in St. Louis, Oregon State University, The University of Alabama, and The University of Puerto Rico on gene annotation projects. The GEP uses a crowdsourcing model, where a large centralized project is broken up into smaller projects that are then made available to participating faculty and their undergraduate students [2,3]. Every annotation project is completed independently at least twice by different students from different colleges/universities, and the results are reconciled by experienced students at a central location (e.g. Washington University in St. Louis, Oregon State University, or The University of Alabama) to ensure reproducibility and quality control.

This extensive collaboration, which includes training for new faculty and ongoing professional development for current faculty, has been supported by grants from the Howard Hughes Medical Institute (HHMI), National Institutes of Health (NIH), and National Science Foundation (NSF). The GEP is one such example of a current educational effort that aims to address the genomics education gap in the undergraduate biology curriculum. GEP-trained faculty utilize common genomics research projects that can easily be implemented in their classrooms and in doing so, create a community of practice [3]. In addition, curriculum modules on a variety of related topics, such as on eukaryotic gene structure, transcription start sites, or sequencing technologies, can be used in conjunction with the research projects or separately are available on the GEP website, many written collaboratively by the GEP faculty. Some of these curriculum modules, including assessment tools, have been published in CourseSource [4,5] (see PART III).

Currently, the GEP has two main projects underway, both of which are discussed in the following section. Each project is headed by a scientific lead who is responsible for determining the overall research question, as well initiating publications based on the results. Publications based on information from student annotation projects include all students and faculty who have contributed completed annotations, and have reviewed, critiqued, and approved the manuscript, as coauthors. In some cases, the number of authors on such publications can exceed 1000 [6].

GEP Project #1: Drosophila F element, or dot chromosome

The dot chromosome in *Drosophila* is a small, heterochromatic chromosome that contains approximately 80 genes; surprisingly, these genes are expressed at levels seen for euchromatic genes, indicating that they have escaped heterochromatic silencing. Students involved in this project work on annotating these euchromatic genes from many different species of *Drosophila*. Of particular interest is the annotation of the transcription start sites of these genes, as this informs the location of the promoter region that is subsequently analyzed for sequence motifs. These analyses shed light on how genes can be expressed even in a heterochromatic environment.

A typical student annotation project would involve examining a ~40 kb contiguous region (contig) using a mirror of the USCS Genome Browser to determine the number of genes and repetitive elements in the contig. Using *Drosophila melanogaster* as a reference, students look for evidence of conservation (results of BLAST searches) in conjunction with the results obtained by ab initio gene finders and data obtained by RNA-seq, ultimately producing a model for the structure of each gene, including its transcription start site (TSS) (Fig. 6.1).

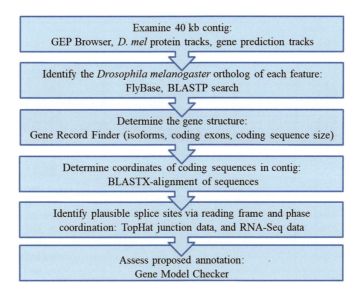

FIG. 6.1

Student workflow for *Drosophila* F element gene annotation projects.

Students must support their hypothesized model with evidence from comparative genomics and alignment tools as well as various evidence tracks in the browser, such as computationally predicted gene models (Fig. 6.2). Students generally report and defend their work before their class or in a poster at an undergraduate research symposium, both of which provide an excellent opportunity to foster public speaking and scientific communication skills. Further, students will often submit their hypothesized models in a standardized format to a central data repository for reconciliation.

Previous student research projects have included "finishing" projects which involved the examination of sequences in *Drosophila* genome assemblies for any errors or gaps. Gaps were corrected, or filled in, where possible through additional sequencing reactions, which students could perform as part of their laboratory course.

Publications stemming from this project have focused on the evolution of the unique properties of the F element in *Drosophila* and the mechanism by which the F element has expanded in some *Drosophila* species [6–8]. The initial student study showed that the fly F element has maintained its special heterochromatic structure over 40 million years of evolution. The heterochromatic environment, which severely curtails recombination, appears to have impacted the evolution of the genes, resulting in very low codon bias [6]. In some cases, the F element has expanded due to high levels of repetitious sequences; in *D. ananassae*, not only is the chromosome larger, but the genes are almost all correspondingly larger, again with very low codon bias [8] (GEP F element project).

GEP Project #2: Evolution of genes in metabolic pathways

The second large project that the GEP is undertaking focuses on annotating specific genes found in pathways of interest in several species. Currently, there are two pathways of interest: the insulin signaling pathway in *Drosophila*, and genes that encode venom proteins in parasitoid wasps. The parasitoid

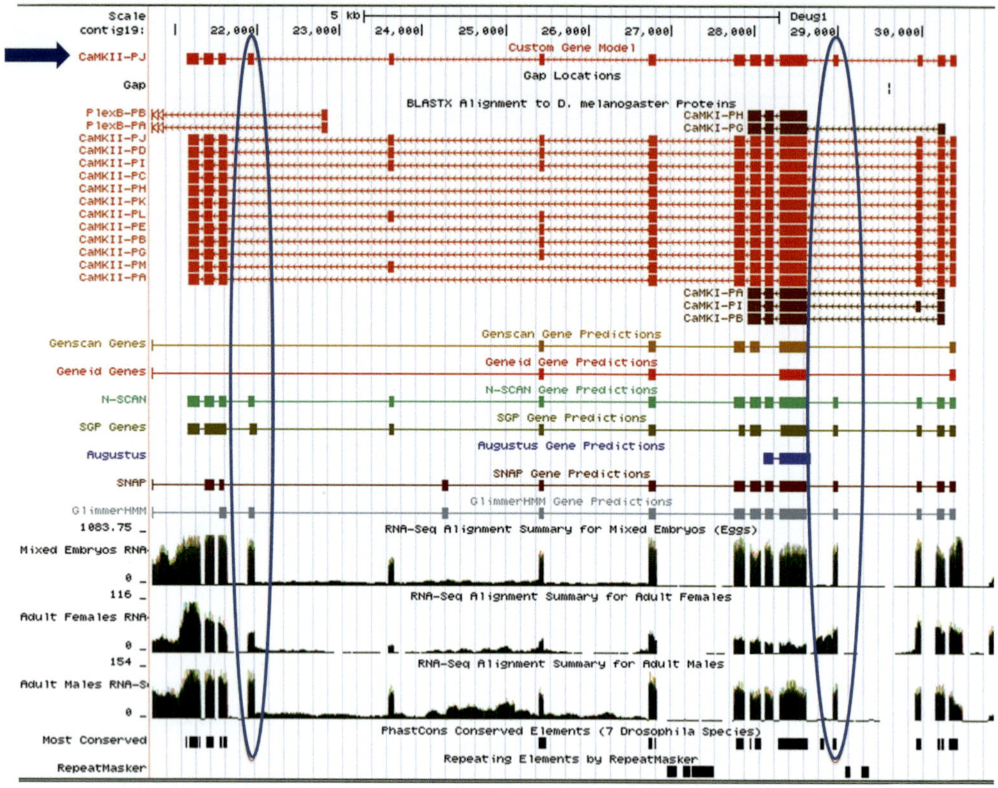

FIG. 6.2

A custom gene model for CaMKII-PJ created by students in a GEP course. This model *(blue arrow)* incorporates two coding exons that were not predicted in the BlastX alignment, but that are supported by RNA-seq data *(blue circles)*.

wasp genomes were newly sequenced and assembled using a genome browser created in collaboration with G-OnRamp [9]. Other similar pathway projects may be added in the near future. The workflow used for these metabolic pathways projects is similar to that of the *Drosophila* F element gene annotation project.

Insulin signaling pathway genes

This project centers on the evolution of genes in the insulin signaling pathway of 27 species of *Drosophila*, representing 40 million years of evolution. Although the insulin signaling pathway is well-characterized and fairly well conserved in *Drosophila* (as well as across all metazoans), some genes in the pathway have undergone duplication while others have been lost within the *Drosophila* genus [10]. Annotation of these genes in this pathway and their TSSs will be used to determine the rates and patterns of evolution of regulatory regions for the genes in this pathway (GEP Pathways project). While this particular project is focused on the insulin signaling pathway, the generalized workflow can be expanded to other pathways in the future.

Parasitoid wasp venom genes

This project focuses on the annotation of genes coding for venom proteins in three species of parasitoid wasps. When wasps infect fly larvae, laying an egg, they also transfer venom proteins to the host, which then manipulates the host physiology [11–14]. These annotation projects, therefore, focus on annotating specific genes involved in venom production in three species of parasitoid wasps, *Leptopilina boulardi, Leptopilina heterotoma,* and *Ganaspis* sp. *1*, with the dual goals of examining the evolution of venom genes and the identification of conserved domains and structural motifs in venom protein coding genes, including regulatory regions (GEP Parasitoid Wasp Project).

GEP material has the distinction of being able to be implemented in a variety of different ways: from the use of only one part of a curriculum module to the entire curriculum on genes and genomes; as part of a genetics course or including full annotation projects worked on over a 16-week semester; or, as part of an independent research project. GEP faculty are equally varied in their positions: some are in 4-year PUIs, others in community colleges, HBCUs, and R1 universities. Students participating in courses that use GEP material, be it curriculum modules or annotation projects, are assessed on their comprehension of gene structure and genomics using pre- and postclass surveys. While more time spent on GEP curriculum leads to stronger outcomes, significant learning gains have been shown, regardless of how GEP material was used in the class [1].

The GEP continues to accept new faculty into the partnership and offers online and regional training for incoming faculty. Annual professional development workshops for current faculty serve to connect faculty from over 100 institutions, review the progress of current projects, and discuss the future direction of the GEP. More information can be found on the GEP's website (GEP Prospective Members).

SEA-PHAGES

HHMI's SEA-PHAGES (Science Education Alliance-Phage Hunters Advancing Genomics Evolutionary Science) program focuses on investigating the diversity of bacteriophages. The SEA-PHAGES curriculum is designed as a two-semester CURE for incoming students with little-to-no laboratory experience to "increase their interest and retention in the biological sciences through immediate immersion in authentic, valuable, yet accessible research" [15]. In the first semester, students isolate, purify, and amplify phages they obtain from soil samples in the hope of discovering novel phages. In the second semester, students annotate the genomes of these phages and, like with GEP projects, identify genes, regulatory elements, and other sequence features. Institutions must apply to the SEA-PHAGES program and demonstrate the capability to offer this rigorous course in a sustainable manner for several years. Once an institution is accepted into the SEA-PHAGES program, faculty from that institution are trained and join a consortium of over 200 institutions. The SEA-PHAGES projects will be discussed in more detail in Chapter 7.

The GEP and SEA-PHAGES programs provide distinct advantages for both faculty and students. Faculty seeking to enhance their understanding and teaching of genomics can join an existing robust network that provides extensive training, resources, professional development, and scholarship opportunities. Students have the opportunity to present their work at conferences organized by each program and be included in publications [8] (GEP publications, SEA-PHAGES publications). Additionally, the data analysis part of these CUREs ensures that the student generated data is reproducible. Each GEP gene annotation project is completed by two or more students and reconciled by a separate group of students. Data generated from the SEA-PHAGES annotation projects is uploaded to a central location for quality control before being submitted to GenBank.

Standalone DNA sequencing activities

The previous section focused on large preexisting research projects that faculty can join. However, some faculty may not want or need to incorporate a CURE in their class and would rather have individual, standalone activities that could be completed in several class periods. We provide three such activities in this section and the resources for implementing these activities (handouts, sequence files, etc.) can be found online.

16S rRNA bacterial genotyping/identification activity
Background
The identification of bacteria remains a critical aspect of microbiology lab education for both science and allied health majors. It is clinically relevant to medical laboratories, and teaches students concepts about microbial cell structure, gene structure, and introduces molecular evolution, taxonomy, and phylogeny. While biochemical identification and serological identification methods remain important, sequence-based methods of identification are becoming more common with the 16S rRNA gene being the most frequently used [16]. The 16S rRNA gene is about 1500 nucleotides long and is a sufficient size for comparative analysis. There are 9 hypervariable regions and 8 conserved regions that allow the design of primers for PCR amplification of the gene prior to sequencing. In addition, the gene is highly resistant to changes that affect folding and function of the RNA, since it is essential to protein synthesis and the health of the cell. By using this gene's conserved regions for primer binding and sequencing of the hypervariable regions, it allows the genotypic identification of organisms that are unable to be cultured in the laboratory, particularly useful in environmental studies [17]. The understanding of basic sequencing practices predicates the understanding of next-generation sequencing techniques (NGS), so undergraduates should be well-versed in traditional methods before introducing NGS.

This sequence analysis activity is appropriate for an introductory nonmajors or majors general biology class, molecular biology class, or a microbiology class, and can be performed completely as a computer-based activity. In this case, identification can be done with provided raw data files (see resources—GitHib Link). Alternatively, this activity can be scaled up into a three-part wet lab activity where students identify unknown bacteria from a pure culture. In this case, the 16S rRNA gene amplification, purification, and sequencing could be performed by the students so they are analyzing sequences of their own samples. Many other examples of 16S rRNA gene-based projects have been implemented and assessed in the education literature, including examples where students perform identification in inquiry-based projects in lecture or lab [18–22]. Inquiry-based personal microbiome activities that use NGS technologies are available as well, but so far the authors have found NGS classroom activities to be cost- and time-prohibitive [23].

For viewing electropherograms and base calling, any sequence viewer software capable of displaying (.ab1) files can be used, but the provided walkthrough is specific to Benchling, a cloud-based all-in-one molecular biology application. The resulting trimmed sequence file from other software can be manually entered into NCBI BLAST site, while in Benchling there is a shortcut BLAST button. The entire analysis takes about 1 h (see timing below). If students are performing PCR and sending their own sequences for analysis, two 3-h lab periods are needed for sample preparation prior to this activity.

Learning objectives
The following learning outcomes should be met upon completion of this activity:

- Describe the activity of DNA polymerase and how the incorporation of dideoxynucleotides (ddNTPs) terminates chain elongation to be used for sequence analysis.
- Trim and tidy sequence data and search 16S rRNA sequence database using BLAST.
- Explain the importance of the 16S rRNA gene in bacterial identification.

Classroom management
Overview
Students should be familiar with DNA replication and central dogma concepts prior to this activity. To understand the use of the 16S rRNA gene in particular, students should understand general ribosome structure and function, the structure of the 16S rRNA gene, and basic concepts in spontaneous mutation and genomic change. This is often achieved through the lecture portion of the course, or prerequisite courses. Preclass readings, questions, and discussions should be planned ahead of the activity if the students are high school or introductory biology students.

Preclass preparation
The following steps are recommended before running the activity in class:

- Ensure that computers with internet access are available for the activity. If computers are limited, students can work in pairs or groups of three for the activity. Students can also work on their own laptops or tablets (as long as they have internet and browser access).
- Ensure that students have access to a data file (the DNA sequences). You can load these sequences onto Benchling, onto each computer, deliver them through the course learning management system, or e-mail them to students. NOTE: If using Benchling, have the students created an account and logged in to join your "organization" before class?
- Make copies and/or distribute electronically the prelab handout. Provide this to students with sufficient time in advance of class/lab to complete prelab readings and handout questions.
- Make copies of the walkthrough handout or distribute electronically for students to follow along during class.

In-class activity
1. Review *Prelab* questions.
2. Discuss slides on Sanger sequencing, dideoxy termination method (solidifies concepts in polymerase function).
3. Start the analysis together.
4. Students work independently or in groups using walkthrough as a guide, instructor circulates to answer any questions and provides guidance.
5. Review *Thought Questions* in walkthrough and discuss results.

Troubleshooting
This activity uses Benchling, a web-based software that updates and adds new features on occasion, which could change the precise appearance of the tools in the software from the student handout.

However, during the 5 years, the authors have been using this software in courses, the overall look has not changed significantly enough to require a complete overhaul of the student handouts, only minor updates have been necessary.

At times, students cannot edit their sequences to trim "N" basecalls or fix improper basecalls. To alleviate this, ensure that files uploaded to Benchling (or other sequence software) are not locked for editing.

Online implementation
This activity can be performed completely online without the wet-lab portion, or using an online wet-lab simulator such as the one available from HHMI Biointeractive ("Bacterial Identification Virtual Lab") [24].

Parts and timing
Prelab (20 min outside of class)
The prelab of this activity functions to help students come into the class with a basic understanding of the concepts behind this activity. Prelab readings and questions aim to:

1. Familiarize students with the vocabulary behind sequencing and use of genomic information for bacterial identification.
2. Generate interest in the topic. If students are interested, they are more likely to value the activity and more effectively learn.

All of the concepts from the prelab will be revisited and then expanded upon in class during the activity, to enhance understanding of the student.

Introductory slideset
The introductory slides reinforce the concepts from the prelab. The slides start with the idea of gene sequence differences being used as a molecular clock. Then, the gene structure of the three bacterial rRNA genes and their general function as components of the ribosome are described. Next, a description details why we use the 16S version, as it is the "goldilocks" gene for taxonomic analysis: it is small enough for easy amplification and sequencing, but large enough to provide enough sequence information for an accurate alignment and identification at the species level. There are limitations, however, as some differences cannot be distinguished, for instance, several *Streptococcus* spp. [25].

Prelab questions addressed
- Why do we need to identify bacteria?
- What are some of the ways bacteria are identified in the lab?
- What is the 16S rRNA gene?
- Why is this gene used for identification of bacteria?

Analysis (5 min guided, 20 min working independently)
Instructors should start with basic guidance on how to navigate the software and find the folder containing individual students' sequence files. Then an overview of the purpose of the analysis helps students envision the end goal—finding a sequence match from the 16S rRNA NCBI database. The instructor could do the entire analysis as an overview demonstration in about 5 min, with the students simply watching the process. Students then can work through the walkthrough sequence protocol on their

individual computers, but since these are parallel projects, students can confer with each other on the common steps of the analysis, which is helpful for guiding struggling students through the analysis.

The first part of the analysis involves setting up and viewing the chromatogram. Students should assess the quality of their sequence data, and whether it is usable overall. Undergraduate students often find it challenging to read electropherograms and identify types of errors in sequence data. Thought-provoking questions ask the students to recall what the colored peaks in the electropherogram represent, and if a peak always represents the exact sequence of the DNA molecule, or if technical artifacts could arise. If the student's sequence failed, they should reflect on how the data might indicate why it failed—Is the signal completely indistinguishable from background noise (all Ns), or are there many base-called peaks with extensive noise that could interfere with the base-called sequence information? Students should reflect on what might have occurred and review their DNA sample's spectrophotometer data for purity and yield to draw some conclusions about why the sequence reaction failed. These students can be given "clean" data to work with, if the instructor wishes. This allows the student to see the analysis through to the end and learn about alignment and genomic database searching.

The next part of the analysis involves the trimming of "N" basecalls at the beginning and end of the sequence, and the fixing of improper basecalls. Thought-provoking questions in the walkthrough prompt students to think about why the ends have messy sequence data and why it is ethical to remove the ends for the purposes of alignment and sequence identification. In addition, students are prompted to consider why sometimes computer basecall issues arise within the sequence, when it is acceptable to fix these basecalls, and when it is not.

Once the sequence is trimmed and tidied, the students perform a BLAST search to determine the closest match to their unknown sequence. The walkthrough clearly delineates the stops to set up the query search and parameters. The results from this BLAST search are captured in screenshots and thought-provoking questions prompt students to consider what each result graphic means and what the overall conclusion is—that is, what is the identity of their unknown organism.

Results discussion and review (10 min)
At the conclusion of the analysis, it is helpful to have a class discussion of the results, where students can report on their overall process and conclusion. This can be done using any reflection and reporting out method the instructor wishes to use. The thought-provoking questions should be discussed as a group so that students can bring up issues or questions they had about their individual sequence, and the important concepts explained and solidified. Students often have some misconceptions that arise during the *Thought Question* discussions as to ethical and unethical treatment of data, the uncertainty in alignments, and how similar, but not identical, hits can arise from the sequence search. Discussions of what the various scores mean in the BLAST results help students interpret this sequence data alignment more completely. See Fig. 6.3 for a complete overview of this activity.

Building and interpreting molecular phylogenies
Background
The ability to build and interpret valid phylogenies is an important and applicable skill in many types of genetic research. Nowadays, many software packages have tree-building capabilities built-in. However, if researchers do not understand the basic principles and settings, the phylogenetic trees that they create may not be robust. Further, it is important to understand that multiple sequence alignments and trees should be constructed and compared using different algorithms or approaches.

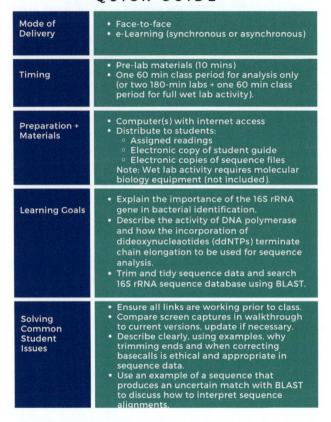

FIG. 6.3

16S Ribosomal RNA sequencing activity quick guide.

Students from a wide array of interests could be engaged in this process by using examples and case studies from a variety of topics. For example, phylogenies are frequently used in biodiversity conservation and to study organism groups of interest [26,27]. Also, recent research surrounding the origin and spread of SARS-CoV-2 could be an excellent case study for students to see how phylogeny construction is used in public health [28]. The mitochondrial gene cytochrome *c* oxidase subunit I (*COX1*) was selected for this exercise because of its highly conserved nature among animal species and the ability to use DNA sequence data instead of protein sequences to infer a phylogeny. *COX1* is also widely considered a universal "barcoding" gene and sequence data exists for almost any eukaryotic organism of interest [29], but the exercise could be tailored to any gene(s) of interest (e.g., 16S rDNA, 18S rDNA).

This activity is intended for upper-division students studying evolution, molecular biology, genetics, or a related course. Students will be guided through the process of constructing a "maximum likelihood" tree and learn about selecting a nucleotide substitution model. An "unknown" sequence is provided for students to demonstrate their ability to complete a basic interpretation of the tree. While this particular exercise is completely computer-based, it could be expanded upon to include wet lab components where students generate their own sequences from species of interest. In the activity as planned, students are given FASTA data files, but they could be asked to gather sequences from databases such as GenBank to allow them to explore those resources.

Objectives
- After completing this activity, students will be able to the following:
- Construct and interpret a basic nucleotide sequence alignment.
- Determine an appropriate substitution model for their alignment.
- Construct a phylogeny using neighbor joining and maximum likelihood methods.
- Compare and interpret the phylogenies made using two different methods.

Classroom management
Overview
A brief project introduction and analysis (individually or in pairs) can be completed in a 75-min class period. The class can take place in a computer lab; however, since no specific software requirements are necessary, all analyses can be completed on a student's personal laptop (PC or Mac) using a web browser. This is a suitable lesson plan for e-learning contexts.

Preclass preparation
The following steps are recommended before running the activity in class:

- Assign the reading and any prelab/homework questions.
- Either ensure that institutional computers with internet access are available for the activity, or require students to bring their personal laptops to class that day (as long as they have internet and browser access). For virtual learning settings, students should have an updated browser and adequate internet connectivity.
- Students can work individually or in pairs, but it is encouraged that all students attempt the work on their own device—even if working in pairs. Some students tend to get overwhelmed with sequence data and working in pairs can help to alleviate that. They also work together to troubleshoot issues which is a valuable skill.
- Upload the necessary sequence files to your learning management system or email to the students.
- Complete the walkthrough handout on your own to ensure all links are still live and that screen captures adequately represent the online resources.
- Distribute the walkthrough handout electronically so that they can fill it in as they work with their answers and the tree image files.

In-class activity
A walk-though worksheet and all necessary sequence files are available online.

Troubleshooting
The flexibility of using free online resources means that these resources could be removed at some point, however a wide range of such resources exist and the activity itself could be modified to use other resources, as necessary. Allowing students to use their own devices also can get into tricky territory

with file types. The activity was developed for PC, and also works on Mac OS; however, the sequence text files need to be edited in the most-simple text editors on either device. Instruct students not to use Microsoft Word or another word processing software and instead only use "Notepad" or similar on PC and "Text Edit" on Mac. Make sure to familiarize yourself with the file formats, too. Additional characters (enters, periods, or spaces) can cause headaches and should be avoided.

Online implementation

While this activity was developed for face-to-face instruction, it could easily be completed by students in an asynchronous or synchronous online format. In an asynchronous format, the instructor could post a brief introductory lecture and walkthroughs of each step for viewing after the student attempts it on their own. In a synchronous format, the instructor provides an introduction, students then work independently with the instructor available for questions. Breakout groups within the online conferencing platform can be employed, with screen-sharing employed with other students and/or the instructor. The instructor could use the interrupted format (similar as described for the face-to-face implementation) and stop the class after certain points in the analysis. Students can upload their completed worksheet and analysis via the institution's learning management system.

Parts and timing

Prelab and introduction: time at home plus approximately 10 min at the start of class. This activity was not meant to be used for a general introduction to phylogenies. Students should already have knowledge of what a phylogenetic tree is, the generalities of how they are built, and how they are interpreted. They should have a basic understanding of genetics and sequencing.

Prior to coming to class, students were assigned reading in their textbook (e.g., chapter 8 in Emlen and Zimmer's *Evolution* textbook [30] gives an excellent introduction to gene trees versus species trees, molecular methods for tree building, and molecular clocks). Depending on the specific class for which this is being used, other textbooks also cover appropriate material, such as the section "How Genomes Evolve" in chapter 4 of Alberts et al. *Molecular Biology of the Cell* textbook. Another open access resource, although somewhat older is "A step by step guide to phylogeny reconstruction" [31]. The beginning of this article discusses the various steps and then goes on to explain how to use two programs. An alternative, free, online resource is available by the University of California Museum of Paleontology through their "Understanding Evolution" website. Specifically, the sections on "Evolution 101" [32] and "Phylogenetic systematics, a.k.a. evolutionary trees" [33] could be assigned, but these are at a much less technical level than the other resources.

At the start of class, instructors should give a quick review of the material from the reading and prior lecture to ensure students are all prepared for the activity. In this review, instructors should discuss the basic steps of phylogeny construction. Depending on the level of the students, the instructor may also wish to review some basic genetics (DNA structure, nucleotide substitutions (transitions and transversions), molecular clocks, etc.).

Prelab questions addressed

- What are the key steps in constructing a phylogenetic tree?
- How can you tell the difference between a nucleotide an amino acid sequence?
- What is a nucleotide substitution model and what do various models take into account?

Analysis (50 min): Students will work through the activity independently or in pairs (but each on their own computer). The activity worksheet walks students through each step in the process and asks questions as they progress through. The instructor could choose to allow the students to work through the activity at their own pace, to completion, or they can use an interrupted format as described in the following section. Alternatively, as the instructor floats around the classroom to facilitate the activity, they can choose organic points to interrupt the class if many students are having the same questions and address them all together.

The first step in the activity requires students to download the sequence files. The instructor should direct students to download them and have them ready on their computer before class to save time. Files can be saved anywhere the student wishes, but you may suggest they save them in a temporary file folder on the Desktop for ease of retrieval. Students should start the class period by opening these files and examining them, and the first few questions of the worksheet ask students to describe the sequence files. If using the interrupted format, the instructor should pause the class at this point, review the steps they have completed, and ensure that students understand they are working with DNA sequences.

The second part of the activity includes selecting an appropriate nucleotide substitution model. The model selection can be done automatically, but this section walks the student through the process. This concept may be challenging for students, but it is an important step. By reviewing some basic genetics or assigning appropriate reading material prior to class, they should recall the differences between transitions and transversions. The goal for this activity is to demonstrate there are different models and the "big picture" ideas behind the various models, but the instructor can certainly expand upon this information if it fits the context of the course and level of the students. At this point if an instructor is using the interrupted format, the instructor should ask students to pause and review the concept of model selection and why it is important.

The third part of the activity walks students though building the phylogeny. First, students will learn some basics of statistical support for the branches using the bootstrap method. They will then reconstruct and compare phylogenies using maximum likelihood and neighbor joining. There is also an "unknown" sequence included and students can interpret the tree to determine which taxa they think the sequence belongs to.

Results discussion and review (15 min): At the conclusion of the activity, the instructor should review some of the questions in the handout, especially the last few questions regarding the interpretation and comparison of the phylogenies. Students should be asked to describe the process they just completed. This could be done as a discussion, a think-pair-share, or as a minute paper and then report out. A summary of the activity is provided in Fig. 6.4.

Bitter tasting ability (PTC) genotyping activity
Background
This activity is ideal for high school or introductory biology or genetics classes in postsecondary settings and can easily be adapted for either audience.

The original observation of differences in bitter tasting ability among humans was made in 1931 in Arthur Fox's lab at the Dupont Chemical Company. As he was pouring some powdered phenylthiocarbimide (PTC) into a bottle, some of the dust accidentally escaped into the air. "Another occupant of the laboratory, Dr. C.R. Noller, complained of the bitter taste of the dust, but the author, who was much closer, observed no taste, and so stated. He even tasted some of the crystals and assured Dr. Noller they were tasteless, but Dr. Noller was equally certain it was the dust he had tasted. He tried some of the crystals and found them extremely bitter" [34].

Each time they tasted the PTC powder, the results were the same—Noller could taste a strong bitter taste while Fox could not taste anything. The gene controlling this phenotype, the TAS2R38 gene, was

BUILDING + INTERPRETING MOLECULAR PHYLOGENIES
QUICK GUIDE

Mode of Delivery	• Face-to-face • e-Learning (synchronous or asynchrounous)
Timing	• Pre-lab materials (10 mins) • One 75 min class period or two 50-min sessions.

Preparation + Materials	• Computer(s) with internet access
	• Distribute to students:
	  ◦ Assigned readings
	  ◦ Electronic copy of student guide
	  ◦ Electronic copies of sequence files

Learning Goals	• Construct and interpret a basic nucleotide sequence alignment. • Determine an appropriate substitution model for their alignment. • Construct a phylogeny using neighbor joining and maximum likelihood methods. • Compare and interpret the phylogenies made using two different methods.
Solving Common Student Issues	• Ensure all links are still "live" prior to class. • Compare screen captures in handout to current versions of software, update if necessary. • Have students only use basic text editor (not MS Word or equivalent) • Interrupted format works well for many students.

FIG. 6.4

Building and interpreting molecular phylogenies quick guide. Overview of the building and interpreting molecular phylogenies activity for classroom implementation.

discovered in 2003, 70 years after the first observation of the phenotype [35]. This historical aspect of this phenotype, as well as the evolution of lab safety protocols, may be of interest to some students and instructors.

TAS2R38 is a simple gene—it is 1002 base pairs long, contains a single exon, and has two alleles, the dominant taster allele and the recessive nontaster allele. The taster and nontaster alleles in humans differ by three polymorphisms located at nucleotides: 145, 785, and 886. This simple Mendelian trait allows for exploration of a wide range of topics spanning beginning to more advanced levels. This lab can be used to introduce beginning students to probability and Punnett square problems and common molecular biology lab techniques such as DNA extraction, PCR, restriction enzyme digests, and agarose gel electrophoresis. More advanced students can use the data generated in this lab to explore genotype-phenotype correlations and Hardy-Weinberg calculations and use the information about the SNPs in TAS2R38 to learn about haplotypes and GWAS studies.

Objectives
After completing this activity, students will be able to:

- Describe the basic process of isolating genomic DNA from cells.
- Describe the three steps in a PCR reaction.
- Describe the properties of restriction enzymes.
- Explain how the technique of gel electrophoresis works to separate DNA fragments based on size.
- Identify sizes of DNA fragments of DNA on an agarose gel.
- Understand how a BLAST search works.
- Understand how E-values are used to interpret the results of a BLAST search.

Classroom management
Overview
The lab is divided into two main sections—a wet lab that allows students to genotype and phenotype themselves and a dry bioinformatic analysis lab that explores the molecular evolution of tasting ability. The first section is further subdivided into four parts: DNA extraction, PCR amplification of part of the TAS2R38 gene, genotyping by restriction enzyme digest, and gel electrophoresis. Several companies sell kits for this experiment, such as Carolina Biological Supply Company and miniPCRbio. Although there are three SNPs that differ between the taster and nontaster alleles, each of these kits genotypes only one of the three SNPs. Carolina's protocol allows for genotyping of the first SNP, a G/C SNP at position 145, whereas miniPCR bio's protocol genotypes the C/T SNP at position 785 [36]. These differences in genotyping have implications in student comprehension (see "Troubleshooting" section).

Since instructor guides and student worksheets are provided by each of these companies and can be found online, we will focus here instead on highlighting conceptual problems students typically encounter during these labs, as well as possible extension activities which could be added.

Parts and timing
The wet lab can be accomplished in two 3-h lab periods (Carolina) or two 45-min class periods (miniPCR bio), making this latter protocol more accessible for high school labs. The bioinformatic analysis can be completed in 60–90 min. An overview of the activity is provided in Fig. 6.5.

Troubleshooting
Genotyping by restriction enzyme digest, Carolina has engineered a recognition site for *HaeIII* into the forward PCR primer. While this may be a common practice in many labs, this presents a conceptual stumbling block for many students and muddies the understanding of the bioinformatic analyses. For example, in the bioinformatic analysis section, students are asked to BLAST the forward and reverse primers in order to explore how conserved the *TAS2R38* gene is across evolution. However, since a restriction site was engineered into the primer, this base appears to be "missing" in all the BLAST hits. For students who are just learning to use BLAST and interpret the finding of any BLAST search, this is an esoteric point that is hard to grasp for many students. miniPCR bio, on the other hand, uses the naturally occurring *Fnu4HI* recognition sequence to distinguish the taster from nontaster alleles at position 785, which eliminates this potential area of confusion.

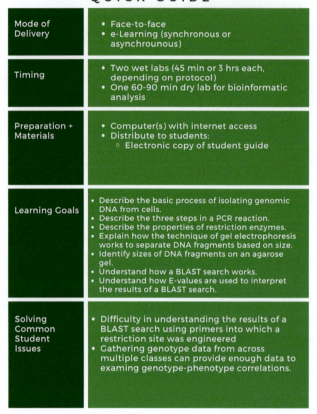

FIG. 6.5

Bitter tasting ability genotyping activity quick guide. Overview of the bitter tasting ability genotyping activity for classroom implementation.

Extension/challenge activities

Genotype-phenotype correlation and hypothesis testing: Collating the data from one class, or across many sections, allows students to examine genotype-phenotype correlations and explore questions of variability in tasting ability. Students typically observe that there is a correlation between having at least one dominant allele and the ability to taste PTC. However, many students will also observe that among those students who can taste PTC, there are those who react very strongly to the taste and those who are weak tasters. The genotype-phenotype correlation becomes less pronounced when taking into consideration strong vs weak tasters. This can lead into interesting discussions about haplotypes and the other polymorphisms' influence on tasting ability.

Hypothesis testing and experimental design: Based on their knowledge of the different alleles that control tasting ability, students can generate hypotheses about tasting ability and food preferences or behaviors. Students can expand upon this and design their own experiments to test their hypotheses. For example, students could hypothesize if tasters or nontasters would be more likely to be smokers, like drinking black coffee, etc.

Evolution of tasting ability: Bioinformatic analyses are particularly well suited for more challenging activities for upper level students. Students can begin by BLASTing the primers or the sequence of the entire human tasting allele to explore how conserved this gene is across evolution. Students can use sequence alignment programs such as Cold Spring Harbor Laboratory's Dolan DNA Learning Center's Bioservers to align the TAS2R38 gene from different species to investigate the evolution of tasting ability in other species (Fig. 6.6).

Hardy-Weinberg: Early investigations across many populations estimated an average frequency of the nontaster allele to be 50%, with a range from 13% to 63% [35]. Students can investigate the frequency of the nontaster allele in their class population and compare it to these earlier studies.

Online implementation: Nothing can replace the hands-on experience that students obtain while being in a lab. However, many of the concepts illustrated in this lab can be completed in an online format. There is no shortage of videos about DNA extraction from cheek cells, PCR amplification, restriction enzyme digest, and gel electrophoresis. During the COVID-19 epidemic, when schools transitioned to

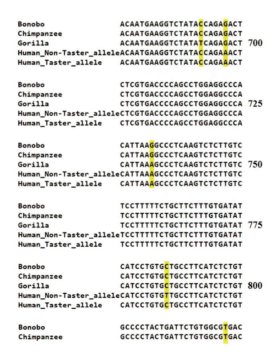

FIG. 6.6

Multiple alignment of the TAS2R38 gene.

remote learning, miniPCR offered a webinar on this very lab, which can be assigned to students. Data from previous semesters can be gathered and presented to students for analysis, whether it is for genotype analysis or Hardy-Weinberg calculations. The bioinformatic analyses can be done synchronously or asynchronously.

Other resources for DNA sequencing and analysis activities

A variety of resources on DNA sequencing activities exists online that are continuously being updated. Rather than summarize specific individual activities, which may change over time, we summarize these organizations/sites and the types of projects included on the sites.

This open-access journal contains peer-reviewed, evidence-based teaching resources for courses in an undergraduate biology curriculum. The resources on CourseSource are organized by course type, for example, Anatomy and Physiology, Genetics, or Immunology. Therefore, the best way to find resources on DNA sequencing is to search the site, as some of these resources can be found in different courses. Learning goals, assessment strategies, PowerPoint slides, and lesson plans are provided for each activity published on CourseSource.

National Center for Case Study Teaching in Science

The National Center for Case Study Teaching in Science (NCCSTS) focuses on publishing case studies that rely on active learning techniques to teach science. Like the materials on CourseSource, the case study activities published on NCCSTS are peer-reviewed and geared so that faculty can incorporate them into their classes with little to no modification. However, the case studies are not evidence based, as the materials on CourseSource are. As with CourseSource, the best way to find DNA sequencing activities on this site is to search the case collection.

Association for Biology Laboratory Education

This website collates tested, peer-reviewed laboratory exercises. The laboratory exercises are published along with a list of materials needed for the lab and instructions for the student and instructor. Since DNA sequencing activities can be found under several different topics, the best way to find such resources is to search the collection. The full pdfs of older activities can be fully accessed without having to be a member of the Association for Biology Laboratory Education (ABLE), while complete articles of more recently published activities require membership, however the membership fee is nominal.

Genetics Society of America

The Genetics Society of America (GSA) is a professional organization that provides two resources for genetics education:

1. Peer-Reviewed Educational Portal. The activities published here are peer-reviewed activities that make use of evidence-based teaching methodology. A range of activities are published on this site from in-class exercises to individual lab protocols, to lab exercises to course-based research projects. Some of the material found on this site is also published in CourseSource.

2. Online Resource Room. This is a collection of nonpeer-reviewed animations, videos, PowerPoint presentations, problem sets, and additional online readings organized by topic. DNA sequencing is one such topic under the "Methods and Tools in Genetics" section. Both Sanger sequencing and some types of next-generation sequencing are featured.

HHMI biointeractive

This site is similar in its offerings to Online Resource Room through the GSA in that the resources, organized by topic, span everything from animations and videos to interactive walkthroughs to complete activities that faculty incorporate into their classes as is. Both Sanger and next-gen sequencing are addressed by HHMI Biointeractive resources.

QUBES

This site provides a space for math and biology educators to share resources and classroom activities, in addition to supporting online communities or faculty mentoring networks. Their Open Education Resources contains a number of DNA sequencing-related activities, including exercises related to barcoding, RNAseq, annotation, SNP analysis, and other modules. This site is geared toward the sharing, dissemination, and discussion of ideas, rather than the sharing of completed classroom activities. The activities that are on this site have not undergone the peer-review process. QUBES differs from the other sites in that it functions as more of an idea sharing platform, whereas sites such as CourseSource provide readymade pedagogical activities, many of which are peer-reviewed and evidence-based, for faculty to use in their classes. Below, we highlight two networks on QUBES that pertain to genotyping and DNA sequencing.

1. Network for Integrating Bioinformatics into Life Sciences Education (NIBLSE). This online community within QUBES is dedicated to bioinformatics education in particular. As with the other online resources, NIBLSE is a centralized repository for bioinformatics education resources. Some of the resources on NIBLSE are published in CourseSource or the GSA Online Resources, but some of the activities are unique to NIBLSE.
2. Build-A-Genome (BAG). The Build-A-Genome CURE was first implemented at Johns Hopkins University in 2007. In 2014, the BAG network was established, with funding from and NSF RCN-UBE Incubator award. The grant enabled several workshops to be held in order to teach interested faculty about synthetic biology and recruit new institutions into the network. Currently, this network is seeking out new institutions to join.

GCAT-SEEK

The education modules on this website mainly use next-gen sequencing data, which will be covered more fully in Section 3 (Chapters 8–11).

Bio-rad cloning and sequencing explorer series

Instructors looking for a way to get started with sequencing in their class will find Bio-Rad's Cloning and Sequencing Explorer Series to be a great way to teach students the essential skills from wet lab to a basic bioinformatic analysis. The kit can be ordered as one unit that takes students through 6–8 weeks of investigation or individual modules can be ordered separately. Students will extract DNA from plants, use PCR to amplify a gene (GAPDH), purify and clone the PCR product, grow and isolate the plasmid DNA, perform a restriction digest, and then analyze the gene sequences.

Although it is a "kit," it has been designed so that students can use this for authentic research. Students could make and test hypotheses about plant phylogeny, gene flow/distribution, and more. When implementing this kit in an undergraduate genetics course, students can generate hypotheses about the plants in the area, sample from different regions or specimens, extract the DNA in class, then analyze the sequences and upload to NCBI GenBank. Implementing this kit is very straightforward for the instructor with student and instructor manuals for each module.

References

[1] J. Buhler, W. Leung, D. Lopatto, S.C.R. Elgin, P. Szauter, J.S. Thompson, M. Wawersik, J. Youngblom, L. Zhou, E.R. Mardis, A course-based research experience: how benefits change with increased investment in instructional time, CBE Life Sci. Educ. 13 (2014) 111–130, https://doi.org/10.1187/cbe-13-08-0152.

[2] S. Elgin, C. Hauser, T. Holzen, The GEP: crowd-sourcing big data analysis with undergraduates, Trends Genet. 33 (2016) 81–85.

[3] E.R. Mardis, D. Lopatto, S.C.R. Elgin, J. Stamm, J.S. Thompson, M. Wawersik, B.A. Wilson, J. Youngblom, W. Leung, J. Buhler, The genomics education partnership: successful integration of research into laboratory classes at a diverse group of undergraduate institutions, CBE Life Sci. Educ. 9 (2010) 55–69, https://doi.org/10.1187/09-11-0087.

[4] A.E. Weisstein, E. Gracheva, Z. Goodwin, Z. Qi, W. Leung, C.D. Shaffer, S.C.R. Elgin, A hands-on introduction to hidden Markov models, CourseSource (2016), https://doi.org/10.24918/cs.2016.8.

[5] M.M. Laakso, L.V. Paliulis, P. Croonquist, B. Derr, E. Gracheva, C. Hauser, C. Howell, C.J. Jones, J.D. Kagey, J. Kennell, S.C. Silver Key, H. Mistry, S. Robic, J. Sanford, M. Santisteban, C. Small, R. Spokony, J. Stamm, M. Van Stry, W. Leung, S.C.R. Elgin, An undergraduate bioinformatics curriculum that teaches eukaryotic gene structure, CourseSource (2017), https://doi.org/10.24918/cs.2017.13.

[6] N.C. Riddle, J. Buhler, E.R. Mardis, S.C.R. Elgin, M. Kaur, M. Semon, D. Serjanov, A. Tooric, C. Wilson, Drosophila Muller F elements maintain a distinct set of genomic properties over 40 million years of evolution, G3: genes, genomes, Genetics 5 (2015) 719–740, https://doi.org/10.1534/g3.114.015966.

[7] S.C.R. Elgin, L. Sabin, A. Shah, A. Sharma, S. Singhal, F. Song, C. Swope, C.B. Wilen, J. Buhler, E.R. Mardis, Evolution of a distinct genomic domain in drosophila: comparative analysis of the dot chromosome in Drosophila melanogaster and Drosophila virilis, Genetics 185 (2010) 1519–1534, https://doi.org/10.1534/genetics.110.116129.

[8] W. Leung, C.D. Shaffer, E.J. Chen, T.J. Quisenberry, Retrotransposons are the major contributors to the expansion of the Drosophila ananassae muller F element, G3: genes, genomes, Genetics 7 (2017) 2439–2460, https://doi.org/10.1534/g3.117.040907.

[9] Y. Liu, L. Sargent, W. Leung, S.C.R. Elgin, J. Goecks, G-OnRamp Goecks: a Galaxy-based platform for collaborative annotation of eukaryotic genomes, Bioinformatics (2019), https://doi.org/10.1093/bioinformatics/btz309. pii: btz309 31070714.

[10] D. Alvarez-Ponce, M. Aguadé, J. Rozas, Network-level molecular evolutionary analysis of the insulin/TOR signal transduction pathway across 12 Drosophila genomes, Genome Res. 19 (2009) 234–242, https://doi.org/10.1101/gr.084038.108.

[11] J. Goecks, N.T. Mortimer, J.A. Mobley, G.J. Bowersock, J. Taylor, T.A. Schlenke, Integrative approach reveals composition of endoparasitoid wasp venoms, PLoS One 8 (2013), https://doi.org/10.1371/journal.pone.0064125.

[12] N.T. Mortimer, Parasitoid wasp virulence: a window into fly immunity, Fly 7 (2013). https://www.landesbioscience.com/journals/fly/2013FLY0020R.pdf.
[13] N.T. Mortimer, J. Goecks, B.Z. Kacsoh, J.A. Mobley, G.J. Bowersock, J. Taylor, T.A. Schlenke, Parasitoid wasp venom SERCA regulates Drosophila calcium levels and inhibits cellular immunity, Proc. Natl. Acad. Sci. USA 110 (2013) 9427–9432, https://doi.org/10.1073/pnas.1222351110.
[14] C. Small, I. Paddibhatla, R. Rajwani, S. Govind, An introduction to parasitic wasps of drosophila and the antiparasite immune response, J. Vis. Exp. (2012) 1, https://doi.org/10.3791/3347.
[15] SEA-PHAGES, 2020. https://seaphages.org/. (Accessed 20 June 2020).
[16] J.E. Clarridge 3rd, Impact of 16S rRNA gene sequence analysis for identification of bacteria, Clin. Microbiol. Rev. 17 (2004) 840–862 (table of contents) https://doi.org/10.1128/CMR.17.4.840-862.2004.
[17] D.M. Ward, R. Weller, M.M. Bateson, 16S rRNA sequences reveal numerous uncultured microorganisms in a natural, Nature 345 (1990) 63–65, https://doi.org/10.1038/345063a0.
[18] S.M. Boomer, D.P. Lodge, B.E. Dutton, Bacterial diversity studies using the 16S rRNA gene provide a powerful, Microbiol. Educ. 3 (2002) 18–25. https://www.ncbi.nlm.nih.gov/pmc/articles/PMC3633119/.
[19] S.J. Rahman, T.C. Charles, P. Kaur, Metagenomic approaches to identify novel organisms from the soil, J. Microbiol. Biol. Educ. 17 (2016) 423–429, https://doi.org/10.1128/jmbe.v17i3.1115.
[20] T.C. Tobin, A. Shade, A town on fire! Integrating 16S rRNA gene amplicon analyses into an, FEMS Microbiol. Lett. 365 (2018), https://doi.org/10.1093/femsle/fny104.
[21] J.T.H. Wang, J.N. Daly, D.L. Willner, J. Patil, R.A. Hall, M.A. Schembri, G.W. Tyson, P. Hugenholtz, Do you kiss your mother with that mouth? An authentic large-scale, J. Microbiol. Biol. Educ. 16 (2015) 50–60, https://doi.org/10.1128/jmbe.v16i1.816.
[22] A.J. Zelaya, N.M. Gerardo, L.S. Blumer, C.W. Beck, The bean beetle microbiome project: a course-based undergraduate research, Front. Microbiol. 11 (2020) 577621, https://doi.org/10.3389/fmicb.2020.577621.
[23] M.R. Hartman, K.T. Harrington, C.M. Etson, M.B. Fierman, D.K. Slonim, D.R. Walt, Personal microbiomes and next-generation sequencing for laboratory-based, FEMS Microbiol. Lett. 363 (2016), https://doi.org/10.1093/femsle/fnw266.
[24] HHMI, Biointeractive, 2020. https://www.biointeractive.org/. (Accessed 15 June 2020).
[25] J.M. Janda, S.L. Abbott, 16S rRNA gene sequencing for bacterial identification in the diagnostic, J. Clin. Microbiol. 45 (2007) 2761–2764, https://doi.org/10.1128/JCM.01228-07.
[26] N.S. Upham, J.A. Esselstyn, W. Jetz, Inferring the mammal tree: species-level sets of phylogenies for questions in ecology, evolution, and conservation, PLoS Biol. 17 (2019) e3000494, https://doi.org/10.1371/journal.pbio.3000494.
[27] T.F. Scheelings, R.J. Moore, T.T.H. Van, M. Klaassen, R.D. Reina, Microbial symbiosis and coevolution of an entire clade of ancient vertebrates: the gut microbiota of sea turtles and its relationship to their phylogenetic history, Animal Microbiome 2 (2020) 17, https://doi.org/10.1186/s42523-020-00034-8.
[28] T. Bedford, R. Neher, Nextstrain: Real-Time Tracking of Pathogen Evolution, 2020. https://nextstrain.org/. (Accessed 30 June 2020).
[29] S. Ratnasingham, P.D.N. Hebert, bold: The barcode of life data system (http://www.barcodinglife.org), Mol. Ecol. Notes 7 (2007) 355–364, https://doi.org/10.1111/j.1471-8286.2007.01678.x.
[30] D.J. Emlen, C. Zimmer, Evolution Making Sense of Life, third ed., MacMillan Learning, New York, NY, 2020.
[31] C. Jill Harrison, J.A. Langdale, A step by step guide to phylogeny reconstruction, Plant J. 45 (2006) 561–572, https://doi.org/10.1111/j.1365-313X.2005.02611.x.
[32] Evolution 101, Understanding Evolution. (2023). https://evolution.berkeley.edu/evolibrary/article/evo_01 (Accessed May 25, 2020).
[33] Phylogenetic Systematics, a.k.a. Evolutionary Trees, Understanding Evolution. (2023). https://evolution.berkeley.edu/evolibrary/article/phylogenetics_01 (Accessed May 25, 2020).

[34] A.L. Fox, The relationship between chemical constitution and taste, PNAS 18 (1932) 115–120, https://doi.org/10.1073/pnas.18.1.115.
[35] S. Wooding, Phenylthiocarbamide: a 75-year adventure in genetics and natural selection, Genetics 172 (2006) 2015–2023. http://www.genetics.org/cgi/reprint/172/4/2015.
[36] miniPCR bio PTC Lab Teacher Guide, 2020. https://www.minipcr.com/wp-content/uploads/miniPCR-PTC-Lab-Teachers-Guide_v1.1.pdf. (Accessed 20 June 2020).

CHAPTER 7

Classroom to career: Implementation considerations for engaging students with meaningful DNA sequencing learning opportunities

Charles Wray

The Jackson Laboratory, Genomic Education, Bar Harbor, ME, United States

Introduction

DNA is one of the truly alluring subjects of science. Watson and Crick's elegant one-page *Nature* paper [1] kicked off seven decades of discovery, hype, hope, and promise in biology, medicine, and now genomics. The *Double Helix* [2] rapidly accelerated public interest in DNA, being both a compelling scientific adventure and subtly objectionable and self-promoting autobiography. Since the discovery of the Sanger di-deoxy chain termination method [3], DNA sequencing has itself become an exciting, extremely fast-paced field. Next generation sequencing methods and technologies—discoveries made possible by the first generation of DNA sequencing—and subsequent applications of nucleic acid sequencing across nearly all subdisciplines of biology have led to significant curricular integration of DNA biology from middle school to medical school and beyond.

Today, nearly all general biology educators need to be able to teach some level of DNA biology. Basic DNA structure lessons are common in middle school; high schools routinely run basic, molecular genetics laboratory exercises; and many college students use PCR to first amplify and subsequently sequence DNA. A growing number of advanced undergraduate students gain experience with high-throughput sequencing, including analysis of RNA-seq data sets. Established pedagogies exist with a plethora of educational DNA sequencing activities to train high school and undergraduate students. The dominant pedagogies [4] used include inquiry-based and project-based learning, built upon active learning strategies and constructivism theory. While numerous learning opportunities exist for integrating DNA sequencing (and its downstream applications) into student learning, this chapter focuses more specifically on best practices and considerations for incorporating DNA sequencing into classroom/training laboratory situation.

It is imperative that undergraduate institutions recognize the varied educational backgrounds and opportunities from which their students hail; while some incoming students may have had access to genetic sequencing technology in high school, others may be completely unfamiliar. As such, offering foundational genetics education to all incoming students is critical to student success, prior to engaging with advanced DNA sequencing technologies. Furthermore, not all DNA sequencing efforts lend themselves to successful student learning. This contribution will briefly describe some of the pressures placed on educators and will explore the sometimes less-than-successful uses of DNA sequencing as a teaching and learning activity. For instance, technological development can create an arms race for institutions to acquire the newest, high-throughput sequencer. In certain circumstances, the desire for new technology outpaces reasonable educational approaches. The recent focus on project-based learning [4] and doing real science in high schools, as well as the growth of course-based research experiences for undergraduates (CURES) [5–7], enriches learning; however, it is essential that the scope and scale of DNA sequencing projects are appropriate for each learner group. Several exemplary research-training projects are described herein.

DNA sequencing at the high school level
Essential content and learning goals

In order to introduce DNA sequencing into classroom and student laboratory settings, students must have content knowledge covering the structure of DNA. Understanding the basic biochemistry, specific molecular structure, and fundamental 5'-3' antiparallel orientation of double-stranded DNA are essential concepts students should master prior to incorporating DNA sequencing or most comparative DNA sequence analysis exercises into classroom settings. Under ordinary *North American* school group circumstances (exclusive of extremely high-performing younger students), middle school students are not taught detailed biochemistry and molecular structure of DNA. Therefore, it is in high school that students may first encounter DNA sequencing technology and DNA sequences as content in applied exercises such as demonstrating the central dogma of molecular biology, highlighting the effects of DNA mutation/variation on amino acid sequence of proteins, or comparative DNA sequence analysis for phylogenetics. While it is rare for middle school students to investigate or generate an actual DNA sequence, there are exceptional 3D models and learning tools available for middle school students, including Dynamic DNA from 3-D Molecular Designs (www.3dmoleculardesigns.com, funded by an NIH Small Business Innovation Research award) and prototype Lego DNA kits developed at MIT (https://edgerton.mit.edu/DNA-proteins-sets). In the 3-D Molecular Designs and MIT DNA kits, the three-dimensional models and associated learning goals are aligned to Next Generation Science Standards (NGSS) [8].

Recent work by Teaching the Genome Generation (TtGG) [9], which supports the integration of modern genetics and genomics into high school biology classes, indicates that the majority of high school teachers are comfortable with content and delivery of DNA biology content. The TtGG team used the Genetics Literacy Assessment Instrument (GLAI) [10] to assess teachers for their genetics content knowledge prior to summer professional development (PD) courses. All GLAI subconcepts were addressed with the exception of *Transmission*. Over 100 teachers scored very well on the genetics knowledge assessment over 4 years, averaging a $92.5 \pm 6.5\%$ correct response rate on all items. The TtGG study indicates that high school biology teachers are fully capable of preparing their students with the foundational content knowledge necessary before undertaking small-scale, traditional DNA sequencing demonstrations or laboratories.

Content knowledge and familiarity with DNA biology is a prerequisite for leading molecular genetics laboratory exercises geared toward DNA sequencing, but this does not imply that high school teachers are prepared to do so. Without previous experience in a research setting, many teachers lack the self-efficacy to launch DNA sequencing efforts with their students. Further, today's biology teachers are required to teach a huge range of subtopics and rarely have significant time in the laboratory with their students. Compounding this, high schools need access to basic equipment and instrumentation such as pipettors, PCR machines, and research-grade centrifuges, all of which are necessary to prepare and amplify nucleic acids. Additionally, access to automated sequencing machines, either Sanger-based, ABI systems, or Illumina next-generation systems normally requires a personal connection to a research facility as well as a budget set aside for purchasing sequencing services. Taken together, most high school science departments and their teachers face significant obstacles.

Despite these obstacles, a number of efforts have introduced DNA sequencing into high school curricula. The majority of efforts have grown out of either large-scale research projects or federally funded science education programs and partnerships. Collaborative efforts that combine large networks of scientists and instructors with high school students and teachers include DNA Barcoding, such as the Canadian School Malaise Trap Program, microbiome sequencing with target high schools (bioSeq at Tufts, http://ase.tufts.edu/chemistry/walt/sepa/index.html), and a small number of high schools that have participated in the HHMI/University of Pittsburgh SEA-PHAGE/Phage Hunter programs. In nearly all these cases, a university, a research institution, or scientific network provides DNA sequencing services and technical support.

The School Malaise Trap Program (SMTP) [11] successfully integrated DNA Barcoding and place-based natural history in a high school context across Canada. SMTP supported student collection of insects on their school campuses. Specimens collected by students were sent to the Center for Biodiversity Genomics at the University at Guelph (Ontario), for sequencing of a portion of the cytochrome c oxidase 1 (CO1) gene. The SMTP program engaged 350 schools and 15,000 students with school lessons that integrated DNA sequences, field collection activities and the importance of metadata, and critical issues of biodiversity. Across the United States, other high schools have successfully integrated DNA barcoding into biology courses and excellent support continues to be provided by the Barcode of Life Data System (BOLD), particularly through the BOLD Student Data Portal (BOLD-SDP, http://v3.boldsystems.org/index.php/SDP_Home).

TtGG is a teacher PD program that seeks to prepare high school students for life in the genomics era. TtGG staff at The Jackson Laboratory (JAX), with partners from the Personal Genetics Education Project (pgEd, Harvard Medical School), deliver a summer PD course that integrates instruction in molecular genetics laboratory techniques, bioinformatics, and bioethics. After summer PD, teachers are empowered to lead molecular genetics laboratories at their high schools. The TtGG program provides the participating schools with customized mobile laboratory kits, extensive curriculum resources, and access to DNA sequencing services. The TtGG project supports the sequencing of anonymous student samples at various loci across the human genome. High school students extract anonymous human DNA samples, amplify the samples using PCR, and prepare samples for DNA sequencing and shipment to JAX. The program uses a highly participatory model, combined with a focus on human genomics, to excite students about genetics and personalized medicine. Over 22,000 high school students have participated in the program by training and empowering teachers, mapping all TtGG lessons and laboratory activities to U.S. Next Generation Science Standards (Fig. 7.1), and providing extensive technical support and resources.

Widely disseminated high school DNA sequencing efforts share five common features. These important characteristics underpin SMTP, TtGG, as well as the Barcode of Life and the more university-focused SEA-PHAGES programs. First, students need to be actively engaged in the project;

Chapter 7 DNA sequencing in the classroom: Implementation considerations for meaningful learning opportunities

Teaching the Genome Generation High School Program: Curricular Mapping to High School Standards	U.S. NGSS Alignment						U.S. AP Biology Alignment								
	HS-LS1-1	HS-LS3-1	HS-LS3-2	HS-LS3-3	HS-LS4-1	HS-LS4-2	3A1	3A3	3C1	3C2	SP 3.3	SP 4.3	SP 5	SP 5.1	SP 7.2
Laboratory Protocol: DNA Extraction	■						■								■
Laboratory Protocol: PCR Amplification	■						■								■
Laboratory Protocol: Restriction Digest	■						■		■						■
Laboratory Protocol: Gel Electrophoresis	■				■				■	■		■			■
Laboratory Protocol: Prep for Sequencing	■						■								■
Laboratory Protocol: Sequence Analysis		■	■	■	■	■						■		■	
Bioinformatics Exercises		■	■	■	■	■					■		■		

FIG. 7.1

High school laboratory exercises mapped to Science Standards. Teaching the Genome Generation laboratory exercises mapped to Next Generation Science Standards(NGSS) and U.S. Advanced Placement Biology standards.

No permission required.

SMTP achieves this through insect trapping, TtGG does this through use of human DNA and students doing their own molecular genetics laboratory work. Second, learning activities need to be simplified and made part of an achievable workflow that students and teachers can execute. Third, developing foundational grade-appropriate content knowledge with teachers and providing PD builds confidence and provides critical technical skill training. Fourth, fostering and nurturing a professional network across schools serves to enrich, engage, and support teachers as they take on challenging new laboratory activities with students. Fifth, all programmatic resources need to be available through user-friendly web resources. These common features, when applied appropriately, are essential to technical success, student engagement and, very likely, effective learning outcomes.

Beyond short-term DNA sequencing activities offered by adjacent institutions, effective integration of DNA sequencing into high school science curricula is essential. While field and lab work actively engage students, significant learning gains are likely when DNA sequencing projects are integrated into theory-based instruction. The five characteristics described above are operational requirements; alignment and integration with course content is also critical. Fortunately, DNA sequence, other molecular data (e.g., restriction fragment driven genotypes, forensic DNA fingerprints, microsatellites length differences), and bioinformatics databases lend themselves to integration across numerous topics in high school biology. DNA sequence activities align with basic molecular biology, the structure of genes, and the central dogma of molecular genetics. Comparative sequence analysis at the nucleotide or amino acid level is widely used in teaching evolution. Distinguishing between sequence variation and mutation across genes and genomes between individuals in a species offers the opportunity to link genotype to phenotype and highlight enormous amounts of sequence diversity that is both neutral and potentially available for natural selection. DNA Barcoding offers an elegant entry point into teaching ecology concepts through biodiversity and conservation genetics. Finally, the technical aspects of Sanger sequencing, Illumina sequencing by synthesis, and Long Read pore-based sequencing align well with biochemistry and engineering learning standards.

Visualizing DNA on an electrophoresis gel or reading a DNA sequence chromatograph trace is a eureka-moment for a high school student. While purchasing a DNA sequencer may not be advisable for a high school, there are unique cases where sequencing in high school or reading a sequence file can ignite a passion for science and become transformational for students. It may appear simple or rudimentary to a research scientist, but finding a heterozygous gene sequence (Fig. 7.2) allows younger

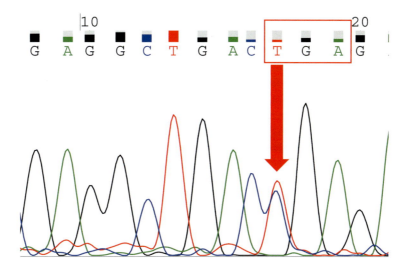

FIG. 7.2

Human ACTN3 DNA sequence from the TtGG program displaying heterozygous individual at the arrow. In this individual, CGA and TGA codons (codon triplet in box at top) code for Arginine and a premature Stop. Visualizing a simple heterozygote in DNA sequence is an important learning moment for high school students.

No permission required.

students to make the important connection between the genetics abstraction of two alleles and an actual heterozygous state in DNA sequence. An exemplary use of DNA sequencing in a high school is the Sacred Heart Academy in Hamden, Connecticut (Box 7.1).

Technology considerations for high schools

The emergence of small, lower-cost sequencing instruments begs the question whether high schools should own and operate their own DNA sequencers. It is enticing for lab-savvy biology teachers and ambitious schools to consider purchase of the minION pore-based platform (Oxford Nanopore Technologies) or iSeq 100 (Illumina Inc.) instrument. While the sample preparation protocols are less challenging and lower cost than they used to be, sample preparation for even these instruments would likely be a challenge for the majority of high school teachers. In most public high schools, the total budget for all laboratory activities across all 9–12 classes can be significantly less than $10,000. Setting aside even $1000–2000 to prepare samples and run a minION or iSeq 100 may not be justifiable relative to budget constraints.

In 2016, the chapter author worked with an independent school to evaluate the purchase and application of a minION for their science program. Two teachers from the school routinely used PCR and downstream techniques with their students; however, the library prep appeared to be both a technical challenge and cost obstacle for this well-resourced school. Moreover, upon discussing data analysis workflows, scale of data produced, the types of samples for sequencing, and what type of targeted sequencing project might fit into course curricula, it became clear that the teachers and school were not prepared to weave high throughput DNA sequencing into their science program. It is certainly possible that a high school could, has, or is in the process of, building a minION sequencing program that includes the five features discussed above: student engagement, simplified learning tools, teacher PD, building a network across schools or teachers, and acquiring resources and a budget. An approach of this type would be a promising step forward worthy of study and replication. For high school students, use of high-throughput (next generation) sequencing platforms would lead to invaluable research training, but for now these learning experiences may be best offered during multiweek summer research internships or intensive, boot-camp style courses. Box 7.1 offers additional examples.

DNA sequencing is now an essential topic and laboratory platform for project-based learning in high school. Students may ask [15] "Why do we have to learn about DNA sequencing?" Driven by high-throughput sequencing, genetics and genomics are subjects with ever-increasing relevance to everyday life. Today's high school students represent the next generation of healthcare consumers, biological scientists, and policy makers. Through well-designed DNA sequencing activities and participation in teaching and learning networks, including Barcode of Life efforts, TtGG, and future programs, high school students will continue to find biology, genetics, and genomics as interesting and relevant, whether or not they pursue STEM careers.

DNA sequencing at the undergraduate level

For numerous reasons, both Sanger-based and high throughput DNA sequencing fits well within undergraduate science education curricula. Most high school applications of DNA sequencing as a learning tool do not require the use of next generation sequence analysis platforms (e.g., Galaxy, Integrative Genomics Viewer, etc.). Through their graduate school training, college and university faculty are prepared to execute a wide range of molecular genetic laboratory procedures. Many faculty manage their own active research laboratories, supervise lab technicians and student researchers, and have access to professionally staffed core facilities. Today, faculty across large research universities, as well as small regional colleges, are encouraged to provide course-based undergraduate research experiences (CUREs) for students. Federal agencies and foundations provide grant support and training networks, within which professors can brush up on specific skills and required bioinformatics workflows [16,17] (https://datascience.nih.gov/bd2k, gcat-seek.weebly.com). If faculty and teaching assistant staff want to launch a new DNA sequence-based course or enhance an existing core course, funds and resources

---

Box 7.1 DNA sequencing at the Sacred Heart Academy.

Sister Mary Jane Paolella was an innovative teacher at the Sacred Heart Academy. Beginning in 1997, she embarked on integrating DNA sequencing into the all-girls catholic school where she spent the majority of her career. With training and support from Yale University School of Medicine, she began innovative sequencing projects with dozens of students at Sacred Heart. Impressed by Sr. Mary Jane's program, Applied BioSystems donated an ABI Prism 310 single capillary sequencer. Over the years, her students contributed more than 25 gene sequences to NCBI/Genbank and published several papers [12–14]. Sr. Mary Jane managed the sequencing instrument, operating as both the sequencing technician/problem solver and teacher/curriculum innovator. A significant number of Sr. Mary Jane's students embarked on STEM careers. The Sacred Heart example suggests that highly motivated teachers, with robust support from their school, and access to sequencing and ancillary instrumentation, can design cutting-edge courses that directly integrate DNA sequencing into high school classes.

can be found from grants or foundations. As a result, undergraduate students encounter a wide range of DNA sequence-based activities across numerous types of courses Fig. 7.3.

Undergraduates may encounter DNA sequencing in their first year or years later in semester-long to year-long senior thesis projects. Whether students engage in first year sequencing as a research experience designed to excite them about being an STEM major, or as an undergraduate research effort, careful alignment of sequencing with learning content or goals remains very important. For many undergraduates, the years spent obtaining a bachelor's degree are a transition from annual education to a career. As a result, undergraduate laboratory training (and DNA sequencing) fulfill a number of different objectives. For each student, incorporation of traditional Sanger or second to third generation DNA sequencing into undergraduate courses can be STEM career skills training, preparation for graduate school, an introduction to hypothesis or discovery-driven research, a step on the path to medical school, or new and expanded content knowledge that improves their genetics, genomics, bioengineering, or overall scientific literacy. Herein, the discussion focuses on three general goals relative to the incorporation of DNA sequencing technology into the undergraduate experience:

1. Engaging and exciting students
2. Promoting research careers
3. Providing forward-looking skills and career training.

These three goals are in no way mutually exclusive of each other and many teaching applications of DNA sequencing simultaneously achieve all three and other objectives.

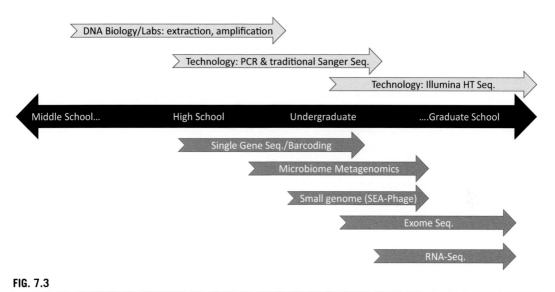

FIG. 7.3

Continuum of DNA sequence-based Learning Opportunities. For middle school through graduate school different levels of DNA sequence biology and different DNA sequencing technologies can be integrated into courses. Technologies appear in light gray, applications of technologies in darker gray.

No permission required.

Engaging and exciting students

By the early 1990s, Sanger sequencing was introduced to first and second year undergraduates as a tool for identifying species, uncovering food fraud, and generally engaging students with cutting edge technologies. Polymerase chain reaction (PCR) amplification of DNA fragments made it possible to design laboratory exercises where students would get an unknown sample and over the course of numerous laboratory periods they would extract DNA, amplify a target locus via PCR, sequence their amplified products and learn basic sequence analysis techniques to identify the sample [18]. At Yale University, undergraduates used this approach to identify the species of origin of meat and poultry products from local grocery stores. These early teaching approaches connected DNA sequencing to everyday life, enabling students to learn skills and produce simple, meaningful species identification outcomes (or in case of sequencing hotdogs, multispecies identifications).

Within 10 years, the research community would collectively launch the Barcode of Life Initiative, which now supports high school [11] and extensive university level DNA sequencing [19–22]. The expanded use of DNA barcoding [23] as a teaching and learning tool is enabled by the excellent Barcode of Life Data System (http://www.boldsystems.org/). At the university-level, DNA Barcoding fosters content knowledge, increases science self-efficacy, improves technical skills, and engages students through ownership of the laboratory (and fieldwork based) experience [24]. An exemplary model of DNA Barcoding occurs as part of separate Ecology and Molecular Biology classes at the University of California, San Diego (UCSD) [22]. Over a 2-week collaborative period, ecology students collect invertebrate specimens in the Scripps Coastal Reserve, gaining experience in fieldwork, collecting habitat information, and documenting and photographing specimens. The ecology students then transfer the invertebrates over to molecular biology students who extract DNA, amplify *cytochrome c oxidase* (COI) using PCR, and send PCR products on for sequencing. Students across both courses work together on consensus sequences comparing sequences to records in the BOLD database system, building phylogenetic trees, and investigating sequence variation. Like Sr. Mary Jane Paolella's students who submit their data to Genbank, UCSD students are excited to contribute new records to the BOLD database. UCSD postcourse evaluation data indicate that students gain self-confidence as young scientists through the barcoding learning experience [19–23].

The core activity of DNA sequencing unknown samples, first introduced more than 25 years ago, still resonates with undergraduates. Students are excited to learn new lab skills and solve a puzzle. Recently, several undergraduate classes have tested the Oxford Nanopore minION, third generation long read sequencing platform [25,26]. Utilizing nonamplified genomic DNA, students at Columbia University and the University of North Carolina (Chapel Hill) sequenced unknown human and Bahamian pupfish samples, respectively. These pioneering undergraduate student uses of third generation DNA sequencing technology did encounter technical and data analysis challenges, including managing large numbers of sequence reads without computer science skills necessary to sort and bin and lack of available analysis tools necessary to analyze their sequence data in a high throughput manner [25]. Nonetheless, problem solving and facing bioinformatics challenges are extremely valuable learning experiences for engaged students. It is likely that long read sequencing technology will be widely used by undergraduates once an applicable research education project or education network optimizes an experimental application and provides technical and pedagogical support to professors and students.

Promoting research careers

The Science Education [27–29] Alliance-Phage Hunters Advancing Genomics and Evolutionary Science (SEA-PHAGE) project [30,31] with nearly 500 trained college and university faculty, is the largest current effort that integrates DNA sequencing into undergraduate classes. Supported by the Howard Hughes Medical Institute and led by Graham Hatfull at the University of Pittsburgh, SEA-PHAGE (seaphages.org, phagesdb.org) is a two-semester undergraduate program that successfully engages and excites students while promoting research careers. Over 20,000 undergraduates across nearly 200 institutions have isolated bacteriophage from soil, extracted and purified DNA, sent DNA for sequencing, and subsequently annotated novel new bacteriophages. Hundreds of undergraduates are authors on peer-reviewed publications, and the SEA-PHAGE program is creating course-based research efforts at research universities and isolated small and tribal colleges across the United States.

The SEA-PHAGES program increases student learning and undergraduate retention in STEM majors [28,29]. After the year-long SEA-PHAGE program, undergraduates score at least as well or better on 20 learning gains when compared to undergraduates who participated in summer-long apprentice-based undergraduate research programs [29] (SURE and CURE programs). The largest gains made by SEA-PHAGE students included improvements in: knowledge construction, solving real problems, and tolerance for obstacles. SEA-PHAGE students also stay in STEM majors and achieve higher semester grades in introductory lecture courses when compared to matched, non-SEA-PHAGE participants at their home institutions. Through thorough PD, laboratory resources provisioning, technical support, networking events (including research symposia for faculty and students), phage DNA sequencing services and an excellent web-based support platform [28,27], the SEA-PHAGE program has become a national model for disseminated research-based undergraduate courses.

While the SEA-PHAGE project continues to promote research experiences with undergraduates, countless individual faculty projects also provide valuable learning and research experience for undergraduates [20,27]. By connecting their own research efforts with upper-level classes and undergraduate independent research projects, faculty greatly enrich student-learning experiences. Effective research-based undergraduate sequencing efforts need to be practical in scale, discovery based, inclusive for multiple students and well linked to or derived from core course material. See Boxes 7.2 and 7.3 for additional examples.

Box 7.2 Case study: demodex mite genetics.

A project undertaken at Bowdoin College in Brunswick, Maine demonstrates a successful integration of faculty research and undergraduate DNA sequencing. In an effort to create a laboratory exercise, Professor of Biology Michael Palopoli initiated a guided inquiry project sequencing *Demodex* mites that live in human hair follicles and sebaceous glands. One goal of the student laboratory was to survey sequence variation in a natural population, linking the effort to learning goals in basic population genetics. At the start of the project very little was known about either the population genetics or the dynamics of the evolutionary association between the *Demodex* mites and humans. The effort was discovery based and used degenerate PCR primers to sequence portions of the mitochondrial genome of *Desmodex folliculorum*. Importantly, the effort was participatory, as undergraduates collected hair follicle mites from themselves, other Bowdoin faculty, classmates, and their family members. The initial data became a laboratory report undergraduates completed for Palopoli's Evolution course. Subsequently, several talented undergraduates used long-range PCR to amplify and sequence the complete mitochondrial genomes of two species of *Demodex*, leading to BMC Genomics and Proceeding of the National Academy of Sciences publications [32,33] where numerous Bowdoin undergraduates were co-authors.

> **Box 7.3 Case study: immersive genomic sequencing.**
>
> Another outstanding example of blending a research program into an undergraduate course is The *Ecological Metagenomics* course led by Elizabeth Dinsdale at San Diego State University (SDSU). Since 2010, SDSU undergraduates have sequenced bacterial genomes, completed numerous metagenomics profiles and sequenced the full genome of the California Sea Lion (*Zalophus californianus*) at a depth of 12× (https://dinsdalelab.sdsu.edu/wordpress/ecological-metagenomics/) [22]. Unlike many other undergraduate laboratory experiences, Dinsdale's SDSU students complete all the wet-bench laboratory (genomic library) preparation work and run the sequencing instrument themselves. The 15-week semester is dedicated to combined lectures and laboratory sessions necessary to convert undergraduates from novices to competent genomic technicians and data analysts. Through access initially to a Roche 454 FLX titanium sequencer, and more recently to other second and third generation high throughput sequencing platforms, *Ecological Metagenomics* is a full immersion course in genomic sequencing. Common factors to the success of this program, as well as most other efforts discussed herein, include a tight connection between the sequencing activity and the course content in marine ecology and genomics, and perhaps more importantly deep levels of student project ownership created by the program.

Not all faculty research projects lend themselves to effective undergraduate learning experiences: when the expertise of the undergraduates and the complexity of the research projects are mismatched, it is more difficult to provide successful learning outcomes for the students. For example, in 2016 while teaching undergraduates a workflow processing millions of FASTQ sequences through a lengthy variant calling exercise, a student asked the author what it meant to "align reads to the reference genome." This question reveals that unless well-versed in genomics, most undergraduates face a steep learning curve when first encountering very large genomic data sets. This is recapitulated time and time again, wherein a professor attempts to merge a moderate scale genomic sequencing project into course content and the students are immediately overwhelmed by the complexity of the experiment and data analysis. Teaching the genomic and technical content thoroughly and well in advance, providing clear glossaries of terms and file types, and starting with small, pilot scale sequencing experiments (e.g., sequencing PCR products) will all help students gain important knowledge and skills in advance of a research scale undergraduate effort.

For instance, one might design an RNA-seq experiment with a model organism with several experimental treatments (e.g., knock-out or morpholino knock-down versus wildtype animals) and three biological replicates of each sample and possibly technical replicates. Suddenly, two or three dozen samples are queued-up for sequencing; the scope and complexity of the project exceeds the capacity of a group of undergraduates, and may exceed the computational resources provided to the professor or to the department. While undergraduates in this type of situation may be getting a real, research-scale experience, the slightest quality control issue or computational obstacle can frustrate the students and minimize positive learning outcomes. The most advanced undergraduate students with particular interest and completed coursework in genomics or bioinformatics will likely be able to handle complex experimental data; however, it is important to match the learner audience with the complexity of the sequencing project.

Prior to considering the blending of a faculty research project with a course, it may be important to consider several factors (Table 7.1). For instance, one must decide who will do the wet-bench molecular biological preparations prior to sequencing. If students do the laboratory work, a range of excellent to poor samples will be a likely result. If a core facility or paid commercial service prepares DNA or genomic libraries, then students have limited ownership of the project. In some instances, it may be most

Table 7.1 Considerations and questions to address before integrating research DNA sequencing into undergraduate courses. Options are provided as you move through the table.

Do students have the lab competencies required to isolate, amplify and clone DNA samples?	Do you have an institutional DNA sequencing service or ongoing contract with a sequencing vendor; do they offer genomic library preparation services?	What is your estimated budget for the course-based research exercise, including consumables and sequencing, and possibly software licenses or access?	Do you have access to Bioinformatics support, through an institutional core or through committed collaborators?	Do students have basic competency using BLAST or BLAT? If yes, are students familiar with Ensembl and UCSC genome browsers?
If no, are there graduate students or qualified teaching or lab assistants available to support your students?	If no, do you have lead time to prepare samples and build a relationship with a sequencing center?	Can you access institutional funds for sequencing or other required supplies?	Do students have computers and data analysis packages or access to packages required to analyze data?	Do you have software licenses or data analysis tools readily available?
If no, consider project with advanced undergraduates. Or develop a single gene sequencing project (Fig. 10.3)	If time is short consider data analysis experience based on downloading public DNA sequence data	If budget is very tight, consider data analysis experience or find a collaborator	If you sense data analysis challenges for you and students, can you start with a bioinformatics course and build capacity?	If no, consider a prerequisite course either in bioinformatics or consider data carpentry training or R/R-Studio training

suitable to have students prepare samples, with duplicate samples prepared by lab technical staff or professional personnel to address potential quality control issues. A simple-sounding solution, however, a dual laboratory work approach is expensive, and students normally figure out that they are simply going through the motions of doing the bench work. A natural tension between wanting high-quality research data and maximizing learning opportunities for students needs balancing prior to deciding if undergraduates will embark on more than an exploratory DNA sequencing project.

Skills and career training

DNA sequencing has become a commodity business provided by core service laboratories and commercial vendors across the world. Traditional Sanger sequencing and next generation sequencing costs are low on a nucleotide base pair basis. While costs appear low to research scientists, many schools, colleges, and universities have extremely limited budgets set aside for demonstrative laboratory exercises such as DNA sequencing. The fact that third-party facilities may do the actual sequencing and the constraints of tight budgets beg the question: is it appropriate to undertake DNA sequencing projects when the goal is student learning?

In all of the examples described above, providing a sense of project ownership by students is vitally important. Additionally, sequencing projects focused on learning gains by students need careful alignment with course content areas. However, if promoting bioinformatics and computational skills training is a focus of student learning, then skipping DNA sequencing altogether is a viable option. At no cost, gigabytes of sequencing data can be downloaded from the NCBI Sequence Read Archive

(SRA) and used as training data sets for undergraduates. Alternatively, large genomic data sets can be generated by a professional laboratory and form the basis of training exercises across numerous undergraduate programs.

The genomics education partnership (GEP) led by Washington University in St. Louis is an example of an excellent undergraduate training program where the students start with raw data generated by a professional sequencing facility [34–36] (https://gep.wustl.edu/curriculum/course_materials_WU/introduction_to_genomics). Since 2006, the GEP has used a crowd-sourced model to engage over 1000 undergraduates from research universities to small private colleges. Via GEP, participating university and college faculty receive training in summer workshops, have access to academic year webinars, and build a collaborative peer network. Faculty embed the GEP undergraduate sequence annotation projects into their own courses, thus providing flexibility to faculty as they add genome annotation and mentored student research into their courses. GEP provides faculty with modular curricular elements that they use to provide preliminary skills training for their students.

In the GEP, undergraduates download scaled packages (~40 kb) of *Drosophila* spp. genome sequences. The majority of GEP projects focus on the dot chromosome (F element) across a range of *Drosophila* species. Students then work to investigate and construct gene models based on other evidence presented in a local instance of the UCSC Genome Browser. With hundreds of students investigating individual and shared portions of the dot chromosome, the GEP effort leverages student numbers to crowd source genome annotation. At the same time, students learn about eukaryotic gene structure, genome evolution, and gain valuable skills in bioinformatics. While students do not become fully invested in the research project through actually doing any wet bench work, they gain ownership in their projects through their work and submission of gene annotations to FlyBase, as well as through frequent peer-reviewed publications with 100 s of undergraduate authors [35].

Full genome or exome sequencing of undergraduate students is imminent. It is possible that an undergraduate course-based research experience has already completed personal genomes of students. Starting in 2012, the Icahn School of Medicine at Mount Sinai in New York began a graduate and medical student course, *Practical Analysis of Your Personal Genome*, wherein participants voluntarily sequenced and annotated their own full genomes [37]. The adult students at the Icahn School of Medicine were able to either voluntarily provide a DNA sample or opt for analyzing an anonymous human genome; 95% ($n=56$) of students chose to have their genomes sequenced. Similar to other efforts wherein postbaccalaureate students analyzed their own genomes [38,39] the Icahn-Mount Sinai participants self-reported enhanced learning and engagement [39]. These postbaccalaureate examples of student participatory genomics suggest that undergraduate participatory genomics may soon be upon us. It is likely that commodity DNA sequencing facilities will be able (or are already able) to provide individual student genomes at a cost-point similar to a technical/advanced course textbook. The responsibility of building pedagogical frameworks and the robust ethical procedures necessary to undertake personal genomics with undergraduates is and will be an exciting challenge for university and college faculty.

It is clear that there is significant and growing demand for computational biologists, bioinformaticians, and biologically focused computer scientists in industry, academia, and research. Workforce demand for computational biologists continues to be substantial. In February 2019, LinkedIn listed over 9000 open positions for Informatics jobs, with over 3500 of these jobs specifically in Bioinformatics. At the same time, GenomeWeb was posting dozens of new bioinformatics positions each week. Skills training for careers in genomics is currently the domain of masters and doctoral degree programs, but looking ahead, undergraduate institutions will have an emerging and important role to play in providing foundational content, skills, and career training necessary to fill thousands of new jobs. Students with solid knowledge of R and or Python and those able to manage and move large data sets and promote reproducibility using clear code should start graduate school or jobs ahead of many of their peers.

Future considerations

There is no doubt that laboratory exercises in molecular genetics promote student confidence, develop important skills, foster understanding of experimental design, increase engagement, create discovery moments, and promote technical understanding of molecular-level phenomena. For these reasons, it is valuable to continue hands-on laboratory practice leading up to DNA sequencing. Moving forward, educators will need to decide whether to have students do the laboratory work necessary to also sequence DNA. DNA Barcoding, professor-led research integrated into courses, and large networks such as SEA-PHAGES will continue to provide critically important training for the next generation of young scientists. At the same time, the emergence of low-cost, genome-scale DNA sequencing provides a range of laboratory-free learning opportunities, particularly for undergraduate students. The GEP program may be an indicator of future learning opportunities wherein students increase their content knowledge, share the experience of a research scientist and create new bioinformatics-based knowledge, while doing computer or "dry-lab" research.

Genomic scale DNA sequence data will also promote teaching and learning in data science. With large-scale DNA and RNA sequence data sets, undergraduate biology education can begin to train students in bioinformatics core competencies [40], skills that will be valuable across many areas of data sciences. Rather than gaining pipetting skills, undergraduates will learn about sequence alignment algorithms, write simple shell scripts, run statistical analyses in R, download omics data, and build networks of genetic interactions. At the same time, it will be possible to promote reproducibility through documentation of bioinformatics workflows and scripts, and provide students with training on the importance of metadata and FAIR data standards (Data that is Findable, Accessible, Interoperable, and Reusable).

Twenty years after the release of the draft sequence of the human genome, genetics and DNA science continue to be exhilarating subjects. The genomics revolution is rapidly making DNA sequence and DNA science a truly personal subject. Direct-to-consumer genomics companies target STEM learners and humanity students alike. Indeed, most students, as current and future healthcare consumers, would be well-served by a required course in genetics and bioethics. DNA sequencing, sequence analysis, and the complexity of genomics is essential content, remains fertile ground for advanced learning, and, importantly, will lead to significant career opportunities.

References

[1] J.D. Watson, F.H.C. Crick, Genetical implications of the structure of deoxyribonucleic acid, Nature 171 (1953) 964–967, https://doi.org/10.1038/171964b0.
[2] J.D. Watson, The Double Helix: A Personal Account of the Discovery of the Structure of DNA, 1968.
[3] F. Sanger, S. Nicklen, A.R. Coulson, DNA sequencing with chain-terminating inhibitors, Proc. Natl. Acad. Sci. (1977) 5463–5467, https://doi.org/10.1073/pnas.74.12.5463.

[4] M. Capraro, J.M. Capraro, STEM Project-Based Learning: An Integrated Science, Technology, Engineering, 2013.
[5] Vision and Change in Undergraduate Biology Education: A Call to Action, 2009.
[6] D. Lopatto, Undergraduate research experiences support science career decisions and active learning, CBE Life Sci. Educ. 6 (2007) 297–306, https://doi.org/10.1187/cbe.07-06-0039.
[7] C.A. Wei, T. Woodin, Undergraduate research experiences in biology: alternatives to the apprenticeship model, CBE Life Sci. Educ. 10 (2011) 123–131, https://doi.org/10.1187/cbe.11-03-0028.
[8] Next Generation Science Standards: For States, By States, 2013.
[9] K. Larue, M. McKernan, K. Bass, C. Wray, Teaching the Genome Generation: bringing modern human genetics into the classroom through teacher professional development, J. STEM Outreach. 1 (2018), https://doi.org/10.15695/jstem/v1i1.12.
[10] B.V. Bowling, E.E. Acra, L. Wang, M.F. Myers, G.E. Dean, G.C. Markle, C.L. Moskalik, C.A. Huether, Development and evaluation of a genetics literacy assessment instrument for undergraduates, Genetics 178 (2008) 15–22, https://doi.org/10.1534/genetics.107.079533.
[11] D. Steinke, V. Breton, P.H. Berzitis, The school malaise trap program: coupling educational outreach with scientific discovery, PLoS Biol. 15 (2017).
[12] W.D. Bonds, M.J. Paolella, Human gene discovery laboratory: a problem-based learning experience, Am. Biol. Teach. 68 (2006) 538–543, https://doi.org/10.2307/4452061.
[13] M.J. Paolella, An honors biotechnology practical approach to learning: sequencing *Bos taurus* conserved regions in five calcium homeostasis-related genes: CALCR, COL1A1, VDR, BMP2, and osteocalcin, journal of bone and mineral, Research 24 (2009) 168–169, https://doi.org/10.1359/jbmr.080811.
[14] M. Watsa, G.A. Erkenswick, A. Pomerantz, S. Prost, Portable sequencing as a teaching tool in conservation and biodiversity research, PLoS Biol. 18 (2020), https://doi.org/10.1371/journal.pbio.3000667.
[15] R.J. Redfield, Why do we have to learn this stuff?—a new genetics for 21st century students, PLoS Biol. 10 (2012).
[16] V.P. Buonaccorsi, M.D. Boyle, D. Grove, C. Praul, E. Sakk, A. Stuart, T. Tobin, J. Hosler, S.L. Carney, M.J. Engle, B.E. Overton, J.D. Newman, M. Pizzorno, J.R. Powell, N. Trun, GCAT-SEEKquence: genome consortium for active teaching of undergraduates through increased faculty access to next-generation sequencing data, CBE Life Sci. Educ. 10 (2011) 342–345, https://doi.org/10.1187/cbe.11-08-0065.
[17] A.M. Campbell, M.L.S. Ledbetter, L.L.M. Hoopes, T.T. Eckdahl, L.J. Heyer, A. Rosenwald, E. Fowlks, S. Tonidandel, B. Bucholtz, G. Gottfried, Genome consortium for active teaching: meeting the goals of BIO2010, CBE Life Sci. Educ. 6 (2007) 109–118, https://doi.org/10.1187/cbe.06-10-0196.
[18] J.T. Millard, A.M. Pilon, Identification of forensic samples via mitochondrial DNA in the undergraduate biochemistry laboratory, J. Chem. Educ. 80 (2003) 444–446, https://doi.org/10.1021/ed080p444.
[19] J. Drew, E. Triplett, Whole genome sequencing in the undergraduate classroom: outcomes and lessons from a pilot course, J. Microbiol. Biol. Educ. 9 (2008) 3–11.
[20] D. Lopatto, C. Alvarez, D. Barnard, C. Chandrasekaran, H.M. Chung, C. Du, T. Eckdahl, A.L. Goodman, C. Hauser, C.J. Jones, O.R. Kopp, G.A. Kuleck, G. McNeil, R. Morris, J.L. Myka, A. Nagengast, P.J. Overvoorde, J.L. Poet, K. Reed, G. Regisford, D. Revie, A. Rosenwald, K. Saville, M. Shaw, G.R. Skuse, C. Smith, M. Smith, M. Spratt, J. Stamm, J.S. Thompson, B.A. Wilson, C. Witkowski, J. Youngblom, W. Leung, C.D. Shaffer, J. Buhler, E. Mardis, S.C.R. Elgin, Undergraduate research: genomics education partnership, Science 322 (2008) 684–685, https://doi.org/10.1126/science.1165351.
[21] J. Marcus, T. Hughes, D. McElroy, R.E. Wyatt, Engaging first-year undergraduates in hands-on research experiences: the upper Green River barcode of life project, J. Coll. Sci. Teach. 39 (2010).
[22] R.A. Edwards, J.M. Haggerty, N. Cassman, J.C. Busch, K. Aguinaldo, S. Chinta, M.H. Vaughn, R. Morey, T.T. Harkins, C. Teiling, K. Fredrikson, E.A. Dinsdale, Microbes, metagenomes and marine mammals: enabling the next generation of scientist to enter the genomic era, BMC Genom. 14 (2013), https://doi.org/10.1186/1471-2164-14-600.

[23] H. Henter, R. Imondi, K. James, D. Spencer, Steinke, DNA barcoding in diverse educational settings: five case studies, Philos. Trans. R. Soc. Lond. Ser. B Biol. Sci. 371 (1702).

[24] V. Savolainen, A. Cowan, G. Vogler, R.L. Roderick, Towards writing the encyclopedia of life: an introduction to DNA barcoding, Philos. Trans. R. Soc. Lond. Ser. B Biol. Sci. 360 (1462) 1805–1811.

[25] S. Zaaijer, Y. Erlich, Using mobile sequencers in an academic classroom, eLife 5 (2016).

[26] Y. Zeng, C.H. Martin, Oxford Nanopore Sequencing in a Research-Based Undergraduate Course, bioRxiv, 2017, https://doi.org/10.1101/227439.

[27] D. Lopatto, C. Hauser, C.J. Jones, D. Paetkau, V. Chandrasekaran, D. Dunbar, C. MacKinnon, J. Stamm, C. Alvarez, D. Barnard, J.E.J. Bedard, A.E. Bednarski, S. Bhalla, J.M. Braverman, M. Burg, H.M. Chung, R.J. DeJong, J.R. DiAngelo, C. Du, T.T. Eckdahl, J. Emerson, A. Frary, D. Frohlich, A.L. Goodman, Y. Gosser, S. Govind, A. Haberman, A.T. Hark, A. Hoogewerf, D. Johnson, L. Kadlec, M. Kaehler, S.C.S. Key, N.P. Kokan, O.R. Kopp, G.A. Kuleck, J. Lopilato, J.C. Martinez-Cruzado, G. McNeil, S. Mel, A. Nagengast, P.J. Overvoorde, S. Parrish, M.L. Preuss, L.D. Reed, E.G. Regisford, D. Revie, S. Robic, J.A. Roecklien-Canfield, A.G. Rosenwald, M.R. Rubin, K. Saville, S. Schroeder, K.A. Sharif, M. Shaw, G. Skuse, C.D. Smith, M. Smith, S.T. Smith, E.P. Spana, M. Spratt, A. Sreenivasan, J.S. Thompson, M. Wawersik, M.J. Wolyniak, J. Youngblom, L. Zhou, J. Buhler, E. Mardis, W. Leung, C.D. Shaffer, J. Threlfall, S.C.R. Elgin, A central support system can facilitate implementation and sustainability of a classroom-based undergraduate research experience (CURE) in genomics, CBE Life Sci. Educ. 13 (2014) 711–723, https://doi.org/10.1187/cbe.13-10-0200.

[28] D.I. Hanauer, M.J. Graham, L. Betancur, A. Bobrownicki, S.G. Cresawn, R.A. Garlena, D. Jacobs-Sera, N. Kaufmann, W.H. Pope, D.A. Russell, W.R. Jacobs, V. Sivanathan, D.J. Asai, G.F. Hatfull, An inclusive research education community (iREC): impact of the SEA-PHAGES program on research outcomes and student learning, Proc. Natl. Acad. Sci. U. S. A. 114 (2017) 13531–13536, https://doi.org/10.1073/pnas.1718188115.

[29] D.I. Hanauer, G. Hatfull, Measuring networking as an outcome variable in undergraduate research experiences, CBE Life Sci. Educ. 14 (2015), https://doi.org/10.1187/cbe.15-03-0061.

[30] T.C. Jordan, S.H. Burnett, S. Carson, S.M. Caruso, K. Clase, R.J. DeJong, J.J. Dennehy, D.R. Denver, D. Dunbar, S.C.R. Elgin, A.M. Findley, C.R. Gissendanner, U.P. Golebiewska, N. Guild, G.A. Hartzog, W.H. Grillo, G.P. Hollowell, L.E. Hughes, A. Johnson, R.A. King, L.O. Lewis, W. Li, F. Rosenzweig, M.R. Rubin, M.S. Saha, J. Sandoz, C.D. Shaffer, B. Taylor, L. Temple, E. Vazquez, V.C. Ware, L.P. Barker, K.W. Bradley, D. Jacobs-Sera, W.H. Pope, D.A. Russell, S.G. Cresawn, D. Lopatto, C.P. Bailey, G.F. Hatfull, A broadly implementable research course in phage discovery and genomics for first-year undergraduate students, mBio 5 (2014), https://doi.org/10.1128/mBio.01051-13.

[31] D.I. Hanauer, D. Jacobs-Sera, M.L. Pedulla, S.G. Cresawn, R.W. Hendrix, G.F. Hatfull, Teaching scientific inquiry, Science 314 (2006) 1880–1881, https://doi.org/10.1126/science.1136796.

[32] M.F. Palopoli, D.J. Fergus, S. Minot, D.T. Pei, W.B. Simison, I. Fernandez-Silva, M.S. Thoemmes, R.R. Dunn, M. Trautwein, Global divergence of the human follicle mite demodex folliculorum: persistent associations between host ancestry and mite lineages, Proc. Natl. Acad. Sci. U. S. A. 112 (2015) 15958–15963, https://doi.org/10.1073/pnas.1512609112.

[33] M.F. Palopoli, S. Minot, D. Pei, A. Satterly, J. Endrizzi, Complete mitochondrial genomes of the human follicle mites demodex brevis and *D. folliculorum*: novel gene arrangement, truncated tRNA genes, and ancient divergence between species, BMC Genomics 15 (2014), https://doi.org/10.1186/1471-2164-15-1124.

[34] S.C.R. Elgin, C. Hauser, T.M. Holzen, C. Jones, A. Kleinschmit, J. Leatherman, The GEP: crowd-sourcing big data analysis with undergraduates, Trends Genet. 33 (2017) 81–85, https://doi.org/10.1016/j.tig.2016.11.004.

[35] W. Leung, C.D. Shaffer, L.K. Reed, S.T. Smith, W. Barshop, W. Dirkes, M. Dothager, P. Lee, J. Wong, D. Xiong, H. Yuan, J.E.J. Bedard, J.F. Machone, S.D. Patterson, A.L. Price, B.A. Turner, S. Robic, E.K. Luippold, S.R. McCartha, T.A. Walji, C.A. Walker, K. Saville, M.K. Abrams, A.R. Armstrong, W. Armstrong, R.J. Bailey, C.R. Barberi, L.R. Beck, A.L. Blaker, C.E. Blunden, J.P. Brand, E.J. Brock, D.W. Brooks, M. Brown,

S.C. Butzler, E.M. Clark, N.B. Clark, A.A. Collins, R.J. Cotteleer, P.R. Cullimore, S.G. Dawson, C.T. Docking, S.L. Dorsett, G.A. Dougherty, K.A. Downey, A.P. Drake, E.K. Earl, T.G. Floyd, J.D. Forsyth, J.D. Foust, S.L. Franchi, J.F. Geary, C.K. Hanson, T.S. Harding, C.B. Harris, J.M. Heckman, H.L. Holderness, N.A. Howey, D.A. Jacobs, E.S. Jewell, M. Kaisler, E.A. Karaska, J.L. Kehoe, H.C. Koaches, J. Koehler, D. Koenig, A.J. Kujawski, J.E. Kus, J.A. Lammers, R.R. Leads, E.C. Leatherman, R.N. Lippert, G.S. Messenger, A.T. Morrow, H.J. NewcombVictoria, S.J.P. Plasman, M.K. Powers, R.M. Reem, J.P. Rennhack, K.R. Reynolds, L.A. Reynolds, D.K. Rhee, A.B. Rivard, A.J. Ronk, M.B. Rooney, L.S. Rubin, L.R. Salbert, R.K. Saluja, T. Schauder, A.R. Schneiter, R.W. Schulz, K.E. Smith, S. Spencer, B.R. Swanson, M.A. Tache, A.A. Tewilliager, A.K. Tilot, E. VanEck, M.M. Villerot, M.B. Vylonis, D.T. Watson, J.A. Wurzler, L.M. Wysocki, M. Yalamanchili, M.A. Zaborowicz, J.A. Emerson, C. Ortiz, F.J. Deuschle, L.A. DiLorenzo, K.L. Goeller, C.R. Macchi, S.E. Muller, B.D. Pasierb, J.E. Sable, J.M. Tucci, M. Tynon, D.A. Dunbar, L.H. Beken, A.C. Conturso, B.L. Danner, G.A. DeMichele, J.A. Gonzales, M.S. Hammond, C.V. Kelley, E.A. Kelly, D. Kulich, C.M. Mageeney, N.L. McCabe, A.M. Newman, L.A. Spaeder, R.A. Tumminello, D. Revie, J.M. Benson, M.C. Cristostomo, P.A. DaSilva, K.S. Harker, J.N. Jarrell, L.A. Jimenez, B.M. Katz, W.R. Kennedy, K.S. Kolibas, M.T. LeBlanc, T.T. Nguyen, D.S. Nicolas, M.D. Patao, S.M. Patao, B.J. Rupley, B.J. Sessions, J.A. Weaver, A.L. Goodman, E.L. Alvendia, S.M. Baldassari, A.S. Brown, I.O. Chase, M. Chen, S. Chiang, A.B. Cromwell, A.F. Custer, T.M. DiTommaso, J. El-Adaimi, N.C. Goscinski, R.A. Grove, N. Gutierrez, R.S. Harnoto, H. Hedeen, E.L. Hong, B.L. Hopkins, V.F. Huerta, C. Khoshabian, K.M. LaForge, C.T. Lee, B.M. Lewis, A.M. Lydon, B.J. Maniaci, R.D. Mitchell, E.V. Morlock, W.M. Morris, P. Naik, N.C. Olson, J.M. Osterloh, M.A. Perez, J.D. Presley, M.J. Randazzo, M.K. Regan, F.G. Rossi, M.A. Smith, E.A. Soliterman, C.J. Sparks, D.L. Tran, T. Wan, A.A. Welker, J.N. Wong, A. Sreenivasan, J. Youngblom, A. Adams, J. Alldredge, A. Bryant, D. Carranza, A. Cifelli, K. Coulson, C. Debow, N. Delacruz, C. Emerson, C. Farrar, D. Foret, E. Garibay, J. Gooch, M. Heslop, S. Kaur, A. Khan, V. Kim, T. Lamb, P. Lindbeck, G. Lucas, E. Macias, D. Martiniuc, L. Mayorga, J. Medina, N. Membreno, S. Messiah, L. Neufeld, S.F. Nguyen, Z. Nichols, G. Odisho, D. Peterson, L. Rodela, P. Rodriguez, V. Rodriguez, J. Ruiz, W. Sherrill, V. Silva, J. Sparks, G. Statton, A. Townsend, I. Valdez, M. Waters, K. Westphal, S. Winkler, J. Zumkehr, R.J. DeJong, A.J. Hoogewerf, C.M. Ackerman, I.O. Armistead, L. Baatenburg, M.J. Borr, L.K. Brouwer, B.J. Burkhart, K.T. Bushhouse, L. Cesko, T.Y.Y. Choi, H. Cohen, A.M. Damsteegt, J.M. Darusz, C.M. Dauphin, Y.P. Davis, E.J. Diekema, M. Drewry, M.E.M. Eisen, H.M. Faber, K.J. Faber, E. Feenstra, I.T. Felzer-Kim, B.L. Hammond, J. Hendriksma, M.R. Herrold, J.A. Hilbrands, E.J. Howell, S.A. Jelgerhuis, T.R. Jelsema, B.K. Johnson, K.K. Jones, A. Kim, R.D. Kooienga, E.E. Menyes, E.A. Nollet, B.E. Plescher, L. Rios, J.L. Rose, A.J. Schepers, G. Scott, J.R. Smith, A.M. Sterling, J.C. Tenney, C. Uitvlugt, R.E. VanDyken, M.V. Vennen, S. Vue, N.P. Kokan, K. Agbley, S.K. Boham, D. Broomfield, K. Chapman, A. Dobbe, I. Dobbe, W. Harrington, M. Ibrahem, A. Kennedy, C.A. Koplinsky, C. Kubricky, D. Ladzekpo, C. Pattison, R.E. Ramirez, L. Wande, S. Woehlke, M. Wawersik, E. Kiernan, J.S. Thompson, R. Banker, J.R. Bartling, C.I. Bhatiya, A.L. Boudoures, L. Christiansen, D.S. Fosselman, K.M. French, I.S. Gill, J.T. Havill, J.L. Johnson, L.J. Keny, J.M. Kerber, B.M. Klett, C.N. Kufel, F.J. May, J.P. Mecoli, C.R. Merry, L.R. Meyer, E.G. Miller, G.J. Mullen, K.C. Palozola, J.J. Pfeil, J.G. Thomas, E.M. Verbofsky, E.P. Spana, A. Agarwalla, J. Chapman, B. Chlebina, I. Chong, I.N. Falk, J.D. Fitzgibbons, H. Friedman, O. Ighile, A.J. Kim, K.A. Knouse, F. Kung, D. Mammo, C.L. Ng, V.S. Nikam, D. Norton, P. Pham, J.W. Polk, S. Prasad, H. Rankin, C.D. Ratliff, V. Scala, N.U. Schwartz, J.A. Shuen, A. Xu, T.Q. Xu, Y. Zhang, A.G. Rosenwald, M.G. Burg, S.J. Adams, M. Baker, B. Botsford, B. Brinkley, C. Brown, S. Emiah, E. Enoch, C. Gier, A. Greenwell, L. Hoogenboom, J.E. Matthews, M. McDonald, A. Mercer, N. Monsma, K. Ostby, A. Ramic, D. Shallman, M. Simon, E. Spencer, T. Tomkins, P. Wendland, A. Wylie, M.J. Wolyniak, G.M. Robertson, S.I. Smith, J.R. DiAngelo, E.D. Sassu, S.C. Bhalla, K.A. Sharif, T. Choeying, J.S. Macias, F. Sanusi, K. Torchon, A.E. Bednarski, C.J. Alvarez, K.C. Davis, C.A. Dunham, A.J. Grantham, A.N. Hare, J. Schottler, Z.W. Scott, G.A. Kuleck, N.S. Yu, M.M. Kaehler, J. Jipp, P.J. Overvoorde, E. Shoop, O. Cyrankowski, B. Hoover, M. Kusner, D. Lin, T. Martinov, J. Misch, G. Salzman, H. Schiedermayer, M. Snavely, S. Zarrasola, S. Parrish, A. Baker, A. Beckett, C. Belella, J. Bryant,

T. Conrad, A. Fearnow, C. Gomez, R.A. Herbstsomer, S. Hirsch, C. Johnson, M. Jones, R. Kabaso, E. Lemmon, C.M. dos Santos Vieira, D. McFarland, C. McLaughlin, A. Morgan, S. Musokotwane, W. Neutzling, J. Nietmann, C. Paluskievicz, J. Penn, E. Peoples, C. Pozmanter, E. Reed, N. Rigby, L. Schmidt, M. Shelton, R. Shuford, T. Tirasawasdichai, B. Undem, D. Urick, K. Vondy, B. Yarrington, T.T. Eckdahl, J.L. Poet, A.B. Allen, J.E. Anderson, J.M. Barnett, J.S. Baumgardner, A.D. Brown, J.E. Carney, R.A. Chavez, S.L. Christgen, J.S. Christie, A.N. Clary, M.A. Conn, K.M. Cooper, M.J. Crowley, S.T. Crowley, J.S. Doty, B.A. Dow, C.R. Edwards, D.D. Elder, J.P. Fanning, B.M. Janssen, A.K. Lambright, C.E. Lane, A.B. Limle, T. Mazur, M.R. McCracken, A.M. McDonough, A.D. Melton, P.J. Minnick, A.E. Musick, W.H. Newhart, J.W. Noynaert, B.J. Ogden, M.W. Sandusky, S.M. Schmuecker, A.L. Shipman, A.L. Smith, K.M. Thomsen, M.R. Unzicker, W.B. Vernon, W.W. Winn, D.S. Woyski, X. Zhu, C. Du, C. Ament, S. Aso, L.S. Bisogno, J. Caronna, N. Fefelova, L. Lopez, L. Malkowitz, J. Marra, D. Menillo, I. Obiorah, E.N. Onsarigo, S. Primus, M. Soos, A. Tare, A. Zidan, C.J. Jones, T. Aronhalt, J.M. Bellush, C. Burke, S. DeFazio, B.R. Does, T.D. Johnson, N. Keysock, N.H. Knudsen, J. Messler, K. Myirski, J. LeaRekai, R.M. Rempe, M.S. Salgado, E. Stagaard, J.R. Starcher, A.W. Waggoner, A.K. Yemelyanova, A.T. Hark, A. Bertolet, C.E. Kuschner, K. Parry, M. Quach, L. Shantzer, M.E. Shaw, M.A. Smith, O. Glenn, P. Mason, C. Williams, S.C.S. Key, T.C.P. Henry, A.G. Johnson, J.X. White, A. Haberman, S. Asinof, K. Drumm, T. Freeburg, N. Safa, D. Schultz, Y. Shevin, P. Svoronos, T. Vuong, J. Wellinghoff, L.L.M. Hoopes, K.M. Chau, A. Ward, E.G.C. Regisford, L.J. Augustine, B. Davis-Reyes, V. Echendu, J. Hales, S. Ibarra, L. Johnson, S. Ovu, J.M. Braverman, T.J. Bahr, N.M. Caesar, C. Campana, D.W. Cassidy, P.A. Cognetti, J.D. English, M.C. Fadus, C.N. Fick, P.J. Freda, B.M. Hennessy, K. Hockenberger, J.K. Jones, J.E. King, C.R. Knob, K.J. Kraftmann, L. Li, L.N. Lupey, C.J. Minniti, T.F. Minton, J.V. Moran, K. Mudumbi, E.C. Nordman, W.J. Puetz, L.M. Robinson, T.J. Rose, E.P. Sweeney, A.S. Timko, D.W. Paetkau, H.L. Eisler, M.E. Aldrup, J.M. Bodenberg, M.G. Cole, K.M. Deranek, M. DeShetler, R.M. Dowd, A.K. Eckardt, S.C. Ehret, J. Fese, A.D. Garrett, A. Kammrath, M.L. Kappes, M.R. Light, A.C. Meier, M.P. AllisonO'Rouke, K. Ramsey, J.R. Ramthun, M.T. Reilly, D. Robinett, N.L. Rossi, M.G. Schueler, E. Shoemaker, K.M. Starkey, A. Vetor, A. Vrable, V. Chandrasekaran, C. Beck, K.R. Hatfield, D.A. Herrick, C.B. Khoury, C. Lea, C.A. Louie, S.M. Lowell, T.J. Reynolds, J. Schibler, A.H. Scoma, M.T. Smith-Gee, S. Tuberty, C.D. Smith, J.E. Lopilato, J. Hauke, J.A. Roecklein-Canfield, M. Corrielus, H. Gilman, S. Intriago, A. Maffa, S.A. Rauf, K. Thistle, M. Trieu, J. Winters, B. Yang, C.R. Hauser, T. Abusheikh, Y. Ashrawi, P. Benitez, L.R. Boudreaux, M. Bourland, M. Chavez, S. Cruz, G.N. Elliott, J.R. Farek, S. Flohr, A.H. Flores, C. Friedrichs, Z. Fusco, Z. Goodwin, E. Helmreich, J. Kiley, J.M. Knepper, C. Langner, M. Martinez, C. Mendoza, M. Naik, A. Ochoa, N. Ragland, E. Raimey, S. Rathore, E. Reza, G. Sadovsky, B.S. Marie-Isabelle, J.E. Smith, A.K. Unruh, V. Velasquez, M.W. Wolski, Y. Gosser, S. Govind, N. Clarke-Medley, L. Guadron, D. Lau, A. Lu, C. Mazzeo, M. Meghdari, S. Ng, B. Pamnani, O. Plante, Y.K.W. Shum, R. Song, D.E. Johnson, M. Abdelnabi, A. Archambault, N. Chamma, S. Gaur, D. Hammett, A. Kandahari, G. Khayrullina, S. Kumar, S. Lawrence, N. Madden, M. Mandelbaum, H. Milnthorp, S. Mohini, R. Patel, S.J. Peacock, E. Perling, A. Quintana, M. Rahimi, K. Ramirez, R. Singhal, C. Weeks, T. Wong, A.T. Gillis, Z.D. Moore, C.D. Savell, R. Watson, S.F. Mel, A.A. Anilkumar, P. Bilinski, R. Castillo, M. Closser, N.M. Cruz, T. Dai, G.F. Garbagnati, L.S. Horton, D. Kim, J.H. Lau, J.Z. Liu, S.D. Mach, T.A. Phan, Y. Ren, K.E. Stapleton, J.M. Strelitz, R. Sunjed, J. Stamm, M.C. Anderson, B. GraceBonifield, D. Coomes, A. Dillman, E.J. Durchholz, A.E. Fafara-Thompson, M.J. Gross, A.M. Gygi, L.E. Jackson, A. Johnson, Z. Kocsisova, J.L. Manghelli, K. McNeil, M. Murillo, K.L. Naylor, J. Neely, E.E. Ogawa, A. Rich, A. Rogers, J.D. Spencer, K.M. Stemler, A.A. Throm, M. VanCamp, K. Weihbrecht, T.A. Wiles, M.A. Williams, M. Williams, K. Zoll, C. Bailey, L. Zhou, D.M. Balthaser, A. Bashiri, M.E. Bower, K.A. Florian, N. Ghavam, E.S. Greiner-Sosanko, H. Karim, V.W. Mullen, C.E. Pelchen, P.M. Yenerall, J. Zhang, M.R. Rubin, S.M. Arias-Mejias, A.G. Bermudez-Capo, G.V. Bernal-Vega, M. Colon-Vazquez, A. Flores-Vazquez, M. Gines-Rosario, I.G. Llavona-Cartagena, J.O. Martinez-Rodriguez, L. Ortiz-Fuentes, E.O. Perez-Colomba, J. Perez-Otero, E. Rivera, L.J. Rodriguez-Giron, A.J. Santiago-Sanabria, A.M. Senquiz-Gonzalez, F.R.S. Valle, D. Vargas-Franco, K.I. Velázquez-Soto, J.D. Zambrana-Burgos, J.C. Martinez-Cruzado, L. Asencio-Zayas, K. Babilonia-Figueroa, F.D. Beauchamp-Pérez,

J. Belén-Rodríguez, L. Bracero-Quiñones, A.P. Burgos-Bula, X.A. Collado-Méndez, L.R. Colón-Cruz, A.I. Correa-Muller, J.L. Crooke-Rosado, J.M. Cruz-García, M. Defendini-Ávila, F.M. Delgado-Peraza, A.J. Feliciano-Cancela, V.M. Gónzalez-Pérez, W. Guiblet, A. Heredia-Negrón, J. Hernández-Muñiz, L.N. Irizarry-González, Á.L. Laboy-Corales, G.A. Llaurador-Caraballo, F. Marín-Maldonado, U. Marrero-Llerena, H.A. Martell-Martínez, I.M. Martínez-Traverso, K.N. Medina-Ortega, S.G. Méndez-Castellanos, K.C. Menéndez-Serrano, C.I. Morales-Caraballo, S. Ortiz-DeChoudens, P. Ortiz-Ortiz, H. Pagán-Torres, D. Pérez-Afanador, E.M. Quintana-Torres, E.G. Ramírez-Aponte, C. Riascos-Cuero, M.S. Rivera-Llovet, I.T. Rivera-Pagán, R.E. Rivera-Vicéns, F. Robles-Juarbe, L. Rodríguez-Bonilla, B.O. Rodríguez-Echevarría, P.M. Rodríguez-García, A.E. Rodríguez-Laboy, S. Rodríguez-Santiago, M.L. Rojas-Vargas, E.N. Rubio-Marrero, A. Santiago-Colón, J.L. Santiago-Ortiz, C.E. Santos-Ramos, J. Serrano-González, A.M. Tamayo-Figueroa, E.P. Tascón-Peñaranda, J.L. Torres-Castillo, N.A. Valentín-Feliciano, Y.M. Valentín-Feliciano, N.M. Vargas-Barreto, M. Vélez-Vázquez, L.R. Vilanova-Vélez, C. Zambrana-Echevarría, C. MacKinnon, H.M. Chung, C. Kay, A. Pinto, O.R. Kopp, J. Burkhardt, C. Harward, R. Allen, P. Bhat, J.H.C. Chang, Y. Chen, C. Chesley, D. Cohn, D. DuPuis, M. Fasano, N. Fazzio, K. Gavinski, H. Gebreyesus, T. Giarla, M. Gostelow, R. Greenstein, H. Gunasinghe, C. Hanson, A. Hay, T.J. He, K. Homa, R. Howe, J. Howenstein, H. Huang, A. Khatri, Y.L. Kim, O. Knowles, S. Kong, R. Krock, M. Kroll, J. Kuhn, M. Kwong, B. Lee, R. Lee, K. Levine, Y. Li, B. Liu, L. Liu, M. Liu, A. Lousararian, J. Ma, A. Mallya, C. Manchee, J. Marcus, S. McDaniel, M.L. Miller, J.M. Molleston, C.M. Diez, P. Ng, N. Ngai, H. Nguyen, A. Nylander, J. Pollack, S. Rastogi, H. Reddy, N. Regenold, J. Sarezky, M. Schultz, J. Shim, T. Skorupa, K. Smith, S.J. Spencer, P. Srikanth, G. Stancu, A.P. Stein, M. Strother, L. Sudmeier, M. Sun, V. Sundaram, N. Tazudeen, A. Tseng, A. Tzeng, R. Venkat, S. Venkataram, L. Waldman, T. Wang, H. Yang, J.Y. Yu, Y. Zheng, M.L. Preuss, A. Garcia, M. Juergens, R.W. Morris, A.A. Nagengast, J. Azarewicz, T.J. Carr, N. Chichearo, M. Colgan, M. Donegan, B. Gardner, N. Kolba, J.L. Krumm, S. Lytle, L. MacMillian, M. Miller, A. Montgomery, A. Moretti, B. Offenbacker, M. Polen, J. Toth, J. Woytanowski, L. Kadlec, J. Crawford, M.L. Spratt, A.L. Adams, B.K. Barnard, M.N. Cheramie, A.M. Eime, K.L. Golden, A.P. Hawkins, J.E. Hill, J.A. Kampmeier, C.D. Kern, E.E. Magnuson, A.R. Miller, C.M. Morrow, J.C. Peairs, G.L. Pickett, S.A. Popelka, A.J. Scott, E.J. Teepe, K.A. TerMeer, C.A. Watchinski, L.A. Watson, R.E. Weber, K.A. Woodard, D.C. Barnard, I. Appiah, M.M. Giddens, G.P. McNeil, A. Adebayo, K. Bagaeva, J. Chinwong, C. Dol, E. George, K. Haltaufderhyde, J. Haye, M. Kaur, M. Semon, D. Serjanov, A. Toorie, C. Wilson, N.C. Riddle, J. Buhler, E.R. Mardis, S.C.R. Elgin, Drosophila Muller F elements maintain a distinct set of genomic properties over 40 million years of evolution, G3: genes, genomes, Genetics 5 (2015) 719–740, https://doi.org/10.1534/g3.114.015966.

[36] C.D. Shaffer, C. Alvarez, C. Bailey, D. Barnard, S. Bhalla, C. Chandrasekaran, V. Chandrasekaran, H.M. Chung, D.R. Dorer, C. Du, T.T. Eckdahl, J.L. Poet, D. Frohlich, A.L. Goodman, Y. Gosser, C. Hauser, L.L.M. Hoopes, D. Johnson, C.J. Jones, M. Kaehler, N. Kokan, O.R. Kopp, G.A. Kuleck, G. McNeil, R. Moss, J.L. Myka, A. Nagengast, R. Morris, P.J. Overvoorde, E. Shoop, S. Parrish, K. Reed, E.G. Regisford, D. Revie, A.G. Rosenwald, K. Saville, S. Schroeder, M. Shaw, G. Skuse, C. Smith, M. Smith, E.P. Spana, M. Spratt, J. Stamm, J.S. Thompson, M. Wawersik, B.A. Wilson, J. Youngblom, W. Leung, J. Buhler, E.R. Mardis, D. Lopatto, S.C.R. Elgin, The genomics education partnership: successful integration of research into laboratory classes at a diverse group of undergraduate institutions, CBE Life Sci. Educ. 9 (2010) 55–69, https://doi.org/10.1187/09-11-0087.

[37] M.D. Linderman, S.C. Sanderson, A. Bashir, G.A. Diaz, A. Kasarskis, R. Zinberg, M. Mahajan, S.A. Suckiel, M. Zweig, E.E. Schadt, Impacts of incorporating personal genome sequencing into graduate genomics education: a longitudinal study over three course years, BMC Med. Genet. 11 (2018), https://doi.org/10.1186/s12920-018-0319-0.

[38] K. Salari, K.J. Karczewski, L. Hudgins, K.E. Ormond, Evidence that personal genome testing enhances student learning in a course on genomics and personalized medicine, PLoS One 8 (2013), https://doi.org/10.1371/journal.pone.0068853.

[39] K.S. Weber, J.L. Jensen, S.M. Johnson, Anticipation of personal genomics data enhances interest and learning environment in genomics and molecular biology undergraduate courses, PLoS One 10 (2015), https://doi.org/10.1371/journal.pone.0133486.

[40] M.A.W. Sayres, C. Hauser, M. Sierk, S. Robic, A.G. Rosenwald, T.M. Smith, E.W. Triplett, J.J. Williams, E. Dinsdale, W.R. Morgan, J.M. Burnette, S.S. Donovan, J.C. Drew, S.C.R. Elgin, E.R. Fowlks, S. Galindo-Gonzalez, A.L. Goodman, N.F. Grandgenett, C.C. Goller, J.R. Jungck, J.D. Newman, W. Pearson, E.F. Ryder, R. Tosado-Acevedo, W. Tapprich, T.C. Tobin, A. Toro-Martínez, L.R. Welch, R. Wright, L. Barone, D. Ebenbach, M. McWilliams, K.C. Olney, M.A. Pauley, Bioinformatics core competencies for undergraduate life sciences education, PLoS ONE 13 (2018), https://doi.org/10.1371/journal.pone.0196878.

SECTION 3

Next-generation sequencing & gene expression

CHAPTER 8

Review of gene expression using microarray and RNA-seq

Ana B. Villaseñor-Altamirano[a], Yalbi Itzel Balderas-Martínez[b], and Alejandra Medina-Rivera[a]

[a]*Laboratorio Internacional de Investigación sobre el Genoma Humano, UNAM, Querétaro, Mexico,* [b]*Instituto Nacional de Enfermedades Respiratorias Ismael Cosio Villegas, Laboratorio de Biología Computacional, Mexico City, Mexico*

Introduction

Many different technologies exist to determine expression levels of genes; this chapter provides an overview discussion of high-throughput techniques for conducting these experiments, as well as touching on key points required for reproducibility. High-throughput techniques refer to approaches that allow for assessment and identification of thousands of elements in a single experiment, for example, determining mRNA expression levels from the whole genome of an organism, as opposed to assessing each gene's expression level individually. This type of whole-genome expression analysis is termed *transcriptomics* and came about with the advent of high-throughput techniques, particularly next-generation sequencing (NGS) methods. Prior to the era of high-throughput transcriptomics, various low-throughput methods permitted analysis of small snapshots of transcriptional expression, which are briefly reviewed later.

Of the low-throughput methods for analyzing RNA expression, Sanger sequencing was an early method used to sequence individual transcripts called expressed sequence tags (ESTs). These ESTs are fragments of mRNA sequences, each one obtained from a cDNA sequence. Using ESTs in 1991, it was possible to obtain the first brain transcriptome with 600 sequences [1]. Later, in 1995, serial analysis of gene expression (SAGE) was published [2], with the advantage of analyzing thousands of transcripts. SAGE works by extracting mRNA from a sample and obtaining cDNA with reverse transcriptase; the cDNA is digested with restriction enzymes to produce "tag fragments," and these fragments are then concatenated and sequenced. Sequenced reads are quantified with computer programs to assess the occurrence of individual tags. Additional RNA expression methods exist, such as northern blotting, nylon membrane arrays, and reverse transcriptase quantitative PCR, providing focused analysis of a specific fraction of the transcriptome [3].

This chapter reviews two major high-throughput techniques applied in transcriptomics: microarrays and sequencing-based methods (Fig. 8.1). Microarrays are a classic technique that measures specific genes based on oligomers attached to a plate, representing the genes of interest; this was the first high-throughput technique for measuring gene expression. Sequencing-based methods rely upon determination of the nucleotide composition of a chain of DNA, where the DNA sequence is derived via reverse transcription of the transcriptome (the organism's RNA), giving rise to complimentary DNA (cDNA).

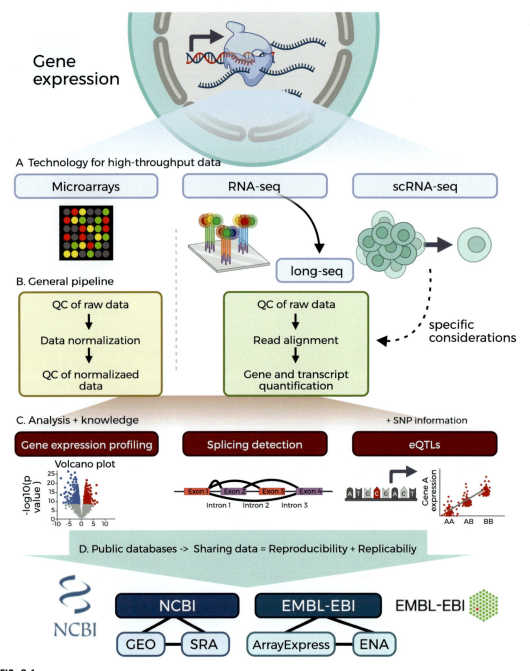

FIG. 8.1

Overview of transcriptomic technologies, their general analysis pipelines, and applications. (A) Microarrays and RNA-seq are the major transcriptomic technologies; RNA-seq has been paired with new approaches to allow for single-cell RNA-seq. (B) General analysis pipelines corresponding to each of the transcriptome technologies. (C) Major applications of transcriptomics to generate biological knowledge. Gene expression profiling aims to identify which genes are relevant for a specific condition. Splicing detection aims to identify the different isoforms present in specific conditions. eQTL analysis aims to identify the relation between genetic variation and gene expression. (D) Published transcriptomic data are available through different public resources so that it can be viewed or repurposed for new studies.

No permission required.

RNA sequencing has wider-reaching applications resulting from the development of new technologies such as long-read sequencing and single-cell sequencing. A critical challenge facing high-throughput experiments is reproducibility, as variation can come from many external variables, including interlaboratory differences, platform utilized (microarray or sequencing), and different analysis pipelines. The following sections review these techniques, the challenges faced by each, and suggest points of consideration for maximizing the impact of one's research by ensuring reproducibility.

High-throughput techniques to assess gene expression
Microarrays—What are they, and how are they conducted?

Microarrays were first developed in the 1990s and continue to be used due to their low cost and method maturity. In general, the technology consists of measuring specific transcripts' abundances through mRNA hybridization with complementary probes in an array. The first historical paper demonstrating the great potential for microarray technology was published in 1995 [4]. Shena and colleagues used 45 complementary cDNA probes of *Arabidopsis thaliana* to hybridize with mRNA simultaneously; the paper demonstrated that using robotic printing made it feasible to scale up the microarray plate preparation process.

Microarrays can cover entire genomes or just a subset of genes specific to a certain event (e.g., apoptosis) or pathway; custom microarrays can be designed. In general, splicing is not considered in a microarray assay, but some versions can detect splice variants (see "Splicing detection" section). Microarrays are most commonly used for exploratory/basic science research and in clinical practice for diagnosis or decision-making [5], but they also have uses in agricultural settings.

Microarrays probes are spotted or directly synthesized onto a glass or silicon surface; presently, it is possible to cover up to 6 million probes in a single chip. Probes are usually localized at the 3′ region of the transcript, requiring a priori knowledge about the genome. Hybridization between the sample cDNA and microarray probes produces a fluorescence intensity that correlates with the abundance of the transcripts. The experiment should have a control (or reference) sample to compare with the experimental sample (e.g., treatment, knockout, disease); this comparison is made using one- or two-color arrays. A one-color array requires a new microarray plate for each sample. A two-color array uses a single microarray plate, but samples are labeled with different fluorescent tags (Cy3, Cy5), permitting comparison between the samples.

Analysis of microarray data usually starts with raw expression measurements. These are expressed as a matrix of numbers that indicate the intensity of the hybridization signal of each sequence probe, which can be translated to gene expression. Data can be analyzed using proprietary manufacturer software, or using an open-source software such as Bioconductor packages [6,7], which has a variety of packages developed for different steps and platforms [8]. The basic steps for analyzing microarray data are outlined in Fig. 8.1B and detailed here.

1. **Quality control of the raw data**: First, it is necessary to perform a quality control analysis of the raw data. This step is critical as introducing poor-quality data will lead to unreliable results. Exploratory analyses such as boxplots of data intensity and principal component analysis (PCA) can highlight issues in the experimental procedure. Depending on these figures, one may choose to exclude certain sample (particularly large outliers) before the normalization step.

2. **Data normalization**: This step is essential to make samples comparable and eliminate technical variation. The normalization will change depending on the platform used.

For Affymetrix chips, RMA (robust-multiarray average) is the most common algorithm used, which implements this strategy:

   (i) Background correction,
   (ii) Quantile normalization, and
   (iii) Summarization with median polish, which results in log2 transformed values [9].

For Agilent chips:

   (i) Background correction is calculated using median and
   (ii) Global loss normalization on individual arrays or quantile normalization between arrays needs to be calculated before performing differential gene expression analysis [10].

3. **Quality control of the normalized data and gene annotation**: After normalization, it is recommended to repeat step 1, comparing output of the normalized data with the raw data. It is important to verify that medians have the same expression level. One should also check if there are batch effects or errors introduced by groups of samples being processed at different times or places (see "Batch effect" section). Annotation files will be required to assign the probe to the respective gene.
4. **Define the contrast model**: After quality control has passed, it is necessary to define which samples (conditions or treatments) will be compared, which can be done using limma [11]. A table describing which samples correspond to which experimental condition or treatment will be required. If there are additional covariates (i.e., age, time, sex), the model can be used to describe them [11,12].
5. **Differential gene expression**: Limma package implements the empirical Bayes model to fit a linear model and moderates the standard errors of the estimated log-fold changes. The output will contain the list of genes with their log-fold change, *P*-value, and adjusted *P*-value, which corrects for false positives by using a multiple testing correction on the *P*-value. Then, adjusted p-values can be used for filtering genes with a defined cutoff, and the log-fold change will allow observing if there are large fold changes in expression. With this list of genes, the functional relevance can be evaluated through a Gene Ontology enrichment analysis using a package like enrichR [13,14].

Disadvantages of microarray analysis

Microarrays have limited sensitivity in the low-expression and high-expression ranges; in these cases, RNA-seq provides more accurate quantifications. Additionally, microarrays suffer from two distinct limitations: technical noise and background hybridization affecting expression measurements.

Technical noise

When analyzing microarray data, the main challenge is technical noise. This noise comes from manipulation errors of the microarray itself which can derive in intensity differences that are not due to gene expression changes. In most studies, there are few replicates to help disentangle this noise, amplifying the problem. Different methods for dealing with technical noise have been proposed for each step of quality control, normalization, and differential expression analyses; most of these have already been implemented on Bioconductor packages (another great reason to use these packages!) [6,7].

Background hybridization
The accuracy of the expression measurements can be affected by background hybridization of transcripts with low abundance or inclusive saturation for those transcripts with high abundance; as mentioned before, this is a limitation of microarrays, as a limited number of probes is available and RNA from highly expressed genes will saturate them quickly, and the rest of the RNA can unspecifically bind the surface or other probes. Probes are different in their hybridization properties, and small mismatches between the probe and the target molecule can exist; improving specificity or removing probes binding to off-target sequences can help mitigate background hybridization, but also risks having information lost. This latter point is particularly relevant in species where genome variation is not established.

Sequencing
Next-generation sequencing
NGS refers to the series of platforms that started being developed in the mid-2000s. These platforms had a significant impact on the Human Genome Project, enabling cost reduction with increased sequencing throughput. Before NGS, the major sequencing technique was Sanger, a technique still considered the gold standard, as it can produce long sequences with low error rates [15]. The technique is still widely used to confirm genetic variants, particularly for disease research or diagnosis.

NGS sequencing platforms are based on parallelization, where thousands or millions of sequencing reactions are happening simultaneously. In the year 2000, massively parallel signature sequencing Lynx technologies launched the first NGS platform, which was then purchased by sequencing giant Illumina. Next, in 2005, a technique based on pyrosequencing, where DNA polymerase activity was measured based on the detection of pyrophosphate [16], was commercialized as the 454 platform. Also during that year, Solexa released a platform with technology based on sequencing by synthesis using a reversible dye terminator. Solexa was acquired by Illumina in 2007. Over the years, different technologies and innovations have been incorporated into NGS techniques. Platforms vary in their cost, throughput, and read length. Several reviews further describe the details of the different chemistries used in commercial sequencing platforms [17–19]; readers are encouraged to review these papers.

Currently, sequencing technologies can be categorized into two major groups: sequencing by ligation and sequencing by synthesis.

Sequencing by ligation
As the name states, this procedure involves ligating labeled probes and anchoring sequences to the cDNA. Labeled probes are small sequences (eight nucleotides) that have a fluorophore attached; this sequence will bind to the cDNA we want to measure, and the attached fluorophore indicates the sequence composition. Later on, a second labeled probe will bind next to the previous one and promote ligation; this reaction will release the label, and this reaction will be repeated to sequence the template cDNA. SOLiD [20] and BGISEq. [21] sequencing platforms are the two most widely accepted sequencing platforms; however, SOLiD has been discontinued, and BGISEQ is still under development and has not been launched for commercialization yet. At the time, no other commercial platform offers technology based on this principle.

Sequencing by synthesis

Sequencing methods that rely on DNA polymerase are considered "sequencing by synthesis." Illumina and Ion Torrent are the major commercial platforms used in NGS based upon this technology. NGS applications are broad; the technology has been adapted to characterize not only an organism's genome [22], but also transcriptome (RNA) [23], or epigenome (regulatory sequences) [24]. A significant advantage of RNA-seq for transcriptomic examinations is that it enables the detection of transcripts without predetermining targets, as is required for microarrays. This unbiased approach has made the detection of new transcripts possible [25], and specifically, it has facilitated the study of isoforms [26], novel noncoding RNAs [27], and genomic variants.

A major drawback of NGS is the high error rate (ranging from 0.1% to 1% of bases being miscalled) and the short read lengths (50–400 bp); new developments allow for long-read sequencing, which will be discussed later in this chapter. **Read length** refers to the number of nucleotides that can be determined from one sequence; the reads generated by a single sequencing assay will be of the same size (length), on average. **Error rate** can be compensated for by increasing the number of sequences, which is known as sequencing depth or coverage—the number of times (or average of times) a unique sequence represents a nucleotide from the genome, i.e., 10 unique sequencing reads can contain a nucleotide from the genome and confirm its identity.

Adaptations have been made to RNA capture protocols to facilitate the identification of transcription start sites; one of these modifications is the cap analysis of gene expression (CAGE) analysis [28]. The aim of CAGE is to select complete transcripts for sequencing. A 5′ methyl-guanosine cap is added to the RNA and retrotranscribed from the poly-A tail. The cap allows for the RNA template to remain complete while indicating, via the cap, the original transcription start site, enabling targeting sequencing to these sites. CAGE has been extensively used in combination with sequencing platforms by the FANTOM consortium [29] to identify transcription start sites across human and mouse tissues and cell types.

The general workflow for RNA-seq data analysis is as follows (detailed in Fig. 8.1B):

1. **Quality control of the sequencing data**. Each sequencing platform will have particular sources of noise. For this reason, it is important to check the general quality of the obtained sequences: base quality, CG content, presence of enriched k-mers (usually, adapters from the sequencing platforms), read length, sequence duplication levels, etc. FastQC [30] is an easy-to-use software that can facilitate this task.
2. **Read alignment**. Each read obtained from sequencing must be "aligned" to the reference genome, to identify its genomic source. Many aligners are available for RNA-seq analysis, and development is ongoing [31]. Take into account that each tool has been designed to help answer a particular question, so be sure to understand which aligner will fit the specific research question being asked. Commonly used aligners include Bowtie2 [32], Kallisto [33], STAR [34], and Salmon [35].
3. **Gene and transcript quantification**. Expression of a gene can be interpreted from the number of times reads align to a gene or transcript. Some aligners, such as Kallisto, Salmon, and STAR, can also perform this task. These counts are usually normalized to allow for comparisons. Some commonly used normalization procedures are reads per kilobase million (RPKM), fragments per kilobase (FPKM), and transcripts per million (TPM). A benchmark on methods to assess the performance of RNA quantification methods is presented by Teng et al. [36].

4. **Differential gene expression**. Once gene expression is quantified, one of the analyses usually performed is comparing expression of genes or transcripts across different samples. Many tools are available to perform these comparisons; a few prominent tools include edgeR [37], DESeq2 [38], and limma [11].
5. **Additional analyses**. RNA-seq data can provide the opportunity to answer many biological questions. Coexpression networks give information regarding gene regulation [39]. Clustering analysis gives insight into shared pathways or mechanisms between genes or samples [40]. Comparing gene expression data across species using orthology information can give insight into key conserved pathways or mechanisms [41].

There are some available pipelines to ease the task of integrating these analysis steps such as RASflow [42], VIPER [43], and BioJupies [44]. Additionally, Galaxy, a web platform, provides the possibility of using RNA-seq analysis pipelines without installing software or updates [45].

Long-read sequencing

Each cell has an enormous diversity of RNA molecules: messenger RNAs can come from different RNA isoforms, and noncoding RNAs can vary in length between short (~10–40 bps) [46] and long (1–10 kbs) [47]. In fact, some messenger RNAs can be 1–2 kilobases (kbs) or longer [48]! Messenger and long noncoding RNAs (lncRNAs) can also undergo alternative splicing (AS), further increasing RNA diversity.

A major limitation of NGS platforms, such as Illumina, is the necessity for RNA to be converted to cDNA and cut it to lengths of around 300 base pairs (bps) to enable library preparation [49]. This has imposed a limit on our capacity to study full transcripts and hence, RNA isoforms.

Studies aiming to identify isoforms using NGS technologies have been enabled through transcriptome analysis tools [50]. Some of these tools are focused on matching reads with similarities, like a puzzle, to identify full transcripts. This is known as transcriptome assembly and has become a key analysis framework to study nonmodel organisms as it does not require having a sequenced reference genome [51]. Nevertheless, there are limitations to these analyses: one must correctly identify the exons that belong together in the same molecule versus separating isoforms with different exons. Long-read sequencing technologies can overcome these challenges.

How does one overcome read length challenges?

The third generation of sequencing technologies refers to the ones that enable long-read sequencing in the range of kilobases. The two primary technologies currently used in this field are PacBio (read lengths of up to 15 kbs) and Oxford Nanopore (reads >30 kbs). PacBio sequencing is based on synthesis; a DNA polymerase is fixed to a zero-mode waveguide (ZMW) well, each nucleotide has a florescent dye attached, the polymerase will replicate the template DNA, and every time a nucleotide is added, the ZMW can measure the fluorescence and identify the base. On the other hand, Oxford Nanopore sequencing operates by measuring electrical disruption as each base passes through the patented Nanopore—in doing so, the technology allows for direct RNA sequencing, without the need to convert to cDNA, reducing biases caused by this step.

A notable drawback of long-read sequencing technologies is the low-throughput nature of the assays. While NGS technologies output millions of reads per run, long-read technologies reach at most 20 gigabases with Oxford Nanopore and a mere 700,000 reads for PacBio sequencing. Furthermore, long-read sequencing still encounters high false positives—PacBio currently reports ~15% error rates compared to ~0.1% observed in Illumina. As such, long-read sequencing is primarily employed for niche applications where such technology is required (e.g., sequencing long tandem repeat tracks).

Toward reducing the high error rate in Oxford Nanopore sequencing, one can add redundancy by sequencing the same chain repeatedly; this is the same rationale behind genome coverage and by seeking reproducibility, improves the accuracy of the sequenced product. One of the strategies used in PacBio is circular consensus sequencing (CCS) [52], where, through circularization of the template, the sequence is read twice, and a consensus is created. However, CCS conveys a reduction in the length of the obtained reads, and new algorithms are being developed to propose new barcode designs that will allow for redundancy without affecting the sequence length [53]. Other computational approaches to reduce error rates in long-read sequencing incorporate data from short read RNA-seq data [54,55].

The baseline protocol to analyze long-read sequencing data includes the following general steps:

1. **Identify full-length transcripts**. Transcripts are identified based on read clustering; in this way, similar reads can be integrated to confirm one transcript. Genome information can be taken into account to correct transcript annotation [55].
2. **Quantify transcripts and genes**. Reads are assigned to transcripts, and the number of reads per transcript is converted to transcript per million to normalize.

Even with their current limitations, long-read sequencing has proven to be useful for sequencing full-length transcripts in human cell lines [56], to detect variation in isoform expression in human primary cells [57], and to survey full-length transcripts in nonmodel organisms [58]. Computational methods continue to be developed to improve data analysis [59]. Nevertheless, RNA-seq based on NGS technology will remain the standard for the coming years, particularly in clinical applications, as the technology has matured, and computational methods for analysis are well established.

Applications

Regardless of the method used to measure gene expression, profiling of RNA molecules can have diverse downstream applications. In the next section, an overview is presented of the most common applications: gene expression profiling, splice site detection, expression quantitative trait loci assessment, and identification of single-cell transcriptional profiles (Fig. 8.1C).

Gene expression profiling

Gene expression profiling refers to the measurement of gene expression activity in thousands of genes simultaneously. This can be done in samples coming from tissues, cell cultures, or other sample types, and it has been primarily used in research, but it has shown to have clinical potential [60].

Depending on the tissue, cell type, treatment, or health/disease condition, cells will have different gene expression profiles, meaning different sets of genes will be expressed or repressed.

Gene expression profiling studies usually aim to identify sets of genes that are changed by disease status, environmental factors, or other variables (tissue type, age, and so on), or that are unique to a particular state. In order to be able to identify changes in gene expression across samples, it is important to properly plan and identify the relevant questions to be answered. High-throughput methods can have high technical variability, and this variability can have an impact on the results and could cause errors in interpretation.

As discussed earlier in this chapter, there are several platforms that can be used to measure gene expression, but each of them has its particular pros and cons, so defining the question to be asked is very important for selecting the most suitable method. For example, the best sequencing set up to detect lncRNAs is not the best for detecting polyadenylated mRNAs that code for proteins; lncRNAs are best detected by using a combination of Illumina sequencing with PacBio, while for the latter, an Illumina approach with poly-A capturing would suffice. For lnRNAs, PacBio allows for full transcript identification, as these genes are poorly annotated, long-read sequencing is required to have this information, and Illumina read can help correct sequencing errors common in PacBio. Protein-coding genes are usually well annotated, and poly-A capture help focusing sequencing only on these genes, optimizing the costs.

Once a platform that best matches the requirements is selected, it is time to plan the experiment or sampling that is required. Here, it is very important again to remember that technical variability can come from the platforms selected. This includes sample processing that can cause batch effects. To this end, it is important to avoid processing samples at different times—all samples should be processed identically at the same time points—or batching together samples that have different biological features. For example, treatment samples and controls should ideally be processed at the same time, on the same day, and in the same location to avoid technical differences that could be interpreted as biological ones. In an ideal world, all experimental work would be conducted so rigorously; however, this is not always possible. In the event that batch effects might be introduced to one's data, there are methods discussed later in this chapter for addressing and minimizing the batch effect.

In general, gene expression profiling analysis has the following steps:

1. **Quality control (QC) of gene expression data**. Depending on the platform used (sequencing or microarray), it is important to evaluate that the obtained measurements can be used to proceed. See earlier discussions of QC.
2. **Gene and transcript quantification**. Each platform will provide a measurement of gene or transcript quantification. Microarrays present this as signal intensity; sequencing outputs present it as number of reads sequenced for a gene or transcript. These measurements constitute the first gene expression profile. However, these have to be corrected for any potential confounders or batch effects (see "Batch effect" section), so they can be correctly interpreted or compared.
3. **Differential gene expression**. Using corrected gene expression profiles, it is possible to compare gene expression across samples.
4. **Visualization**. Heat maps are one of the classic displays used to show gene expression profiles. A heat map is a matrix that contains expression values per gene per sample/condition. Volcano plot is another graphic representation of gene expression that has on the x-axis the logFC, and the y-axis the p-adjusted value in $-\log 10$ scale. To visualize only one gene, box plots are useful, and the x-axis will have the categorical variable (e.g., treatment, or disease stage), and the y-axis gene expression.

Reproducibility can be a challenge in gene expression profiling analysis. As described before, batch effects are a common downside to these experiments, and this can affect reproducibility. Nevertheless, metaanalysis approaches can be used to account for variations that come from the experimental origin (i.e., different laboratory and/or year), allowing a comparison of results that come from different studies (for a more detailed description of these procedures, see "Batch effect" section).

Gene expression profiling has been widely standardized, allowing it to be used to inform clinical practice [61], and it has become one of the baseline analyses in transcriptomics, allowing scientists to assess biological and clinical questions. Currently, gene expression profiling has been included as an standard analysis in major databases like the UK biobank [62] and the Cancer Genome Atlas (TCGA) [63] and has been used to achieve a better understanding of the biology of human traits and diseases. Moreover, gene expression profile data are widely available for other organisms, which has opened the door for comparative genomics analysis, where conservation of gene expression levels is assessed. Comparative genomics gene expression has shown that in general, mRNA patterns are shared across species in a given tissue [64]; however, when splicing patterns are compared, it is possible to observe that there is organism-specific variation [41].

Splicing detection

Genome sequencing revealed that cellular complexity has poor correlation with the surprisingly low number of protein-coding genes. It was subsequently determined that alternative splicing (AS) is used as a mechanism to create diversity, which may lead to increased complexity [65]. AS is a biological mechanism that generates multiple RNA isoforms from a single gene by selecting different exons, and sometimes retaining introns. This mechanism allows for protein diversity and even opposing functions to be encoded within the same genomic locus [66].

Multiple isoforms from the same gene have been recognized in eukaryotic genomes such as vertebrates, plants, and fungi, even unicellular organisms (i.e., *Sphaeroforma arctica*) [67]. This process is highly conserved and has been tracked to the last eukaryotic common ancestor (LECA) [68]. In humans, AS plays a crucial role in proteome diversity: 92%–94% of protein-coding genes in humans go through this process [69]; it is tissue-specific [70,71], relevant in organ development [72], and has clinical applicability in human diseases, such as cancer [73]. A large and historic database to visually search for AS variations in humans is SpliceSeq, a tool with RNA-seq data that facilitates visualization graphs. More recently, the same research team developed a resource with TCGA data to explore splicing events using a web-based tool [74].

Classification of AS events vary in definition; however, the most common categories of AS include: inclusion or skipping of exons, intron retention, alternative 5′ or 3′ splice site events, differentially untranslated regions (UTRs), and alternative first and last exons. These classifications are furthered described by Wang and colleagues [69,75].

Exon skipping isoforms are particularly relevant in animals and have been shown to participate in regulating protein–protein interactions. Similarly, intron retentions have been associated with downregulation by controlling nonsense-mediated decay (NMD), a process that regulates mRNA levels with premature stop codons [72].

Current research based on the identification of AS has been applied to cancer data in the TCGA consortium [73]. Using annotations from the SpliceSeq databases, Zhang and colleagues were able to predict prognostic signatures for 31 cancer types using AS, this means that prognosis for patients with certain types of cancers can be predicted by AS patterns [73]. This shows the impact of AS in disease mechanisms and how it varies across individuals.

Alternative splicing in microarrays

Evaluating AS events using high-throughput microarray technology was swiftly taken up by the scientific community. Affymetrix developed microarrays with specific probes that recognized spliced exons, introns, and exonic junctions. While this technology allows for rapid screening, it is limited by probe design, as the probes are based on current reference genome annotations, and thus may not capture the full complement of spliced variants. Affymetrix tried to compensate for this by predicting possible new splicing regions to help de novo discoveries, and as consequence, probe sets where classified into three groups [76]:

- Core (based on RefSeq and GenBank annotation),
- Extended (core probes plus EST and partial mRNA annotation), and
- Full (extended plus ab-initio predictions).

In addition to the limitations highlighted earlier, microarrays use short probes, which are between 25 and 60 bp depending on the platform [77]; short probes are less specific as the small sequence can be found with higher probability by chance, in particular, across very similar isoforms. In some cases, unique matches in small exons were challenging to design and had [78] a validation rate of 33%–86% [79]. Despite its disadvantages, microarrays have successfully been used to uncover AS events in different conditions, such as tissue-specific variants [80], cancer [81], and dioxin exposure, [82] among others.

The general pipeline for analyzing AS with microarrays is similar to the analysis of microarrays for gene expression (see earlier) with the following changes:

1. Quality control of raw data.
2. Normalization.
3. Quality control for normalized data.
4. AS analysis. Different methods have been developed to detect AS events using microarrays. Packages include Splicing Index [78], FIRMA [83], MADS[84], MiDAS [85], ARH [86], and many others.

Alternative splicing event detection using RNA sequencing

RNA sequencing improved AS analysis dramatically; it has been used to confirm alternative isoforms and find new isoforms as well. This relatively unbiased technology gives the possibility to analyze transcriptome data without needing prior knowledge to design probes. While de novo isoform discoveries have been made possible by RNA-seq, challenges to applying the technology exist:

- Aligning ambiguous reads from 100 to 150 bp in some genomic regions is problematic because of nonuniqueness of the sequence, particularly repetitive sequence [87].
- Transcript isoform abundance also partakes in AS detection accuracy, as highly expressed isoforms have comparable accuracy across different methods, but this decreases with low abundance transcripts [88].

Overall, RNA-seq enables better analysis and understanding of AS events compared with microarrays (for a thorough comparison of these factors, see tables in Refs [3,89]), but experimenters must take into account biases highlighted earlier when studying differential splicing.

The following is the general pipeline to analyze AS events with RNA-seq data:

1. Quality control of sequencing data.
2. Read alignment. RNA-seq read alignment can be classified depending on the method employed:

a. **Reference genome**. Able to detect novel transcripts, computationally expensive. Reads are mapped to a reference genome (i.e., TopHat, STAR, HISAT2), and then another program is used to quantify transcript abundance (i.e., Cufflinks) [90].
   b. **Alignment-free or pseudoalignment**. Faster than aligning to a reference genome, but only recognizes known transcripts (i.e., Kallisto, Salmon)[33].
   c. **De novo transcript assembly**. Helpful for unreliable reference genomes, unexisting references, and distinguishing new transcripts (i.e., Trinity, Oases, DiffSplice) [91,92].
3. Gene and transcript quantification.
4. Differential AS analysis. After calculating read counts, differential analysis is performed to discover AS variants across different conditions. Comparison of the methods can be reviewed in Refs. [91,92]. The following are the two methods suggested for this step:
   d. **Isoform-based** methods reconstruct full-length transcripts and then quantify the abundance for comparison among groups (Cufflinks [93], DiffSplice [87], Sleuth [94]).
   e. **Count-based** methods include *exon-based*, which measures reads falling into an exon or junction region (limma [11], edgeR [37], and DESeq2 [38]), and *event-based* analysis, which calculate the fraction of spliced event types (spliced exons, retained introns) and then compare ratios between groups (rMATS [95], SUPPA [96]).

Aside from taking into account if a reference genome or transcriptome is available, experimental design must be considered when selecting an AS quantification methods [91], in particular, when the design requires comparing more than two groups of samples (i.e., time series experimental designs). Different benchmark studies have compared these methods [36,50,91,92]. The experimentalist is advised to understand all parameters used by each method and adapt them accordingly, as choosing default values may not model experimental design adequately, notably altering results [97].

Representing AS events visually

Representing AS events with a visualization plot can be challenging because of the complex nature of this information. Heatmaps are one of the classic displays used to show gene expression profiles, but this is not useful for accurate AS representation. In AS, the most common visualization strategy is to use sashimi plots [98] and box plots with expression per exon. Strobelt and colleagues have also developed a visual analysis tool for AS exploration with different alternative visualization graphs [99].

With single-cell data (discussed later), new methods are under development [100], and tools such as Salmon, RSEM, Kallisto, among others, have shown similar performance in scRNA-seq and bulk RNA-seq [101]. Technical obstacles remain to be addressed to study alternative isoforms in single-cell data [102].

Expression quantitative trait loci assessment

Genetic variation refers to the loci in a genome that can differ across members of a population. Two humans selected by chance from the population will have genomes differing by approximately 0.1% (only identical twins share the same genome). This means that about 1 in every 1000 bps will differ between any two individuals [103]. Genetic variation comprises single nucleotide polymorphisms (SNPs—changes of one nucleotide between two individual, i.e., A>T); insertions and deletions of nucleotides (indels); and copy number variations (CNVs), which is the repetition in tandem of a chain of nucleotides. Variation can be assessed either by microarrays or by whole-genome sequencing.

Genetic variation is largely predictive of health and disease status. This is widely addressed by Genome Wide Associations Studies (GWAS) that establish an association between genetic variants and human traits [104]. It has been reported that more than 90% of the SNPs that have been associated with traits or diseases in humans through GWAS are located in noncoding regions [105]. These variants have been related to changes in gene expression, probably affecting transcriptional regulation, as they are commonly found in open chromatin regions where regulatory proteins tend to bind [106].

An SNP in a noncoding region with a potential regulatory function associated with gene expression is known as expression quantitative trait loci (eQTL) [107]. In order to establish the relation between an SNP and gene expression phenotype, we require a direct association test between genetic markers and gene expression levels. This analysis typically requires tens or hundreds of individuals.

eQTLs lay a connection between the genetic background and cell function, so as gene expression can be tissue-specific, so also can eQTLs be tissue- and cell-type specific (GTEx Consortium, 2020). Moreover, eQTLs can also be population-specific; thus, ancestry has to be taken into account. For example, a study performed in colon samples of a Han Chinese population found 5940 eQTLs, from which only 21.4% had been previously reported in other populations [108].

The Genotype-Tissue Expression (GTEx) project launched in 2013 [109,110] is the biggest resource of eQTL mapping in the human genome, with data for 15,201 RNA-sequencing samples from 49 tissues and 838 postmortem donors. GTEx measures tissue-specific gene expression and identifies eQTLs in humans. These results have been harnessed in GWAS research to establish links between diseases and tissues [111]. Moreover, eQTLs can be leveraged to establish biological mechanisms related to GWAS-associated variants [112].

In general, an eQTL mapping study requires three things:

1. <u>Gene RNA levels</u>: transcriptome data from tens, hundreds, or thousands of participants [113–115]. Gene expression analysis can be performed as described previously.
2. <u>Genotypes</u>: SNPs in a genome of the same individuals from where transcriptome data were obtained. The data have to undergo its own quality control; methods will differ depending on whether it was obtained from genotyping arrays of whole-genome sequencing. This subject is out of the scope of the chapter.
3. <u>Covariates</u>: Information from participants such as sex, age, ancestry, health status, etc.

This information will be used to determine associations between SNPs and genetic expression and then later interpreted as a molecular trait. This association is possible based on the fact that gene expression, measured as RNA abundance, is heritable [116].

To establish the relation of an SNP loci to gene expression, the correlation between the expression of a gene and the alleles present in a group of people has to be assessed. While this sounds easy, it is important to take into account the genetic composition of the population (i.e., local ancestry and global ancestry) [117] and the established covariates that could bias the analysis. Several tools have been designed to perform this analysis, e.g., Matrix eQTL [118], QTL Tools [119], and FastQTL [120].

There are some considerations to be taken into account when performing eQTL mapping that could lead to false discoveries. We will mention the most relevant ones:

1. Allele frequency in the population can bias results. This means the number of individuals should be higher if the variant is present in low frequency.
2. Another common issue is the *winner's curse*—the true genetic effect of an eQTL is lower than the estimation obtained from the sample population, which can lead to replication problems [121].

As stated before, eQTLs are variants associated with gene expression. Nevertheless, this association does not imply causality, and the identification of the causal variant requires fine mapping. eQTL variants are proxies of variants that are in linkage disequilibrium, meaning that variants within a loci are inherited together more times than expected by chance. There are different tools to infer causal variants, such as CaVEMaN [122], CAVIAR [123], and DAP-G [124].

eQTLs are a useful resource to identify the plausible regulatory mechanisms affecting the expression of a gene in a given tissue or cell type. With the lowering cost of genotyping and RNA sequencing, this tool has become broadly accessible and will enable a better understanding of the interplay between genetics and function.

Single-cell RNA-seq

Single-cell RNA sequencing (scRNA-seq) technology enables the experimentalist to obtain transcriptome profiles per cell, permitting improved resolution of the molecular events inside a cell (as compared with bulk RNA sequencing). The first experiment was used in a four-cell stage blastomere [125]; a decade later, numerous methods have been developed and standardized, which will be expanded upon later.

Single-cell RNA-seq is applied to resolve cell type subpopulations, evaluate the heterogeneity of the cells, or understand the dynamics of biological processes. In general, they share some steps: (1) isolation of single cells, (2) reverse transcription, (3) cDNA amplification, and (4) sequencing library preparation [126].

A crucial step in scRNA-seq is the isolation of the cells because it is essential to maintain RNA integrity. RNA isolation could be done with different experimental methods, and the most used are enzymatic treatment, laser capture microdissection, and patch clamping.

Different numbers of cells can be used; the range will depend on the type of research, e.g., if the focus is to study the development of an embryo, we can start with less than a hundred cells; and if our interest is to study a human biopsy, we could have 10,000 cells per sample. Micropipetting or patch-seq are used to capture less than 50 cells [127]. Fluorescence-activated cell sorting and microfluidic approaches can be used to obtain a higher number of cells (1000–10,000) [128,129]. It is also possible to perform scRNA-seq of nuclei using FACS. In this case, it is obtained from unprocessed mRNA, which is less compared to the mRNA of the whole cell. The big challenge with this experiment is to remove the cytoplasm.

Once RNA is obtained, then reverse transcription can be used to synthesize cDNA. Usually, the protocol uses oligo-dT priming to avoid ribosomal RNA and select mRNAs, and some noncoding RNAs. cDNA can be amplified using SMART technology that switches the mechanism at the 5′ end of the RNA template. Another option could be ligating the 5′ end of cDNA with poly(A) or poly(C) to build common adaptors for PCR amplification. Here, the problem could be the generation of making short amplicons shorter and with less G-C content. To solve this, CEL-sEq. [130] or MARS-sEq. [131] method can be used to perform in vitro transcription.

After this step, there are different methods of sequencing: (1) those based on full length where the full coverage is obtained and (2) those based on 5′ or 3′ tags. scRNA-seq methods based on tags can be combined with unique molecular identifiers (UMIs) to quantify transcript molecules, but reads obtained through this method cannot be used for splicing detection. So, depending on the biological question, one method could be better than the other. For projects interested in discovering cell types and knowing cell composition, a method based on tags could be enough and can help to reduce costs. When comparing methods, Smart-seq2 is better in sensitivity and reproducibility, but for a large number of cells, a technology like Drop-seq could detect up to 4000 genes [126].

When analyzing single-cell data, there are some challenges that need to be considered. First, the matrix with the genes in rows and cells in columns is going to be full of zeros, and high cell-to-cell variability that can be caused by technical and biological noise. Particularly, a cell can be in a specific cell cycle, in the transition to a different cellular state, have a specific size, and have a stochastic gene expression. Like other experiments, it is possible to have batch effects that can be detected through some methods such as PCA and can be corrected using multiple batch correction methods. It is also possible to add spike-in RNA standards of known abundance to the endogenous samples (e.g., spike-ins of the RNA Control Consortium ERCC), which can be used for quality control or normalization.

Public databases
Databases

Public biological databases are helpful to share data and promote open science by organizing and making data available. The following are the main public databases storing transcriptome data:

1. The National Center for Biotechnology Information (NCBI)
2. The European Molecular Biology Laboratory's European Bioinformatics Institute (EMBL-EBI)
3. The DNA Databank of Japan (DDBJ)

These databases often share data between them, provide links to interconnect information, and store different genetic, transcriptomic, proteomic, and other biological information. This has led to reorganizing each database into more specialized repositories [132] that can store raw and preprocessed high-throughput data.

The particular repositories to store functional genomic projects from high-throughput experiments (i.e., microarrays and sequencing data) are **Gene Expression Omnibus (GEO)** [133,134] from NCBI and **ArrayExpress** from EMBL-EBI [135,136]. With the development of NGS technologies, the amount of data has increased exponentially, and specific databases for efficiently storing raw data and alignment information had to be developed. As a consequence, the **Sequence Read Archive (SRA)** [136] and its analogous form in EMBL-EBI [135], the **European Nucleotide Archive** [137,138], were created to meet this need (Fig. 8.1D).

Both databases, SRA and ENA, contain high-throughput DNA and RNA sequencing data of different organisms, metagenome studies, and environmental impact surveys. It is essential to highlight the importance of raw data for reproducibility, as results replication highly depend on this. We will further discuss this issue in section "Key point reproducibility, replicability, robustness, generalizability."

NCBI
GEO-NCBI

GEO has been a useful free resource to archive genomic high-throughput data. We can find gene expression experiments using microarrays or RNA-seq, but also methylation, genetic variation (SNPs), noncoding microarrays or sequencing, as well as immunoprecipitation studies for analyzing chromatin accessibility and protein analysis, among others. This repository not only aims to store raw and/or normalized data, but also to be up-to-date for the scientific community, and to provide an interface that allows to query, detect and download data.

GEO can be overwhelming for new users due to how the data are stored, the amount of data it contains, and the lack of uniformity to annotate the information. GEO organizes the information by using three main records, and each of them has its own ID. The following are the three records:

a. Samples (GSMxxx): This record corresponds to a unique sample element, and it is used to describe the background, conditions, experimental details, and specific descriptors of the biological sample, such as disease status, treatment, age, sex, culture media, temperature, etc. This ID contains preprocessed data, and the normalized method should be indicated. Optionally, a supplementary file with raw data can be included. The identification number is constructed by GSM plus numbers (i.e., GSM4565413) that will be connected to only one platform (e.g., Affymetrix) and can be part of multiple experiments (or Series).
b. Series (GSExxx): This record links samples, and it is used to give a global description of the experiment, such as the summary, analysis, and finding of each study, as well as the aim, general methodology, number of samples used in each comparative group, person in charge and contact information, submission date, study location, etc. This ID is unique and can contain raw data, meta information from all samples, or any additional file required for the experiment. Typically, each series will group samples from the sample platform and later assign another GSE ID. This record is called a SuperSeries, and it can indicate an experiment that uses different platforms (i.e., when a sample was measured using gene expression arrays and methylation arrays, check GSE56342 as example). However, there are no rules on how to use SuperSeries identifiers.
c. Platform (GPLxxx): A GPL ID corresponds to a unique identification for a platform used to measure high-throughput data in a sample (i.e., an specific Affymetrix array). It contains a description of the technology used, number of measured features, as well as manufacturer information, such as company, year, etc. and can have tables with probe information. It is particularly useful for getting back gene names from array probes.

In addition to these records, there is another type which tracks GSE IDs (or series experiments) that have been curated by the NCBI team and corresponds to GDS ID. These types of IDs can be used for tools provided by NCBI, such as gene expression profile charts and DataSet clusters, but this advantage for using additional tools is limited to specific curated data sets.

It is important to understand how public databases store high-throughput data, but it is also relevant to know how to download all this information. The following are some useful ways to do this process:

a. Website: The normalized data will be accessible for either microarrays or RNA-seq and raw data for microarrays can be found as supplementary data either as an experiment using GSE or per sample using GSM. However, be aware that some experiments will not provide that information. It is also possible to use the utility GEO2R for a quick analysis to search the expression of specific genes or probes and obtain a general behavior of data; however, this analysis is lacking the visual inspection of the chips, an important quality step. Raw data for RNA-seq can be downloaded from SRA (see later).
b. *E*-Utils: A tool for downloading data through a programmatic access [139].
c. ftp: Data can be downloaded using an ftp structure such as ftp://ftp.ncbi.nlm.nih.gov/geo/series/GSE1nnn/GSE1000/suppl/GSE1000_RAW.tar; for more information, check the tutorial [139].

d. GEOquery: A package from Bioconductor that will download GEO data into R [140] environment.
   e. GEOparse: A python package analogous to GEOquery to download GEO data into python [141].

Sequence read archive (SRA)

NCBI has a special repository for raw high-throughput sequencing data called the Sequence Read Archive (SRA) [136]. The aim of SRA is to make data available for reproducibility and facilitate collaborative work [142,143].

For downloading SRA data, you can use:

   a. Website: SRA website contains a browser that facilitates exploration of data. This information is linked to GEO, so it is easy to move across the two databases. It is possible to use this interface to download the desired raw data files and meta information.
   b. SRA toolkit: Series of commands that enable reading and downloading of sequencing files from SRA.
   c. SRAdb: A Bioconductor package to download SRA data into R [144].
   d. pysradb: A python package to download SRA and ENA data into python [145].

EMBL-EBI

ArrayExpress

ArrayExpress contains high-throughput data from genomic experiments in different organisms, is part of the EMBL-EBI service, and is connected to the GEO database. ArrayExpress, as well as GEO, storages description and metadata information, raw data files, and preprocessed data, and all this information is publicly available. The data contained in ArrayExpress can be either imported from GEO or directly submitted to the repository, ergo it has some GEO experiments and unique submissions [135,136].

The datasets directly submitted into ArrayExpress are manually curated to verify the minimum requirements to enable reproducibility of the results [146]. This process follows specific guidelines that are Minimum Information About a Microarray Experiment (MIAME) [147,148] and Minimum Information about a high-throughput SEQuencing Experiment (MINSEQE) [149]. These standards make ArrayExpress a consistent repository.

Data can be downloaded by:

   a. Website: A searchable website that allows the usage of keywords to facilitate exploration. Direct links to processed and raw data are available.
   b. ftp: Raw data are stored in the European Nucleotide Archive (ENA) database that is directly linked in ArrayExpress. In each project (i.e., experiment) website, a table will be displayed with the available files, and direct download options are displayed. ArrayExpress is connected directly to the Galaxy platform [150] (https://usegalaxy.org/), which is a web-based analysis resource highly recommended for new NGS users that wish to analyze their own data.

ENA

Similar to SRA, the ENA stores raw data, assembly information, and functional annotation to facilitate data accessibility. ENA does not provide specialized tools to download data, as links are available and can be plugged into wget commands in a Unix terminal.

Public data-based resources

Public databases have been used to reanalyze and collect available data such as Recount [151], COLOMBOS [152,153], VESPUCCI [154], and PulmonDB [155] with different aims but all taking advantage of published stored data. Moreover, different tools allow you to reanalyze public data through specific programming languages (i.e., GEOquery, SRAdb, pysradb), and other tools have created more user-friendly and web-based applications such as GREIN [156], ImaGEO [157], GEOprofiles [158], Galaxy [150], and GIANT [159]. It is important to note that not all experiments may be available, and it is focused generally in human, rat, mouse, and other model organisms.

Reproducibility across studies

Part of the scientific process is to have a systematic methodology allowing other groups to replicate results. This is the very basis of how a scientific community creates knowledge. Therefore, a researcher's responsibility is to publish clear, honest methods and experimental results by minimizing potential errors and noise so that another researcher can replicate the findings. Particularly with computational experiments, the results should be recreated by using the same data and code, and this is considered to be reproducibility [160,161].

Reproducibility and **replicability** are oft-confused terms. Generally, replicability refers to recreating an entire study from an independent investigator with new data and similar algorithms. In contrast, reproducibility is the ability to produce the same results using the same data. Plesser discussed these concepts in detail and reviewed different meanings used in the literature [160].

Key point reproducibility, replicability, robustness, generalizability

The Turing Way community has created a great resource that describes these terms and created guidelines for reproducibility [162]:

> Reproducibility: Same data, same analysis.
> Replicability: Different data, same analysis.
> Robustness: Same data, different analysis.
> Generalizability: Different data, different analysis.

In this section, we focus on computational reproducibility and replicability in high-throughput techniques to assess gene expression. We will also briefly discuss methods that allow us to summarize information from different studies such as batch effect correction and metaanalysis.

When RNA-seq started to be used for measuring gene expression, scientists questioned if microarrays and RNA-seq were generalizable results. Marioni and colleagues [163] proved microarrays can be comparable with RNA-seq by using the same samples in different platforms with technical replicates per technology. The authors found RNA-seq has small changes in technical replicates, and high correlation with microarray results since both platforms identified similar differentially expressed genes.

Moreover, several publications have evaluated the generalizability of high-throughput transcriptomic platforms by comparing microarrays and RNA-seq. For example, using data from TCGA, 11,120 transcripts were found to be correlated across the two technologies [164]. Nevertheless, authors reported that transcripts with too high or low expression levels were more discrepant across methods.

Batch effect

In all experiments, there will be noise but depending on the experimental design, that noise may interfere with the analysis and lead to wrong conclusions. This unwanted heterogeneity can be identified and corrected. As discussed earlier, it is ideal to have all samples tested at the same time by the same person to avoid differences of nonbiological relevance. However, it is not always feasible and is necessary to add this variation into our analysis, so the experiment needs to be planned carefully for being able to identify the technical variation.

When samples cannot be processed together, the experiment can be designed in batches, but it is crucial to always process the samples with your interested condition (i.e., disease, mutation, temperature, drug) mixed together with samples from the control group. It does not matter the technology or platform, experimental design is key, and it will help answer the desired scientific question. Some guidelines with good experimental design points can be found in Refs [165–168] and for scRNA-Seq [169].

Running the analysis in batches can lead to technical noise, and exploratory analyses are highly recommended before performing the downstream analyses [166]. One of the most widely used exploratory analyses is dimensional reduction algorithms, such as PCA and t-Distributed Stochastic Neighbor (tSNE). It is a good practice to always visualize your data before differential expression analysis to detect potential batch effects. If the experiment has additional variables (also called covariates such as age, sex, day of processing, and library size) that can create batches, these covariates can be described to fit the differential expression model design [170,171].

If an experiment has not been designed properly, there is slim room for improvement. Nevertheless, there are methods for combining different experiments. Adding more samples into the analysis will increase power; however, raw data from different experiments cannot be merged directly, even if they have the same hypothesis, because they are run by different laboratories and most of the time using different platforms.

There are some approaches to integrate datasets with similar hypotheses but different origins; some authors refer to these methods as cross-platform [172] when multiple experiments are integrated into one dataset. These methods can estimate the technical variance and correct the expression before a downstream analysis, such as surrogate variable analysis (SVA)[173,174]. A benchmarking comparison for different batch effect correction methods in microarray data (without considering SVA) concluded that ComBat outperformed over other methods [175]. However, in RNA-seq experiments, ComBat showed overestimation of batch effect correction and SVA had better performance [176].

Metaanalysis

Another approach to summarize different results is a metaanalysis, which helps to identify generalizable results. A metaanalysis is a statistical approach that helps to integrate results from different studies to obtain global conclusions. There are two main models for a metaanalysis: (i) fixed-effect model, which assumes a conditional inference and estimates the effect size only using the studies included in the metaanalysis and (ii) random-effect model, which assumes studies are a random sample, and calculates a theoretical overall population that provides an unconditional inference [177,178].

A metaanalysis is not restricted to high-throughput data, and it has been used for gene expression [177,179]. R packages for conducting a metaanalysis [180] include metafor [178] and rmeta [181]).

There are web-based tools to help researchers to do a metaanalysis using public gene expression data such as ImaGEO [157] and ExAtlas [182] (for a complete list of softwares and websites available for microarray metaanalysis, see Ref. [172]).

Conclusion and remarks

High-throughput techniques have transformed and improved transcriptomic analysis since they can identify gene expression levels ranging from thousands of elements to the whole genome of an organism. Microarrays were the first technology of this category and are still in use today, but different RNA-seq methodologies emerged to enable unbiased profiling gene expression, and over the years, RNA-seq has become a popular, cheap, and versatile technology. More recently, scRNA-seq has been used for characterizing gene expression from a unique cell.

In this chapter, we introduced general pipelines for analyzing transcriptomic data depending on the technology used to measure gene expression. These pipelines are key to avoid systematic errors that can drive the analysis to wrong conclusions or misleading results. Therefore, a mandatory practice in transcriptomic analysis is to check data quality and remove noise. Other standard practices during preprocessing data are to (i) annotate the used parameters and additional files, such as the reference genome used for alignment in RNA-seq, (ii) visualize raw data and preprocessed data, and (iii) identify potential outliers and/or batch effects.

Multiple analyses can be performed after preprocessing transcriptomic data. A frequent analysis is to compare gene expression profiling, which determines differential expression genes that change across different conditions. At the same time, high-throughput transcriptomic data (usually RNA-seq) can evaluate splicing events and investigate eQTLs if SNP information is available.

This chapter also highlighted the most common analysis for gene expression data; however, there are other analyses that can be computed. For example, allele expression in which RNA-seq data can be used to identify allele-specific expressions that together with eQTLs can be integrated to determine the regulatory effect on each allele [183].

A key factor for ensuring reproducibility and replicability is rigorous reporting of methods, and data sharing—practicing Open Science (within the restrictions of research ethics board approvals for human studies and patient health information). This chapter summarized two massive public repositories for gene expression, GEO from NCBI and ArrayExpress from EMBL-EBI, and briefly reviewed repositories for sequencing data (i.e., SRA and ENA), offering procedures for downloading data.

Code can, and should, be shared using platforms as GitHub (https://github.com/). Where the documentation of an analysis and the code used to perform, it can be posted for other people to access and follow it. One great asset of GitHub is that is allows for recognition of this work to the original author of the code. These open-source resources help improve analysis pipelines while creating a community and facilitating collaboration.

Acknowledgments

YIBM was supported by CF-2023-I-1653 Conahcyt. AMR was supported by CONACYT-FORDECYT-PRONACES grant no. [11311], and Programa de Apoyo a Proyectos de Investigación e Innovación Tecnológica–Universidad Nacional Autónoma de México (PAPIIT-UNAM) grant IA203021 and IN218023.

References

[1] M.D. Adams, J.M. Kelley, J.D. Gocayne, M. Dubnick, M.H. Polymeropoulos, H. Xiao, et al., Complementary DNA sequencing: expressed sequence tags and human genome project, Science 252 (1991) 1651–1656.

[2] V.E. Velculescu, L. Zhang, B. Vogelstein, K.W. Kinzler, Serial analysis of gene expression, Science 270 (1995) 484–487.

[3] R. Lowe, N. Shirley, M. Bleackley, S. Dolan, T. Shafee, Transcriptomics technologies, PLoS Comput. Biol. 13 (2017) e1005457.

[4] M. Schena, D. Shalon, R.W. Davis, P.O. Brown, Quantitative monitoring of gene expression patterns with a complementary DNA microarray, Science 270 (1995) 467–470.

[5] R. Govindarajan, J. Duraiyan, K. Kaliyappan, M. Palanisamy, Microarray and its applications, J. Pharm. Bioallied Sci. 4 (2012) S310–S312.

[6] W. Huber, V.J. Carey, R. Gentleman, S. Anders, M. Carlson, B.S. Carvalho, et al., Orchestrating high-throughput genomic analysis with bioconductor, Nat. Methods 12 (2015) 115–121.

[7] R.C. Gentleman, V.J. Carey, D.M. Bates, B. Bolstad, M. Dettling, S. Dudoit, et al., Bioconductor: open software development for computational biology and bioinformatics, Genome Biol. 5 (2004) R80.

[8] Arrays, Bioconductor, [cited 14 Jul 2023]. Available at: https://www.bioconductor.org/packages/release/workflows/html/arrays.html.

[9] R.A. Irizarry, B. Hobbs, F. Collin, Y.D. Beazer-Barclay, K.J. Antonellis, U. Scherf, et al., Exploration, normalization, and summaries of high density oligonucleotide array probe level data, Biostatistics 4 (2003) 249–264.

[10] M.E. Ritchie, J. Silver, A. Oshlack, M. Holmes, D. Diyagama, A. Holloway, et al., A comparison of background correction methods for two-colour microarrays, Bioinformatics 23 (2007) 2700–2707.

[11] M.E. Ritchie, B. Phipson, D. Wu, Y. Hu, C.W. Law, W. Shi, et al., Limma powers differential expression analyses for RNA-sequencing and microarray studies, Nucleic Acids Res. 43 (2015) e47.

[12] X. Wang, Y. Lin, C. Song, E. Sibille, G.C. Tseng, Detecting disease-associated genes with confounding variable adjustment and the impact on genomic meta-analysis: with application to major depressive disorder, BMC Bioinform. 13 (2012) 52.

[13] M.V. Kuleshov, M.R. Jones, A.D. Rouillard, N.F. Fernandez, Q. Duan, Z. Wang, et al., Enrichr: a comprehensive gene set enrichment analysis web server 2016 update, Nucleic Acids Res. 44 (2016) W90–W97.

[14] E.Y. Chen, C.M. Tan, Y. Kou, Q. Duan, Z. Wang, G.V. Meirelles, et al., Enrichr: interactive and collaborative HTML5 gene list enrichment analysis tool, BMC Bioinform. 14 (2013) 128.

[15] F. Sanger, S. Nicklen, A.R. Coulson, DNA sequencing with chain-terminating inhibitors, Proc. Natl. Acad. Sci. U. S. A. 74 (1977) 5463–5467.

[16] A. Ahmadian, M. Ehn, S. Hober, Pyrosequencing: history, biochemistry and future, Clin. Chim. Acta 363 (2006) 83–94.

[17] J. Zhang, R. Chiodini, A. Badr, G. Zhang, The impact of next-generation sequencing on genomics, J. Genet. Genomics. 38 (2011) 95–109.

[18] S.E. Levy, R.M. Myers, Advancements in next-generation sequencing, Annu. Rev. Genomics Hum. Genet. 17 (2016) 95–115.

[19] S. Goodwin, J.D. McPherson, W.R. McCombie, Coming of age: ten years of next-generation sequencing technologies, Nat. Rev. Genet. 17 (2016) 333–351.

[20] A. Valouev, J. Ichikawa, T. Tonthat, J. Stuart, S. Ranade, H. Peckham, et al., A high-resolution, nucleosome position map of *C. elegans* reveals a lack of universal sequence-dictated positioning, Genome Res. 18 (2008) 1051–1063.

[21] J. Huang, X. Liang, Y. Xuan, C. Geng, Y. Li, H. Lu, et al., A reference human genome dataset of the BGISEQ-500 sequencer, Gigascience 6 (2017) 1–9.

[22] J.R. Lupski, J.G. Reid, C. Gonzaga-Jauregui, D. Rio Deiros, D.C.Y. Chen, L. Nazareth, et al., Whole-genome sequencing in a patient with Charcot-Marie-Tooth neuropathy, N. Engl. J. Med. 362 (2010) 1181–1191.
[23] Z. Wang, M. Gerstein, M. Snyder, RNA-seq: a revolutionary tool for transcriptomics, Nat. Rev. Genet. 10 (2009) 57–63.
[24] D.S. Johnson, A. Mortazavi, R.M. Myers, B. Wold, Genome-wide mapping of in vivo protein-DNA interactions, Science 316 (2007) 1497–1502.
[25] T. Weirick, G. Militello, R. Müller, D. John, S. Dimmeler, S. Uchida, The identification and characterization of novel transcripts from RNA-seq data, Brief. Bioinform. 17 (2016) 678–685.
[26] S.A. Hardwick, A. Joglekar, P. Flicek, A. Frankish, H.U. Tilgner, Getting the entire message: progress in isoform sequencing, Front. Genet. 10 (2019) 709.
[27] X. Shi, M. Sun, H. Liu, Y. Yao, Y. Song, Long non-coding RNAs: a new frontier in the study of human diseases, Cancer Lett. 339 (2013) 159–166.
[28] T. Shiraki, S. Kondo, S. Katayama, K. Waki, T. Kasukawa, H. Kawaji, et al., Cap analysis gene expression for high-throughput analysis of transcriptional starting point and identification of promoter usage, Proc. Natl. Acad. Sci. U. S. A. 100 (2003) 15776–15781.
[29] The FANTOM Consortium and the RIKEN PMI and CLST (DGT), A promoter-level mammalian expression atlas, Nature 507 (2014) 462–470.
[30] S. Andrews, Others., FastQC: A Quality Control Tool for High Throughput Sequence Data, Babraham Bioinformatics, Babraham Institute, Cambridge, United Kingdom, 2010.
[31] S. Arora, S.S. Pattwell, E.C. Holland, H. Bolouri, Variability in estimated gene expression among commonly used RNA-seq pipelines, Sci. Rep. 10 (2020) 2734.
[32] B. Langmead, S.L. Salzberg, Fast gapped-read alignment with Bowtie 2, Nat. Methods 9 (2012) 357–359.
[33] N.L. Bray, H. Pimentel, P. Melsted, L. Pachter, Near-optimal probabilistic RNA-seq quantification, Nat. Biotechnol. 34 (2016) 525–527.
[34] A. Dobin, C.A. Davis, F. Schlesinger, J. Drenkow, C. Zaleski, S. Jha, et al., STAR: ultrafast universal RNA-seq aligner, Bioinformatics 29 (2013) 15–21.
[35] R. Patro, G. Duggal, M.I. Love, R.A. Irizarry, C. Kingsford, Salmon provides fast and bias-aware quantification of transcript expression, Nat. Methods 14 (2017) 417–419.
[36] M. Teng, M.I. Love, C.A. Davis, S. Djebali, A. Dobin, B.R. Graveley, et al., A benchmark for RNA-seq quantification pipelines, Genome Biol. 17 (2016) 74.
[37] M.D. Robinson, D.J. McCarthy, G.K. Smyth, edgeR: a bioconductor package for differential expression analysis of digital gene expression data, Bioinformatics 26 (2010) 139–140.
[38] M.I. Love, W. Huber, S. Anders, Moderated estimation of fold change and dispersion for RNA-seq data with DESeq2, Genome Biol. 15 (2014) 550.
[39] T. Moerman, S. Aibar Santos, C. Bravo González-Blas, J. Simm, Y. Moreau, J. Aerts, et al., GRNBoost2 and Arboreto: efficient and scalable inference of gene regulatory networks, Bioinformatics 35 (2019) 2159–2161.
[40] L. Zhao, H. Zhao, H. Yan, Gene expression profiling of 1200 pancreatic ductal adenocarcinoma reveals novel subtypes, BMC Cancer 18 (2018) 603.
[41] N.L. Barbosa-Morais, M. Irimia, Q. Pan, H.Y. Xiong, S. Gueroussov, L.J. Lee, et al., The evolutionary landscape of alternative splicing in vertebrate species, Science 338 (2012) 1587–1593.
[42] X. Zhang, I. Jonassen, RASflow: an RNA-seq analysis workflow with Snakemake, BMC Bioinform. 21 (2020) 110.
[43] M. Cornwell, M. Vangala, L. Taing, Z. Herbert, J. Köster, B. Li, et al., VIPER: visualization pipeline for RNA-seq, a Snakemake workflow for efficient and complete RNA-seq analysis, BMC Bioinform. 19 (2018) 135.
[44] D. Torre, A. Lachmann, Ma'ayan A., BioJupies: automated generation of interactive notebooks for RNA-seq data analysis in the cloud, Cell Syst. 7 (2018) 556–561.e3.

[45] J. Taylor, I. Schenck, D. Blankenberg, A. Nekrutenko, Using galaxy to perform large-scale interactive data analyses, Curr. Protoc. Bioinform. (2007). Chapter 10: Unit 10.5.
[46] S.D. Boyd, Everything you wanted to know about small RNA but were afraid to ask, Lab. Investig. 88 (2008) 569–578.
[47] A. Zampetaki, A. Albrecht, K. Steinhofel, Corrigendum: long non-coding RNA structure and function: is there a link? Front. Physiol. 10 (2019) 1127.
[48] J. Harrow, A. Frankish, J.M. Gonzalez, E. Tapanari, M. Diekhans, F. Kokocinski, et al., GENCODE: the reference human genome annotation for the ENCODE project, Genome Res. 22 (2012) 1760–1774.
[49] J. Pease, R. Sooknanan, A rapid, directional RNA-seq library preparation workflow for Illumina® sequencing, Nat. Methods 9 (2012) i–ii.
[50] G.A. Merino, A. Conesa, E.A. Fernández, A benchmarking of workflows for detecting differential splicing and differential expression at isoform level in human RNA-seq studies, Brief. Bioinform. 20 (2019) 471–481.
[51] M. Hölzer, M. Marz, De novo transcriptome assembly: a comprehensive cross-species comparison of short-read RNA-seq assemblers, Gigascience (2019) 8, https://doi.org/10.1093/gigascience/giz039.
[52] A. Rhoads, K.F. Au, PacBio sequencing and its applications, Genomics Proteomics Bioinform. 13 (2015) 278–289.
[53] J. Ezpeleta, F.J. Krsticevic, P. Bulacio, E. Tapia, Designing robust watermark barcodes for multiplex long-read sequencing, Bioinformatics 33 (2017) 807–813.
[54] H.A. Chowdhury, D.K. Bhattacharyya, J.K. Kalita, Differential expression analysis of RNA-seq reads: overview, taxonomy, and tools, IEEE/ACM Trans. Comput. Biol. Bioinform. 17 (2020) 566–586.
[55] D. Wyman, A. Mortazavi, TranscriptClean: variant-aware correction of indels, mismatches and splice junctions in long-read transcripts, Bioinformatics 35 (2019) 340–342.
[56] H. Tilgner, D. Raha, L. Habegger, M. Mohiuddin, M. Gerstein, M. Snyder, Accurate identification and analysis of human mRNA isoforms using deep long read sequencing, G3 (Bethesda) 3 (2013) 387–397.
[57] A. Byrne, A.E. Beaudin, H.E. Olsen, M. Jain, C. Cole, T. Palmer, et al., Nanopore long-read RNAseq reveals widespread transcriptional variation among the surface receptors of individual B cells, Nat. Commun. 8 (2017) 16027.
[58] J. Ye, S. Cheng, X. Zhou, Z. Chen, S.U. Kim, J. Tan, et al., A global survey of full-length transcriptome of *Ginkgo biloba* reveals transcript variants involved in flavonoid biosynthesis, Ind. Crop. Prod. 139 (2019) 111547.
[59] D. Wyman, A. Mortazavi, ENCODE long read RNA-seq analysis protocol for human samples (v3.0), [cited 14 Jul 2023]. Available at: https://www.encodeproject.org/documents/81af563b-5134-4f78-9bc4-41cb42cc6a48/@@download/attachment/ENCODE%20Long%20Read%20RNA-Seq%20Analysis%20Pipeline%20v3%20%28Human%29.pdf.
[60] C.M. Claussen, H. Lee, J.J. Shah, T. Richards, N. Shah, K. Patel, et al., Gene expression profiling predicts clinical outcomes in newly diagnosed multiple myeloma patients in a standard of care setting, Blood 128 (2016) 5628.
[61] R. Szalat, H. Avet-Loiseau, N.C. Munshi, Gene expression profiles in myeloma: ready for the real world? Clin. Cancer Res. 22 (2016) 5434–5442.
[62] C. Sudlow, J. Gallacher, N. Allen, V. Beral, P. Burton, J. Danesh, et al., UK biobank: an open access resource for identifying the causes of a wide range of complex diseases of middle and old age, PLoS Med. 12 (2015) e1001779.
[63] L.B. Alexandrov, J. Kim, N.J. Haradhvala, M.N. Huang, A.W. Tian Ng, Y. Wu, et al., The repertoire of mutational signatures in human cancer, Nature 578 (2020) 94–101.
[64] D. Brawand, M. Soumillon, A. Necsulea, P. Julien, G. Csárdi, P. Harrigan, et al., The evolution of gene expression levels in mammalian organs, Nature 478 (2011) 343–348.
[65] B.J. Blencowe, Alternative splicing: new insights from global analyses, Cell 126 (2006) 37–47.

[66] H.-D. Li, R. Menon, G.S. Omenn, Y. Guan, The emerging era of genomic data integration for analyzing splice isoform function, Trends Genet. 30 (2014) 340–347.
[67] X. Grau-Bové, I. Ruiz-Trillo, M. Irimia, Origin of exon skipping-rich transcriptomes in animals driven by evolution of gene architecture, Genome Biol. 19 (2018) 135.
[68] M. Csuros, I.B. Rogozin, E.V. Koonin, A detailed history of intron-rich eukaryotic ancestors inferred from a global survey of 100 complete genomes, PLoS Comput. Biol. 7 (2011) e1002150.
[69] E.T. Wang, R. Sandberg, S. Luo, I. Khrebtukova, L. Zhang, C. Mayr, et al., Alternative isoform regulation in human tissue transcriptomes, Nature 456 (2008) 470–476.
[70] E.F. Modafferi, D.L. Black, Combinatorial control of a neuron-specific exon, RNA 5 (1999) 687–706.
[71] S.-J. Noh, K. Lee, H. Paik, C.-G. Hur, TISA: tissue-specific alternative splicing in human and mouse genes, DNA Res. 13 (2006) 229–243.
[72] F.E. Baralle, J. Giudice, Alternative splicing as a regulator of development and tissue identity, Nat. Rev. Mol. Cell Biol. 18 (2017) 437–451.
[73] Y. Zhang, L. Yan, J. Zeng, H. Zhou, H. Liu, G. Yu, et al., Pan-cancer analysis of clinical relevance of alternative splicing events in 31 human cancers, Oncogene 38 (2019) 6678–6695.
[74] M. Ryan, W.C. Wong, R. Brown, R. Akbani, X. Su, B. Broom, et al., TCGASpliceSeq a compendium of alternative mRNA splicing in cancer, Nucleic Acids Res. 44 (2016) D1018–D1022.
[75] Y. Wang, J. Liu, B.O. Huang, Y.-M. Xu, J. Li, L.-F. Huang, et al., Mechanism of alternative splicing and its regulation, Biomed. Rep. 3 (2015) 152–158.
[76] S. Subbaram, M. Kuentzel, D. Frank, C.M. Dipersio, S.V. Chittur, Determination of alternate splicing events using the Affymetrix Exon 1.0 ST arrays, Methods Mol. Biol. 632 (2010) 63–72.
[77] R. Jaksik, M. Iwanaszko, J. Rzeszowska-Wolny, M. Kimmel, Microarray experiments and factors which affect their reliability, Biol. Direct 10 (2015) 46.
[78] K. Srinivasan, L. Shiue, J.D. Hayes, R. Centers, S. Fitzwater, R. Loewen, et al., Detection and measurement of alternative splicing using splicing-sensitive microarrays, Methods 37 (2005) 345–359.
[79] M.J. Moore, P.A. Silver, Global analysis of mRNA splicing, RNA 14 (2008) 197–203.
[80] T.A. Clark, A.C. Schweitzer, T.X. Chen, M.K. Staples, G. Lu, H. Wang, et al., Discovery of tissue-specific exons using comprehensive human exon microarrays, Genome Biol. 8 (2007) R64.
[81] A. Lapuk, H. Marr, L. Jakkula, H. Pedro, S. Bhattacharya, E. Purdom, et al., Exon-level microarray analyses identify alternative splicing programs in breast cancer, Mol. Cancer Res. 8 (2010) 961–974.
[82] A.B. Villaseñor-Altamirano, J.D. Watson, S.D. Prokopec, C.Q. Yao, P.C. Boutros, R. Pohjanvirta, et al., 2,3,7,8-Tetrachlorodibenzo-p-dioxin modifies alternative splicing in mouse liver, PLoS One 14 (2019) e0219747.
[83] E. Purdom, K.M. Simpson, M.D. Robinson, J.G. Conboy, A.V. Lapuk, T.P. Speed, FIRMA: a method for detection of alternative splicing from exon array data, Bioinformatics 24 (2008) 1707–1714.
[84] Y. Xing, P. Stoilov, K. Kapur, A. Han, H. Jiang, S. Shen, et al., MADS: a new and improved method for analysis of differential alternative splicing by exon-tiling microarrays, RNA 14 (2008) 1470–1479.
[85] A. GeneChip, Exon Array Whitepaper Collection, "Alternative Transcript Analysis Methods for Exon Arrays," rev. Oct. 11, 2005. ver. 1.1.
[86] A. Rasche, R. Herwig, ARH: predicting splice variants from genome-wide data with modified entropy, Bioinformatics 26 (2010) 84–90.
[87] Y. Hu, Y. Huang, Y. Du, C.F. Orellana, D. Singh, A.R. Johnson, et al., DiffSplice: the genome-wide detection of differential splicing events with RNA-seq, Nucleic Acids Res. 41 (2013) e39.
[88] A. Kanitz, F. Gypas, A.J. Gruber, A.R. Gruber, G. Martin, M. Zavolan, Comparative assessment of methods for the computational inference of transcript isoform abundance from RNA-seq data, Genome Biol. 16 (2015) 150.
[89] X. Li, S. Teng, RNA sequencing in schizophrenia, Bioinform. Biol. Insights 9 (2015) 53–60.

[90] S. Ghosh, C.-K.K. Chan, Analysis of RNA-seq data using TopHat and cufflinks, Methods Mol. Biol. 1374 (2016) 339–361.
[91] A. Mehmood, A. Laiho, M.S. Venäläinen, A.J. McGlinchey, N. Wang, L.L. Elo, Systematic evaluation of differential splicing tools for RNA-seq studies, Brief. Bioinform. (2019), https://doi.org/10.1093/bib/bbz126.
[92] S.M.E. Sahraeian, M. Mohiyuddin, R. Sebra, H. Tilgner, P.T. Afshar, K.F. Au, et al., Gaining comprehensive biological insight into the transcriptome by performing a broad-spectrum RNA-seq analysis, Nat. Commun. 8 (2017) 59.
[93] C. Trapnell, A. Roberts, L. Goff, G. Pertea, D. Kim, D.R. Kelley, et al., Differential gene and transcript expression analysis of RNA-seq experiments with TopHat and cufflinks, Nat. Protoc. 7 (2012) 562–578.
[94] H. Pimentel, N.L. Bray, S. Puente, P. Melsted, L. Pachter, Differential analysis of RNA-seq incorporating quantification uncertainty, Nat. Methods 14 (2017) 687–690.
[95] S. Shen, J.W. Park, Z.-X. Lu, L. Lin, M.D. Henry, Y.N. Wu, et al., rMATS: robust and flexible detection of differential alternative splicing from replicate RNA-seq data, Proc. Natl. Acad. Sci. U. S. A. 111 (2014) E5593–E5601.
[96] G.P. Alamancos, A. Pagès, J.L. Trincado, N. Bellora, E. Eyras, Leveraging transcript quantification for fast computation of alternative splicing profiles, RNA 21 (2015) 1521–1531.
[97] G. Baruzzo, K.E. Hayer, E.J. Kim, B. Di Camillo, G.A. FitzGerald, G.R. Grant, Simulation-based comprehensive benchmarking of RNA-seq aligners, Nat. Methods 14 (2017) 135–139.
[98] D. Garrido-Martín, E. Palumbo, R. Guigó, A. Breschi, ggsashimi: Sashimi plot revised for browser- and annotation-independent splicing visualization, PLoS Comput. Biol. 14 (2018) e1006360.
[99] H. Strobelt, B. Alsallakh, J. Botros, B. Peterson, M. Borowsky, H. Pfister, et al., Vials: visualizing alternative splicing of genes, IEEE Trans. Vis. Comput. Graph. 22 (2016) 399–408.
[100] Y. Huang, G. Sanguinetti, BRIE: transcriptome-wide splicing quantification in single cells, Genome Biol. 18 (2017) 123.
[101] J. Westoby, M.S. Herrera, A.C. Ferguson-Smith, M. Hemberg, Simulation-based benchmarking of isoform quantification in single-cell RNA-seq, Genome Biol. 19 (2018) 191.
[102] J. Westoby, P. Artemov, M. Hemberg, A. Ferguson-Smith, Obstacles to detecting isoforms using full-length scRNA-seq data, Genome Biol. 21 (2020) 74.
[103] National Institutes of Health (US), Biological Sciences Curriculum Study, Understanding Human Genetic Variation, NIH Curriculum Supplement Series, NIH, 2007.
[104] T.A. Manolio, Genomewide association studies and assessment of the risk of disease, N. Engl. J. Med. 363 (2010) 166–176.
[105] J. MacArthur, E. Bowler, M. Cerezo, L. Gil, P. Hall, E. Hastings, et al., The new NHGRI-EBI catalog of published genome-wide association studies (GWAS catalog), Nucleic Acids Res. 45 (2017) D896–D901.
[106] J. Vierstra, J. Lazar, R. Sandstrom, J. Halow, K. Lee, D. Bates, et al., Global reference mapping of human transcription factor footprints, Nature 583 (2020) 729–736.
[107] A.C. Nica, E.T. Dermitzakis, Expression quantitative trait loci: present and future, Philos. Trans. R. Soc. Lond. Ser. B Biol. Sci. 368 (2013) 20120362.
[108] C.C. Guo, N. Wei, S.H. Liang, B.L. Wang, S.M. Sha, K.C. Wu, Population-specific genome-wide mapping of expression quantitative trait loci in the colon of Han Chinese, J. Dig. Dis. 17 (2016) 600–609.
[109] GTEx Consortium, The Genotype-Tissue Expression (GTEx) project, Nat. Genet. 45 (6) (2013), https://doi.org/10.1038/ng.2653.
[110] GTEx Consortium, The GTEx Consortium atlas of genetic regulatory effects across human tissues, Science 369 (6509) (2020), https://doi.org/10.1126/science.aaz1776.
[111] E.R. Gamazon, A.V. Segrè, M. van de Bunt, X. Wen, H.S. Xi, F. Hormozdiari, et al., Using an atlas of gene regulation across 44 human tissues to inform complex disease- and trait-associated variation, Nat. Genet. 50 (2018) 956–967.

[112] U.M. Marigorta, L.A. Denson, J.S. Hyams, K. Mondal, J. Prince, T.D. Walters, et al., Transcriptional risk scores link GWAS to eQTLs and predict complications in Crohn's disease, Nat. Genet. 49 (2017) 1517–1521.

[113] S.C.L. Lock, Functional Data Analysis and QTL Detection across Time within the Circadian Clock, mscresearch, University of York, 2017. Available at https://etheses.whiterose.ac.uk/19451/.

[114] M. Pala, Z. Zappala, M. Marongiu, X. Li, J.R. Davis, R. Cusano, et al., Population- and individual-specific regulatory variation in Sardinia, Nat. Genet. 49 (2017) 700–707.

[115] D.V. Zhernakova, P. Deelen, M. Vermaat, M. van Iterson, M. van Galen, W. Arindrarto, et al., Identification of context-dependent expression quantitative trait loci in whole blood, Nat. Genet. 49 (2017) 139–145.

[116] K.G. Ouwens, R. Jansen, M.G. Nivard, J. van Dongen, M.J. Frieser, J.-J. Hottenga, et al., A characterization of cis- and trans-heritability of RNA-seq-based gene expression, Eur. J. Hum. Genet. 28 (2020) 253–263.

[117] N.R. Gay, M. Gloudemans, M.L. Antonio, N.S. Abell, B. Balliu, Y. Park, et al., Impact of admixture and ancestry on eQTL analysis and GWAS colocalization in GTEx, Genome Biol. 21 (2020) 233.

[118] A.A. Shabalin, Matrix eQTL: ultra fast eQTL analysis via large matrix operations, Bioinformatics 28 (2012) 1353–1358.

[119] O. Delaneau, H. Ongen, A.A. Brown, A. Fort, N.I. Panousis, E.T. Dermitzakis, A complete tool set for molecular QTL discovery and analysis, Nat. Commun. 8 (2017) 15452.

[120] H. Ongen, A. Buil, A.A. Brown, E.T. Dermitzakis, O. Delaneau, Fast and efficient QTL mapper for thousands of molecular phenotypes, Bioinformatics 32 (2016) 1479–1485.

[121] Q.Q. Huang, S.C. Ritchie, M. Brozynska, M. Inouye, Power, false discovery rate and Winner's curse in eQTL studies, Nucleic Acids Res. 46 (2018) e133.

[122] A.A. Brown, A. Viñuela, O. Delaneau, T.D. Spector, K.S. Small, E.T. Dermitzakis, Predicting causal variants affecting expression by using whole-genome sequencing and RNA-seq from multiple human tissues, Nat. Genet. 49 (2017) 1747–1751.

[123] F. Hormozdiari, S. Gazal, B. van de Geijn, H. Finucane, C.J.-T. Ju, P.-R. Loh, et al., Leveraging molecular QTL to understand the genetic architecture of diseases and complex traits, bioRxiv. (2017) 203380, https://doi.org/10.1101/203380.

[124] X. Wen, R. Pique-Regi, F. Luca, Integrating molecular QTL data into genome-wide genetic association analysis: probabilistic assessment of enrichment and colocalization, PLoS Genet. 13 (2017) e1006646.

[125] F. Tang, C. Barbacioru, Y. Wang, E. Nordman, C. Lee, N. Xu, et al., mRNA-seq whole-transcriptome analysis of a single cell, Nat. Methods 6 (2009) 377–382.

[126] E. Hedlund, Q. Deng, Single-cell RNA sequencing: technical advancements and biological applications, Mol. Asp. Med. 59 (2018) 36–46.

[127] C.R. Cadwell, A. Palasantza, X. Jiang, P. Berens, Q. Deng, M. Yilmaz, et al., Electrophysiological, transcriptomic and morphologic profiling of single neurons using patch-seq, Nat. Biotechnol. 34 (2016) 199–203.

[128] X. Liao, M. Makris, X.M. Luo, Fluorescence-activated cell sorting for purification of plasmacytoid dendritic cells from the mouse bone marrow, J. Vis. Exp. (2016), https://doi.org/10.3791/54641.

[129] W.-M. Zhou, Y.-Y. Yan, Q.-R. Guo, H. Ji, H. Wang, T.-T. Xu, et al., Microfluidics applications for high-throughput single cell sequencing, J Nanobiotechnol. 19 (2021) 312.

[130] T. Hashimshony, F. Wagner, N. Sher, I. Yanai, CEL-seq: single-cell RNA-seq by multiplexed linear amplification, Cell Rep. 2 (2012) 666–673.

[131] J.R. Dobson, D. Hong, A.R. Barutcu, H. Wu, A.N. Imbalzano, J.B. Lian, et al., Identifying nuclear matrix-attached DNA across the genome, J. Cell. Physiol. 232 (2017) 1295–1305.

[132] Genomic Data Resources: Curation, Databasing, and Browsers. [cited 21 Sep 2020]. Available at: https://www.nature.com/scitable/topicpage/genomic-data-resources-challenges-and-promises-743721/.

[133] R. Edgar, M. Domrachev, A.E. Lash, Gene Expression Omnibus: NCBI gene expression and hybridization array data repository, Nucleic Acids Res. 30 (1) (2002) 207–210, https://doi.org/10.1093/nar/30.1.207.

[134] T. Barrett, S.E. Wilhite, P. Ledoux, C. Evangelista, I.F. Kim, M. Tomashevsky, K.A. Marshall, K.H. Phillippy, P.M. Sherman, M. Holko, A. Yefanov, H. Lee, N. Zhang, C.L. Robertson, N. Serova, S. Davis, A. Soboleva, NCBI GEO: archive for functional genomics data sets—update, Nucleic Acids Res. 41 (Database issue) (2013) D991–D995, https://doi.org/10.1093/nar/gks1193.

[135] W. Li, A. Cowley, M. Uludag, T. Gur, H. McWilliam, S. Squizzato, Y.M. Park, N. Buso, R. Lopez, The EMBL-EBI bioinformatics web and programmatic tools framework, Nucleic Acids Res. 43 (W1) (2015) W580–W584, https://doi.org/10.1093/nar/gkv279.

[136] R. Leinonen, H. Sugawara, M. Shumway, International Nucleotide Sequence Database Collaboration, The sequence read archive, Nucleic Acids Res. 39 (Database issue) (2011) D19–D21, https://doi.org/10.1093/nar/gkq1019.

[137] R. Leinonen, R. Akhtar, E. Birney, L. Bower, A. Cerdeno-Tárraga, Y. Cheng, I. Cleland, N. Faruque, N. Goodgame, R. Gibson, G. Hoad, M. Jang, N. Pakseresht, S. Plaister, R. Radhakrishnan, K. Reddy, Siamak Sobhany, P. Ten Hoopen, R. Vaughan, V. Zalunin, G. Cochrane, The European Nucleotide Archive, Nucleic Acids Res. 39 (Database issue) (2011) D28–D31, https://doi.org/10.1093/nar/gkq967.

[138] J. Burgin, A. Ahamed, C. Cummins, R. Devraj, K. Gueye, D. Gupta, V. Gupta, M. Haseeb, M. Ihsan, E. Ivanov, S. Jayathilaka, V. Balavenkataraman Kadhirvelu, M. Kumar, A. Lathi, R. Leinonen, M. Mansurova, J. McKinnon, Colman O'Cathail, Joana Paupério, S. Pesant, N. Rahman, G. Rinck, S. Selvakumar, S. Suman, S. Vijayaraja, Z. Waheed, P. Woollard, D. Yuan, A. Zyoud, T. Burdett, G. Cochrane, The European Nucleotide Archive in 2022, Nucleic Acids Res. 51 (D1) (2023) D121–D125, https://doi.org/10.1093/nar/gkac1051.

[139] geo, Programmatic Access to GEO, [cited 21 Oct 2020]. Available at: https://www.ncbi.nlm.nih.gov/geo/info/geo_paccess.html.

[140] S. Davis, P.S. Meltzer, GEOquery: a bridge between the gene expression omnibus (GEO) and BioConductor, Bioinformatics (2007) 1846–1847, https://doi.org/10.1093/bioinformatics/btm254.

[141] Welcome to GEOparse's Documentation!—GEOparse 1.2.0 Documentation. [cited 21 Oct 2020]. Available at: https://geoparse.readthedocs.io/en/latest/index.html.

[142] Home—SRA—NCBI. [cited 15 Oct 2020]. Available at: https://www.ncbi.nlm.nih.gov/sra.

[143] R. Leinonen, H. Sugawara, M. Shumway, International nucleotide sequence database collaboration. The sequence read archive, Nucleic Acids Res. 39 (2011) D19–D21.

[144] Y. Zhu, R.M. Stephens, P.S. Meltzer, S.R. Davis, SRAdb: query and use public next-generation sequencing data from within R, BMC Bioinform. 14 (2013) 19.

[145] S. Choudhary, pysradb: a Python package to query next-generation sequencing metadata and data from NCBI Sequence Read Archive, F1000Res 8 (2019) 532, https://doi.org/10.12688/f1000research.18676.1.

[146] EMBL-EBI. What is Bioinformatics? [cited 20 May 2022]. Available at: https://www.ebi.ac.uk/training/online/courses/bioinformatics-terrified/what-bioinformatics/.

[147] FGED: MIAME. [cited 19 Oct 2020]. Available at: http://www.fged.org/projects/miame/.

[148] T.F. Rayner, P. Rocca-Serra, P.T. Spellman, H.C. Causton, A. Farne, E. Holloway, R.A. Irizarry, J. Liu, D.S. Maier, M. Miller, K. Petersen, J. Quackenbush, G. Sherlock, C.J. Stoeckert Jr., J. White, P.L. Whetzel, F. Wymore, H. Parkinson, U. Sarkans, C.A. Ball, A. Brazma, A simple spreadsheet-based, MIAME-supportive format for microarray data: MAGE-TAB, BMC Bioinf. 7 (2006) 489, https://doi.org/10.1186/1471-2105-7-489.

[149] N. Abeygunawardena, MINSEQE-Workgroups-FGED, 2008, Available at: http://www.fged.org/projects/minseqe/.

[150] E. Afgan, D. Baker, B. Batut, M. van den Beek, D. Bouvier, M. Cech, et al., The galaxy platform for accessible, reproducible and collaborative biomedical analyses: 2018 update, Nucleic Acids Res. 46 (2018) W537–W544.

[151] L. Collado-Torres, A. Nellore, A.E. Jaffe, Recount workflow: accessing over 70,000 human RNA-seq samples with Bioconductor, F1000Res 6 (2017) 1558.

[152] M. Moretto, P. Sonego, N. Dierckxsens, M. Brilli, L. Bianco, D. Ledezma-Tejeida, et al., COLOMBOS v3.0: leveraging gene expression compendia for cross-species analyses, Nucleic Acids Res. 44 (2016) D620–D623.
[153] K. Engelen, Q. Fu, P. Meysman, A. Sánchez-Rodríguez, R. De Smet, K. Lemmens, et al., COLOMBOS: access port for cross-platform bacterial expression compendia, PLoS One 6 (2011) e20938.
[154] M. Moretto, P. Sonego, S. Pilati, G. Malacarne, L. Costantini, L. Grzeskowiak, et al., VESPUCCI: exploring patterns of gene expression in grapevine, Front. Plant Sci. 7 (2016) 633.
[155] A.B. Villaseñor-Altamirano, M. Moretto, M. Maldonado, A. Zayas-Del Moral, A. Munguía-Reyes, Y. Romero, et al., PulmonDB: a curated lung disease gene expression database, Sci. Rep. 10 (2020) 514.
[156] N.A. Mahi, M.F. Najafabadi, M. Pilarczyk, M. Kouril, M. Medvedovic, GREIN: an interactive web platform for re-analyzing GEO RNA-seq data, Sci. Rep. 9 (2019) 7580.
[157] D. Toro-Domínguez, J. Martorell-Marugán, R. López-Domínguez, A. García-Moreno, V. González-Rumayor, M.E. Alarcón-Riquelme, et al., ImaGEO: integrative gene expression meta-analysis from GEO database, Bioinformatics 35 (2019) 880–882.
[158] T. Barrett, T.O. Suzek, D.B. Troup, S.E. Wilhite, W.-C. Ngau, P. Ledoux, et al., NCBI GEO: mining millions of expression profiles—database and tools, Nucleic Acids Res. 33 (2005) D562–D566.
[159] J. Vandel, C. Gheeraert, B. Staels, J. Eeckhoute, P. Lefebvre, J. Dubois-Chevalier, GIANT: galaxy-based tool for interactive analysis of transcriptomic data, Sci. Rep. 10 (2020) 19835.
[160] H.E. Plesser, Reproducibility vs., Replicability: a brief history of a confused terminology, Front. Neuroinform. 11 (2017) 76.
[161] N.P. Rougier, K. Hinsen, F. Alexandre, T. Arildsen, L.A. Barba, F.C.Y. Benureau, et al., Sustainable computational science: the ReScience initiative, PeerJ Comput Sci. 3 (2017) e142.
[162] T.T.W. Community, B. Arnold, L. Bowler, S. Gibson, P. Herterich, R. Higman, et al., The Turing Way: A Handbook for Reproducible Data Science, Zenodo, 2019, https://doi.org/10.5281/ZENODO.3233853.
[163] J.C. Marioni, C.E. Mason, S.M. Mane, M. Stephens, Y. Gilad, RNA-seq: an assessment of technical reproducibility and comparison with gene expression arrays, Genome Res. (2008) 1509–1517, https://doi.org/10.1101/gr.079558.108.
[164] M. Chen, B. Liu, J. Xiao, Y. Yang, Y. Zhang, A novel seven-long non-coding RNA signature predicts survival in early stage lung adenocarcinoma, Oncotarget 8 (2017) 14876–14886.
[165] A. Conesa, P. Madrigal, S. Tarazona, D. Gomez-Cabrero, A. Cervera, A. McPherson, et al., A survey of best practices for RNA-seq data analysis, Genome Biol. 17 (2016) 13.
[166] J.T. Leek, R.B. Scharpf, H.C. Bravo, D. Simcha, B. Langmead, W.E. Johnson, et al., Tackling the widespread and critical impact of batch effects in high-throughput data, Nat. Rev. Genet. 11 (2010) 733–739.
[167] L.M. McIntyre, K.K. Lopiano, A.M. Morse, V. Amin, A.L. Oberg, L.J. Young, et al., RNA-seq: technical variability and sampling, BMC Genomics 12 (2011) 293.
[168] N.J. Schurch, P. Schofield, M. Gierliński, C. Cole, A. Sherstnev, V. Singh, et al., How many biological replicates are needed in an RNA-seq experiment and which differential expression tool should you use? RNA 22 (2016) 839–851.
[169] S.C. Hicks, F.W. Townes, M. Teng, R.A. Irizarry, Missing data and technical variability in single-cell RNA-sequencing experiments, Biostatistics 19 (2018) 562–578.
[170] M.I. Love, S. Anders, W. Huber, Analyzing RNA-seq Data with DESeq2, 2020, [cited 2 Dec 2020]. Available at: https://bioconductor.org/packages/release/bioc/vignettes/DESeq2/inst/doc/DESeq2.html.
[171] S. Mostafavi, A. Battle, X. Zhu, A.E. Urban, D. Levinson, S.B. Montgomery, et al., in: P.V. Benos (Ed.), Normalizing RNA-Sequencing Data by Modeling Hidden Covariates with Prior Knowledge, 2013, https://doi.org/10.1371/journal.pone.0068141.
[172] C.J. Walsh, P. Hu, J. Batt, C.C.D. Santos, Microarray meta-analysis and cross-platform normalization: integrative genomics for robust biomarker discovery, Microarrays (Basel) 4 (2015) 389–406.

[173] J.T. Leek, Svaseq: removing batch effects and other unwanted noise from sequencing data, Nucleic Acids Res. (2014) e161, https://doi.org/10.1093/nar/gku864.
[174] J.T. Leek, W.E. Johnson, H.S. Parker, A.E. Jaffe, J.D. Storey, The sva package for removing batch effects and other unwanted variation in high-throughput experiments, Bioinformatics 28 (2012) 882–883.
[175] C. Chen, K. Grennan, J. Badner, D. Zhang, E. Gershon, L. Jin, et al., Removing batch effects in analysis of expression microarray data: an evaluation of six batch adjustment methods, PLoS One 6 (2011) e17238.
[176] Q. Liu, M. Markatou, Evaluation of methods in removing batch effects on RNA-seq data, Infect. Dis. Transl. Med. 2 (2016) 3–9.
[177] T.E. Sweeney, W.A. Haynes, F. Vallania, J.P. Ioannidis, P. Khatri, Methods to increase reproducibility in differential gene expression via meta-analysis, Nucleic Acids Res. 45 (2017) e1.
[178] W. Viechtbauer, Conducting meta-analyses in R with the meta for package, J. Stat. Softw. 36 (2010) 1–48.
[179] D. Reinhold, J.D. Morrow, S. Jacobson, J. Hu, B. Ringel, M.A. Seibold, et al., Meta-analysis of peripheral blood gene expression modules for COPD phenotypes, PLoS One 12 (2017) e0185682.
[180] J.R. Polanin, E.A. Hennessy, E.E. Tanner-Smith, A review of meta-analysis packages in R, J. Educ. Behav. Stat. 42 (2017) 206–242.
[181] T. Lumley, rmeta: Meta-Analysis. R Package Version, 2009, p. 2.
[182] A.A. Sharov, D. Schlessinger, M.S.H. Ko, ExAtlas: an interactive online tool for meta-analysis of gene expression data, J. Bioinform. Comput. Biol. 13 (2015) 1550019.
[183] S.E. Castel, F. Aguet, P. Mohammadi, GTEx Consortium, K.G. Ardlie, T. Lappalainen, A vast resource of allelic expression data spanning human tissues, Genome Biol. 21 (2020) 234.

CHAPTER 9

Guidelines and important considerations for 'omics-level studies

Francesca Luca[a,b,c] and Athma A. Pai[d]

[a]Center for Molecular Medicine and Genetics, Wayne State University, Detroit, MI, United States,
[b]Department of Obstetrics and Gynecology, Wayne State University, Detroit, MI, United States,
[c]Department of Biology, University of Rome "Tor Vergata", Rome, Italy, [d]RNA Therapeutics Institute, University of Massachusetts Chan Medical School, Worcester, MA, United States

The advent of high-throughput sequencing has made it possible to conduct unbiased genome-wide measurements of a large range of molecular phenotypes. While these techniques are increasingly easy to implement in all labs familiar with standard molecular biology experimentation the complexity of these datasets motivates many important considerations when undertaking 'omics-level studies. This chapter reviews several best experimental and analytical practices with a focus on RNA-sequencing studies for gene expression and mRNA splicing.

RNA sequencing (RNA-seq) refers to high-throughput sequencing of either the entire population or subsets of cellular RNA. This chapter focuses on the most common implementation of RNA-seq: complementary DNA (cDNA) sequencing after capture and reverse-transcription of polyadenylated messenger RNA (mRNA). This technique allows for the quantification of steady-state levels of mature mRNA molecules (fully transcribed, capped and polyadenylated), but often is not powered to analyze lowly expressed, transient, or quickly degraded mRNA species. RNA-seq data can be used not only for the quantification of gene expression levels, but also for identifying the composition of expressed genes or mRNA isoforms, quantification of specific mRNA isoforms, and evaluation of the relative inclusion of specific exons or splice sites in mRNA molecules. Thus, RNA-seq provides insight into molecular processes ranging from transcription, mRNA splicing and processing, and mRNA degradation.

Before high-throughput sequencing approaches, mRNA profiling was performed with microarrays spotted with a set number of probes in the 3' untranslated regions (UTRs) of genes or alternatively spliced exons. Following hybridization of cellular cDNA, fluorescence intensity was used to quantify gene expression or exon usage. These methods were dependent on prior knowledge of gene and isoform structure, prohibiting discovery of novel transcribed regions, isoforms, or alternatively spliced exons. Furthermore, if the annotation used to create the probe set was incorrect, measurements could bias biological insight into the gene expression or splicing. Finally, microarray approaches only allowed for limited genome-wide analyses, since the number of probes was limited by the microarray scaffold. RNA-seq approaches overcome these difficulties and allow for unbiased sequencing of cDNA that allow for simultaneous mRNA discovery, annotation, and quantification.

Overview of the RNA-sequencing experimental protocol

RNA-seq library preparation protocols not only share several steps with other high-throughput sequencing (HTS) protocols, but also pose unique challenges due to the delicate nature of the starting material. Here, we will review the major steps of the protocol to prepare libraries from polyadenylated mRNA, with a special focus on the steps that can be modified/adjusted for specific applications (Fig. 9.1A). One of the challenges for researchers that are getting started on RNA-sequencing is the length and complexity of the experimental protocol. However, aside from a few individual steps, the majority of the steps require basic molecular biology techniques.

It is crucial that RNase-free consumables and reagents are used at all times. This includes standard reagents like water, which should be molecular grade and nuclease-free for all steps. Whenever possible, RNA-seq protocols should be performed in a dedicated laboratory space with partition between pre- and post-PCR steps to decrease the risk of contamination.

RNA isolation is the first step in the protocol. Most RNA-seq applications use total RNA as starting material, followed by either mRNA capture or rRNA depletion. rRNA depletion is preferred when the study aims to profile the entire RNA population, including small RNAs and other RNA molecules that may not be polyadenylated. When limited material is available, it is possible to capture mRNA directly from cell lysate, thus skipping the total RNA isolation step and increasing RNA yield. This is usually a safe stopping point in the protocol and isolated RNA can be preserved at −80°C.

The RNA is then fragmented and primed for cDNA synthesis. cDNA synthesis is performed with random primers; however, particular applications may require using targeted primers. At this step, it is also possible to add Unique Molecular Identifiers (UMI) to allow for direct counting of RNA molecules (Fig. 9.1B). Following first and second strand cDNA synthesis, the library preparation protocol can be stopped and the cDNA stored at −20°C for up to 1 week.

The cDNA is cleaned with SPRI beads, followed by end repair and adapter ligation. This is a critical step as the adapter design depends on the library preparation kit and sequencing strategy of choice. During this step, or the subsequent PCR step, short oligonucleotides (indexes) can be added to each library to identify them when pooling multiple libraries in a single sequencing run. Ideally, dual indexing should be used to maximize the probability that reads are correctly assigned to their sample of origin despite sequencing errors in the indexes (Fig. 9.1B). This is particularly important when a large number of samples are pooled together in the same sequencing run and for certain sequencing platforms.

At this point, it is necessary to enrich fragments within a specific size range in the library. This step serves two purposes: (1) Removing potential adapter dimers that formed during ligation because of imbalances in the cDNA/adapter ratio, which may occur if cDNA synthesis had suboptimal efficiency and (2) removing large cDNA fragments, which could create steric interference on the sequencer during cluster generation. The correct beads-to-DNA ratio is used to efficiently remove very short fragments that are preferentially sequenced and could drastically reduce the sequencing yield.

Deciding the upper limit for size selection depends on the scientific question. Short reads provide adequate information for most applications; however, splicing and allele-specific analyses require longer reads (i.e.; 300 cycles) to increase the probability of sampling the splice junction or polymorphic site, respectively. Special consideration should be paid to both the size-selection and bead clean-up steps, as the techniques used are specific to HTS protocols. Beads are used to bind the nucleic acid (i.e., cDNA) and, depending on the beads:nucleic acid ratio, they can also be used to select a specific range of fragment lengths. Generally, size selection is most efficient for large fragment sizes (right side of

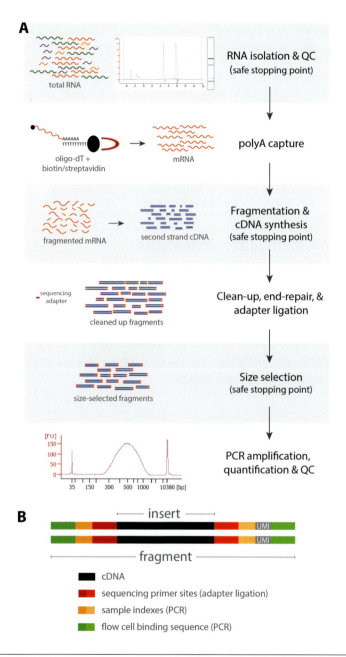

FIG. 9.1

RNA-seq experimental protocol. (A) Protocol steps with examples of Agilent Bioanalyzer results for a high quality RNA sample (RIN > 9, *top*) and for a final library (*bottom*). The average fragment size of the library is 500 bp, which is ideal for splicing and ASE analysis, in addition to gene expression quantification and differential analyses. This protocol is usually executed over 3 days, safe stop points indicated. (B) Final library construct design, including dual indexes (*yellow*) and UMIs (*gray*).

the fragment size distribution) as opposed to removing short fragments (left side of the fragment size distribution). It is advisable to practice bead clean-up and size selection prior to performing the first RNA-seq experiment.

The adapter-ligated DNA is ready to undergo PCR enrichment, followed by quality control of the library and determination of optimal loading concentration. The quality control step aims to confirm that the library produced is within the concentration range recommended for sequencing. It is advisable to use a capillary electrophoresis approach to QC the library, rather than an agarose gel or spectrophotometer to visualize and quantify the library, respectively. A library with low concentration may be indicative of low quality/concentration RNA, inefficient cDNA synthesis, issues with bead clean up steps or failed adapter ligation. Additionally, this step confirms that the prepared library is of the expected size and that adapter dimers were efficiently removed during size selection.

Study design considerations—Definitions

The key to a successful 'omics study is integration of experimental and data analysis considerations at an early study design stage. Many experimental details or protocol decisions that seem purely technical in nature are actually deeply connected to the biological question being asked. Ignoring these considerations may often lead to false-positive results. Designing a well-powered study begins with access to high-quality samples and is strongly dependent on the technical steps that constitute the experimental execution of the project. Here, we will focus on key experimental aspects that bear a significant weight in the overall success of a study and discuss how to integrate them in the data analysis considerations, which will be further expanded in the next section.

RNA quality

In some cases, a project starts with the investigator having full control over the sample collection. In this best case scenario, it is of critical importance to optimize the sample preparation so that the RNA integrity is preserved. However, it is more often the case that the researcher only has access to previously collected samples, thus having limited control over the RNA quality. RNA integrity is crucial to the success of an RNA-sequencing experiment. It is important to note that RNA-sequencing is a less-forgiving experiment than quantitative real-time PCR.

It is standard in the field to report RNA quality using an RNA Integrity Number (RIN), which varies on a scale of 1–10 (10 being best and 1 being worse). While the RIN of RNA isolated from cell cultures or model organisms in the laboratory is usually very high (9 or greater), the RIN can be quite low for RNA isolated from other sources. Nevertheless, RNA-sequencing of postmortem samples in humans has yielded high-quality libraries with minimal RIN values of 6 [1]. When RIN varies greatly across samples, it is advisable to include it as a technical variable/confounder in statistical models for RNA-seq analysis [2]. RNA quality also affects library complexity and data quality. A library with low complexity results in a high proportion of reads from PCR duplicates, which are identical copies of independent cDNA fragments but do not contribute independent information to the overall measure of gene expression.

To limit the occurrence of PCR duplicates, RNA-seq protocols minimize the number of cycles during the PCR enrichment step. However, when low concentration or low-quality RNA is used as input, even a low number of PCR cycles may result in a high proportion of PCR duplicates or to library preparation failure. Another feature of RNA-seq data that is influenced by RNA quality is the proportion of reads mapping to exons. When this number is low, it may indicate genomic DNA contamination, RNA degradation and/or low efficiency in the polyA RNA capture or rRNA depletion. Thus, the proportion of reads mapping to exons is another important technical variable to consider when modeling gene expression during data analysis.

Confounders

Technical variables for RNA quality are an example of experimental confounders that may complicate the analysis of RNA-seq data. The most well-known experimental confounders are batch effects, which is an umbrella term for any grouping of experimental samples that may result in features of the data that are shared within the group and do not have direct biological relevance for the research question. Batch effects can be introduced at any step along the experimental protocol, starting with sample collection, RNA isolation, library preparation, and sequencing. Working in batches is inevitable in functional genomics experiments, where the number of samples and experimental conditions considered is usually large. Therefore, while batch effects are unavoidable, it is critical to limit batch effects with careful study design, to record confounders for further consideration during data analysis, and to avoid batch effects that are confounded with the research question.

In particular, when designing a study, batches should be balanced for all variables relevant to the research question (Fig. 9.2). When relevant variables are too numerous, they can be randomized across batches. For example, in a study aimed at identifying gene expression differences across multiple cell types, it may be difficult to have each cell type represented in a batch because of different growth rates. In this case, the cell types included in each batch could be randomized to avoid batches that contain exclusively or mostly one cell type.

Finally, it is crucial to carefully define the research question(s) prior to beginning sample collection. This will ensure that sample groupings in batches do not confound group contrasts. The example in Fig. 9.2 presents a hypothetical study aimed at identifying gene expression differences between cases and controls for a pathological condition, thus disease status is the variable of interest. Let us imagine that after the data are collected the investigator decides to use this dataset to also study the effect of medication on gene expression. If the batches are designed only considering disease status as the variable of interest, it is likely that some batches only contain medicated subjects, whereas others are overrepresented in unmedicated subjects (Fig. 9.2; Batch 4). Therefore it would be impossible to investigate differences in gene expression due to medication. However, if the batches are balanced also by medication status, they can be used to investigate this additional research question. It is also possible that changes in gene expression associated with disease are also correlated to differences due to medication status, so the best practice would be to control for this variable regardless of the additional research question.

Replicates

Replicability is one of the key concepts in science. A rigorous finding is based on results that have been replicated in independent experiments or sample groups. Because of the complexity of genomics experiments, replication may take several forms and occur at multiple steps along the experimental

194 Chapter 9 Important considerations for 'omics-level studies

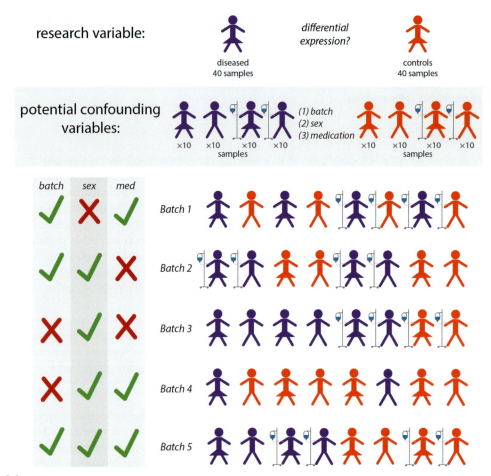

FIG. 9.2

Study design and confounders. Example of a study design to identify differential gene expression between diseased and control patients. The sample size is 40 for each group (balanced sample size). Potential confounding variables are the patient's sex and medication status. The bottom panel presents different batch designs. For each batch design, the left column indicates if the study design is balanced (*green check*) or if the study variable (disease status) is confounded with any of the other variables (sex, medication status, batch) (*red cross*). In a well-designed study, the variable of interest (disease status) is not confounded with any additional variable.

process. Depending on which segment of an experiment is replicated, replication may have caveats or yield conservative results. Researchers across scientific fields have not reached a consensus on the definition of technical and biological replicates. Intuitively, these two terms are based on the sources of variability that may contribute to a lack of reproducibility. If the source of variability is biological, repeating the experimental step to replicate the sampling of this biological variability would be considered a biological replicate. If the source of variability is technical—does not originate from a biological process but is inherent to the experimental procedure—replicating the relevant experimental step would result in a technical replication.

Some examples may help illustrate these points. In the example shown in Fig. 9.2, experiments performed on samples from different individual donors from the same disease status group are considered biological replicates. If multiple aliquots of the same cell pellet are used to independently isolate RNA and prepare RNA-seq libraries, these would be technical replicates of the RNA isolation and library preparation steps. However, other types of replication experiments are possible and their definition becomes more controversial. For example, for some researchers, independent growth of cell cultures from the same donor would be considered biological replicates of the culturing process, while for other researchers they are technical replicates of the experimental procedure to grow cells. Ideally, one would perform replicates of different experimental steps to identify the most relevant sources of variation and focus on those for all subsequent experiments. However, this strategy may not always be feasible and researchers often prefer full biological replicates, when possible.

Sequencing depth

When designing an 'omics experiment, the experimenter must also consider the final characteristics and amount of data needed to answer specific biological questions. For high-throughput sequencing experiments, it is important to think about the number of short reads needed to estimate biological parameters at genome-wide scale. This is referred to as *sequencing depth*, defined as the number of reads sequenced from a library. Importantly, the depth is determined by the experimenter, through two main choices. First, the total read output differs between sequencing instruments. For Illumina short-read sequencing, the number of reads sequenced on a single run can range between 4 million on an iSeq 100 instrument to 2.6 billion on a NovaSeq 6000. Second, users can determine the proportion of reads from a lane or flowcell that are allocated for each library by multiplexing libraries. *Multiplexing* refers to the practice of mixing multiple different libraries to sequence them together. This is achieved by using unique library-specific indexes during library preparation (usually during the PCR step as described earlier) in Fig. 9.1B, which are then used to disambiguate libraries while mapping (see in the following text).

If multiplexing, it is critical not to combine two libraries with the same indexes and most ideal to use a set of indexes that all have at least two nucleotide differences from another index. Indexes are generally 6- or 8-mers. While there is no limit to the number of libraries that can be mixed together, higher multiplexing reduces the number of reads obtained from each library and can present challenges for the downstream computational separation of reads from individual libraries.

The ideal sequencing depth of a library is determined by the anticipated complexity of a library and genomic coverage. Library *complexity* refers to the number of unique molecular fragments within a library, driven by the amount of starting material and the target molecular enrichment (i.e., mRNA, small RNAs, protein binding peaks, etc.), see earlier. Libraries prepared with very low amounts of material or targeting low-frequency genomic regions are likely to have lower complexity, leading to saturation of unique biological information with lower numbers of reads. Standard RNA-seq libraries are usually extremely complex, owing to the large diversity and dynamic range of mRNAs in most cells. However, there are exceptions—for instance, RNA-seq libraries from red blood cells generally have low complexity as red blood cells are enriched for hemoglobin mRNA [3,4].

Genomic *coverage* refers to the number and distribution of reads across all expected genomic regions. Higher sequencing depth usually results in higher coverage—specifically, a more even distribution of reads across all regions (not just highly expressed genes) and enough reads at each region to provide power to make biological conclusions. For DNA sequencing experiments, coverage is usually a primary consideration, where Nx coverage refers to N reads overlapping every base or region on average.

Decisions about how deeply to sequence a library are usually driven by a balance between statistical power and experimental cost for replicate samples. Too few reads would lead to lower coverage and thus prohibit robust and reproducible quantification of genes or isoforms [5]. This would particularly affect the quantification of low- to medium-expressed genes and result in low confidence for exon-specific analyses in these genes, which rely on rare splice junction reads. On the other hand, it is expensive to achieve higher sequencing depths for each library and potentially not needed. For highly complex RNA-seq libraries, studies have found that gene expression can be robustly estimated using 10–15 million fragments (10–15 million single end reads or 20–30 million paired end reads) and isoform or splicing analyses can be conducted with 30–50 million fragments on average. A similarly high sequencing depth (40–100 million fragments or more) is required also for allele-specific expression analysis, to confidently calculate allelic ratios. If preliminary analyses indicate that more reads are necessary, it is always possible to re-sequence the same library to obtain more reads, but it is necessary to ensure that potential confounds are taken into account before combining data from two runs (described earlier).

Read length and type

Two important sequencing parameters are the length of the reads and the number of reads sequenced from each independent fragment. Read length is determined by the number of sequencing cycles run on the sequencer and can range between 30 and 300 nucleotides on an Illumina machine. Note that specific sequencing kits must be used for higher cycle numbers, with different instruments being able to sustain longer sequencing runs (i.e., MiSeq will sequence 300 nt reads). Fragments in the library can either be sequenced from only one end (single end read) or from both ends (paired end read). When choosing read length and single vs paired end reads, it is important to consider the average *insert size* of the library, defined as the length of the cDNA fragment between the two adapters (fragment length—total adapter length; Fig. 9.2B).

Read lengths longer than the insert size will result in sequencing into the adapter, which can be avoided with shorter reads. Similarly, if the insert size is shorter than the sum of the paired end reads, the read pair will be providing redundant information, artificially increasing the coverage at the overlapping region. If aiming for a specific read length for single or paired end reads, then it is necessary to prepare the library with an appropriate fragment length (see the earlier description). After sequencing, software tools like Picard [6] can be used to infer the insert length from the data to determine whether adapter trimming might be necessary or paired ends reads are likely to be overlapping.

Decisions about read length and pairing influence the robustness of the downstream analyses. First, these parameters influence the ability to confidently map or align reads to the genome. Longer reads are more likely to map to unique positions in the genome. Similarly, paired end reads can help to position reads since both reads can be used to disambiguate between similar regions. For instance, if one read maps to a homologous or repeat region, the second read may be used to properly position the entire fragment if mapped to a unique region. Second, longer reads and paired end reads aid with isoform quantification and splicing analyses. Longer reads are more likely to overlap an informative splice junction, providing direct evidence of a splicing event. Analogously, long reads are more likely to overlap one or multiple heterozygous site(s) for allele-specific analysis purposes. Paired end reads can be used to quantify splicing of an exon even when both reads are within flanking exons by inferring whether the pair of reads could have arisen from an unspliced versus spliced transcript given the average insert length of the library.

Analysis of RNA-seq data

Similar to other 'omics applications, data generation is just the first step. A complex and careful data analysis plan is necessary to preprocess and QC the data and to test the hypothesis or identify patterns of biological interest. Importantly, data analysis should not be seen as independent from data generation, as they are two highly interconnected steps of the overall RNA-seq experiment. Ideally the experimentalist and data analyst are either the same person or work very closely in designing the experiment, recording potential confounders, and interpreting the data. This section presents key concepts and approaches to RNA-seq data analysis, many of which are also relevant for other 'omics applications.

Quality control

Before initiating any biological analyses, it is important to ensure that high-throughput sequencing data is of high quality. As described earlier, there are several experimental and technical considerations that could influence data quality and bias gene expression levels. Experimental considerations like unhealthy or dying cells and poor RNA quality must be assessed prior to sequencing a library, since they are impossible to assess in the final data but can have a large impact on biological interpretation. In contrast, sample preparation issues resulting in DNA, rRNA, or adapter contamination in the final library should be avoided but can be assessed in the final library as described in the following text.

When such sample preparation issues are identified, reads can either be filtered or libraries can be re-prepared after DNase digestion, further rRNA removal or polyA selection, and selection for larger fragment sizes to correct the issues listed earlier, respectively. It is customary to run several quality control checks on raw sequencing data before proceeding with mapping and quantifying mRNA levels. These quality control checks include examining the distribution of read quality scores (for confidence in base calls), evaluating overrepresented sequences (to assess library complexity), adapter reads, adapter contaminated reads (at the ends), and read quality near the ends of reads. There are many packages (i.e., fastQC, RNA-seQC [7,8]) that streamline the implementation of these checks and provide visual analyses for quick evaluation of data quality.

Mapping of reads

The first step in any high-throughput sequencing analysis workflow is to "map" reads to genomic coordinates or align reads to reference genome or transcriptome sequences. While there are dozens of mapping software that have been written for short read sequencing data, each has specific properties that are important to consider based on the biological question. Here, we focus on two considerations

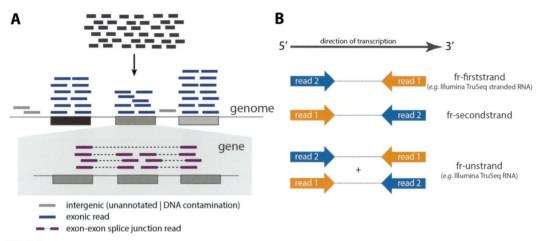

FIG. 9.3

Mapping and strandedness. (A) RNA-seq reads include intergenic (DNA contamination), exonic, and exon-exon splice junction reads. A splicing-aware software allows for proper mapping of exon-exon splice junction reads and quantification of DNA contamination. (B) Libraries can be prepared using stranded or unstranded protocols. The details of the protocol are important to infer correct strand information from mapped data or input into downstream quantification or read counting algorithms.

that are specifically crucial for RNA-seq analyses. First, when mapping RNA-seq data, it is important to use a splicing-aware mapper (i.e., TopHat [9], STAR [10], HiSAT2 [11]) to map splice junction reads, which contain large genomic gaps that cannot be handled by standard genome mappers (Fig. 9.3A). Though it is possible to use a nonsplicing-aware mapper to map RNA-seq reads to a transcriptome reference sequence instead of the genome, it is more advisable to map to a genome reference sequence. Mapping to the genome with a splicing aware mapper allows for quantification of novel transcribed regions or splice junctions after scaffolding on known splicing events and allows for the quantification of DNA contamination (reads mapping to intergenic or intronic regions) (Fig. 9.3A).

However, when sequencing RNA from a species that has no reference genome or a poorly annotated one, it may be advisable to use the RNA-seq data to assemble a reference transcriptome, scaffolded on the genome or transcriptome of a closely related species. Second, for genetic analyses (e.g., allele-specific expression), it is important to avoid mapping artifacts that lead to a biased allelic representation among the mapped reads. Since mapping software rely on a reference genome, reads are more likely to be mapped properly when they exactly match the reference alleles. This can be alleviated using a haplotype or genotype aware mapper (i.e., HISAT2), which accounts for both allelic options during mapping. Furthermore, this problem can be exacerbated when there are experimentally driven substitutions in the data, such as in SLAM-seq experiments where uridines labeled with 4sU appear as T>C substitutions in the sequencing data [12]. In these cases, it is useful to downweigh known substitutions using a mapper such as NextGenMap [13] (not splicing aware) or HISAT3n [14] (splicing aware). Lastly, following mapping, it is worth looking at raw data on a genome viewer (such as iGV [15]) to identify any issues and evaluate your expectations.

Note that read alignment software will assign reads to the strand that directly matches the read sequence. However, the assigned strand may not reflect the strand from which the parent transcript was transcribed. If the library was not prepared with steps to maintain strand information, reads have an equal probability of mapping to either strand. If a stranded library preparation was used, the transcriptional strand can be identified. Libraries can be prepared using one of two strandedness protocols, which result in the first read matching either the sense or antisense strand (Fig. 9.3B). Thus, it is important to identify the experimental specifics to infer correct strand information from mapped data or input into downstream quantification or read counting algorithms.

Quantifying gene and isoform expression

When using RNA-seq data to quantify mRNA expression levels and compare the expression levels across genes or samples, it is crucial to account for two parameters. First, read counts must be normalized by the total number of reads sequenced. If sample 1 has been sequenced to a depth of 50 million reads and sample 2 only has a total of 10 million reads, each gene is likely to have 5 times more reads in sample 1 than sample 2 even if the abundance of all mRNAs is equivalent between samples. Second, it is necessary to account for the total length of a gene or isoform. If gene 1 is longer than gene 2, there is a higher probability of sequencing reads from gene 1 independent of gene expression levels. Thus, mRNA expression levels are generally computed with one of two metrics that accounts for these confounding variables: Reads per kilobase per million (RPKM) and Transcripts per million (TPM).

RPKM: A straightforward metric that simply divides read counts by the total number of reads sequenced for the library (in millions) and then by the length of the transcribed region from which reads are counted (in kilobases) [16]. A variation of this is fragments per kilobase per million (FPKM), which is applied to paired end reads and uses the count of read pairs rather than individual reads, since paired reads are not statistically independent from each other.

TPM: A similar metric, where read counts are first divided by length and then by the total number of reads sequenced [17]. While this seems like an inconsequential mathematical change, the difference in normalization order greatly improves the ability to compare relative gene expression levels across samples. The second normalization by the total number of million reads forces the sum of TPMs within each sample to be 10^6 (not true for RPKM or FPKM metrics), which standardizes the proportional levels across samples. Importantly, the need to normalize by library-specific parameters makes it only possible to quantify relative expression levels (rather than absolute) for standard RNA-seq libraries. For these reasons, TPM has become the preferred metric to quantify gene expression from RNA-seq data.

Either of these quantification metrics can be calculated in several ways. To estimate RPKM or TPM by hand, the researcher first needs to count reads for each exon or gene feature (using software such as ht-seq [18] or featureCounts [19]) and then use the formulas described earlier to calculate the desired metric. These read counts, either using only uniquely mapping or including reads that map to multiple locations, can also be used as input for differential expression analyses, for which statistical models are run on raw rather than normalized counts (see in the following sections). A more robust method of quantification involves using a statistical maximum likelihood model to probabilistically assign multimapping reads across the multiple locations to which they map. This approach is implemented in software like RSEM, which uses an expectation-maximization algorithm to calculate maximum likelihood abundance estimates from mapped reads scaffolded on known transcriptome annotations [17].

This approach will output: (1) adjusted read counts that accounts for the expected read count based on both uniquely and multimapping reads and (2) TPM values using these adjusted read counts. Finally, a more extreme implementation of the expectation-maximization algorithm is used in reference-free pseudoalignment-based isoform and gene quantification approaches (i.e., Kallisto [20] or Salmon [21]), which do not rely on an initial alignment step but match k-mers within reads to compatible transcripts to obtain maximum likelihood isoform or gene abundance estimates. By circumventing the mapping step, this reference-free approach alleviates issues arising from mapping biases of individual regions and instead focuses solely on quantification of known isoform sequences.

Pseudoalignment quantification approaches are extremely fast and result in TPM levels for isoforms, however they do not provide a list of mapped regions with their genomic coordinates. Since both maximum likelihood approaches quantify gene expression using a full cohort of reads rather than only those mapping to unique regions, they are more likely to provide robust abundance estimates. Furthermore, these approaches enable isoform level quantification, which is harder to do with read counts alone where exon features can be shared across isoforms. It is important to note that however isoform-level quantification always relies on known annotations and specifically known junctions between exons. When these annotations are incorrect or incomplete, they can bias isoform TPMs. However, it is always possible to add the isoform TPMs from a single gene to get a gene-level TPM, which is more robust to annotation biases since it relies only on knowledge of exonic regions rather than specific splicing patterns.

Quantifying absolute mRNA abundance

All quantification methods within samples highlighted earlier are inherently relative measurements, since they must account for differences in sequencing depth and other library-specific properties. Specifically, metrics such as TPMs are designed to describe the relative proportion of mRNA that arises from a given gene—a gene with TPM=1 represents 1 millionth of the total mRNA population, while TPM=1000 represents a thousandth of the total population. Thus, an increase in TPM between samples supports an increase in the relative proportion of mRNA represented by that gene but does not allow any conclusions to be drawn about the absolute amount of mRNA produced from that gene.

To perform absolute quantification, libraries must be designed to account for differences in cell count and total RNA yield between samples, the loss of material during RNA extraction and library prep, and any biases that occur during sequencing. The most popular method to do this is to spike a pool of exogenous RNA into a sample at the beginning of RNA extraction or library preparation. This "spike-in" RNA can be from a different species (preferably far enough diverged from the species of interest to allow abundant sequence dissimilarity) or a population of synthetic RNA designed to account for different transcript lengths and sequence compositions (i.e., the controls designed by the External RNA Controls Consortium (ERCC)) [41].

Furthermore, if a polyA selection is performed, the exogenous RNA must have polyA tails. A small amount of exogenous RNA is spiked in relative to the number of cells (or a similarly relevant parameter like the weight of tissue) used to extract RNA from the sample of interest, which allows calibration for differences in total RNA yield and thus absolute RNA abundance between samples [22]. This addition should happen as early as possible in sample preparation (i.e., cell lysis) so that the spiked-in RNA goes through the entire sample preparation process simultaneously with the sample of interest. Sequencing reads from a library with spiked-in RNA should be mapped to a genome where the spike-in sequences or genome of the spike-in species have been combined with the genome of interest as distinct and labeled chromosomes. To obtain an abundance estimate that accounts for absolute differences in RNA abundance, TPMs from the samples of interest can then be normalized by the proportion of reads that map to the spike-in samples.

Comparing gene expression between groups

A common application of RNA-seq is to compare expression levels between two groups of samples and identify differentially expressed genes. This general framework applies to several research questions, for example identifying gene expression changes induced by a treatment in vitro or in vivo or comparing expression in treated and untreated samples. Another common scenario is the comparison of gene expression between cases and control groups to identify genes that are differentially expressed in a disease state. Finally, gene expression levels are also commonly compared across developmental stages and tissue or cell types. Generally, the researcher will identify a dichotomous variable of interest, which will be used to partition the samples into two groups. Because of the confounders discussed earlier, the variable of interest must be defined before collecting the RNA samples and performing the experiment, thus ensuring that any potential batch effect is not confounded with the variable of interest.

Read length is not a critical factor for these studies, as even the shortest reads (75 cycles split over pair-end reads on the Illumina NextSeq500) are sufficient for accurate gene expression quantification and comparison between groups. Many studies use a sequencing depth of 20–40 million paired-end reads per sample, however studies have demonstrated that even <10 million reads per sample are sufficient to determine whether the variable of interest is associated with gene expression differences between the two groups (e.g., Ref. [23]). If the researcher is interested in a detailed characterization of gene expression changes between two groups, greater sequencing depth is necessary to capture smaller effects.

Prior to performing formal statistical tests for differential gene expression, a few data processing/visualization approaches are helpful to identify major axes of variation in the dataset, determine if the data present any unexpected structure (due for example to an unmeasured confounder), and confirm which confounding variables should be included in the statistical model. Principal component analysis of the gene counts and a scatter plot of the first few (1–4) PCs can be used to visualize if the samples form unexpected clusters. Correlation analysis of the PCs with the variable of interest and potential confounders can also be used to quantitatively determine the main axes of variation in the data. Pairwise correlation between all samples, followed by visualization, is a complementary method to visualize the data structure and identify unexpected clustering of the samples. This can be further investigated through hierarchical clustering on the correlation matrix.

To identify genes differentially expressed between the two groups, a linear model is used to model expression of each gene as a function of the variable of interest, of measured confounders and error. Common packages such as EdgeR [24] and DESEQ2 [25] implement a fixed effect linear model, which does not allow random effects between individuals to be modeled. This is generally acceptable for most study designs; however if multiple measurements of the same individual are included (i.e., in time-series experiments or experimental replicates), a random effect linear model or a mixed effect linear model are more appropriate, as they will account for the correlation structure that exists in the data due to multiple measurements from the same individual. LIMMA [26] is one of the most common packages used to analyze RNA-seq data using a random effect linear model.

One key difference between these two classes of methods is also the data preprocessing. Methods such as DESEQ can be applied directly on the count matrix and use a negative binomial model to estimate the overdispersion parameters in the data, thus not requiring an additional normalization step. The LIMMA package includes a few different normalization approaches that are performed prior to testing for differential gene expression. In both cases, the output will be a measure of the expression change for each gene, usually indicated as average log2(fold change) and the statistical significance (P-value).

To account for the large number of tests conducted, it is important to perform multiple test correction. Two common multiple test correction approaches are the Benjamini-Hochberg P-adjusted to control for the false discovery rate [27] and Storey's q-value [28]. The final list of differentially expressed genes will be defined at a certain false discovery rate (q-value or P-adjusted) and potentially also after setting a threshold on the fold change in gene expression if the researcher is interested in filtering effects based on the magnitude of the change. To ensure that the tests are well calibrated, it is common to create a qq-plot to compare the observed P-value distribution to the expected uniform distribution. Volcano plots are also used to visualize the relationship between fold change and P-value and set the aforementioned thresholds.

While gene expression comparisons between two discrete groups is a common framework for many research questions, sometimes the variable of interest may not be dichotomous (e.g., presence/absence of treatment) but varies continuously in the entire sample. Examples of continuous variables are time/age (e.g., in time-series experiments), dosage (e.g., in dose-response experiments), serum analytes levels and many others. In these cases, there are two options. First, the variable can be encoded as continuous and a linear model is used to test for an association between the variable of interest and gene expression over a continuous scale. Alternatively, the variable can be discretized by creating arbitrary breaks that may reflect experimental or biologically meaningful categories (e.g., low, intermediate, high level of serum analyte). In both cases, the underlying assumption is a linear relationship between gene expression and the variable of interest. However, this assumption may not always be true. For example, in a dose-response curve, the expression of a gene may increase initially and then plateau or even decrease, depending on the underlying regulatory mechanism and potential feedback loops. A simple but limited approach to this problem is to reduce these types of datasets to multiple pairwise comparisons and compare each data point to the previous in the series or to the baseline. While this is still an ongoing area of statistical development, clustering approaches have also been used effectively to identify patterns of gene expression in series of measurements.

Studying the genetic determinants of gene expression

An application of RNA-seq that has become widely popular in the past decade is expression quantitative trait locus (eQTL) mapping (Fig. 9.4). With the decrease in sequencing costs and the ability to perform RNA-seq on a large number of samples, researchers have paired RNA-seq data with genotype data to identify associations between genotype at a locus and expression of a nearby (*cis*-eQTL) or distant (trans-eQTL) gene. Here, we will discuss eQTL mapping in general with a focus on *cis*-eQTLs since those are the easiest to identify, but many of the considerations presented are

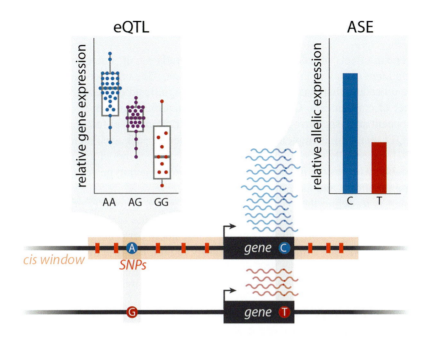

FIG. 9.4

Methods to characterize genetic regulation of gene expression. In the eQTL example (*left*), each dot of the boxplot represents an individual. The median gene expression is indicated by the line in the box and is calculated across individuals within each genotype class. A significant association is present between genotype at the noncoding SNP and gene expression across individuals, with higher expression associated with increasing number of copies of the A allele. The ASE example (*right*) shows the distribution of reads at a coding heterozygous site within one individual sample. A significant allelic imbalance is present between the C and T alleles, with higher expression of the C allele. In the absence of additional information, the eQTL has a higher probability of being the true causal variant compared to the coding SNP.

also applicable to *trans*-eQTL mapping. Furthermore, the general QTL mapping approach can also be used to map loci associated with other molecular phenotypes, and this can be achieved by combining genotype data with other types of 'omics data (e.g., splicing QTLs, chromatin accessibility QTLs, etc.).

Generating RNA-seq data for eQTL mapping presents unique challenges compared to the other applications discussed here. First, these studies need a much larger sample size. There is good consensus in the scientific community, supported by both power calculations and large scale studies (e.g., GTEx), that a sample size of at least 70 individuals is sufficient to detect large genetic effects on gene expression for a large number of genes. Though eQTL mapping studies have also used thousands of samples, they are still quite smaller than the hundreds of thousands of samples used for genome-wide association studies. Working in batches for eQTL mapping is inevitable; therefore, all the study design considerations discussed earlier apply. Furthermore, if the goal is to compare genetic effects on gene expression across conditions (response-eQTL mapping, reQTL), the different conditions should be processed in parallel within the same batch. Sequencing depths and read lengths appropriate for the comparisons between groups are also appropriate for eQTL mapping.

Because genetic effects on gene expression are generally small; unmeasured and measured confounders have a large impact in this application of RNA-seq. The effects of confounders are removed via an iterative procedure that maximizes the number of significant eQTLs by progressively removing a larger number of confounders. Confounders can be quantified through several methods, including principal component analysis, surrogate variable analysis [29] and peer factor analysis [30]. The statistical test for eQTL mapping is implemented in a few different packages, including FastQTL [31] and MatrixQTL [32]. Essentially these are different flavors of a linear model where gene expression is the dependent variable, whereas the independent variables are the genotype dosage (0, 1, or 2 number of copies of the minor allele), and the confounders.

The first step is to define the *cis*-regulatory region, which is usually arbitrary defined between 100 kb and 1 Mb. The RNA-seq data can be normalized using the same procedure as in the LIMMA preprocessing (commonly Voom normalization). Association between genotype and expression is tested for each gene/SNP pair within the *cis*-regulatory region. Because the tests are not independent, due to linkage disequilibrium between SNPs in the same *cis*-regulatory region, the P-value distribution is often inflated. Permutations are used to correct the P-value distribution, followed by Storey's q-value method for multiple test correction.

Identifying response eQTLs, which are genetic loci associated with a difference in gene expression levels upon stimulation or cellular perturbation, is more challenging and there is limited consensus on the methodological approach. The traditional approach uses the same statistical framework used for eQTL mapping but with expression change as the dependent variable. However, this approach has limited power and can only capture the most extreme cases, when the association between genotype and gene expression is only present in one condition or has opposite sign in the two conditions considered. In most cases, the genetic effect on gene expression undergoes more subtle changes across conditions, thus more sophisticated methods are needed. Bayesian methods have been successfully applied to this problem, and have demonstrated to have the potential to identify condition/context-specific genetic effects across several groups and beyond the two groups' comparison [33].

An alternative and complementary approach to QTL mapping is allele-specific analysis (Fig. 9.4). As for QTL mapping approaches, allele-specific analysis can be performed on different types of 'omics data. Allele-specific expression (ASE) analysis is used to identify genes with regulatory variants from RNA-seq data. ASE analysis is performed only on genes that contain heterozygous variants and assumes that the true causal site is a variant in the regulatory region, which is also heterozygous in the sample under consideration. Under the null hypothesis of absence of regulatory variation at a gene locus, the two alleles at the heterozygous coding site should be represented by a similar number of reads in the RNA-seq data. ASE is defined as a departure from this 50:50 allelic ratio, and is formally tested with a beta-binomial model that accounts for overdispersion in the allelic read counts. It should be noted that in rare cases, ASE may result from technical or biological processes that are independent of gene regulatory variants (e.g., imprinting). ASE analysis can be performed in a single individual sample, which is an advantage over eQTL mapping. Furthermore, trans-effects are fully controlled because allelic effects are compared within the same individual. However, detection of ASE for a gene is limited by the number of polymorphic sites in the coding region and by their heterozygosity.

Analysis of alternative splicing events

While some analyses of mRNA splicing patterns can be conducted by using isoform-level TPM values, it is often more precise to delve into individual exon-level changes to understand alternative splicing changes. There are several types of exon-level changes that are often studied, which can be broken down into three main categories: (1) alternative transcription start sites, including alternative first exons, (2) alternative splice sites, including alternative 3′ or 5′ splice sites, cassette or skipped exons, mutually exclusive exons, and retained introns, and (3) alternative polyadenylation sites, including alternative last exons and tandem 3′ untranslated regions (Fig. 9.5).

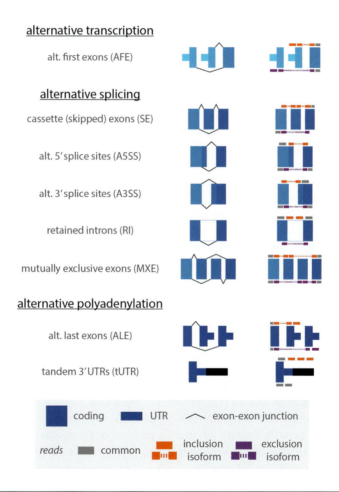

FIG. 9.5

Alternative splicing and other exon-level changes. Examples of alternative transcription, splicing, and polyadenylation events that can be quantified in RNA-seq data. The left column indicates the splicing isoforms and the right column shows the reads that are informative for quantifying the usage of inclusion (*top, orange*) or exclusion (*bottom, purple*) isoforms.

While the first and the third are usually quantified using read coverage in the relevant exons, analyzing alternative splice site usage relies on combined exonic and junction read coverage. Popular metrics to quantify alternative splicing include percent spliced in (PSI as in Ref. [34]) or percent alternative usage (PAU as in Ref. [35]), which broadly calculate the percentage of reads that support a defined inclusion isoform (i.e., inclusion of a cassette exon) relative to the total informative reads in the region.

Quantification software generally takes one of two approaches: (1) local read analysis or (2) isoform-anchored local event analysis. In the first approach, exonic and junction reads specific to a particular event are used to assign an PSI value, where the definition of alternative events and inclusion versus exclusion isoforms is based on known annotations [34,36]. Since this method relies heavily on junction reads, many software packages have options to initially discover novel events in a dataset and incorporate these into downstream quantification [37,38]. Crucially, all quantification occurs only at a local level, agnostic to what happens upstream or downstream of the specific event of interest and thus avoiding any biases caused by long-range isoform annotation. However, since the approach gains power through a high density of informative reads within a small region, many events do not have sufficient read density to allow for quantification.

The second approach uses isoform-level TPM quantification (as described earlier) to infer a quantification of local event usage by aggregating information from isoforms that include versus exclude a given exon or site [39]. Since the definition of isoforms is necessary, this approach often precludes de novo identification and quantification of events. Additionally, the reliance on long-range annotations could bias exon-specific quantification if isoform structures are incorrect or incomplete. However, isoform-anchored local event analysis is often much quicker and easier to implement, making it a good option for a first pass analysis. Furthermore, since isoform level TPM quantification uses reads across the entire isoform, these approaches often have more power to quantify alternative splicing in lower expressed genes. Finally, these approaches are often better suited for measuring terminal exon usage of either alternative first or last exons since those event types may not have informative junction reads [35,40]. Regardless of the method used to quantify PSI or PAU values, these values can be used for splicing QTL analyses with the same considerations and statistical approaches described earlier.

Summary

In this chapter, we summarize the experimental and computational considerations for high-throughput RNA-seq libraries. We focus primarily on the most common polyA RNA-seq libraries, which are enriched for mature mRNA molecules. However, there are many other types of RNA-seq libraries that either enrich for different RNA species (i.e., total RNA population, small RNAs) or target RNA at different stages of their lifecycle (i.e., chromatin associated RNA, nuclear or cytoplasmic RNA, ribosome-associated RNA). Here, we walk through the specific protocol and important considerations for polyA library preparation. Importantly, we go through several best practices for the study design of these experiments, including measuring the quality of RNA prior to starting libraries, potential biological and technical confounders, replicates necessary for statistical analyses, and sequencing parameters. Designing a well-controlled and well-calibrated study is important for any high-throughput experiment, so most of these considerations apply to all high-throughput sequencing studies. Furthermore, they will likely all continue to be important regardless of future experimental or computational developments in the field. In our last section, we outline steps for analyzing RNA-seq datasets, with a focus on initial QC, gene expression measurements, differential expression and splicing, and using these data to study the influences of genetic variation on mRNA levels and compositions.

Finally, it is important to note that there is constant development of new high-throughput sequencing protocols to sequence different subsets of RNA, with many new technologies being published and widely adopted each year. Similarly, statistical analysis methods, software, and study design standards are an area of active research and new techniques may correct previous biases or improve biological interpretability of results. Thus, it is crucial to survey the literature for new developments prior to undertaking an RNA-seq experiment to ensure conformation with the latest best practices. Our hope is that the considerations presented in this chapter can provide a useful guide to identify important and relevant aspects when keeping up to date in the field.

References

[1] GTEx Consortium, The genotype-tissue expression (GTEx) project, Nat. Genet. 45 (6) (2013) 580–585.
[2] G. Romero, A.A. Irene, J.T. Pai, Y. Gilad, RNA-Seq: impact of RNA degradation on transcript quantification, BMC Biol. 12 (May) (2014) 42.
[3] C.A. Harrington, S.S. Fei, J. Minnier, L. Carbone, R. Searles, B.A. Davis, K. Ogle, S.R. Planck, J.T. Rosenbaum, D. Choi, RNA-Seq of human whole blood: evaluation of globin RNA depletion on ribo-zero library method, Sci. Rep. (2020), https://doi.org/10.1038/s41598-020-62801-6.
[4] F. Uellendahl-Werth, M. Wolfien, A. Franke, O. Wolkenhauer, D. Ellinghaus, A benchmark of hemoglobin blocking during library preparation for mRNA-sequencing of human blood samples, Sci. Rep. 10 (1) (2020) 5630.
[5] D. Sims, I. Sudbery, N.E. Ilott, A. Heger, C.P. Ponting, Sequencing depth and coverage: key considerations in genomic analyses, Nat. Rev. Genet. 15 (2) (2014) 121–132.
[6] P. Toolkit, Picard Toolkit, Broad Institute, Github Repository, 2019.
[7] S. Andrews, et al., FastQC: A Quality Control Tool for High Throughput Sequence Data 2010, Babraham Bioinformatics, 2017. https://www.bioinformatics.babraham.ac.uk/projects/fastqc/.
[8] A. Graubert, F. Aguet, A. Ravi, K.G. Ardlie, G. Getz, RNA-SeQC 2: efficient RNA-Seq quality control and quantification for large cohorts, Bioinformatics (2021), https://doi.org/10.1093/bioinformatics/btab135. March.
[9] C. Trapnell, L. Pachter, S.L. Salzberg, TopHat: discovering splice junctions with RNA-Seq, Bioinformatics 25 (9) (2009) 1105–1111.
[10] A. Dobin, C.A. Davis, F. Schlesinger, J. Drenkow, C. Zaleski, S. Jha, P. Batut, M. Chaisson, T.R. Gingeras, STAR: ultrafast universal RNA-Seq aligner, Bioinformatics 29 (1) (2013) 15–21.
[11] D. Kim, J.M. Paggi, C. Park, C. Bennett, S.L. Salzberg, Graph-based genome alignment and genotyping with HISAT2 and HISAT-genotype, Nat. Biotechnol. 37 (8) (2019) 907–915.
[12] V.A. Herzog, B. Reichholf, T. Neumann, P. Rescheneder, P. Bhat, T.R. Burkard, W. Wlotzka, A. von Haeseler, J. Zuber, S.L. Ameres, Thiol-linked alkylation of RNA to assess expression dynamics, Nat. Methods 14 (12) (2017) 1198–1204.
[13] F.J. Sedlazeck, P. Rescheneder, A. von Haeseler, NextGenMap: fast and accurate read mapping in highly polymorphic genomes, Bioinformatics 29 (21) (2013) 2790–2791.
[14] Y. Zhang, C. Park, C. Bennett, M. Thornton, D. Kim, Rapid and accurate alignment of nucleotide conversion sequencing reads with HISAT-3N, Genome Res. (2021), https://doi.org/10.1101/gr.275193.120. June.
[15] J.T. Robinson, H. Thorvaldsdóttir, W. Winckler, M. Guttman, E.S. Lander, G. Getz, J.P. Mesirov, Integrative genomics viewer, Nat. Biotechnol. 29 (1) (2011) 24–26.

[16] A. Mortazavi, B.A. Williams, K. McCue, L. Schaeffer, B. Wold, Mapping and quantifying mammalian transcriptomes by RNA-Seq, Nat. Methods (2008), https://doi.org/10.1038/nmeth.1226.
[17] B. Li, C.N. Dewey, RSEM: accurate transcript quantification from RNA-Seq data with or without a reference genome, BMC Bioinform. 12 (August) (2011) 323.
[18] S. Anders, P.T. Pyl, W. Huber, HTSeq—a Python Framework to work with high-throughput sequencing data, Bioinformatics (2015), https://doi.org/10.1093/bioinformatics/btu638.
[19] Y. Liao, G.K. Smyth, W. Shi, featureCounts: an efficient general purpose program for assigning sequence reads to genomic features, Bioinformatics 30 (7) (2014) 923–930.
[20] N.L. Bray, H. Pimentel, P. Melsted, L. Pachter, Erratum: near-optimal probabilistic RNA-Seq quantification, Nat. Biotechnol. 34 (8) (2016) 888.
[21] R. Patro, G. Duggal, M.I. Love, R.A. Irizarry, C. Kingsford, Salmon provides fast and bias-aware quantification of transcript expression, Nat. Methods 14 (4) (2017) 417–419.
[22] J. Lovén, D.A. Orlando, A.A. Sigova, C.Y. Lin, P.B. Rahl, C.B. Burge, D.L. Levens, T.I. Lee, R.A. Young, Revisiting global gene expression analysis, Cell 151 (3) (2012) 476–482.
[23] G.A. Moyerbrailean, A.L. Richards, D. Kurtz, C.A. Kalita, G.O. Davis, C.T. Harvey, A. Alazizi, et al., High-throughput allele-specific expression across 250 environmental conditions, Genome Res. (2016), https://doi.org/10.1101/gr.209759.116. October.
[24] M.D. Robinson, D.J. McCarthy, G.K. Smyth, edgeR: a bioconductor package for differential expression analysis of digital gene expression data, Bioinformatics 26 (1) (2010) 139–140.
[25] M.I. Love, W. Huber, S. Anders, Moderated estimation of fold change and dispersion for RNA-seq data with DESeq2, Genome Biol. 15 (12) (2014) 550.
[26] M.E. Ritchie, B. Phipson, W. Di, H. Yifang, C.W. Law, W. Shi, G.K. Smyth, Limma powers differential expression analyses for RNA-sequencing and microarray studies, Nucleic Acids Res. 43 (7) (2015) e47.
[27] Y. Benjamini, Y. Hochberg, Controlling the false discovery rate: a practical and powerful approach to multiple testing, J. R. Stat. Soc. 57 (1) (1995) 289–300.
[28] J.D. Storey, The positive false discovery rate: a Bayesian interpretation and the Q-value, Ann. Stat. 31 (6) (2003) 2013–2035.
[29] J.T. Leek, J.D. Storey, Capturing heterogeneity in gene expression studies by surrogate variable analysis, PLoS Genet. 3 (9) (2007) 1724–1735.
[30] O. Stegle, L. Parts, R. Durbin, J. Winn, A Bayesian framework to account for complex non-genetic factors in gene expression levels greatly increases power in eQTL studies, PLoS Comput. Biol. 6 (5) (2010) e1000770.
[31] H. Ongen, A. Buil, A.A. Brown, E.T. Dermitzakis, O. Delaneau, Fast and efficient QTL mapper for thousands of molecular phenotypes, Bioinformatics 32 (10) (2016) 1479–1485.
[32] A.A. Shabalin, Matrix eQTL: ultra fast eQTL analysis via large matrix operations, Bioinformatics 28 (10) (2012) 1353–1358.
[33] S.M. Urbut, G. Wang, P. Carbonetto, M. Stephens, Flexible statistical methods for estimating and testing effects in genomic studies with multiple conditions, Nat. Genet. 51 (1) (2019) 187–195.
[34] Y. Katz, E.T. Wang, E.M. Airoldi, C.B. Burge, Analysis and design of RNA sequencing experiments for identifying isoform regulation, Nat. Methods 7 (12) (2010) 1009–1015.
[35] K.C.H. Ha, B.J. Blencowe, Q. Morris, QAPA: a new method for the systematic analysis of alternative polyadenylation from RNA-Seq data, Genome Biol. 19 (1) (2018) 45.
[36] S. Shen, J.W. Park, L. Zhi-Xiang, L. Lin, M.D. Henry, W. Ying Nian, Q. Zhou, Y. Xing, rMATS: robust and flexible detection of differential alternative splicing from replicate RNA-Seq data, Proc. Natl. Acad. Sci. USA 111 (51) (2014) E5593–E5601.
[37] Y.I. Li, D.A. Knowles, J.K. Pritchard, LeafCutter: annotation-free quantification of RNA splicing, bioRxiv (2016). March.

[38] J. Vaquero-Garcia, A. Barrera, M.R. Gazzara, J. González-Vallinas, N.F. Lahens, J.B. Hogenesch, K.W. Lynch, Y. Barash, A new view of transcriptome complexity and regulation through the lens of local splicing variations, elife 5 (February) (2016) e11752.

[39] G.P. Alamancos, A. Pagès, J.L. Trincado, N. Bellora, E. Eyras, SUPPA: a super-fast pipeline for alternative splicing analysis from RNA-Seq, bioRxiv (2014), https://doi.org/10.1101/008763.

[40] R. Goering, K.L. Engel, A.E. Gillen, N. Fong, D.L. Bentley, J. Matthew Taliaferro, LABRAT reveals association of alternative polyadenylation with transcript localization, RNA binding protein expression, transcription speed, and cancer survival, BMC Genom. (2020), https://doi.org/10.1101/2020.10.05.326702.

[41] External RNA Controls—The Joint Initiative for Metrology in Biology." n.d. Accessed November 21, 2021. https://jimb.stanford.edu/ercc/.

CHAPTER 10

Rigor and reproducibility of RNA sequencing analyses

Dominik Buschmann[a], Tom Driedonks[a], Yiyao Huang[a,b], Juan Pablo Tosar[c,d], Andrey Turchinovich[e,f], and Kenneth W. Witwer[a]

[a]Department of Molecular and Comparative Pathobiology, Johns Hopkins University School of Medicine, Baltimore, MD, United States, [b]Department of Laboratory Medicine, Nanfang Hospital, Southern Medical University, Guangzhou, Guangdong, China, [c]Nuclear Research Center, School of Science, Universidad de la República, Montevideo, Uruguay, [d]Functional Genomics Laboratory, Institut Pasteur de Montevideo, Montevideo, Uruguay, [e]Division of Cancer Genome Research, German Cancer Research Center (DKFZ) and German Cancer Consortium (DKTK), Heidelberg, Germany, [f]Heidelberg Biolabs GmbH, Heidelberg, Germany

Rigor and reproducibility of RNA sequencing analyses

Introduction to RNA sequencing

High-throughput sequencing (HTS) of nucleic acids has revolutionized the biological sciences and ushered in new prospects for molecular medicine. Also referred to as next-generation sequencing (NGS) or massive(ly) parallel sequencing (MPS), HTS has in part superseded techniques such as Sanger sequencing and hybridization or amplification microarrays. As inputs into the HTS workflow, both DNA and RNA can be used, but the latter must be converted first into complementary DNA (cDNA). HTS for RNA (RNA sequencing or RNA-Seq) allows the analysis of the transcriptome at single-nucleotide resolution, identifying RNA sequences without a priori knowledge of genomic DNA, splicing, or posttranscriptional modifications. The technique can thus complement genomic DNA sequencing by capturing a "snapshot" of the transcriptomic output within an organism, tissue, cell, or even extracellular environment.

A current frontier of RNA-Seq is low-input applications, perhaps most consequentially those directed at extracellular RNA (exRNA). Both living and dying cells release exRNA, offering an unparalleled opportunity to assess the health status of releasing cells and their tissues by relatively noninvasive "liquid biopsy": harvesting and analyzing biological fluids that retain an RNA imprint of disease. Unfortunately, RNA is labile because of its chemical structure (prone to hydrolysis relative to DNA) and its susceptibility to breakdown by ubiquitous RNA degrading enzymes (RNases). Nevertheless, this unlikely survivor can be protected from degradation inside extracellular vesicles (EVs), in association with lipoprotein particles (LPPs), by RNA-binding proteins (RBPs), chemical modifications, and self-folding into digestion-refractory structures. However, the maxim of "garbage in, garbage out" is particularly true for low-input RNA-Seq, in which quality and quantity of input material influence results.

To realize the promise of exRNA sequencing, we must first recognize and accommodate the possible errors and biases that may compromise our conclusions. Some of these hurdles are common to all

RNA-Seq measurements, while others may be unique to low-input workflows. In this chapter, we review the entire lifecycle of an RNA-Seq experiment, from sample collection through data analysis. We discuss common sources of bias at different steps and provide tips for setting up reliable experiments and comparisons. Along the way, special considerations of exRNA profiling are our guiding light.

Extracellular RNA: A case study for the challenges of low-input RNA-Seq

exRNA has been studied extensively in settings ranging from basic biology to clinical applications. For exRNA in biofluids such as serum, plasma, urine, and cerebrospinal fluid (CSF), most studies have focused on the role of noncoding RNAs (ncRNAs), which have been implicated in cancer progression [1], seem to drive inflammatory responses [2], and are touted as biomarkers for maladies including but not limited to Alzheimer's disease [3,4], various cancers [5], organ damage, and metabolic disease [6]. Illustrating the interest in exRNA analysis, an effort to unravel basic concepts of exRNA biology, as well as potential utility for clinical purposes, was undertaken by the Extracellular RNA Communication Consortium (ERCC) [7]. Launched in 2013 by the US National Institutes of Health, the program, which is now in its second iteration, aims at cataloging exRNAs in various biofluids, establishing robust methods for their isolation and analysis, and developing tools and resources that aid the growing community of exRNA researchers [8]. Specific RNA carriers have also been addressed. For example, the International Society for Extracellular Vesicles (ISEV) has fostered progress and standardization efforts for the EV research community, including RNA-focused initiatives that have provided guidance on topics ranging from sample collection to bioinformatic analysis [9–12].

Although most studies have historically focused on miRNAs, other classes of ncRNAs including long noncoding RNA (lncRNA), ribosomal RNA (rRNA), transfer RNA (tRNA), small nuclear RNA (snRNA), small nucleolar RNA (snoRNA), and Y RNA, as well as protein-coding RNA, can be detected outside of cells [13]. A common feature of these RNAs is that they are protected from degradation by RNases, which abound in many biofluids. This protection can be conveyed by several mechanisms including association with lipoproteins [14], argonaute proteins [15–17], and ribonucleoprotein complexes [18], as well as encapsulation in EVs. For some RNAs, the conventional concept of needing a carrier entity for protection does not seem to apply, as demonstrated in recent experiments that discovered how specific tRNA halves form dimers that are themselves resistant to RNase digestion [19]. Importantly, these fragments are formed by nucleases outside of cells, hinting at an impact of extracellular RNA processing on exRNA profiles that we are just beginning to understand [20,21]. exRNA abundance is thus a function of both release and differential extracellular stability, and studies on exRNAs outside of EVs are likely to capture a processed snapshot of the extracellular RNAome rather than the complete set of released RNA. Supporting this hypothesis, specific tRNA halves are enriched in nonvesicular biofluid fractions compared with EV fractions [22].

EVs are nanosized vesicles secreted or shed from cells that feature a lipid bilayer and a hydrophilic core. They are thought to play important roles in intercellular communication by exchanging signaling molecules, nutrients, and other biomolecular information between specific cell populations. Ever since the discovery that EVs contain RNA that is functional in recipient cells [23], an avalanche of studies on EV-mediated exRNA shuttling was performed, although with varying conclusions. While there is little doubt that EVs carry a certain amount of nucleic acids, including mitochondrial and genomic DNA [24], their contribution to miRNA signaling is less than clear, and, in contrast to initial beliefs, the majority of plasma miRNA might not be associated with EVs [16,17].

Complicated by technical challenges and complex matrix effects, among others, unraveling the profile of exRNA in biofluids and their association with different carriers is an active and rapidly evolving area of research. Whenever the exRNA associated with a specific carrier is to be profiled, highly pure separation of the said carrier is the first step toward reliable results. Despite the availability of various commercial and in-house separation methods for many of the exRNA carriers mentioned above, high-purity isolation without sacrificing yield remains an elusive goal in many scenarios. EVs, which can be separated by methods ranging from differential ultracentrifugation to highly sophisticated microfluidics, are particularly prone to contamination with other exRNA carriers such as lipoproteins, with which they share physicochemical features including size and density [25]. Quantitatively assessing both successful enrichment of the exRNA carrier of interest and depletion of potential contaminants is thus an essential part of pre-analytical sample characterization [26]. For in vitro exRNA studies, assessing and reporting cell viability is another important quality control. Large numbers of RNA-containing apoptotic bodies are quickly released under stressful conditions, contaminating the profile of genuinely secreted exRNAs [27]. A control condition that includes purposefully inducing apoptosis and comparing resulting exRNA profiles to those of the experimental samples might thus help achieve more reliable results.

Exclusively associating nucleic acid detected in bulk EV preparations with specific carriers remains difficult; additionally, nucleic acids might be genuinely enclosed by vesicles or associated with their surface, which is thought to display a corona rich with loosely and tightly attached proteins [28,29]. Specific experiments to assess nucleic acid topology can therefore be useful. For instance, nuclease protection assays have been used to analyze the association of both DNA and RNA in EV samples [24,30]. Treating samples with proteinases and nucleases before analysis will eliminate most nucleic acids not protected by EVs. Detergent treatments can then be used to liberate genuinely EV-encapsulated nucleic acids [31].

Experimental design and preanalytical variables
The measurement process

Transcriptome analysis by RNA-Seq is based on the identification and quantitation of all RNAs in a sample. However, sequencing may not reveal the entire RNA content. In low-input experiments in particular, the complexity of simultaneously measuring thousands of highly diluted RNAs makes this approximation highly error-prone. Errors have multiple causes. Some are intrinsic to the measurement process, while others are challenges of transcriptome analysis in general. We will examine errors according to the step of the measurement process in which they manifest Fig. 10.1.

Suppose we wish to analyze the small RNA content of EVs in the blood of a cancer patient before and after surgery. *Sampling* is taking a representative portion of the material to be studied, say, 5 mL of venous blood. This *sample* corresponds to roughly 1/1000 of the patient's blood, and its contents should be a reasonable representation of circulating contents. Sampling is followed by sample preservation, aiming to minimize physical or chemical transformations of the sample. Next is *sample preparation*: manipulations to obtain molecules that can be directly analyzed (i.e., chemical analysis) using our instrument of choice. The final step is *data analysis* [32]. In our example of a low-input RNA-Seq project, the "sample preparation" steps include addition of anticoagulants; centrifugation and separation of plasma; separation of EVs by differential centrifugation, density gradients, or other methods; RNA

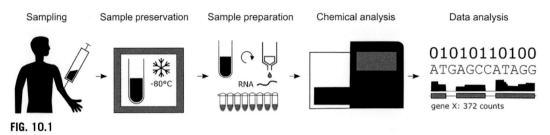

FIG. 10.1

The RNA-Seq workflow. Sample collection, preservation, and processing are followed by chemical analysis (sequencing) and data analysis.

extraction; and then sequencing library preparation steps such as adapter ligation, reverse transcription (RT), polymerase chain reaction (PCR) amplification, and size-selection. Some steps are intertwined. Anticoagulant addition may be part of both sample preparation and sample preservation. EVs might be frozen at −80°C after separation and before RNA extraction.

Sample preservation: What is the original state?

The goal of RNA sequencing is usually to determine the identities and quantities of RNA species present at the time of sampling. However, during and after the sampling procedure, chemical, biochemical, and biological processes may continue or begin within the sample, altering the "original state" of the RNA. In this context, "sample preservation" refers to steps taken to preserve that original state. Where possible, quality controls should be included to assess sample preservation. We will now give some examples of how pre-analytical variables may affect preservation of the initial RNA state.

Example 1: Effects of cell harvest procedures

If live cells are present in the sampled material, they may continue to make or release RNA, and this output may also be affected by conditions. Take the case of adherent cells in culture. Lysis buffer might be added directly to the cells after removal of culture medium, effectively "fixing" the state of cellular RNA at that moment. However, if cells are first removed from the substrate and centrifuged, they might sense the loss of anchorage and respond with an altered transcriptional program [33]. Similarly, imagine six flasks of cells, three treated with a drug and three left untreated. Researcher one removes medium and lyses cells in one flask at a time. Researcher two removes medium from all flasks, then lyses cells in one group before moving to the next group. What might be the consequence of one group being left without medium for a longer period of time? Such questions cannot always be answered practically or cheaply, so it is a best practice to ensure that all samples are harvested in the same way, or in a staggered or randomized order to minimize introduction of batch effects.

Example 2: Effect of extracellular ribonucleases on exRNA profiles

The majority of exRNA studies have been focused on EV-associated RNA; however, most microRNAs (miRNAs) circulating in human plasma, as well as multiple other RNA classes, are not associated with EVs (see "Extracellular RNA: A case study for the challenges of low-input RNA-Seq" section). Remarkably, addition of ribonuclease inhibitors to extracellular samples, or depletion of extracellular RNases, induces dramatic changes in vesicles-free exRNA profiles [20,21], including apparent depletion of tRNA-derived fragments and emergence of their full-length precursors. Thus, extracellular RNases shape the extracellular RNAome by degrading multiple nonvesicular RNAs. As a result, fragments showing a higher extracellular stability tend to accumulate over time [19,20] which results in highly dynamic exRNA profiles. Therefore, the results of exRNA analysis can vary dramatically, depending on pre-analytical variables such as processing time and temperature or the presence of RNase inhibitors.

Example 3: Microbiological decomposition

Human tissues, particularly from deceased individuals, may be subject to the influence of microbes. For example, in a study of lung tissue from two decedents [34], one sample contained about 1500 reads per million that mapped to the genus *Lactobacillus* (representing 60% of all exogenous, nonmammalian sequences) (Tosar et al. unpublished analysis). Although we cannot discard pre-mortem bacterial infection, facultative anaerobic bacteria such as *Lactobacillus* predominate in organ tissues with short post-mortem intervals [35]. Recently, the Genotype-Tissue Expression (GTEx) Consortium (gtexportal.org/) released its final datasets [36], including 15,201 RNA-sequencing samples from 49 tissues of 838 postmortem donors. Looking at earlier releases, others found bacterial reads in all GTEx tissues except testis, adrenal gland, heart, brain, and nerve [37]. However, it is not clear from these analyses whether those reads represent the natural microbiome of each tissue or rather result from contamination or the onset of microbial decomposition of postmortem tissues. Of course, tissue decomposition that begins before RNA extraction can add bacterial sequences but can also affect host RNA.

Contamination introduced before or during library preparation

Contamination is the inclusion or addition of unwanted sequences prior to sequencing. Contaminating sequences might be intrinsic or extrinsic. Intrinsic contamination results from inefficient separation protocols. For instance, it might refer to RNA contributed by lymphocytes remaining in a population of purified macrophages [38] or by lipoproteins in a preparation of blood plasma EVs [39]. Extrinsic contamination occurs when foreign sequences enter the sample of interest. This might occur when samples from related organisms are inadvertently mixed (in which case the "foreign" nature of the RNA might not be identifiable) or when RNA from different species appears. Note that the presence of microbiome signals might not be considered contamination.

Sources of extrinsic contamination are numerous. In cell culture, fetal bovine serum (FBS, also called fetal calf serum, FCS) contributes contaminating RNAs to exRNA preparations, even when EV-depleted FBS is used [40,41]. Even defined (i.e., serum-free) medium formulations can introduce contaminating RNAs. For example, an enzyme included in one such formulation was found to contain miR-122 and miR-451a as contaminants [42]. Labware and reagents might also contribute contaminating nucleic acids. In 2013, a seemingly novel virus was identified in patients with seronegative hepatitis, a disease with no known etiological agent [43]. However, this virus was found by other investigators to be associated with algae used to generate the matrix for silica spin columns [44]. Although this is a story of DNA rather than RNA contamination, it illustrates the extent to which unidentified contamination events can affect interpretation of results. Indeed, RNA contamination can also occur, for example, from exogenous small RNAs in spin columns [45]. We have identified sequences from bivalves in small RNA-Seq data from a variety of human tissues sequenced by different groups. Presumably, the origin of these enigmatic sequences is molecular biology-grade glycogen, often used for RNA precipitation and derived from mussels [41]. We also found *Trypanosoma* sequences in human samples and could trace these to a bottle of TRIzol that had been used for a Trypanosome project a year earlier. These sequences dropped off when we used a new, noncontaminated bottle of TRIzol [41]. What about cross-contamination of samples in, say, a pathology laboratory, where tissues may be present from many individuals of the same species? Here, tissue-specific transcripts not expected in the tissue under study might be used to assess cross-sample contamination [46,47].

Cross-sample contamination can also occur during sequencing library preparation. In a sequencing facility where technicians were blinded to the experiment, after processing a low-input human RNA sample next to a high-input tissue sample from turtles, a measurable number of turtle-derived reads were found in the human sample [48]. The percentage of turtle-derived reads increased when the human RNA was previously treated with sodium periodate. This reagent renders most RNAs refractory to standard sequencing; exceptions are those with 2′-o-methylation at the 3′ end (like plant miRNAs and human PIWI-associated RNAs (piRNAs)). That is, by diminishing the amount of "sequenceable" RNA in a sample, the effects of contamination become more apparent, with obvious implications for low-input RNA-Seq. Indeed, serial dilutions of environmental samples greatly increased the diversity of identifiable microbial sequences [49].

Another illustrative example is the story of diet-derived xenomiRs. In 2012, rice-derived miR-168a was reportedly found in the sera of Chinese patients [50]. Evidence was presented that diet-derived miRNA could regulate mammalian gene expression in the liver. However, independent groups failed to replicate these findings [51–54]. Of note, the plant RNA had been identified only after treatment with sodium periodate [55], which, as stated, allows better detection of low-abundance contaminants [48]. Interestingly, we identified the same rice-derived miRNAs in small RNA sequencing data obtained previously by the same authors from lancelets [48]. Since lancelets do not eat rice, this observation strongly indicates contamination, not dietary uptake, as the explanation for the analysis results from the original dataset.

Quality control

The starting material for RNA-Seq, usually a total RNA extract, should be accurately quantified and checked for purity and integrity. High-input RNA from tissues or bulk cells can be quantified and its purity assessed by spectrophotometry. As nucleic acids and common impurities specifically absorb light at different wavelengths, absorbance ratios are frequently used to assess RNA purity. For pure RNA, the 260 nm vs. 280 nm absorbance ratio is around 2.0, whereas lower ratios indicate contamination with proteins or phenolic compounds, most likely derived from RNA extraction reagents. Similarly, the 260 nm vs. 230 nm ratio of pure RNA ranges between 2.0 and 2.2 and may be lowered by contaminants with strong absorbance at 230 nm. For low-input samples such as exRNA, the utility of spectrophotometry is questionable due to its sensitivity limits and the strong relative impact of contaminant absorbance when nucleic acid concentrations are low. These samples usually require more sensitive fluorescence methods or amplification-based quantification of several RNA species.

Integrity of RNA can be estimated by taking a "fingerprint" of abundance of distinct RNA species across the size spectrum and/or comparing ratios of specific RNAs, for example, specific rRNAs. Commercially available metrics for estimating integrity such as RNA Integrity Number (RIN) are based on these approaches. However, RIN and similar metrics are often of limited or no value for exRNA-Seq studies, since longer RNAs used to calculate these ratios are mostly absent or degraded in these samples. Furthermore, exRNA samples are often reported to be enriched with small RNA species, which violates the criteria that quality metric algorithms developed for cellular RNA are based on. For these samples, evaluating RNA size distributions by capillary electrophoresis using fluorescent dyes may be still helpful, as may establishing the presence and quantity of specific target RNAs prior to RNA-Seq. For longer exRNAs, integrity might be assessed by using reverse transcription quantitative PCR (RT-qPCR) to quantify the relative expression of amplicons located at their 3′ and 5′ ends, with increasing ratios indicating degradation [56,57]. Several caveats apply, including that degradation kinetics might vary for different transcripts and, importantly, that detecting fragments of longer transcripts may reflect the true state of an exRNA profile rather than low integrity induced by poor sample preservation and handling.

As results of RNA-Seq experiments are directly impacted by the quality and quantity of starting material for library preparation, input normalization is an important consideration for high-input and low-input samples alike. While high-input samples that lend themselves to precise RNA quantification are usually normalized to the same amount of total RNA, exRNA samples might require different approaches. Normalizing to biofluid volume, i.e., using the entire yield from the same starting volume for all samples, is a common method, as is normalizing to particle count, e.g., constructing libraries from the same number of EVs irrespective of the biofluid volume they were isolated from. Alternative strategies include normalizing to the same amount of a specific transcript previously quantified using amplification-based methods or to the same amount of a particular RNA carrier, although both require a priori knowledge about the biological system at hand.

Experimental design and validation

Important aspects of experimental design include number of biological and technical replicates, library preparation technique, sequencing mode, sequencing depth, bioinformatics workflows, and whether and how results will be validated. Sequencing reads may be as short as 50 base pairs (bp) or as long as several hundred bp (although some technologies can achieve even longer reads). Shorter reads are usually used for small RNA species and referred to as small RNA-Seq, while longer reads are needed for longer RNA species or poorly annotated transcriptomes.

Before beginning the RNA-Seq protocol, experiments should be designed to be statistically rigorous, with a number of replicates decided by a sample size calculation if possible. RNA-Seq replicates can be technical or biological. Regarding technical replicates, RNA-Seq has less technical variability compared with microarray platforms, so technical replicates may be unnecessary and not worth the cost [58]. A minimum of three to four biological replicates is needed to assess biological variability accurately [59], but this number could also be much higher depending on variance, the expected effect size, and the scientific question. In a study of whole blood RNA, for example, the probability of detecting a twofold expression difference was estimated at 98% when using five replicates, while 10 replicates were needed to detect a 1.5-fold difference with 91% probability [60]. Another study found that at least 12 biological replicates were needed [61]. Unfortunately, many RNA-Seq studies are underpowered due to poor planning or financial constraints, and some seemingly significant results may not reflect a biological reality [62].

Sequencing depth is the total number of reads obtained per sample. Shallow sequencing may miss low-abundance transcripts and obtain too few reads for accurate differential expression (DE) analysis. In contrast, sequencing too deeply may be resource-inefficient, especially if sequencing depth is achieved at the expense of statistical power [63]. For general DE experiments for eukaryotic genes, an average of 10 to 30 million reads per sample have been recommended, although as few as one million reads per sample can provide accurate estimates for highly expressed genes [58]. For miRNAs, as few as 1.5 million mapped reads may accurately represent sample composition, but five million mapped reads are advantageous for DE analysis [64,65]. However, these numbers are, at best, rules of thumb, and sequencing depth should be carefully decided, based on the knowledge of expected abundance of target RNAs, peculiarities of library preparation methods, and other available data for each experiment. A good practice is to estimate read coverage across all transcripts and samples in a pilot RNA-Seq experiment to determine whether additional sequencing is likely to increase the dynamic range of the detected genes as well as sensitivity of DE analysis. Related to sequence depth, multiplexing is when cDNA molecules from individual samples are tagged with a specific barcode (index) to allow sequencing of multiple samples in the same sequencing lane. This can help avoid batch effects, especially in large and multi-condition experiments where technical and biological factors can be confounded [66,67]. The number of samples that can be multiplexed is determined by the total output of the sequencing run, the desired number of reads per sample, and the availability of unique indices.

Despite its overall accuracy and reliability, RNA-Seq can introduce biases [68]. Sequencing artifacts include GC content bias, overrepresentation of certain sequences, and even erroneous "detection" of transcripts not present in the original RNA [69,70]. RNA-Seq results are therefore commonly validated by independent methods such as RT-qPCR and digital droplet PCR (ddPCR), although some researchers question the need for validation, especially for RNA-Seq of large numbers of replicates. In many studies ranging from transcriptome profiling to single-cell sequencing, expression data from RNA-Seq and RT-qPCR correlate very well (Pearson's $r > .8$) [71–73]. Still, since both RNA-Seq and RT-qPCR rely on RT and subsequent cDNA analysis, RT-induced biases in validation data sets cannot be excluded, and orthogonal methods such as probe-based counting may be advantageous [69,74]. Regardless of which method is used, it may be useful to include independent biological samples instead of or in addition to the previously sequenced samples.

The impact of library preparation methods

After RNA isolation, quantification and quality control, library preparation converts RNA into a form that is compatible with the sequencing system. The core steps of library preparation include RNA fragmentation, RT into cDNA, and attachment of sequencing adapters (Fig. 10.2). In this section, we discuss how library preparation affects sequencing outcome and how to minimize biases. For clarity, we follow the workflow laid out in a commonly used total RNA library preparation kit and discuss biases associated with each step. Although the order of some steps might be reversed for small RNA library preparation, these biases still apply and need to be taken into account when settling on a particular library construction strategy.

RNA and cDNA fragmentation

RNA fragmentation must be done for longer RNAs (e.g., messenger RNA (mRNA), lncRNA) to obtain inserts within the size limitation of the sequencing platform (e.g., < 600 bp on Illumina sequencers, <400 bp on Ion Torrent, etc.). RNA can be fragmented by hydrolysis, nebulization, or enzymatic digestion. Hydrolysis uses divalent cations (Zn^{2+}, $MgCl_2^+$, etc.) to catalyze RNA hydrolysis at an elevated

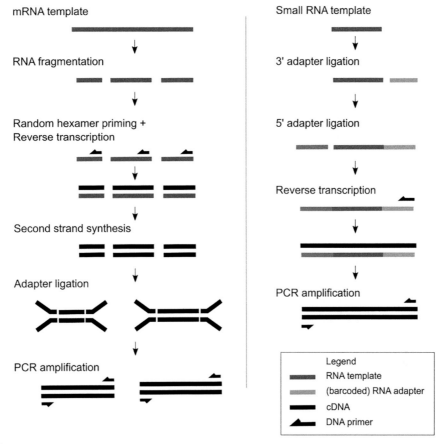

FIG. 10.2

Comparison of RNA-Seq library preparation workflows. Examples of RNA-Seq library preparation protocols for coding RNAs *(left)* and small RNAs *(right)*. Note that some widely used library preparation workflows introduce modifications to these schemes.

temperature to mitigate the effect of RNA secondary structure on fragmentation. The process achieves random cleavage, leaving a 5′ hydroxyl and a 2′,3′ cyclic phosphate ester [75,76]. Without sequence or structure specificity, the method can produce uniform read coverage across a transcript. It has been widely used in many library preparation methods (e.g., the Illumina TruSeq Stranded mRNA kit and the RNA Fragmentation and Sequencing (RF-Seq) method) [77]. Enzyme digestion often uses RNase III (e.g., Ion Total RNA-Seq Kit) since it produces fragments with 5′ phosphate and 3′ OH, allowing subsequent adapter ligation [78,79]. However, RNase III is known to cleave double-stranded RNA (dsRNA) in a sequence- and secondary structure-determined manner, which may lead to heterogeneous cleavage and uneven coverage. A recent study [76,80] showed that although RNase III digestion did not affect mRNA abundance in the library, it underestimated some ncRNAs, including tRNAs and rRNAs, when compared with Zinc cleavage.

Fragmentation can also be done after cDNA production by shear force or DNase I treatment. The original deoxy-UTP (dUTP) protocol [81] and the NuGEN RNA-Seq kit fragment cDNA by sonication, while other kits, such as the SMARTer Ultra Low RNA Kit (Clontech), use enzyme-based methods. Furthermore, the recently developed transposon-based tagmentation method (used in Nextera XT library preparation by Illumina) can cleave and tag cDNA with adapters at both ends in a single step (reviewed in [82]). Still, similar to other enzyme-based methods, the ratio between samples and transposase will need to be optimized in case of under- or overtagmentation. cDNA fragmentation is reportedly associated with higher 3':5' bias and poor uniform sequence coverage compared with a controlled RNA hydrolysis fragmentation method [83]. Consequently, RNA fragmentation remains the more prevalent approach.

In summary, nonuniform fragmentation methods produce bias in transcript reassembly. Technical factors such as operation temperature, time, and enzyme concentration can influence fragment size. Analysis of RNA fragments via, e.g., capillary electrophoresis, is recommended to verify the product amount and size distribution before moving to the next step.

Reverse transcription

Reverse transcription of RNA is a required step for RNA library preparation sequence, as most RNA-Seq experiments are carried out on technically mature instruments used for DNA sequencing, with the notable exception of direct RNA or cDNA sequencing using Oxford Nanopore technology [84]. The template RNA converts to first strand cDNA in the presence of primers, reverse transcriptase (RTase), and dNTPs. The first strand cDNA can then be used in PCR to synthesize and amplify double-stranded DNA (dsDNA), facilitating detection of low abundance samples.

RT primers are a key factor in cDNA synthesis. Oligo-dT primers consist of 12–20 deoxythymidines and can directly anneal to poly(A) tails of mRNAs [85]. Oligo-dT primers complete mRNA enrichment and RT in a single step, but they induce high 3' bias since poly(A) tails are primarily located at the 3' end of RNAs [82]. Because poly(A) sequences are sometimes found internally, the method could also produce truncated cDNAs [86]. Finally, oligo-dT-based methods are restricted to poly(A)+mRNAs and are not useful for whole transcriptome analysis, which includes both coding and noncoding RNAs. They are also not the best choice when working with partially degraded samples including those derived from clinically relevant specimens stored in formalin-fixed paraffin-embedded blocks. Also note that some specific mRNAs do not contain poly(A) tails, such as those coding for metazoan replication-dependent histones [87].

Random hexamer primers are an alternative, and they start RT across the entire length of a transcript, regardless of polyadenylation status. However, ribosomal RNA (rRNA) depletion may be needed when using random primers to avoid making libraries that consist mostly of rRNA. This can be achieved by pulling out rRNA from the solution with rRNA-specific probes coupled to magnetic beads or through RNase H-mediated degradation of rRNA/DNA probe hybrids [88]. As an alternative, not-so-random (NSR) primers were developed to exclude rRNA-binding sequences and thus prepare libraries directly from total RNA without rRNA depletion [89,90]. Although random primers are the most prevalent method to initiate cDNA synthesis, priming bias and mispriming have been reported [91–93]. Since methods to overcome primer bias have yet to be developed, bioinformatics methods are often used to mitigate the impact of biases [91,94].

The fidelity and capability of RTases are also critical in cDNA synthesis. Commonly used RTases are from Murine Leukemia Virus (MMLV) and Avian Myeloblastosis Virus (AMV) [95]. Both have polymerase and RNase H activity and are active at 37°C. Only AMV RTases have DNA $3' \rightarrow 5'$ endonuclease activity [95]. The base misincorporation rates of AMV and MLV RTases are low enough to guarantee the fidelity of cDNA synthesis from small RNAs [96,97]. To increase cDNA synthesis sensitivity and reproducibility, some engineered, thermostable RTases and group II intron RTases [98] can work at higher temperatures to process through even highly structured RNAs such as tRNAs [99,100] reduce nonspecific primer binding, and reduce the overrepresentation of RNA fragments (Fig. 10.3). Group II intron RTases can also ligate adapters via template switching activity, avoiding biases introduced by phage ligases [101]. A systematic comparison of 11 commercially available RTases [102] found high performance for all RTases, but reported that Maxima H and Superscript II RTases exhibited better efficiency, reproducibility, and sensitivity. Another study showed that combining MMLV RTases with other thermostable DNA polymerases and DNA/RNA helicases improved sensitivity compared with RTase alone [103]. More comparisons may be needed to evaluate the processivity, efficiency, accuracy, and sensitivity of RTases.

The preservation of RNA direction information is important for identifying antisense RNA and novel RNA species (as reviewed in [82]). Antisense transcripts that complement sense transcripts are known to regulate gene expression in eukaryotes [104,105]. However, the original hexamer primer-based method based on end-repair DNA synthesis does not indicate which DNA strand corresponds to the RNA template, so strand-specific information was lost. To maintain strand directionality, the cDNA second strand can be chemically modified. A well-known example used in several commercial kits (Illumina TruSeq Stranded mRNA kit and NuGEN Universal RNA-Seq with NuQuant, etc.) is incorporating 2′-Deoxyuridine, 5′-Triphosphate (dUTP) in replace of dTTP during the second strand synthesis. The dUTP-containing strand is a poor template for thermostable polymerase and can be degraded before PCR amplification with uracil DNA glycosylase (UDG) [81,82,106]. Thus, only the first strand of cDNA, conserving RNA direction information, will be amplified and preserved.

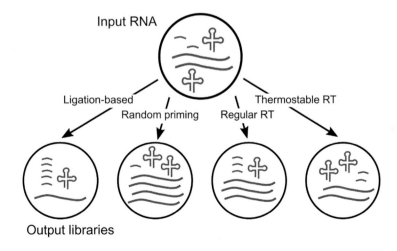

FIG. 10.3

Library preparation-induced biases in RNA-Seq data. Variables that may introduce biases in RNA-Seq results.

Another solution is based on the directional switching mechanism at the 5′ end of the RNA transcript (SMART) to facilitate cDNA synthesis from RNA templates [107–110]. The SMART method harnesses the intrinsic template switching activity of MMLV RTases to add a few nontemplated nucleotides (mostly deoxycytidine (dC)) to the 3′ end of the first strand of cDNA when the RTase reaches the 5′ end of the template RNA. A template switch oligonucleotide (TSO) with dG sequence can then pair with a nontemplate dC anchoring site to attach to the 3′ terminus of the cDNA. By doing so, the completed cDNA will contain the entire full-length transcripts with both 3′ and 5′ ends in a sequence-independent manner. This approach was now used in Low Input RNA Kits for Sequencing (Clontech) to generate double-stranded full-length cDNA from the total RNA. In studies systematically comparing several strand-specific methods [93,111], the SMART method showed better performance on low-quantity samples with good library complexity but may induce more 3′ bias and GC content bias. However, the dUTP method serves as the leading protocol as it better balances strand specificity, library complexity, coverage evenness, and accuracy. Other solutions based on adapter attachment with known orientation to 3′ and 5′ to RNA/cDNA, which is mainly used for small RNA library preparation, will be discussed in the adapter attachment section below.

Adapter attachment

Adapters attached to the ends of RNA fragments/cDNA contain functional elements such as flow-cell/bead binding sequences, sequencing primer sites, sample indices, and unique molecular identifiers (UMIs). These sequences may be required for sample hybridization and sequencing on specific platforms and may also reveal strand-specific information. In the following, we review several approaches to adapter attachment.

For mRNA library preparation, adapters are added to double-stranded cDNA using platform-specific library construction kits. Ligation-based methods (e.g., Ion Xpress Plus Fragment Library Preparation and Illumina TruSeq Stranded mRNA kit) include sequential enzymatic steps of end repair, phosphorylation, polyadenylation, and adapter ligation (as reviewed in [112]). Adapters can also be directly ligated to the ends of unknown RNA molecules using T4 RNA ligases to allow RT priming and subsequent PCR [97,113]. Adapters used on different platforms include double-stranded adapters (SOLiD sequencing platform), 3′ adenylated DNA and 5′ RNA adapters (Illumina, sequential ligation, small RNA only), and hybrid RNA–DNA adapters (Ion Torrent, one-step ligation). The ligation-based method, originally developed for miRNAs [113,114], makes use of T4 RNA ligases including T4 Rn1 and T4 Rn2, which are key enzymes commonly involved in sequential ligation steps. Taking the Illumina protocol of adding adapters in a sequential manner as an example, the pre-adenylated 3′ adapter hybridized with the complementary RT primers is ligated to the 3′ end of RNA template by Rn2, followed by the 5′ adapter being added to another end of RNA in the presence of ATP and Rn1. The RTase then synthesizes the first strand of cDNA. Some biases in read distribution are expected because the substrate sequence as well as structure specificities of enzymes will lead to uneven and noncontinuous coverage of RNA libraries [115–117].

The RNA termini modification also impacts the enzymatical steps since the effective ligation of adapters requires the formation of a 3′- to 5′-phosphodiester bond between a 3′-hydroxyl and a 5′-phosphate group (as reviewed in [97]). Some small RNA species, such as piRNAs or plant miRNAs with the 3′ methyl modification, are ligated less efficiently and under-represented in the libraries [118]. Instead of sequential ligation of 3′ and 5′ adapters before the RT, the 5′ adapter can also be ligated to the first strand of cDNA instead of the template RNA after the 3′ adapter ligation-initiated RT [93,119].

This method then does not depend on the 5′ monophosphate of RNAs. Alternatively, transposase-based methods fragment cDNA and attach adapters in one step [120,121]. Called "tagmentation" and also discussed under RNA fragmentation, this method is now used in Illumina Nextera DNA Sample Preparation Kits and Low Input Library Prep Kits. To minimize bias induced by the ligation step, strategies such as optimization of reaction time and temperature [118], adapter pooling [116,122], and randomized adapters [97] have been developed to generate more accurate small RNA profiles.

Ligation-free methods have also been developed for small RNA library preparation. Based on the SMART principle used for mRNA library construction, capture, and amplification by tailing and switching (CATS) protocol was developed for small RNA (<150 bp). After polyadenylation of the 3′ end of using a poly(A) polymerase [123], cDNA synthesis starts with an anchored poly (dT) primer incorporating 3′ adapter sequences. The terminal transferase activity of the RTase adds additional dC upon reaching the 5′ end of the RNA, then 5′ adapters with rG nucleotides are added by template switching to start second strand cDNA synthesis. Thus, both 3′ and 5′ adapters are added to the cDNA independent of RNA template sequences and without ligation. In addition to reducing the bias introduced by ligation steps, CATS does not require additional purification before cDNA pre-amplification since the PCR primers are not complementary to either 5′ nucleotides or poly (dT) primers.

This method was commercialized by Diagenode as the CATS Small RNA-Seq Kit and the more recent D-Plex Small RNA-eq Library Prep Kit. The sensitivity of this method was demonstrated by preparing small RNA libraries from trace amounts of blood RNA [124]. However, as the protocol begins with a poly(A) tailing step, it may obscure the detection of isomiRs containing 3′-(A) nucleotides after the 3′ trim [124]. Also, the d(T) priming method has the potential to cause 3′ bias [82]. Clontech now also offers a similar ligation-independent SMARTer smRNA-Seq Kit.

Library amplification and unique molecular identifiers

For low-input samples, PCR pre-amplification is widely used to increase the amount of material for sequencing. However, PCR may unequally amplify different templates, or even introduce sequence alterations that may appear to be novel sequences [125,126]. To identify and correct these biases, UMIs or molecular barcodes (MBC), a mixture of random, degenerate bases, are used to label individual cDNA fragments prior to PCR amplification [127–129]. After sequencing, the number of starting molecules can be determined by counting each UMI once to normalize the data. Use of randomized UMIs has been reported to improve mRNA sequencing accuracy for protocols including dUTP strand-specific libraries [130], SMART mRNA libraries [129], tagmentation [131], and Illumina TrueSeq libraries [132], as well as small RNA libraries prepared by the Diagenode D-Plex Small RNA-Seq Kit. However, while several studies have advocated UMIs in small RNA-Seq [70,101,129,130,133], others suggest that UMIs are unnecessary since amplification step biases are minimal in small RNA-Seq [115,116,134]. In contrast, the use of UMIs is widespread in other low-input applications, such as single-cell RNA sequencing [135]. UMI design and diversity are also important. A recent study showed that UMIs of eight nucleotides (nt) in length may not be diverse enough to tag all RNAs uniquely and may thus distort the data [136].

Impact of size selection on RNA fragment distribution and detected RNA biotypes

After library construction, size selection may be needed to remove adapter dimers or enrich target transcripts of a specific length. This can be done by polyacrylamide gel electrophoresis or by automated systems (such as Pippin Prep). Small RNA-Seq library preparation kits designed for miRNA sequencing (e.g., TruSeq and NEBNext), may recommend size ranges specific for miRNAs (18–35 nt). Nevertheless, inserts of this size may also include fragments of other small ncRNAs or longer RNAs. A wealth of literature now describes RNA fragments formed from small ncRNAs by enzymatic cleavage; however, some apparent RNA fragments may also be artifacts that arise during library preparation. After mapping to the genome, these fragments could be counted as if the full-length transcript were present. Size selection influences the abundance of biological or artefactual RNA fragments. Thus, speculation about the biological function of RNA fragments should be tempered with caution. In this section, we will highlight how RNA fragments are formed and how size selection affects the composition of RNA-Seq libraries.

RNA fragments: Biological entities or artifacts?

Many types of RNA are known to be cleaved into smaller fragments [137]. Fragmented RNAs may have functions that differ from their full-length counterparts. For example, fragments of Vault RNA and tRNA can regulate the translation of certain mRNAs [138,139]. Nevertheless, RNA fragments may be formed only under specific circumstances. tRNAs are cleaved by angiogenin in severely stressed cells (UV-irradiation, arsenite treatment) [140], and Y-RNA fragments are generated by RNase L after the dsRNA receptor TLR-3 is triggered [141]. exRNAs in conditioned culture medium and biofluids may also be fragmented by RNase 1 [20,21].

However, apparent RNA fragments in sequencing data may also be artifacts from library preparation. Studies of purified EVs and their parent cells have identified fragments and speculated on their potential function in intercellular communication [22,142–145]. While RNA fragments may be apparent in coverage plots, this does not necessarily mean that these fragments were present in the sample at similar levels. For example, tRNA and Y-RNA from dendritic cells appeared to be fragmented in RNA-Seq coverage plots, but predominantly full-length sequences were found by Northern blot [142]. Conventional small RNA-Seq kits may thus be biased toward fragments. Subsequent selection for short fragments will further increase the abundance of ncRNA fragments over their full-length counterparts. Furthermore, some RNA subtypes may generate more RNA fragments than others. Exemplifying this, 15–40 nt sized RNA-Seq libraries differed in the abundance of tRNA and Y-RNA subtypes compared with 40–100 nt libraries [144]. miRNA-sized RNA sequencing libraries may contain reads from other ncRNA types such as Y-RNA and tRNA. The relative abundance of these reads may not entirely reflect the abundance of full-length species. Thus, although size selection can be advantageous to enrich miRNAs in sequencing libraries, it may introduce bias in the representation of other small ncRNA types.

How to interpret RNA fragmentation?

Taken together, RNA fragments, some of them introduced by incomplete RT, may be overrepresented in libraries selected for small insert size. Thermostable RTases, used at high temperatures, can remediate formation of artefactual RNA fragments or yield more reliable fragment-to-precursor ratios (Fig. 10.3). However, representation of full-length precursors can still suffer from the presence of posttranscriptional modifications that can provoke abortive RT products, as are frequently found in tRNAs [146,147]. If one is interested in ncRNAs beyond miRNAs, we recommend including longer inserts (up to 300 nt) in sequencing libraries to obtain better coverage of full-length transcripts. Finally, Northern blots can validate conclusions about fragmentation.

Data analysis

With proliferation of sequencing analysis software and pipelines [58,60,148], the chosen analysis approach can influence outcomes via differences in read quantification, normalization, and DE analysis, among others [149–151]. In choosing a pipeline, one should consider (1) the type of RNA sequencing (mRNA-Seq, total RNA-Seq, small RNA-Seq or single-cell RNA-Seq; other types of sequencing are not covered here [152–154]); (2) the biological question; (3) the source of RNA; and (4) available computational resources. Whatever pipeline is chosen, analysis of RNA-Seq data follows the four steps: (1) preprocessing of raw reads; (2) read alignment; (3) read quantification; and (4) downstream processing/analysis, most commonly normalization and DE analysis (Fig. 10.4).

Preprocessing

Files containing base-called sequencing reads, usually in FASTQ format [155], are the starting point for HTS analyses. Raw reads may be processed for quality trimming, size-selection, and to remove adapter sequences. The *FastQC* software (Babraham Bioinformatics) is the most popular tool for quality monitoring of raw and preprocessed reads in a FASTQ file, including calculating the overall duplication rate (percentage of reads having the same sequence), visualizing the distribution of read

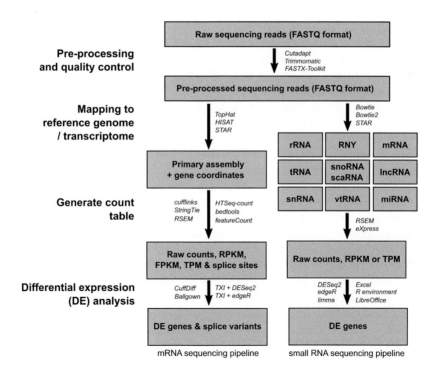

FIG. 10.4

Schematic of RNA-Seq data analysis workflow. Steps in RNA-Seq data analysis.

length and nucleotide content, calculating and visualizing GC distribution, and extracting overrepresented sequences. Quality trimming entails discarding low-quality reads or poor-quality bases, unless accurate identification of 3′ ends is a desired output of the analysis, as is often the case in small RNA-Seq. Size selection is optional, but it is highly recommended to remove reads shorter than 15 nt to avoid ambiguous alignments. Adapter trimming is necessary only when adapters are present in a significant percentage of reads, which in turn depends on the average insert length and sequencing length. For small RNA-Seq, adapters are usually present and must be trimmed. However, if library preparation includes size selection for longer RNAs, trimming might be obviated since sequencing length will likely be shorter than the length of most inserts. Currently, *cutadapt* is the most widely used software for removal of adapters (or other unwanted sequences), discarding low-quality reads, and performing final size-selection. Other packages (such as *Trimmomatic* and *FASTX*-Toolkit) are also available [156].

Read mapping

In read mapping, preprocessed (trimmed and size-selected) reads are aligned to a reference genome or transcriptome or both. Mapping to a reference genome is most commonly done using *TopHat*, *HISAT*, or *STAR* aligners, since these tools can perform a so-called "spliced alignment" that allows for gaps in the reads corresponding to intron-interrupted exons [157–159]. Transcriptome mapping, on the other hand, is usually performed by *Bowtie* and *Bowtie2* aligners, which offer accuracy and speed but lack the ability to identify intron-interrupted sequences [160,161]. However, both *HISAT* and *STAR* can use transcriptomes as a reference. Likewise, *Bowtie* and *Bowtie2* can be used to map reads to a reference genome. Mapping to the transcriptome can be significantly faster than genome mapping but does not allow de novo transcript discovery.

The primary indicator of mapping quality is the percentage of reads mapped to a corresponding reference. Mapping statistics can reveal the accuracy of sequencing, correctness of the raw read preprocessing, and presence of contaminating DNA in the original RNA sample. In a typical, successful human RNA-Seq experiment, one might expect 50%–95% of reads to map to the genome depending on the alignment software and the stringency of mapping parameters. However, optimal mapping percentages are expected to vary by species, depending on the reference genomes or transcriptomes. Uniformity of read coverage across the transcript is another important quality parameter. If reads primarily accumulate at the 3′ end of transcripts in poly(A)-selected samples, this might indicate RNA degradation [60]. Finally, the GC content of mapped reads may reveal biases introduced during library preparation. The most popular tools for analyzing coverage and GC content include *Qualimap*, *Picard tools*, and *RSeQC* [162,163].

A fraction of reads may be assigned to more than one location regardless of whether a reference genome or transcriptome is used. These "multi-mapping reads" or "multi-reads" [164] arise primarily because of repetitive sequences, similar exons within gene paralogs, and pseudogenes within introns and intergenic regions, and occur more commonly with alignment to transcriptomes, which include different transcript isoforms. To reduce the incidence of multireads in the latter case, a curated transcriptome reference containing only one (canonic) isoform per gene can be utilized. How to properly handle ambiguously mapped reads remains currently unresolved [58]. However, most researchers either randomly assign the multi-reads to one of the best alignments or split the multi-reads equally between all alignments.

Alternatively, multi-reads could be ignored or counted once for each possible alignment [164]. However, discarding multi-mapping reads will underestimate certain gene sets and biotypes, and can dramatically reduce the dynamic range of DE analysis, especially in the case of total RNA-Seq or small RNA-Seq. Likewise, counting all valid alignments for a read will lead to overestimations, since reads counted more than once will inflate the number of molecules that appear to have been sequenced [164]. In contrast, uniformly distributing the multireads by keeping a single random alignment or by splitting the count between each alignment will ensure that every read is counted only once, giving a better representation of the proportion of each RNA biotype [164].

Overlapping annotations and database quality

In order to make biological inferences, mapped reads subsequently need to be matched with genome or transcriptome annotations, which may emphasize certain RNA biotypes and is not a trivial issue even for well-annotated organisms. Researchers interested in miRNAs might map to mature miRNA sequences downloaded from miRBase (mirbase.org). Although this procedure is straightforward, its outcome depends on the accuracy of the reference database, and information on the genomic context of each mapped read is lost. Databases that are not well curated contain entries that overlap with other RNA categories. miRbase, for example, includes many fragments of tRNAs, rRNAs, snoRNAs, and others that are incorrectly tagged as miRNAs. These errors might even be clear from sequencing coverage plots that are inconsistent with hallmarks of miRNA biogenesis [165]. As an example, human miR-1246 has been widely studied in cancer as an extracellularly enriched miRNA. However, it is one of the most abundant miRNAs in FBS ([40], see "Experimental design and validation" section). While hsa-miR-1246 is present in a single genomic locus on chromosome 2, the mature miRbase sequence annotated is only 19 nt long and maps to additional loci, including 100% overlap with the U2 small nuclear RNA. The reported oncogenic properties of putative miR-1246, arrived at in part by silencing experiments [166] could potentially be due instead to silencing of U2, a core component of the splicing machinery. Of note, there are no entries for miR-1246 in MirGeneDB (mirgenedb.org), a curated metazoan miRNA database [167].

Annotation issues also affect piRNAs, which silence transposable elements and regulate gene expression in the germline and gonad somatic cells by guiding PIWI-clade Argonaute proteins to transcripts in a sequence-specific manner. Although they function similarly to si/miRNAs, their biogenesis is quite different. Most piRNA database retrieve sequences from the publication that first identified piRNAs in mouse testes [168]. Although cloned sequences mapping to tRNAs and rRNAs were removed, ~1% of mammalian piRNA database entries correspond to miRNAs or ncRNA fragments (tRNA, rRNA, YRNA, snoRNA) [169]. Strikingly, virtually all papers describing piRNAs in nongonadal mammalian tissues (where piRNA expression is not expected) or in human blood plasma or serum report sequences in this small subset of likely false positives [169,170]. Interestingly, when these contaminating sequences are removed, aberrant expression of PIWI-clade proteins in cancer does not seem to reactivate piRNA expression [171,172]. Thus, the quest to find somatic piRNAs in mammals may have produced false positives due to overlapping annotations and lack of manual curation in reference databases.

As a take-home message, we strongly encourage readers to identify genomic loci that gave rise to identified sequences. Additionally, it is important to assess read coverage in these loci, judge compatibility with known RNA biogenesis mechanisms, and perform a BLAST search to identify alternative annotations.

Read quantification

Estimation of gene expression remains the primary application of RNA-Seq and requires quantification of reads that map to each transcript and/or RNA class. The approach depends on whether genome or transcriptome references are used for the initial alignment. For genome mapping, the most frequently used approach is to aggregate transcripts and extract raw (or normalized) counts of mapped reads using precise genomic coordinates of genes, exons, and/or different RNA classes (usually in the formats of *GTF* or *GFF* files). The latter procedure is usually performed using *cufflinks*, *StringTie*, *RSEM*, *HTSeq-count*, *featureCounts* or *bedtools* software [173–177]. In contrast, quantification of reads after transcriptome mapping does not rely on genomic coordinate files and is commonly performed by *RSEM* or *eXpress* software packages [174,178]. The ultimate output of read quantification procedures will depend on the quantification software used and may include: (1) raw count matrices; (2) normalized count matrices; and (3) other transcript assembly and abundance metrics, which can be used for downstream DE analysis and/or detection of differentially spliced mRNA isoforms. Some software packages (such as *Salmon*, *Sailfish*, and *Kallisto*) allow read quantification directly, without the need for prior mapping and rely on k-mer counting within the reads [179–181]. While these algorithms have accurately counted long and highly abundant transcripts, they are less successful with short and low-abundance transcripts [182].

Data quality control

As explained in the paragraph on size selection, miRNA sequencing libraries will often contain other small ncRNA transcripts, which may be fragmented or whose length is not captured by the RNA sequencing protocol. These short RNA fragments will be mapped to their parental genes and will be counted as such when generating count tables. For example, a lncRNA may appear to be highly abundant in a count table even if only a minor fragment is present in the sequencing library. Coverage plots can identify coverage patterns between conditions and replicates, including presence of full-length or degradation products or transcriptional noise.

How to make a coverage plot

Coverage plots can be made from various file types, most commonly BAM files. These can be loaded into the UCSC Genome Browser [183] online or locally into the Integrated Genome Viewer (IGV) [184].

Coverage plots in UCSC Genome browser
    Loading data:
- Ensure BAM file availability on an online FTP server—Go to https://genome.ucsc.edu/—Click "My Data," followed by "Custom Tracks"—Select the appropriate genome in the drop-down menu—Copy-paste the URL containing your files in the text field. Click Submit—The Genome Browser appears, showing additional tracks containing your RNA-Seq coverage data

Coverage plots in IGV genome viewer
- Place sorted BAM (.bam) and BAM index (.bam.bai) files in same local folder—Download and install IGV from http://software.broadinstitute.org/software/igv/—Select the appropriate genome in the top-left drop-down menu—Click "File," and "Load from File," select all datasets you want to load, and click "Open"—Your datasets appear as individual tracks in the genome browser

Advantages and disadvantages of methods

Both methods come with advantages and disadvantages. The genomes in IGV show annotations only for mRNAs, while the UCSC Genome Browser contains annotation data of RefSeq and other ncRNA databases, including expression data from functional genomic consortia. This makes the UCSC Genome Browser more convenient to use for small ncRNA data. Navigating through the genome is slightly easier in IGV, since UCSC must refresh after every move. Comparing large numbers of samples side-by-side is more convenient with UCSC, since one can scroll from top to bottom through all tracks, while IGV displays all tracks in one nonscrollable window. Using UCSC requires space for data storage on an external FTP server. Once a track is built in UCSC, it can be accessed from multiple computers and can be shared between users. IGV can be run independently on any computer but will require more local storage space, making sharing of coverage data more challenging.

Normalization and differential expression analysis

Distribution of reads is affected by sequencing depth, sequencing biases, and transcript lengths, so raw counts are insufficient to compare expression levels among samples. Raw read counts are thus often converted into RPKM (reads per kilobase per million reads), FPKM (fragments per kilobase per million mapped reads), TPM (transcripts per million), or simply CPM (counts per million mapped reads). All but CPM remove library size bias and feature-length bias [83,185,186]. Correcting for gene length is not necessary when expression is compared between the same genes across multiple samples. However, correction for gene length may introduce a bias in per-gene variance of some genes (particularly lowly expressed ones) [187,188]. As a result, some algorithms for DE analysis (such as *edgeR*) work with count values and then perform their own internal normalization under certain assumptions (see next paragraph) [189,190]. Log-transformed RPKM, FPKM, TPM, or CPM are also often plotted at this step to assess sequencing quality between technical and biological replicates. In addition, the magnitude of gene expression differences can be visualized via supervised or unsupervised clustering in principal component analysis (PCA) and multidimensional scaling (MDS) plots showing similarities and dissimilarities between samples [190,191].

DE analysis can be performed in standard spreadsheet environments using RPKM, FPKM, TPM, or CPM tables generated by transcript abundance quantifiers (e.g., *RSEM* or *eXpress*) or "manually" from raw count matrices. However, R/Bioconductor packages such as *edgeR*, *DESeq2*, *limma-voom*, *DSS*, *EBSeq,* and *baySeq* provide more sophisticated statistical analysis, unsupervised clustering of samples, and visualization options. These options, though, may expect input data in the form of raw (un-normalized) count matrices. It should be mentioned that the R/Bioconductor *tximport* package can be used to create gene-level raw count matrices from the quantification data generated by various quantifiers of transcript abundance, including *StringTie* and *RSEM* [192]. In addition, R/Bioconductor packages introduce options for additional "inter-sample" normalization of log-transformed values by computing a scaling factor for each replicate/sample based on total gene expression across the analyzed datasets (e.g., TMM method in *edgeR*). However, these secondary normalization strategies are based on the assumptions that (1) most genes are not differentially expressed and (2) for those differentially expressed, over- and under-expression are approximately balanced [193]. Therefore, such augmented data normalization is not suitable for all RNA-Seq experiments, including those in which RNA profiles are compared between cellular and extracellular fractions.

Functional analysis of RNA-Seq data

The principle of gene set enrichment analysis (GSEA) is to search for sets of genes that are significantly over-represented in a priori gene sets grouped by biological function, chromosomal location, or regulation [194,195]. The ultimate goal of GSEA is to determine if a gene set shows statistically significant differences between two conditions. Predefined gene sets are found at the Molecular Signatures Database (MSigDB) [196], and software developed by the Broad Institute (http://software.broadinstitute.org/gsea/) is commonly used for identifying significantly enriched or depleted genes [197]. A comprehensive description of the GSEA algorithm can be found in the original paper [195].

Interpretation of differential expression patterns may involve analysis of putative molecular and functional pathways to which the DE genes purportedly contribute. These classifications usually involve tests such as "GO enrichment analysis" and "gene set enrichment analysis "(GSEA), heavily relying on published functional annotation data for a given organism. Importantly, adequate functional annotation databases are currently available only for protein-coding genes and some well-studied small RNA classes such as miRNA. Gene Ontology (GO) enrichment analysis is a system for hierarchically classifying genes or gene products into terms organized in a graph structure depending on functional characteristics [198]. GO terms are grouped into three categories: (1) molecular function of a gene product (e.g., a receptor), (2) biological process to which the gene product contributes (e.g., apoptosis), and (3) cellular component or location of the gene product (e.g., plasma membrane) [199]. Each gene can be annotated with more than one GO term. "Enrichment" means that a GO term appears more frequently than would be expected by chance. The Kyoto Encyclopedia of Genes and Genomes (KEGG) is another enrichment database [200,201].

To date, multiple GO analysis software and online tools have been developed, but there is little consensus on which is most reliable. Since GO analysis resources depend on up-to-date annotations and continued maintenance, caution should be applied when using online tools and software. We recommend *Panther* (offered directly from the GO website) [202] and *GOrilla* [203], which are frequently updated web-based tools for GO analysis and visualization.

Conclusions

RNA-Seq represents a powerful technology for transcriptomic analyses in various contexts. Its single-nucleotide resolution combined with high accuracy and ever-falling per-reaction costs have turned RNA-Seq into a widely used method. While utilizing it to profile intracellular gene expression or extracellular RNA profiles, detecting both novel transcripts and isoforms of previously known sequences, and opening new avenues for clinical use, technological advances fostered even more sophisticated applications. Having progressed from being able to analyze tissues or bulk cells to single cells and cell-free samples, RNA-Seq is now used to understand tissue heterogeneity, establish novel liquid biopsies, and devise personalized treatment strategies for complex diseases. However, this powerful method must be utilized appropriately and cautiously, as various steps spanning the entire RNA-Seq workflow from sampling and sequencing to in silico data processing and analysis are prone to induce errors and biases.

As we stressed in this chapter, RNA-Seq experiments should be planned out carefully before even touching a pipette. The maxim of *"garbage in, garbage out"* applies to RNA-Seq studies. Experimental design has a strong impact on the final outcome, both at the quantitative level (e.g. statistical power) and in the types of RNA molecules that will be analyzed and those that will be left behind. Even for high-quality data, analysis and interpretation need to be performed carefully and thoughtfully. If RNA-Seq is operated as a black box, chances of misinterpreting the resulting data are higher. Sequencing facilities and commercial companies now offer efficient standardized workflows for library preparation, sequencing and data analysis that might, however, be suboptimal for specific research questions.

The optimal RNA-Seq data analysis pipeline may not yet have been established and remains of great demand, especially for small RNA research or in situations where basic assumptions of the most popular differential expression algorithms are not satisfied. Even for identical samples, the choice of different analysis pipelines can have a profound impact on the final outcome [204].

Two important take-home messages are:

(i) detailed methodological reporting is crucial due to the reasons mentioned above and
(ii) conclusions should be made considering the limitations imposed by the experimental design and intrinsic technical biases, as with any other experiment.

Readers should not infer that the output of RNA-Seq is equivalent to a reliable snapshot of the RNA composition of a cell or an exRNA sample [205]. Fortunately, most technical biases tend to be systematic, and relative comparisons of gene expression between samples (i.e., how transcript A changes in comparison to transcript B between sample X and sample Y) are more trustworthy than within-sample gene expression rankings [101].

Although most of the points raised above apply to any RNA-Seq experiment, they become critical when RNA quantities are small. Examples include single-cell RNA-Seq (current challenges reviewed in [206]) and the entire exRNA field. The bright side is that the more we understand technical biases and RNA-Seq limitations, the broader the horizon of available methods to overcome or mitigate these limitations. As a consequence, these are exciting times for RNA and exRNA biology, where conventional assumptions and established concepts are being updated based on data derived from ever-refined sequencing methods. Quoting the brilliant molecular biologist Sydney Brenner (1927–2019): *"Progress in science depends on new techniques, new discoveries and new ideas, probably in that order."*

References

[1] J.S. Redzic, L. Balaj, K.E. van der Vos, X.O. Breakefield, Extracellular RNA mediates and marks cancer progression, Semin. Cancer Biol. 28 (2014) 14–23, https://doi.org/10.1016/j.semcancer.2014.04.010.

[2] A.-K. Elsemüller, V. Tomalla, U. Gärtner, K. Troidl, S. Jeratsch, J. Graumann, N. Baal, H. Hackstein, M. Lasch, E. Deindl, K.T. Preissner, S. Fischer, Characterization of mast cell-derived rRNA-containing microvesicles and their inflammatory impact on endothelial cells, FASEB J. 33 (2019) 5457–5467, https://doi.org/10.1096/fj.201801853RR.

[3] D. Siedlecki-Wullich, J. Català-Solsona, C. Fábregas, I. Hernández, J. Clarimon, A. Lleó, M. Boada, C.A. Saura, J. Rodríguez-Álvarez, A.J. Miñano-Molina, Altered microRNAs related to synaptic function as potential plasma biomarkers for Alzheimer's disease, Alzheimers Res. Ther. 11 (2019) 46, https://doi.org/10.1186/s13195-019-0501-4.

[4] Y. Zhao, Y. Zhang, L. Zhang, Y. Dong, H. Ji, L. Shen, The potential markers of circulating microRNAs and long non-coding RNAs in Alzheimer's disease, Aging Dis. 10 (2019) 1293–1301, https://doi.org/10.14336/AD.2018.1105.

[5] J.F. Quinn, T. Patel, D. Wong, S. Das, J.E. Freedman, L.C. Laurent, B.S. Carter, F. Hochberg, K. Van Keuren-Jensen, M. Huentelman, R. Spetzler, M.Y.S. Kalani, J. Arango, P.D. Adelson, H.L. Weiner, R. Gandhi, B. Goilav, C. Putterman, J.A. Saugstad, Extracellular RNAs: development as biomarkers of human disease, J. Extracell. Vesicles 4 (2015) 27495, https://doi.org/10.3402/jev.v4.27495.

[6] R. Shah, V. Murthy, M. Pacold, K. Danielson, K. Tanriverdi, M.G. Larson, K. Hanspers, A. Pico, E. Mick, J. Reis, S. de Ferranti, E. Freinkman, D. Levy, U. Hoffmann, S. Osganian, S. Das, J.E. Freedman, Extracellular RNAs are associated with insulin resistance and metabolic phenotypes, Diabetes Care 40 (2017) 546–553, https://doi.org/10.2337/dc16-1354.

[7] A.M. Ainsztein, P.J. Brooks, V.G. Dugan, A. Ganguly, M. Guo, T.K. Howcroft, C.A. Kelley, L.S. Kuo, P.A. Labosky, R. Lenzi, G.A. McKie, S. Mohla, D. Procaccini, M. Reilly, J.S. Satterlee, P.R. Srinivas, E.S. Church, M. Sutherland, D.A. Tagle, J.M. Tucker, S. Venkatachalam, The NIH extracellular RNA communication consortium, J. Extracell. Vesicles 4 (2015) 27493, https://doi.org/10.3402/jev.v4.27493.

[8] S. Das, K.M. Ansel, M. Bitzer, X.O. Breakefield, A. Charest, D.J. Galas, M.B. Gerstein, M. Gupta, A. Milosavljevic, M.T. McManus, T. Patel, R.L. Raffai, J. Rozowsky, M.E. Roth, J.A. Saugstad, K. Van Keuren-Jensen, A.M. Weaver, L.C. Laurent, The extracellular RNA communication consortium: establishing foundational knowledge and technologies for extracellular RNA research, Cell 177 (2019) 231–242, https://doi.org/10.1016/j.cell.2019.03.023.

[9] A.F. Hill, D.M. Pegtel, U. Lambertz, T. Leonardi, L. O'Driscoll, S. Pluchino, D. Ter-Ovanesyan, E.N.M., Nolte-'t Hoen, ISEV position paper: extracellular vesicle RNA analysis and bioinformatics, J. Extracell. Vesicles 2 (2013), https://doi.org/10.3402/jev.v2i0.22859.

[10] K.W. Witwer, E.I. Buzás, L.T. Bemis, A. Bora, C. Lässer, J. Lötvall, E.N. Nolte-'t Hoen, M.G. Piper, S. Sivaraman, J. Skog, C. Théry, M.H. Wauben, F. Hochberg, Standardization of sample collection, isolation and analysis methods in extracellular vesicle research, J. Extracell. Vesicles 2 (2013), https://doi.org/10.3402/jev.v2i0.20360.

[11] B. Mateescu, E.J.K. Kowal, B.W.M. van Balkom, S. Bartel, S.N. Bhattacharyya, E.I. Buzás, A.H. Buck, P. de Candia, F.W.N. Chow, S. Das, T.A.P. Driedonks, L. Fernández-Messina, F. Haderk, A.F. Hill, J.C. Jones, K.R. Van Keuren-Jensen, C.P. Lai, C. Lässer, I. di Liegro, T.R. Lunavat, M.J. Lorenowicz, S.L.N. Maas, I. Mäger, M. Mittelbrunn, S. Momma, K. Mukherjee, M. Nawaz, D.M. Pegtel, M.W. Pfaffl, R.M. Schiffelers, H. Tahara, C. Théry, J.P. Tosar, M.H.M. Wauben, K.W. Witwer, E.N. Nolte-'t Hoen, Obstacles and opportunities in the functional analysis of extracellular vesicle RNA – an ISEV position paper, J. Extracell. Vesicles 6 (2017) 1286095, https://doi.org/10.1080/20013078.2017.1286095.

[12] C. Soekmadji, A.F. Hill, M.H. Wauben, E.I. Buzás, D. Di Vizio, C. Gardiner, J. Lötvall, S. Sahoo, K.W. Witwer, Towards mechanisms and standardization in extracellular vesicle and extracellular RNA studies: results of a worldwide survey, J. Extracellul. Vesicles 7 (2018) 1535745, https://doi.org/10.1080/20013078.2018.1535745.

[13] A. Turchinovich, O. Drapkina, A. Tonevitsky, Transcriptome of extracellular vesicles: state-of-the-art, Front. Immunol. 10 (2019) 202, https://doi.org/10.3389/fimmu.2019.00202.

[14] K.C. Vickers, B.T. Palmisano, B.M. Shoucri, R.D. Shamburek, A.T. Remaley, MicroRNAs are transported in plasma and delivered to recipient cells by high-density lipoproteins, Nat. Cell Biol. 13 (2011) 423–433, https://doi.org/10.1038/ncb2210.

[15] A. Turchinovich, B. Burwinkel, Distinct AGO1 and AGO2 associated miRNA profiles in human cells and blood plasma, RNA Biol. 9 (2012) 1066–1075, https://doi.org/10.4161/rna.21083.

[16] J.D. Arroyo, J.R. Chevillet, E.M. Kroh, I.K. Ruf, C.C. Pritchard, D.F. Gibson, P.S. Mitchell, C.F. Bennett, E.L. Pogosova-Agadjanyan, D.L. Stirewalt, J.F. Tait, M. Tewari, Argonaute2 complexes carry a population of circulating microRNAs independent of vesicles in human plasma, Proc. Natl. Acad. Sci. U. S. A. 108 (2011) 5003–5008, https://doi.org/10.1073/pnas.1019055108.

[17] A. Turchinovich, L. Weiz, A. Langheinz, B. Burwinkel, Characterization of extracellular circulating microRNA, Nucleic Acids Res. 39 (2011) 7223–7233, https://doi.org/10.1093/nar/gkr254.

[18] Z. Wei, A.O. Batagov, S. Schinelli, J. Wang, Y. Wang, R. El Fatimy, R. Rabinovsky, L. Balaj, C.C. Chen, F. Hochberg, B. Carter, X.O. Breakefield, A.M. Krichevsky, Coding and noncoding landscape of extracellular RNA released by human glioma stem cells, Nat. Commun. 8 (2017) 1145, https://doi.org/10.1038/s41467-017-01196-x.

[19] J.P. Tosar, F. Gámbaro, L. Darré, S. Pantano, E. Westhof, A. Cayota, Dimerization confers increased stability to nucleases in 5′ halves from glycine and glutamic acid tRNAs, Nucleic Acids Res. 46 (2018) 9081–9093, https://doi.org/10.1093/nar/gky495.

[20] J.P. Tosar, M. Segovia, M. Castellano, F. Gámbaro, Y. Akiyama, P. Fagúndez, Á. Olivera, B. Costa, T. Possi, M. Hill, P. Ivanov, A. Cayota, Fragmentation of extracellular ribosomes and tRNAs shapes the extracellular RNAome, Nucleic Acids Res. 48 (2020) 12874–12888, https://doi.org/10.1093/nar/gkaa674.

[21] G. Nechooshtan, D. Yunusov, K. Chang, T.R. Gingeras, Processing by RNase 1 forms tRNA halves and distinct Y RNA fragments in the extracellular environment, Nucleic Acids Res. 48 (2020) 8035–8049, https://doi.org/10.1093/nar/gkaa526.

[22] J.P. Tosar, F. Gámbaro, J. Sanguinetti, B. Bonilla, K.W. Witwer, A. Cayota, Assessment of small RNA sorting into different extracellular fractions revealed by high-throughput sequencing of breast cell lines, Nucleic Acids Res. 43 (2015) 5601–5616, https://doi.org/10.1093/nar/gkv432.

[23] H. Valadi, K. Ekström, A. Bossios, M. Sjöstrand, J.J. Lee, J.O. Lötvall, Exosome-mediated transfer of mRNAs and microRNAs is a novel mechanism of genetic exchange between cells, Nat. Cell Biol. 9 (2007) 654–659.

[24] E. Lázaro-Ibáñez, C. Lässer, G.V. Shelke, R. Crescitelli, S.C. Jang, A. Cvjetkovic, A. García-Rodríguez, J. Lötvall, DNA analysis of low- and high-density fractions defines heterogeneous subpopulations of small extracellular vesicles based on their DNA cargo and topology, J. Extracell. Vesicles 8 (2019) 1656993, https://doi.org/10.1080/20013078.2019.1656993.

[25] N. Karimi, A. Cvjetkovic, S.C. Jang, R. Crescitelli, M.A. Hosseinpour Feizi, R. Nieuwland, J. Lötvall, C. Lässer, Detailed analysis of the plasma extracellular vesicle proteome after separation from lipoproteins, Cell. Mol. Life Sci. CMLS 75 (2018) 2873–2886, https://doi.org/10.1007/s00018-018-2773-4.

[26] C. Théry, K.W. Witwer, E. Aikawa, M.J. Alcaraz, J.D. Anderson, R. Andriantsitohaina, A. Antoniou, T. Arab, F. Archer, G.K. Atkin-Smith, D.C. Ayre, J.-M. Bach, D. Bachurski, H. Baharvand, L. Balaj, S. Baldacchino, N.N. Bauer, A.A. Baxter, M. Bebawy, C. Beckham, A.B. Zavec, A. Benmoussa, A.C. Berardi, P. Bergese, E. Bielska, C. Blenkiron, S. Bobis-Wozowicz, E. Boilard, W. Boireau, A. Bongiovanni, F.E. Borràs, S. Bosch, C.M. Boulanger, X. Breakefield, A.M. Breglio, M.Á. Brennan, D.R. Brigstock, A. Brisson, M.L. Broekman, J.F. Bromberg, P. Bryl-Górecka, S. Buch, A.H. Buck, D. Burger, S. Busatto, D. Buschmann, B. Bussolati, E.I. Buzás, J.B. Byrd, G. Camussi, D.R. Carter, S. Caruso, L.W. Chamley, Y.-T. Chang, C. Chen, S. Chen, L. Cheng, A.R. Chin, A. Clayton, S.P. Clerici, A. Cocks, E. Cocucci, R.J. Coffey, A. Cordeiro-da-Silva, Y. Couch, F.A. Coumans, B. Coyle, R. Crescitelli, M.F. Criado, C. D'Souza-Schorey, S. Das, A.D. Chaudhuri, P. de Candia, E.F. De Santana, O. De Wever, H.A. Del Portillo, T. Demaret, S. Deville, A. Devitt, B. Dhondt, D. Di Vizio, L.C. Dieterich, V. Dolo, A.P.D. Rubio, M. Dominici, M.R. Dourado, T.A. Driedonks, F.V. Duarte, H.M. Duncan, R.M. Eichenberger, K. Ekström, S. El Andaloussi, C. Elie-Caille, U. Erdbrügger, J.M. Falcón-Pérez, F. Fatima, J.E. Fish, M. Flores-Bellver, A. Försönits, A. Frelet-Barrand, F. Fricke, G. Fuhrmann, S. Gabrielsson, A. Gámez-Valero, C. Gardiner, K. Gärtner, R. Gaudin, Y.S. Gho, B. Giebel, C. Gilbert, M. Gimona, I. Giusti, D.C. Goberdhan, A. Görgens, S.M. Gorski, D.W. Greening, J.C. Gross, A. Gualerzi, G.N. Gupta, D. Gustafson, A. Handberg, R.A. Haraszti, P. Harrison, H. Hegyesi, A. Hendrix, A.F. Hill, F.H. Hochberg, K.F. Hoffmann, B. Holder, H. Holthofer, B. Hosseinkhani, G. Hu, Y. Huang, V. Huber, S. Hunt, A.G.-E. Ibrahim, T. Ikezu, J.M. Inal, M. Isin, A. Ivanova, H.K. Jackson, S. Jacobsen, S.M. Jay, M. Jayachandran, G. Jenster, L. Jiang, S.M. Johnson, J.C. Jones, A. Jong, T. Jovanovic-Talisman, S. Jung, R. Kalluri, S.-I. Kano, S. Kaur, Y. Kawamura, E.T. Keller, D. Khamari, E. Khomyakova, A. Khvorova, P. Kierulf, K.P. Kim, T. Kislinger, M. Klingeborn, D.J. Klinke, M. Kornek, M.M. Kosanović, Á.F. Kovács, E.-M. Krämer-Albers, S. Krasemann, M. Krause, I.V. Kurochkin, G.D. Kusuma, S. Kuypers, S. Laitinen, S.M.

Langevin, L.R. Languino, J. Lannigan, C. Lässer, L.C. Laurent, G. Lavieu, E. Lázaro-Ibáñez, S. Le Lay, M.-S. Lee, Y.X.F. Lee, D.S. Lemos, M. Lenassi, A. Leszczynska, I.T. Li, K. Liao, S.F. Libregts, E. Ligeti, R. Lim, S.K. Lim, A. Linē, K. Linnemannstöns, A. Llorente, C.A. Lombard, M.J. Lorenowicz, Á.M. Lörincz, J. Lötvall, J. Lovett, M.C. Lowry, X. Loyer, Q. Lu, B. Lukomska, T.R. Lunavat, S.L. Maas, H. Malhi, A. Marcilla, J. Mariani, J. Mariscal, E.S. Martens-Uzunova, L. Martin-Jaular, M.C. Martinez, V.R. Martins, M. Mathieu, S. Mathivanan, M. Maugeri, L.K. McGinnis, M.J. McVey, D.G. Meckes, K.L. Meehan, I. Mertens, V.R. Minciacchi, A. Möller, M.M. Jørgensen, A. Morales-Kastresana, J. Morhayim, F. Mullier, M. Muraca, L. Musante, V. Mussack, D.C. Muth, K.H. Myburgh, T. Najrana, M. Nawaz, I. Nazarenko, P. Nejsum, C. Neri, T. Neri, R. Nieuwland, L. Nimrichter, J.P. Nolan, E.N. Nolte-'t Hoen, N.N. Hooten, L. O'Driscoll, T. O'Grady, A. O'Loghlen, T. Ochiya, M. Olivier, A. Ortiz, L.A. Ortiz, X. Osteikoetxea, O. Østergaard, M. Ostrowski, J. Park, D.M. Pegtel, H. Peinado, F. Perut, M.W. Pfaffl, D.G. Phinney, B.C. Pieters, R.C. Pink, D.S. Pisetsky, E.P. von Strandmann, I. Polakovicova, I.K. Poon, B.H. Powell, I. Prada, L. Pulliam, P. Quesenberry, A. Radeghieri, R.L. Raffai, S. Raimondo, J. Rak, M.I. Ramirez, G. Raposo, M.S. Rayyan, N. Regev-Rudzki, F.L. Ricklefs, P.D. Robbins, D.D. Roberts, S.C. Rodrigues, E. Rohde, S. Rome, K.M. Rouschop, A. Rughetti, A.E. Russell, P. Saá, S. Sahoo, E. Salas-Huenuleo, C. Sánchez, J.A. Saugstad, M.J. Saul, R.M. Schiffelers, R. Schneider, T.H. Schøyen, A. Scott, E. Shahaj, S. Sharma, O. Shatnyeva, F. Shekari, G.V. Shelke, A.K. Shetty, K. Shiba, P.R.-M. Siljander, A.M. Silva, A. Skowronek, O.L. Snyder, R.P. Soares, B.W. Sódar, C. Soekmadji, J. Sotillo, P.D. Stahl, W. Stoorvogel, S.L. Stott, E.F. Strasser, S. Swift, H. Tahara, M. Tewari, K. Timms, S. Tiwari, R. Tixeira, M. Tkach, W.S. Toh, R. Tomasini, A.C. Torrecilhas, J.P. Tosar, V. Toxavidis, L. Urbanelli, P. Vader, B.W. van Balkom, S.G. van der Grein, J. Van Deun, M.J. van Herwijnen, K. Van Keuren-Jensen, G. van Niel, M.E. van Royen, A.J. van Wijnen, M.H. Vasconcelos, I.J. Vechetti, T.D. Veit, L.J. Vella, É. Velot, F.J. Verweij, B. Vestad, J.L. Viñas, T. Visnovitz, K.V. Vukman, J. Wahlgren, D.C. Watson, M.H. Wauben, A. Weaver, J.P. Webber, V. Weber, A.M. Wehman, D.J. Weiss, J.A. Welsh, S. Wendt, A.M. Wheelock, Z. Wiener, L. Witte, J. Wolfram, A. Xagorari, P. Xander, J. Xu, X. Yan, M. Yáñez-Mó, H. Yin, Y. Yuana, V. Zappulli, J. Zarubova, V. Žėkas, J.-Y. Zhang, Z. Zhao, L. Zheng, A.R. Zheutlin, A.M. Zickler, P. Zimmermann, A.M. Zivkovic, D. Zocco, E.K. Zuba-Surma, Minimal information for studies of extracellular vesicles 2018 (MISEV2018): a position statement of the International Society for Extracellular Vesicles and update of the MISEV2014 guidelines, J. Extracell. Vesicles 7 (2018) 1535750, https://doi.org/10.1080/20013078.2018.1535750.

[27] R. Crescitelli, C. Lässer, T.G. Szabó, A. Kittel, M. Eldh, I. Dianzani, E.I. Buzás, J. Lötvall, Distinct RNA profiles in subpopulations of extracellular vesicles: apoptotic bodies, microvesicles and exosomes, J. Extracell. Vesicles 2 (2013), https://doi.org/10.3402/jev.v2i0.20677.

[28] M. Palviainen, M. Saraswat, Z. Varga, D. Kitka, M. Neuvonen, M. Puhka, S. Joenväärä, R. Renkonen, R. Nieuwland, M. Takatalo, P.R.M. Siljander, Extracellular vesicles from human plasma and serum are carriers of extravesicular cargo-implications for biomarker discovery, PLoS One 15 (2020) e0236439, https://doi.org/10.1371/journal.pone.0236439.

[29] J.B. Simonsen, R. Münter, Pay attention to biological nanoparticles when studying the protein corona on nanomedicines, Angew. Chem. (Int. Ed. in Engl.) 59 (2020) 12584–12588, https://doi.org/10.1002/anie.202004611.

[30] G.V. Shelke, C. Lässer, Y.S. Gho, J. Lötvall, Importance of exosome depletion protocols to eliminate functional and RNA-containing extracellular vesicles from fetal bovine serum, J. Extracell. Vesicles 3 (2014), https://doi.org/10.3402/jev.v3.24783.

[31] X. Osteikoetxea, B. Sódar, A. Németh, K. Szabó-Taylor, K. Pálóczi, K.V. Vukman, V. Tamási, A. Balogh, Á. Kittel, É. Pállinger, E.I. Buzás, Differential detergent sensitivity of extracellular vesicle subpopulations, Org. Biomol. Chem. 13 (2015) 9775–9782, https://doi.org/10.1039/c5ob01451d.

[32] S. Mitra, R. Brukh, Sample preparation techniques in analytical, Chemistry 162 (2003), https://doi.org/10.1002/0471457817.ch1.

[33] E. Della Bella, M.J. Stoddart, Cell detachment rapidly induces changes in noncoding RNA expression in human mesenchymal stromal cells, BioTechniques 67 (2019) 286–293, https://doi.org/10.2144/btn-2019-0038.

[34] M.A. Faghihi, M. Zhang, J. Huang, F. Modarresi, M.P. Van der Brug, M.A. Nalls, M.R. Cookson, G. St-Laurent, C. Wahlestedt, Evidence for natural antisense transcript-mediated inhibition of microRNA function, Genome Biol. 11 (2010) R56, https://doi.org/10.1186/gb-2010-11-5-r56.

[35] I. Can, G.T. Javan, A.E. Pozhitkov, P.A. Noble, Distinctive thanatomicrobiome signatures found in the blood and internal organs of humans, J. Microbiol. Methods 106 (2014) 1–7, https://doi.org/10.1016/j.mimet.2014.07.026.

[36] The GTEx Consortium, The GTEx Consortium atlas of genetic regulatory effects across human tissues, Science (New York, N.Y.) 369 (2020) 1318–1330, https://doi.org/10.1126/science.aaz1776.

[37] S. Mangul, H.T. Yang, N. Strauli, F. Gruhl, H.T. Porath, K. Hsieh, L. Chen, T. Daley, S. Christenson, A. Wesolowska-Andersen, R. Spreafico, C. Rios, C. Eng, A.D. Smith, R.D. Hernandez, R.A. Ophoff, J.R. Santana, E.Y. Levanon, P.G. Woodruff, E. Burchard, M.A. Seibold, S. Shifman, E. Eskin, N. Zaitlen, ROP: dumpster diving in RNA-sequencing to find the source of 1 trillion reads across diverse adult human tissues, Genome Biol. 19 (2018) 36, https://doi.org/10.1186/s13059-018-1403-7.

[38] U. Schleicher, A. Hesse, C. Bogdan, Minute numbers of contaminant CD8+ T cells or CD11b+CD11c+ NK cells are the source of IFN-gamma in IL-12/IL-18-stimulated mouse macrophage populations, Blood 105 (2005) 1319–1328.

[39] B.W. Sódar, Á. Kittel, K. Pálóczi, K.V. Vukman, X. Osteikoetxea, K. Szabó-Taylor, A. Németh, B. Sperlágh, T. Baranyai, Z. Giricz, Z. Wiener, L. Turiák, L. Drahos, É. Pállinger, K. Vékey, P. Ferdinandy, A. Falus, E.I. Buzás, Low-density lipoprotein mimics blood plasma-derived exosomes and microvesicles during isolation and detection, Sci. Rep. 6 (2016) 24316, https://doi.org/10.1038/srep24316.

[40] Z. Wei, A.O. Batagov, D.R.F. Carter, A.M. Krichevsky, Fetal bovine serum RNA interferes with the cell culture derived extracellular RNA, Sci. Rep. 6 (2016) 31175, https://doi.org/10.1038/srep31175.

[41] J.P. Tosar, A. Cayota, E. Eitan, M.K. Halushka, K.W. Witwer, Ribonucleic artefacts: are some extracellular RNA discoveries driven by cell culture medium components? J. Extracell. Vesicles 6 (2017) 1272832, https://doi.org/10.1080/20013078.2016.1272832.

[42] M. Auber, D. Fröhlich, O. Drechsel, E. Karaulanov, E.-M. Krämer-Albers, Serum-free media supplements carry miRNAs that co-purify with extracellular vesicles, J. Extracell. Vesicles 8 (2019) 1656042, https://doi.org/10.1080/20013078.2019.1656042.

[43] B. Xu, N. Zhi, G. Hu, Z. Wan, X. Zheng, X. Liu, S. Wong, S. Kajigaya, K. Zhao, Q. Mao, N.S. Young, Hybrid DNA virus in Chinese patients with seronegative hepatitis discovered by deep sequencing, Proc. Natl. Acad. Sci. U. S. A. 110 (2013) 10264–10269, https://doi.org/10.1073/pnas.1303744110.

[44] S.N. Naccache, A.L. Greninger, D. Lee, L.L. Coffey, T. Phan, A. Rein-Weston, A. Aronsohn, J. Hackett, E.L. Delwart, C.Y. Chiu, The perils of pathogen discovery: origin of a novel parvovirus-like hybrid genome traced to nucleic acid extraction spin columns, J. Virol. 87 (2013) 11966–11977, https://doi.org/10.1128/JVI.02323-13.

[45] A. Heintz-Buschart, D. Yusuf, A. Kaysen, A. Etheridge, J.V. Fritz, P. May, C. de Beaufort, B.B. Upadhyaya, A. Ghosal, D.J. Galas, P. Wilmes, Small RNA profiling of low biomass samples: identification and removal of contaminants, BMC Biol. 16 (2018) 52, https://doi.org/10.1186/s12915-018-0522-7.

[46] T.O. Nieuwenhuis, S.Y. Yang, R.X. Verma, V. Pillalamarri, D.E. Arking, A.Z. Rosenberg, M.N. McCall, M.K. Halushka, Consistent RNA sequencing contamination in GTEx and other data sets., nature, Communications 11 (2020) 1933, https://doi.org/10.1038/s41467-020-15821-9.

[47] B.A. Haider, A.S. Baras, M.N. McCall, J.A. Hertel, T.C. Cornish, M.K. Halushka, A critical evaluation of microRNA biomarkers in non-neoplastic disease, PLoS One 9 (2014) e89565, https://doi.org/10.1371/journal.pone.0089565.

[48] J.P. Tosar, C. Rovira, H. Naya, A. Cayota, Mining of public sequencing databases supports a non-dietary origin for putative foreign miRNAs: underestimated effects of contamination in NGS, RNA (New York, N.Y.) 20 (2014) 754–757, https://doi.org/10.1261/rna.044263.114.

[49] S.J. Salter, M.J. Cox, E.M. Turek, S.T. Calus, W.O. Cookson, M.F. Moffatt, P. Turner, J. Parkhill, N.J. Loman, A.W. Walker, Reagent and laboratory contamination can critically impact sequence-based microbiome analyses, BMC Biol. 12 (2014) 87, https://doi.org/10.1186/s12915-014-0087-z.

[50] L. Zhang, D. Hou, X. Chen, D. Li, L. Zhu, Y. Zhang, J. Li, Z. Bian, X. Liang, X. Cai, Y. Yin, C. Wang, T. Zhang, D. Zhu, D. Zhang, J. Xu, Q. Chen, Y. Ba, J. Liu, Q. Wang, J. Chen, J. Wang, M. Wang, Q. Zhang, J. Zhang, K. Zen, C.-Y. Zhang, Exogenous plant MIR168a specifically targets mammalian LDLRAP1: evidence of cross-kingdom regulation by microRNA, Cell Res. 22 (2012) 107–126, https://doi.org/10.1038/cr.2011.158.

[51] B. Dickinson, Y. Zhang, J.S. Petrick, G. Heck, S. Ivashuta, W.S. Marshall, Lack of detectable oral bioavailability of plant microRNAs after feeding in mice, Nat. Biotechnol. 31 (2013) 965–967, https://doi.org/10.1038/nbt.2737.

[52] K.W. Witwer, M.A. McAlexander, S.E. Queen, R.J. Adams, Real-time quantitative PCR and droplet digital PCR for plant miRNAs in mammalian blood provide little evidence for general uptake of dietary miRNAs: limited evidence for general uptake of dietary plant xenomiRs, RNA Biol. 10 (2013) 1080–1086, https://doi.org/10.4161/rna.25246.

[53] J.W. Snow, A.E. Hale, S.K. Isaacs, A.L. Baggish, S.Y. Chan, Ineffective delivery of diet-derived microRNAs to recipient animal organisms, RNA Biol. 10 (2013) 1107–1116, https://doi.org/10.4161/rna.24909.

[54] W. Kang, C.H. Bang-Berthelsen, A. Holm, A.J.S. Houben, A.H. Müller, T. Thymann, F. Pociot, X. Estivill, M.R. Friedländer, Survey of 800+ data sets from human tissue and body fluid reveals xenomiRs are likely artifacts, RNA (New York, N.Y.) 23 (2017) 433–445, https://doi.org/10.1261/rna.059725.116.

[55] X. Chen, K. Zen, C.-Y. Zhang, Reply to lack of detectable oral bioavailability of plant microRNAs after feeding in mice, Nat. Biotechnol. 31 (2013) 967–969, https://doi.org/10.1038/nbt.2741.

[56] T. Nolan, R.E. Hands, S.A. Bustin, Quantification of mRNA using real-time RT-PCR, Nat. Protoc. 1 (2006) 1559–1582.

[57] J.V. Die, Á. Obrero, C.I. González-Verdejo, B. Román, Characterization of the 3′:5′ ratio for reliable determination of RNA quality, Anal. Biochem. 419 (2011) 336–338, https://doi.org/10.1016/j.ab.2011.08.012.

[58] R. Stark, M. Grzelak, J. Hadfield, RNA sequencing: the teenage years, Nat. Rev. Genet. 20 (2019) 631–656, https://doi.org/10.1038/s41576-019-0150-2.

[59] S. Lamarre, P. Frasse, M. Zouine, D. Labourdette, E. Sainderichin, G. Hu, V. Le Berre-Anton, M. Bouzayen, E. Maza, Optimization of an RNA-Seq differential gene expression analysis depending on biological replicate number and library size, Front. Plant Sci. 9 (2018) 108, https://doi.org/10.3389/fpls.2018.00108.

[60] A. Conesa, P. Madrigal, S. Tarazona, D. Gomez-Cabrero, A. Cervera, A. McPherson, M.W. Szcześniak, D.J. Gaffney, L.L. Elo, X. Zhang, A. Mortazavi, A survey of best practices for RNA-seq data analysis, Genome Biol. 17 (2016) 13, https://doi.org/10.1186/s13059-016-0881-8.

[61] N.J. Schurch, P. Schofield, M. Gierliński, C. Cole, A. Sherstnev, V. Singh, N. Wrobel, K. Gharbi, G.G. Simpson, T. Owen-Hughes, M. Blaxter, G.J. Barton, How many biological replicates are needed in an RNA-seq experiment and which differential expression tool should you use? RNA (New York, N.Y.) 22 (2016) 839–851, https://doi.org/10.1261/rna.053959.115.

[62] K.D. Hansen, Z. Wu, R.A. Irizarry, J.T. Leek, Sequencing technology does not eliminate biological variability, Nat. Biotechnol. 29 (2011) 572–573, https://doi.org/10.1038/nbt.1910.

[63] T. Ching, S. Huang, L.X. Garmire, Power analysis and sample size estimation for RNA-Seq differential expression, RNA (New York, N.Y.) 20 (2014) 1684–1696, https://doi.org/10.1261/rna.046011.114.

[64] R.P.R. Metpally, S. Nasser, I. Malenica, A. Courtright, E. Carlson, L. Ghaffari, S. Villa, W. Tembe, K. Van Keuren-Jensen, Comparison of analysis tools for miRNA high throughput sequencing using nerve crush as a model, Front. Genet. 4 (2013) 20, https://doi.org/10.3389/fgene.2013.00020.

[65] J.D. Campbell, G. Liu, L. Luo, J. Xiao, J. Gerrein, B. Juan-Guardela, J. Tedrow, Y.O. Alekseyev, I.V. Yang, M. Correll, M. Geraci, J. Quackenbush, F. Sciurba, D.A. Schwartz, N. Kaminski, W.E. Johnson, S. Monti, A. Spira, J. Beane, M.E. Lenburg, Assessment of microRNA differential expression and detection in multiplexed small RNA sequencing data, RNA (New York, N.Y.) 21 (2015) 164–171, https://doi.org/10.1261/rna.046060.114.

[66] C. Hartl, Y. Gao, Clarifying the effect of library batch on extracellular RNA sequencing, Proc. Natl. Acad. Sci. U. S. A. 117 (2020) 1849–1850, https://doi.org/10.1073/pnas.1916312117.

[67] M.G. Ross, C. Russ, M. Costello, A. Hollinger, N.J. Lennon, R. Hegarty, C. Nusbaum, D.B. Jaffe, Characterizing and measuring bias in sequence data, Genome Biol. 14 (2013) R51, https://doi.org/10.1186/gb-2013-14-5-r51.

[68] D. Buschmann, A. Haberberger, B. Kirchner, M. Spornraft, I. Riedmaier, G. Schelling, M.W. Pfaffl, Toward reliable biomarker signatures in the age of liquid biopsies – how to standardize the small RNA-Seq workflow, Nucleic Acids Res. 44 (2016) 5995–6018, https://doi.org/10.1093/nar/gkw545.

[69] P.M. Godoy, A.J. Barczak, P. DeHoff, S. Srinivasan, A. Etheridge, D. Galas, S. Das, D.J. Erle, L.C. Laurent, Comparison of reproducibility, accuracy, sensitivity, and specificity of miRNA quantification platforms, Cell Rep. 29 (2019) 4212–4222.e5, https://doi.org/10.1016/j.celrep.2019.11.078.

[70] C. Wright, A. Rajpurohit, E.E. Burke, C. Williams, L. Collado-Torres, M. Kimos, N.J. Brandon, A.J. Cross, A.E. Jaffe, D.R. Weinberger, J.H. Shin, Comprehensive assessment of multiple biases in small RNA sequencing reveals significant differences in the performance of widely used methods, BMC Genomics 20 (2019) 513, https://doi.org/10.1186/s12864-019-5870-3.

[71] U. Nagalakshmi, Z. Wang, K. Waern, C. Shou, D. Raha, M. Gerstein, M. Snyder, The transcriptional landscape of the yeast genome defined by RNA sequencing, Science (New York, N.Y.) 320 (2008) 1344–1349, https://doi.org/10.1126/science.1158441.

[72] Y. Li, L. Zhang, R. Li, M. Zhang, Y. Li, H. Wang, S. Wang, Z. Bao, Systematic identification and validation of the reference genes from 60 RNA-Seq libraries in the scallop Mizuhopecten yessoensis, BMC Genomics 20 (2019) 288, https://doi.org/10.1186/s12864-019-5661-x.

[73] A.R. Wu, N.F. Neff, T. Kalisky, P. Dalerba, B. Treutlein, M.E. Rothenberg, F.M. Mburu, G.L. Mantalas, S. Sim, M.F. Clarke, S.R. Quake, Quantitative assessment of single-cell RNA-sequencing methods, Nat. Methods 11 (2014) 41–46, https://doi.org/10.1038/nmeth.2694.

[74] E. Speranza, L.A. Altamura, K. Kulcsar, S.L. Bixler, C.A. Rossi, R.J. Schoepp, E. Nagle, W. Aguilar, C.E. Douglas, K.L. Delp, T.D. Minogue, G. Palacios, A.J. Goff, J.H. Connor, Comparison of transcriptomic platforms for analysis of whole blood from ebola-infected cynomolgus macaques, Sci. Rep. 7 (2017) 14756, https://doi.org/10.1038/s41598-017-15145-7.

[75] V.M. Shelton, J.R. Morrow, Catalytic transesterification and hydrolysis of RNA by zinc(II) complexes, Inorg. Chem. 30 (1991) 4295–4299. https://pubs.acs.org/doi/10.1021/ic00023a003.

[76] M. Wery, M. Descrimes, C. Thermes, D. Gautheret, A. Morillon, Zinc-mediated RNA fragmentation allows robust transcript reassembly upon whole transcriptome RNA-Seq, Methods (San Diego Calif.) 63 (2013) 25–31, https://doi.org/10.1016/j.ymeth.2013.03.009.

[77] Y. Veeranagouda, A. Remaury, J.-C. Guillemot, M. Didier, RNA fragmentation and sequencing (RF-Seq): cost-effective, time-efficient, and high-throughput 3' mRNA sequencing library construction in a single tube, Curr. Protocols Mol. Biol. 129 (2019) e109, https://doi.org/10.1002/cpmb.109.

[78] D. Faktorová, R.E.R. Nisbet, J.A.F. Robledo, E. Casacuberta, L. Sudek, A.E. Allen, M. Ares, C. Aresté, C. Balestreri, A.C. Barbrook, P. Beardslee, S. Bender, D.S. Booth, F.-Y. Bouget, C. Bowler, S.A. Breglia, C. Brownlee, G. Burger, H. Cerutti, R. Cesaroni, M.A. Chiurillo, T. Clemente, D.B. Coles, J.L. Collier, E.C. Cooney, K. Coyne, R. Docampo, C.L. Dupont, V. Edgcomb, E. Einarsson, P.A. Elustondo, F. Federici, V. Freire-Beneitez, N.J. Freyria, K. Fukuda, P.A. García, P.R. Girguis, F. Gomaa, S.G. Gornik, J. Guo, V. Hampl, Y. Hanawa, E.R. Haro-Contreras, E. Hehenberger, A. Highfield, Y. Hirakawa, A. Hopes, C.J. Howe, I. Hu, J. Ibañez, N.A.T. Irwin, Y. Ishii, N.E. Janowicz, A.C. Jones, A. Kachale, K. Fujimura-Kamada, B.

Kaur, J.Z. Kaye, E. Kazana, P.J. Keeling, N. King, L.A. Klobutcher, N. Lander, I. Lassadi, Z. Li, S. Lin, J.-C. Lozano, F. Luan, S. Maruyama, T. Matute, C. Miceli, J. Minagawa, M. Moosburner, S.R. Najle, D. Nanjappa, I.C. Nimmo, L. Noble, A.M.G. Novák Vanclová, M. Nowacki, I. Nuñez, A. Pain, A. Piersanti, S. Pucciarelli, J. Pyrih, J.S. Rest, M. Rius, D. Robertson, A. Ruaud, I. Ruiz-Trillo, M.A. Sigg, P.A. Silver, C.H. Slamovits, G. Jason Smith, B.N. Sprecher, R. Stern, E.C. Swart, A.D. Tsaousis, L. Tsypin, A. Turkewitz, J. Turnšek, M. Valach, V. Vergé, P. von Dassow, T. von der Haar, R.F. Waller, L. Wang, X. Wen, G. Wheeler, A. Woods, H. Zhang, T. Mock, A.Z. Worden, J. Lukeš, Genetic tool development in marine protists: emerging model organisms for experimental cell biology, Nat. Methods 17 (2020) 481–494, https://doi.org/10.1038/s41592-020-0796-x.

[79] J.J. Dunn, RNase III cleavage of single-stranded RNA. Effect of ionic strength on the fideltiy of cleavage, J. Biol. Chem. 251 (1976) 3807–3814.

[80] E.L. van Dijk, H. Auger, Y. Jaszczyszyn, C. Thermes, Ten years of next-generation sequencing technology, Trends Genet. 30 (2014) 418–426, https://doi.org/10.1016/j.tig.2014.07.001.

[81] D. Parkhomchuk, T. Borodina, V. Amstislavskiy, M. Banaru, L. Hallen, S. Krobitsch, H. Lehrach, A. Soldatov, Transcriptome analysis by strand-specific sequencing of complementary DNA, Nucleic Acids Res. 37 (2009) e123, https://doi.org/10.1093/nar/gkp596.

[82] R. Hrdlickova, M. Toloue, B. Tian, RNA-Seq methods for transcriptome analysis, Wiley Interdisc. Rev. RNA 8 (2017), https://doi.org/10.1002/wrna.1364.

[83] A. Mortazavi, B.A. Williams, K. McCue, L. Schaeffer, B. Wold, Mapping and quantifying mammalian transcriptomes by RNA-Seq, Nat. Methods 5 (2008) 621–628, https://doi.org/10.1038/nmeth.1226.

[84] C. Soneson, Y. Yao, A. Bratus-Neuenschwander, A. Patrignani, M.D. Robinson, S. Hussain, A comprehensive examination of nanopore native RNA sequencing for characterization of complex transcriptomes., nature, Communications 10 (2019) 3359, https://doi.org/10.1038/s41467-019-11272-z.

[85] A.G. Hunt, R. Xu, B. Addepalli, S. Rao, K.P. Forbes, L.R. Meeks, D. Xing, M. Mo, H. Zhao, A. Bandyopadhyay, L. Dampanaboina, A. Marion, C. Von Lanken, Q.Q. Li, Arabidopsis mRNA polyadenylation machinery: comprehensive analysis of protein-protein interactions and gene expression profiling, BMC Genomics 9 (2008) 220, https://doi.org/10.1186/1471-2164-9-220.

[86] D.K. Nam, S. Lee, G. Zhou, X. Cao, C. Wang, T. Clark, J. Chen, J.D. Rowley, S.M. Wang, Oligo(dT) primer generates a high frequency of truncated cDNAs through internal poly(A) priming during reverse transcription, Proc. Natl. Acad. Sci. U. S. A. 99 (2002) 6152–6156.

[87] M. Dávila López, T. Samuelsson, Early evolution of histone mRNA 3' end processing, RNA (New York, N.Y.) 14 (2008) 1–10.

[88] Z.T. Herbert, J.P. Kershner, V.L. Butty, J. Thimmapuram, S. Choudhari, Y.O. Alekseyev, J. Fan, J.W. Podnar, E. Wilcox, J. Gipson, A. Gillaspy, K. Jepsen, S.S. BonDurant, K. Morris, M. Berkeley, A. LeClerc, S.D. Simpson, G. Sommerville, L. Grimmett, M. Adams, S.S. Levine, Cross-site comparison of ribosomal depletion kits for Illumina RNAseq library construction, BMC Genomics 19 (2018) 199, https://doi.org/10.1186/s12864-018-4585-1.

[89] C.D. Armour, J.C. Castle, R. Chen, T. Babak, P. Loerch, S. Jackson, J.K. Shah, J. Dey, C.A. Rohl, J.M. Johnson, C.K. Raymond, Digital transcriptome profiling using selective hexamer priming for cDNA synthesis, Nat. Methods 6 (2009) 647–649, https://doi.org/10.1038/nmeth.1360.

[90] O. Arnaud, F. Le Loarer, F. Tirode, BAFfling pathologies: alterations of BAF complexes in cancer, Cancer Lett. 419 (2018) 266–279, https://doi.org/10.1016/j.canlet.2018.01.046.

[91] K.D. Hansen, S.E. Brenner, S. Dudoit, Biases in Illumina transcriptome sequencing caused by random hexamer priming, Nucleic Acids Res. 38 (2010) e131, https://doi.org/10.1093/nar/gkq224.

[92] T.P. van Gurp, L.M. McIntyre, K.J.F. Verhoeven, Consistent errors in first strand cDNA due to random hexamer mispriming, PLoS One 8 (2013) e85583, https://doi.org/10.1371/journal.pone.0085583.

[93] J.Z. Levin, M. Yassour, X. Adiconis, C. Nusbaum, D.A. Thompson, N. Friedman, A. Gnirke, A. Regev, Comprehensive comparative analysis of strand-specific RNA sequencing methods, Nat. Methods 7 (2010) 709–715, https://doi.org/10.1038/nmeth.1491.

[94] W. Zheng, L.M. Chung, H. Zhao, Bias detection and correction in RNA-sequencing data, BMC Bioinformatics 12 (2011) 290, https://doi.org/10.1186/1471-2105-12-290.
[95] J.D. Roberts, B.D. Preston, L.A. Johnston, A. Soni, L.A. Loeb, T.A. Kunkel, Fidelity of two retroviral reverse transcriptases during DNA-dependent DNA synthesis in vitro, Mol. Cell. Biol. 9 (1989) 469–476.
[96] L. Menéndez-Arias, Mutation rates and intrinsic fidelity of retroviral reverse transcriptases, Viruses 1 (2009) 1137–1165, https://doi.org/10.3390/v1031137.
[97] F. Zhuang, R.T. Fuchs, G.B. Robb, Small RNA expression profiling by high-throughput sequencing: implications of enzymatic manipulation, J. Nucleic Acids 2012 (2012) 360358, https://doi.org/10.1155/2012/360358.
[98] H. Xu, J. Yao, D.C. Wu, A.M. Lambowitz, Improved TGIRT-seq methods for comprehensive transcriptome profiling with decreased adapter dimer formation and bias correction, Sci. Rep. 9 (2019) 7953, https://doi.org/10.1038/s41598-019-44457-z.
[99] Y. Qin, J. Yao, D.C. Wu, R.M. Nottingham, S. Mohr, S. Hunicke-Smith, A.M. Lambowitz, High-throughput sequencing of human plasma RNA by using thermostable group II intron reverse transcriptases, RNA (New York, N.Y.) 22 (2016) 111–128, https://doi.org/10.1261/rna.054809.115.
[100] M.J. Shurtleff, J. Yao, Y. Qin, R.M. Nottingham, M.M. Temoche-Diaz, R. Schekman, A.M. Lambowitz, Broad role for YBX1 in defining the small noncoding RNA composition of exosomes, Proc. Natl. Acad. Sci. U. S. A. 114 (2017) E8987–E8995, https://doi.org/10.1073/pnas.1712108114.
[101] M.D. Giraldez, R.M. Spengler, A. Etheridge, P.M. Godoy, A.J. Barczak, S. Srinivasan, P.L. De Hoff, K. Tanriverdi, A. Courtright, S. Lu, J. Khoory, R. Rubio, D. Baxter, T.A.P. Driedonks, H.P.J. Buermans, E.N.M. Nolte-'t Hoen, H. Jiang, K. Wang, I. Ghiran, Y.E. Wang, K. Van Keuren-Jensen, J.E. Freedman, P.G. Woodruff, L.C. Laurent, D.J. Erle, D.J. Galas, M. Tewari, Comprehensive multi-center assessment of small RNA-seq methods for quantitative miRNA profiling, Nat. Biotechnol. 36 (2018) 746–757, https://doi.org/10.1038/nbt.4183.
[102] D. Zucha, P. Androvic, M. Kubista, L. Valihrach, Performance comparison of reverse transcriptases for single-cell studies, Clin. Chem. 66 (2020) 217–228, https://doi.org/10.1373/clinchem.2019.307835.
[103] H. Okano, Y. Katano, M. Baba, A. Fujiwara, R. Hidese, S. Fujiwara, I. Yanagihara, T. Hayashi, K. Kojima, T. Takita, K. Yasukawa, Enhanced detection of RNA by MMLV reverse transcriptase coupled with thermostable DNA polymerase and DNA/RNA helicase, Enzym. Microb. Technol. 96 (2017) 111–120, https://doi.org/10.1016/j.enzmictec.2016.10.003.
[104] Y. He, B. Vogelstein, V.E. Velculescu, N. Papadopoulos, K.W. Kinzler, The antisense transcriptomes of human cells, Science (New York, N.Y.) 322 (2008) 1855–1857, https://doi.org/10.1126/science.1163853.
[105] J.D. Mills, Y. Kawahara, M. Janitz, Strand-specific RNA-Seq provides greater resolution of transcriptome profiling, Curr. Genomics 14 (2013) 173–181, https://doi.org/10.2174/1389202911314030003.
[106] T. Borodina, J. Adjaye, M. Sultan, A strand-specific library preparation protocol for RNA sequencing, Methods Enzymol. 500 (2011) 79–98, https://doi.org/10.1016/B978-0-12-385118-5.00005-0.
[107] M. Matz, D. Shagin, E. Bogdanova, O. Britanova, S. Lukyanov, L. Diatchenko, A. Chenchik, Amplification of cDNA ends based on template-switching effect and step-out PCR, Nucleic Acids Res. 27 (1999) 1558–1560.
[108] D. Ramsköld, S. Luo, Y.-C. Wang, R. Li, Q. Deng, O.R. Faridani, G.A. Daniels, I. Khrebtukova, J.F. Loring, L.C. Laurent, G.P. Schroth, R. Sandberg, Full-length mRNA-Seq from single-cell levels of RNA and individual circulating tumor cells, Nat. Biotechnol. 30 (2012) 777–782.
[109] R. Wellenreuther, I. Schupp, A. Poustka, S. Wiemann, SMART amplification combined with cDNA size fractionation in order to obtain large full-length clones, BMC Genomics 5 (2004) 36.
[110] Y.Y. Zhu, E.M. Machleder, A. Chenchik, R. Li, P.D. Siebert, Reverse transcriptase template switching: a SMART approach for full-length cDNA library construction, BioTechniques 30 (2001) 892–897.
[111] X. Adiconis, D. Borges-Rivera, R. Satija, D.S. DeLuca, M.A. Busby, A.M. Berlin, A. Sivachenko, D.A. Thompson, A. Wysoker, T. Fennell, A. Gnirke, N. Pochet, A. Regev, J.Z. Levin, Comparative analysis of RNA sequencing methods for degraded or low-input samples, Nat. Methods 10 (2013) 623–629, https://doi.org/10.1038/nmeth.2483.

[112] S.R. Head, H.K. Komori, S.A. LaMere, T. Whisenant, F. Van Nieuwerburgh, D.R. Salomon, P. Ordoukhanian, Library construction for next-generation sequencing: overviews and challenges, BioTechniques 56 (2014) 61–64. 66, 68, passim https://doi.org/10.2144/000114133.

[113] M. Hafner, P. Landgraf, J. Ludwig, A. Rice, T. Ojo, C. Lin, D. Holoch, C. Lim, T. Tuschl, Identification of microRNAs and other small regulatory RNAs using cDNA library sequencing, Methods (San Diego, Calif.) 44 (2008) 3–12.

[114] S. Pfeffer, M. Lagos-Quintana, T. Tuschl, Chapter 26: Cloning of small RNA molecules, in: Current Protocols in Molecular Biology, 2005, https://doi.org/10.1002/0471142727.mb2604s72. Unit 26.4.

[115] M. Hafner, N. Renwick, M. Brown, A. Mihailović, D. Holoch, C. Lin, J.T.G. Pena, J.D. Nusbaum, P. Morozov, J. Ludwig, T. Ojo, S. Luo, G. Schroth, T. Tuschl, RNA-ligase-dependent biases in miRNA representation in deep-sequenced small RNA cDNA libraries, RNA (New York, N.Y.) 17 (2011) 1697–1712, https://doi.org/10.1261/rna.2799511.

[116] A.D. Jayaprakash, O. Jabado, B.D. Brown, R. Sachidanandam, Identification and remediation of biases in the activity of RNA ligases in small-RNA deep sequencing, Nucleic Acids Res. 39 (2011) e141, https://doi.org/10.1093/nar/gkr693.

[117] K. Sorefan, H. Pais, A.E. Hall, A. Kozomara, S. Griffiths-Jones, V. Moulton, T. Dalmay, Reducing ligation bias of small RNAs in libraries for next generation sequencing, Silence 3 (2012) 4, https://doi.org/10.1186/1758-907X-3-4.

[118] D.B. Munafó, G.B. Robb, Optimization of enzymatic reaction conditions for generating representative pools of cDNA from small RNA, RNA (New York, N.Y.) 16 (2010) 2537–2552, https://doi.org/10.1261/rna.2242610.

[119] J. Pak, A. Fire, Distinct populations of primary and secondary effectors during RNAi in *C. elegans*, Science (New York, N.Y.) 315 (2007) 241–244.

[120] A. Adey, H.G. Morrison, X. Asan, J.O. Xun, E.H. Kitzman, B. Turner, A.P. Stackhouse, N.C. MacKenzie, X. Caruccio, J.S. Zhang, Rapid, low-input, low-bias construction of shotgun fragment libraries by high-density in vitro transposition, Genome Biol. 11 (2010) R119, https://doi.org/10.1186/gb-2010-11-12-r119.

[121] S. Picelli, A.K. Björklund, B. Reinius, S. Sagasser, G. Winberg, R. Sandberg, Tn5 transposase and tagmentation procedures for massively scaled sequencing projects, Genome Res. 24 (2014) 2033–2040, https://doi.org/10.1101/gr.177881.114.

[122] G. Sun, X. Wu, J. Wang, H. Li, X. Li, H. Gao, J. Rossi, Y. Yen, A bias-reducing strategy in profiling small RNAs using Solexa, RNA (New York, N.Y.) 17 (2011) 2256–2262, https://doi.org/10.1261/rna.028621.111.

[123] A. Turchinovich, H. Surowy, A. Serva, M. Zapatka, P. Lichter, B. Burwinkel, Capture and amplification by tailing and switching (CATS). An ultrasensitive ligation-independent method for generation of DNA libraries for deep sequencing from picogram amounts of DNA and RNA, RNA Biol. 11 (2014) 817–828, https://doi.org/10.4161/rna.29304.

[124] M. Pirritano, T. Fehlmann, T. Laufer, N. Ludwig, G. Gasparoni, Y. Li, E. Meese, A. Keller, M. Simon, Next generation sequencing analysis of total small noncoding RNAs from low input RNA from dried blood sampling, Anal. Chem. 90 (2018) 11791–11796, https://doi.org/10.1021/acs.analchem.8b03557.

[125] S.G. Acinas, R. Sarma-Rupavtarm, V. Klepac-Ceraj, M.F. Polz, PCR-induced sequence artifacts and bias: insights from comparison of two 16S rRNA clone libraries constructed from the same sample, Appl. Environ. Microbiol. 71 (2005) 8966–8969.

[126] T. Kanagawa, Bias and artifacts in multitemplate polymerase chain reactions (PCR), J. Biosci. Bioeng. 96 (2003) 317–323.

[127] B.E. Miner, R.J. Stöger, A.F. Burden, C.D. Laird, R.S. Hansen, Molecular barcodes detect redundancy and contamination in hairpin-bisulfite PCR, Nucleic Acids Res. 32 (2004) e135.

[128] S. Islam, A. Zeisel, S. Joost, G. La Manno, P. Zajac, M. Kasper, P. Lönnerberg, S. Linnarsson, Quantitative single-cell RNA-seq with unique molecular identifiers, Nat. Methods 11 (2014) 163–166, https://doi.org/10.1038/nmeth.2772.

[129] T. Kivioja, A. Vähärautio, K. Karlsson, M. Bonke, M. Enge, S. Linnarsson, J. Taipale, Counting absolute numbers of molecules using unique molecular identifiers, Nat. Methods 9 (2011) 72–74, https://doi.org/10.1038/nmeth.1778.

[130] Y. Fu, P.-H. Wu, T. Beane, P.D. Zamore, Z. Weng, Elimination of PCR duplicates in RNA-seq and small RNA-seq using unique molecular identifiers, BMC Genomics 19 (2018) 531, https://doi.org/10.1186/s12864-018-4933-1.

[131] J. Cao, J.S. Packer, V. Ramani, D.A. Cusanovich, C. Huynh, R. Daza, X. Qiu, C. Lee, S.N. Furlan, F.J. Steemers, A. Adey, R.H. Waterston, C. Trapnell, J. Shendure, Comprehensive single-cell transcriptional profiling of a multicellular organism, Science (New York, N.Y.) 357 (2017) 661–667, https://doi.org/10.1126/science.aam8940.

[132] J. Hong, D. Gresham, Incorporation of unique molecular identifiers in TruSeq adapters improves the accuracy of quantitative sequencing, BioTechniques 63 (2017) 221–226, https://doi.org/10.2144/000114608.

[133] O.R. Faridani, I. Abdullayev, M. Hagemann-Jensen, J.P. Schell, F. Lanner, R. Sandberg, Single-cell sequencing of the small-RNA transcriptome, Nat. Biotechnol. 34 (2016) 1264–1266, https://doi.org/10.1038/nbt.3701.

[134] C.D. Belair, T. Hu, B. Chu, J.W. Freimer, M.R. Cooperberg, R.H. Blelloch, High-throughput, efficient, and unbiased capture of small RNAs from low-input samples for sequencing, Sci. Rep. 9 (2019) 2262, https://doi.org/10.1038/s41598-018-38458-7.

[135] W. Chen, Y. Li, J. Easton, D. Finkelstein, G. Wu, X. Chen, UMI-count modeling and differential expression analysis for single-cell RNA sequencing, Genome Biol. 19 (2018) 70, https://doi.org/10.1186/s13059-018-1438-9.

[136] K. Saunders, A.G. Bert, B.K. Dredge, J. Toubia, P.A. Gregory, K.A. Pillman, G.J. Goodall, C.P. Bracken, Insufficiently complex unique-molecular identifiers (UMIs) distort small RNA sequencing, Sci. Rep. 10 (2020) 14593, https://doi.org/10.1038/s41598-020-71323-0.

[137] A.C. Tuck, D. Tollervey, RNA in pieces, Trends Genet. 27 (2011) 422–432, https://doi.org/10.1016/j.tig.2011.06.001.

[138] H.K. Kim, G. Fuchs, S. Wang, W. Wei, Y. Zhang, H. Park, B. Roy-Chaudhuri, P. Li, J. Xu, K. Chu, F. Zhang, M.-S. Chua, S. So, Q.C. Zhang, P. Sarnow, M.A. Kay, A transfer-RNA-derived small RNA regulates ribosome biogenesis, Nature 552 (2017) 57–62, https://doi.org/10.1038/nature25005.

[139] H. Persson, A. Kvist, J. Vallon-Christersson, P. Medstrand, A. Borg, C. Rovira, The non-coding RNA of the multidrug resistance-linked vault particle encodes multiple regulatory small RNAs, Nat. Cell Biol. 11 (2009) 1268–1271, https://doi.org/10.1038/ncb1972.

[140] P. Ivanov, M.M. Emara, J. Villen, S.P. Gygi, P. Anderson, Angiogenin-induced tRNA fragments inhibit translation initiation, Mol. Cell 43 (2011) 613–623, https://doi.org/10.1016/j.molcel.2011.06.022.

[141] J. Donovan, S. Rath, D. Kolet-Mandrikov, A. Korennykh, Rapid RNase L-driven arrest of protein synthesis in the dsRNA response without degradation of translation machinery, RNA (New York, N.Y.) 23 (2017) 1660–1671, https://doi.org/10.1261/rna.062000.117.

[142] T.A.P. Driedonks, S.G. van der Grein, Y. Ariyurek, H.P.J. Buermans, H. Jekel, F.W.N. Chow, M.H.M. Wauben, A.H. Buck, P.A.C. Hoen, E.N.M. Nolte-'t Hoen, Immune stimuli shape the small non-coding transcriptome of extracellular vesicles released by dendritic cells, Cell. Mol. Life Sci. 75 (2018) 3857–3875, https://doi.org/10.1007/s00018-018-2842-8.

[143] E.N.M. Nolte-'t Hoen, H.P.J. Buermans, M. Waasdorp, W. Stoorvogel, M.H.M. Wauben, P.A.C. Hoen, Deep sequencing of RNA from immune cell-derived vesicles uncovers the selective incorporation of small non-coding RNA biotypes with potential regulatory functions, Nucleic Acids Res. 40 (2012) 9272–9285, https://doi.org/10.1093/nar/gks658.

[144] L. Vojtech, S. Woo, S. Hughes, C. Levy, L. Ballweber, R.P. Sauteraud, J. Strobl, K. Westerberg, R. Gottardo, M. Tewari, F. Hladik, Exosomes in human semen carry a distinctive repertoire of small non-coding RNAs with potential regulatory functions, Nucleic Acids Res. 42 (2014) 7290–7304, https://doi.org/10.1093/nar/gku347.

[145] B.W.M. van Balkom, A.S. Eisele, D.M. Pegtel, S. Bervoets, M.C. Verhaar, Quantitative and qualitative analysis of small RNAs in human endothelial cells and exosomes provides insights into localized RNA processing, degradation and sorting, J. Extracell. Vesicles 4 (2015) 26760, https://doi.org/10.3402/jev.v4.26760.

[146] A.E. Cozen, E. Quartley, A.D. Holmes, E. Hrabeta-Robinson, E.M. Phizicky, T.M. Lowe, ARM-seq: AlkB-facilitated RNA methylation sequencing reveals a complex landscape of modified tRNA fragments, Nat. Methods 12 (2015) 879–884, https://doi.org/10.1038/nmeth.3508.

[147] G. Zheng, Y. Qin, W.C. Clark, Q. Dai, C. Yi, C. He, A.M. Lambowitz, T. Pan, Efficient and quantitative high-throughput tRNA sequencing, Nat. Methods 12 (2015) 835–837, https://doi.org/10.1038/nmeth.3478.

[148] S.M.E. Sahraeian, M. Mohiyuddin, R. Sebra, H. Tilgner, P.T. Afshar, K.F. Au, N. Bani Asadi, M.B. Gerstein, W.H. Wong, M.P. Snyder, E. Schadt, H.Y.K. Lam, Gaining comprehensive biological insight into the transcriptome by performing a broad-spectrum RNA-seq analysis, Nat. Commun. 8 (2017) 59, https://doi.org/10.1038/s41467-017-00050-4.

[149] L.A. Corchete, E.A. Rojas, D. Alonso-López, J. De Las Rivas, N.C. Gutiérrez, F.J. Burguillo, Systematic comparison and assessment of RNA-seq procedures for gene expression quantitative analysis, Sci. Rep. 10 (2020) 19737, https://doi.org/10.1038/s41598-020-76881-x.

[150] L. Tong, P.-Y. Wu, J.H. Phan, H.R. Hassazadeh, W. Tong, M.D. Wang, Impact of RNA-seq data analysis algorithms on gene expression estimation and downstream prediction, Sci. Rep. 10 (2020) 17925, https://doi.org/10.1038/s41598-020-74567-y.

[151] P.-Y. Wu, M.D. Wang, The selection of quantification pipelines for illumina RNA-seq data using a subsampling approach, in: IEEE-EMBS International Conference on Biomedical and Health Informatics. IEEE-EMBS International Conference on Biomedical and Health Informatics, 2016, pp. 78–81, https://doi.org/10.1109/BHI.2016.7455839.

[152] J. Hör, S.A. Gorski, J. Vogel, Bacterial RNA biology on a genome scale, Mol. Cell 70 (2018) 785–799, https://doi.org/10.1016/j.molcel.2017.12.023.

[153] S. Schwartz, Y. Motorin, Next-generation sequencing technologies for detection of modified nucleotides in RNAs, RNA Biol. 14 (2017) 1124–1137, https://doi.org/10.1080/15476286.2016.1251543.

[154] E.T. Wang, R. Sandberg, S. Luo, I. Khrebtukova, L. Zhang, C. Mayr, S.F. Kingsmore, G.P. Schroth, C.B. Burge, Alternative isoform regulation in human tissue transcriptomes, Nature 456 (2008) 470–476, https://doi.org/10.1038/nature07509.

[155] P.J.A. Cock, C.J. Fields, N. Goto, M.L. Heuer, P.M. Rice, The sanger FASTQ file format for sequences with quality scores, and the Solexa/Illumina FASTQ variants, Nucleic Acids Res. 38 (2010) 1767–1771, https://doi.org/10.1093/nar/gkp1137.

[156] A.M. Bolger, M. Lohse, B. Usadel, Trimmomatic: a flexible trimmer for Illumina sequence data, Bioinformatics (Oxford, England) 30 (2014) 2114–2120, https://doi.org/10.1093/bioinformatics/btu170.

[157] A. Dobin, C.A. Davis, F. Schlesinger, J. Drenkow, C. Zaleski, S. Jha, P. Batut, M. Chaisson, T.R. Gingeras, STAR: ultrafast universal RNA-seq aligner, Bioinformatics (Oxford, England) 29 (2013) 15–21, https://doi.org/10.1093/bioinformatics/bts635.

[158] D. Kim, B. Langmead, S.L. Salzberg, HISAT: a fast spliced aligner with low memory requirements, Nat. Methods 12 (2015) 357–360, https://doi.org/10.1038/nmeth.3317.

[159] D. Kim, G. Pertea, C. Trapnell, H. Pimentel, R. Kelley, S.L. Salzberg, TopHat2: accurate alignment of transcriptomes in the presence of insertions, deletions and gene fusions, Genome Biol. 14 (2013) R36, https://doi.org/10.1186/gb-2013-14-4-r36.

[160] B. Langmead, S.L. Salzberg, Fast gapped-read alignment with Bowtie 2, Nat. Methods 9 (2012) 357–359, https://doi.org/10.1038/nmeth.1923.

[161] B. Langmead, C. Trapnell, M. Pop, S.L. Salzberg, Ultrafast and memory-efficient alignment of short DNA sequences to the human genome, Genome Biol. 10 (2009) R25, https://doi.org/10.1186/gb-2009-10-3-r25.

[162] F. García-Alcalde, K. Okonechnikov, J. Carbonell, L.M. Cruz, S. Götz, S. Tarazona, J. Dopazo, T.F. Meyer, A. Conesa, Qualimap: evaluating next-generation sequencing alignment data, Bioinformatics (Oxford, England) 28 (2012) 2678–2679, https://doi.org/10.1093/bioinformatics/bts503.

[163] L. Wang, S. Wang, W. Li, RSeQC: quality control of RNA-seq experiments, Bioinformatics (Oxford, England) 28 (2012) 2184–2185, https://doi.org/10.1093/bioinformatics/bts356.

[164] G. Deschamps-Francoeur, J. Simoneau, M.S. Scott, Handling multi-mapped reads in RNA-seq, Comput. Struct. Biotechnol. J. 18 (2020) 1569–1576, https://doi.org/10.1016/j.csbj.2020.06.014.

[165] B. Fromm, T. Billipp, L.E. Peck, M. Johansen, J.E. Tarver, B.L. King, J.M. Newcomb, L.F. Sempere, K. Flatmark, E. Hovig, K.J. Peterson, A uniform system for the annotation of vertebrate microRNA genes and the evolution of the human microRNAome, Annu. Rev. Genet. 49 (2015) 213–242, https://doi.org/10.1146/annurev-genet-120213-092023.

[166] P. Du, Y.-H. Lai, D.-S. Yao, J.-Y. Chen, N. Ding, Downregulation of microRNA-1246 inhibits tumor growth and promotes apoptosis of cervical cancer cells by targeting thrombospondin-2, Oncol. Lett. 18 (2019) 2491–2499, https://doi.org/10.3892/ol.2019.10571.

[167] B. Fromm, D. Domanska, E. Høye, V. Ovchinnikov, W. Kang, E. Aparicio-Puerta, M. Johansen, K. Flatmark, A. Mathelier, E. Hovig, M. Hackenberg, M.R. Friedländer, K.J. Peterson, MirGeneDB 2.0: the metazoan microRNA complement, Nucleic Acids Res. 48 (2020) D132–D141, https://doi.org/10.1093/nar/gkz885.

[168] A. Girard, R. Sachidanandam, G.J. Hannon, M.A. Carmell, A germline-specific class of small RNAs binds mammalian Piwi proteins, Nature 442 (2006) 199–202.

[169] J.P. Tosar, C. Rovira, A. Cayota, Non-coding RNA fragments account for the majority of annotated piRNAs expressed in somatic non-gonadal tissues, Commun. Biol. 1 (2018) 2, https://doi.org/10.1038/s42003-017-0001-7.

[170] J.P. Tosar, M.R. García-Silva, A. Cayota, Circulating SNORD57 rather than piR-54265 is a promising biomarker for colorectal cancer: common pitfalls in the study of somatic piRNAs in cancer, RNA (New York, N.Y.) 27 (2021) 403–410, https://doi.org/10.1261/rna.078444.120.

[171] P. Genzor, S.C. Cordts, N.V. Bokil, A.D. Haase, Aberrant expression of select piRNA-pathway genes does not reactivate piRNA silencing in cancer cells, Proc. Natl. Acad. Sci. U. S. A. 116 (2019) 11111–11112, https://doi.org/10.1073/pnas.1904498116.

[172] S. Shi, Z.-Z. Yang, S. Liu, F. Yang, H. Lin, PIWIL1 promotes gastric cancer via a piRNA-independent mechanism, Proc. Natl. Acad. Sci. U. S. A. 117 (2020) 22390–22401, https://doi.org/10.1073/pnas.2008724117.

[173] C. Trapnell, A. Roberts, L. Goff, G. Pertea, D. Kim, D.R. Kelley, H. Pimentel, S.L. Salzberg, J.L. Rinn, L. Pachter, Differential gene and transcript expression analysis of RNA-seq experiments with TopHat and cufflinks, Nat. Protoc. 7 (2012) 562–578, https://doi.org/10.1038/nprot.2012.016.

[174] B. Li, C.N. Dewey, RSEM: accurate transcript quantification from RNA-Seq data with or without a reference genome, BMC Bioinformatics 12 (2011) 323, https://doi.org/10.1186/1471-2105-12-323.

[175] S. Anders, P.T. Pyl, W. Huber, HTSeq- -a Python framework to work with high-throughput sequencing data, Bioinformatics (Oxford, England) 31 (2015) 166–169, https://doi.org/10.1093/bioinformatics/btu638.

[176] Y. Liao, G.K. Smyth, W. Shi, featureCounts: an efficient general purpose program for assigning sequence reads to genomic features, Bioinformatics (Oxford, England) 30 (2014) 923–930, https://doi.org/10.1093/bioinformatics/btt656.

[177] A.R. Quinlan, BEDTools: the Swiss-Army tool for genome feature analysis, Curr. Protoc. Bioinformatics 47 (2014) 11.12.1-34, https://doi.org/10.1002/0471250953.bi1112s47.

[178] A. Roberts, L. Pachter, Streaming fragment assignment for real-time analysis of sequencing experiments, Nat. Methods 10 (2013) 71–73, https://doi.org/10.1038/nmeth.2251.

[179] R. Patro, G. Duggal, M.I. Love, R.A. Irizarry, C. Kingsford, Salmon provides fast and bias-aware quantification of transcript expression, Nat. Methods 14 (2017) 417–419, https://doi.org/10.1038/nmeth.4197.

[180] R. Patro, S.M. Mount, C. Kingsford, Sailfish enables alignment-free isoform quantification from RNA-seq reads using lightweight algorithms, Nat. Biotechnol. 32 (2014) 462–464, https://doi.org/10.1038/nbt.2862.

[181] N.L. Bray, H. Pimentel, P. Melsted, L. Pachter, Near-optimal probabilistic RNA-seq quantification, Nat. Biotechnol. 34 (2016) 525–527, https://doi.org/10.1038/nbt.3519.

[182] D.C. Wu, J. Yao, K.S. Ho, A.M. Lambowitz, C.O. Wilke, Limitations of alignment-free tools in total RNA-seq quantification, BMC Genomics 19 (2018) 510, https://doi.org/10.1186/s12864-018-4869-5.

[183] W.J. Kent, C.W. Sugnet, T.S. Furey, K.M. Roskin, T.H. Pringle, A.M. Zahler, D. Haussler, The human genome browser at UCSC, Genome Res. 12 (2002) 996–1006.

[184] J.T. Robinson, H. Thorvaldsdóttir, W. Winckler, M. Guttman, E.S. Lander, G. Getz, J.P. Mesirov, Integrative genomics viewer, Nat. Biotechnol. 29 (2011) 24–26, https://doi.org/10.1038/nbt.1754.

[185] C. Trapnell, B.A. Williams, G. Pertea, A. Mortazavi, G. Kwan, M.J. van Baren, S.L. Salzberg, B.J. Wold, L. Pachter, Transcript assembly and quantification by RNA-Seq reveals unannotated transcripts and isoform switching during cell differentiation, Nat. Biotechnol. 28 (2010) 511–515, https://doi.org/10.1038/nbt.1621.

[186] C.W. Law, M. Alhamdoosh, S. Su, X. Dong, L. Tian, G.K. Smyth, M.E. Ritchie, RNA-seq analysis is easy as 1-2-3 with limma, Glimma and edgeR, F1000Research 5 (2016), https://doi.org/10.12688/f1000research.9005.3.

[187] A. Oshlack, M.J. Wakefield, Transcript length bias in RNA-seq data confounds systems biology, Biol. Direct 4 (2009) 14, https://doi.org/10.1186/1745-6150-4-14.

[188] X. Li, G.N. Brock, E.C. Rouchka, N.G.F. Cooper, D. Wu, T.E. O'Toole, R.S. Gill, A.M. Eteleeb, L. O'Brien, S.N. Rai, A comparison of per sample global scaling and per gene normalization methods for differential expression analysis of RNA-seq data, PLoS One 12 (2017) e0176185, https://doi.org/10.1371/journal.pone.0176185.

[189] M.D. Robinson, D.J. McCarthy, G.K. Smyth, edgeR: a Bioconductor package for differential expression analysis of digital gene expression data, Bioinformatics (Oxford, England) 26 (2010) 139–140, https://doi.org/10.1093/bioinformatics/btp616.

[190] M.I. Love, W. Huber, S. Anders, Moderated estimation of fold change and dispersion for RNA-seq data with DESeq2, Genome Biol. 15 (2014) 550.

[191] X. Chen, B. Zhang, T. Wang, A. Bonni, G. Zhao, Robust principal component analysis for accurate outlier sample detection in RNA-Seq data, BMC Bioinformatics 21 (2020) 269, https://doi.org/10.1186/s12859-020-03608-0.

[192] C. Soneson, M.I. Love, M.D. Robinson, Differential analyses for RNA-seq: transcript-level estimates improve gene-level inferences, F1000Research 4 (2015) 1521, https://doi.org/10.12688/f1000research.7563.2.

[193] M.-A. Dillies, A. Rau, J. Aubert, C. Hennequet-Antier, M. Jeanmougin, N. Servant, C. Keime, G. Marot, D. Castel, J. Estelle, G. Guernec, B. Jagla, L. Jouneau, D. Laloë, C. Le Gall, B. Schaëffer, S. Le Crom, M. Guedj, F. Jaffrézic, A comprehensive evaluation of normalization methods for Illumina high-throughput RNA sequencing data analysis, Brief. Bioinform. 14 (2013) 671–683, https://doi.org/10.1093/bib/bbs046.

[194] C. Simillion, R. Liechti, H.E.L. Lischer, V. Ioannidis, R. Bruggmann, Avoiding the pitfalls of gene set enrichment analysis with SetRank, BMC Bioinformatics 18 (2017) 151, https://doi.org/10.1186/s12859-017-1571-6.

[195] A. Subramanian, P. Tamayo, V.K. Mootha, S. Mukherjee, B.L. Ebert, M.A. Gillette, A. Paulovich, S.L. Pomeroy, T.R. Golub, E.S. Lander, J.P. Mesirov, Gene set enrichment analysis: a knowledge-based approach for interpreting genome-wide expression profiles, Proc. Natl. Acad. Sci. U. S. A. 102 (2005) 15545–15550.

[196] A. Liberzon, C. Birger, H. Thorvaldsdóttir, M. Ghandi, J.P. Mesirov, P. Tamayo, The molecular signatures database (MSigDB) hallmark gene set collection, Cell Syst. 1 (2015) 417–425.

[197] J. Reimand, R. Isserlin, V. Voisin, M. Kucera, C. Tannus-Lopes, A. Rostamianfar, L. Wadi, M. Meyer, J. Wong, C. Xu, D. Merico, G.D. Bader, Pathway enrichment analysis and visualization of omics data using g:profiler, GSEA, Cytoscape and EnrichmentMap, Nat. Protoc. 14 (2019) 482–517, https://doi.org/10.1038/s41596-018-0103-9.

[198] The gene ontology project in 2008, Nucleic Acids Res. 36 (2008) D440–D444.
[199] S.Y. Rhee, V. Wood, K. Dolinski, S. Draghici, Use and misuse of the gene ontology annotations, Nat. Rev. Genet. 9 (2008) 509–515, https://doi.org/10.1038/nrg2363.
[200] M. Kanehisa, S. Goto, KEGG: kyoto encyclopedia of genes and genomes, Nucleic Acids Res. 28 (2000) 27–30.
[201] M. Kanehisa, Y. Sato, KEGG Mapper for inferring cellular functions from protein sequences, Protein Sci. 29 (2020) 28–35, https://doi.org/10.1002/pro.3711.
[202] H. Mi, A. Muruganujan, J.T. Casagrande, P.D. Thomas, Large-scale gene function analysis with the PANTHER classification system, Nat. Protoc. 8 (2013) 1551–1566, https://doi.org/10.1038/nprot.2013.092.
[203] E. Eden, R. Navon, I. Steinfeld, D. Lipson, Z. Yakhini, GOrilla: a tool for discovery and visualization of enriched GO terms in ranked gene lists, BMC Bioinformatics 10 (2009) 48, https://doi.org/10.1186/1471-2105-10-48.
[204] C. Robert, M. Watson, Errors in RNA-Seq quantification affect genes of relevance to human disease, Genome Biol. 16 (2015) 177, https://doi.org/10.1186/s13059-015-0734-x.
[205] N.F. Lahens, I.H. Kavakli, R. Zhang, K. Hayer, M.B. Black, H. Dueck, A. Pizarro, J. Kim, R. Irizarry, R.S. Thomas, G.R. Grant, J.B. Hogenesch, IVT-seq reveals extreme bias in RNA sequencing, Genome Biol. 15 (2014) R86, https://doi.org/10.1186/gb-2014-15-6-r86.
[206] D. Lähnemann, J. Köster, E. Szczurek, D.J. McCarthy, S.C. Hicks, M.D. Robinson, C.A. Vallejos, K.R. Campbell, N. Beerenwinkel, A. Mahfouz, L. Pinello, P. Skums, A. Stamatakis, C.S.-O. Attolini, S. Aparicio, J. Baaijens, M. Balvert, B. de Barbanson, A. Cappuccio, G. Corleone, B.E. Dutilh, M. Florescu, V. Guryev, R. Holmer, K. Jahn, T.J. Lobo, E.M. Keizer, I. Khatri, S.M. Kielbasa, J.O. Korbel, A.M. Kozlov, T.-H. Kuo, B.P.F. Lelieveldt, I.I. Mandoiu, J.C. Marioni, T. Marschall, F. Mölder, A. Niknejad, L. Raczkowski, M. Reinders, J. de Ridder, A.-E. Saliba, A. Somarakis, O. Stegle, F.J. Theis, H. Yang, A. Zelikovsky, A.C. McHardy, B.J. Raphael, S.P. Shah, A. Schönhuth, Eleven grand challenges in single-cell data science, Genome Biol. 21 (2020) 31, https://doi.org/10.1186/s13059-020-1926-6.

CHAPTER 11

Validation of gene expression by quantitative PCR

Arundhati Das, Debojyoti Das, and Amaresh C. Panda
Institute of Life Sciences, Bhubaneswar, Odisha, India

Introduction

The development of genome-wide sequencing technologies revealed that ~80% of the human genome is transcribed into RNA [1,2]. Transcriptomic studies led to many surprising discoveries that changed the perception of genes in the genome and their regulation. Intriguingly, only ~2% of the human genome contributes to RNA species coding for functional proteins, while the majority of the RNAs are nonprotein-coding in nature [1–5]. The transcripts within cells are broadly categorized into two types, messenger RNAs (mRNAs), those coding for proteins, and noncoding RNAs (ncRNAs) [3]. Depending on their size and function, there are several types of ncRNAs, including ribosomal RNA (rRNAs), transfer RNAs (tRNAs), long ncRNAs (lncRNAs), small nuclear RNAs (snRNAs), microRNAs (miRNAs), and newly discovered circular RNAs (circRNAs) [4–8]. Several ncRNAs play many critical roles in the regulation of transcription, splicing, export, stability, and translation of mRNA into protein [9]. Several approaches have been developed for the expression analysis of mRNAs and ncRNAs such as RNA-sequencing, microarray, serial analysis of gene expression, northern blotting, and reverse transcription followed by quantitative PCR (RT-qPCR) [10–14].

RT-qPCR is one of the most widely used methods for quantification of gene expression [14–16]. The main advantages of RT-qPCR over other approaches include quicker analysis, simplicity, sensitivity, specificity, reproducibility, and affordability. Various modifications have been made to the RT-qPCR method to quantify the expression level of all kinds of RNAs, including mRNA, lncRNA, miRNA, and circRNA. RT-qPCR technique involves the reverse transcription of the target RNA species to cDNA, followed by qPCR using specific primer sets for target RNA amplification. The expression level of the target RNA is measured by the amount of fluorescence signal generated by the PCR amplicons. Currently, many DNA-binding dyes or fluorescent probes are used in qPCR to detect the target gene's expression. In this chapter, we present a detailed protocol for the validation of gene expression by RT-qPCR that includes different reverse transcription strategies, tools for designing primers, and qPCR conditions for long mRNAs, lncRNAs, circRNAs, and small miRNAs.

Materials

Caution: The workplace and reagents should be clean and nuclease-free. All the reagents and equipment must be handled delicately to isolate good quality RNA.

RNA extraction

1. Cells
2. 1.7 mL microcentrifuge tubes
3. Refrigerated microcentrifuges
4. TRIzol (Thermo Fisher Scientific)
5. Ethanol
6. Chloroform
7. Isopropanol
8. Vortex mixer
9. Thermomixer
10. Turbo DNase

Assessment of RNA

1. Multiskan sky microplate spectrophotometer (Thermo Fisher Scientific)
2. Qubit 4.0 fluorometer
3. Qubit RNA BR assay kit
4. Vortex mixer
5. Benchtop microfuge
6. Agarose
7. Nuclease-free water
8. E-gel snap viewer

cDNA synthesis and qPCR

1. 1.7 mL microcentrifuge tubes
2. Total RNA purified from cells
3. Mir-X miRNA First-Strand Synthesis Kit (Takara Bio)
4. High-capacity cDNA reverse transcription kit (Thermo Fisher Scientific)
5. Murine RNase inhibitor (NEB)
6. dNTPs mix (dATP, dCTP, dGTP, dTTP)
7. Random hexamer
8. Nuclease-free water
9. Thermal cycler
10. 96 well PCR plates
11. MicroAmp optical adhesive film
12. 2× PowerUp SYBR Green PCR master mix
13. QuantStudio 3 real-time PCR system
14. Primers for target RNAs

Method

The following protocol describes a detailed workflow of expression analysis of mRNA, lncRNA, miRNA, and circRNA (Fig. 11.1).

Total RNA isolation

The presence of RNases in the tissues as well as in the environment and the instability of RNA, create great challenges in the process of RNA isolation. Hence, precaution should be taken during RNA preparation (see Note 1). Here, we describe the classical total RNA isolation procedure using TRIzol [17] (see Note 2).

1. Take a 10 cm dish of ~50% confluent adherent cell and discard the culture media.
2. Add 1 mL of TRIzol reagent to it followed by incubation on a rocker for 5 min at room temperature.
3. Harvest the cell lysate in a microcentrifuge tube by pipetting vigorously to get a clear suspension from the viscous one.
4. Add 200 μL of chloroform to the 1 mL of TRIzol homogenized solution and vortex the mixture vigorously for 10 s.
5. Centrifuge at 15000g for 10 min at 4°C.
6. Transfer 400 μL of the aqueous phase into another microcentrifuge tube and add 400 μL of isopropanol and mix by inverting the tube a few times.
7. Incubate the tube for 10 min at room temperature followed by centrifugation at 15000g for 10 min at 4°C.

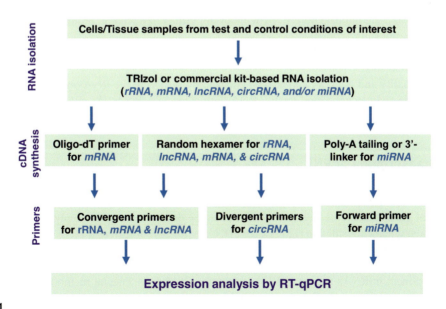

FIG. 11.1

Flowchart of quantification of gene expression by RT-qPCR analysis.

8. Discard the supernatant and add 1 mL of 70% ethanol into the tube containing RNA pellet.
9. Vortex for a few seconds and centrifuge at 15000g for 10 min at room temperature.
10. Discard the supernatant completely and let the pellet air-dry for 2–3 min.
11. Add 50 µL of nuclease-free water to the pellet and incubate it at 55°C on a thermomixer for 5 min.
12. Immediately place the RNA tubes on ice.
13. *(Optional) DNase treatment*—Add 6 µL of 10× TURBO DNase I buffer, 1 µL of TURBO DNase I, 3 µL of nuclease-free water to 50 µL of RNA, followed by incubation for 30 min at 37°C in a thermomixer. Add 6 µL (0.1 volumes) of DNase inactivation reagent to the reaction mixture and incubate at room temperature for 5 min. Centrifuge at 10,000g for 1 min at room temperature. Transfer the supernatant into a new microcentrifuge tube.

Various column-based commercial RNA extraction kits are currently available for total RNA isolation (see Note 3). The column-based RNA isolation may be coupled with on-column DNase digestion to purify DNA-free total RNA [18,19]. Although column-based kits are widely used, most commercial kits can isolate long RNAs with more than 200 nucleotides in length efficiently. A few kits can isolate the miRNAs along with long RNAs. Hence, great care should be taken to choose the correct total RNA isolation kit, if the researcher is interested in miRNA analysis along with mRNAs.

Analysis of RNA quantity, purity, and integrity

The quality, purity, and yield of RNA are determined through the use of a Multiskan sky microplate spectrophotometer and Qubit 4 fluorometer.

- Measure the RNA quantity and quality on a microplate spectrophotometer [20]. The $A_{260/280}$ ratio of 2 indicates good quality RNA without DNA contamination. Also, the $A_{260/230}$ ratio ranges between 2.0 and 2.2 indicate pure RNA without impurities like phenol, protein, polysaccharides, and guanidinium salts.
- In addition concentration of RNA can be measured with a Qubit 4 fluorometer using the Qubit RNA assay reagents.
- The RNA quality and integrity can also be checked by resolving the RNA on a denaturing formaldehyde agarose gel. Good quality RNA with clean 28s rRNA, 18s rRNA, and 5s rRNA/tRNA bands on a gel without any smear can be taken for downstream reverse transcription.

Reverse transcription of RNA

Many commercial reverse transcription kits are available for the preparation of cDNA from total RNA. The amount of total RNA required for cDNA synthesis can vary depending on the sample quantity and the number of genes that are needed to be analyzed in the downstream qPCR analysis. However, care should be taken when choosing the right kit for cDNA synthesis for different kinds of RNAs. Most of the long RNAs, including mRNAs, lncRNAs, and circRNAs, can be reverse transcribed with random hexamer primers, while oligo-dT primers can only be used for cDNA synthesis of mRNA/lncRNA with poly-A tails. Since, miRNAs are small RNAs without poly-A tails, special methods are required for cDNA synthesis that add a poly-A tail or a 3′ linker or a stem-loop oligo that hybridize miRNAs followed by cDNA synthesis using appropriate reverse transcription primer (Fig. 11.2) [21].

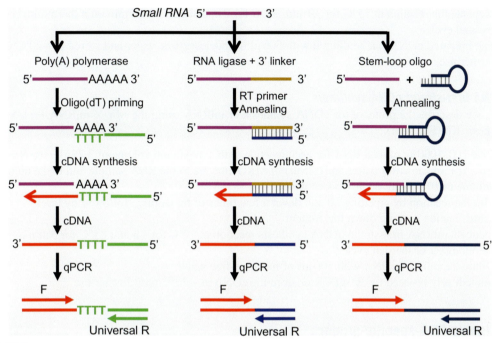

FIG. 11.2

Schematic illustration of cDNA synthesis for small RNAs using different approaches. The RT-qPCR analysis of miRNA is performed using the forward primer targeting the miRNA and the universal reverse primer.

cDNA synthesis of mRNAs, lncRNAs, and circRNA

Total RNA in the cell contains various types of RNAs, including rRNA, mRNA, lncRNA, miRNA, and circRNAs. Among these, the mRNAs and some lncRNAs containing poly-A tails can be reverse transcribed using oligo-dT primers. Importantly, cDNA prepared with oligo-dT can only be used for expression analysis of RNAs with poly-A tails and not for other RNAs. However, cDNA synthesis with a random hexamer oligo can reverse-transcribe all longer RNA species together. Here, we describe the cDNA synthesis protocol with random hexamer oligos using the High Capacity cDNA synthesis kit.

1. Take 1 μg of total RNA for cDNA synthesis reaction using the High Capacity cDNA synthesis kit following the manufacturer's protocol (see Note 4).
2. Add 2 μL of 10× RT buffer, 2 μL of random hexamer oligo, 1 μL of dNTPs, 0.5 μL of Murine RNase Inhibitor, and 1 μL of MultiScribe reverse transcriptase enzyme to the RNA, and make up the volume to 20 μL with nuclease-free water.
3. For no-RT controls, prepare an identical 20 μL reaction where the reverse transcriptase enzyme is replaced with 1 μL of nuclease-free water.
4. Mix the above reaction thoroughly by tapping or pipetting, followed by a short spin to bring the reaction mixture to the bottom of the tube.

5. Incubate the reaction at 25°C for 10 min, 37°C for 2 h, and 85°C for 5 min on a thermomixer or thermal cycler.
6. The prepared cDNA can be diluted with 500 μL of nuclease-free water and stored at −20°C until further use.

miRNA first-strand cDNA synthesis

Here, we describe the protocol for cDNA synthesis of miRNA using the Mir-X miRNA First-Strand Synthesis Kit (Fig. 11.2).

1. Take 200–1000 ng of total RNA for each reaction with a maximum volume of 3.75 μL (see Note 4).
2. Add 5 μL of 2× MRQ buffer and 1.25 μL of MRQ enzyme to the RNA sample and make up the volume to 10 μL with nuclease-free water.
3. Mix the reaction by pipetting up and down a few times or by tapping, followed by brief centrifugation to bring down the content.
4. Incubate miRNA first-strand cDNA synthesis reaction at 37°C for 1 h and 85°C for 5 min on a thermomixer or thermal cycler.
5. Dilute the miRNA cDNA with 500 μL of nuclease-free water.
6. Immediately proceed for RT-qPCR or store the cDNA at −20°C until further use.

Designing RNA-specific primers

Different RNAs require different type of primer sets for qPCR validation. Long RNAs can be validated by forward and reverse primers targeting the RNA of interest, while miRNAs can be validated by a forward primer targeting the miRNA sequence of interest and a universal reverse primer. Here, we describe different tools and strategies for designing primers for different kinds of RNAs (Fig. 11.3).

Primers for long RNAs, including mRNA and lncRNAs

For mRNAs and lncRNAs, forward and reverse primers can be designed in such a way that they do not amplify the genomic DNA during qPCR. Placing the primers on different exons or across the exon-exon junction reduces false-positive genomic DNA amplification and increases sensitivity. There are several tools available for primer design such as Primer 3 and NCBI primer blast, which we have used with excellent success. Here, we describe the designing the primer using NCBI primer-BLAST (http://www.ncbi.nlm.nih.gov/tools/primer-blast/).

1. Search the species-specific gene name or transcript ID for a particular mRNA or lncRNA in the NCBI nucleotide website (https://www.ncbi.nlm.nih.gov/nucleotide/) and then select the gene of interest.
2. The new page shows the complete information about the transcript of interest along with the sequence.
3. Click the "pick primer" link on the right side of the web page.
4. The new primer design page has various options, which users can choose before designing the forward and reverse primer pair.
5. Set the PCR product size between 120 and 180 bp (a longer amplicon may amplify splice variants and reduce accuracy of the quantification). For better specificity select the option "primer must span exon-exon junction," and click on *get primers*.

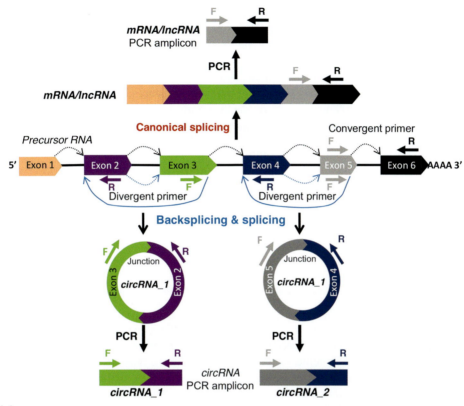

FIG. 11.3

Schematic representation of biogenesis of mature mRNA/lncRNA (top) and circRNA (bottom) from the precursor RNA by splicing and backsplicing, respectively. Schematic illustration of the position of the convergent and divergent primers are used in the qPCR analysis of mRNA and circular RNAs, respectively.

6. The new page gives 5 pairs of primers along with the position of the primers on the target transcript.
7. Select one pair of specific primer for synthesis from a company (e.g., SIGMA).

Alternatively, Primer3web (http://primer3.ut.ee/) can be used to design primers [22]. Paste the target DNA sequence in the "source sequence" window and select the "product size" between 120 and 180 bp. Click the "get primer," which opens a new page with different primer sets for qPCR. Select the best one and order for synthesis.

Divergent primer design for circRNA

Given that circRNAs are generated from linear RNA by backsplicing, the sequence of the circRNA is the same as the parental mRNA. However, the backsplicing generates the head-to-tail joining of circRNA, giving a unique backsplice junction sequence for specific detection of circRNAs. Here, the forward and reverse primers are designed across the junction sequence, making the primers face outward on the linear RNA and are called divergent primers. Here, we describe the method for designing divergent primers for circRNA validation by qPCR (Fig. 11.3) [23].

1. Get the mature circRNA sequence from RNA sequencing data analysis or circRNA database using circRNA ID or by using UCSC genome browser (https://genome.ucsc.edu/), where a mature sequence can be formed by joining the intervening exon sequence present between the genome coordinates of backsplicing sites.
2. Prepare a 200 nucleotides long PCR template by pasting the last 100 nucleotides of the circRNA sequence in front of the first nucleotide 100 nucleotides sequences.
3. For circRNAs less than 200 nucleotides, divide the entire sequence into two halves and then paste the 3′ half of the sequence at the 5′ end of the 5′ half, making the complete template.
4. Use the above backsplice junction sequence PCR template for divergent circRNA primer design using Primer3web (http://primer3.ut.ee/) [22].
5. Select a primer pair with PCR product size between 120 and 180 bp.

Alternatively, divergent primer pairs for human circRNAs with circBase ID can be designed through Circinteractome web tool using the divergent primer design option (https://circinteractome.nia.nih.gov/divergent_primers.html) [24,25].

Primer for microRNA

Since miRNAs are 19–21 nucleotides long, only the forward primer is designed to target specific miRNAs, while the reverse primer is the universal one depending on the kit used for miRNA cDNA synthesis.

1. Use miRBase (http://www.mirbase.org/index.shtml) to get the sequence of the mature miRNA of a particular species by selecting the species name and entering the appropriate miRNA name [26].
2. Design the forward primer by replacing the "U"s present in the miRNA sequence with "T."
3. It must be noted that some miRNAs are members of large miRNA families with similar sequences. The PCR condition must be set according to the primer Tm to avoid nonspecific amplification of other miRNAs of the same family.

Validation of expression by quantitative (q)PCR

Here, we describe the RT-qPCR assay using SYBR Green dye for validation of gene expression (see Note 5).

Quantitative PCR setup for mRNA, lncRNA, and circRNA

1. Prepare 100 μM primer stock with nuclease-free water.
2. Prepare a 1 μM working forward and reverse primer mix by mixing 10 μL each of forward and reverse primer in 980 μL of nuclease-free water.
3. Use housekeeping genes such as *18s* rRNA, *GAPDH*, or *ACTB* primers as loading controls for each sample, along with primers for target RNA.
4. Thaw the cDNA prepared from different RNA samples, primer mix, and 2× PowerUp SYBR green master mix.
5. Prepare qPCR mix of 20 μL in a 96 well plate using 10 μL of 2× PowerUp SYBR green master mix, 5 μL of primer mix, and 5 μL of cDNA. Each reaction should be prepared in triplicates.
6. Seal the plates and vortex slightly to mix the reaction properly, followed by a short spin to bring down the reaction.

7. Set up the qPCR protocol using PowerUp SYBR Green master mix and run the following program on a QuantStudio 3 real-time system: 95°C for 2 min, 40 cycles of 95°C for 2 s, and 60°C for 20 s.
8. Include the melting curve analysis in the qPCR protocol to calculate the melting temperature (Tm) of the amplified PCR product. Single Tm represents specific PCR product amplification, which is suitable for downstream data analysis (see Note 6).
9. The PCR products may be resolved on a 2% Agarose gel to verify the specific amplification of the target PCR amplicon (see Note 7).
10. Calculate the average cycle threshold (C_T) value for each condition from the triplicate reactions. Calculate the relative change in expression of the gene or RNA of interest using the comparative delta C_T method. Use housekeeping genes such as *18s*, *GAPDH*, and *ACTB* as loading controls [15] (see Note 8).

qPCR assay for microRNAs

1. Prepare 500 μL of working solution of 1 μM miRNA primer mix using 5 μL of 100 μM miRNA-specific forward primer and 50 μL of 10 μM MRQ 3′ primer (universal reverse primer from Mir-X miRNA First-Strand Synthesis Kit), and 445 μL of nuclease-free water. Vortex the primer mix followed by a short spin.
2. Use U1 and U6 (snRNA) as controls for each sample along with specific target miRNA primers. Thaw miRNA cDNA samples and primer mix on ice.
3. Setup the real-time PCR assay of 20 μL in a 96 well plate using 10 μL of 2× PowerUp SYBR Green master mix, 5 μL of miRNA Primer mix, and 5 μL of miRNA cDNA.
4. Seal the plate using an optical adhesive film followed by a gentle vortex to mix the reaction.
5. Centrifuge for a few seconds to settle the reaction at the bottom of the well.
6. Use the following program to run the qPCR on QuantStudio 3 real-time platform; 95°C for 2 min, followed by 40 cycles of 95°C for 2 s, and 60°C for 10 s.
7. Analyze the relative expression of target miRNAs in different samples with the comparative delta C_T method using U6 or U1 snRNA as loading controls [15] (see Note 8).

Technical notes

1. Sample preparation area, all reagents, and instruments must be clean and RNase-free for isolation of good quality RNA for downstream analysis.
2. RNA isolation with TRIzol uses hazardous reagents. Working in a fume hood is recommended.
3. The commercial kits for RNA isolation must be selected with great care considering the downstream analysis. The majority of the column-based kits are good for isolating RNAs longer than 200 nucleotides efficiently. Some kits such as the miRNAeasy mini kit (Qiagen) and Direct-zol RNA kit (Zymo) isolate miRNAs along with long RNAs. For the isolation of small RNAs, PureLink miRNA isolation kit (Thermo Scientific) or other miRNA isolation kit may be used.
4. Reverse transcription with a higher amount of total RNA is better for qPCR. Using an equal amount of RNA for all samples is recommended for cDNA synthesis.

5. Unlike SYBR Green-based qPCR, the fluorochrome attached probes targeting the RNA of interest are highly specific to the target RNA. Using the labeled probes such as TaqMan, FRET, and molecular beacons overcome the nonspecific amplification and primer-dimer amplification issues in SYBR Green-based qPCR quantification.
6. Multiple "Tm" in the melting curve analysis suggests the amplification of multiple products, which may not be considered for further analysis. Also, a Tm lower than 70°C indicates primer dimer amplification. Usually, No-RT or NTC samples amplify primer dimers. However, primer concentration must be reduced, if primer-dimers are amplified in cDNA samples.
7. The PCR products amplified with divergent primers must be verified by Sanger sequencing to confirm the specific amplification of target circRNA.
8. The average Ct values for the target gene and the housekeeping genes should be obtained from the real-time PCR system. The differential expression of the gene of interest can be calculated using $2^{-\Delta\Delta Ct}$, where $\Delta\Delta Ct$ represents "ΔCt test" minus "ΔCt control." Calculate the average Ct value of the triplicate reactions of the target gene and housekeeping gene in the test and control samples. ΔCt represents the difference between the Ct of the target gene minus the Ct of the housekeeping reference gene.

Acknowledgments

This work was supported by the intramural funding from the Institute of Life Sciences and the Wellcome Trust/DBT India Alliance Fellowship [IA/I/18/2/504017] awarded to Amaresh C. Panda. We thank other laboratory members for proofreading the article.

Conflict of interest

The authors declare no conflict of interest.

References

[1] The ENCODE Project Consortium, An integrated encyclopedia of DNA elements in the human genome, Nature 489 (2012) 57–74, https://doi.org/10.1038/nature11247.
[2] Y. Ruan, B. Wold, P. Carninci, R. Guig, T.R. Gingeras, T. Hubbard, A. Reymond, S.E. Antonarakis, G. Hannon, M.C. Giddings, Landscape of transcription in human cells, Nature 489 (2012) 101–108, https://doi.org/10.1038/nature11233.
[3] M. Kellis, B. Paten, A. Reymond, M.L. Tress, P. Flicek, B. Aken, J.S. Choudhary, M. Gerstein, R. Guigó, T.J.P. Hubbard, GENCODE reference annotation for the human and mouse genomes, Nucleic Acids Res. 47 (2019) D766–D773, https://doi.org/10.1093/nar/gky955.
[4] A.F. Palazzo, E.S. Lee, Non-coding RNA: what is functional and what is junk? Front. Genet. 6 (2015) 2, https://doi.org/10.3389/fgene.2015.00002.
[5] C.A. Davis, R. Shiekhattar, T.R. Gingeras, T.J. Hubbard, C. Notredame, J. Harrow, R. Guigó, L. Lipovich, J.M. Gonzalez, M. Thomas, The GENCODE v7 catalog of human long noncoding RNAs: analysis of their gene structure, evolution, and expression, Genome Res. 22 (2012) 1775–1789, https://doi.org/10.1101/gr.132159.111.

[6] M. Esteller, Non-coding RNAs in human disease, Nat. Rev. Genet. 12 (2011) 861–874, https://doi.org/10.1038/nrg3074.
[7] A. Keller, E. Meese, U. Fischer, C. Backes, V. Galata, M. Minet, M. Hart, M. Abu-Halima, F.A. Grässer, H.P. Lenhof, An estimate of the total number of true human miRNAs, Nucleic Acids Res. 47 (2019) 3353–3364, https://doi.org/10.1093/nar/gkz097.
[8] A.C. Panda, I. Grammatikakis, R. Munk, M. Gorospe, K. Abdelmohsen, Emerging roles and context of circular RNAs, Wiley Interdiscip. Rev. RNA 8 (2017), https://doi.org/10.1002/wrna.1386.
[9] V.S. Patil, R. Zhou, T.M. Rana, Gene regulation by non-coding RNAs, Crit. Rev. Biochem. Mol. Biol. 49 (2014) 16–32, https://doi.org/10.3109/10409238.2013.844092.
[10] Z. Wang, M. Gerstein, M. Snyder, RNA-Seq: a revolutionary tool for transcriptomics, Nat. Rev. Genet. 10 (2009) 57–63, https://doi.org/10.1038/nrg2484.
[11] M. Schena, D. Shalon, R.W. Davis, P.O. Brown, Quantitative monitoring of gene expression patterns with a complementary DNA microarray, Science 270 (1995) 467–470, https://doi.org/10.1126/science.270.5235.467.
[12] J.C. Alwine, D.J. Kemp, G.R. Stark, Method for detection of specific RNAs in agarose gels by transfer to diazobenzyloxymethyl-paper and hybridization with DNA probes, Proc. Natl. Acad. Sci. USA 74 (1977) 5350–5354, https://doi.org/10.1073/pnas.74.12.5350.
[13] V.E. Velculescu, L. Zhang, B. Vogelstein, K.W. Kinzler, Serial analysis of gene expression, Science 270 (1995) 484–487, https://doi.org/10.1126/science.270.5235.484.
[14] T. Nolan, R.E. Hands, S.A. Bustin, Quantification of mRNA using real-time RT-PCR, Nat. Protoc. 1 (2006) 1559–1582, https://doi.org/10.1038/nprot.2006.236.
[15] K.J. Livak, T.D. Schmittgen, Analysis of relative gene expression data using real-time quantitative PCR and the 2-$\Delta\Delta$CT method, Methods 25 (2001) 402–408, https://doi.org/10.1006/meth.2001.1262.
[16] S.N. Peirson, J.N. Butler, R.G. Foster, Experimental validation of novel and conventional approaches to quantitative real-time PCR data analysis, Nucleic Acids Res. 31 (2003) e73, https://doi.org/10.1093/nar/gng073.
[17] D.C. Rio, M. Ares Jr., G.J. Hannon, T.W. Nilsen, Purification of RNA using TRIzol (TRI reagent), Cold Spring Harb. Protoc. (2010), https://doi.org/10.1101/pdb.prot5439. pdb.prot5439-pdb.prot5439.
[18] T. Lucelia, P.M. Alves, R.B. Ferreira, C.N. Santos, Comparison of different methods for DNA-free RNA isolation from SK-N-MC neuroblastoma, BMC Res. Note. (2011) 3, https://doi.org/10.1186/1756-0500-1-140.
[19] D.C. Rio, M.J. Ares, G.J. Hannon, T.W. Nilsen, Guidelines for the use of RNA purification kits, Cold Spring Harb. Protoc. 5 (2010), https://doi.org/10.1101/pdb.ip79.
[20] K.L. Manchester, Use of UV methods for measurement of protein and nucleic acid concentrations, BioTechniques 20 (1996) 968–970, https://doi.org/10.2144/96206bm05.
[21] D.A. Forero, Y. González-Giraldo, L.J. Castro-Vega, G.E. Barreto, qPCR-based methods for expression analysis of miRNAs, BioTechniques 67 (2019) 192–199, https://doi.org/10.2144/btn-2019-0065.
[22] K. Triinu, L. Maarja, K. Lauris, R. Kairi, A. Reidar, R. Maido, H. John, Primer3_masker: integrating masking of template sequence with primer design software, Bioinformatics (2018) 1937–1938, https://doi.org/10.1093/bioinformatics/bty036.
[23] P. Amaresh, G. Myriam, Detection and analysis of circular RNAs by RT-PCR, Bio-Protocol (2018), https://doi.org/10.21769/BioProtoc.2775.
[24] P. Glažar, P. Papavasileiou, N. Rajewsky, CircBase: a database for circular RNAs, RNA 20 (2014) 1666–1670, https://doi.org/10.1261/rna.043687.113.
[25] A.C. Panda, D.B. Dudekula, K. Abdelmohsen, M. Gorospe, Analysis of circular RNAs using the web tool CircInteractome, in: Methods in Molecular Biology, Humana Press, United States, 2018, pp. 43–56, https://doi.org/10.1007/978-1-4939-7562-4_4.
[26] A. Kozomara, M. Birgaoanu, S. Griffiths-Jones, MiRBase: from microRNA sequences to function, Nucleic Acids Res. 47 (2019) D155–D162, https://doi.org/10.1093/nar/gky1141.

SECTION 4

Epigenetic analyses

CHAPTER 12

Best practices for epigenome-wide DNA modification data collection and analysis

Joseph Kochmanski[a,b] and Alison I. Bernstein[c,d]

[a]Rancho BioSciences, San Diego, CA, United States, [b]Department of Translational Neuroscience, Michigan State University, East Lansing, MI, United States, [c]Department of Pharmacology and Toxicology, Rutgers University, Piscataway, NJ, United States, [d]Environmental and Occupational Health Science Institute, Rutgers University, Piscataway, NJ, United States

Introduction

In response to a series of analyses that revealed very low reproducibility of studies in life sciences, many initiatives have been launched to assist, establish, and improve standards for rigor and reproducibility in life sciences research [1]. In 2012, Landis and colleagues published a core set of guidelines to establish standards for reporting methods and data [2]. These include standards for randomization, blinding, sample-size planning, and data handling. In response to these guidelines, the FAIR Guiding Principles for data management and stewardship were published to clearly define good data management practices [3], and new repositories have been launched to facilitate sharing of data, analysis methods, and codes, including https://fairsharing.org [4]. In addition, as discussed later in this chapter, preregistration is becoming more common to prevent statistical negligence [5]. These issues are particularly relevant to the field of epigenetics, where adherence to these principles is made more difficult by the underlying biology of epigenetic marks, rapidly evolving methods, and statistical challenges. As this field continues to grow, it is critical that researchers adhere to existing recommendations as they plan, execute, and report their studies. In this chapter, we provide an overview of a specific class of epigenetic marks, DNA modifications. We discuss why these marks are important, how to measure them, and best practices for ensuring rigor and reproducibility in DNA modification research, with a focus on epigenome-wide association studies (EWAS).

DNA modifications

Epigenetics generally refers to a set of mechanisms that affect gene expression without altering the DNA sequence, including DNA modifications, histone modifications, noncoding RNAs, and alterations in chromatin accessibility [6–8]. The initiation and maintenance of marks play critical roles in many aspects of mammalian development, including stem cell self-renewal, stem cell differentiation, and neurodevelopment [9–11]. Dysregulation of epigenetic mechanisms has been identified in a variety of disorders and it is thought that these mechanisms, in particular DNA modifications, are especially important for the development and function of the central nervous system (CNS) [10–16]. In addition, epigenetic marks are sensitive to the environment, established during cellular differentiation, and regulate gene expression throughout the lifespan [17,18]. Given these unique traits, the epigenome has been recognized as a mediator of the relationship between environmental exposures and disease [6,19].

Over the past two decades, a rapidly growing number of epigenome-wide association studies have investigated the associations between DNA modifications and disease. As this field has exploded, researchers have begun to realize the need to ensure rigor and reproducibility in EWAS, specifically related to the biological features of DNA modifications, as well as the technical and statistical aspects of data analysis. In this chapter, we provide a brief overview of DNA modifications and the methods used to measure them and then focus on issues related to rigor and reproducibility in EWAS.

DNA modifications are covalent modifications of DNA that do not alter the genetic sequence itself. The most widely studied is methylation of the fifth position of cytosine (5-methylcytosine, 5-mC), which most commonly occurs at cytosine-phospho-guanine (CpG) dinucleotides. However, DNA methylation can also occur at non-CpG sites, including cytosine-phospho-adenine (CpA), cytosine-phospho-cytosine (CpC), and cytosine-phospho-thymine (CpT) [15]. Methylation of cytosines is catalyzed by DNA methyltransferase (DNMT) enzymes, which either carry out de novo methylation or maintain methylation through DNA replication [20]. Once established, 5-mC is recognized by methylation-specific DNA-binding proteins, including methyl-CpG binding protein 2 (MeCp2) and the methyl-CpG binding domain proteins 1-6 (MDB1-6) [21,22]. DNA methylation has been demonstrated to play an important role in transcriptional regulation, maintenance of chromatin structure, genomic imprinting, X-chromosome inactivation, and genomic stability [9,23,24].

While 5-mC is a fairly stable modification, it can be enzymatically oxidized to 5-hydroxymethylcytosine (5-hmC) and eventually to 5-formylcytosine (5-fC) and 5-carboxylcytosine (5-caC) through the activity of ten-eleven translocation (Tet) enzymes [25–36] (Fig. 12.1).

DNA modifications in health and disease

In the past two decades, an explosion of research has investigated the potential role of DNA modifications in human health and disease. These studies have documented clear associations between DNA modifications and disease risk in multiple contexts [37–41]. This is most well-established in the cancer field, such that for some cancers, research findings have been translated to clinical practice using DNA modifications as clinical biomarkers and treatments [42–49].

DNA modifications are particularly relevant to work on neurodegenerative diseases [38,50–52], as a growing body of research supports a critical role for DNA modifications, particularly 5-hydroxymethylcytosine (5-hmC), in neurological function and in the response to environmental exposures [38,40,53–55]. The specific enrichment and the temporal and the spatial patterns of 5-mC and 5-hmC in the mammalian brain suggest that these modifications are critical for brain development and function [32,56–58]. Changes in DNA methylation and hydroxymethylation have been associated with a range of neurodevelopmental and neurological disorders, including but not limited to Rett syndrome,

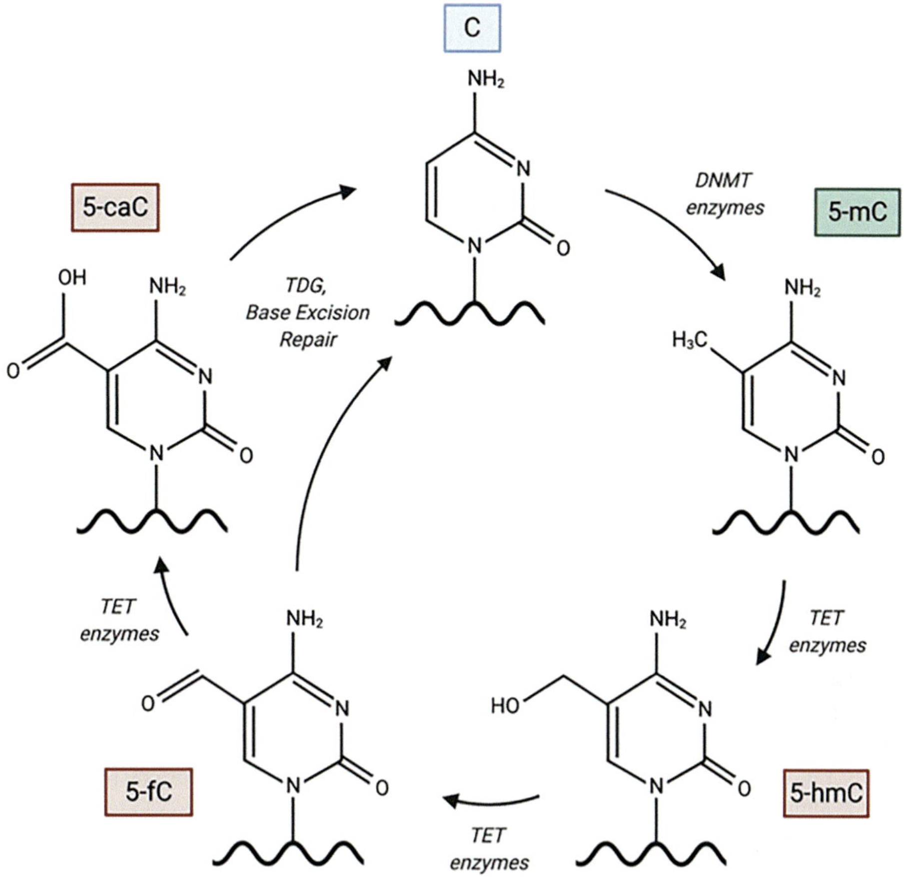

FIG. 12.1

DNA modification pathways. Cytosine (C) is methylated to 5-methylcytosine (5-mC) through activity of DNA methyltransferase (DNMT) enzymes. From there, 5-mC can be further oxidized to 5-hydroxymethylcytosine (5-hmC), 5-formylcytosine (5-fC), and 5-carboxylcytosine (5-caC) through the activity of ten-eleven translocation (TET) enzymes. 5-fC and 5-caC can then be converted back to an unmodified cytosine through thymine DNA glycosylase (TDG)-mediated base excision repair.

autism spectrum disorders, schizophrenia, Alzheimer's disease, Huntington's disease, and Parkinson's disease [38,40,48,52,59–64].

In addition, the epigenome is recognized as a potential mediator of the relationship between exposures and disease. Epigenetic marks are sensitive to the environment, established during cellular differentiation, and regulate gene expression throughout the lifespan [17,18]. Of note, epigenetic marks can change in response to lifestyle changes, diet, and environmental exposures, both during early development and into adulthood [6,65–72]. Given this, DNA modifications have emerged as an important area of research in understanding the mechanisms of disease etiology for a wide range of human diseases.

Methods for detection of DNA modifications

Over the past two decades, studies have interrogated genome-wide DNA modifications using multiple methods to measure modifications. These methods operate at three different genomic resolutions: global, regional, and base-pair. The pros and cons of these methods are comprehensively reviewed [73–75]. In general, the field is moving away from global and regional methods, as they generate data that is difficult to interpret in functional terms. Given this trend, for purposes of this chapter, we will focus only on the widely used base-pair resolution methods that are suitable for EWAS. EWAS are used to find associations between genome-wide epigenetic marks and phenotypes, much as genome-wide association studies are used to identify SNPs associated with phenotypes.

Bisulfite conversion methods

Base-pair resolution methods quantify DNA modifications at individual cytosine bases. Traditionally, these methods have relied on the bisulfite (BS) conversion of DNA. BS deaminates unmodified cytosines, 5-caC, and 5-fC to uracil, but not to 5-mC or 5-hmC [76–78]. Converted uracils are then read as thymine during PCR amplification, sequencing, or detection on a methylation array. From this data, the ratio of cytosines to thymines at a given cytosine is used to estimate a beta value, which represents the proportion of cytosines at a given locus that are modified. All of the methods described below produce beta values as their output; in EWAS, these beta values are then compared between groups.

The gold standard method in the field is whole-genome bisulfite sequencing (WGBS), which pairs BS conversion with next-generation sequencing to provide single base-pair resolution beta values at cytosines across the entire sequenced genome [79–81]. While WGBS produces comprehensive data, cost remains a huge barrier for most labs, even with declining sequencing prices [82]. Two common genome-wide alternatives with lower costs are reduced representation bisulfite sequencing (RRBS) and Illumina methylation arrays.

RRBS reduces the number of nucleotides sequenced by incorporating a methylation-insensitive restriction enzyme digestion step prior to BS conversion [83–85]. Typically, *Msp*I is used, which cuts at C/CGG recognition sites, regardless of the methylation status. When combined with size selection of restriction fragments, this generates a "reduced representation" of the genome, enriching for fragments containing potentially modified cytosines. RRBS data typically produces beta value data for approximately 2–4 million cytosines.

Illumina DNA methylation arrays combine sodium bisulfite treatment, whole-genome amplification, and sequence-specific probe hybridization using BeadChip technology to estimate DNA methylation at a preselected set of cytosines across the genome [86]. The current version of this technology, the Illumina Methylation EPIC array, covers over 850,000 CpG sites. Unfortunately, as of this writing, these arrays are only available for human samples, so their application to animal models is limited. Readers interested in detailed comparisons of RRBS, methylation arrays, and WGBS should refer to publications on the topic [74,75,87].

Another, newer option for reducing cost is capture hybridization [88], which allows for targeted selection of specific sequences prior to BS conversion and sequencing. A disadvantage of this method is that it requires prior knowledge of which regions to select, and has not been widely adopted as of this writing.

A major drawback of BS-based methods is that BS treatment does not differentiate 5-mC from 5-hmC [89]. Both 5-mC and 5-hmC are protected from standard BS-mediated deamination and read the same in BS-based sequencing or array output [90].

Alternatives to BS-based methods

Oxidative BS (oxBS) conversion allows for the detection of only 5-mC without 5-hmC [91]. In oxBS, the first step is the selective, chemical-mediated oxidation of 5-hmC to 5-fC by potassium perruthenate ($KuRO_4$). The oxidized 5-fC is then sensitive to BS treatment and is converted like an unmodified cytosine in a BS conversion step to uracil [91–96].

In addition to comeasurement of 5-mC and 5-hmC, methods have also been developed to address the high input requirements for BS-based methods and to allow interrogation of DNA modifications from low amounts of input DNA. One of these methods, enzymatic methyl-sequencing (EM-Seq), has been commercialized in the NEBNext Enzymatic Methyl-seq Kit (New England BioLabs). This is a two-step enzymatic conversion with no chemical conversion step, allowing for successful detection of DNA modifications from low input DNA [97,98]. In this method, TET2 and T4-phage β glucosyltransferase (T4-βGT) convert 5-mC and 5-hmC to 5-caC and 5-(β-glucosyloxymethyl) cytosine (5-gmC), respectively. Then APOBEC3A deaminates unmodified cytosines to uracils. EM-seq output mimics the output of BS conversion (Fig. 12.2).

Other methods, like apolipoprotein B mRNA editing enzyme, catalytic polypeptide-like 3A (APOBEC)-coupled epigenetic sequencing (ACE-seq), and TET-assisted pyridine borane sequencing (TAPS), also remove the BS conversion step. Unlike EM-seq, these alternative methods each generate slightly different output, detecting either only 5-hmC (ACE-Seq) or both marks (TAPS) (summarized in Refs. [90,99]).

Challenges to reproducibility in DNA modification EWAS research

While methodological advancements provide exciting opportunities in the field, the unique combination of complicated biology and the massive scale of modern sequencing data present a number of challenges for EWAS. These challenges fall into several categories: biology, methodology, statistics, experimental planning, and data and method reporting. Overcoming these challenges for epigenome-wide data hinges upon rigorous and reproducible methodology and analysis. Here, we highlight the challenges listed above and present some solutions for addressing them in EWAS research.

Biology

Tissue heterogeneity and cell-type specificity

A major challenge standing in the way of producing rigorous, reproducible epigenome-wide results is the fact that different cell types exhibit distinct epigenetic profiles [100–102]. As a result, studies that use tissue comprised of a mixture of cell types introduce noise into their data, washing out potentially important results with unwanted variability. Recent work has begun to address this issue using both experimental and computational tools [103,104]. For example, bioinformatics tools have been developed to estimate cell type composition from reference, bulk-tissue epigenetic data, allowing for correction during statistical analysis [105]. Some methods even allow for reference-free cell type estimation, enabling researchers to estimate

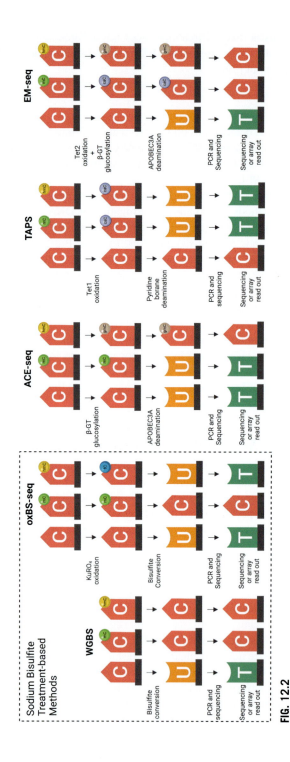

FIG. 12.2

Conversion methods to measure base-pair resolution DNA modifications. Over the past two decades, the gold standard method to measure base-pair resolution DNA methylation has been whole-genome bisulfite sequencing (WGBS). While reliable, this method involves a sodium bisulfite conversion step, which damages DNA, and does not distinguish between 5-mC and 5-hmC. To avoid these issues, researchers have used a variety of different enzymatic and chemical treatments to develop improved methods. In some cases, these methods allow for specific measurement of either 5-mC (oxBS-seq) or 5-hmC (ACE-seq), or simply improve on the damaging effects of sodium bisulfite conversion through the use of enzymatic deamination via APOBEC3A enzymes (TAPS and EM-seq). In the diagram, different cytosine modifications are indicated by colored bubbles; *green*, 5mC; *yellow*, 5-hmC; *blue*, 5-fC; *purple*, 5-caC; *gray*, 5-(β-glucosyloxymethyl)cytosine (5-gmC). Abbreviations: *WGBS*, whole-genome bisulfite sequencing; *oxBS-seq*, oxidative bisulfite sequencing; *ACE-seq*, APOBEC-coupled epigenetic sequencing; *TAPS*, TET-assisted pyridine borane sequencing; *EM-seq*, enzymatic methyl-sequencing; *KuRO4*, potassium perruthenate; *PCR*, polymerase chain reaction; *β-GT*, b-glucosyltranferase; *Tet1/2*, ten-eleven translocase ½; *APOBEC3A*, apolipoprotein B mRNA editing enzyme, catalytic polypeptide-like 3A.

cell types without reference data from their tissue of interest [106,107]. Meanwhile, at the benchtop, cell sorting methods, including flow-assisted cell sorting and magnetic-assisted cell sorting [108,109] can be used to enrich specific cell populations prior to isolating DNA for downstream epigenetic follow-up.

Another enrichment method is laser capture microdissection (LCM), which allows for targeted dissection of individual cells or tissue regions prior to DNA isolation and measurement of DNA modifications [110]. The LCM method is less applicable to genome-wide methods that require high DNA inputs, as is often the case for DNA modifications, but as technology improves, this technology will become more feasible to pair with EWAS. These types of laboratory-based enrichment tools are critical for reducing random variability in EWAS studies, allowing for not only improved biological interpretability but also increased reproducibility. The issue of cell-type heterogeneity is well-recognized in epigenetics research, and in an ideal world, all EWAS studies would include techniques to isolate specific cell populations for analysis. However, universal implementation of cell enrichment approaches will require improved compatibility with low amounts of input DNA. Furthermore, to ensure reproducibility, researchers must not only report their cell type enrichment and/or estimation methods, but also share any code or protocols related to these efforts, as they will benefit the entire field.

Temporal changes in DNA modifications
DNA modifications can change with age [111–115], and preliminary evidence suggests that early-life exposures can modify rates of epigenetic aging [116]. As a result, the age at which tissue collection takes place could affect DNA modification signatures, complicating interpretation of EWAS results. Luckily, there are some simple ways to account for this variability during experimental planning:

- In human studies, researchers can control for the effects of age on DNA modifications by either age-matching human samples or including age as a covariate in statistical modeling.
- In animal studies, where greater experimental control is available, animals should always be age-matched during tissue collection.

Sex-related changes in DNA modifications
In addition to age, sex affects levels of DNA modification across the genome [117–119]. Of note, evidence shows that sex is not only associated with differences in baseline DNA methylation at specific genes, but also the epigenetic response to the environment [51,120–123]. As a result, studies in humans and animals that include both males and females must account for sex in their analysis. This can be done in a similar fashion to age:

- In human studies, researchers can control for the effects of sex on DNA modifications by either sex-matching human samples or including sex as a covariate in statistical modeling.
- In animal studies, where there is greater experimental control, statistical analyses should be stratified by sex.

Methodology
Another major challenge facing the field is the ever-changing landscape of methods to interrogate the epigenome [73,74,87,124]. Technologies are changing and updating so quickly that grants often include methods that are outdated by the time grants are funded. In addition to new conversion methods,

there has also been a flood of new, improved bioinformatics tools for data processing and analysis [125–127]. While these new methods and tools present exciting new possibilities in the field, the selection of the correct analysis pipeline is not always clear, as there are subtle differences between many of the statistical approaches employed by these methods.

There are also differences in the structure and scale of EWAS data that affect downstream statistical inference. A useful example is a comparison of the analysis pipelines for RRBS data and Illumina EPIC array data. These two methods cover similar numbers of CpG sites across the genome (RRBS: ~2–4 million CpGs; EPIC array: ~850,000 CpGs). At first glance, one might think that similar methods could be used to analyze both of these datasets. However, due to methodological differences, the pipelines used to analyze RRBS and EPIC data are vastly different. RRBS produces sequencing reads in the form of .fastq files, which must be aligned to the genome using specific software settings prior to analysis [128]. Meanwhile, the EPIC array produces probe-based red and green channel.idat files, which require a specialized quality control, processing, and normalization pipeline [129]. Furthermore, when running statistical tests on these large datasets, the effect of controlling for multiple testing will vary based on the number of sites being analyzed. As a result, one chromosomal position that shows a false discovery rate < 0.05 in an EPIC dataset may not in an RRBS dataset, even if the effect size and sample size are the same. This is because more comparisons are being performed in the RRBS dataset, therefore limiting the power to detect changes when correcting for multiple hypothesis testing. These methodological and statistical differences between EWAS datasets lead to a lack of reproducibility in the field. While some of these differences are inherent to a study's data structure, epigenetics researchers must work to establish standard, method-specific pipelines for genome-wide 5-mC and 5-hmC data analysis. Readers interested in the specific recommendations regarding methodology in EWAS research, including ways to avoid bias in tissue processing and statistical analysis, can find additional information in a number of recent articles [82,130–133]. By incorporating existing methodological guidelines, the field of epigenetics research could improve its rigor and reproducibility, overcoming the uncertainty that has plagued its data for decades.

Statistics

Sample size planning and power analysis

The most important statistical consideration for any EWAS is sample-size planning. This process should be performed during the initial planning stages of an experiment, and involves analysis of statistical power, or the probability that a statistical test will correctly reject the null hypothesis.

Statistical power analysis requires four main parameters:

1. Estimated effect size.
2. Estimated data variability.
3. Sample size.
4. Selected alpha significance level (e.g., $P < .05$), and the desired statistical power (e.g., 0.8)

As such, for a scientist to estimate the required sample size for an experiment at a given power, they will need to input their estimated effect size, data variability, and alpha significance level. This is a foundational power analysis as it is typically applied to a simple statistical test; for example, a Student's T-test.

However, when analyzing genome-wide data, researchers are typically conducting tests on thousands of gene regions at once. In this larger-scale scenario, power analysis must consider additional parameters, including the false discovery rate to correct for multiple testing, the distribution of read counts, and the dispersion coefficients of analyzed genes [134]. To address these additional issues, recent work has used existing RNA-sequencing data as a frame of reference to create bioinformatics tools to perform power analysis for EWAS. Examples include the *RNASeqPower* [134], *PROPER* [135], *powsimR* [136], and *RnaSeqSampleSize* [137], among others. The sequencing data produced in RNA-sequencing experiments are similar in structure to those generated by RRBS or WGBS, meaning the tools developed for RNA-seq experiments can provide some guidance for EWAS studies. However, the data dispersion and statistical assumptions used to analyze RRBS or WGBS data are not the same as those used for RNA-seq, meaning it would be better to use specialized techniques.

Unfortunately, as of this writing, there are only a few existing tools to specifically performs power analysis on WGBS or Illumina Beadchip array data. For WGBS data, power analysis can be performed using the *Bisulfite sequencing simulator/power analysis* shiny app developed by the Tung Lab [138]. Meanwhile, for Illumina Beadchip arrays, there are two existing tools in the literature: *pwrEWAS*, an R shiny app [139], and an interactive web application specifically for conducting power calculations for Illumina EPIC array datasets [130]. Along with this web application, the authors of the second tool also provide a suggested P-value significance cutoff for EPIC array data that accounts for both multiple testing and correlation between probes: $P < 9 \times 10^{-8}$. They also provide recommendations regarding sample size for studies that utilize this methodology. Based on their simulations, at a sample size of $N=200$ ($n=100$ case, $n=100$ control), approximately 85% of analyzed sites will have >80% power to detect a beta value difference of 0.05 or greater [130]. Utilization of the highlighted sample size planning tools is critical for ensuring statistical rigor in future EPIC array studies.

Human epidemiological studies

Human epidemiological cohorts are heterogeneous and provide little control over environmental and lifestyle factors that can affect the epigenome, including diet, stress, chemical exposures, and physical activity level [18,69,70,140–144]. As such, it can be difficult to determine the effect of a single variable, e.g., exposure to a toxicant, on DNA modifications in a human cohort. Generally, epidemiological studies attempt to control for potential confounders by including them as covariates in statistical modeling. In this way, it is possible to account for the heterogeneity of human populations while examining DNA modifications as a primary outcome measure. There is certainly value in this approach, but variable selection must be performed carefully. Adding too many additional covariates can inflate model estimates and lead to multicollinearity, thereby impacting the ability of a model to accurately detect significant changes [145]. Of note, these issues are exacerbated by small sample sizes, as often occurs in EWAS research. As a result, careful model building is incredibly important for producing reliable results for human EWAS. Strategies for variable selection are varied, but two of the most common approaches are testing the significance of covariate coefficients or testing whether the magnitude of changes in the outcome due to covariate inclusion are biologically meaningful (e.g., above a predetermined threshold) [145]. Unfortunately, specific recommendations are difficult to come by, as selection of covariate inclusion criteria will vary based on study design and sample size. Despite this uncertainty, it should be emphasized that EWAS investigators must always consider variable selection during model building, as inclusion/exclusion of covariates will have a significant effect on results.

Hypothesis testing in EWAS

While rigorous and reproducible analysis of epigenome-wide datasets is certainly possible, it is not simple or straightforward, and epigenome-wide association studies from different labs typically use very different methods to test their hypotheses, making it difficult to reconcile data between groups. As of now, there are some existing guidelines from the NIH Roadmap Epigenomics Program regarding methodology, data reporting, and analysis for epigenome-wide association studies [146]. Unfortunately, these guidelines provide only general suggestions regarding data handling and reporting, e.g., investigators report all data handling steps prior to read mapping, and do not give researchers guidance regarding specific analysis parameters. Other large epigenomics consortia, including the NIEHS TaRGET II Consortium [147], provide protocol files for their bioinformatics pipelines, which is extremely helpful for the field. However, the establishment of best practice guidelines for EWAS analysis is not straightforward and represents an area of continuing development.

A major issue in establishing standardized analysis practices is that both 5-mC and 5-hmC beta values are heteroscedastic in their data distributions, with most cytosines showing either very high or very low beta values [96]. Given this unique distribution, statistical tests to test for differential DNA modifications must utilize statistical methods that can account for this unique beta value distribution. Although the field has not settled on a single best practice to address this heteroscedasticity, there are several methods that are commonly used during hypothesis testing in EWAS.

1. *Logit transformation*
   a. One of the most common method uses logit transformation to generate m-values from the beta values [148]. Under this approach, m-values exhibit a homoscedastic distribution and can be analyzed using standard linear regression methods.
   b. This type of analysis is often implemented in pipelines that rely upon the *limma* and *BSmooth* R packages [127]. Importantly, linear regression requires logit transformation to m-values prior to modeling because linear regression assumes homoscedasticity.
   c. A weakness of this method is that m-values are less interpretable than beta values and are not recommended for data visualization [148,149]. As such, we only recommend using this method, if investigators have no intention of visualizing their data after detection of differential DNA modifications.
2. *Regression modeling*
   a. There are several analysis approaches that do not rely upon the generation of m-values; these include beta regression, rank-based regression, and beta-binomial regression modeling [138,150–152].
   b. These types of alternative regression methods, which are more appropriate for analyzing raw beta values, are implemented in a number of existing R packages, including *DSS, BiSeq, MethylSig,* and *MOABS,* among others [127].
   c. The exact choice of method will depend on data distribution and the desired modeling flexibility, but in general, we recommend using one of these alternative regression-based methods when investigators want to keep their untransformed beta values for improved data interpretation and visualization (Box 12.1).

Even when using these more appropriate analysis methods, the validity and sensitivity of genome-wide analyses of DNA modifications can break down when test assumptions do not match data distributions. **To ensure that investigators are using the most appropriate methods for their data, they should**

> **Box 12.1 Heteroscedasticity**
>
> Although, it sounds like a difficult concept to understand, heteroscedasticity (or "different dispersion") is actually relatively simple. It refers to data in which the variability of an outcome variable (i.e., DNA modification beta value) is unequal across the range of values for a second, independent variable (i.e., experimental treatment). Homoscedasticity, meanwhile, refers to data in which the variability of an outcome variable is equal across the range of values for a second variable. Linear regression assumes homoscedastic data, but DNA modification data often shows inherently unequal variance across its own distribution, with the majority of values close to 0 or 1. As such, it is important to account for the potential effects of heteroscedasticity when applying regression methods to DNA modification data.

always double-check their modeling output to determine the rigor of their experiment. For example, when conducting a large number of hypotheses tests in EWAS, the authors recommend visualization of the P-value distribution to diagnose potential issues with error rates [153]. By comparing an experimental P-value distribution shape to the expected uniform distribution under the null hypothesis, it is possible to determine whether an analysis is problematic. When P-value histograms do not show a peak near zero and/or broadly deviate from uniformity, investigators should consider adjusting their model, stratifying data to control for uncontrolled confounders, or revisiting power analysis to determine whether their data is adequately testing the question of interest. While it is difficult to provide specific recommendations given that each experiment will vary in its hypothesis testing, we strongly recommend examining P-value histograms for all EWAS research.

Reconciling paired 5-mC and 5-hmC data

Recent technological advances have made it possible to measure both 5-mC and 5-hmC from the same tissues, providing opportunities to tease apart associations between these two marks and gene regulation. However, there are not yet standard statistical methods for co-analyzing 5-mC and 5-hmC data. Recent studies have either examined the distribution of 5-hmC alone across the genome [93,154,155] or treated 5-mC and 5-hmC as independent outcomes, analyzing each as a separate epigenetic mark [94,156,157]. While these approaches are not wrong, they ignore the biological interdependence of 5-mC and 5-hmC.

In a recent publication, the authors proposed a new statistical approach for reconciling combined 5-mC and 5-hmC data [96]. In this streamlined analysis pipeline, rather than treating 5-mC and 5-hmC as separate outcomes, a mixed-effects (ME) modeling approach is proposed that treats the two epigenetic marks as "repeated" measures of a single outcome variable. An interaction term is used to determine whether the relationship between the outcome (i.e., beta value) and experimental treatment varies by DNA modification. One limitation of this approach is that it is specific to sites with measurable beta values for both 5-mC and 5-hmC; for those sites that show only 5-mC or 5-hmC, separate analyses for each mark are still needed. Implementation of mixed modeling for 5-mC and 5-hmC data is still in its infancy, and as of now, there is no specialized software to perform only this analysis method. However, it can be achieved using mixed-effects regression modeling in the existing *gamlss* R package [158]. To improve the co-analysis of paired 5-mC and 5-hmC data, future bioinformatics tools should work to streamline this process, implementing a mixed-effects modeling approach on a genome-wide scale into an R package or shiny application. We also hope that other groups will take up this challenge and develop additional methods for coanalyzing 5-mC and 5-hmC data.

Experimental planning & reporting: Methods, code, and data

Preregistration

Another way to improve experimental planning in general is through preregistration. This is particularly important in epigenetics, as methodologies and analysis pipelines are continually changing. **Preregistration refers to the process of defining one's research questions and proposed analyses prior to data collection** [5]. By preregistering a research project, it is possible to distinguish data generated as part of predictive hypothesis testing as opposed to exploratory analyses.

A variety of services are available online to preregister studies; as one example, the Open Science Framework (OSF) is a free, open-access service that provides workflows for preregistration [159]. This service is maintained by the Center for Open Science, a nonprofit organization committed to improving rigor and reproducibility in scientific research. By completing a preregistration, researchers can avoid hindsight bias and P-value hacking, common analysis pitfalls that can occur when working with large-scale datasets that present many possible avenues for statistical analysis.

In vivo animal colony reporting

A crucial component of in vivo EWAS studies is careful reporting of animal breeding schemes. Many EWAS studies utilize animal models to examine the link between developmental exposures and the later-life epigenome in offspring. This type of intergenerational and/or transgenerational research is critical to improving our understanding of the Developmental Origins of Health and Disease (DOHaD) hypothesis [19,160,161], but the field has been hampered by a lack of reproducibility [162]. One of the major reasons for the poor reproducibility in these types of studies is a lack of transparency regarding animal breeding schemes. In particular, many studies fail to account for litter effects [163,164].

To address these persistent deficiencies, researchers recently released the animal research: reporting of in vivo experiments (ARRIVE) guidelines, a checklist of items to include in animal model publications [165]. These guidelines include reporting suggestions for study design, sample size, inclusion/exclusion criteria, randomization, and more. It is recommended that investigators utilize these guidelines both during experimental planning and reporting. By adhering to these criteria for publication, scientists who use animals in their EWAS research can improve the transparency of their projects, thereby increasing rigor and reproducibility.

Sharing data through repositories

Another major area of continued improvement is in data publication. Recently, there has been a significant push toward data publication in public, online repositories. Some well-known examples include the Gene Expression Omnibus (GEO), Sequence Read Archives (SRA), and BioStudies, which are the repositories maintained by funding agencies. More recently, independent repositories have been created, including Dryad and Mendeley Data. Specific to epigenetics data, the National Center for Biotechnology Information (NCBI) recently established an Epigenomics database [166]. **Given these readily available resources, no published EWAS should ever use the term "data not shown."** Rather, all data, both raw and processed, should be provided in an online repository as part of a manuscript's publication. Furthermore, all data tables resulting from statistical analysis should be shared as supplementary tables. This will ensure that all aspects of a study's results are reported, allowing for verification of findings by other researchers.

Code sharing

Given the evolving EWAS methodology and analysis tools highlighted earlier in this chapter, **code sharing is critical for ensuring rigor and reproducibility** in EWAS. Many research groups use their own custom pipelines to analyze data, yet they do not publish code alongside their publications. There is nothing inherently wrong with using an in-house analysis pipeline, especially given the number of choices and methods available, but doing so limits the interoperability of data, decreasing the potential for shared procedures and results. In essence, an anticollaborative culture is created not by intention but by necessity. Groups only know how to process their own data, which reduces the chances for cooperation that might spur innovation in the field.

To combat this trend, it is critical for researchers to share their code along with their publications so that their analysis is clear. When shared, code should be usable and editable; preferably in a format that can be easily rerun. Related to this, it is recommended that researchers avoid simply sharing raw code, and instead provide annotations or comments that explain the parameter choices that were made during initial analysis. In many cases, default software settings are not used, so the selection of specific parameters must be clear for users. We also recommend that all quality control figures and metrics produced in data processing be shared publicly. The exact implementation of these recommendations will vary based on the selected analysis software, but they could be implemented through the Github platform or through the use of R Notebooks. This will increase a future user's ability to recreate past analyses, thereby improve the reproducibility of a bioinformatics pipeline.

FAIR principles for data sharing

Data repositories and code sharing are certainly steps in the right direction, but data files and code provided by the authors are often difficult to interpret, with obtuse, nondescriptive file names, and little annotation available to recreate analyses. In response to this lack of standardization, a diverse group of scientists recently came together to design the **FAIR Data Principles**, which act as a set of guidelines for improving the reusability of data after publication [3]. Under these principles, publicly hosted data are suggested to be four things: Findable, Accessible, Interoperable, and Reusable (FAIR). Under this framework,

> **Findable**—refers to registration in a searchable resource along with rich metadata, including a unique identifier variable.
> **Accessible**—refers to the data being retrievable through a free, user-friendly communications portal or protocol.
> **Interoperable**—refers to the inclusion formal, shared language (e.g., code), clear references to related data, and a consistent vocabulary. Finally,
> **Reusable**—refers to providing a rich metadata and linking data to a usage license.

By adhering to these principles, scientists can not only increase the believability of their results, but also provide clear opportunities for rigorous data validation.

Existing data repositories in the life sciences have variable implementations of FAIR principles, putting much of the responsibility for accurate and complete data reporting on users [167,168]. Furthermore, the lack of standardized language in these repositories can make it difficult to utilize publicly available data. For example, specific information regarding experimental samples may be lacking, including tissue heterogeneity, age, and collection methods. To help investigators identify their best options for data sharing, https://fairsharing.org is a publicly available resource that provides data and metadata standards as well as ratings for existing repositories [4]. To improve the interoperability of public EWAS data, investigators must aim to adhere to FAIR principles when submitting their data to a repository.

Conclusion

In this chapter, we describe a number of existing methods for analyzing DNA modification data. We also presented challenges related to modern sequencing datasets, including biology, methodology, statistics, experimental planning, and data and method reporting. Overcoming these challenges for epigenome-wide data hinges on rigorous and reproducible methodology and analysis. Related to this point, we highlighted a number of specific recommendations for EWAS research.

To address the biological heterogeneity of the epigenome across cell types, we recommend cell type enrichment methods (e.g., flow-assisted cell sorting) and/or computational estimation. For statistical analysis, we recommend specific tools for analysis planning and power analysis in EWAS, including *pwrEWAS*. We also discuss covariate inclusion criteria in human EWAS, methods for addressing beta value heteroscedasticity, *P*-value histograms as a diagnostic tool, and mixed effects regression methods for coanalyzing paired 5-mC and 5-hmC data. After that, we provided recommendations regarding reporting in EWAS research. In particular, we highlight areas of potential improvement in the field: preregistration to avoid p-hacking, adherence to the ARRIVE principles for in vivo studies, data reporting via public repositories, code and parameter sharing, and the FAIR principles for data sharing.

Through careful implementation of the provided recommendations, we believe the field can begin to produce epigenome-wide association studies that are both rigorous and reproducible.

References

[1] M. Baker, D. Penny, Is there a reproducibility crisis? Nature 533 (2016) 452–454, https://doi.org/10.1038/533452A.

[2] S.C. Landis, S.G. Amara, K. Asadullah, C.P. Austin, R. Blumenstein, E.W. Bradley, R.G. Crystal, R.B. Darnell, R.J. Ferrante, H. Fillit, R. Finkelstein, M. Fisher, H.E. Gendelman, R.M. Golub, J.L. Goudreau, R.A. Gross, A.K. Gubitz, S.E. Hesterlee, D.W. Howells, J. Huguenard, K. Kelner, W. Koroshetz, D. Krainc, S.E. Lazic, M.S. Levine, M.R. MacLeod, J.M. McCall, R.T.M. Iii, K. Narasimhan, L.J. Noble, S. Perrin, J.D. Porter, O. Steward, E. Unger, U. Utz, S.D. Silberberg, A call for transparent reporting to optimize the predictive value of preclinical research, Nature 490 (2012) 187–191, https://doi.org/10.1038/nature11556.

[3] M.D. Wilkinson, M. Dumontier, I.J. Aalbersberg, G. Appleton, M. Axton, A. Baak, N. Blomberg, J.W. Boiten, L.B. da Silva Santos, P.E. Bourne, J. Bouwman, A.J. Brookes, T. Clark, M. Crosas, I. Dillo, O. Dumon, S. Edmunds, C.T. Evelo, R. Finkers, A. Gonzalez-Beltran, A.J.G. Gray, P. Groth, C. Goble, J.S. Grethe, J. Heringa, P.A.C. 't Hoen, R. Hooft, T. Kuhn, R. Kok, J. Kok, S.J. Lusher, M.E. Martone, A. Mons, A.L. Packer, B. Persson, P. Rocca-Serra, M. Roos, R. van Schaik, S.A. Sansone, E. Schultes, T. Sengstag, T. Slater, G. Strawn, M.A. Swertz, M. Thompson, J. Van Der Lei, E. Van Mulligen, J. Velterop, A. Waagmeester, P. Wittenburg, K. Wolstencroft, J. Zhao, B. Mons, Comment: The FAIR guiding principles for scientific data management and stewardship, Sci. Data 3 (2016), https://doi.org/10.1038/sdata.2016.18.

[4] S.A. Sansone, P. McQuilton, P. Rocca-Serra, A. Gonzalez-Beltran, M. Izzo, A.L. Lister, M. Thurston, FAIRsharing as a community approach to standards, repositories and policies, Nat. Biotechnol. 37 (2019) 358–367, https://doi.org/10.1038/s41587-019-0080-8.

[5] B.A. Nosek, C.R. Ebersole, A.C. DeHaven, D.T. Mellor, The preregistration revolution, Proc. Natl. Acad. Sci. USA 115 (2018) 2600–2606, https://doi.org/10.1073/pnas.1708274114.

[6] A.J. Bernal, R.L. Jirtle, Epigenomic disruption: the effects of early developmental exposures, Birth Defects Res. A Clin. Mol. Teratol. 88 (2010) 938–944, https://doi.org/10.1002/bdra.20685.

[7] B.E. Bernstein, A. Meissner, E.S. Lander, The mammalian epigenome, Cell 128 (2007) 669–681, https://doi.org/10.1016/j.cell.2007.01.033.

[8] G. Egger, G. Liang, A. Aparicio, P.A. Jones, Epigenetics in human disease and prospects for epigenetic therapy, Nature 429 (2004) 457–463, https://doi.org/10.1038/nature02625.

[9] E. Li, Chromatin modification and epigenetic reprogramming in mammalian development, Nat. Rev. Genet. 3 (2002) 662–673, https://doi.org/10.1038/nrg887.

[10] Z.D. Smith, A. Meissner, DNA methylation: roles in mammalian development, Nat. Rev. Genet. 14 (2013) 204–220, https://doi.org/10.1038/nrg3354.

[11] B. Yao, P. Jin, Cytosine modifications in neurodevelopment and diseases, Cell Mol. Life Sci. 71 (2014) 405–418, https://doi.org/10.1007/s00018-013-1433-y.

[12] J. Feng, G. Fan, The role of DNA methylation in the central nervous system and neuropsychiatric disorders, Int. Rev. Neurobiol. 89 (2009), https://doi.org/10.1016/S0074-7742(09. 89004-1.

[13] K. Gapp, B.T. Woldemichael, J. Bohacek, I.M. Mansuy, Epigenetic regulation in neurodevelopment and neurodegenerative diseases, Neuroscience 264 (2014) 99–111, https://doi.org/10.1016/j.neuroscience.2012.11.040.

[14] M. Jakovcevski, S. Akbarian, Epigenetic mechanisms in neurological disease, Nat. Med. 18 (2012) 1194–1204, https://doi.org/10.1038/nm.2828.

[15] H.S. Jang, W.J. Shin, J.E. Lee, J.T. Do, CpG and non-CpG methylation in epigenetic gene regulation and brain function, Genes 8 (2017) 2–20, https://doi.org/10.3390/genes8060148.

[16] A. Rudenko, L.H. Tsai, Epigenetic modifications in the nervous system and their impact upon cognitive impairments, Neuropharmacology 80 (2014) 70–82, https://doi.org/10.1016/j.neuropharm.2014.01.043.

[17] C.D. Allis, T. Jenuwein, The molecular hallmarks of epigenetic control, Nat. Rev. Genet. 17 (2016) 487–500, https://doi.org/10.1038/nrg.2016.59.

[18] C. Faulk, D.C. Dolinoy, Timing is everything: the when and how of environmentally induced changes in the epigenome of animals, Epigenetics 6 (2011) 791–797, https://doi.org/10.4161/epi.6.7.16209.

[19] T. Bianco-Miotto, J.M. Craig, Y.P. Gasser, S.J. Van Dijk, S.E. Ozanne, Epigenetics and DOHaD: from basics to birth and beyond, J. Dev. Orig. Health Dis. 8 (2017) 513–519, https://doi.org/10.1017/S2040174417000733.

[20] H. Gowher, A. Jeltsch, Mammalian DNA methyltransferases: new discoveries and open questions, Biochem. Soc. Trans. 46 (2018) 1191–1202, https://doi.org/10.1042/BST20170574.

[21] O. Bogdanović, G.J.C. Veenstra, DNA methylation and methyl-CpG binding proteins: developmental requirements and function, Chromosoma 118 (2009) 549–565, https://doi.org/10.1007/s00412-009-0221-9.

[22] Q. Du, P.L. Luu, C. Stirzaker, S.J. Clark, Methyl-CpG-binding domain proteins: readers of the epigenome, Epigenomics 7 (2015) 1051–1073, https://doi.org/10.2217/epi.15.39.

[23] S. Gopalakrishnan, B.O. Van Emburgh, K.D. Robertson, DNA methylation in development and human disease, Mutat. Res. Fundam. Mol. Mech. Mutagen. 647 (2008) 30–38, https://doi.org/10.1016/j.mrfmmm.2008.08.006.

[24] M.M. Suzuki, A. Bird, DNA methylation landscapes: provocative insights from epigenomics, Nat. Rev. Genet. 9 (2008) 465–476, https://doi.org/10.1038/nrg2341.

[25] S. Ito, L. Shen, Q. Dai, S.C. Wu, L.B. Collins, J.A. Swenberg, C. He, Y. Zhang, Tet proteins can convert 5-methylcytosine to 5-formylcytosine and 5-carboxylcytosine, Science 333 (2011) 1300–1303, https://doi.org/10.1126/science.1210597.

[26] R.M. Kohli, Y. Zhang, TET enzymes, TDG and the dynamics of DNA demethylation, Nature 502 (2013) 472–479, https://doi.org/10.1038/nature12750.

[27] M. Tahiliani, K.P. Koh, Y. Shen, W.A. Pastor, H. Bandukwala, Y. Brudno, S. Agarwal, L.M. Iyer, D.R. Liu, L. Aravind, A. Rao, Conversion of 5-methylcytosine to 5-hydroxymethylcytosine in mammalian DNA by MLL partner TET1, Science 324 (2009) 930–935, https://doi.org/10.1126/science.1170116.

[28] H. Wu, Y. Zhang, Mechanisms and functions of Tet proteinmediated 5-methylcytosine oxidation, Genes Dev. 25 (2011) 2436–2452, https://doi.org/10.1101/gad.179184.111.

[29] M. Bachman, S. Uribe-Lewis, X. Yang, M. Williams, A. Murrell, S. Balasubramanian, 5-Hydroxymethylcytosine is a predominantly stable DNA modification, Nat. Chem. 6 (2014) 1049–1055, https://doi.org/10.1038/nchem.2064.

[30] M.A. Hahn, P.E. Szabó, G.P. Pfeifer, 5-Hydroxymethylcytosine: a stable or transient DNA modification? Genomics 104 (2014) 314–323, https://doi.org/10.1016/j.ygeno.2014.08.015.

[31] C.X. Song, C. He, Potential functional roles of DNA demethylation intermediates, Trends Biochem. Sci. 38 (2013) 480–484, https://doi.org/10.1016/j.tibs.2013.07.003.

[32] Y. Chen, N.P. Damayanti, J. Irudayaraj, K. Dunn, F.C. Zhou, Diversity of two forms of DNA methylation in the brain, Front. Genet. 5 (2014), https://doi.org/10.3389/fgene.2014.00046.

[33] T. Khare, S. Pai, K. Koncevicius, M. Pal, E. Kriukiene, Z. Liutkeviciute, M. Irimia, P. Jia, C. Ptak, M. Xia, R. Tice, M. Tochigi, S. Moréra, A. Nazarians, D. Belsham, A.H.C. Wong, B.J. Blencowe, S.C. Wang, P. Kapranov, R. Kustra, V. Labrie, S. Klimasauskas, A. Petronis, 5-hmC in the brain is abundant in synaptic genes and shows differences at the exon-intron boundary, Nat. Struct. Mol. Biol. 19 (2012) 1037–1044, https://doi.org/10.1038/nsmb.2372.

[34] C.G. Spruijt, F. Gnerlich, A.H. Smits, T. Pfaffeneder, P.W.T.C. Jansen, C. Bauer, M. Münzel, M. Wagner, M. Müller, F. Khan, H.C. Eberl, A. Mensinga, A.B. Brinkman, K. Lephikov, U. Müller, J. Walter, R. Boelens, H. Van Ingen, H. Leonhardt, T. Carell, M. Vermeulen, Dynamic readers for 5-(hydroxy)methylcytosine and its oxidized derivatives, Cell 152 (2013) 1146–1159, https://doi.org/10.1016/j.cell.2013.02.004.

[35] H. Stroud, S. Feng, S. Morey Kinney, S. Pradhan, S.E. Jacobsen, 5-Hydroxymethylcytosine is associated with enhancers and gene bodies in human embryonic stem cells, Genome Biol. 12 (2011), https://doi.org/10.1186/gb-2011-12-6-r54.

[36] A.A. Sérandour, S. Avner, F. Oger, M. Bizot, F. Percevault, C. Lucchetti-Miganeh, G. Palierne, C. Gheeraert, F. Barloy-Hubler, C.L. Péron, T. Madigou, E. Durand, P. Froguel, B. Staels, P. Lefebvre, R. Métivier, J. Eeckhoute, G. Salbert, Dynamic hydroxymethylation of deoxyribonucleic acid marks differentiation-associated enhancers, Nucleic Acids Res. 40 (2012) 8255–8265, https://doi.org/10.1093/nar/gks595.

[37] M.J. Armstrong, Y. Jin, E.G. Allen, P. Jin, Diverse and dynamic DNA modifications in brain and diseases, Hum. Mol. Genet. 28 (2019) R241–R253, https://doi.org/10.1093/hmg/ddz179.

[38] Y. Cheng, A. Bernstein, D. Chen, P. Jin, 5-Hydroxymethylcytosine: a new player in brain disorders? Exp. Neurol. 268 (2015) 3–9, https://doi.org/10.1016/j.expneurol.2014.05.008.

[39] Z. Jin, Y. Liu, DNA methylation in human diseases, Genes Dis. 5 (2018) 1–8, https://doi.org/10.1016/j.gendis.2018.01.002.

[40] R. Lardenoije, E. Pishva, K. Lunnon, D.L. van den Hove, Neuroepigenetics of aging and age-related neurodegenerative disorders, in: Progress in Molecular Biology and Translational Science, Elsevier B.V, Netherlands, 2018, pp. 49–82, https://doi.org/10.1016/bs.pmbts.2018.04.008.

[41] A. Portela, M. Esteller, Epigenetic modifications and human disease, Nat. Biotechnol. 28 (2010) 1057–1068, https://doi.org/10.1038/nbt.1685.

[42] N. Ahuja, A.R. Sharma, S.B. Baylin, Epigenetic therapeutics: a new weapon in the war against cancer, Annu. Rev. Med. 67 (2016) 73–89, https://doi.org/10.1146/annurev-med-111314-035900.

[43] K. Kamińska, E. Nalejska, M. Kubiak, J. Wojtysiak, Ł. Żołna, J. Kowalewski, M.A. Lewandowska, Prognostic and predictive epigenetic biomarkers in oncology, Mol. Diagn. Ther. 23 (2019) 83–95, https://doi.org/10.1007/s40291-018-0371-7.

[44] M. Kulis, M. Esteller, DNA methylation and cancer, Spain (2010), https://doi.org/10.1016/B978-0-12-380866-0.60002-2.

[45] W.J. Locke, D. Guanzon, C. Ma, Y.J. Liew, K.R. Duesing, K.Y.C. Fung, J.P. Ross, DNA methylation cancer biomarkers: translation to the clinic, Front. Genet. 10 (2019), https://doi.org/10.3389/fgene.2019.01150.

[46] A. Nebbioso, F.P. Tambaro, C. Dell'Aversana, L. Altucci, Cancer epigenetics: moving forward, PLoS Genet. 14 (2018), https://doi.org/10.1371/journal.pgen.1007362.

[47] A. Roberti, A.F. Valdes, R. Torrecillas, M.F. Fraga, A.F. Fernandez, Epigenetics in cancer therapy and nanomedicine, Clin. Epigenetics 11 (2019), https://doi.org/10.1186/s13148-019-0675-4.

[48] M.L. Thomas, P. Marcato, Epigenetic modifications as biomarkers of tumor development, therapy response, and recurrence across the cancer care continuum, Cancer 10 (2018), https://doi.org/10.3390/cancers10040101.

[49] S. Virani, S. Virani, J.A. Colacino, J.H. Kim, L.S. Rozek, Cancer epigenetics: a brief review, ILAR J. 53 (2012) 359–369, https://doi.org/10.1093/ilar.53.3-4.359.

[50] A.I. Bernstein, Y. Lin, R.C. Street, L. Lin, Q. Dai, L. Yu, H. Bao, M. Gearing, J.J. Lah, P.T. Nelson, C. He, A.I. Levey, J.G. Mullé, R. Duan, P. Jin, 5-Hydroxymethylation-associated epigenetic modifiers of Alzheimer's disease modulate Tau-induced neurotoxicity, Hum. Mol. Genet. 25 (2016) 2437–2450, https://doi.org/10.1093/hmg/ddw109.

[51] J. Kochmanski, S.E. Vanoeveren, J.R. Patterson, A.I. Bernstein, Developmental dieldrin exposure alters DNA methylation at genes related to dopaminergic neuron development and Parkinson's disease in mouse midbrain, Toxicol. Sci. 169 (2019) 593–607, https://doi.org/10.1093/toxsci/kfz069.

[52] J. Kochmanski, N.C. Kuhn, A.I. Bernstein, Parkinson's disease-associated, sex-specific changes in DNA methylation at PARK7 (DJ-1), SLC17A6 (VGLUT2), PTPRN2 (IA-2β), and NR4A2 (NURR1) in cortical neurons, NPJ Parkinsons Dis. 8 (1) (2022) 120, https://doi.org/10.1038/s41531-022-00355-2.

[53] T. Dao, R.Y.S. Cheng, M.P. Revelo, W. Mitzner, W.Y. Tang, Hydroxymethylation as a novel environmental biosensor, Curr. Environ. Health Rep. 1 (2014), https://doi.org/10.1007/s40572-013-0005-5.

[54] O.A. Efimova, A.S. Koltsova, M.I. Krapivin, A.V. Tikhonov, A.A. Pendina, Environmental epigenetics and genome flexibility: focus on 5-hydroxymethylcytosine, Int. J. Mol. Sci. 21 (2020), https://doi.org/10.3390/ijms21093223.

[55] J. Kochmanski, A.I. Bernstein, The impact of environmental factors on 5-hydroxymethylcytosine in the brain, Curr. Environ Health Rep. 7 (2020) 109–120, https://doi.org/10.1007/s40572-020-00268-3.

[56] A. Madrid, P. Chopra, R.S. Alisch, Species-specific 5 mC and 5 hmC genomic landscapes indicate epigenetic contribution to human brain evolution, Front. Mol. Neurosci. 11 (2018), https://doi.org/10.3389/fnmol.2018.00039.

[57] H. Spiers, E. Hannon, L.C. Schalkwyk, N.J. Bray, J. Mill, 5-Hydroxymethylcytosine is highly dynamic across human fetal brain development, BMC Genomics 18 (2017), https://doi.org/10.1186/s12864-017-4091-x.

[58] L. Wen, X. Li, L. Yan, Y. Tan, R. Li, Y. Zhao, Y. Wang, J. Xie, Y. Zhang, C. Song, M. Yu, X. Liu, P. Zhu, X. Li, Y. Hou, H. Guo, X. Wu, C. He, R. Li, F. Tang, J. Qiao, Whole-genome analysis of 5-hydroxymethylcytosine and 5-methylcytosine at base resolution in the human brain, Genome Biol. 15 (2014), https://doi.org/10.1186/gb-2014-15-3-r49.

[59] S. Kriaucionis, A. Bird, DNA methylation and Rett syndrome, Hum. Mol. Genet. 12 (2003) R221–R227, https://doi.org/10.1093/hmg/ddg286.

[60] S. Marques, T.F. Outeiro, Epigenetics in Parkinson's and Alzheimer's diseases, Subcell. Biochem. 61 (2013) 507–525, https://doi.org/10.1007/978-94-007-4525-4_22.

[61] E. Miranda-Morales, K. Meier, A. Sandoval-Carrillo, J. Salas-Pacheco, P. Vázquez-Cárdenas, O. Arias-Carrión, Implications of DNA methylation in Parkinson's disease, Front. Mol. Neurosci. 10 (2017), https://doi.org/10.3389/fnmol.2017.00225.

[62] L.K. Pries, S. Gülöksüz, G. Kenis, DNA methylation in schizophrenia, in: Advances in Experimental Medicine and Biology, Springer, New York LLC, Netherlands, 2017, pp. 211–236, https://doi.org/10.1007/978-3-319-53889-1_12.

[63] S. Rangasamy, S.R. D'Mello, V. Narayanan, Epigenetics, autism spectrum, and neurodevelopmental disorders, Neurotherapeutics 10 (2013) 742–756, https://doi.org/10.1007/s13311-013-0227-0.

[64] E.A. Thomas, DNA methylation in Huntington's disease: implications for transgenerational effects, Neurosci. Lett. 625 (2016) 34–39, https://doi.org/10.1016/j.neulet.2015.10.060.

[65] O.S. Anderson, M.S. Nahar, C. Faulk, T.R. Jones, C. Liao, K. Kannan, C. Weinhouse, L.S. Rozek, D.C. Dolinoy, Epigenetic responses following maternal dietary exposure to physiologically relevant levels of bisphenol A, Environ. Mol. Mutagen. 53 (2012) 334–342, https://doi.org/10.1002/em.21692.

[66] M.J. Essex, W. Thomas Boyce, C. Hertzman, L.L. Lam, J.M. Armstrong, S.M.A. Neumann, M.S. Kobor, Epigenetic vestiges of early developmental adversity: childhood stress exposure and DNA methylation in adolescence, Child Dev. 84 (2013) 58–75, https://doi.org/10.1111/j.1467-8624.2011.01641.x.

[67] J.J. Kochmanski, E.H. Marchlewicz, R.G. Cavalcante, B.P.U. Perera, M.A. Sartor, D.C. Dolinoy, Longitudinal effects of developmental bisphenol a exposure on epigenome-wide DNA hydroxymethylation at imprinted loci in mouse blood, Environ. Health Perspect. 126 (2018), https://doi.org/10.1289/EHP3441.

[68] F.A.D. Leenen, C.P. Muller, J.D. Turner, DNA methylation: conducting the orchestra from exposure to phenotype? Clin. Epigenetics 8 (2016), https://doi.org/10.1186/s13148-016-0256-8.

[69] C.J. Marsit, Influence of environmental exposure on human epigenetic regulation, J. Exp. Biol. 218 (2015) 71–79, https://doi.org/10.1242/jeb.106971.

[70] B.P.U. Perera, C. Faulk, L.K. Svoboda, J.M. Goodrich, D.C. Dolinoy, The role of environmental exposures and the epigenome in health and disease, Environ. Mol. Mutagen. 61 (2020) 176–192, https://doi.org/10.1002/em.22311.

[71] G. Singh, V. Singh, Z.X. Wang, G. Voisin, F. Lefebvre, J.M. Navenot, B. Evans, M. Verma, D.W. Anderson, J.S. Schneider, Effects of developmental lead exposure on the hippocampal methylome: influences of sex and timing and level of exposure, Toxicol. Lett. 290 (2018) 63–72, https://doi.org/10.1016/j.toxlet.2018.03.021.

[72] R.O. Wright, J. Schwartz, R.J. Wright, V. Bollati, L. Tarantini, S.K. Park, H. Hu, D. Sparrow, P. Vokonas, A. Baccarelli, Biomarkers of lead exposure and DNA methylation within retrotransposons, Environ. Health Perspect. 118 (2010) 790–795, https://doi.org/10.1289/ehp.0901429.

[73] M.F. Fraga, M. Esteller, DNA methylation: a profile of methods and applications, BioTechniques 33 (2002) 632–649, https://doi.org/10.2144/02333rv01.

[74] S. Kurdyukov, M. Bullock, DNA methylation analysis: choosing the right method, Biology 5 (2016), https://doi.org/10.3390/biology5010003.

[75] W.S. Yong, F.M. Hsu, P.Y. Chen, Profiling genome-wide DNA methylation, Epigenetics Chromatin 9 (2016), https://doi.org/10.1186/s13072-016-0075-3.

[76] S. Beck, V.K. Rakyan, The methylome: approaches for global DNA methylation profiling, Trends Genet. 24 (2008) 231–237, https://doi.org/10.1016/j.tig.2008.01.006.

[77] S.J. Clark, A. Statham, C. Stirzaker, P.L. Molloy, M. Frommer, DNA methylation: bisulphite modification and analysis, Nat. Protoc. 1 (2006) 2353–2364, https://doi.org/10.1038/nprot.2006.324.

[78] T. Rein, M.L. DePamphilis, H. Zorbas, Identifying 5-methylcytosine and related modifications in DNA genomes, Nucleic Acids Res. 26 (1998) 2255–2264, https://doi.org/10.1093/nar/26.10.2255.

[79] S.J. Cokus, S. Feng, X. Zhang, Z. Chen, B. Merriman, C.D. Haudenschild, S. Pradhan, S.F. Nelson, M. Pellegrini, S.E. Jacobsen, Shotgun bisulphite sequencing of the Arabidopsis genome reveals DNA methylation patterning, Nature 452 (2008) 215–219, https://doi.org/10.1038/nature06745.

[80] R. Lister, M. Pelizzola, R.H. Dowen, R.D. Hawkins, G. Hon, J. Tonti-Filippini, J.R. Nery, L. Lee, Z. Ye, Q.M. Ngo, L. Edsall, J. Antosiewicz-Bourget, R. Stewart, V. Ruotti, A.H. Millar, J.A. Thomson, B. Ren, J.R. Ecker, Human DNA methylomes at base resolution show widespread epigenomic differences, Nature 462 (2009) 315–322, https://doi.org/10.1038/nature08514.

[81] M. Suzuki, W. Liao, F. Wos, A.D. Johnston, J. DeGrazia, J. Ishii, T. Bloom, M.C. Zody, S. Germer, J.M. Greally, Whole-genome bisulfite sequencing with improved accuracy and cost, Genome Res. 28 (2018) 1364–1371, https://doi.org/10.1101/gr.232587.117.

[82] V.K. Rakyan, T.A. Down, D.J. Balding, S. Beck, Epigenome-wide association studies for common human diseases, Nat. Rev. Genet. 12 (2011) 529–541, https://doi.org/10.1038/nrg3000.

[83] A. Akalin, F.E. Garrett-Bakelman, M. Kormaksson, J. Busuttil, L. Zhang, I. Khrebtukova, T.A. Milne, Y. Huang, D. Biswas, J.L. Hess, C.D. Allis, R.G. Roeder, P.J.M. Valk, B. Löwenberg, R. Delwel, H.F. Fernandez, E. Paietta, M.S. Tallman, G.P. Schroth, C.E. Mason, A. Melnick, M.E. Figueroa, Base-pair resolution DNA methylation sequencing reveals profoundly divergent epigenetic landscapes in acute myeloid leukemia, PLoS Genet. 8 (2012), https://doi.org/10.1371/journal.pgen.1002781.

[84] H. Gu, Z.D. Smith, C. Bock, P. Boyle, A. Gnirke, A. Meissner, Preparation of reduced representation bisulfite sequencing libraries for genome-scale DNA methylation profiling, Nat. Protoc. 6 (2011) 468–481, https://doi.org/10.1038/nprot.2010.190.

[85] A. Meissner, A. Gnirke, G.W. Bell, B. Ramsahoye, E.S. Lander, R. Jaenisch, Reduced representation bisulfite sequencing for comparative high-resolution DNA methylation analysis, Nucleic Acids Res. 33 (2005) 5868–5877, https://doi.org/10.1093/nar/gki901.

[86] R. Pidsley, E. Zotenko, T.J. Peters, M.G. Lawrence, G.P. Risbridger, P. Molloy, S. Van Djik, B. Muhlhausler, C. Stirzaker, S.J. Clark, Critical evaluation of the illumina methylationEPIC BeadChip microarray for whole-genome DNA methylation profiling, Genome Biol. 17 (2016), https://doi.org/10.1186/s13059-016-1066-1.

[87] D. Barros-Silva, C.J. Marques, R. Henrique, C. Jerónimo, Profiling DNA methylation based on next-generation sequencing approaches: new insights and clinical applications, Genes 9 (2018), https://doi.org/10.3390/genes9090429.

[88] J. Wang, H. Jiang, G. Ji, F. Gao, M. Wu, J. Sun, H. Luo, J. Wu, R. Wu, X. Zhang, High resolution profiling of human exon methylation by liquid hybridization capture-based bisulfite sequencing, BMC Genomics 12 (2011), https://doi.org/10.1186/1471-2164-12-597.

[89] S.G. Jin, S. Kadam, G.P. Pfeifer, Examination of the specificity of DNA methylation profiling techniques towards 5-methylcytosine and 5-hydroxymethylcytosine, Nucleic Acids Res. 38 (2010) e125, https://doi.org/10.1093/nar/gkq223.

[90] E.K. Schutsky, J.E. Denizio, P. Hu, M.Y. Liu, C.S. Nabel, E.B. Fabyanic, Y. Hwang, F.D. Bushman, H. Wu, R.M. Kohli, Nondestructive, base-resolution sequencing of 5-hydroxymethylcytosine using a DNA deaminase, Nat. Biotechnol. 36 (2018) 1083–1090, https://doi.org/10.1038/nbt.4204.

[91] M.J. Booth, T.W.B. Ost, D. Beraldi, N.M. Bell, M.R. Branco, W. Reik, S. Balasubramanian, Oxidative bisulfite sequencing of 5-methylcytosine and 5-hydroxymethylcytosine, Nat. Protoc. 8 (2013) 1841–1851, https://doi.org/10.1038/nprot.2013.115.

[92] Z. Xu, J.A. Taylor, Y.K. Leung, S.M. Ho, L. Niu, OxBS-MLE: an efficient method to estimate 5-methylcytosine and 5-hydroxymethylcytosine in paired bisulfite and oxidative bisulfite treated DNA, Bioinformatics 32 (2016) 3667–3669, https://doi.org/10.1093/bioinformatics/btw527.

[93] J.R. Hernandez Mora, M. Sanchez-Delgado, P. Petazzi, S. Moran, M. Esteller, I. Iglesias-Platas, D. Monk, Profiling of oxBS-450K 5-hydroxymethylcytosine in human placenta and brain reveals enrichment at imprinted loci, Epigenetics 13 (2018) 182–191, https://doi.org/10.1080/15592294.2017.1344803.

[94] A.R. Smith, R.G. Smith, E. Pishva, E. Hannon, J.A.Y. Roubroeks, J. Burrage, C. Troakes, S. Al-Sarraj, C. Sloan, J. Mill, D.L. Van Den Hove, K. Lunnon, Parallel profiling of DNA methylation and hydroxymethylation highlights neuropathology-associated epigenetic variation in Alzheimer's disease, Clin. Epigenetics 11 (2019), https://doi.org/10.1186/s13148-019-0636-y.

[95] S.K. Stewart, T.J. Morris, P. Guilhamon, H. Bulstrode, M. Bachman, S. Balasubramanian, S. Beck, OxBS-450K: a method for analysing hydroxymethylation using 450K BeadChips, Methods 72 (2015) 9–15, https://doi.org/10.1016/j.ymeth.2014.08.009.

[96] J. Kochmanski, C. Savonen, A.I. Bernstein, A novel application of mixed effects models for reconciling base-pair resolution 5-methylcytosine and 5-hydroxymethylcytosine data in neuroepigenetics, Front. Genet. 10 (2019), https://doi.org/10.3389/fgene.2019.00801.

[97] A. Hoppers, L. Williams, V.K.C. Ponnaluri, B. Sexton, L. Saleh, M. Campbell, K. Marks, M. Samaranayake, L. Ettwiller, S. Guan, H. Church, B. Langhorst, Z. Sun, T.C. Evans, R. Vaisvila, E. Dimalanta, F. Stewart, Enzymatic methyl-seq: next generation methylomes, J. Biomol. Tech. 31 (2020) S15.

[98] R. Vaisvila, V.K. Chaithanya Ponnaluri, Z. Sun, B.W. Langhorst, L. Saleh, S. Guan, N. Dai, M.A. Campbell, B. Sexton, K. Marks, M. Samaranayake, J.C. Samuelson, H.E. Church, E. Tamanaha, I.R. Corrêa, S. Pradhan, E.T. Dimalanta, T.C. Evans, L. Williams, T.B. Davis, EM-seq: detection of DNA methylation at single base resolution from picograms of DNA, bioRxiv (2019), https://doi.org/10.1101/2019.12.20.884692.

[99] Y. Liu, P. Siejka-Zielińska, G. Velikova, Y. Bi, F. Yuan, M. Tomkova, C. Bai, L. Chen, B. Schuster-Böckler, C.X. Song, Bisulfite-free direct detection of 5-methylcytosine and 5-hydroxymethylcytosine at base resolution, Nat. Biotechnol. 37 (2019) 424–429, https://doi.org/10.1038/s41587-019-0041-2.

[100] H. Ji, L.I.R. Ehrlich, J. Seita, P. Murakami, A. Doi, P. Lindau, H. Lee, M.J. Aryee, R.A. Irizarry, K. Kim, D.J. Rossi, M.A. Inlay, T. Serwold, H. Karsunky, L. Ho, G.Q. Daley, I.L. Weissman, A.P. Feinberg, Comprehensive methylome map of lineage commitment from haematopoietic progenitors, Nature 467 (2010) 338–342, https://doi.org/10.1038/nature09367.

[101] L.E. Reinius, N. Acevedo, M. Joerink, G. Pershagen, S.E. Dahlén, D. Greco, C. Söderhäll, A. Scheynius, J. Kere, Differential DNA methylation in purified human blood cells: implications for cell lineage and studies on disease susceptibility, PLoS One 7 (2012), https://doi.org/10.1371/journal.pone.0041361.

[102] B. Zhang, Y. Zhou, N. Lin, R.F. Lowdon, C. Hong, R.P. Nagarajan, J.B. Cheng, D. Li, M. Stevens, H.J. Lee, X. Xing, J. Zhou, V. Sundaram, G. Elliott, J. Gu, T. Shi, P. Gascard, M. Sigaroudinia, T.D. Tlsty, T. Kadlecek, A. Weiss, H. O'Geen, P.J. Farnham, C.L. Maire, K.L. Ligon, P.A.F. Madden, A. Tam, R. Moore, M. Hirst, M.A. Marra, B. Zhang, J.F. Costello, T. Wang, Functional DNA methylation differences between tissues, cell types, and across individuals discovered using the M&M algorithm, Genome Res. 23 (2013) 1522–1540, https://doi.org/10.1101/gr.156539.113.

[103] S.A. Smallwood, H.J. Lee, C. Angermueller, F. Krueger, H. Saadeh, J. Peat, S.R. Andrews, O. Stegle, W. Reik, G. Kelsey, Single-cell genome-wide bisulfite sequencing for assessing epigenetic heterogeneity, Nat. Methods 11 (2014) 817–820, https://doi.org/10.1038/nmeth.3035.

[104] A.J. Titus, R.M. Gallimore, L.A. Salas, B.C. Christensen, Cell-type deconvolution from DNA methylation: a review of recent applications, Hum. Mol. Genet. 26 (2017) R216–R224, https://doi.org/10.1093/hmg/ddx275.

[105] E. Rahmani, R. Schweiger, B. Rhead, L.A. Criswell, L.F. Barcellos, E. Eskin, S. Rosset, S. Sankararaman, E. Halperin, Cell-type-specific resolution epigenetics without the need for cell sorting or single-cell biology, Nat. Commun. 10 (2019), https://doi.org/10.1038/s41467-019-11052-9.

[106] E.A. Houseman, M.L. Kile, D.C. Christiani, T.A. Ince, K.T. Kelsey, C.J. Marsit, Reference-free deconvolution of DNA methylation data and mediation by cell composition effects, BMC Bioinform. 17 (2016), https://doi.org/10.1186/s12859-016-1140-4.

[107] M. Scherer, P.V. Nazarov, R. Toth, S. Sahay, T. Kaoma, V. Maurer, N. Vedeneev, C. Plass, T. Lengauer, J. Walter, P. Lutsik, Reference-free deconvolution, visualization and interpretation of complex DNA methylation data using DecompPipeline, MeDeCom and FactorViz, Nat. Protoc. 15 (2020) 3240–3263, https://doi.org/10.1038/s41596-020-0369-6.

[108] S. Basu, H.M. Campbell, B.N. Dittel, A. Ray, Purification of specific cell population by fluorescence activated cell sorting (FACS), J. Vis. Exp. (2010), https://doi.org/10.3791/1546.

[109] B. Schmitz, A. Radbruch, T. Kümmel, C. Wickenhauser, H. Korb, M.L. Hansmann, J. Thiele, R. Fischer, Magnetic activated cell sorting (MACS)—a new immunomagnetic method for megakaryocytic cell isolation: comparison of different separation techniques, Eur. J. Haematol. 52 (1994) 267–275, https://doi.org/10.1111/j.1600-0609.1994.tb00095.x.

[110] L. Hackler, T. Masuda, V.F. Oliver, S.L. Merbs, D.J. Zack, Use of laser capture microdissection for analysis of retinal mRNA/miRNA expression and DNA methylation, Methods Mol. Biol. 884 (2012) 289–304, https://doi.org/10.1007/978-1-61779-848-1_21.

[111] S. Horvath, K. Raj, DNA methylation-based biomarkers and the epigenetic clock theory of ageing, Nat. Rev. Genet. 19 (2018) 371–384, https://doi.org/10.1038/s41576-018-0004-3.

[112] A.A. Johnson, K. Akman, S.R.G. Calimport, D. Wuttke, A. Stolzing, J.P. De Magalhães, The role of DNA methylation in aging, rejuvenation, and age-related disease, Rejuvenation Res. 15 (2012) 483–494, https://doi.org/10.1089/rej.2012.1324.

[113] N.D. Johnson, L. Huang, R. Li, Y. Li, Y. Yang, H.R. Kim, C. Grant, H. Wu, E.A. Whitsel, D.P. Kiel, A.A. Baccarelli, P. Jin, J.M. Murabito, K.N. Conneely, Age-related DNA hydroxymethylation is enriched for gene expression and immune system processes in human peripheral blood, Epigenetics 15 (2020) 294–306, https://doi.org/10.1080/15592294.2019.1666651.

[114] M.J. Jones, S.J. Goodman, M.S. Kobor, DNA methylation and healthy human aging, Aging Cell 14 (2015) 924–932, https://doi.org/10.1111/acel.12349.

[115] J. Kochmanski, E.H. Marchlewicz, R.G. Cavalcante, M.A. Sartor, D.C. Dolinoy, Age-related epigenome-wide DNA methylation and hydroxymethylation in longitudinal mouse blood, Epigenetics 13 (2018) 779–792, https://doi.org/10.1080/15592294.2018.1507198.

[116] J. Kochmanski, L. Montrose, J.M. Goodrich, D.C. Dolinoy, Environmental deflection: the impact of toxicant exposures on the aging epigenome, Toxicol. Sci. 156 (2017) 325–335, https://doi.org/10.1093/toxsci/kfx005.

[117] E. Hall, P. Volkov, T. Dayeh, J.L.S. Esguerra, S. Salö, L. Eliasson, T. Rönn, K. Bacos, C. Ling, Sex differences in the genome-wide DNA methylation pattern and impact on gene expression, microRNA levels and insulin secretion in human pancreatic islets, Genome Biol. 15 (2014) 522, https://doi.org/10.1186/s13059-014-0522-z.

[118] J. Liu, M. Morgan, K. Hutchison, V.D. Calhoun, A study of the influence of sex on genome wide methylation, PLoS One 5 (2010) e10028, https://doi.org/10.1371/journal.pone.0010028.

[119] H. Xu, F. Wang, Y. Liu, Y. Yu, J. Gelernter, H. Zhang, Sex-biased methylome and transcriptome in human prefrontal cortex, Hum. Mol. Genet. 23 (2014) 1260–1270, https://doi.org/10.1093/hmg/ddt516.

[120] S.W. Curtis, S.A. Gerkowicz, D.O. Cobb, V. Kilaru, M.L. Terrell, M.E. Marder, D.B. Barr, C.J. Marsit, M. Marcus, K.N. Conneely, A.K. Smith, Sex-specific DNA methylation differences in people exposed to polybrominated biphenyl, Epigenomics 12 (2020) 757–770, https://doi.org/10.2217/epi-2019-0179.

[121] R. Massart, Z. Nemoda, M.J. Suderman, S. Sutti, A.M. Ruggiero, A.M. Dettmer, S.J. Suomi, M. Szyf, Early life adversity alters normal sex-dependent developmental dynamics of DNA methylation, Dev. Psychopathol. 28 (2016) 1259–1272.

[122] E.W. Tobi, L.H. Lumey, R.P. Talens, D. Kremer, H. Putter, A.D. Stein, P.E. Slagboom, B.T. Heijmans, DNA methylation differences after exposure to prenatal famine are common and timing- and sex-specific, Hum. Mol. Genet. 18 (2009) 4046–4053, https://doi.org/10.1093/hmg/ddp353.

[123] K. Wang, S. Liu, L.K. Svoboda, C.A. Rygiel, K. Neier, T.R. Jones, J.A. Colacino, D.C. Dolinoy, M.A. Sartor, Tissue- and sex-specific DNA methylation changes in mice perinatally exposed to lead (Pb), Front. Genet. 11 (2020) 840, https://doi.org/10.3389/fgene.2020.00840.

[124] P.W. Laird, Principles and challenges of genome-wide DNA methylation analysis, Nat. Rev. Genet. 11 (2010) 191–203, https://doi.org/10.1038/nrg2732.

[125] I. Rauluseviciute, F. Drabløs, M.B. Rye, DNA methylation data by sequencing: experimental approaches and recommendations for tools and pipelines for data analysis, Clin. Epigenetics 11 (2019), https://doi.org/10.1186/s13148-019-0795-x.

[126] A. Shafi, C. Mitrea, T. Nguyen, S. Draghici, A survey of the approaches for identifying differential methylation using bisulfite sequencing data, Brief. Bioinform. 19 (2018) 737–753, https://doi.org/10.1093/bib/bbx013.

[127] K. Wreczycka, A. Gosdschan, D. Yusuf, B. Grüning, Y. Assenov, A. Akalin, Strategies for analyzing bisulfite sequencing data, J. Biotechnol. 261 (2017) 105–115, https://doi.org/10.1016/j.jbiotec.2017.08.007.

[128] X. Sun, Y. Han, L. Zhou, E. Chen, B. Lu, Y. Liu, X. Pan, A.W. Cowley, M. Liang, Q. Wu, Y. Lu, P. Liu, A comprehensive evaluation of alignment software for reduced representation bisulfite sequencing data, Bioinformatics 34 (2018) 2715–2723, https://doi.org/10.1093/bioinformatics/bty174.

[129] J.P. Fortin, T.J. Triche, K.D. Hansen, Preprocessing, normalization and integration of the Illumina Human MethylationEPIC array with minfi, Bioinformatics 33 (2017) 558–560, https://doi.org/10.1093/bioinformatics/btw691.

[130] G. Mansell, T.J. Gorrie-Stone, Y. Bao, M. Kumari, L.S. Schalkwyk, J. Mill, E. Hannon, Guidance for DNA methylation studies: statistical insights from the Illumina EPIC array, BMC Genomics 20 (2019), https://doi.org/10.1186/s12864-019-5761-7.

[131] K.B. Michels, A.M. Binder, S. Dedeurwaerder, C.B. Epstein, J.M. Greally, I. Gut, E.A. Houseman, B. Izzi, K.T. Kelsey, A. Meissner, A. Milosavljevic, K.D. Siegmund, C. Bock, R.A. Irizarry, Recommendations for the design and analysis of epigenome-wide association studies, Nat. Methods 10 (2013) 949–955, https://doi.org/10.1038/nmeth.2632.

[132] J. Mill, B.T. Heijmans, From promises to practical strategies in epigenetic epidemiology, Nat. Rev. Genet. 14 (2013) 585–594, https://doi.org/10.1038/nrg3405.

[133] C.L. Relton, G. Davey Smith, Epigenetic epidemiology of common complex disease: prospects for prediction, prevention, and treatment, PLoS Med. 7 (2010) e1000356, https://doi.org/10.1371/journal.pmed.1000356.

[134] S.N. Hart, T.M. Therneau, Y. Zhang, G.A. Poland, J.P. Kocher, Calculating sample size estimates for RNA sequencing data, J. Comput. Biol. 20 (2013) 970–978, https://doi.org/10.1089/cmb.2012.0283.

[135] H. Wu, C. Wang, Z. Wu, PROPER: comprehensive power evaluation for differential expression using RNA-seq, Bioinformatics 31 (2015) 233–241, https://doi.org/10.1093/bioinformatics/btu640.

[136] B. Vieth, C. Ziegenhain, S. Parekh, W. Enard, I. Hellmann, powsimR: power analysis for bulk and single cell RNA-seq experiments, Bioinformatics (Oxford, England) 33 (2017) 3486–3488, https://doi.org/10.1093/bioinformatics/btx435.

[137] S. Zhao, C.I. Li, Y. Guo, Q. Sheng, Y. Shyr, RnaSeqSampleSize: real data based sample size estimation for RNA sequencing, BMC Bioinform. 19 (2018), https://doi.org/10.1186/s12859-018-2191-5.

[138] A.J. Lea, T.P. Vilgalys, P.A.P. Durst, J. Tung, Maximizing ecological and evolutionary insight in bisulfite sequencing data sets, Nat. Ecol. Evol. 1 (2017) 1074–1083, https://doi.org/10.1038/s41559-017-0229-0.

[139] S. Graw, R. Henn, J.A. Thompson, D.C. Koestler, PwrEWAS: a user-friendly tool for comprehensive power estimation for epigenome wide association studies (EWAS), BMC Bioinform. 20 (2019), https://doi.org/10.1186/s12859-019-2804-7.

[140] O.S. Anderson, K.E. Sant, D.C. Dolinoy, Nutrition and epigenetics: an interplay of dietary methyl donors, one-carbon metabolism and DNA methylation, J. Nutr. Biochem. 23 (2012) 853–859, https://doi.org/10.1016/j.jnutbio.2012.03.003.

[141] J. Bakusic, W. Schaufeli, S. Claes, L. Godderis, Stress, burnout and depression: a systematic review on DNA methylation mechanisms, J. Psychosom. Res. 92 (2017) 34–44, https://doi.org/10.1016/j.jpsychores.2016.11.005.

[142] E. Grazioli, I. Dimauro, N. Mercatelli, G. Wang, Y. Pitsiladis, L. Di Luigi, D. Caporossi, Physical activity in the prevention of human diseases: role of epigenetic modifications, BMC Genomics 18 (2017), https://doi.org/10.1186/s12864-017-4193-5.

[143] N. Matosin, C. Cruceanu, E.B. Binder, Preclinical and clinical evidence of DNA methylation changes in response to trauma and chronic stress, Chronic Stress 1 (2017), https://doi.org/10.1177/2470547017710764.

[144] S. Voisin, N. Eynon, X. Yan, D.J. Bishop, Exercise training and DNA methylation in humans, Acta Physiol. 213 (2015) 39–59, https://doi.org/10.1111/apha.12414.

[145] S. Greenland, N. Pearce, Statistical foundations for model-based adjustments, Annu. Rev. Public Health 36 (2015) 89–108, https://doi.org/10.1146/annurev-publhealth-031914-122559.

[146] L.H. Chadwick, The NIH roadmap epigenomics program data resource, Epigenomics 4 (2012) 317–324, https://doi.org/10.2217/epi.12.18.

[147] T. Wang, E.C. Pehrsson, D. Purushotham, D. Li, X. Zhuo, B. Zhang, H.A. Lawson, M.A. Province, C. Krapp, Y. Lan, C. Coarfa, T.A. Katz, W.Y. Tang, Z. Wang, S. Biswal, S. Rajagopalan, J.A. Colacino, Z.T.Y. Tsai, M.A. Sartor, K. Neier, D.C. Dolinoy, J. Pinto, R.B. Hamanaka, G.M. Mutlu, H.B. Patisaul, D.L. Aylor, G.E. Crawford, T. Wiltshire, L.H. Chadwick, C.G. Duncan, A.E. Garton, K.A. McAllister, M.S. Bartolomei, C.L. Walker, F.L. Tyson, The NIEHS TaRGET II Consortium and environmental epigenomics, Nat. Biotechnol. 36 (2018) 225–227, https://doi.org/10.1038/nbt.4099.

[148] P. Du, X. Zhang, C.C. Huang, N. Jafari, W.A. Kibbe, L. Hou, S.M. Lin, Comparison of beta-value and M-value methods for quantifying methylation levels by microarray analysis, BMC Bioinform. 11 (2010), https://doi.org/10.1186/1471-2105-11-587.

[149] C. Xie, Y.K. Leung, A. Chen, D.X. Long, C. Hoyo, S.M. Ho, Differential methylation values in differential methylation analysis, Bioinformatics 35 (2019) 1094–1097, https://doi.org/10.1093/bioinformatics/bty778.

[150] E. Dolzhenko, A.D. Smith, Using beta-binomial regression for high-precision differential methylation analysis in multifactor whole-genome bisulfite sequencing experiments, BMC Bioinform. 15 (2014), https://doi.org/10.1186/1471-2105-15-215.

[151] M. Saadati, A. Benner, Statistical challenges of high-dimensional methylation data, Stat. Med. 33 (2014) 5347–5357, https://doi.org/10.1002/sim.6251.

[152] W.J. Seow, A.C. Pesatori, E. Dimont, P.B. Farmer, B. Albetti, A.S. Ettinger, V. Bollati, C. Bolognesi, P. Roggieri, T.I. Panev, T. Georgieva, D.F. Merlo, P.A. Bertazzi, A.A. Baccarelli, Urinary benzene biomarkers and DNA methylation in bulgarian petrochemical workers: study findings and comparison of linear and beta regression models, PLoS One 7 (2012), https://doi.org/10.1371/journal.pone.0050471.

[153] P. Breheny, A. Stromberg, J. Lambert, P-value histograms: inference and diagnostics, High-Throughput 7 (2018), https://doi.org/10.3390/HT7030023.

[154] B.B. Green, E.A. Houseman, K.C. Johnson, D.J. Guerin, D.A. Armstrong, B.C. Christensen, C.J. Marsit, Hydroxymethylation is uniquely distributed within term placenta, and is associated with gene expression, FASEB J. 30 (2016) 2874–2884, https://doi.org/10.1096/fj.201600310R.

[155] K.C. Johnson, E.A. Houseman, J.E. King, K.M. Von Herrmann, C.E. Fadul, B.C. Christensen, 5-Hydroxymethylcytosine localizes to enhancer elements and is associated with survival in glioblastoma patients, Nat. Commun. 7 (2016), https://doi.org/10.1038/ncomms13177.

[156] W.K. Glowacka, H. Jain, M. Okura, A. Maimaitiming, Y. Mamatjan, R. Nejad, H. Farooq, M.D. Taylor, K. Aldape, P. Kongkham, 5-Hydroxymethylcytosine preferentially targets genes upregulated in isocitrate dehydrogenase 1 mutant high-grade glioma, Acta Neuropathol. 135 (2018) 617–634, https://doi.org/10.1007/s00401-018-1821-3.

[157] Z. Xue, C. Xiaoting, M.T. Weirauch, X. Zhang, J.D. Burleson, E.B. Brandt, H. Ji, Diesel exhaust and house dust mite allergen lead to common changes in the airway methylome and hydroxymethylome, Environ. Epigen. (2018), https://doi.org/10.1093/eep/dvy020.

[158] R.A. Rigby, D.M. Stasinopoulos, Generalized additive models for location, scale and shape (with discussion), J. R. Stat. Soc.: Ser. C: Appl. Stat. 54 (2005) 507–554, https://doi.org/10.1111/j.1467-9876.2005.00510.x.

[159] M.E.D. Foster, M.A. Deardorff, Open science framework (OSF), J. Med. Libr. Assoc. (2017), https://doi.org/10.5195/jmla.2017.88.

[160] J.J. Heindel, L.N. Vandenberg, Developmental origins of health and disease: a paradigm for understanding disease cause and prevention, Curr. Opin. Pediatr. 27 (2015) 248–253, https://doi.org/10.1097/MOP.0000000000000191.

[161] B. Janis, J. Chandni, B. Mary, F. Caroline, H. Mark, H. Nicholas, I. Hazel, K. Kalyanaraman, C. Cyrus, Developmental origins of health and disease: a lifecourse approach to the prevention of non-communicable diseases, Healthcare 14 (2017), https://doi.org/10.3390/healthcare5010014.

[162] J. Bohacek, I.M. Mansuy, A guide to designing germline-dependent epigenetic inheritance experiments in mammals, Nat. Methods 14 (2017) 243–249, https://doi.org/10.1038/nmeth.4181.

[163] S.E. Lazic, L. Essioux, Improving basic and translational science by accounting for litter-to-litter variation in animal models, BMC Neurosci. 14 (2013), https://doi.org/10.1186/1471-2202-14-37.

[164] I. Plewis, Pesticides and transgenerational inheritance of pathologies: designing, analysing and reporting rodent studies, PLoS One 15 (2020), https://doi.org/10.1371/journal.pone.0228762.

[165] N.P. du Sert, V. Hurst, A. Ahluwalia, S. Alam, M.T. Avey, M. Baker, W.J. Browne, A. Clark, I.C. Cuthill, U. Dirnagl, M. Emerson, P. Garner, S.T. Holgate, D.W. Howells, N.A. Karp, S.E. Lazic, K. Lidster, C.J. MacCallum, M. Macleod, E.J. Pearl, O.H. Petersen, F. Rawle, P. Reynolds, K. Rooney, E.S. Sena, S.D. Silberberg, T. Steckler, H. Würbel, The arrive guidelines 2.0: updated guidelines for reporting animal research, PLoS Biol. 18 (2020), https://doi.org/10.1371/journal.pbio.3000410.

[166] I.M. Fingerman, L. McDaniel, X. Zhang, W. Ratzat, T. Hassan, Z. Jiang, R.F. Cohen, G.D. Schuler, N.C.B.I. Epigenomics, A new public resource for exploring epigenomic data sets, Nucleic Acids Res. 39 (2011) D908–D912, https://doi.org/10.1093/nar/gkq1146.

[167] A. Ammar, S. Bonaretti, L. Winckers, J. Quik, M. Bakker, D. Maier, I. Lynch, J. van Rijn, E. Willighagen, A semi-automated workflow for fair maturity indicators in the life sciences, Nanomaterials 10 (2020) 1–14, https://doi.org/10.3390/nano10102068.

[168] M. Corpas, N.V. Kovalevskaya, A. McMurray, F.G.G. Nielsen, A FAIR guide for data providers to maximise sharing of human genomic data, PLoS Comput. Biol. 14 (2018), https://doi.org/10.1371/journal.pcbi.1005873.

CHAPTER 13

Best practices for the ATAC-seq assay and its data analysis

Haibo Liu[a], Rui Li[a], Kai Hu[a], Jianhong Ou[b], Magnolia Pak[a], Michael R. Green[a], and Lihua Julie Zhu[a,c]

[a]Department of Molecular, Cell and Cancer Biology, University of Massachusetts Chan Medical School, Worcester, MA, United States, [b]Department of Cell Biology, Duke University School of Medicine, Duke Regeneration Center, Duke University, Durham, NC, United States, [c]Department of Molecular Medicine, Program in Bioinformatics and Integrative Biology, University of Massachusetts Chan Medical School, Worcester, MA, United States

An overview of ATAC-seq

With nucleosomes as its basic building blocks, eukaryotic nuclear DNA is hierarchically packaged into chromatin, and different regions across the genome are condensed at different levels [1,2]. Based on its compactness, chromatin is categorized into two major functional states: lightly packed, transcriptionally active euchromatin, and highly condensed, transcriptionally inert heterochromatin (Fig. 13.1) [3,4]. A subset of euchromatin regions, devoid of histones and other chromatin-associated factors and thus accessible by transcriptional machinery, are called accessible chromatin regions. While accessible chromatin DNA only encompasses a small fraction of the genome in any given cell type, it contains the majority of gene regulatory elements, such as promoters, enhancers, insulators, silencers, and locus control regions, which collectively regulate gene expression to maintain cellular identity and functions [5,6]. For instance, across Tiers 1 and 2 cell types that the encyclopedia of DNA elements (ENCODE) consortium surveyed, accessible DNA comprises an average of 1% of each cell type's genome sequence and in total only 3.9% of the human genome. Meanwhile, it represents above 90% of binding regions of transcription factors (TFs) [5,7]. The landscape of chromatin accessibility not only varies among different cell types but also dynamically changes in response to internal and external stimuli, as well as cell cycle progression [5,8–11]. Importantly, abnormal alterations in chromatin accessibility have been implicated in a wide range of diseases [12–16]. Therefore, genome-wide chromatin accessibility profiling can offer important biological insights into the regulatory landscape of the genome during normal cell differentiation, development, and disease development.

Several strategies have been developed to profile genome-wide chromatin accessibility, including FAIRE-seq [17], DNase-seq [18], MNase-seq [19], NOMe-seq [20], and assay of transposase accessible chromatin sequencing (ATAC-seq) [21]. These methods have been comprehensively reviewed elsewhere [22,23]. In general, they use either enzymatic or physical approaches to obtain accessible chromatin and then leverage next-generation sequencing (NGS) for quantification. Among these methods, ATAC-seq has gained popularity due to several advantages [21]. First, the ATAC-seq experimental protocol is relatively simple (consisting of three major steps) and fast (taking from 2 to 3 hours). Second, it is highly sensitive and can be carried out using substantially less starting material than other methods.

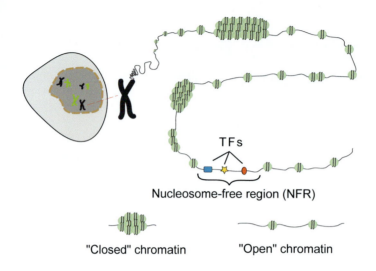

FIG. 13.1

Chromatin accessibility. In eukaryotic cells, DNA of ~146 bp wraps around a histone octamer to form a nucleosome core particle (NCP). Neighboring NCPs are connected by linker DNA of ~20–90 bp to form arrays of nucleosomes, which could be further coiled and stacked together into tightly packed chromatin. Some chromatin regions are loosely packed or even not occupied by histone octamers at all, and thus they are accessible to the transcriptional machinery. These regions are called "open" chromatin. Most transcription factors (TF) predominantly bind to the DNA in nucleosome-free regions. However, a majority of the genome is tightly packed in a less or even not accessible form called "closed" chromatin.

A typical ATAC-seq workflow is illustrated in Fig. 13.2. Briefly, intact nuclei are first isolated from cells or tissue samples. Then, accessible chromatin is preferentially fragmented and simultaneously tagged through a process known as "tagmentation" using hyperactive Tn5 transposase (Tnp) preloaded in vitro with sequencing adapters. A subsequent PCR step enables limited fragment amplification and simultaneous incorporation of adapters for sequencing and sample indexing. The resulting product is then purified and sequenced on NGS platforms. ATAC-seq has been successfully applied to different cell/tissue types with tailored nuclei preparation protocols for various species [11,24–27]. Notably, compared to other ATAC-seq protocols, Omni-ATAC [27] offers improved cell permeabilization while dramatically depleting mitochondria DNA contamination and boosting the signal-to-noise ratio. Thus, it significantly increases ATAC-seq efficacy and broadens its scope to include nuclei harvested from difficult-to-prepare sources such as snap-frozen tissues and refractory cell types.

Since its inception in 2013, ATAC-seq has been adopted by numerous individual labs [28,29] and large consortia, such as the ENCODE [30], the Functional Annotation of Animal Genomes (FAANG) [31,32], and the Cancer Genome Atlas (TCGA) [33], to generate atlases of chromatin accessibility in arrays of cell types and tissues of different species, including tumor samples. ATAC-seq has also been used to study differential chromatin accessibility during normal cell differentiation, external stimuli, and cancer evolution to reveal underlying molecular mechanisms [8–11]. In addition to the accessibility landscape, ATAC-seq data also provide positioning information of nucleosomes flanking nucleosome-free regions (NFRs) and reveal footprints for actively bound TFs [21], which allow us to reconstruct

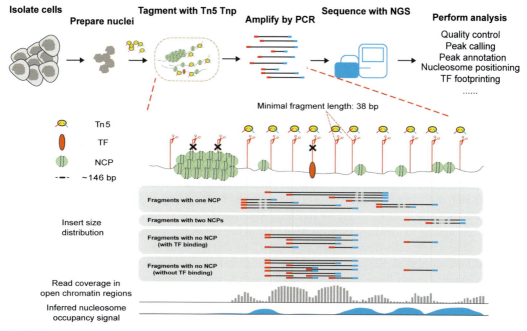

FIG. 13.2

ATAC-seq and its applications. The overall workflow of ATAC-seq is illustrated on the top panel. The enlarged bottom panel shows the details of the "tagmentation" process, the resulting characteristic fragments, the read coverage, and the nucleosome occupancy.

gene regulatory networks (GRNs). Furthermore, the recently developed method of single-cell ATAC-seq (scATAC-seq) allows one to study the cellular heterogeneity in chromatin accessibility [34–37].

Notably, although ATAC-seq experimental protocols are well established, the lack of standardized bioinformatics pipelines often confounds ATAC-seq analysis. This chapter aims to describe the current best practices for designing and conducting ATAC-seq experiments and recommend bioinformatics software for comprehensive data analysis.

Generating high-quality ATAC-seq data
Experimental design

Though the ATAC-seq wet-lab protocol is relatively straightforward to implement [21], to ensure a successful ATAC-seq experiment, valid experimental design and step-by-step quality control (QC) are essential. Biological replicates allow one to identify robust accessible chromatin regions and control for sample-specific biases. Although the majority of published ATAC-seq experiments used two biological replicates per biological condition, more biological replicates help to identify chromatin regions of low accessibility and moderate differences in chromatin accessibility between biological conditions. To be cost-effective, control experiments for ATAC-seq assays are usually not performed. Nevertheless, one can use deproteinized genomic DNA fragmented by Tn5 tagmentation as an optional control to adjust for potential artifacts such as sequencing or mapping biases. Along with replication, guidelines for the design of RNA-seq experiments such as randomization and blocking are applicable to ATAC-seq experimental design to enable one to remove batch effects and other confounding factors [38].

Nuclei preparation and quality control

Different ATAC-seq protocols vary mainly in cell breakage, nuclei preparation, and the setup of the transposition reaction [11,27,34,39–41]. Methods for preparing single-cell suspension and nuclei should be optimized based on species, tissue types, and cell contexts. When working with cultured animal cells as the initial material, preparing single-cell suspension from nonadherent cells is straightforward, but it becomes essential to employee protease treatment when dealing with adherent cells. Fresh animal tissue samples require cold-active proteases to disaggregate the tissue pieces into single-cell suspension [27,42]. In all cases, we recommend removing dead cells and cell-free, genomic DNA through flow sorting or density centrifugation [27]. After obtaining a homogeneous cell suspension, one can utilize single cells directly for Omni-ATAC, or disrupt them to isolate high-purity nuclei, as described by Corces et al. [27] Additionally, it is feasible to isolate single nuclei directly from tissue pieces or 50-μm sections through physical means, such as grinding tissue samples into fine powder in liquid N_2 with a mortar and pestle [43], disrupting tissues using a dounce homogenizer [27], or finely slicing tissues with a sharp razor blade [44]. Subsequently, nuclei are purified by iodixanol density gradient centrifugation [27].

For unicellular or filamentous fungi, the initial step involves preparing protoplasts through enzymatic degradation of cell walls [45–50]. Subsequently, protoplasts can be used for nuclei isolation like mammalian cells, or they can be directly used for Omni-ATAC. For nonfilamentous, multicellular fungi, cells can be broken up by physical means to prepare nuclei [51,52].

Applying ATAC-seq to plant tissues is more challenging due to the rigid plant cell walls and abundant mitochondrial and chloroplast DNA. To isolate plant nuclei, plant cell walls have to be first broken up mechanically or enzymatically, depending on the composition of the walls [44,53]. Subsequently, nuclei can be purified using methods such as iodixanol density gradient centrifugation [54], the INTACT method (Isolation of Nuclei Tagged in Specific Cell Types) that isolates nuclei from individual cell types with biotinylated nuclear envelope [55], or fluorescent activated nuclei sorting (FANS) [40]. The latter two methods produce higher-quality ATAC-seq data because they significantly reduce organelle DNA contamination and improve the signal-to-noise ratio. However, the INTACT method depends on special transgenes and is thus not broadly applicable. Theoretically, it might be feasible to apply Omni-ATAC to protoplasts prepared from young plant tissues such as leaves and seedlings.

After nuclei are extracted, it is recommended to stain a small aliquot with DAPI [44,56] or Hoechst [24] and assess the nuclei's quality through morphological examination, employing a phase-contrast microscope and a fluorescent microscope. It is expected to observe well-dispersed, intact nuclei exhibiting a round or oval shape [24]. For troubleshooting, it can be beneficial to check the copy number of the carried-over organelle DNA relative to that of the nuclear genomic DNA using qPCR, with the total DNA from the unprocessed tissue/cells as the control.

In summary, it is critical to prepare high-purity, intact nuclei containing chromatin as native as possible to preserve the structure of chromatin-associated proteins for successful and effective ATAC-seq assays. Empirically, fresh samples are the best for ATAC-seq assays, followed by cryopreserved samples, snap-frozen samples [27], and then briefly cross-linked samples [26,34].

Tagmentation, PCR amplification, and quality control

The number of nuclei used for an ATAC-seq assay depends on the sample availability and the amount of open chromatin per nucleus, which is a function of the genome size, tissue, and cell type [34,40,57]. For mammals, the number of nuclei ranges from ~500 to 50,000, with more cells producing libraries with higher complexity [21,58]. Given the variations in the amount of open chromatin and organelle DNA contamination, it is usually necessary to determine the optimal ratio of Tn5 Tnp to nuclei by performing a serial of tagmentation assays. These assays should, for a given duration, use either a fixed number of nuclei alongside a serial dilution of Tn5 Tnp, or a fixed dosage of Tn5 Tnp alongside a serial dilution of nuclei. For a given number of cells, over-transposition caused by excessive Tnp increases background noise, whereas under-transposition due to insufficient Tnp leads to a low resolution [34].

The transposition product is initially extended and preamplified for three to five cycles of PCR. This number of cycles allows the efficient addition of Illumina sequencing adapters and sample indices [34]. To limit the size and GC biases introduced during PCR amplification, the number of additional PCR cycles is determined by qPCR with 10% of the preamplified product as templates [34]. The C_t value, where the fluorescence intensity reaches one-third of the maximum, equals the required number of additional PCR cycles for further amplification of the tagmentation product [34]. Usually, it is amplified by at least three, but no more than 11, cycles in total [34]. If multiple replicates from one or more conditions are to be sequenced, primers with different indices should be used during PCR amplification. This would allow multiplexing for sequencing, thus controlling potential lane effect and/or reducing sequencing cost. It is worth mentioning that through tagmentation and PCR, adapter sequences of 128 bp are added to the ends of genomic DNA fragments for single-indexed ATAC-seq libraries [34].

One of the important QC steps is to determine the size distribution of PCR products using a microfluidics-based electrophoresis system, such as the Agilent TapeStation or Fragment Analyzer system (Agilent Technologies, Inc., Santa Clara, CA). For an optimal ATAC-seq library, the gel electrophoresis-like image should exhibit a distinctive DNA laddering pattern consisting of well-defined bands. These bands should progressively decrease in intensities from small fragments (~200 ± 30 bp) to large fragments (~900 ± 30 bp). These fragments result from NFRs and regions occupied by one to four nucleosomes [21,34]. At a minimum, the electropherogram of an acceptable ATAC-seq library should display bands for fragments from NFRs and mononucleosome-occupied regions (~200 bp).

If the electropherogram reveals significant fractions of primer dimers (<110 bp) or fragments exceeding 1000 bp, it is essential to remove them from the amplified library. This can be achieved through purification methods such as PAGE gels, AMpure magnetic beads or their equivalents [59]. Another effective option is the Sage BluePippin size selection system (Sage Science, Inc. Beverly, MA, USA). This process is crucial for optimizing the sequencing efficiency. Importantly when opting for magnetic bead-based size selection, it is critical to use an optimal bead-to-DNA solution volume ratio to select fragments in the 150 to 1000 bp range without apparent bias [34]. The Sage BluePippin system is recommended for automatic, precise size selection [60]. To quantitate the library's concentration following size selection, qPCR is preferred to absorbance-based measurements, such as Qubit (Thermo Fisher Scientific, Waltham, MA, USA) fluorometric quantification. This preference arises from the relatively wide range of the library's fragment sizes (from 150 to 1000 bp) [34].

To further validate the library's quality, qPCR can be used to assess fragment enrichment in a few known positive regions and negative regions. This assessment is performed using the total DNA extracted from the unprocessed tissue or nuclei as a reference for normalization [61]. For example, proximal promoter regions—centered on transcription start sites (TSSs) of highly expressed house-keeping genes such as GAPDH and ACTB—can be used as positive controls, while a few genomic regions in gene deserts can be used as negative controls. In addition, a few organelle DNA regions can be checked for organelle DNA contamination. A normalized relative enrichment fold for each positive control region, calculated using the method described by Lacazette [62], should be no smaller than ten to ensure a high signal-to-noise ratio. Meanwhile, the normalized enrichment fold for mitochondrial regions may vary depending on the source of biomaterials and ATAC-seq protocols. For mammals with the genome size similar to humans, this fold should typically not exceed sixty-four if the original ATAC-seq protocol [21] is employed. This value is roughly equivalent to approximately 50% of the reads being mitochondrial reads, taking into account that 40% of the reads are attributed to genomic background and assuming each cell contains 5000 copies of mitochondrial DNA molecules [63]. Notably, ATAC Primer Tool is recommended for ATAC-qPCR primer design [64].

Sequencing

Given that a significant fraction of fragments in ATAC-seq libraries contain DNA inserts from 40 to 100 bp, 2×40-bp paired-end sequencing is sufficient and cost-effective [21]. Longer reads (up to 2×100-bp) may slightly improve the unique mapping rate, but many reads would contain adapter sequences due to the insert sizes of many fragments being less than the targeted read length. Besides improving unique mapping rate, paired-end reads also allow for more accurate removal of PCR duplicates and better characterization of the accessible chromatin regions, and enable the inference of the fragment length and the grouping of reads into bins of nucleosome-free, and mono-, di-, and tri-nucleosomal regions [21]. It is difficult to predetermine the desired sequencing depth due to the varying efficacy of different ATAC-seq protocols, a variable percentage of organelle DNA contamination [65], various levels of chromatin accessibility in different libraries [66], diverse genome sizes, and different types of downstream analyses of interest. The ENCODE consortium recommends 25 million (M) nonduplicate, nonmitochondria, uniquely aligned fragments for (differential) open chromatin analysis in humans and mice [34,67], which is consistent with a study by Karabacak Calviello et al [41]. Sequencing saturation analyses should be performed retrospectively; further sequencing should be performed if more open chromatin regions are expected and the complexity of the libraries are sufficiently high. To profile nucleosome positioning at a single-base-pair resolution, higher read coverage is typically required [46], though no general recommendation on sequencing depth exists. In a study by Schep et al., 20 M reads were required to reach saturation in *Saccharomyces cerevisiae*, but the number of detected nucleosomes kept increasing even at 80 M reads in the human sample [46]. Similarly, there is no general guideline on sequencing depth needed for effective footprinting analysis, as the desired coverage depends on multiple factors, including TF properties and the factors mentioned earlier [41,67–69]. DNase-seq experiments indicate that the number of detected footprints is highly correlated with sequencing depth and saturates only at very high sequencing depth (~400 M for human samples) [70,71]. Buenrostro et al. successfully identified CTCF footprints using the standard ATAC-seq protocol with 100 M paired, nonduplicate, nonmitochondria, uniquely aligned fragments sequenced for human samples [34]. On the other hand, Li et al. suggested Omni-ATAC data with moderate (~50 M) or even shallow (~25 M) sequencing depth is useful for footprinting analysis due to the improved performance of the protocol [67].

Analyzing ATAC-seq data: From stringent data quality control to comprehensive data mining

Raw read quality control and preprocessing

The quality of raw reads is usually checked using software, such as FastQC [72], before any formal analyses. In an ATAC-seq experiment with multiple samples, it is helpful to use MultiQC [73] to generate a single interactive QC report from the multiple FastQC outputs. Important QC metrics include per-base sequence quality, per-sequence quality score, per-sequence GC content, and adapter content. Typically, for reads generated on an Illumina HiSeq series, the base quality is slightly low at the beginning, increases, and levels off in the middle, and may decrease as the bases approach the ends if the read is longer than 50 bp. If the per-base sequence quality of the reads is acceptable (i.e., the median base quality score is above 20), quality-based trimming is optional because DNA aligners such as Bowtie [74], Bowtie2 [75], and BWA [76] can perform soft clipping of the unalignable read ends. If necessary, 3′-end bases of very low quality (quality score < 10) and sequencing adapters can be trimmed with tools such as Cutadapt [77], Trimmomatic [78], and fastp [79], which support the trimming of paired-end reads. However, trimming at the 5′-end is not recommended even if the quality of 5′ bases is poor because the 5′-ends of reads represent the exact transposition sites, which are needed for TF footprinting analysis. Additionally, the library GC-content distributions of different biological replicates should be very similar.

Alignment, postalignment processing, and quality control

Next, reads are aligned to a reference genome using BWA-mem [76], Bowtie [74] (for ≤50-bp reads), or Bowtie2 [75] (for >50-bp reads). The alignment files can be processed by using the SAMtools [80] and/or Picard Tools to generate sorted, duplicate-marked BAM [80] files and basic mapping statistics, such as unique mapping rate, duplication rate, and mitochondrial read rate. BAM files should then be filtered with the SAMtools [80] or custom scripts [65] to remove duplicates, organelle reads, and reads that are unmapped, multimapped, mapped with low-quality scores, or improperly paired. Improperly paired reads refer to read pairs that are discordantly mapped or have inferred fragment sizes shorter than 38 or longer than 2000 bp [81]. Low mapping quality scores are those that lie at the lower tail of the empirical score distribution. Additionally, alignments that appear in the genome-specific blacklist should be removed [69]. Blacklists can be downloaded for model organisms [82] or created using the tool BlackList [83] or PeakPass [84] for nonmodel organisms. The tool BlackList requires a large set of independent, whole-genome sequencing data and genome mappability information. For organisms with very limited whole-genome sequencing data, a greylist can be created using the GreyListChIP package [85]. It is important to emphasize that the filtered BAM file should be used to determine the unbiased insert size distribution for an ATAC-seq library.

Tn5 Tnp cleaves the double-stranded DNA at the transposition site and creates a 9-bp overhang at the 5′-end [86], to which the transposition adaptors are appended. Thus, for peak calling and TF footprinting analysis, the coordinates of reads mapping to the positive and negative strands are shifted by +4 and −5, respectively, so that the read start sites correspond to the middle of transposition events [21,86]. In the process of identifying open chromatin regions and inferring nucleosome positioning, the shifted reads are segregated into different bins based on their fragment sizes, namely bins for reads from putative NFRs and those located within regions occupied by one to three nucleosomes [21].

TSSs of active or poised promoters are typically found within NFRs, which are surrounded by well-positioned nucleosomes [23,87]. Therefore, high-quality ATAC-seq data should show enriched signals around TSSs. TSS enrichment scores and density plots displaying the distribution of fragments from NFRs and of fragments bound by mononucleosomes around TSSs (TSS $\pm$ 1 kb) can be used as an important QC metric. In addition to open chromatin regions found within active and poised proximal promoter regions, the majority of accessible regions are distal, nonpromoter *cis*-regulatory elements, such as enhancers and insulators [5,39]. Therefore, it is also important to use a genome browser like the integrative genomics viewer (IGV) [88] to assess data quality thoroughly. To achieve this, one should examine the fragment distribution across tens of large genomic regions containing both transcriptionally active and inactive genes, tailored to the specific tissues or cells being investigated. Visualizing signal and noise distribution along these large genomic regions provides a visual gauge of data quality, complementing the TSS enrichment analysis.

Genomic regions tightly bound by nonhistone proteins, such as transcription factors, are locally protected from Tn5 Tnp tagmentation. This protection results in distinct narrow regions characterized by decreased cutting frequency within open chromatin regions of high cutting frequency, commonly referred to as "footprints" [21,89]. With high-quality ATAC-seq data, one should be able to detect footprints of high-affinity DNA binding factors [21,90]. Thus, the aggregated footprints for CTCF or other stably bound TFs also serve as a valuable metric indicating the quality of ATAC-seq data [65].

Oftentimes, it is necessary to determine whether the sequencing is saturated and whether a library is complex enough for further sequencing [91,92]. Sequencing saturation analyses can be performed using a peak-based method with peaks generated from the filtered BAM files (see "Peak calling" section) [58,92], while the library complexity should be estimated with the unfiltered BAM files [65].

If an experiment includes multiple ATAC-seq assays, it is expected that if there are any condition-specific effects, libraries from biological replicates should be more similar to one another than those from different conditions. Sample similarity can be determined by hierarchical clustering analysis (HCA) [93] or principal component analysis (PCA) [94]. For both HCA and PCA, a count matrix including all the samples of an experiment is constructed using a sliding window-based or consensus peak-based approach, followed by normalization (see "Peak calling" and "Differential peak analysis" sections) and a regularized log (rlog) transformation or variance stabilizing transformation (vst) [93]. Euclidean distances among samples calculated with the normalized, transformed count matrix are used for the HCA, while the normalized, transformed count matrix is directly used for the PCA. Cautiously, a correlation coefficient alone is not sufficient measure of reproducibility [95].

The ATACseqQC package [65] provides convenient functions for all postalignment quality assessments and preprocessing such as alignment shifting, read binning based on inferred fragment size and genomic origin, ATAC-seq-specific QC metrics and visualization (insert size distribution, TSS enrichment, signal and noise distribution along broad genomic regions, and aggregated TF footprint visualization), and sequencing saturation and library complexity analyses [65]. Fig. 13.3 shows the postalignment QC plots, the distributions of insert size, and TSS enrichment, for an ATAC-seq dataset from the lymphoblastoid cell line GM12878 (SRR891269) [21] (Fig. 13.3A and B) and an ENCODE mouse liver ATAC-seq dataset (ENCSR609OHJ) [96] (Fig. 13.3C and D). Based on the insert size distribution plots, the library quality of the former was higher than the latter. Readers can refer to Ou et al. [65] for additional QC plots of libraries with various qualities. Besides the ATACseqQC package [65], the ChIPQC package [97] can also provide additional important functionality to quantify ATAC-seq data quality, such

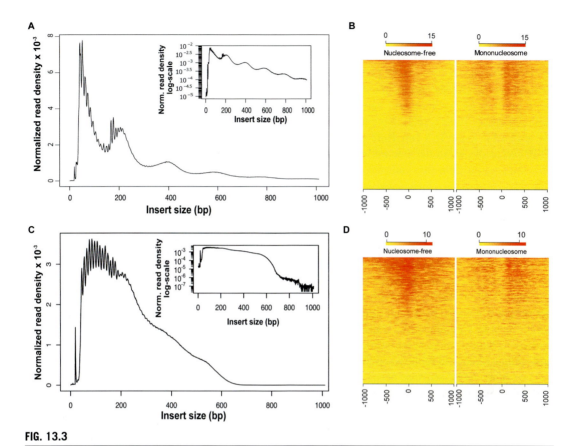

FIG. 13.3

Postalignment ATAC-seq QC. The ATAC-seq data from the lymphoblastoid cell line GM12878 (SRR891269) and a mouse liver sample (ENCSR609OHJ, two replicates) were mapped to the human and mouse reference genome using BWA-mem, respectively. The BAM files were fully filtered as recommended in this chapter. The package ATACseqQC was used for postalignment QC. (A and C) Density plots showing the insert size distributions for datasets SRR891269 and ENCSR609OHJ, respectively. (B and D) Heatmaps displaying the aggregate, nucleosome-free, mononucleosome signal enrichment around TSSs for datasets SRR891269 and ENCSR609OHJ, respectively.

as calculating the fraction of reads in peaks (FRiP) [98]. A recent ATAC-seq QC tool, ataqv, can provide dynamic, comparative visualization of some ATACseqQC metrics plus some other functionalities [99].

Peak calling

Peak calling represents a fundamental step in the analysis of ATAC-seq data, essential for detecting accessible chromatin regions. Tools used for ATAC-seq peak calling can be roughly classified into general-purpose and ATAC-seq-specific peak callers. The general-purpose peak callers were historically developed for calling peaks with ChIP-seq data. Briefly, signal pileups of ChIP samples in sliding windows are compared against matching windows in input samples and/or local background regions, and adjacent windows with statistically higher signals in the ChIP sample are merged and called peaks. For most published ATAC-seq data, peaks have been called using general-purpose peak callers [69,100], especially MACS2 [101], which has been adopted by the ENCODE ATAC-seq pipeline [67]. Many bioinformatics pipelines/guidelines use MACS2 with the BAMPE mode, where paired-end reads are transformed into coverage signals of full-length fragments from the 5' end of Read 1 (R1) to the 5' end of Read 2 (R2). Note that this assumes that all regions covered by the whole fragments are accessible to the Tn5 Tnp. Here we call it the "MACS2.PE" method. This setting is appropriate for ChIP-seq analysis, but the assumption does not hold for ATAC-seq data, because, for ATAC-seq data, only the transposition sites indicate accessibility. To consider that fragment ends represent transposition sites in ATAC-seq data, some improved pipelines recommend using MACS2 for ATAC-seq peak calling with parameters --shift -s and --extend $2s$ to center the pileup signals on the actual transposase cleavage sites, where s is the number of bases by which the coordinates of reads are shifted toward the 5' direction [16]. We refer this approach as the "MACS2.SITE" method.

Recently, two ATAC-seq-specific peak callers have been developed, one of which is Genrich [102]. Genrich enables users to define the 5' extremities of both R1 and R2 of paired-end reads to be the centered Tn5 insert sites, to create insert site signal tracks in the bedGraph format, and to call open chromatin regions based on the actual transposition sites. To smooth signals and improve peak calling performance, Genrich by default extends the insert site signal by 50bp on both sides, with the extension length tunable. Genrich works similarly for single-end (SE) ATAC-seq data. In addition, Genrich has options to remove mitochondrial reads, eliminate PCR duplicates, make use of multimapping reads, and include biological replicates in its model.

The other ATAC-seq-specific peak caller HMMRATAC uses a machine-learning approach [103]. HMMRATAC first decomposes ATAC-seq reads into four signal tracks: signal tracks for NFRs and mono-, di-, and tri-nucleosome according to their fragment sizes. Based on the combinatory patterns of these four signal tracks, HMMRATAC builds a Hidden Markov Model (HMM), which is used to classify genomic regions into the background (low in all signals, State $E0$), nucleosome (low in NFRs, high in nucleosome signals, State $E1$), and center (high in all signals, State $E2$) regions. Then, HMMRATAC combines nucleosome and center regions to output open chromatin regions in the gappedPeak format. This "decomposition and integration" idea could potentially increase peak calling performance if the training dataset is large and diverse enough. In the current implementation, the training dataset used by HMMRATAC consists of 1000 "good" peaks selected internally such that their fold changes over the background fall within a certain range. As a result, HMMRATAC tends to favor high/medium-signal peaks and miss low-coverage ones.

We compared the performance of the four peak callers with the ATAC-seq dataset from the lymphoblastoid cell line GM12878 (SRR891268–SRR891271) [21], using the default settings of each peak caller. Descriptive summary statistics of peaks detected by each caller are shown in Fig. 13.4A–C. For this dataset with four replicates, Genrich detected the largest number of peaks, while HMMRATAC detected the least number of peaks (Fig. 13.4A). Peaks detected by HMMRATAC tend to have a wider range of width than those detected by other peak callers (Fig. 13.4B). Consensus peaks are of mid/high ATAC-seq signals, while peak caller-specific peaks are of lower signals (Fig. 13.4C). Peaks called by Genrich, HMMRATAC, and MACS.SITE have well-defined boundaries, with lower signals clearly observed right before peak start sites (PSSs) and after peak end sites (PESs) (Fig. 13.4D). Notably, peaks uniquely detected by MACS2.PE showed a weird enrichment pattern with two small peaks at the edges separated by a shallow valley (Fig. 13.4E). This is most likely due to the improper signal representation by MACS2.PE.

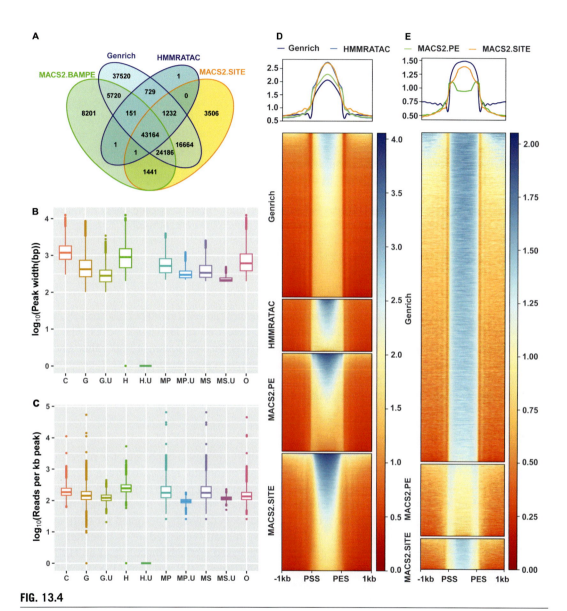

FIG. 13.4

Evaluation of ATAC-seq peak callers. The ATAC-seq dataset with four biological replicates from the lymphoblastoid cell line GM12878 (SRR891268–SRR891271) was used to evaluate the four peak callers, MACS2.PE, MACS2.SITE, Genrich, and HMMRATAC. The paired-end reads in the FASTQ format were mapped to the human reference genome with BWA-mem, and sorted with SAMtools. Duplicated reads were removed using Picard Tools. Then, aligned reads in the BAM format were further filtered to remove (1) alignments with estimated fragment length not in the range from 38 to 2000 bp, (2) alignment with mapping quality lower than 20, (3) alignments hitting on the mitochondrial genome (chrM), and (4) alignments which are not properly paired.

(Continued)

FIG. 13.4, CONT'D
The resulting four BAM files were used as input for Genrich as biological replicates, and pooled before peak calling using HMMRATAC, MACS2.BAMPE, and MACS2.SITE. Default parameter settings and an FDR cutoff of 0.05 were applied for all peak calling. Heatmaps were generated using deeptools, with the bedGraph output from Genrich as input, to extract extended Tnp cleavage signals (signals over cleavage sites ±50 bp). (A) Venn diagram displaying the number of peaks called by each peak caller. (B) Box plots showing the width (bp) distributions of peaks called by different tools. (C) Box plots exhibiting the distributions of the number of fragments per kb within all peaks called by each tool and within peaks uniquely detected by a given peak caller. (D) Heatmaps displaying extended Tnp cleavage signals in all peak regions scaled across PSSs (peak start sites) and PESs (peak end sites), and their unscaled flanking regions (peak regions ±1 kb adjacent regions) by each peak caller. (E) Heatmaps showing extended Tnp cleavage signals in peak regions and their flanking regions (peak regions ±1 kb adjacent regions) uniquely detected by each peak caller. Here, H, MP, and MS represent peaks called by Genrich, HMMRATAC, MACS2.PE, and MACS2.SITE, respectively. C represents consensus peaks called by all four peak callers. G.U, H.U, MP.U, and MS.U are for unique peaks each called by only one peak caller. O represents peaks that fall into other categories in the Venn diagram that are called by two or three peak callers.

In summary, mid/high-signal peaks maintain their robustness across a variety of peak callers, whereas the choice of peak callers can significantly impact the detection of low-signal peaks. Specifically, HMMRATAC is the most conservative peak caller. Genrich is the most sensitive peak caller when there are multiple replicates. Conversely, when there is only a single replicate, MACS2.SITE emerges as the most sensitive choice (data not shown). Taken together, we recommend using Genrich when biological replicates are available, while we advise against using MACS2.PE for ATAC-seq peak calling.

Differential peak analysis

When well-designed ATAC-seq experiments have multiple biological replicates for more than one condition, researchers can identify differentially accessible chromatin regions between conditions via differential peak analysis. Although a set of tools have been developed for differential peak analyses of ChIP-seq data [69,104,105], no tools have been developed specifically for ATAC-seq differential peak analysis. Therefore, researchers have to rely on the tools originally designed for ChIP-seq data to perform differential peak analysis. Yan et al. [69] have recently reviewed a few potential tools for ATAC-seq differential peak analysis. The sliding window-based methods make use of an HMM such as ChIPDiff [106], or a negative binomial model such as diffReps [107] and csaw [108] to evaluate all sliding windows. These methods are more sensitive in detecting local chromatin accessibility changes than consensus peak-based methods [69], such as DiffBind [109]. Thus, they are more appropriate for differential peak analysis of ATAC-seq data because ATAC-seq peaks have complex patterns that contain both read count information and the distribution shape profile. Of note, the sensitive sliding window-based tools tend to detect many false positives; this needs to be controlled by applying stringent filtering criteria, such as a small FDR and a relatively high fold change threshold. Differential peaks of interest can be validated using qPCR with primers designed by ATAC Primer Tool [64].

Notably, ATAC-seq assays for differential accessibility analyses are prone to technical biases, such as variabilities in organelle DNA contamination [65], transposition reaction efficiency, and library complexity [65]. Additionally, the levels of chromatin accessibility between conditions of interest may vary significantly [66]. To have a meaningful biological interpretation, proper normalization of ATAC-seq data is critical for differential accessibility analysis [110]. Differences in genome-level chromatin accessibilities between conditions can be experimentally determined by the ATAC-see method [26]. If no significant differences are observed between conditions, one can consider using a complexity-based normalization method via a stochastic subsampling process to equalize library complexity among conditions [110]. Following that, widely accepted, count-based RNA-seq normalization methods such as TMM [111] and DESeq [93], can be applied. Otherwise, the complexity-based normalization method should be avoided and the condition-aware one, YARN [112], can be used.

Annotation and functional analysis

The ATAC-seq peaks contain a list of regulatory regions, including promoters, enhancers, and insulators. The width of ATAC-seq peaks varies from extremely narrow (a few hundred base pairs) to extremely broad [21], and the total number of peaks ranges from tens of thousands to hundreds of thousands [29], depending on peak callers, biological samples, and the implementation of ATAC-seq assays. The variations in ATAC-seq peak sizes and numbers complicate peak annotation and functional analysis. Additionally, only a minority of ATAC-seq peaks are within 1-kb regions flanking TSSs. This introduces more challenges to ATAC-seq peak annotations.

To gain biological insights from ATAC-seq peaks, ATAC-seq peaks can be associated with their potential target genes via a process similar to ChIP-seq peak annotation. Thus, software designed for ChIP-seq peak annotation such as ChIPpeakAnno [113,114], ANNOVAR [115], GREAT [116], UROPA [117], ChIPseeker [118], HOMER [119], BEDTools [120], and geneXtendeR [121], can be directly applied to ATAC-seq peak annotation. Notably, most of the annotation tools simply assign the peaks to the closest, overlapping, or nearby features.

However, it is known that regulatory elements can influence the expression of genes tens to thousands of kilobases away. Simply annotating peaks to nearby genes will miss important features of long-range regulation. To fully annotate ATAC-seq peaks, it is necessary to integrate other omics data such as long-range interaction data generated by technologies such as Hi-C [122], 3C-Seq [123], Next Generation Capture-C [124], and HiChIP [125], RNA-seq [126], histone modification ChIP-seq [127], and PRO-seq [128]. Currently, the function findEnhancers in the ChIPpeakAnno package [113] can make use of available chromatin interaction data to assign distal enhancers to their target genes.

For experiments aimed at identifying regulatory elements of differential accessibility between conditions, once peaks are assigned to their putative target genes, gene ontology [129] or pathway analysis [130] can be performed using the functions getEnrichedGO and getEnrichedPATH from the ChIPpeakAnno package [113].

Visualization

Like other genomic data, ATAC-seq data are usually displayed as signal tracks in a genome browser, along with a variety of genomic annotations, such as gene models, CpG islands, and SNPs. ATAC-seq data in various formats, including the BAM [80], bedGraph [131], and bigWig [131] ones, can be used for visualization. Among them, bigWig is the most widely used format for dense and continuous data, which can be converted from BAM files with the bamCoverage function of deepTools [132], or bwtool [133].

Several genome browsers and other visualization tools have been developed and successfully used for visualizing the ATAC-seq and other types of genomic data, such as the offline browsers trackViewer [134], and IGV [88], and the online browsers the University of California Santa Cruz (UCSC) Genome Browser [135], UCSC Cancer Genomics Browser [136], WashU Epigenome Browser [137], STAR [138], and Epiviz [139]. The UCSC Genome Browser is a well-known and easy-to-use tool that can output high-quality figures with multiple genomic tracks. Users can create custom tracks by uploading their own data or the corresponding URLs for integrative visualization along with other public genomic data in UCSC Genome Browser. For efficient access to large datasets, users are recommended to create track hubs [140], which can be programmatically generated using trackhub (https://github.com/daler/trackhub). Bioconductor [141]-based offline genome browsers such as rtracklayer [142], Gviz [143], and trackViewer [134], make use of the flexible R graphics system to display large sets of numeric data. The package rtracklayer also provides an interface to online genome browsers and associated annotation tracks, while the trackViewer package can be used for interactive visualization of ATAC-seq tracks and for producing high-quality plots with multiomics data. In addition, the IGVSnapshot function in the ATACseqQC package [65] can streamline the visualization of many genomic regions of interest using IGV [88].

Nucleosome positioning

As mentioned earlier, nucleosomes not only play a critical role in organizing chromatin but also regulate the underlying DNA accessibility to DNA-binding proteins. Numerous genome-wide studies have demonstrated that nucleosome positioning along the genome is not random; instead, nucleosomes are enriched at some positions while depleted at others [6,144]. The best-known canonical pattern comes from studying *S. cerevisiae*, where a clearly defined NFR, surrounded by two-phased nucleosomes, is detected near the TSSs [145,146]. However, this phenomenon is less obvious in multicellular eukaryotes [7,147]. Nucleosome positioning is also highly dynamic. Even cells with the same genome sequence can exhibit distinct nucleosome positioning patterns depending on their specific states and tissue contexts [148]. Determinants of nucleosome positioning, including TFs, have been reviewed elsewhere [149].

By nature, ATAC-seq is biased toward loosely packed, "open" chromatin regions since Tnp can barely access the "closed" chromatin regions. As a result, most fragments come from NFRs and are short. Long fragments spanning mono- or oligo-nucleosomes are most likely from the regions flanking NFRs, which makes it feasible to infer nucleosome positions for these regions. Though over 25 tools are currently available for nucleosome positioning analysis [150], most of which are designed for MNase-seq, such as DANPOS2 [151], iPNS [152], nucleR [153], and NucTools [154]. Although some of these tools such as DANPOS2 [151], iNPS [152], and NucTools [154] can be adapted for ATAC-seq data after the removal of nucleosome-free fragments [69], all of them except NucleoATAC fail to consider the unique ATAC-seq features, such as the Tnp transposition site preference [67].

To date, only NucleoATAC is specifically designed for nucleosome positioning analysis of ATAC-seq data based on the unique ATAC-seq fragmentation pattern around nucleosomes depicted by V-plots [46]. It also accounts for Tn5 Tnp transposition site bias and has been shown to outperform other nucleosome positioning tools designed for non-ATAC-seq data [46]. For input, NucleoATAC requires a sorted and indexed BAM file, a BED file containing open chromatin regions, and a genome reference sequence in the FASTA format along with its .fai index file. By default, NucleoATAC generates multiple outputs, including tracks for nucleosome occupancy scores, NFRs, and nucleosome dyad centers. For illustration purposes, we analyzed the ATAC-seq dataset from the lymphoblastoid cell line GM12878 (SRR891268–SRR891271) [21] using NucleoATAC with default parameter settings and presented the results in Fig. 13.5. The called dyad centers and the NRFs are displayed in tracks 3 and 4, respectively. An aggregated V-plot showing the distribution of midpoints of fragments of different insert sizes relative to nucleosome dyads is displayed in Fig. 13.5B.

Changes in nucleosome positioning play a crucial role in modulating numerous cellular processes, including DNA replication and gene expression. For dynamic analysis of nucleosome positioning, tools originally developed for MNase-seq including DANPOS2 [151], Dimnp [155], and the most recent Nucleosome Dynamics [156] can be considered, although none of them account for Tn5 Tnp transposition bias.

TF occupancy inference

Aside from a small number of pioneer TFs [157], almost all other TFs exclusively bind accessible chromatin by recognizing specific DNA sequences called TF binding sites (TFBSs) [5]. To date, consensus motif sequences of a large array of TFs from a variety of species have been derived experimentally or computationally. They have been compiled into many databases, such as JASPAR [158], CIS-BP [159], GTRD [160], and TRANSFAC [161] for eukaryotic TFs, HOCOMOCO [162] for human and mouse TFs only, OnTheFly [163] and FlyFactorSurvey [164] specifically for fruit fly TFs. FootprintDB [165] is an integrated database compiled and updated with the information of 19 other databases for eukaryotic TFs and their DNA binding motifs, including JASPAR [158] and HOCOMOCO [162]. The motif information is kept in different formats in different databases, such as the plain text, HTML, and XML formats. Position weight matrices (PWM) and position frequency matrices (PFM) are commonly used text formats.

Interactions between TFs and their TFBSs could interfere with the cutting efficiency of Tn5 Tnp or DNase I in the regions bound by the TFs and their immediate vicinities, thus leaving small footprints with low read coverage that can be detected by ATAC-seq [21] or DNase-seq [166] assays. These footprints can be used to infer TF occupancy and thus reveal the underlying regulatory mechanisms of gene expression. Karabacak Calviello et al. [41] systematically compared ATAC-seq and DNase-seq on inferring TF footprints. They concluded that the positions of TF footprints inferred by the two methods are largely concordant, though the footprints' shapes are method-specific and the performance of these two methods is TF-dependent [41].

However, a few technical challenges hinder TF occupancy inference. First, the established ATAC-seq and DNase-seq protocols allow one to detect footprints of stably bound TFs preferentially, even when the library is deeply sequenced [90]. Seeking to increase the likelihood of detecting footprints for transiently bound TFs, Oh et al. [167] modified the ATAC-seq protocol: they performed a brief cross-link of the cells with formaldehyde prior to the standard ATAC-seq assay [21,34] and a reversal of cross-link after tagmentation. Unfortunately, for unknown reasons, the modified protocol impaired the ATAC-seq-based footprint detection for transiently bound TFs, even though the same modifications successfully improved DNase-seq-based footprint signals of the same TFs [167]. To date, Omni-ATAC remains the most efficient ATAC-seq protocol for footprinting analysis [67].

Second, the homo-dimer Tn5 Tnp has sequence-dependent transposition site bias [67,168]. To distinguish authentic footprints from Tn5 Tnp-disfavored sites, good footprinting analysis tools should

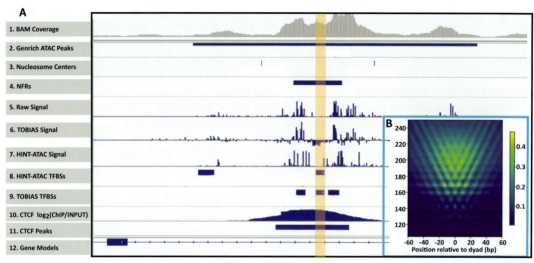

FIG. 13.5

Snapshot showing nucleosomes inferred by NucleoATAC and footprints identified by HINT-ATAC and TOBIAS. The ATAC-seq data from the lymphoblastoid cell line GM12878 (SRR891268-SRR891271) were mapped to the human reference genome using BWA-mem. The BAM files were first filtered as recommended in this chapter and combined. Then open chromatin regions were identified using Genrich with default parameter settings and an FDR cutoff of 0.05, followed by the removal of the human genome blacklist. The remaining open chromatin regions were extended by 600 bp at both ends with the bedtools slop module and then used for nucleosome positioning analysis with NucleoATAC. The un-extended open chromatin regions were used for footprinting analysis with TOBIAS and HINT-ATAC separately. Motifs of 283 TFs (HOCOMOCO V11 Core human collection) expressed in the same cell line were used for TF occupancy inference by both tools. The predicted TFBSs were filtered to remove those that overlap nucleosome-occupied regions (nucleosome center ±73 bp) inferred by NucleoATAC. The remaining TFBSs of 33 TFs were validated using ENCODE ChIP-seq data (ENCFF274BBE, ENCFF701QXK, ENCFF926LHG, ENCFF768VSH, ENCFF773OQL, ENCFF196JGP, ENCFF980VOD, ENCFF958GXF, ENCFF830BRO, ENCFF223MUF, ENCFF677QUK, ENCFF510NDO, ENCFF942MDT, ENCFF873DJD, ENCFF186AWV, ENCFF432AQP, ENCFF270NAL, ENCFF199HGX, ENCFF193POQ, ENCFF960ZGP, ENCFF652BRY, ENCFF249SVT, ENCFF514SWA, ENCFF603BID, ENCFF766WWB, ENCFF946ACA, ENCFF152RNE, ENCFF786YYI, ENCFF335ADU, ENCFF971VHK, ENCFF091YID, ENCFF948CPI, and ENCFF370ZNL) of the same 33 TFs from the same cell line. (A) Visualization of ATAC-seq data and analysis results. The raw BAM coverage and narrowPeak tracks are displayed in tracks 1 and 2, respectively. The centers of nucleosomes and nucleosome-free regions inferred by NucleoATAC are displayed in tracks 3 and 4, respectively. With the default parameter settings, NucleoATAC called 113,467 unique nucleosomes. The raw signals of transposition sites and the signals corrected for Tn5 transposition bias by TOBIAS and HINT-ATAC are shown in tracks 5–7, respectively. TF occupancy inferred by HINT-ATAC and TOBIAS are shown in tracks 8 and 9, respectively. A CTCF binding site, predicted by both footprinters and highlighted in light brown, is located in an NFR and overlaps the center of a CTCF ChIP-seq peak (tracks 10 and 11). (B) A characteristic V-plot showing the distribution of midpoints of fragments of different insert sizes relative to nucleosome dyads.

be able to correct for the bias [67,169]. To our knowledge, msCentipede [170], HINT-ATAC [67], and TOBIAS [169] are the only full-featured tools for ATAC-seq footprinting analysis. msCentipede is a motif-centric footprinting method for both DNase-seq and ATAC-seq data. Although msCentipede does not perform direct bias correction, it builds hierarchical multiscale models with heterogeneous Poisson processes to account for spatial variation in enzymatic cleavage patterns around putative TFBSs for different TFs.

The required input for msCentipede is BAM files for biological replicates and a list of precomputed putative TFBSs of a given TF, including TFBS' genomic locations, strand, and PWM score [170,171]. A two-component mixture model is used to fit the cleavage frequencies around putative TFBSs. One component is for bound motifs and the other for unbound motifs (the background model) using the aforementioned hierarchical multiscale model with different parameter distributions for each component. msCentipede classifies each putative TFBS as bound or unbound based on its posterior odds [170]. By building background models (msCentipede-flexbg) with sequencing data from naked DNA tagmented by Tn5 Tnp, one can slightly improve the accuracy of msCentipede [170]. Of note, msCentipede has to fit complicated models for each TF for every run, which demands a high-performance computing environment, though it performs well in general [169].

Unlike msCentipede, HINT-ATAC [67] and TOBIAS [169] are de novo footprinting methods specifically developed for ATAC-seq data analysis. They can predict footprint sites across predefined open chromatin regions by making use of the unique "peak-dip-peak" feature of typical footprints after explicitly correcting for Tn5 transposition bias. HINT-ATAC, part of the Regulatory Genomics Toolbox (RGT), utilizes a position dependency model (PDM) with a word size of eight to correct for strand-specific Tn5 transposition bias. It predicts footprints using an HMM with strand-specific, nucleosome-size decomposed, bias-corrected signals. The HMM is trained in a semisupervised way with TFBSs derived from high-quality ChIP-seq data of at least one TF for a given organism. Following the footprinting analysis, the HINT-ATAC-pipeline infers TF occupancy using a motif matching tool implemented in the RGT, which is based on the MOODS C++ library [172].

By contrast, TOBIAS [169] first builds a dinucleotide weight matrix (DWM) [173], which can capture inter-base dependencies, with reads mapped to nonopen chromatin regions. It then uses this weight matrix to correct for Tn5 transposition bias in open chromatin regions. The bias-corrected signals are used to estimate footprint scores across open chromatin regions. Next, TOBIAS infers TF occupancy by matching TF motifs to the predicted footprints using the same MOODS library. The single nucleotide frequencies from the input peak regions are used by MOODS to build a background model for calculating the log odds ratio of each TFBS. TFBSs are classified into bound and unbound sites with an empirical TF-specific footprint score threshold. The threshold is set at the upper tail's critical point of a normal distribution that models the background footprint scores such that the significance level is the user-defined P-value cutoff.

Using ATAC-seq data with matching TF ChIP-seq results from the ENCODE project as the ground truth, Bentsen et al. showed that TOBIAS has similar accuracy to msCENTIPEDE [169]. However, due to multithreading, TOBIAS is more efficient than msCENTIPEDE [169]. TOBIAS also outperforms HINT-ATAC [67] in Tn5 bias correction, although no TF occupancy comparison between the two has been published [169]. Thus, we compared TOBIAS with HINT-ATAC using the ATAC-seq dataset from the lymphoblastoid cell line GM12878 (SRR891268–SRR891271) with default parameter settings [21].

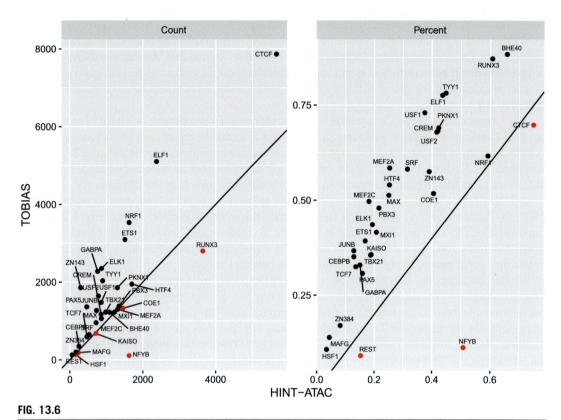

FIG. 13.6

TOBIAS outperformed HINT-ATAC in TF occupancy inference. The ATAC-seq dataset from the lymphoblastoid cell line GM12878 (SRR891268–SRR891271) was analyzed as described in the legend of Fig. 13.5. The number and percentage of inferred TFBSs that overlap the corresponding TF's ChIP-seq peak centers (summit ± 100 bp) are shown here for TOBIAS and HINT-ATAC. TOBIAS performed better than HINT-ATAC in terms of the number and percentage of ChIP-seq-concordant TFBSs for most TFs, except a few ones colored in red.

After removing TFBSs that overlap NucleoATAC-called putative nucleosomes and the blacklist regions of the human reference genome [82], we obtained 1,720,772 and 1,147,883 TFBSs for 283 expressed TFs (> 0.1 TPM (Transcripts Per Million)) with HINT-ATAC and TOBIAS, respectively. Tracks 8 and 9 in Fig. 13.5 display TFBSs detected by the two methods, respectively. The two tools detected 216,789 TFBSs in common. We validated the TFBSs detected by TOBIAS or HINT-ATAC using the ENCODE ChIP-seq data of 33 TFs for the same GM12878 cell line. We consider a predicted TFBS correct if it overlaps with a peak center region (summit ± 100 bp) from the corresponding TF's ChIP-seq assay in the same cell line. Fig. 13.6 shows the number and percentage of correctly predicted TFBSs by TOBIAS and HINT-ATAC. For most TFs, TOBIAS has higher prediction accuracy than HINT-ATAC. Our results are consistent with Bentsen et al [169]. Based on these findings, we recommend TOBIAS for TF occupancy inference with ATAC-seq data.

Motif mapping

Given that TF footprinting analysis is not efficient for transiently bound TFs, motif mapping analysis for NFRs is a useful alternative to studying gene transcription regulation. Popular motif mappers for open chromatin regions detected by ATAC-seq are command line tools such as HOMER [119], Cluster-Buster [174], and MOODS [172], Bioconductor packages TFBStools [175], motifmatchr [176], and universalmotif [177], and web servers PWMScan [171], RSAT [178], MotifMap [179], and FIMO [180] in the MEME suite [181]. Systematical comparison of the performance of these tools have been performed [182,183]. In summary, FIMO is one of the most stringent motif mappers, which has a relatively low false positive rate (FPR), but low sensitivity [182,183]. By contrast, MOODS is one of the most lenient motif mappers, which outputs 10 times more TFBSs than FIMO and has the highest FPR [182,183]. An ensemble motif mapping approach that combines top-scored motifs from different tools, such as FIMO and Cluster-Buster, has been suggested to improve both sensitivity and accuracy [183]. Of note, large-scale motif search produces many false positives because motif sequences are short and degenerative. Thus, stringent filtering steps as follows are recommended to reduce the FPR:

(1) restrict the search space by eliminating nucleosome-occupied regions from the open chromatin regions;
(2) only search for motifs of TFs that have a detectable expression in the target tissue or cell type by using existing gene expression data if available or that have putative maternal effect genes in the target sample [169];
(3) apply a stringent statistical threshold, such as FDR < 0.05;
(4) filter putative TFBSs based on additional sources of evidence, such as gene co-regulation [184] and evolutionary sequence conservation [185], though not all functional binding sites are evolutionarily conserved [186].

Differential TF binding activity analysis

TF binding changes the flanking chromatin accessibility by recruiting other factors such as chromatin modelers [187–190], and may leave footprints. Any changes that affect the TFs' binding activities [191] or expression levels [192] can potentially cause changes in chromatin accessibility in the vicinity of their binding sites and changes the depth of their footprints. TF binding activity can be indirectly measured by its aggregated footprint depth (FPD) and/or the accessibility of the footprints' flanking regions (for short called flanking accessibility, FA) [67,169,193]. This type of score reflects TF binding activity better than that which uses TFBS entity alone [193,194].

DiffTF [195] infers TFs with binding activity changes using differences in flanking accessibility (ΔFA) as a surrogate; it calculates ΔFA for each putative TF binding site in open chromatin regions under different conditions. DiffTF also classifies TFs as activators or repressors when matching RNA-seq data are available [195]. By contrast, BaGFoot [193], HINT-ATAC [67], and TOBIAS [169] make use of both ΔFA and difference in footprint depth (ΔFPD) to infer changes in TF binding activity between conditions after Tn5 transposition bias correction and sequencing depth normalization. BagFoot uses k-mer frequencies (by default $k=6$) to correct for Tn5 transposition bias and FIMO for finding putative TFBSs of all known motifs across ATAC-seq peak regions. For each condition, genome-wide FPD and FA of each TF are calculated and normalized. Then, between-condition ΔFPD and ΔFA are calculated by subtracting the corresponding values of one condition from those of the other condition. TFs with significant ΔFPD and/or ΔFA are considered as candidates with significant differential binding activity, with *P*-values determined by a bivariate outlier analysis [193].

BagFoot is robust to the choice of peak callers, the sequencing depth, and the way the k-mer frequencies for bias correction are calculated [193]. Unlike BagFoot, both HINT-ATAC and TOBIAS identify footprints after bias correction and discover putative TFBSs within footprints with the motif mapper MOODS. For each TF of a given condition, HINT-ATAC adds the means of both FDP and FA to calculate an activity score. TFs with significant changes in the normalized activity scores between two conditions are considered to have differential binding activities. By contrast, for each putative TFBS, TOBIAS uses the FPD as the footprint score. A log_2 fold change of footprint score (log_2FC) per TFBS between conditions is calculated. A background distribution of log_2FC is derived based on the random sampling of one value of cut site frequency per 200-bp intervals from the peak regions of every condition with the same sampling seed. The mean log_2FC for each TF is compared to that of the corresponding background distribution, and the significance is determined with the one-sample Student's *t*-test. TFs with significant differential binding scores are considered to have differential binding activity [169]. Currently, the performances of DiffTF, HINT-ATAC, and TOBIAS in differential TF binding activity analysis have not been compared.

Due to the technical limitations of the footprinting analysis mentioned in the previous section ("TF occupancy inference" section), footprint-independent motif enrichment analysis can be considered as a supplementary method for inferring TF binding activity. Differential motif enrichments in open chromatin regions under different conditions may indicate alterations in TF binding activity [193]. After identifying the locations and frequencies of TFBSs in open chromatin regions for a condition through motif analyses, one can compare the locations and frequencies to those in a random background or in open chromatin regions under another condition. Often, differential motif analyses are performed on differentially accessible regions between conditions of interest. HOMER [119], AME [196] and CentriMO [197] in the MEME suite [181], and DAStk [198] are the best tools for this purpose. However, this type of analysis is not sensitive because only regions with significant accessibility changes are included [193]. In addition, differentially enriched motifs only provide indirect evidence for inferring TFs' differential binding activity [193]. Therefore, the results should be interpreted cautiously and validated with orthogonal methods.

Reconstruction of gene regulatory network

A GRN describes how each gene is regulated under a given condition by its *cis*-regulatory elements, including its promoters and enhancers via direct and/or indirect binding of trans-regulatory factors. Constructing conditional GRNs and comparing GRNs under different conditions are fundamental to understand the underlying mechanisms of cell differentiation, development, and pathogenesis [199–202]. ATAC-seq allows one to identify genome-wide *cis*-regulatory elements and TFBSs within these elements through TF footprinting analysis and motif mapping. In addition, the *cis*-regulatory elements can be linked to their target genes through integration with long-range interaction data, as described earlier (see "Annotation and functional analysis" section). TOBIAS can build TF-TF regulatory networks based on the inferred TF occupancy over the regulatory elements of all expressing TFs [169], whereas paired expression and chromatin accessibility (PECA) modeling [203] can infer the GRN by regressing target gene expression on the TF expression, the chromatin remodeler expression, and the chromatin accessibility of their *cis*-regulatory elements. Thus, it is possible to build directional GRNs by integrating ATAC-seq data with multiple layers of other omics data [204,205].

Summary

ATAC-seq has become one of the most popular methods for profiling chromatin accessibility since its debut in 2013. Over the years, the original ATAC-seq protocol has been optimized to adapt to different types and sources of biological samples. Nevertheless, comprehensive guidelines for ATAC-seq assays and data analysis are not available. This chapter systematically described all steps of ATAC-seq assays for different species and ATAC-seq data analysis. URLs and references to relevant tutorials, protocols, and computation tools are available in Table 13.1.

When designing an ATAC-seq experiment, at least two biological replicates should be considered for each condition, and blocking and randomization should be conducted to discern batch effect and

Table 13.1 Tutorials, protocols, and tools for best practices of the ATAC-seq assay and its data analysis.

	URL	Reference
General tutorials for ATAC-seq: from experimental design to data analysis		
Niu's tutorial	https://yiweiniu.github.io/blog/2019/03/ATAC-seq-data-analysis-from-FASTQ-to-peaks/	[206]
Illumina's tutorial	https://www.illumina.com/techniques/popular-applications/epigenetics/atac-seq-chromatin-accessibility.html	
Gaspar's tutorial	https://informatics.fas.harvard.edu/atac-seq-guidelines.html	[207]
Galaxy tutorial	https://training.galaxyproject.org/training-material/topics/epigenetics/tutorials/atac-seq/tutorial.html	[208]
Protocols for ATAC-seq assays		
Buenrostro's protocol	https://www.ncbi.nlm.nih.gov/pmc/articles/PMC4374986	[34]
Plant ATAC-seq	https://www.ncbi.nlm.nih.gov/pmc/articles/PMC5693289	[209]
Cold Spring Harbor's protocol	http://cshprotocols.cshlp.org/content/2019/1/pdb.prot098327	[61]
Omni-ATAC	https://www.nature.com/articles/nmeth.4396	[27]
Fungi ATAC-seq	https://bmcbiol.biomedcentral.com/articles/10.1186/s12915-021-01114-0	[210]
Tools for ATAC-seq data quality control		
FastQC	https://www.bioinformatics.babraham.ac.uk/projects/fastqc/	[72]
MultiQC	https://multiqc.info/	[73]

Continued

Table 13.1 Tutorials, protocols, and tools for best practices of the ATAC-seq assay and its data analysis—cont'd

	URL	Reference
ATACseqQC	https://bioconductor.org/packages/ATACseqQC/	[65]
ATAC-seq QC tutorial	https://haibol2016.github.io/ATACseqQCWorkshop	[211]
	https://www.youtube.com/watch?v=VZFUu_cJxyI	
ataqv	https://github.com/ParkerLab/ataqv/	[99]
ChIPQC	https://bioconductor.org/packages/ChIPQC/	[97]
Trimming tools		
cutadapt	https://cutadapt.readthedocs.io/	[77]
Trimmomatic	http://www.usadellab.org/cms/?page=trimmomatic	[78]
fastp	https://github.com/OpenGene/fastp	[79]
Alignment tools		
BWA	http://bio-bwa.sourceforge.net/	[76]
Bowtie	http://bowtie-bio.sourceforge.net/index.shtml	[74]
Bowtie2	http://bowtie-bio.sourceforge.net/bowtie2/index.shtml	[75]
SAMtools	http://www.htslib.org/	[80]
Picard tools	https://broadinstitute.github.io/picard/	
Tools for blacklist or graylist generation		
BlackList	https://github.com/Boyle-Lab/Blacklist	[83]
PeakPass	https://github.com/ewimberley/peakPass	[212]
GreyListChIP	https://bioconductor.org/packages/GreyListChIP/	[213]
Tools for peak calling		
MACS2	https://github.com/macs3-project/MACS	[101]
Genrich	https://github.com/jsh58/Genrich	[214,215]
	https://bigmonty12.github.io/peak-calling-benchmark	
HMMRATAC	https://github.com/LiuLabUB/HMMRATAC	[103]

Table 13.1 Tutorials, protocols, and tools for best practices of the ATAC-seq assay and its data analysis—cont'd

	URL	Reference
Tools for differential peak analysis and validation		
DESeq2	https://bioconductor.org/packages/csaw/	[216]
DiffBind	https://bioconductor.org/packages/DiffBind/	[109]
diffReps	https://github.com/shenlab-sinai/diffreps	[107]
csaw	https://bioconductor.org/packages/csaw/	[108]
YARN	https://github.com/QuackenbushLab/yarn	[112]
APT	https://github.com/ChangLab/ATACPrimerTool	[64]
Tools for annotation and functional analysis		
ChIPpeakAnno	https://bioconductor.org/packages/ChIPpeakAnno/	[113]
ChIPpeakAnno tutorial	https://hukai916.github.io/IntegratedChIPseqWorkshop/	[217,218]
	https://www.youtube.com/watch?v=EAxFrz_F4bg	
ChIPseeker	https://bioconductor.org/packages/ChIPseeker/	[118]
UROPA	https://github.com/loosolab/UROPA	[117]
GREAT	http://great.stanford.edu/public/html/	[116]
Visualization tools		
trackViewer	https://bioconductor.org/packages/trackViewer/	[134]
trackViewer tutorial	https://jianhong.github.io/workshop2020/articles/trackViewer.html	[218,219]
	https://www.youtube.com/watch?v=EAxFrz_F4bg (starts at 23:16)	
IGV	https://software.broadinstitute.org/software/igv/	[88]
UCSC Genome Browser	https://genome.ucsc.edu/	[135]
WashU Epigenome Browser	https://epigenomegateway.wustl.edu/	[220]

Continued

Table 13.1 Tutorials, protocols, and tools for best practices of the ATAC-seq assay and its data analysis—cont'd

	URL	Reference
Tool for nucleosome positioning		
NulceoATAC	https://github.com/GreenleafLab/NucleoATAC	[46]
Tool for TF occupancy inference		
msCentipede	https://github.com/rajanil/msCentipede	[170]
HINT-ATAC	http://www.regulatory-genomics.org/rgt	[67]
TOBIAS	https://github.com/loosolab/TOBIAS	[169]
Motif mapping tools		
FIMO	https://meme-suite.org/meme/tools/fimo	[180]
MOODS	https://github.com/jhkorhonen/MOODS	[172]
Tools for differential TF binding activity analysis		
DiffTF	https://git.embl.de/grp-zaugg/diffTF	[195]
BaGFoot	https://sourceforge.net/projects/bagfootr	[193]
HINT-ATAC	http://www.regulatory-genomics.org/rgt	[67]
TOBIAS	https://github.com/loosolab/TOBIAS	[169]
HOMER	http://homer.ucsd.edu/homer	[119]
AME	https://meme-suite.org/meme/doc/ame.html	[196]
CentriMO	https://meme-suite.org/meme/doc/centrimo.html	[197]
DAStk	https://github.com/Dowell-Lab/DAStk	[198]
Tools for reconstruction of gene regulatory networks		
TOBIAS	https://github.com/loosolab/TOBIAS	[169]
PECA	https://github.com/SUwonglab/PECA	[203]

other confounding factors. As for nuclei preparation, it is critical to isolate high-quality nuclei with fresh or cryopreserved tissues or cells, as these are the preferred starting materials. Background DNA from medium or dead cells should be removed, and organelle DNA should be depleted as much as possible. Nuclei preparation should be checked using microscopy.

Whenever possible, the Omni-ATAC protocol should be adopted to ensure high-quality data. The ratio of the amount of Tn5 enzyme to the number of nuclei should be optimized. The number of PCR cycles for amplifying the tagmentation product should be experimentally determined to reduce biases introduced by over-amplification. The size distribution of the resulting library should be examined by microfluidic electrophoresis. If necessary, size selection can be applied to remove primer dimers and/or fragments longer than 1000 bp. The final library's concentration should be determined using qPCR. The library should be sequenced on an NGS platform in a paired-end mode to generate reads with lengths between 40 and 100 bp. The sequencing depth per library depends on the experimental goal and species, with at least 50 M filtered aligned reads for open chromatin analysis and at least 200 M filtered aligned reads for footprinting analysis for human samples.

As for ATAC-seq data analysis, FastQC and MultiQC are recommended for raw read QC. Optional read trimming can be performed, using fastp, Trimmomatic, or Cutadapt, before alignment using BWA, Bowtie, or Bowtie2, depending on the read length. The ATACseqQC package is highly recommended for postalignment QC. MACS2.SITE and Genrich are preferred for peak calling with nonreplicated experiments, while Genrich is the recommended peak caller when multiple biological replicates exist. The aligned ATAC-seq data and peaks can be conveniently visualized with UCSC genome browser, trackViewer, or IGV. Peak regions can be annotated using the ChIPpeakAnno package and GREAT. For differential peak analysis, we recommend diffReps or csaw. We also recommend FIMO for TF motif detection, HOMER for motif enrichment, NucleoATAC for nucleosome positioning analysis, and TOBIAS for TF occupancy inference. It is possible to reconstruct GRNs by integrating multiple layers of omics data, especially long-range interaction data, and RNA-seq data, with ATAC-seq data using the PECA method.

The guidelines in this chapter will help researchers perform successful ATAC-seq assays and analysts extract meaningful insights from the data, prioritizing methodological rigor and data reproducibility. As novel tools for ATAC-seq data analysis emerge, some tools recommended here may be replaced by more sophisticated ones in the future.

Acknowledgments

We thank Serena J. Han of Harvard College and Sara Deibler at University of Massachusetts for editorial assistance.

References

[1] R.D. Kornberg, Chromatin structure: a repeating unit of histones and DNA, Science 184 (4139) (1974) 868–871.
[2] K. Luger, J.C. Hansen, Nucleosome and chromatin fiber dynamics, Curr. Opin. Struct. Biol. 15 (2) (2005) 188–196.
[3] N. Saksouk, E. Simboeck, J. Déjardin, Constitutive heterochromatin formation and transcription in mammals, Epigenetics Chromatin 8 (2015) 3.
[4] Y. Murakami, Heterochromatin and euchromatin, in: W. Dubitzky, O. Wolkenhauer, K.-H. Cho, H. Yokota (Eds.), Encyclopedia of Systems Biology, Springer New York, New York, NY, 2013, pp. 881–884.
[5] R.E. Thurman, E. Rynes, R. Humbert, et al., The accessible chromatin landscape of the human genome, Nature 489 (7414) (2012) 75–82.
[6] C.-K. Lee, Y. Shibata, B. Rao, B.D. Strahl, J.D. Lieb, Evidence for nucleosome depletion at active regulatory regions genome-wide, Nat. Genet. 36 (8) (2004) 900–905.

[7] 2 Chromatin patterns at transcription factor binding sites, Nature (2019).
[8] H. Zhou, Y. Xiang, M. Hu, et al., Chromatin accessibility is associated with the changed expression of miRNAs that target members of the Hippo pathway during myoblast differentiation, Cell Death Dis. 11 (2) (2020) 148.
[9] S. Schick, D. Fournier, S. Thakurela, S.K. Sahu, A. Garding, V.K. Tiwari, Dynamics of chromatin accessibility and epigenetic state in response to UV damage, J. Cell Sci. 128 (23) (2015) 4380–4394.
[10] Y. Zhao, D. Zheng, A. Cvekl, Profiling of chromatin accessibility and identification of general cis-regulatory mechanisms that control two ocular lens differentiation pathways, Epigenetics Chromatin 12 (1) (2019) 27.
[11] M.R. Corces, J.D. Buenrostro, B. Wu, et al., Lineage-specific and single-cell chromatin accessibility charts human hematopoiesis and leukemia evolution, Nat. Genet. 48 (10) (2016) 1193–1203.
[12] S.K. Denny, D. Yang, C.-H. Chuang, et al., Nfib promotes metastasis through a widespread increase in chromatin accessibility, Cell 166 (2) (2016) 328–342.
[13] K. Qu, L.C. Zaba, A.T. Satpathy, et al., Chromatin accessibility landscape of cutaneous T cell lymphoma and dynamic response to HDAC inhibitors, Cancer Cell 32 (1) (2017) 27–41.e24.
[14] Y. Wang, X. Zhang, Q. Song, et al., Characterization of the chromatin accessibility in an Alzheimer's disease (AD) mouse model, Alzheimers Res. Ther. 12 (1) (2020) 29.
[15] S.A. McClymont, P.W. Hook, A.I. Soto, et al., Parkinson-associated SNCA enhancer variants revealed by open chromatin in mouse dopamine neurons, Am. J. Hum. Genet. 103 (6) (2018) 874–892.
[16] J. Wang, C. Zibetti, P. Shang, et al., ATAC-Seq analysis reveals a widespread decrease of chromatin accessibility in age-related macular degeneration, Nat. Commun. 9 (1) (2018) 1364.
[17] P.G. Giresi, J. Kim, R.M. McDaniell, V.R. Iyer, J.D. Lieb, FAIRE (Formaldehyde-Assisted Isolation of Regulatory Elements) isolates active regulatory elements from human chromatin, Genome Res. 17 (6) (2007) 877–885.
[18] G.E. Crawford, I.E. Holt, J. Whittle, et al., Genome-wide mapping of DNase hypersensitive sites using massively parallel signature sequencing (MPSS), Genome Res. 16 (1) (2006) 123–131.
[19] K. Cui, K. Zhao, Genome-wide approaches to determining nucleosome occupancy in metazoans using MNase-Seq, Methods Mol. Biol. 833 (2012) 413–419.
[20] T.K. Kelly, Y. Liu, F.D. Lay, G. Liang, B.P. Berman, P.A. Jones, Genome-wide mapping of nucleosome positioning and DNA methylation within individual DNA molecules, Genome Res. 22 (12) (2012) 2497–2506.
[21] J.D. Buenrostro, P.G. Giresi, L.C. Zaba, H.Y. Chang, W.J. Greenleaf, Transposition of native chromatin for fast and sensitive epigenomic profiling of open chromatin, DNA-binding proteins and nucleosome position, Nat. Methods 10 (12) (2013) 1213–1218.
[22] M. Tsompana, M.J. Buck, Chromatin accessibility: a window into the genome, Epigenetics Chromatin 7 (1) (2014) 33.
[23] S.L. Klemm, Z. Shipony, W.J. Greenleaf, Chromatin accessibility and the regulatory epigenome, Nat. Rev. Genet. 20 (4) (2019) 207–220.
[24] P. Milani, R. Escalante-Chong, B.C. Shelley, et al., Cell freezing protocol suitable for ATAC-Seq on motor neurons derived from human induced pluripotent stem cells, Sci. Rep. 6 (2016) 25474.
[25] S. Fujiwara, S. Baek, L. Varticovski, S. Kim, G.L. Hager, High quality ATAC-Seq data recovered from cryopreserved breast cell lines and tissue, Sci. Rep. 9 (1) (2019) 516.
[26] X. Chen, Y. Shen, W. Draper, et al., ATAC-see reveals the accessible genome by transposase-mediated imaging and sequencing, Nat. Methods 13 (12) (2016) 1013–1020.
[27] M.R. Corces, A.E. Trevino, E.G. Hamilton, et al., An improved ATAC-seq protocol reduces background and enables interrogation of frozen tissues, Nat. Methods 14 (10) (2017) 959–962.
[28] J.F. Fullard, M.E. Hauberg, J. Bendl, et al., An atlas of chromatin accessibility in the adult human brain, Genome Res. 28 (8) (2018) 1243–1252.
[29] C. Liu, M. Wang, X. Wei, et al., An ATAC-seq atlas of chromatin accessibility in mouse tissues, Scientific Data. 6 (1) (2019) 65.

[30] ENCODE ATAC-seq Experiment Mattrix. https://www.encodeproject.org/matrix/?type=Experiment&status=released&assay_title=ATAC-seq. (Accessed 27 April 2020).
[31] E. Giuffra, C.K. Tuggle, F. Consortium, Functional Annotation of Animal Genomes (FAANG): current achievements and roadmap, Annu. Rev. Anim. Biosci. 7 (1) (2019) 65–88.
[32] S. Foissac, S. Djebali, K. Munyard, et al., Multi-species annotation of transcriptome and chromatin structure in domesticated animals, BMC Biol. 17 (1) (2019) 108.
[33] M.R. Corces, J.M. Granja, S. Shams, et al., The chromatin accessibility landscape of primary human cancers, Science 362 (6413) (2018) eaav1898.
[34] J.D. Buenrostro, B. Wu, H.Y. Chang, W.J. Greenleaf, ATAC-seq: a method for assaying chromatin accessibility genome-wide, in: Curr. Protoc. Mol. Biol., John Wiley & Sons, Inc., 2015.
[35] A.T. Satpathy, J.M. Granja, K.E. Yost, et al., Massively parallel single-cell chromatin landscapes of human immune cell development and intratumoral T cell exhaustion, Nat. Biotechnol. 37 (8) (2019) 925–936.
[36] A. Mezger, S. Klemm, I. Mann, et al., High-throughput chromatin accessibility profiling at single-cell resolution, Nat. Commun. 9 (1) (2018) 3647.
[37] D.A. Cusanovich, R. Daza, A. Adey, et al., Multiplex single-cell profiling of chromatin accessibility by combinatorial cellular indexing, Science 348 (6237) (2015) 910–914.
[38] P.L. Auer, R.W. Doerge, Statistical design and analysis of RNA sequencing data, Genetics 185 (2) (2010) 405–416.
[39] M. Tannenbaum, A. Sarusi-Portuguez, R. Krispil, et al., Regulatory chromatin landscape in *Arabidopsis thaliana* roots uncovered by coupling INTACT and ATAC-seq, Plant Methods 14 (1) (2018) 113.
[40] Z. Lu, B.T. Hofmeister, C. Vollmers, R.M. DuBois, R.J. Schmitz, Combining ATAC-seq with nuclei sorting for discovery of *cis*-regulatory regions in plant genomes, Nucleic Acids Res. 45 (6) (2016) e41.
[41] A. Karabacak Calviello, A. Hirsekorn, R. Wurmus, D. Yusuf, U. Ohler, Reproducible inference of transcription factor footprints in ATAC-seq and DNase-seq datasets using protocol-specific bias modeling, Genome Biol. 20 (1) (2019) 42.
[42] M. Adam, A.S. Potter, S.S. Potter, Psychrophilic proteases dramatically reduce single-cell RNA-seq artifacts: a molecular atlas of kidney development, Dev. (Cambr., Engl.) 144 (19) (2017) 3625–3632.
[43] S. Preissl, R. Fang, H. Huang, et al., Single-nucleus analysis of accessible chromatin in developing mouse forebrain reveals cell-type-specific transcriptional regulation, Nat. Neurosci. 21 (3) (2018) 432–439.
[44] Y.-J. Lee, P. Chang, J.-H. Lu, P.-Y. Chen, C.-J.R. Wang, Assessing chromatin accessibility in maize using ATAC-seq, bioRxiv (2019) 526079.
[45] L. Huang, X. Li, L. Dong, B. Wang, L. Pan, Profiling of chromatin accessibility across *Aspergillus* species and identification of transcription factor binding sites in the *Aspergillus* genome using filamentous fungi ATAC-seq, bioRxiv (2019) 857284.
[46] A.N. Schep, J.D. Buenrostro, S.K. Denny, K. Schwartz, G. Sherlock, W.J. Greenleaf, Structured nucleosome fingerprints enable high-resolution mapping of chromatin architecture within regulatory regions, Genome Res. 25 (11) (2015) 1757–1770.
[47] N.S. Patil, J.P. Jadhav, Penicillium ochrochloron MTCC 517 chitinase: an effective tool in commercial enzyme cocktail for production and regeneration of protoplasts from various fungi, Saudi J. Biol. Sci. 22 (2) (2015) 232–236.
[48] P.F. Hamlyn, R.E. Bradshaw, F.M. Mellon, C.M. Santiago, J.M. Wilson, J.F. Peberdy, Efficient protoplast isolation from fungi using commercial enzymes, Enzym. Microb. Technol. 3 (4) (1981) 321–325.
[49] M. Nakamura, K. Nomura, A. J-iP, Y. Degawa, M. Kakishima, A simple method for isolation of nuclei from *Basidiobolus ranarum* (*Zygomycota*), Mycoscience 50 (6) (2009) 448.
[50] M. Hsiang, R.D. Cole, The isolation of nuclei from fungi, in: G. Stein, J. Stein, L.J. Kleinsmith (Eds.), Method Cell Biol., vol. 16, Academic Press, 1977, pp. 113–124 (Chapter 6).
[51] M.A. Gealt, G. Sheir-Neiss, N.R. Morris, The isolation of nuclei from the filamentous fungus *Aspergillus nidulans*, Microbiology 94 (1) (1976) 204–210.

[52] C. Yang, L. Ma, D. Xiao, Z. Ying, X. Jiang, Y. Lin, Integration of ATAC-Seq and RNA-Seq identifies key genes in light-induced primordia formation of *Sparassis latifolia*, Int. J. Mol. Sci. 21 (1) (2019) 185.

[53] M.R. Davey, P. Anthony, J.B. Power, K.C. Lowe, Plant protoplasts: status and biotechnological perspectives, Biotechnol. Adv. 23 (2) (2005) 131–171.

[54] J.M. Graham, Isolation of nuclei and nuclear membranes from animal tissues, Curr. Protocols Cell Biol. 12 (1) (2001). 3.10.11–13.10.19.

[55] R.B. Deal, S. Henikoff, A simple method for gene expression and chromatin profiling of invidual cell types within a tissue, Dev. Cell 18 (6) (2010) 1030–1040.

[56] B.I. Tarnowski, F.G. Spinale, J.H. Nicholson, DAPI as a useful stain for nuclear quantitation, Biotech Histochem. 66 (6) (1991) 297–302.

[57] D.G. Hendrickson, I. Soifer, B.J. Wranik, et al., A new experimental platform facilitates assessment of the transcriptional and chromatin landscapes of aging yeast, elife 7 (2018) e39911.

[58] H. Hu, Y.-R. Miao, L.-H. Jia, Q.-Y. Yu, Q. Zhang, A.-Y. Guo, AnimalTFDB 3.0: a comprehensive resource for annotation and prediction of animal transcription factors, Nucleic Acids Res. 47 (D1) (2018) D33–D38.

[59] I. Grbesa, M. Tannenbaum, A. Sarusi-Portuguez, M. Schwartz, O. Hakim, Mapping genome-wide accessible chromatin in primary human T lymphocytes by ATAC-Seq, J. Visual. Exp.—JoVE 129 (2017) 56313.

[60] M.A. Quail, H. Swerdlow, D.J. Turner, Improved protocols for the illumina genome analyzer sequencing system, Curr. Protocols Hum. Genet. (2009), https://doi.org/10.1002/0471142905.hg1802s62 (Chapter 18).

[61] A.R. Bright, G.J.C. Veenstra, Assay for transposase-accessible chromatin-sequencing using *Xenopus* embryos, Cold Spring Harbor Protocols 2019 (1) (2019) pdb.prot098327.

[62] E. Lacazette, A laboratory practical illustrating the use of the ChIP-qPCR method in a robust model: estrogen receptor alpha immunoprecipitation using Mcf-7 culture cells, Biochem. Mol. Biol. Educ. 45 (2) (2017) 152–160.

[63] E.D. Robin, R. Wong, Mitochondrial DNA molecules and virtual number of mitochondria per cell in mammalian cells, J. Cell. Physiol. 136 (3) (1988) 507–513.

[64] K.E. Yost, A.C. Carter, J. Xu, U. Litzenburger, H.Y. Chang, ATAC Primer Tool for targeted analysis of accessible chromatin, Nat. Methods 15 (5) (2018) 304–305.

[65] J. Ou, H. Liu, J. Yu, et al., ATACseqQC: a bioconductor package for post-alignment quality assessment of ATAC-seq data, BMC Genomics 19 (1) (2018) 169.

[66] L. Song, Z. Zhang, L.L. Grasfeder, et al., Open chromatin defined by DNaseI and FAIRE identifies regulatory elements that shape cell-type identity, Genome Res. 21 (10) (2011) 1757–1767.

[67] Z. Li, M.H. Schulz, T. Look, M. Begemann, M. Zenke, I.G. Costa, Identification of transcription factor binding sites using ATAC-seq, Genome Biol. 20 (1) (2019) 45.

[68] B. Quach, T.S. Furey, DeFCoM: analysis and modeling of transcription factor binding sites using a motif-centric genomic footprinter, Bioinform. (Oxford, Engl.) 33 (7) (2017) 956–963.

[69] F. Yan, D.R. Powell, D.J. Curtis, N.C. Wong, From reads to insight: a hitchhiker's guide to ATAC-seq data analysis, Genome Biol. 21 (1) (2020) 22.

[70] I. Barozzi, P. Bora, M.J. Morelli, Comparative evaluation of DNase-seq footprint identification strategies, Front. Genet. 5 (278) (2014).

[71] S. Neph, J. Vierstra, A.B. Stergachis, et al., An expansive human regulatory lexicon encoded in transcription factor footprints, Nature 489 (7414) (2012) 83–90.

[72] S. Andrews, FASTQC. http://www.bioinformatics.babraham.ac.uk/projects/fastqc/. (Accessed 3 April 2020).

[73] P. Ewels, M. Magnusson, S. Lundin, M. Käller, MultiQC: summarize analysis results for multiple tools and samples in a single report, Bioinformatics 32 (19) (2016) 3047–3048.

[74] B. Langmead, Aligning short sequencing reads with Bowtie, Curr. Protoc. Bioinform. (2010). Chapter 11:Unit 11 17.

[75] B. Langmead, S.L. Salzberg, Fast gapped-read alignment with Bowtie 2, Nat. Methods 9 (4) (2012) 357–359.

[76] H. Li, R. Durbin, Fast and accurate short read alignment with Burrows-Wheeler transform, Bioinformatics 25 (14) (2009) 1754–1760.
[77] M. Martin, Cutadapt removes adapter sequences from high-throughput sequencing reads, EMBnet J. 17 (1) (2011) 3.
[78] A.M. Bolger, M. Lohse, B. Usadel, Trimmomatic: a flexible trimmer for Illumina sequence data, Bioinformatics 30 (15) (2014) 2114–2120.
[79] S. Chen, Y. Zhou, Y. Chen, J. Gu, fastp: an ultra-fast all-in-one FASTQ preprocessor, Bioinformatics 34 (17) (2018) i884–i890.
[80] H. Li, B. Handsaker, A. Wysoker, et al., The sequence alignment/Map format and SAMtools, Bioinformatics 25 (16) (2009) 2078–2079.
[81] A. Adey, H.G. Morrison, Asan, et al., Rapid, low-input, low-bias construction of shotgun fragment libraries by high-density *in vitro* transposition, Genome Biol. 11 (12) (2010) R119.
[82] Boyle-Lab Blacklist. https://github.com/Boyle-Lab/Blacklist/tree/master/lists. (Accessed 31 March 2020).
[83] H.M. Amemiya, A. Kundaje, A.P. Boyle, The ENCODE blacklist: identification of problematic regions of the genome, Sci. Rep. 9 (1) (2019) 9354.
[84] C.E. Wimberley, S. Heber, PeakPass: automating ChIP-Seq blacklist creation, J. Comput. Biol. 27 (2) (2019) 259–268.
[85] G. Brown, GreyListChIP: Grey Lists: Mask Artefact Regions Based on ChIP Inputs, R Package Version 1.18.0, 2019.
[86] D.K. Nag, U. DasGupta, G. Adelt, D.E. Berg, IS50-mediated inverse transposition: specificity and precision, Gene 34 (1) (1985) 17–26.
[87] S. Baldi, Nucleosome positioning and spacing: from genome-wide maps to single arrays, Essays Biochem. 63 (1) (2019) 5–14.
[88] J.T. Robinson, H. Thorvaldsdóttir, W. Winckler, et al., Integrative genomics viewer, Nat. Biotechnol. 29 (1) (2011) 24.
[89] D.J. Galas, A. Schmitz, DNAse footprinting: a simple method for the detection of protein-DNA binding specificity, Nucleic Acids Res. 5 (9) (1978) 3157–3170.
[90] M.-H. Sung, S. Baek, G.L. Hager, Genome-wide footprinting: ready for prime time? Nat. Methods 13 (3) (2016) 222–228.
[91] T. Daley, A.D. Smith, Predicting the molecular complexity of sequencing libraries, Nat. Methods 10 (4) (2013) 325–327.
[92] P. Hansen, J. Hecht, D.M. Ibrahim, A. Krannich, M. Truss, P.N. Robinson, Saturation analysis of ChIP-seq data for reproducible identification of binding peaks, Genome Res. 25 (9) (2015) 1391–1400.
[93] S. Anders, W. Huber, Differential expression analysis for sequence count data, Genome Biol. 11 (2010).
[94] C.D. Scharer, A.P.R. Bally, B. Gandham, J.M. Boss, Cutting edge: chromatin accessibility programs CD8 T cell memory, J. Immunol. (2017) 1602086.
[95] M. Teng, M.I. Love, C.A. Davis, et al., A benchmark for RNA-seq quantification pipelines, Genome Biol. 17 (1) (2016) 74.
[96] Experiment Summary for ENCSR609OHJ. https://www.encodeproject.org/experiments/ENCSR609OHJ/. (Accessed 31 March 2020).
[97] T.S. Carroll, Z. Liang, R. Salama, R. Stark, I. de Santiago, Impact of artifact removal on ChIP quality metrics in ChIP-seq and ChIP-exo data, Front. Genet. 5 (2014) 75.
[98] S.G. Landt, G.K. Marinov, A. Kundaje, et al., ChIP-seq guidelines and practices of the ENCODE and modENCODE consortia, Genome Res. 22 (9) (2012) 1813–1831.
[99] P. Orchard, Y. Kyono, J. Hensley, J.O. Kitzman, S.C.J. Parker, Quantification, dynamic visualization, and validation of bias in ATAC-Seq data with ataqv, Cell Syst. 10 (3) (2020) 298–306.e294.
[100] M. Garber, M.G. Grabherr, M. Guttman, C. Trapnell, Computational methods for transcriptome annotation and quantification using RNA-seq, Nat. Methods 8 (2011).

[101] Y. Zhang, T. Liu, C.A. Meyer, et al., Model-based analysis of ChIP-Seq (MACS), Genome Biol. 9 (9) (2008) R137.
[102] J.M. Gaspar, Genrich. https://github.com/jsh58/Genrich. (Accessed 31 March 2020).
[103] E.D. Tarbell, T. Liu, HMMRATAC: a hidden Markov modeler for ATAC-seq, Nucleic Acids Res. 47 (16) (2019) e91.
[104] S. Steinhauser, N. Kurzawa, R. Eils, C. Herrmann, A comprehensive comparison of tools for differential ChIP-seq analysis, Brief. Bioinform. 17 (6) (2016) 953–966.
[105] S. Tu, Z. Shao, An introduction to computational tools for differential binding analysis with ChIP-seq data, Quant. Biol. 5 (3) (2017) 226–235.
[106] H. Xu, C.-L. Wei, F. Lin, W.-K. Sung, An HMM approach to genome-wide identification of differential histone modification sites from ChIP-seq data, Bioinformatics 24 (20) (2008) 2344–2349.
[107] L. Shen, N.-Y. Shao, X. Liu, I. Maze, J. Feng, E.J. Nestler, diffReps: detecting differential chromatin modification sites from ChIP-seq data with biological replicates, PLoS One 8 (6) (2013) e65598.
[108] A.T.L. Lun, G.K. Smyth, csaw: a bioconductor package for differential binding analysis of ChIP-seq data using sliding windows, Nucleic Acids Res. 44 (5) (2016) e45.
[109] C.S. Ross-Innes, R. Stark, A.E. Teschendorff, et al., Differential oestrogen receptor binding is associated with clinical outcome in breast cancer, Nature 481 (7381) (2012) 389–393.
[110] J.J. Reske, M.R. Wilson, R.L. Chandler, ATAC-seq normalization method can significantly affect differential accessibility analysis and interpretation, Epigenetics Chromatin 13 (1) (2020) 22.
[111] M.D. Robinson, A. Oshlack, A scaling normalization method for differential expression analysis of RNA-seq data, Genome Biol. 11 (3) (2010) R25.
[112] J.N. Paulson, C.Y. Chen, C.M. Lopes-Ramos, et al., Tissue-aware RNA-Seq processing and normalization for heterogeneous and sparse data, BMC Bioinform. 18 (1) (2017) 437.
[113] L.J. Zhu, C. Gazin, N.D. Lawson, et al., ChIPpeakAnno: a bioconductor package to annotate ChIP-seq and ChIP-chip data, BMC Bioinform. 11 (1) (2010) 237.
[114] L.J. Zhu, Integrative analysis of ChIP-chip and ChIP-seq dataset, Methods Mol. Biol. 1067 (2013) 105–124.
[115] K. Wang, M. Li, H. Hakonarson, ANNOVAR: functional annotation of genetic variants from high-throughput sequencing data, Nucleic Acids Res. 38 (16) (2010) e164.
[116] C.Y. McLean, D. Bristor, M. Hiller, et al., GREAT improves functional interpretation of cis-regulatory regions, Nat. Biotechnol. 28 (5) (2010) 495–501.
[117] M. Kondili, A. Fust, J. Preussner, C. Kuenne, T. Braun, M. Looso, UROPA: a tool for Universal RObust Peak Annotation, Sci. Rep. 7 (1) (2017) 2593.
[118] G. Yu, L.-G. Wang, Q.-Y. He, ChIPseeker: an R/Bioconductor package for ChIP peak annotation, comparison and visualization, Bioinformatics 31 (14) (2015) 2382–2383.
[119] S. Heinz, C. Benner, N. Spann, et al., Simple combinations of lineage-determining transcription factors prime cis-regulatory elements required for macrophage and B cell identities, Mol. Cell 38 (4) (2010) 576–589.
[120] A.R. Quinlan, I.M. Hall, BEDTools: a flexible suite of utilities for comparing genomic features, Bioinformatics 26 (6) (2010) 841–842.
[121] B. Khomtchouk, W. Koehler, D. Van Booven, C. Wahlestedt, Optimized functional annotation of ChIP-seq data, F1000Research 8 (612) (2019).
[122] N.L. van Berkum, E. Lieberman-Aiden, L. Williams, et al., Hi-C: a method to study the three-dimensional architecture of genomes, J. Visual. Exp.—JoVE 39 (2010) 1869.
[123] H. Tanizawa, K.-i. Noma, Unravelling global genome organization by 3C-seq, Semin. Cell Dev. Biol. 23 (2) (2012) 213–221.
[124] J.O.J. Davies, J.M. Telenius, S.J. McGowan, et al., Multiplexed analysis of chromosome conformation at vastly improved sensitivity, Nat. Methods 13 (1) (2016) 74–80.
[125] M.R. Mumbach, A.J. Rubin, R.A. Flynn, et al., HiChIP: efficient and sensitive analysis of protein-directed genome architecture, Nat. Methods 13 (11) (2016) 919–922.
[126] A.M. Ackermann, Z. Wang, J. Schug, A. Naji, K.H. Kaestner, Integration of ATAC-seq and RNA-seq identifies human alpha cell and beta cell signature genes, Mol Metab. 5 (3) (2016) 233–244.

[127] G. Robertson, M. Hirst, M. Bainbridge, et al., Genome-wide profiles of STAT1 DNA association using chromatin immunoprecipitation and massively parallel sequencing, Nat. Methods 4 (8) (2007) 651–657.
[128] D.B. Mahat, H. Kwak, G.T. Booth, et al., Base-pair-resolution genome-wide mapping of active RNA polymerases using precision nuclear run-on (PRO-seq), Nat. Protoc. 11 (8) (2016) 1455–1476.
[129] The Gene Ontology Consortium, M. Ashburner, C.A. Ball, et al., Gene ontology: tool for the unification of biology, Nat. Genet. 25 (1) (2000) 25–29.
[130] P. Khatri, M. Sirota, A.J. Butte, Ten years of pathway analysis: current approaches and outstanding challenges, PLoS Comput. Biol. 8 (2) (2012) e1002375.
[131] W.J. Kent, A.S. Zweig, G. Barber, A.S. Hinrichs, D. Karolchik, BigWig and BigBed: enabling browsing of large distributed datasets, Bioinformatics 26 (2010).
[132] F. Ramírez, F. Dündar, S. Diehl, B.A. Grüning, T. Manke, deepTools: a flexible platform for exploring deep-sequencing data, Nucleic Acids Res. 42 (Web Server Issue) (2014) W187–W191.
[133] A. Pohl, M. Beato, bwtool: a tool for bigWig files, Bioinformatics 30 (11) (2014) 1618–1619.
[134] J. Ou, L.J. Zhu, trackViewer: a bioconductor package for interactive and integrative visualization of multi-omics data, Nat. Methods 16 (6) (2019) 453–454.
[135] D. Karolchik, R. Baertsch, M. Diekhans, et al., The UCSC genome browser database, Nucleic Acids Res. 31 (1) (2003) 51–54.
[136] J. Zhu, J.Z. Sanborn, S. Benz, et al., The UCSC cancer genomics browser, Nat. Methods 6 (4) (2009) 239–240.
[137] X. Zhou, R.F. Lowdon, D. Li, et al., Exploring long-range genome interactions using the WashU epigenome browser, Nat. Methods 10 (5) (2013) 375–376.
[138] T. Wang, J. Liu, L. Shen, et al., STAR: an integrated solution to management and visualization of sequencing data, Bioinformatics (2013) btt558.
[139] F. Chelaru, L. Smith, N. Goldstein, H.C. Bravo, Epiviz: interactive visual analytics for functional genomics data, Nat. Methods 11 (9) (2014) 938–940.
[140] B.J. Raney, T.R. Dreszer, G.P. Barber, et al., Track data hubs enable visualization of user-defined genome-wide annotations on the UCSC Genome Browser, Bioinform. (Oxford, Engl.) 30 (7) (2014) 1003–1005.
[141] M. Reimers, V.J. Carey, Bioconductor: an open source framework for bioinformatics and computational biology, in: Methods in Enzymology, vol. 411, Academic Press, 2006, pp. 119–134.
[142] M. Lawrence, R. Gentleman, V. Carey, rtracklayer: an R package for interfacing with genome browsers, Bioinformatics 25 (14) (2009) 1841–1842.
[143] F. Hahne, S. Durinck, R. Ivanek, A. Mueller, S. Lianoglou, G. Tan, Gviz: plotting data and annotation information along genomic coordinates, R Package Version 1 (4) (2013).
[144] W. Lee, D. Tillo, N. Bray, et al., A high-resolution atlas of nucleosome occupancy in yeast, Nat. Genet. 39 (10) (2007) 1235–1244.
[145] O. Flores, Ö. Deniz, M. Soler-López, M. Orozco, Fuzziness and noise in nucleosomal architecture, Nucleic Acids Res. 42 (8) (2014) 4934–4946.
[146] Ö. Deniz, O. Flores, M. Aldea, M. Soler-López, M. Orozco, Nucleosome architecture throughout the cell cycle, Sci. Rep. 6 (1) (2016) 19729.
[147] R.L. Martin, J. Maiorano, G.J. Beitel, J.F. Marko, G. McVicker, Y.N. Fondufe-Mittendorf, A comparison of nucleosome organization in *Drosophila* cell lines, PLoS One 12 (6) (2017) e0178590.
[148] A.J. Andrews, K. Luger, Nucleosome structure(s) and stability: Variations on a theme, Annu. Rev. Biophys. 40 (1) (2011) 99–117.
[149] K. Struhl, E. Segal, Determinants of nucleosome positioning, Nat. Struct. Mol. Biol. 20 (3) (2013) 267–273.
[150] V.B. Teif, Nucleosome positioning: resources and tools online, Brief. Bioinform. 17 (5) (2015) 745–757.
[151] K. Chen, Y. Xi, X. Pan, et al., DANPOS: dynamic analysis of nucleosome position and occupancy by sequencing, Genome Res. 23 (2) (2013) 341–351.
[152] W. Chen, Y. Liu, S. Zhu, C.D. Green, G. Wei, J.-D.J. Han, Improved nucleosome-positioning algorithm iNPS for accurate nucleosome positioning from sequencing data, Nat. Commun. 5 (1) (2014) 4909.

[153] O. Flores, M. Orozco, nucleR: a package for non-parametric nucleosome positioning, Bioinformatics 27 (15) (2011) 2149–2150.
[154] Y. Vainshtein, K. Rippe, V.B. Teif, NucTools: analysis of chromatin feature occupancy profiles from high-throughput sequencing data, BMC Genomics 18 (1) (2017) 158.
[155] L. Liu, J. Xie, X. Sun, K. Luo, Z.S. Qin, H. Liu, An approach of identifying differential nucleosome regions in multiple samples, BMC Genomics 18 (1) (2017) 135.
[156] D. Buitrago, L. Codó, R. Illa, et al., Nucleosome dynamics: a new tool for the dynamic analysis of nucleosome positioning, Nucleic Acids Res. 47 (18) (2019) 9511–9523.
[157] A. Mayran, J. Drouin, Pioneer transcription factors shape the epigenetic landscape, J. Biol. Chem. 293 (36) (2018) 13795–13804.
[158] O. Fornes, J.A. Castro-Mondragon, A. Khan, et al., JASPAR 2020: update of the open-access database of transcription factor binding profiles, Nucleic Acids Res. 48 (D1) (2019) D87–D92.
[159] M.T. Weirauch, A. Yang, M. Albu, et al., Determination and inference of eukaryotic transcription factor sequence specificity, Cell 158 (6) (2014) 1431–1443.
[160] I. Yevshin, R. Sharipov, S. Kolmykov, Y. Kondrakhin, F. Kolpakov, GTRD: a database on gene transcription regulation-2019 update, Nucleic Acids Res. 47 (D1) (2019) D100–D105.
[161] V. Matys, O.V. Kel-Margoulis, E. Fricke, et al., TRANSFAC and its module TRANSCompel: transcriptional gene regulation in eukaryotes, Nucleic Acids Res. 34 (Database issue) (2006) D108–D110.
[162] I.V. Kulakovskiy, I.E. Vorontsov, I.S. Yevshin, et al., HOCOMOCO: towards a complete collection of transcription factor binding models for human and mouse via large-scale ChIP-Seq analysis, Nucleic Acids Res. 46 (D1) (2018) D252–D259.
[163] S. Shazman, H. Lee, Y. Socol, R.S. Mann, B. Honig, OnTheFly: a database of *Drosophila melanogaster* transcription factors and their binding sites, Nucleic Acids Res. 42 (Database issue) (2014) D167–D171.
[164] L.J. Zhu, R.G. Christensen, M. Kazemian, et al., FlyFactorSurvey: a database of *Drosophila* transcription factor binding specificities determined using the bacterial one-hybrid system, Nucleic Acids Res. 39 (Database issue) (2011) D111–D117.
[165] A. Sebastian, B. Contreras-Moreira, footprintDB: a database of transcription factors with annotated cis elements and binding interfaces, Bioinformatics 30 (2) (2013) 258–265.
[166] J.R. Hesselberth, X. Chen, Z. Zhang, et al., Global mapping of protein-DNA interactions in vivo by digital genomic footprinting, Nat. Methods 6 (4) (2009) 283–289.
[167] K.-S. Oh, J. Ha, S. Baek, M.-H. Sung, XL-DNase-seq: improved footprinting of dynamic transcription factors, Epigenetics Chromatin 12 (1) (2019) 30.
[168] B. Ason, W.S. Reznikoff, DNA sequence bias during Tn5 transposition, J. Mol. Biol. 335 (5) (2004) 1213–1225.
[169] M. Bentsen, P. Goymann, H. Schultheis, et al., ATAC-seq footprinting unravels kinetics of transcription factor binding during zygotic genome activation, Nat. Commun. 11 (1) (2020) 4267.
[170] A. Raj, H. Shim, Y. Gilad, J.K. Pritchard, M. Stephens, msCentipede: modeling heterogeneity across genomic sites and replicates improves accuracy in the inference of transcription factor binding, PLoS One 10 (9) (2015) e0138030.
[171] G. Ambrosini, R. Groux, P. Bucher, PWMScan: a fast tool for scanning entire genomes with a position-specific weight matrix, Bioinformatics 34 (14) (2018) 2483–2484.
[172] J. Korhonen, P. Martinmäki, C. Pizzi, P. Rastas, E. Ukkonen, MOODS: fast search for position weight matrix matches in DNA sequences, Bioinform. (Oxford, Engl.) 25 (23) (2009) 3181–3182.
[173] R. Siddharthan, Dinucleotide weight matrices for predicting transcription factor binding sites: generalizing the position weight matrix, PLoS One 5 (3) (2010) e9722.
[174] M.C. Frith, M.C. Li, Z. Weng, Cluster-Buster: finding dense clusters of motifs in DNA sequences, Nucleic Acids Res. 31 (13) (2003) 3666–3668.
[175] G. Tan, B. Lenhard, TFBSTools: an R/bioconductor package for transcription factor binding site analysis, Bioinform. (Oxford, Engl.) 32 (10) (2016) 1555–1556.

[176] A. Schep, motifmatchr: Fast Motif Matching in R, R Package Version 1.8.0 https://www.bioconductor.org/packages/release/bioc/html/motifmatchr.html. (Accessed 7 April 2020).
[177] B.J. Tremblay, universalmotif: Import, Modify, and Export Motifs with R, R package version 1.4.9 https://github.com/bjmt/universalmotif. (Accessed 7 April 2020).
[178] N.T.T. Nguyen, B. Contreras-Moreira, J.A. Castro-Mondragon, et al., RSAT 2018: regulatory sequence analysis tools 20th anniversary, Nucleic Acids Res. 46 (W1) (2018) W209–W214.
[179] K. Daily, V.R. Patel, P. Rigor, X. Xie, P. Baldi, MotifMap: integrative genome-wide maps of regulatory motif sites for model species, BMC Bioinform. 12 (2011) 495.
[180] C.E. Grant, T.L. Bailey, W.S. Noble, FIMO: scanning for occurrences of a given motif, Bioinformatics 27 (7) (2011) 1017–1018.
[181] T.L. Bailey, J. Johnson, C.E. Grant, W.S. Noble, The MEME suite, Nucleic Acids Res. 43 (W1) (2015) W39–W49.
[182] N. Jayaram, D. Usvyat, R. Martin AC., Evaluating tools for transcription factor binding site prediction, BMC Bioinform. 17 (1) (2016) 547.
[183] S.R. Kulkarni, D.M. Jones, K. Vandepoele, Enhanced maps of transcription factor binding sites improve regulatory networks learned from accessible chromatin data, Plant Physiol. 181 (2) (2019) 412–425.
[184] S.R. Kulkarni, D. Vaneechoutte, J. Van de Velde, K. Vandepoele, TF2Network: predicting transcription factor regulators and gene regulatory networks in *Arabidopsis* using publicly available binding site information, Nucleic Acids Res. 46 (6) (2018) e31.
[185] B. Tokovenko, R. Golda, O. Protas, M. Obolenskaya, A. El'skaya, COTRASIF: conservation-aided transcription-factor-binding site finder, Nucleic Acids Res. 37 (7) (2009) e49.
[186] E.T. Dermitzakis, A.G. Clark, Evolution of transcription factor binding sites in mammalian gene regulatory regions: conservation and turnover, Mol. Biol. Evol. 19 (7) (2002) 1114–1121.
[187] J.L. Gutiérrez, M. Chandy, M.J. Carrozza, J.L. Workman, Activation domains drive nucleosome eviction by SWI/SNF, EMBO J. 26 (3) (2007) 730–740.
[188] T.C. Voss, G.L. Hager, Dynamic regulation of transcriptional states by chromatin and transcription factors, Nat. Rev. Genet. 15 (2) (2014) 69–81.
[189] H.H. He, C.A. Meyer, M.W. Chen, V.C. Jordan, M. Brown, X.S. Liu, Differential DNase I hypersensitivity reveals factor-dependent chromatin dynamics, Genome Res. 22 (6) (2012) 1015–1025.
[190] E.E. Swinstead, T.B. Miranda, V. Paakinaho, et al., Steroid receptors reprogram FoxA1 occupancy through dynamic chromatin transitions, Cell 165 (3) (2016) 593–605.
[191] I. Goldstein, S. Baek, D.M. Presman, V. Paakinaho, E.E. Swinstead, G.L. Hager, Transcription factor assisted loading and enhancer dynamics dictate the hepatic fasting response, Genome Res. 27 (3) (2017) 427–439.
[192] E.G. Gusmao, M. Allhoff, M. Zenke, I.G. Costa, Analysis of computational footprinting methods for DNase sequencing experiments, Nat. Methods 13 (4) (2016) 303–309.
[193] S. Baek, I. Goldstein, G.L. Hager, Bivariate genomic footprinting detects changes in transcription factor activity, Cell Rep. 19 (8) (2017) 1710–1722.
[194] R.I. Sherwood, T. Hashimoto, C.W. O'Donnell, et al., Discovery of directional and nondirectional pioneer transcription factors by modeling DNase profile magnitude and shape, Nat. Biotechnol. 32 (2) (2014) 171–178.
[195] I. Berest, C. Arnold, A. Reyes-Palomares, et al., Quantification of differential transcription factor activity and multiomics-based classification into activators and repressors: diffTF, Cell Rep. 29 (10) (2019). 3147–3159.e3112.
[196] R.C. McLeay, T.L. Bailey, Motif enrichment analysis: a unified framework and an evaluation on ChIP data, BMC Bioinform. 11 (2010) 165.
[197] T. Lesluyes, J. Johnson, P. Machanick, T.L. Bailey, Differential motif enrichment analysis of paired ChIP-seq experiments, BMC Genomics 15 (1) (2014) 752.
[198] I.J. Tripodi, M.A. Allen, R.D. Dowell, Detecting differential transcription factor activity from ATAC-Seq data, Molecules 23 (5) (2018) 1136.

[199] F. Emmert-Streib, M. Dehmer, B. Haibe-Kains, Gene regulatory networks and their applications: understanding biological and medical problems in terms of networks, Front. Cell Dev. Biol. 2 (38) (2014).
[200] M. Levine, E.H. Davidson, Gene regulatory networks for development, Proc. Natl. Acad. Sci. U. S. A. 102 (14) (2005) 4936–4942.
[201] S. Okawa, S. Nicklas, S. Zickenrott, J.C. Schwamborn, A. Del Sol, A generalized gene-regulatory network model of stem cell differentiation for predicting lineage specifiers, Stem Cell Rep. 7 (3) (2016) 307–315.
[202] S. Chatterjee, N. Ahituv, Gene regulatory elements, major drivers of human disease, Annu. Rev. Genomics Hum. Genet. 18 (1) (2017) 45–63.
[203] Z. Duren, X. Chen, R. Jiang, Y. Wang, W.H. Wong, Modeling gene regulation from paired expression and chromatin accessibility data, Proc. Natl. Acad. Sci. 114 (25) (2017) E4914–E4923.
[204] R.N. Ramirez, N.C. El-Ali, M.A. Mager, D. Wyman, A. Conesa, A. Mortazavi, Dynamic gene regulatory networks of human myeloid differentiation, Cell Syst. 4 (4) (2017). 416–429.e413.
[205] E.R. Miraldi, M. Pokrovskii, A. Watters, et al., Leveraging chromatin accessibility for transcriptional regulatory network inference in T Helper 17 cells, Genome Res. 29 (3) (2019) 449–463.
[206] Y. Niu, ATAC-seq Data Analysis: From FASTQ to Peaks. https://yiweiniu.github.io/blog/2019/03/ATAC-seq-data-analysis-from-FASTQ-to-peaks/. (Accessed 26 June 2023).
[207] J.M. Gaspar, ATAC-seq Guidelines. https://informatics.fas.harvard.edu/atac-seq-guidelines.html. (Accessed 26 June 2023).
[208] L. Delisle, M. Doyle, F. Heyl, ATAC-Seq Data Analysis. https://training.galaxyproject.org/training-material/topics/epigenetics/tutorials/atac-seq/tutorial.html. (Accessed 26 June 2023).
[209] M. Bajic, K.A. Maher, R.B. Deal, Identification of open chromatin regions in plant genomes using ATAC-Seq, Methods Mol. Biol. 1675 (2018) 183–201.
[210] L. Huang, X. Li, L. Dong, B. Wang, L. Pan, Profiling of chromatin accessibility identifies transcription factor binding sites across the genome of *Aspergillus* species, BMC Biol. 19 (1) (2021) 189.
[211] H. Liu, J. Ou, K. Hu, L.J. Zhu, H. Liu, Workshop 200: Best Practices for ATAC seq QC and Data Analysis. https://www.youtube.com/watch?v=VZFUu_cJxyI. (Accessed 28 June 2023).
[212] C.E. Wimberley, S. Heber, PeakPass: Automating ChIP-Seq Blacklist Creation, 2019. Cham.
[213] G. Brown, GreyListChIP: Grey Lists—Mask Artefact Regions Based on ChIP Inputs, version 1320, 2023.
[214] J.M. Gaspar, Genrich: Detecting Sites of Genomic Enrichment. https://github.com/jsh58/Genrich. (Accessed 28 June 2023).
[215] A. Montgomery, Benchmarking ATAC-seq Peak Calling. https://bigmonty12.github.io/peak-calling-benchmark. (Accessed 28 June 2023).
[216] M.I. Love, W. Huber, S. Anders, Moderated estimation of fold change and dispersion for RNA-seq data with DESeq2, Genome Biol. 15 (12) (2014) 550.
[217] K. Hu, J. Ou, R. Li, H. Liu, L.J. Zhu, Integrated ChIP-seq Data Analysis Workshop. https://hukai916.github.io/IntegratedChIPseqWorkshop/. (Accessed 28 June 2023).
[218] K. Hu, L.J. Zhu, H. Liu, J. Ou, K. Hu, J. Zhu, H. Liu, J. Ou, Workshop 200: Integrated ChIP Seq Data Analysis. https://www.youtube.com/watch?v=EAxFrz_F4bg. (Accessed 28 June 2023).
[219] J. Ou, L.J. Zhu, trackViewer_workshop2020. https://jianhong.github.io/workshop2020/articles/trackViewer.html. (Accessed 28 June 2023).
[220] X. Zhou, B. Maricque, M. Xie, et al., The human epigenome browser at Washington University, Nat. Methods 8 (12) (2011) 989–990.

CHAPTER 14

Best practices for ChIP-seq and its data analysis

Huayun Hou, Matthew Hudson, and Minggao Liang
Genetics and Genome Biology, SickKids Research Institute, Toronto, ON, Canada

Introduction

The precise spatial and temporal regulation of eukaryotic gene expression is essential for development and homeostasis. It is a complex process regulated by the interplay between *cis*-regulatory factors, such as promoters and enhancers, and *trans*-factors, such as transcription factors (TFs) and cofactors. TFs play a central role in gene regulation. Many act as "master regulators" and drive the differentiation of specific cell lineages. TFs exert their function by directly or indirectly binding to the DNA. In eukaryotic cells, DNA is wrapped around histone proteins to form chromatin. Chemical modifications to the tails of histone proteins (hereby referred to as histone modifications) render the DNA more or less accessible to other nuclear proteins, and specific combinations of histone modifications have been associated with different genomics functions [1,2]. For example, trimethylation of lysine 4 of histone H3 (H3K4me3) is associated with increased transcription activity at gene promoters, and acetylation of lysine 27 of histone H3 (H3K27ac) is associated with active enhancer regions [3,4]. Thus, identifying genomic regions bound by TFs or marked by histone modifications is a fundamental step in advancing our understanding of gene regulation and cellular functions.

Chromatin immunoprecipitation and sequencing (ChIP-seq) [5] is a technique to characterize regions of the genome that are bound by the targeted proteins, including TFs and histone modifications. ChIP-seq uses antibodies raised against specific nuclear proteins (i.e., TFs) or histone modifications to pull down protein-bound DNA fragments for sequencing in order to identify protein-bound regions genome-wide. This chapter describes how best to design a ChIP-seq experiment based on your cell or tissue type of choice, your protein target of choice, and how much material you are starting with, and follows with the best practices in downstream data analysis after sequencing.

ChIP-seq is a versatile technique when it comes to sample preparation and can be performed on cultured or sorted cells, harvested tissues from animal models, formalin-fixed paraffin-preserved (FFPE) banked human tissues, and fresh or frozen samples. However, certain ChIP-seq techniques are better suited to certain sample preparation methods or to the particular protein of interest. The two most important factors to consider when designing a ChIP-seq experiment are (i) how to fragment the genome for immunoprecipitation and (ii) whether to cross-link proteins in a "snapshot" of their current state or to keep the genome in its native conformation. Each technique has its advantages and disadvantages, and many protocols have been designed with specific situations in mind. Importantly, the analysis of ChIP-seq data is also versatile, requiring careful choices of statistical models and methods based on the type of data and the biological questions. This chapter will outline why these factors are important, how one can identify which protocol and analytical method will best suit their needs, and how will they provide researchers with the knowledge of adjusting a protocol or analysis pipeline according to their needs.

Crucial considerations for a rigorous ChIP-seq experiment
To fix or not to fix?

The most commonly used tissue preparation method for ChIP-seq is formaldehyde fixation of tissue. Dissociated cells or finely minced tissues are fixed by incubation in dilute formaldehyde to cross-link disulfide bonds within and between proteins to preserve protein-protein and protein-DNA interactions. Fixation is a must for any ChIP-seq experiment targeting TFs or other unstable or transient chromatin interactions. A benefit of formaldehyde fixation is that fixed chromatin is quite stable at a range of temperatures—it can be frozen for future use and remains complexed at physiological temperatures (i.e., 37°C), meaning it can be easily manipulated in the lab without concern for loss of sample integrity. Importantly, once formaldehyde-fixed chromatin for ChIP-seq has been immunoprecipitated, the cross-links must be reversed by a high-temperature incubation (~70°C) in a concentrated salt solution to release the DNA from the protein for ChIP-seq library construction [6].

There are some concerns associated with formaldehyde fixation. First, there is a risk of over-fixing chromatin with either too high concentration or too long incubation and vice versa with under-fixing, whereby either too much of nonspecific binding will occur or not enough specific binding will occur, ultimately yielding a ChIP-seq library of low complexity (i.e., overrepresentation of certain genomic regions) [7]. Second, not all protein-DNA interactions can be fixed with formaldehyde; some researchers have resorted to using a double-fixation technique with disuccinimidyl glutarate for such proteins, like NF-kB [8]. Third, precipitation of formaldehyde-fixed chromatin tends to be inefficient, and thus such experiments generally require at least 20 million cells, which is not always feasible for a given cell or tissue type. Finally, formaldehyde is also associated with certain toxicity and may activate DNA damage and apoptosis pathways if its concentration and incubation time are not well controlled [6].

While more unstable and transient protein-DNA interactions are better captured with a cross-linking ChIP-seq protocol, more stable interactions—like those between DNA and histones—do not require fixation for precipitation. The nucleosome can be an abundant source of information owing to its various possible histone modifications. These are frequently used in genomics to complement transcription-factor ChIP-seq, ATAC-seq, and RNA-seq. "Native" ChIP-seq protocols simply skip the fixation steps in cross-linking protocols, and immunoprecipitation is performed on nuclei in their native conformations [9,10]. As most antibodies are raised to recognize epitopes of unfixed or native antigens, an important benefit of native ChIP-seq over fixed ChIP-seq is that a given antibody to a protein of interest is frequently more efficient in native chromatin than in the fixed chromatin, thus yielding better immunoprecipitation yield.

Nuclear isolation for ChIP-seq

Nuclear isolation is strongly recommended for any ChIP-seq experiment, as cytoplasmic proteins can compete for antibodies during immunoprecipitation and fragmentation may be impeded if nuclei are not isolated. Another benefit of isolating nuclei before ChIP-seq is that it ensures chromatin is kept in a more native conformation within the nuclear membrane, which may increase the likelihood of finding biologically relevant results. Nuclei are commonly isolated for ChIP-seq through the use of lysis buffers and mechanical homogenization, which are both easy and inexpensive methods. However, mechanical homogenization in particular may not always be reproducible between cell types, and researchers and may not be suitable for more high-throughput experiments. The author recommends the NEXSON technique (Nuclei Extraction by Sonication), which actually utilizes a light sonication step on whole cells to disrupt the cell membrane. This technique "shakes out" the nuclei with minimal damage to nuclear membranes, allowing the purification of high-quality nuclei in numbers down to a few thousands [11].

Fragmenting chromatin for ChIP-seq

Chromatin needs to be fragmented into smaller pieces to make both immunoprecipitation and library preparation more efficient. Chromatin fragmentation is known as one of the more difficult aspects of ChIP-seq, as there is no "one-size-fits-all" method for every cell type and experiment. Every ChIP-seq experiment requires optimization of fragmentation before proceeding to immunoprecipitation. Generally, the preferred DNA fragment sizes for ChIP-seq are 250–350 bp, and it is important to ascertain the best fragmentation conditions for your experiment [12] (Fig. 14.1). The two methods employed for chromatin fragmentation for ChIP-seq are enzymatic digestion and sonication, both of which are effective regardless of whether chromatin is fixed or not.

Fragmentation by enzymatic digestion is usually done with micrococcal nuclease (MNase), which will digest exposed DNA [13]. MNase digestion creates consistently small fragment sizes in a laddered nucleosomal pattern, thereby increasing the resolution of ChIP-seq and making it a favorite among researchers by examining histone modifications. MNase digestion is a relatively simpler and faster technique than sonication, requiring less optimization: generally, MNase works best at 37°C and over a period of 5–7 min for most applications. However, MNase is known to preferentially cut at AT-rich sites, leading to a potential bias during fragmentation [14].

Sonication is the most commonly used method of fragmentation, whereby chromatin is sheared by bursts of ultrasonic vibration using either a probe inserted directly into a sample or through sample submersion in a water bath. Sonication is preferred by many researchers as it generates unbiased fragment distributions through the randomness of DNA breakpoints, but it can also require significant optimization in order to achieve the desired fragment distribution. Every sonicator will have its own instructions on how to shear chromatin to the desired length based on the company's own quality-control experiments, but it is important to test these conditions yourself to see if they work for your cell type of interest. Sonication conditions are adjusted based on the amplitude or strength of vibrations, length of bursts of sonication and rest periods, and the number of cycles. Depending on the degree of chromatin condensation, these parameters may need to be adjusted. As sonication generates heat, it is important to keep samples cool either in an ice bath (as is usually the case when using a probe) or in a water bath (as is usually the case when using a multisample sonicator) to avoid any dissociation between the DNA and your protein of interest. Low concentrations of sodium dodecyl sulfate (SDS; ~0.1%) should also be used in sonication buffers to increase efficiency [15].

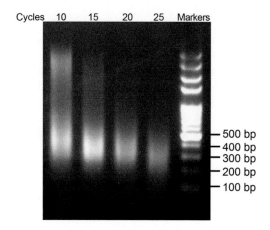

FIG. 14.1

Sonication effect on DNA fragmentation. DNA fragment distribution on an agarose gel after sonicating chromatin from HEK293T cells at 10, 15, 20, and 25 cycles (10s on, 20s off). In this gel, 15 cycles appears to be the best condition as it shows the strongest 300-bp band.

Modified from He, L., Yu, W., Zhang, W., Zhang, L., 2021. An optimized two-step chromatin immunoprecipitation protocol to quantify the associations of two separate proteins and their common target DNA. STAR Protocols 2(2), 100504. https://doi.org/10.1016/j.xpro.2021.100504.

Regardless of the fragmentation method, it is crucial to assess the fragmentation efficiency before proceeding any further with ChIP-seq. This can be done easily by taking a sample of fragmented chromatin, separating the DNA from the proteins (which may require reversing cross-links if using fixed chromatin) and running the DNA on an agarose gel to view the fragment size distribution.

Single-cell and low-input ChIP-seq

While a significant limitation to ChIP-seq in the past has been its need for a large amount of starting material, there are several published low-input [9,16–18] and even single-cell methods [19–21]. In these protocols, single cells or pools of ~20 cells are placed in wells with lysate and MNase for fragmentation, with each well receiving its own unique genetic barcode (see Library Preparation). While these techniques expand the realm of possible cell and tissue types for ChIP-seq, they tend to work best when examining histone modifications due to the nature of the fragmentation technique and the limited sample size, as there may be too few TFs of interest spread throughout one cell's genome to acquire usable data from [22].

Appropriate controls in ChIP-seq

As ChIP-seq experiment is a measure of the enrichment of one particular factor in the genome, be it a histone modification or a TF, it is absolutely necessary to include an unenriched reference sample with which to compare to. There are two types of controls that are used in ChIP-seq: input libraries and immunoglobulin G (IgG)-enriched libraries. An input library is made from a sample of fragmented chromatin before incubation with ChIP antibodies (usually around 10% of the sample that will be immunoprecipitated), which will allow a researcher to examine the "normal" chromatin landscape under whichever fragmentation conditions were used. An IgG-enriched library is made from fragmented chromatin incubated with simple IgG antibodies, which theoretically should not bind to any specific nuclear proteins and therefore should not show any particular enrichment. However, input libraries tend to be preferred over IgG-enriched libraries, as IgG can sometimes introduce artifacts and inputs are considered as a more unbiased control [23].

Antibody incubation, chromatin washing, and elution

Many companies produce "ChIP grade" primary antibodies against popular target proteins and histone modifications (i.e., H3K27ac, H3K27me3, H3K4me3, RNA polymerases, CTCF, etc.), which have been specifically validated for ChIP-seq experiments and are generally monoclonal. Monoclonal antibodies are preferred over polyclonal antibodies to ensure antibody-antigen specificity and minimize cross-reactivity with other proteins or epitopes [24]. Once the preferred primary antibody has been identified, the preferred means of collecting antibody-bound chromatin needs to be considered. The two most popular methods are incubation with agarose beads (i.e., Sepharose) and magnetic beads coated with a secondary antibody (Protein A, Protein G, or streptavidin). Magnetic beads tend to be preferred over agarose beads as they are less time-consuming to use and require samples to simply be held against a magnet to isolate antibody-bound chromatin instead of pelleting beads by centrifugation. If the primary antibody is biotinylated, then a streptavidin-coated bead can be used for immunoprecipitation; otherwise, Protein A- or G-coated beads (or a mixture of the two) will be needed to be used depending on the species and heavy chain of the primary antibody. Proteins A and G are proteins found in streptococcal bacteria that readily bind to IgG: Protein A has higher binding affinity for rabbit, pig, dog, cat, and guinea pig IgG, whereas Protein G has higher binding affinity for mouse, human, rat, cow, and horse IgG [25]. Primary antibodies should be incubated with beads separately (2–4 h, 4°C) in radioimmunoprecipitation (RIPA) buffer before adding to the chromatin to ensure binding between primary and secondary antibodies. Chromatin can also be preincubated with beads alone (no primary antibody) to block any nonspecific interactions between chromatin and beads. Chromatin-bead incubations can be carried out at either room temperature for 2–4 h or overnight (~12 h) at 4°C, depending on the protocol and time restraints.

After antibody incubation, chromatin-bead complexes should be washed with RIPA buffer over an increasing salt gradient to remove any excess unbound antibodies and to break any weak interactions between antibodies and off-target proteins or epitopes. If using magnetic beads, the sample is held against a magnetic rack, the supernatant is removed and the wash buffer is added gently and slowly above the bead pellet to avoid directly disturbing it; this process is repeated several times. If using agarose beads, the beads are pelleted at low speed, the supernatant is removed, and the pellet is resuspended in wash buffer. In either case, it is important to not let the beads dry out, as this will decrease the efficiency of precipitation.

After washing, the chromatin can be eluted off of the beads and cross-links can be reversed (if working with fixed chromatin). This is generally done with heat (65°C) and a high-salt buffer.

Analysis of ChIP-eluted chromatin
Library preparation

Immunoprecipitated chromatin needs to undergo several steps before it can become a ChIP-seq library ready for sequencing: fragment end repair, ligation of sequencing adapters, and amplification (Fig. 14.2). As chromatin shearing will not always lead to blunt-ended fragments, DNA overhangs need to be "filled in" with complementary nucleotides or removed, using T4 DNA polymerase, Klenow

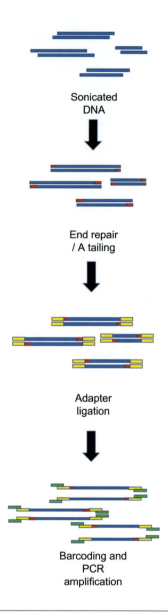

FIG. 14.2

Creating a sequencing library from immunoprecipitated DNA. Sonicated DNA is first repaired to create blunt ends and dATP is added to the ends for adapter ligation. Double-stranded sequencing adapters are then added to the blunt ends. Finally, barcoded primers complementary to the adapters are incorporated into fragments via PCR, giving each PCR reaction its own unique nucleotide barcode for identification after sequencing.

No permission required.

polymerase (*Escherichia coli* DNA Pol I large fragment), and T4 polynucleotide kinase. If using Illumina adapters (which are most common), fragments will need to be A-tailed for ligation to T-tailed adapters using Klenow polymerase and excess dATP. After adapter ligation, each library undergoes PCR with a specific barcoded primer (with usually a 6- to 8-nt code) for deconvolution of sequencing results. Finally, libraries need to be size-selected to remove large fragments (generally >450 bp) which will not amplify efficiently and small fragments (<200 bp, generally dimerized adapters) which may be overrepresented if amplified during PCR. Size selection can be performed using SPRI beads (i.e., AMPure beads; Thermo Scientific), simple excision from an agarose gel, or one of many commercially available machines (i.e., PippinPrep; Sage Biosciences). Finally, once libraries have been size-selected, they can be amplified by PCR using a high-fidelity DNA polymerase. If using a library preparation kit, the manufacturer will supply PCR conditions according to the estimated amount of DNA that was used to make the library, which are usually sufficient. If desired, however, a sample of preamplified DNA can be used in a qPCR reaction to determine the optimal cycle number, as in ATAC-seq (see Chapter 13).

Sequencing a ChIP-seq experiment

It is important to choose a sequencing experimental design appropriate for the type of ChIP-seq experiment, and one that is cost-efficient. Some of the key parameters include sequencing depth, single- or paired-end (PE) sequencing, and read length. The sequencing depth required for ChIP-seq experiments largely depends upon the type of factor profiled as well as the species. Point-source factors, for example, TFs and their cofactors, localize to specific regions of the genome, thus requiring lower sequencing depth. Broad-source factors, including most histone modifications and factors like RNA polymerase II, cover broader regions across the genome and will require higher sequencing depth. Specifically, the encyclopedia of DNA elements (ENCODE) consortium suggests a minimum of 10 million reads for point-source factors for mammalian genomes (4 million for fly and worm genomes), and a minimum of 20 million reads for mammalian broad marks (10 million for fly and worm) [26]. While studies have systematically evaluated the impact of sequencing depth on ChIP-seq experiments [27,28] and provided standard guidelines, it remains important to review one's specific experimental design. **To find the optimal sequencing depth, saturation analysis can also be performed**. For example, a sequenced library can be randomly downsampled to evaluate the impact of different sequencing depths on called peaks.

Single- or paired-end sequencing can also have an impact on the output of ChIP-seq experiments. Single-end (SE) sequencing generates sequencing reads from one end of the DNA fragments, whereas PE sequencing reads a DNA fragment from both ends. PE sequencing can increase the accuracy of read mapping, especially to regions with low complexity, and allows the detection of structural rearrangements like insertions and deletions. Zhang et al. showed that PE sequencing improves peak calling and the accuracy of allele-specific binding detection [28]. With the advance of sequencing technologies, the cost of PE sequencing is becoming comparable with SE sequencing. However, PE sequencing analysis can be more expensive computationally and require analysis with peak callers designed specifically to handle PE sequencing data.

As with any experiment, ChIP experiments should be designed with batch effects in mind, ideally with the inclusion of samples from each condition, control, and input controls per round of sample and library preparation. Given the known impact of sequencing parameters on ChIP-seq data analysis, it is also important that libraries from the same experiment, especially when different conditions are being compared, should be sequenced with the same sequencing platform and to similar depths. If this is not possible, bioinformatics processing such as read downsampling, or read trimming, could be applied to mitigate the impact of sequencing-related technical variations.

Sequencing read preprocessing and alignment

Raw sequencing reads are obtained from the sequencer and demultiplexed. The preprocessing and alignment of sequencing reads from ChIP-seq experiments should follow typical guidelines for any NGS experiments. Briefly, the quality of sequencing reads should be examined (reviewed in the "Quality Control" section). Next, reads will be filtered and trimmed for (1) bases with poor base-calling quality as determined by the sequencer and (2) adaptor sequences that get sequenced at the end of the reads when the read length is longer than the insert size (length of the DNA fragments). Trimming results in a set of highly confident reads and increases mapping efficiency.

Trimmed reads will then be mapped against a reference genome assembly of choice. Short-read alignment is an intriguing computational problem. The large number of short reads obtained from NGS experiments and the relatively large search space of the whole genome make it a game of efficiency. Meanwhile, due to errors in sequencing as well as naturally occurring genetic variants, mismatches should be allowed, which further increases the complexity of the problem. Different strategies of short-read alignment and algorithms available have been extensively reviewed [29,30]. The most popular short-read aligners include Bowtie2 [31] and BWA [32], both of which use Burrows-Wheeler transformation-based indexing to reduce memory requirement and increase the speed.

Additional issues to be considered relate to the mappability of the genome: some reads may be mapped to multiple positions in the genome. Ambiguous read mapping affects peak detection because the total read number can be over- or underrepresented depending on how the specific aligner handles multimapping reads. Consequently, only uniquely mappable reads are typically used for downstream ChIP analyses, unless repeats are the focus of the study, in which case, multiple mapping reads should be carefully processed. Inappropriate read counting at repeat regions could result in false biological conclusions such as overestimation of TF binding to repeat elements.

Furthermore, NGS sequencing has certain technical biases that result in uniformly highly enriched signals regardless of cell type and conditions. These have been systematically characterized, and a "blacklist" of regions giving rise to these artifacts is compiled for each species using the ENCODE datasets [33]. It is advised to exclude reads mapped to these "blacklist" regions before downstream analysis.

Peak calling

After read alignment, peak calling is performed to identify genomic regions with statistically significant enrichment of reads in the experimental samples compared to controls. Most frequently, data from input or IgG-enriched libraries are used as controls. The identified peaks are a proxy for protein binding sites or regions with enriched histone modifications and are used in subsequent analysis.

Numerous peak callers have been developed for ChIP-seq experiments. There are several issues to consider when choosing a peak caller. First, the statistical model used to model read distribution largely affects the outcome. ChIP-seq reads are often summarized in windows across the genome. Due to the sparse nature of the count data, many methods, such as **SICER** and **CCAT** [34,35], model the data with a Poisson distribution which assumes that mean and variance are the same. Other methods model the data with a negative binomial distribution, which accounts for overdispersion, or when the variance is larger than what's assumed by Poisson distribution [36]. Alternatively, **MACS** uses a local Poisson model to account for local chromatin complexity [37]. Another feature for peak callers to model is the shape of the peaks. For point-source factors such as TFs, a monomodal peak is expected and a summit is often estimated as the most probable site of the TF-DNA interaction. Broad-source factors, such as histones PTM, on the other hand, tend to occupy a much broader region. H3K27me3-marked domains, for example, can be dozens of kilobases long. For these factors, it is important to choose peak callers that are developed to specifically identify domains of diffuse signals, such as SICER, CCAT, and MUSIC [34,35,38]. Instead of looking for local enrichment of reads density, these tools aim to identify boundaries of domains where reads cluster by scanning the whole genome in bins. Alternatively, other tools allow the calling of broader domains by making modifications to their original algorithm. MACS, for example, uses a subsequent step to stitch peaks together. A recent review surveyed 30 methods and identified features that distinguish the performances of different peak callers [39]. Using both simulated data and real ChIP-seq datasets, the authors showed that input signal should be used only for ranking peaks but not for identifying peaks; and that methods based on Poisson distribution are more sensitive than the ones based on negative binomial. Such studies could serve as guidelines for choosing a peak caller based on data types as well as the aim of the study.

Quality control (QC)

Careful quality control of ChIP-seq data is required at each step of the analysis. The ENCODE consortium published a comprehensive guideline for evaluating ChIP-seq experiments [26]. The detailed QC standards for ChIP-seq data can be found at https://www.encodeproject.org/data-standards/. The ENCODE-proposed methods for quality assessments and quality cutoffs have been widely adopted by the research community. ChIP-seq QC pipelines have been developed and packaged into ready-to-use packages. For example, preseq [40] introduced measurements of library complexity; and ChIPQC [41] provided a comprehensive assessment report. Overall, the following aspects should be considered:

1. **Sequencing quality**: Similar to RNA-seq, library quality should be examined upon the acquisition of sequencing data. FastQC [42] is the most widely used tool for examining sequencing quality. It provides an overview of base calling quality, GC content, duplication rate, adaptor content, overrepresented sequences, etc. Library failure or bias can be detected and addressed prior to any downstream analysis. This step should be repeated after reads are trimmed for quality, adaptors, and/or filtered for contaminations.
2. **Library complexity**: A failed ChIP-seq experiment may result in very few unique DNA fragments, which upon PCR amplification could be sequenced and result in a library with low complexity. Many factors, such as antibodies, sonication, cross-linking, and overamplification, could result in low-library complexity. It can be measured by PCR bottleneck coefficient (PBC) as recommended by ENCODE.
3. **Signal-to-noise ratio (SNR)**: A successful ChIP-seq experiment should feature high SNR, suggesting that immunoprecipitation (IP) is successful. SNR can be intuitively assessed by performing peak calling, with more peaks indicating better signal enrichment. Meanwhile, a significant proportion of reads should be located within peak regions. This is measured by the fraction of reads in peaks (Frip). In addition, SNR can be measured without peak calling to

avoid biases introduced by different peak callers. For example, "plotFingerpring" function from the **deeptools2** package plots the accumulation of reads across all genomic bins [43]. If reads are found to be highly enriched in a small proportion of the genome, the ChIP enrichment is likely robust. Another method, strand cross-correlation analysis is ideal for TF ChIP-seq data [44]. Because only the beginning of each DNA fragment is sequenced, by shifting reads on one strand, the forward and strand read signals should overlap. This method calculates the correlation between forward and reverse strand signals as the function of strand shift to estimate the average fragment size. Failure to identify an optimal K indicates a lack of signal enrichment.

4. **Replicate numbers and reproducibility**: The ENCODE guideline recommended that at least two biological replicates should be performed for the identification of protein-binding sites. However, for quantitative analyses, for example, to identify differential TF binding, two replicates are not sufficient. The choice of replicate numbers depends on many factors including technical variations, biological variations, and the desired statistical power. For example, samples from human donors will be more heterogeneous than those from model organisms with the same genetic background. Similarly, complex tissues will be more heterogeneous than cultured cell lines. Schurch et al. suggested that for RNA-seq differential analysis, at least six biological replicates should be used to properly control for FDR with the majority of commonly used tools. As ChIP-seq differential analysis often uses the same toolkit and the technical variations of ChIP-seq tend to be higher than RNA-seq, it is recommended to obtain at least the same number of biological replicates as required for RNA-seq. Overall, more biological replicates will increase the sensitivity and specificity of the assay. Careful considerations should be given to balance the benefits and the cost and the availability of biological samples.

The reproducibility between replicates is often measured by calculating the correlation of ChIP signals across the genome. Genomes could be binned into equal intervals, and the number of reads mapped to each bin will be calculated and used for correlation analysis. Reproducibility could also be measured at the level of peaks. Irreproducible discovery rate (IDR) ranks peaks called for both replicates and separates them into reproducible and irreproducible groups with the assumption that peaks ranked higher (i.e., more significant) are more likely to be consistent [26]. IDR not only evaluates replicates reproducibility but also provides a list of robust peaks for downstream analysis.

Visualization of ChIP-seq data

VisualizingChIP-seq signal in its genomic context can greatly facilitate quality assessment and biological interpretation of the data. This is often achieved using genome browsers, which are visual representations of the genome of interest. Genome browsers support the display of a wide variety of genomics data. ChIP-seq data are primarily visualized by showing read density across the genome as well as peak coordinates. Other metrics commonly used (i.e., by ENCODE) include signal P-value and fold-change over input. Web-based genome browsers provide comprehensive sets of genomic annotations, including gene annotations, repeats, conservation scores, and many more. They also offer extensive public datasets including ChIP-seq, ATAC-seq, RNA-seq, single-cell RNA-seq, and chromatin interaction data, from large consortia such as ENCODE and GTEx, as well as individual studies. On genome browsers, user-supplied data can be easily compared with these public resources, allowing an in-depth understanding of the genomic context of the ChIP-seq signal. Moreover, web-based genome browsers allow effortless sharing of browser sessions, making it a convenient option for publishing data and collaborating with others.

The most widely used genome browsers include the **UCSC Genome Browser** [45], the **WashU Epigenome Browser** [46], and the **Integrative Genomics Viewer** (IGV) [47]. The UCSC genome browser features an intuitive user interface and one of the most comprehensive collections of public data resources. The WashU Epigenome Browser is particularly designed for better visualization of the epigenome data. Furthermore, it recently introduced several innovative panels enabling the visualization of 3D genome structure and the animation of dynamic time series data. While a web-based version is available, IGV is commonly used as a desktop application. It is lightweight, flexible, and intuitive to use. IGV is especially convenient for visualizing data from a custom genome.

In addition to genome browsers, **several R packages** are available for visualizing ChIP-seq data along genomic coordinates. **Gviz**, for example, provides flexible plotting options for multiple types of data [48]. With Gviz, users can use predefined genome and annotation data packages in bioconductor or easily extract data from public databases such as ENSEMBL and UCSC. Using R packages facilitates the production of publication-quality figures and allows the generation of large numbers of plots automatically.

Another common approach is to visualize ChIP-seq data as heatmaps and aggregate profiles. In the heatmap, each row represents a genomic region (i.e., peaks, genes, certain distance up/downstream from TSS), color represents ChIP-seq signal (i.e., normalized read counts in 10 base pair bins), and each panel is a sample (i.e., different condition or factor). Signal across a set of genomic regions can then be summarized (i.e., mean, median, sum) to generate an aggregate plot. Heatmaps and aggregate plots are best suited to show the enrichment of protein binding across genomic regions of interest. For example, in Fig. 14.3, panel B shows the enrichment of H3K4me1 at promoter regions and H3K36me3 across gene bodies, generated using **deepTools** [43].

Peak annotation and functional enrichment analysis

The first step to understanding the biological context of protein binding sites identified in ChIP-seq experiments is often peak annotation. This process aims at assigning peaks to functionally annotated genomic regions, including promoters, gene bodies, intergenic regions, etc., providing an overview of the genomic distribution of the peaks. Peak annotation can be done in customized ways such as generating a file of genomic region annotations and then overlapping peaks with these regions. Alternatively, many tools or bioconductor packages such as **ChIPpeakAnno** [49] or **ChIPseeker** [50] can perform peak annotation using either built-in or customized genome annotations and provide other helpful functions such as comparisons between different factors and visualizations of the results.

It is often of interest to examine if the protein-binding sites are associated with certain biological pathways. One common method is to first associate peaks with genes. This is frequently done based on the proximity of ChIP-seq peaks to gene transcription start sites (TSS). Peaks can be associated with the nearest TSS regardless of distance from the TSS or based on a customized distance cutoff. While there are no fixed standards, within 3 kb from TSS is typically considered a cutoff. Once the peaks are associated with genes, pathway enrichment tools such as **ClusterProfiler** [51] can be used for functional enrichment analysis. However, this approach has several limitations. For example, it won't be able to accurately determine the target gene when the binding site is near multiple TSSs and will not annotate distal intergenic regions to genes.

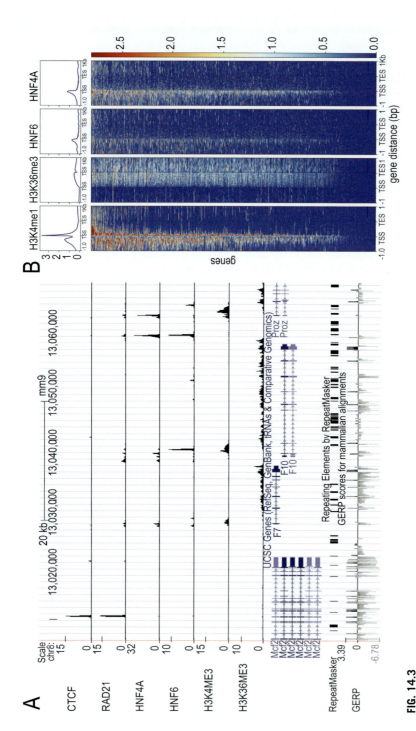

FIG. 14.3

ChIP-seq data visualization. (A) UCSC genome browser screenshot showing ChIP-seq signal of TFs and histone modifications. UCSC gene annotation, RepeatMasker, and GERP score tracks are also shown. (B) Heatmaps showing normalized ChIP-seq signal of different factors across gene bodies and 1 kb up/downstream of genes. 1000 selected mouse liver-expressed genes are shown. Gene bodies are scaled to the same length and ChIP-seq signal is summarized in 10 bp bins.

Datasets used in this figure from Uuskula-Reimand, L., Hou, H., Samavarchi-Tehrani, P., Rudan, M.V., Liang, M., Medina-Rivera, A., Mohammed, H., Schmidt, D., Schwalie, P., Young, E. J., Reimand, J., Hadjur, S., Gingras, A.C., Wilson, M.D., 2016. Topoisomerase II beta interacts with cohesin and CTCF at topological domain borders. Genome Biol. 17(1). https://doi.org/10.1186/s13059-016-1043-8.

The **Genomic Regions Enrichment of Annotations Tool (GREAT)** was developed to address this issue [52]. It also provides a more streamlined analysis by taking peak regions and performing enrichment analysis directly. GREAT is advantageous in two aspects: (1) the association between peaks and genes is more flexible, allowing the inclusion of distal intergenic peaks and (2) it performs both a hypergeometric test over genes and a binomial test over genomic regions. These two types of statistical tests compensate for each other's biases to give a robust set of pathway enrichment.

It is known that enhancers can regulate genes regardless of their distance from the target genes through the formation of long-range chromatin interactions by TFs, cohesin, and other factors [53]. Chromatin interactions can be measured by experimental methods including HiC [54], ChIA-PET [55], and more recently, HiChIP [56] or predicted by computational models such as the activity-by-contact model [57]. If long-range interactions have been mapped or predicted in the tissue or cell type of interest, it is recommended to leverage such information in order to accurately identify target genes of the ChIP-ed factor of interest.

Differential binding analysis

Comparing protein binding changes between different conditions is a common task in ChIP-seq data analysis and is essential for understanding mechanisms underlying development and disease. Overlapping different peak sets identified in different conditions can provide a qualitative measure of differences in protein binding. However, this method is generally not advised as it will not be able to properly account for sample variations or quantify changes at shared protein binding regions. A more biologically important question is: what are the genomic regions showing differences in signal intensity (in this case measures protein-occupancy) across conditions? Compared to other genomics data types such as RNA-seq, answering this question with ChIP-seq data can be particularly challenging for several reasons: (1) regions of interest in the analysis will need to be defined first; (2) ChIP-seq data is typically more variable than RNA-seq due to the technical nature of the method (i.e., different antibody pull-down efficiencies); (3) ChIP-seq profiles can differ more drastically between conditions as it has been shown that enhancer landscapes can be more dynamic than gene expression [58]; (4) ChIP-seq data of different types of factors (i.e., point-source vs. broad source) can differ substantially.

One common approach for ChIP-seq differential analysis is to take advantage of methods initially developed for RNA-seq analysis. In this approach, regions of interest are first defined by peak calling. Peaks called for different conditions are then merged to create a consensus peak set. Next, a count matrix is generated by counting ChIP-seq reads mapped to these peaks. This is similar to RNA-seq data where read counts for each gene are modeled and compared between conditions. The **R package DiffBind** provides a framework for this approach with a series of functions for peak manipulation, read counting, normalization, and performing differential analysis using DESeq2 or EdgeR [59,60]. In addition, DiffBind also presents a comprehensive list of normalization methods including normalizing using the full-library size, using only reads within peaks, against the background, or against spike-ins when available. In its vignette, the authors discussed the impact of different normalization methods on the final interpretation of the data and made suggestions on the best normalization strategy.

However, peak-based methods are challenging for broad-source factors that do not present as narrow, sharp peaks. Many "peak-free" methods, including **diffReps** [61] and **csaw** [62], were developed to address this challenge. Typically, read counts are summarized with a sliding window of user-defined size across the genome. These methods are better suited for identifying differential histone PTM domains.

Recent work by Steinhauser et al. evaluated 14 tools that are developed to identify differential enrichment between conditions [63]. This work is further extended by Eder and Grebien [64] where 33 published tools and custom approaches were evaluated. Both studies used a combination of simulated and real ChIP-seq data for different factors to evaluate the consistency, accuracy, and biological significance of differential binding results. They provide suggestions and decision trees to help guide the choice of tools depending on the type of factor and the biological scenario. Overall, careful considerations need to be taken prior to choosing a method for ChIP-seq differential analysis. For example, how regions of focus are defined (peaks or sliding windows), and what are the statistical presumptions? Practically, it is often beneficial to perform differential analyses using multiple suitable methods to obtain a consensus set of results.

Biological replicates are crucial for reliable ChIP-seq differential binding analysis and are required by most of the tools. However, replicates might not be available in certain research settings, for example, clinical samples from rare disease patients. In these cases, using a tool that is specifically developed for comparison without biological replicates is more beneficial than binary overlapping of peaks. One such tool, **MAnorm** [65], performs signal normalization under the assumption that the majority of the binding sites are shared between conditions. With caution, results obtained from only one replicate could still be informative, especially when combined with other lines of evidence, for example, gene expression changes.

Motif analysis

A common downstream goal of ChIP-seq is to characterize regulatory DNA sequences. This is typically done through the analysis of *motifs*, referring to short (6–20 bp) DNA sequences that encode regulatory information such as binding recognition by sequence-specific DNA binding proteins, such as TFs. Unlike protein coding sequences, regulatory motifs are typically degenerate, meaning that individual positions can be variable in sequence. Motifs are typically represented as *position weight matrices* (PWMs) to reflect this, where each position contains information on both the identity of the sequence nucleotide and its frequency of occurrence, for all possible residues (Fig. 14.4). Motifs can also be visually represented as *sequence logos* where the height of each nucleotide is scaled according to its frequency and the total height represents the information content at a given position and corresponds to positions of high-sequence constraint. Both PWMs and logos may sometimes be represented as *frequencies* rather than in counts and/or information content.

Several types of motif analyses are commonly performed downstream of ChIP experiments. Motif scanning aims to find the occurrence of known motifs in one more multiple DNA sequences, for example, ChIP-seq peaks. Motif enrichment analysis evaluates the enrichment of known motifs within regions of interest, compared to a background model or control sequences. In the case of ChIP-seq for a TF with a known DNA recognition sequence, motif enrichment for the expected target sequence can be used as a QC metric to ascertain whether the peaks detected are specific for the target of interest. Motif analysis can also be employed in the case of non-TF ChIPs to reveal potential regulatory mechanisms, for example, enhancer/promoter modifications. Conversely, motif discovery is employed to assemble de novo motifs and can be used downstream of ChIP for an uncharacterized DNA binding factor as a

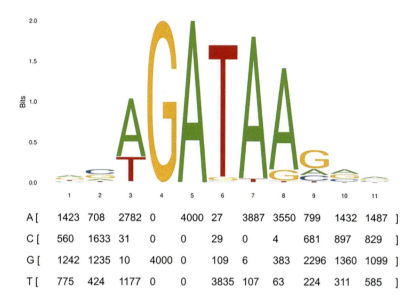

FIG. 14.4

Visual presentation of motifs. The mouse Gata1 TF recognition sequence depicted in both sequence logo (top) and position weight matrices (PWM, bottom) format.

Logo and PWM information obtained for MA0035.2 from the JASPAR 2020 database from Fornes, O., Castro-Mondragon, J. A., Khan, A., Van Der Lee, R., Zhang, X., Richmond, P. A., Modi, B. P., Correard, S., Gheorghe, M., Baranašić, D., Santana-Garcia, W., Tan, G., Chèneby, J., Ballester, B., Parcy, F., Sandelin, A., Lenhard, B., Wasserman, W.W., Mathelier, A., 2020. JASPAR 2020: update of the open-access database of transcription factor binding profiles. Nucleic Acids Res. 48(1), D87–D92. https://doi.org/10.1093/nar/gkz1001.

means to identify DNA-binding sequences. The discovered motifs are often compared with databases of known motifs in order to provide further biological insights into the factor.

A large number of published tools are available for motif-based analyses, for example, the MEME suite [66], HOMER [67], and the Regulatory Sequence Analysis Tools [68]. Each tool relies on slightly different algorithmic implementations and is subject to its own biases. The algorithms and features of several web-based motif analysis tools have been carefully summarized and compared [69]. Overall, a good practice is to run the same analyses using multiple tools and select a consensus result.

Key take-away

By following the abovementioned suggestions and guidelines, researchers can generate robust and meaningful data that advances our understanding of gene regulation in a rigorous and reproducible manner. Key points to remember:

- Formaldehyde fixation is preferred for ChIP-seq experiments targeting TFs, while native ChIP-seq is preferred for histone modifications.
- Nuclear isolation is recommended before chromatin fragmentation.

- Chromatin fragmentation needs to be optimized for every experiment, as it will differ according to cell/tissue type and conditions.
- Always prepare a control ChIP-seq library, preferably an input library.
- If possible, use monoclonal instead of polyclonal antibodies.

For ChIP-seq data analysis:

- Rigorous QC steps can detect problems early and ensure accurate downstream analysis.
- Many bioinformatics tools are available for different steps of ChIP-seq data analysis, including peak calling, differential analysis, and motif analysis. Choosing an appropriate tool often requires considering the type of targeted factor and the biological question at hand.
- Visualizing ChIP-seq data on a genome browser or with R is not only helpful for exploring or presenting the data but also serves as an effective QC measure.

Conclusion

ChIP-seq investigates protein-DNA interactions genome-wide and is an essential technique for studying transcriptional regulation and epigenetic mechanisms. The process involves cross-linking proteins to DNA in cells, fragmenting the chromatin, followed by immunoprecipitation of the protein of interest using an antibody. DNA fragments associated with the target protein are purified and sequenced, followed by a series of data analysis steps to identify protein binding sites, to compare binding across sample conditions, to discover protein binding motifs, and many more.

This chapter reviewed the critical steps in both experimental procedure and data analysis of ChIP-seq. By highlighting common pitfalls and offering suggestions for best practices, we hope this chapter could assist researchers in obtaining high-quality ChIP-seq data, performing robust data analysis, and archiving meaningful biological findings.

Over the past decade, ChIP-seq has been widely employed by the research community. Several consortia, including ENCODE (ENCODE Project Consortium, 2012) [70], Roadmap Epigenomics Consortium [71], and the International Human Epigenome Consortium (IHEC) [72], had systematically profiled TF binding and histone modifications across different tissues and cell lines, for human and other model organisms. These large datasets, together with the increasing number of other publicly available datasets, present an invaluable opportunity for researchers to generate and test novel hypotheses, leading to new discoveries.

One area of ongoing challenge is the integration of different ChIP-seq datasets, with other types of genomics data, such as ATAC-seq [73], RNA-seq, and 3C-based methods [74]. Joint analysis of different types of genomic data can enable a simultaneous survey of different layers of the gene regulatory machinery, facilitating the discovery of interplay between different factors [75].

For example, ChIP-seq data of different TFs and histone modifications can be integrated to identify chromatin states. ATAC-seq, which identifies accessible chromatin regions, can be integrated with ChIP-seq data to study the interaction between TF binding and chromatin accessibility. ChIP-seq and RNA-seq performed on paired samples can be integrated to identify target genes of TFs or discover epigenetic changes underlying gene expression differences. Using chromatin interaction data, such as Hi-C and Hi-ChIP, TFs can be more accurately associated with their potential target genes or coregulators.

Overall, it is advised to combine ChIP-seq with complementary functional genomics assays when possible in order to gain a deeper understanding of the relationship between DNA-regulatory protein interactions and changes in global patterns of expression and/or genome structure.

References

[1] J. Ernst, M. Kellis, Discovery and characterization of chromatin states for systematic annotation of the human genome, Nat. Biotechnol. 28 (2010) 817–825, https://doi.org/10.1038/nbt.1662.

[2] B.D. Strahl, C.D. Allis, The language of covalent histone modifications, Nature 403 (2000) 41–45, https://doi.org/10.1038/47412.

[3] A. Barski, S. Cuddapah, K. Cui, T.Y. Roh, D.E. Schones, Z. Wang, G. Wei, I. Chepelev, K. Zhao, High-resolution profiling of histone methylations in the human genome, Cell 129 (2007) 823–837, https://doi.org/10.1016/j.cell.2007.05.009.

[4] M.P. Creyghton, A.W. Cheng, G.G. Welstead, T. Kooistra, B.W. Carey, E.J. Steine, J. Hanna, M.A. Lodato, G.M. Frampton, P.A. Sharp, L.A. Boyer, R.A. Young, R. Jaenisch, Histone H3K27ac separates active from poised enhancers and predicts developmental state, Proc. Natl. Acad. Sci. U. S. A. 107 (2010) 21931–21936, https://doi.org/10.1073/pnas.1016071107.

[5] D.S. Johnson, A. Mortazavi, R.M. Myers, B. Wold, Genome-wide mapping of in vivo protein-DNA interactions, Science 316 (2007) 1497–1502, https://doi.org/10.1126/science.1141319.

[6] V. Orlando, Mapping chromosomal proteins in vivo by formaldehyde-crosslinked-chromatin immunoprecipitation, Trends Biochem. Sci. 25 (2000) 99–104, https://doi.org/10.1016/S0968-0004(99)01535-2.

[7] L. Baranello, F. Kouzine, S. Sanford, D. Levens, ChIP bias as a function of cross-linking time, Chromosom. Res. 24 (2016) 175–181, https://doi.org/10.1007/s10577-015-9509-1.

[8] B. Tian, J. Yang, A.R. Brasier, Two-step cross-linking for analysis of protein-chromatin interactions, Methods Mol. Biol. 809 (2012) 105–120, https://doi.org/10.1007/978-1-61779-376-9_7.

[9] J. Brind'Amour, S. Liu, M. Hudson, C. Chen, M.M. Karimi, M.C. Lorincz, An ultra-low-input native ChIP-seq protocol for genome-wide profiling of rare cell populations, Nat. Commun. 6 (2015), https://doi.org/10.1038/ncomms7033.

[10] L.P. O'Neill, M.D. VerMilyea, B.M. Turner, Epigenetic characterization of the early embryo with a chromatin immunoprecipitation protocol applicable to small cell populations, Nat. Genet. 38 (2006) 835–841, https://doi.org/10.1038/ng1820.

[11] L. Arrigoni, A.S. Richter, E. Betancourt, K. Bruder, S. Diehl, T. Manke, U. Bönisch, Standardizing chromatin research: a simple and universal method for ChIP-seq, Nucleic Acids Res. 44 (2016) e67, https://doi.org/10.1093/nar/gkv1495.

[12] B.L. Kidder, G. Hu, K. Zhao, ChIP-seq: technical considerations for obtaining high-quality data, Nat. Immunol. 12 (2011) 918–922, https://doi.org/10.1038/ni.2117.

[13] D.E. Schones, K. Cui, S. Cuddapah, T.Y. Roh, A. Barski, Z. Wang, G. Wei, K. Zhao, Dynamic regulation of nucleosome positioning in the human genome, Cell 132 (2008) 887–898, https://doi.org/10.1016/j.cell.2008.02.022.

[14] R.V. Chereji, T.D. Bryson, S. Henikoff, Quantitative MNase-seq accurately maps nucleosome occupancy levels, Genome Biol. 20 (2019) 198, https://doi.org/10.1186/s13059-019-1815-z.

[15] N.A. Pchelintsev, P.D. Adams, D.M. Nelson, M. Wu, Critical parameters for efficient sonication and improved chromatin immunoprecipitation of high molecular weight proteins, PLoS One 11 (2016) e0148023, https://doi.org/10.1371/journal.pone.0148023.

[16] Z. Cao, C. Chen, B. He, K. Tan, C. Lu, A microfluidic device for epigenomic profiling using 100 cells, Nat. Methods 12 (2015) 959–962, https://doi.org/10.1038/nmeth.3488.

[17] C. Schmidl, A.F. Rendeiro, N.C. Sheffield, C. Bock, ChIPmentation: fast, robust, low-input ChIP-seq for histones and transcription factors, Nat. Methods 12 (2015) 963–965, https://doi.org/10.1038/nmeth.3542.
[18] P.J. Skene, J.G. Henikoff, S. Henikoff, Targeted in situ genome-wide profiling with high efficiency for low cell numbers, Nat. Protoc. 13 (2018) 1006–1019, https://doi.org/10.1038/nprot.2018.015.
[19] S. Ai, H. Xiong, C.C. Li, Y. Luo, Q. Shi, Y. Liu, X. Yu, C. Li, A. He, Profiling chromatin states using single-cell itChIP-seq, Nat. Cell Biol. 21 (2019) 1164–1172, https://doi.org/10.1038/s41556-019-0383-5.
[20] A. Rotem, O. Ram, N. Shoresh, R.A. Sperling, A. Goren, D.A. Weitz, B.E. Bernstein, Single-cell ChIP-seq reveals cell subpopulations defined by chromatin state, Nat. Biotechnol. 33 (2015) 1165–1172, https://doi.org/10.1038/nbt.3383.
[21] Q. Wang, H. Xiong, S. Ai, X. Yu, Y. Liu, J. Zhang, A. He, CoBATCH for high-throughput single-cell epigenomic profiling, Mol. Cell 76 (2019) 206–216.e7, https://doi.org/10.1016/j.molcel.2019.07.015.
[22] M. Fosslie, A. Manaf, M. Lerdrup, K. Hansen, G.D. Gilfillan, J.A. Dahl, Going low to reach high: Small-scale ChIP-seq maps new terrain, WIREs Syst. Biol. Med. 12 (2020), https://doi.org/10.1002/wsbm.1465.
[23] C.A. Meyer, X.S. Liu, Identifying and mitigating bias in next-generation sequencing methods for chromatin biology, Nat. Rev. Genet. 15 (2014) 709–721, https://doi.org/10.1038/nrg3788.
[24] M. Busby, C. Xue, C. Li, Y. Farjoun, E. Gienger, I. Yofe, A. Gladden, C.B. Epstein, E.M. Cornett, S.B. Rothbart, C. Nusbaum, A. Goren, Systematic comparison of monoclonal versus polyclonal antibodies for mapping histone modifications by ChIP-seq, Epigenet. Chromatin 9 (2016), https://doi.org/10.1186/s13072-016-0100-6.
[25] J.B. Fishman, E.A. Berg, Protein A and protein G purification of antibodies, Cold Spring Harb. Protoc. 2019 (2019) 82–84, https://doi.org/10.1101/pdb.prot099143.
[26] S.G. Landt, G.K. Marinov, A. Kundaje, P. Kheradpour, F. Pauli, S. Batzoglou, B.E. Bernstein, P. Bickel, J.B. Brown, P. Cayting, Y. Chen, G. DeSalvo, C. Epstein, K.I. Fisher-Aylor, G. Euskirchen, M. Gerstein, J. Gertz, A.J. Hartemink, M.M. Hoffman, V.R. Iyer, Y.L. Jung, S. Karmakar, M. Kellis, P.V. Kharchenko, Q. Li, T. Liu, X.S. Liu, L. Ma, A. Milosavljevic, R.M. Myers, P.J. Park, M.J. Pazin, M.D. Perry, D. Raha, T.E. Reddy, J. Rozowsky, N. Shoresh, A. Sidow, M. Slattery, J.A. Stamatoyannopoulos, M.Y. Tolstorukov, K.P. White, S. Xi, P.J. Farnham, J.D. Lieb, B.J. Wold, M. Snyder, ChIP-seq guidelines and practices of the ENCODE and modENCODE consortia, Genome Res. 22 (2012) 1813–1831, https://doi.org/10.1101/gr.136184.111.
[27] Y.L. Jung, L.J. Luquette, J.W.K. Ho, F. Ferrari, M. Tolstorukov, A. Minoda, R. Issner, C.B. Epstein, G.H. Karpen, M.I. Kuroda, P.J. Park, Impact of sequencing depth in ChIP-seq experiments, Nucleic Acids Res. 42 (2014) e74, https://doi.org/10.1093/nar/gku178.
[28] Q. Zhang, X. Zeng, S. Younkin, T. Kawli, M.P. Snyder, S. Keleş, Systematic evaluation of the impact of ChIP-seq read designs on genome coverage, peak identification, and allele-specific binding detection, BMC Bioinform. 17 (2016), https://doi.org/10.1186/s12859-016-0957-1.
[29] S. Canzar, S.L. Salzberg, Short read mapping: an algorithmic tour, Proc. IEEE 105 (2017) 436–458, https://doi.org/10.1109/JPROC.2015.2455551.
[30] N.A. Fonseca, J. Rung, A. Brazma, J.C. Marioni, Tools for mapping high-throughput sequencing data, Bioinformatics 28 (2012) 3169–3177, https://doi.org/10.1093/bioinformatics/bts605.
[31] B. Langmead, S.L. Salzberg, Fast gapped-read alignment with Bowtie 2, Nat. Methods 9 (2012) 357–359, https://doi.org/10.1038/nmeth.1923.
[32] H. Li, R. Durbin, Fast and accurate short read alignment with Burrows-Wheeler transform, Bioinformatics 25 (2009) 1754–1760, https://doi.org/10.1093/bioinformatics/btp324.
[33] H.M. Amemiya, A. Kundaje, A.P. Boyle, The ENCODE blacklist: identification of problematic regions of the genome, Sci. Rep. 9 (2019), https://doi.org/10.1038/s41598-019-45839-z.
[34] H. Xu, L. Handoko, X. Wei, C. Ye, J. Sheng, C.L. Wei, F. Lin, W.K. Sung, A signal-noise model for significance analysis of ChIP-seq with negative control, Bioinformatics 26 (2010) 1199–1204, https://doi.org/10.1093/bioinformatics/btq128.

[35] C. Zang, D.E. Schones, C. Zeng, K. Cui, K. Zhao, W. Peng, A clustering approach for identification of enriched domains from histone modification ChIP-seq data, Bioinformatics 25 (2009) 1952–1958, https://doi.org/10.1093/bioinformatics/btp340.

[36] N.U. Rashid, P.G. Giresi, J.G. Ibrahim, W. Sun, J.D. Lieb, ZINBA integrates local covariates with DNA-seq data to identify broad and narrow regions of enrichment, even within amplified genomic regions, Genome Biol. 12 (2011), https://doi.org/10.1186/gb-2011-12-7-r67.

[37] Y. Zhang, T. Liu, C.A. Meyer, J. Eeckhoute, D.S. Johnson, B.E. Bernstein, C. Nussbaum, R.M. Myers, M. Brown, W. Li, X.S. Shirley, Model-based analysis of ChIP-seq (MACS), Genome Biol. 9 (2008), https://doi.org/10.1186/gb-2008-9-9-r137.

[38] A. Harmanci, J. Rozowsky, M. Gerstein, MUSIC: identification of enriched regions in ChIP-seq experiments using a mappability-corrected multiscale signal processing framework, Genome Biol. 15 (2014) 474, https://doi.org/10.1186/s13059-014-0474-3.

[39] R. Thomas, S. Thomas, A.K. Holloway, K.S. Pollard, Features that define the best ChIP-seq peak calling algorithms, Brief. Bioinform. 18 (2017) 441–450, https://doi.org/10.1093/bib/bbw035.

[40] T. Daley, A.D. Smith, Predicting the molecular complexity of sequencing libraries, Nat. Methods 10 (2013) 325–327, https://doi.org/10.1038/nmeth.2375.

[41] T.S. Carroll, Z. Liang, R. Salama, R. Stark, I. de Santiago, Impact of artifact removal on ChIP quality metrics in ChIP-seq and ChIP-exo data, Front. Genet. 5 (2014), https://doi.org/10.3389/fgene.2014.00075.

[42] Andrews, FastQC: A Quality Control Tool for High Throughput Sequence Data, Avaliable Online, 2010.

[43] F. Ramírez, D.P. Ryan, B. Grüning, V. Bhardwaj, F. Kilpert, A.S. Richter, S. Heyne, F. Dündar, T. Manke, deepTools2: a next generation web server for deep-sequencing data analysis, Nucleic Acids Res. 44 (2016) W160–W165, https://doi.org/10.1093/NAR/GKW257.

[44] H. Anzawa, H. Yamagata, K. Kinoshita, Theoretical characterisation of strand cross-correlation in ChIP-seq, BMC Bioinform. 21 (2020), https://doi.org/10.1186/s12859-020-03729-6.

[45] W. James Kent, C.W. Sugnet, T.S. Furey, K.M. Roskin, T.H. Pringle, A.M. Zahler, D. Haussler, The human genome browser at UCSC, Genome Res. 12 (2002) 996–1006, https://doi.org/10.1101/gr.229102. Article published online before print in May 2002.

[46] D. Li, D. Purushotham, J.K. Harrison, S. Hsu, X. Zhuo, C. Fan, S. Liu, V. Xu, S. Chen, J. Xu, S. Ouyang, A.S. Wu, T. Wang, WashU epigenome browser update 2022, Nucleic Acids Res. 50 (2022) W774–W781, https://doi.org/10.1093/nar/gkac238.

[47] J.T. Robinson, H. Thorvaldsdóttir, W. Winckler, M. Guttman, E.S. Lander, G. Getz, J.P. Mesirov, Integrative genomics viewer, Nat. Biotechnol. 29 (2011) 24–26, https://doi.org/10.1038/nbt.1754.

[48] F. Hahne, R. Ivanek, Visualizing genomic data using Gviz and bioconductor, in: Methods in Molecular Biology, Humana Press Inc., 2016, pp. 335–351, https://doi.org/10.1007/978-1-4939-3578-9_16. undefined.

[49] L.J. Zhu, C. Gazin, N.D. Lawson, H. Pagès, S.M. Lin, D.S. Lapointe, M.R. Green, ChIPpeakAnno: a bioconductor package to annotate ChIP-seq and ChIP-chip data, BMC Bioinform. 11 (2010), https://doi.org/10.1186/1471-2105-11-237.

[50] G. Yu, L.G. Wang, Q.Y. He, ChIP seeker: an R/bioconductor package for ChIP peak annotation, comparison and visualization, Bioinformatics 31 (2015) 2382–2383, https://doi.org/10.1093/bioinformatics/btv145.

[51] G. Yu, L.G. Wang, Y. Han, Q.Y. He, ClusterProfiler: an R package for comparing biological themes among gene clusters, OMICS J. Integr. Biol. 16 (2012) 284–287, https://doi.org/10.1089/omi.2011.0118.

[52] C.Y. McLean, D. Bristor, M. Hiller, S.L. Clarke, B.T. Schaar, C.B. Lowe, A.M. Wenger, G. Bejerano, GREAT improves functional interpretation of cis-regulatory regions, Nat. Biotechnol. 28 (2010) 495–501, https://doi.org/10.1038/nbt.1630.

[53] G. Li, X. Ruan, R.K. Auerbach, K.S. Sandhu, M. Zheng, P. Wang, H.M. Poh, Y. Goh, J. Lim, J. Zhang, H.S. Sim, S.Q. Peh, F.H. Mulawadi, C.T. Ong, Y.L. Orlov, S. Hong, Z. Zhang, S. Landt, D. Raha, G. Euskirchen, C.L. Wei, W. Ge, H. Wang, C. Davis, K.I. Fisher-Aylor, A. Mortazavi, M. Gerstein, T. Gingeras, B. Wold, Y. Sun, M.J. Fullwood, E. Cheung, E. Liu, W.K. Sung, M. Snyder, Y. Ruan, Extensive promoter-centered

chromatin interactions provide a topological basis for transcription regulation, Cell 148 (2012) 84–98, https://doi.org/10.1016/j.cell.2011.12.014.

[54] E. Lieberman-Aiden, N.L. Van Berkum, L. Williams, M. Imakaev, T. Ragoczy, A. Telling, I. Amit, B.R. Lajoie, P.J. Sabo, M.O. Dorschner, R. Sandstrom, B. Bernstein, M.A. Bender, M. Groudine, A. Gnirke, J. Stamatoyannopoulos, L.A. Mirny, E.S. Lander, J. Dekker, Comprehensive mapping of long-range interactions reveals folding principles of the human genome, Science 326 (2009) 289–293, https://doi.org/10.1126/science.1181369.

[55] M.J. Fullwood, M.H. Liu, Y.F. Pan, J. Liu, H. Xu, Y.B. Mohamed, Y.L. Orlov, S. Velkov, A. Ho, P.H. Mei, E.G.Y. Chew, P.Y.H. Huang, W.J. Welboren, Y. Han, H.S. Ooi, P.N. Ariyaratne, V.B. Vega, Y. Luo, P.Y. Tan, P.Y. Choy, K.D.S.A. Wansa, B. Zhao, K.S. Lim, S.C. Leow, J.S. Yow, R. Joseph, H. Li, K.V. Desai, J.S. Thomsen, Y.K. Lee, R.K.M. Karuturi, T. Herve, G. Bourque, H.G. Stunnenberg, X. Ruan, V. Cacheux-Rataboul, W.K. Sung, E.T. Liu, C.L. Wei, E. Cheung, Y. Ruan, An oestrogen-receptor-α-bound human chromatin interactome, Nature 462 (2009) 58–64, https://doi.org/10.1038/nature08497.

[56] M.R. Mumbach, A.J. Rubin, R.A. Flynn, C. Dai, P.A. Khavari, W.J. Greenleaf, H.Y. Chang, HiChIP: efficient and sensitive analysis of protein-directed genome architecture, Nat. Methods 13 (2016) 919–922, https://doi.org/10.1038/nmeth.3999.

[57] C.P. Fulco, J. Nasser, T.R. Jones, G. Munson, D.T. Bergman, V. Subramanian, S.R. Grossman, R. Anyoha, B.R. Doughty, T.A. Patwardhan, T.H. Nguyen, M. Kane, E.M. Perez, N.C. Durand, C.A. Lareau, E.K. Stamenova, E.L. Aiden, E.S. Lander, J.M. Engreitz, Activity-by-contact model of enhancer-promoter regulation from thousands of CRISPR perturbations, Nat. Genet. 51 (2019) 1664–1669, https://doi.org/10.1038/s41588-019-0538-0.

[58] M. Kasowski, S. Kyriazopoulou-Panagiotopoulou, F. Grubert, J.B. Zaugg, A. Kundaje, Y. Liu, A.P. Boyle, Q.C. Zhang, F. Zakharia, D.V. Spacek, J. Li, D. Xie, A. Olarerin-George, L.M. Steinmetz, J.B. Hogenesch, M. Kellis, S. Batzoglou, M. Snyder, Extensive variation in chromatin states across humans, Science 342 (2013) 750–752, https://doi.org/10.1126/science.1242510.

[59] C.S. Ross-Innes, R. Stark, A.E. Teschendorff, K.A. Holmes, H.R. Ali, M.J. Dunning, G.D. Brown, O. Gojis, I.O. Ellis, A.R. Green, S. Ali, S.F. Chin, C. Palmieri, C. Caldas, J.S. Carroll, Differential oestrogen receptor binding is associated with clinical outcome in breast cancer, Nature 481 (2012) 389–393, https://doi.org/10.1038/nature10730.

[60] R. Stark, G. Brown, DiffBind: Differential Binding Analysis of ChIP-Seq Peak Data, 2011. http://Bioconductor.Org/Packages/Release/Bioc/Vignettes/DiffBind/Inst/Doc/DiffBind.Pdf.

[61] L. Shen, N.Y. Shao, X. Liu, I. Maze, J. Feng, E.J. Nestler, diffReps: detecting differential chromatin modification sites from ChIP-seq data with biological replicates, PLoS One 8 (2013), https://doi.org/10.1371/journal.pone.0065598.

[62] A.T.L. Lun, G.K. Smyth, Csaw: a bioconductor package for differential binding analysis of ChIP-seq data using sliding windows, Nucleic Acids Res. 44 (2015) e45, https://doi.org/10.1093/nar/gkv1191.

[63] S. Steinhauser, N. Kurzawa, R. Eils, C. Herrmann, A comprehensive comparison of tools for differential ChIP-seq analysis, Brief. Bioinform. 17 (2016) bbv110, https://doi.org/10.1093/bib/bbv110.

[64] T. Eder, F. Grebien, Comprehensive assessment of differential ChIP-seq tools guides optimal algorithm selection, Genome Biol. 23 (2022), https://doi.org/10.1186/s13059-022-02686-y.

[65] Z. Shao, Y. Zhang, G.C. Yuan, S.H. Orkin, D.J. Waxman, MAnorm: a robust model for quantitative comparison of ChIP-seq data sets, Genome Biol. 13 (2012), https://doi.org/10.1186/gb-2012-13-3-r16.

[66] T.L. Bailey, M. Boden, F.A. Buske, M. Frith, C.E. Grant, L. Clementi, J. Ren, W.W. Li, W.S. Noble, MEME suite: tools for motif discovery and searching, Nucleic Acids Res. 37 (2009) W202–W208, https://doi.org/10.1093/nar/gkp335.

[67] S. Heinz, C. Benner, N. Spann, E. Bertolino, Y.C. Lin, P. Laslo, J.X. Cheng, C. Murre, H. Singh, C.K. Glass, Simple combinations of lineage-determining transcription factors prime cis-regulatory elements required for macrophage and B cell identities, Mol. Cell 38 (2010) 576–589, https://doi.org/10.1016/j.molcel.2010.05.004.

[68] W. Santana-Garcia, J.A. Castro-Mondragon, M. Padilla-Gálvez, N.T.T. Nguyen, A. Elizondo-Salas, N. Ksouri, F. Gerbes, D. Thieffry, P. Vincens, B. Contreras-Moreira, J. Van Helden, M. Thomas-Chollier, A. Medina-Rivera, RSAT 2022: regulatory sequence analysis tools, Nucleic Acids Res. 50 (2022) W670–W676, https://doi.org/10.1093/nar/gkac312.

[69] N.T.L. Tran, C.H. Huang, A survey of motif finding web tools for detecting binding site motifs in ChIP-seq data, Biol. Direct 9 (2014), https://doi.org/10.1186/1745-6150-9-4.

[70] I. Dunham, A. Kundaje, S.F. Aldred, P.J. Collins, C.A. Davis, F. Doyle, C.B. Epstein, S. Frietze, J. Harrow, R. Kaul, J. Khatun, B.R. Lajoie, S.G. Landt, B.K. Lee, F. Pauli, K.R. Rosenbloom, P. Sabo, A. Safi, A. Sanyal, N. Shoresh, J.M. Simon, L. Song, N.D. Trinklein, R.C. Altshuler, E. Birney, J.B. Brown, C. Cheng, S. Djebali, X. Dong, J. Ernst, T.S. Furey, M. Gerstein, B. Giardine, M. Greven, R.C. Hardison, R.S. Harris, J. Herrero, M.M. Hoffman, S. Iyer, M. Kellis, P. Kheradpour, T. Lassmann, Q. Li, X. Lin, G.K. Marinov, A. Merkel, A. Mortazavi, S.C.J. Parker, T.E. Reddy, J. Rozowsky, F. Schlesinger, R.E. Thurman, J. Wang, L.D. Ward, T.W. Whitfield, S.P. Wilder, W. Wu, H.S. Xi, K.Y. Yip, J. Zhuang, B.E. Bernstein, E.D. Green, C. Gunter, M. Snyder, M.J. Pazin, R.F. Lowdon, L.A.L. Dillon, L.B. Adams, C.J. Kelly, J. Zhang, J.R. Wexler, P.J. Good, E.A. Feingold, G.E. Crawford, J. Dekker, L. Elnitski, P.J. Farnham, M.C. Giddings, T.R. Gingeras, R. Guigó, T.J. Hubbard, W.J. Kent, J.D. Lieb, E.H. Margulies, R.M. Myers, J.A. Stamatoyannopoulos, S.A. Tenenbaum, Z. Weng, K.P. White, B. Wold, Y. Yu, J. Wrobel, B.A. Risk, H.P. Gunawardena, H.C. Kuiper, C.W. Maier, L. Xie, X. Chen, T.S. Mikkelsen, S. Gillespie, A. Goren, O. Ram, X. Zhang, L. Wang, R. Issner, M.J. Coyne, T. Durham, M. Ku, T. Truong, M.L. Eaton, A. Dobin, A. Tanzer, J. Lagarde, W. Lin, C. Xue, B.A. Williams, C. Zaleski, M. Röder, F. Kokocinski, R.F. Abdelhamid, T. Alioto, I. Antoshechkin, M.T. Baer, P. Batut, I. Bell, K. Bell, S. Chakrabortty, J. Chrast, J. Curado, T. Derrien, J. Drenkow, E. Dumais, J. Dumais, R. Duttagupta, M. Fastuca, K. Fejes-Toth, P. Ferreira, S. Foissac, M.J. Fullwood, H. Gao, D. Gonzalez, A. Gordon, C. Howald, S. Jha, R. Johnson, P. Kapranov, B. King, C. Kingswood, G. Li, O.J. Luo, E. Park, J.B. Preall, K. Presaud, P. Ribeca, D. Robyr, X. Ruan, M. Sammeth, K.S. Sandhu, L. Schaeffer, L.H. See, A. Shahab, J. Skancke, A.M. Suzuki, H. Takahashi, H. Tilgner, D. Trout, N. Walters, H. Wang, Y. Hayashizaki, A. Reymond, S.E. Antonarakis, G.J. Hannon, Y. Ruan, P. Carninci, C.A. Sloan, K. Learned, V.S. Malladi, M.C. Wong, G.P. Barber, M.S. Cline, T.R. Dreszer, S.G. Heitner, D. Karolchik, V.M. Kirkup, L.R. Meyer, J.C. Long, M. Maddren, B.J. Raney, L.L. Grasfeder, P.G. Giresi, A. Battenhouse, N.C. Sheffield, K.A. Showers, D. London, A.A. Bhinge, C. Shestak, M.R. Schaner, S.K. Kim, Z.Z. Zhang, P.A. Mieczkowski, J.O. Mieczkowska, Z. Liu, R.M. McDaniell, Y. Ni, N.U. Rashid, M.J. Kim, S. Adar, Z. Zhang, T. Wang, D. Winter, D. Keefe, V.R. Iyer, M. Zheng, P. Wang, J. Gertz, J. Vielmetter, E.C. Partridge, K.E. Varley, C. Gasper, A. Bansal, S. Pepke, P. Jain, H. Amrhein, K.M. Bowling, M. Anaya, M.K. Cross, M.A. Muratet, K.M. Newberry, K. McCue, A.S. Nesmith, K.I. Fisher-Aylor, B. Pusey, G. DeSalvo, S.L. Parker, S. Balasubramanian, N.S. Davis, S.K. Meadows, T. Eggleston, J.S. Newberry, S.E. Levy, D.M. Absher, W.H. Wong, M.J. Blow, A. Visel, L.A. Pennachio, H.M. Petrykowska, A. Abyzov, B. Aken, D. Barrell, G. Barson, A. Berry, A. Bignell, V. Boychenko, G. Bussotti, C. Davidson, G. Despacio-Reyes, M. Diekhans, I. Ezkurdia, A. Frankish, J. Gilbert, J.M. Gonzalez, E. Griffiths, R. Harte, D.A. Hendrix, T. Hunt, I. Jungreis, M. Kay, E. Khurana, J. Leng, M.F. Lin, J. Loveland, Z. Lu, D. Manthravadi, M. Mariotti, J. Mudge, G. Mukherjee, C. Notredame, B. Pei, J.M. Rodriguez, G. Saunders, A. Sboner, S. Searle, C. Sisu, C. Snow, C. Steward, E. Tapanari, M.L. Tress, M.J. Van Baren, S. Washietl, L. Wilming, A. Zadissa, Z. Zhang, M. Brent, D. Haussler, A. Valencia, N. Addleman, R.P. Alexander, R.K. Auerbach, S. Balasubramanian, K. Bettinger, N. Bhardwaj, A.P. Boyle, A.R. Cao, P. Cayting, A. Charos, Y. Cheng, C. Eastman, G. Euskirchen, J.D. Fleming, F. Grubert, L. Habegger, M. Hariharan, A. Harmanci, S. Iyengar, V.X. Jin, K.J. Karczewski, M. Kasowski, P. Lacroute, H. Lam, N. Lamarre-Vincent, J. Lian, M. Lindahl-Allen, R. Min, B. Miotto, H. Monahan, Z. Moqtaderi, X.J. Mu, H. O'Geen, Z. Ouyang, D. Patacsil, D. Raha, L. Ramirez, B. Reed, M. Shi, T. Slifer, H. Witt, L. Wu, X. Xu, K.K. Yan, X. Yang, K. Struhl, S.M. Weissman, L.O. Penalva, S. Karmakar, R.R. Bhanvadia, A. Choudhury, M. Domanus, L. Ma, J. Moran, A. Victorsen, T. Auer, L. Centanin, M. Eichenlaub, F. Gruhl, S. Heermann, B. Hoeckendorf, D. Inoue, T. Kellner, S. Kirchmaier, C. Mueller, R. Reinhardt, L. Schertel, S. Schneider, R.

Sinn, B. Wittbrodt, J. Wittbrodt, G. Jain, G. Balasundaram, D.L. Bates, R. Byron, T.K. Canfield, M.J. Diegel, D. Dunn, A.K. Ebersol, T. Frum, K. Garg, E. Gist, R.S. Hansen, L. Boatman, E. Haugen, R. Humbert, A.K. Johnson, E.M. Johnson, T.V. Kutyavin, K. Lee, D. Lotakis, M.T. Maurano, S.J. Neph, F.V. Neri, E.D. Nguyen, H. Qu, A.P. Reynolds, V. Roach, E. Rynes, M.E. Sanchez, R.S. Sandstrom, A.O. Shafer, A.B. Stergachis, S. Thomas, B. Vernot, J. Vierstra, S. Vong, H. Wang, M.A. Weaver, Y. Yan, M. Zhang, J.M. Akey, M. Bender, M.O. Dorschner, M. Groudine, M.J. MacCoss, P. Navas, G. Stamatoyannopoulos, K. Beal, A. Brazma, P. Flicek, N. Johnson, M. Lukk, N.M. Luscombe, D. Sobral, J.M. Vaquerizas, S. Batzoglou, A. Sidow, N. Hussami, S. Kyriazopoulou-Panagiotopoulou, M.W. Libbrecht, M.A. Schaub, W. Miller, P.J. Bickel, B. Banfai, N.P. Boley, H. Huang, J.J. Li, W.S. Noble, J.A. Bilmes, O.J. Buske, A.D. Sahu, P.V. Kharchenko, P.J. Park, D. Baker, J. Taylor, L. Lochovsky, An integrated encyclopedia of DNA elements in the human genome, Nature 489 (2012) 57–74, https://doi.org/10.1038/nature11247.

[71] Roadmap Epigenomics Consortium, A. Kundaje, W. Meuleman, J. Ernst, M. Bilenky, A. Yen, A. Heravi-Moussavi, P. Kheradpour, Z. Zhang, J. Wang, M.J. Ziller, V. Amin, J.W. Whitaker, M.D. Schultz, L.D. Ward, A. Sarkar, G. Quon, R.S. Sandstrom, M.L. Eaton, Y.C. Wu, A.R. Pfenning, X. Wang, M. Claussnitzer, Y. Liu, C. Coarfa, R.A. Harris, N. Shoresh, C.B. Epstein, E. Gjoneska, D. Leung, W. Xie, R.D. Hawkins, R. Lister, C. Hong, P. Gascard, A.J. Mungall, R. Moore, E. Chuah, A. Tam, T.K. Canfield, R.S. Hansen, R. Kaul, P.J. Sabo, M.S. Bansal, A. Carles, J.R. Dixon, K.H. Farh, S. Feizi, R. Karlic, A.R. Kim, A. Kulkarni, D. Li, R. Lowdon, G. Elliott, T.R. Mercer, S.J. Neph, V. Onuchic, P. Polak, N. Rajagopal, P. Ray, R.C. Sallari, K.T. Siebenthall, N.A. Sinnott-Armstrong, M. Stevens, R.E. Thurman, J. Wu, B. Zhang, X. Zhou, A.E. Beaudet, L.A. Boyer, P.L. De Jager, P.J. Farnham, S.J. Fisher, D. Haussler, S.J.M. Jones, W. Li, M.A. Marra, M.T. McManus, S. Sunyaev, J.A. Thomson, T.D. Tlsty, L.H. Tsai, W. Wang, R.A. Waterland, M.Q. Zhang, L.H. Chadwick, B.E. Bernstein, J.F. Costello, J.R. Ecker, M. Hirst, A. Meissner, A. Milosavljevic, B. Ren, J.A. Stamatoyannopoulos, T. Wang, M. Kellis, Integrative analysis of 111 reference human epigenomes, Nature 518 (2015) 317–329, https://doi.org/10.1038/nature14248.

[72] H.G. Stunnenberg, S. Abrignani, D. Adams, M. de Almeida, L. Altucci, V. Amin, I. Amit, S.E. Antonarakis, S. Aparicio, T. Arima, L. Arrigoni, R. Arts, V. Asnafi, M.E. Badosa, J.B. Bae, K. Bassler, S. Beck, B. Berkman, B.E. Bernstein, M. Bilenky, A. Bird, C. Bock, B. Boehm, G. Bourque, C.E. Breeze, B. Brors, D. Bujold, O. Burren, M.J. Bussemakers, A. Butterworth, E. Campo, E. Carrillo-de-Santa-Pau, L. Chadwick, K.M. Chan, W. Chen, T.H. Cheung, L. Chiapperino, N.H. Choi, H.R. Chung, L. Clarke, J.M. Connors, P. Cronet, J. Danesh, M. Dermitzakis, G. Drewes, P. Durek, S. Dyke, T. Dylag, C.J. Eaves, P. Ebert, R. Eils, J. Eils, C.A. Ennis, T. Enver, E.A. Feingold, B. Felder, A. Ferguson-Smith, J. Fitzgibbon, P. Flicek, R.S.Y. Foo, P. Fraser, M. Frontini, E. Furlong, S. Gakkhar, N. Gasparoni, G. Gasparoni, D.H. Geschwind, P. Glažar, T. Graf, F. Grosveld, X.Y. Guan, R. Guigo, I.G. Gut, A. Hamann, B.G. Han, R.A. Harris, S. Heath, K. Helin, J.G. Hengstler, A. Heravi-Moussavi, K. Herrup, S. Hill, J.A. Hilton, B.C. Hitz, B. Horsthemke, M. Hu, J.Y. Hwang, N.Y. Ip, T. Ito, B.M. Javierre, S. Jenko, T. Jenuwein, Y. Joly, S.J.M. Jones, Y. Kanai, H.G. Kang, A. Karsan, A.K. Kiemer, S.C. Kim, B.J. Kim, H.H. Kim, H. Kimura, S. Kinkley, F. Klironomos, I.U. Koh, M. Kostadima, C. Kressler, R. Kreuzhuber, A. Kundaje, R. Küppers, C. Larabell, P. Lasko, M. Lathrop, D.H.S. Lee, S. Lee, H. Lehrach, E. Leitão, T. Lengauer, Å. Lernmark, R.D. Leslie, G.K.K. Leung, D. Leung, M. Loeffler, Y. Ma, A. Mai, T. Manke, E.R. Marcotte, M.A. Marra, J.H.A. Martens, J.I. Martin-Subero, K. Maschke, C. Merten, A. Milosavljevic, S. Minucci, T. Mitsuyama, R.A. Moore, F. Müller, A.J. Mungall, M.G. Netea, K. Nordström, I. Norstedt, H. Okae, V. Onuchic, F. Ouellette, W. Ouwehand, M. Pagani, V. Pancaldi, T. Pap, T. Pastinen, R. Patel, D.S. Paul, M.J. Pazin, P.G. Pelicci, A.G. Phillips, J. Polansky, B. Porse, J.A. Pospisilik, S. Prabhakar, D.C. Procaccini, A. Radbruch, N. Rajewsky, V. Rakyan, W. Reik, B. Ren, D. Richardson, A. Richter, D. Rico, D.J. Roberts, P. Rosenstiel, M. Rothstein, A. Salhab, H. Sasaki, J.S. Satterlee, S. Sauer, C. Schacht, F. Schmidt, G. Schmitz, S. Schreiber, C. Schröder, D. Schübeler, J.L. Schultze, R.P. Schulyer, M. Schulz, M. Seifert, K. Shirahige, R. Siebert, T. Sierocinski, L. Siminoff, A. Sinha, N. Soranzo, S. Spicuglia, M. Spivakov, C. Steidl, J.S. Strattan, M. Stratton, P. Südbeck, H. Sun, N. Suzuki, Y. Suzuki, A. Tanay, D. Torrents, F.L. Tyson, T. Ulas, S. Ullrich, T. Ushijima, A. Valencia, E. Vellenga, M.

Vingron, C. Wallace, S. Wallner, J. Walter, H. Wang, S. Weber, N. Weiler, A. Weller, A. Weng, S. Wilder, S.M. Wiseman, A.R. Wu, Z. Wu, J. Xiong, Y. Yamashita, X. Yang, D.Y. Yap, K.Y. Yip, S. Yip, J.I. Yoo, D. Zerbino, G. Zipprich, M. Hirst, The international human epigenome consortium: a blueprint for scientific collaboration and discovery, Cell 167 (2016) 1145–1149, https://doi.org/10.1016/j.cell.2016.11.007.

[73] J.D. Buenrostro, B. Wu, H.Y. Chang, W.J. Greenleaf, ATAC-seq: a method for assaying chromatin accessibility genome-wide, Curr. Protocols Mol. Biol. 2015 (2015) 21.29.1–21.29.9, https://doi.org/10.1002/0471142727.mb2129s109.

[74] I. Jerković, G. Cavalli, Understanding 3D genome organization by multidisciplinary methods, Nat. Rev. Mol. Cell Biol. 22 (2021) 511–528, https://doi.org/10.1038/s41580-021-00362-w.

[75] S. Jiang, A. Mortazavi, Integrating ChIP-seq with other functional genomics data, Brief. Funct. Genomics 17 (2018) 104–115, https://doi.org/10.1093/bfgp/ely002.

CHAPTER 15

A practical guide for essential analyses of Hi-C data

Yu Liu* and Erica M. Hildebrand*
Department of Systems Biology, University of Massachusetts Chan Medical School, Worcester, MA, United States

A brief summary of Hi-C and an example analysis pipeline

Human chromosomes are organized in 3D structures. Microscopy has revealed the morphology of interphase nuclei and mitotic chromosomes; however, chromosome organization at the molecular level was not described until the emergence of chromosome conformation capture (3C) and the related technologies [1]. 3C technology measures physical interactions of distant genomic loci in relatively close 3D proximity and demonstrates the organization of chromosomes in 3D. Based on 3C, many other methods have been established, including 4C, 5C, micro-C, and Hi-C [2–5]. Hi-C reveals genome-wide interactions and has been widely applied in chromosome structure studies. In a typical Hi-C experiment, chromosomes are first fixed and then digested using restriction enzymes into short fragments, which are held together by protein-protein and protein-DNA crosslinks due to the fixation [6]. Fragments in close spatial proximity are next ligated, regardless of linear genomic distance. The contact loci can thus be joined together to form chimeric DNA fragments, which are subjected to enrichment and high-throughput sequencing (Fig. 15.1).

In this chapter, we will describe the computational tools that we use for basic Hi-C analysis of different types of chromosome structures. However, we note that this is not a comprehensive comparison or review of all Hi-C analysis software, and many of these analyses have been implemented in other Hi-C analysis packages as well. While specific computational methods may differ, we hope that the description of the types of basic analysis that can be performed and the interpretation of these results will be useful for all who are new to Hi-C analysis, regardless of which analysis package is used.

The key step for Hi-C data processing is to identify chimeric DNA fragments between loci that are in close 3D proximity. This is done by mapping pairs of interacting fragments making up one Hi-C sequencing read to the genome and storing these reads as a matrix of pairwise interactions. There are many different software packages and file formats that have been developed to process Hi-C data [7–9].

In our research, we use the cooler file format to store pairwise interaction data in a sparse format, as storing interaction counts using dense matrices in text files results in very large file sizes, especially with higher-resolution data sets [7,10]. Cooler also provides an interface to access the data and is used in the software package cooltools for analysis of the Hi-C data [11]. We also note that it is possible to

---

*These authors contributed equally as co-corresponding authors.

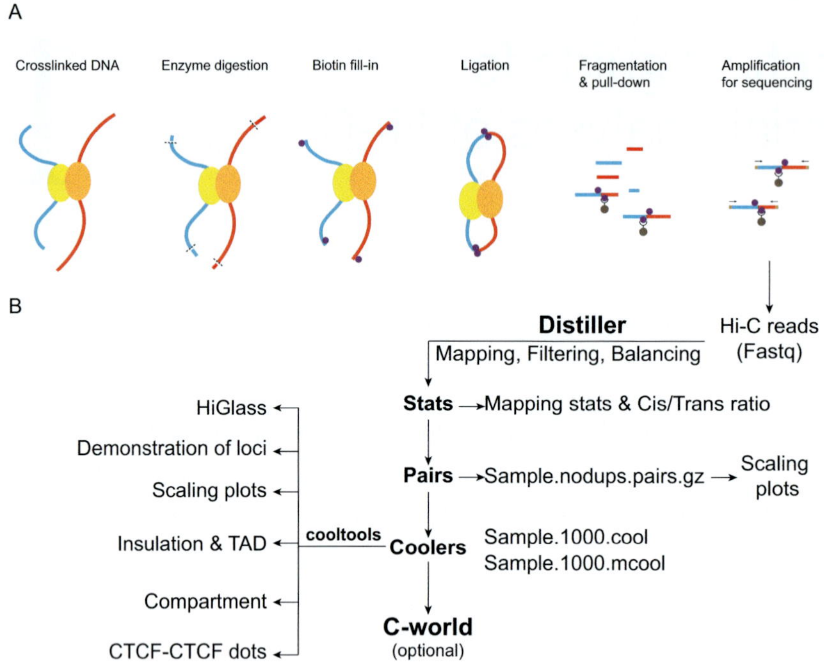

344   Chapter 15 A practical guide for essential analyses of Hi-C data

FIG. 15.1

Introduction to Hi-C method and example analytical pipeline. (A) A brief summary of Hi-C. After genomic DNA is fixed, restriction enzyme is added to cut DNA. The ends of the DNA are then labeled with biotin and rejoined by DNA ligase. Genomic DNA is fragmented, and biotin-labeled DNA fragments are enriched using streptavidin beads and amplified to construct sequencing libraries. (B) Illustration of an example Hi-C processing pipeline. Hi-C reads in fastq format are obtained using the Illumina sequencing platform and processed by the distiller mapping and processing pipeline into the cooler file format. Distiller first maps Hi-C reads onto a reference genome and then generates interaction pairs. After filtering and balancing, distiller constructs Hi-C contact matrices and stores them in cooler files. In the output folder, distiller generates three main folders, Stats, Pairs, and Coolers (single and multiresolution), which contain information and files for quality control and subsequent analysis. The files and their useage are summarized.

No permission required.

convert mapped Hi-C data between different file formats in many cases, which can be useful if a particular analysis tool is only available for a different file format than was created by the initial mapping pipeline.

In this chapter, we will focus on the cooler format and analysis using the related cooltools package, since it generates and operates on cooler files that will balance scalable data storage and data query at different resolutions [7]. Cooltools is written in Python, making many BioConda packages available for integrative analysis [12]. We will introduce the basic usage of cooltools and share our experience using each analytical module in a specific context. We include published examples to show our interpretation of these results, aiming to provide guidance to help explore the biological meaning of chromosome organization.

Hi-C data processing and quality control

Enriched Hi-C fragments are sequenced to generate sequence reads in fastq format. We use the distiller mapping pipeline which processes these Hi-C reads to generate cooler files for subsequent analysis [13]. Distiller includes three main modules: mapping, parsing, and binning. For mapping, distiller integrates BWA-MEM to align reads onto a reference genome and generates SAM files [14]. Then distiller parses SAM files to generate contact pairs, from which spurious reads such as PCR duplicates are filtered out. The remaining pairs are aggregated, binned in different resolutions, and stored in single resolution (.cool) and multiresolution (.mcool) cooler files.

Distiller generates three main output files: Stats, Pairs, and Coolers (Fig. 15.1B). Stats files contain quality control (QC) information of Hi-C libraries and Pairs contain all Hi-C interaction fragments and can be used to calculate the frequency of interaction with genomic separation, called scaling plots. In the Coolers folder, single-resolution cooler files at 1000 bp resolution and multiresolution cooler files starting at 1000 bp resolution of each sample can be found and these files are used for most downstream analysis, including topologically associating domains (TADs), dots, and compartments (discussed below).

In our experience, we use a minimal resolution of 20 kb with sequencing depth of <100 million paired-end reads. Distiller can be downloaded from: https://github.com/open2c/distiller-nf. Distiller can also be set to balance output cooler files using iterative correction [15]. This step is to normalize Hi-C matrices based on the assumption that all the chromosomal loci have the same accessibility to contact each other. In theory, this allows comparison of libraries of different total read numbers on the same scale. However, as library complexity also increases with sequencing depth, we generally also down-sample libraries within the same experiment to the same total read number before balancing, to obtain similar library complexity between the samples. This can be accomplished in the cooler format using cooltools random-sample, or if coolers are converted to dense matrix files for use with other analysis tools such as c-world (the c-world scaling matrix command (scaleMatrix.pl) can also be used for the same purpose [8]).

For subsequent analysis using cooler files, cooltools calculates expected interaction matrices, which are based on the distance decay (further explained in the Scaling Plots section), and uses the ratio of observed/expected signals to examine chromosome organization changes, such as compartment strength and dot calling.

The most important QC factor for Hi-C data is cis percent or cis/trans ratio. Cis percent calculates the percent of cis contacts in all the contacts, including both cis and trans pairs. Cis and trans contacts result from intra- and interchromosome interactions, respectively. Low cis percent indicates high interchromosomal interactions that may result from random ligation of loosened fragments after enzyme digestion. While Hi-C resolution is ultimately determined by the size of the digestion fragments resulting from the initial digestion step, smaller fragments can also lead to higher noise, if the protocol is not adjusted accordingly. For example, digestion enzymes like *Dpn*II and *Mbo*I recognize and cut 4 bp long sequence (4 bp cutter), thus generating more loose fragments than with a 6 bp cutter like *Hin*dIII, as 4 bp sequences occur more frequently throughout the genome than 6 bp sequences. These free fragments can be ligated to create more trans reads and lower cis percent. From our experience, cis percent could be as low as 20% using the original Hi-C protocol with HindIII for some cell types like T cells [16]. However, Hi-C 2.0 optimized the Hi-C protocol to dramatically increase the cis percent [17].

Using Hi-C 2.0, a minimal cis percent of 50% is usually accepted as good Hi-C libraries using *Dpn*II, and is often much higher, depending on cell cycle state or cell type, whereas cis percent of Hi-C libraries using *Hin*dIII is always beyond 75%. Although there are higher trans contacts using 4 cutter enzymes, more looping interactions (short-range interactions) can be detected, as interactions can be pinpointed to smaller regions of the genome. Recently, Akgol Oksuz et al. systematically examined the effects of 4-,6-cutter and varied cross-linkers on Hi-C data quality to detect critical features of chromosome structures [6]. As expected, the cis percent of Hi-C libraries using 6-cutter restriction enzymes is higher than that using 4-cutter enzymes. More importantly, this work showed that the addition of the cross-linker Di(N-succinimidyl) glutarate (DSG) to the cross-linker formaldehyde (FA), which had been used in previous Hi-C protocols, can significantly increase the cis percent of Hi-C libraries using 4-cutter enzymes, thus balancing loop detection and the cis/trans ratio. This important analysis has led to the development of Hi-C 3.0, a new protocol to obtain high-quality Hi-C libraries for both loop and compartment analysis [6,18]. We will discuss the benefits from this new protocol for loop and compartment analysis later in this chapter.

Visualization of Hi-C data

Visual inspection of Hi-C results is an important step to identify features of interest, and to guide quantitative analysis. Many Hi-C analytical tools can generate contact heatmaps that reveal the organization of whole chromosomes or specific loci [19,20]. However, in practice it can be time-consuming to generate separate heatmaps for visual exploration of many loci, or for high-level overviews of the data.

We have found that the HiGlass browser is a useful tool for direct visualization of Hi-C results in a genome-browser-type interface, which works with multiresolution cooler files, as obtained from the distiller mapping pipeline [21]. Linked views within HiGlass allow comparison of chromosome structures at the same loci either at different resolutions within the same data set, or from different samples, as shown for an example of publicly available Human Foreskin Fibroblast Clone 6 (HFFc6) FA *Dpn*II or FA + DSG DpnII HiC data [6] (Fig. 15.2). Moreover, this software can integrate TAD caller tracks, compartment tracks, loop locations, and other types of genomic data. (Fig. 15.2B). Documentation for HiGlass is available at https://docs.higlass.io/.

HiGlass can be run locally or on a server, and importantly, local HiGlass instances can use data saved on remote servers, without copying the data to the local machine, using the –no-upload option. This is essential for visualization of large data sets, and greatly facilitates getting started with HiGlass, particularly if the data is saved on a managed shared computer cluster, which often do not allow the use of Docker, which is required for running HiGlass.

The biggest limitation of HiGlass currently is that it is difficult to make publication quality figures due to autosizing of the heatmap panels, which tend to result in rectangular but not square heatmaps. We find HiGlass to be most useful for visual inspection of contact data, and identifying regions of interest. These regions can then be plotted with more customization by fetching dense matrices of regions of interest using cooler.fetch_matrix, and plotting the heatmaps using the matplotlib package [22]. This allows plotting of specific regions of chromosomes at a chosen resolution, as well as addition of 1D genomic tracks such as insulation, compartment eigenvectors, or ChIP-seq results (Fig. 15.2C). Results can then be saved as publication quality high-resolution PNG or PDF files. A similar functionality can

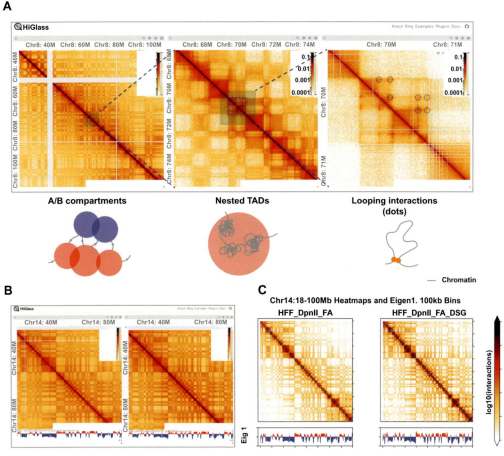

FIG. 15.2

Visualization of Hi-C results and chromosome structures. (A) HiGlass snapshot showing different resolutions of the same Hi-C data set (HFFc6-*Dpn*II-FA from GSE163666). The shaded region in the lower resolution panels shows the area that is depicted at a higher resolution (lower bin size) in the next panel to the right. Compartment interactions are shown in the 250 kb resolution panel (left), TADs are shown in the 25 kb resolution panel (middle), and dots/stripes are shown in the 5 kb resolution panel (right). The dashed box and circles in the middle and right panels highlight TADs and dots, respectively. The cartoons on the lower panel indicate how chromatin is organized in these three structures. (B) HiGlass snapshot of two different data sets (HFFc6-*Dpn*II-FA and HFFc6-DpnII-FA+DSG from GSE163666), both Hi-C contact maps and compartment eigenvectors are shown. Note that location, zoom level, and color scale can be locked between the two data sets. (C) Publication ready Hi-C heatmaps with corresponding eigenvectors made using dense matrices exported from cooler format and Matplotlib of the same Hi-C libraries as in panel B.

Data from B. Akgol Oksuz, L. Yang, S. Abraham, S.V. Venev, N. Krietenstein, K.M. Parsi, H. Ozadam, M.E. Oomen, A. Nand, H. Mao, R.M.J. Genga, R. Maehr, O.J. Rando, L.A. Mirny, J.H. Gibcus, J. Dekker, Systematic evaluation of chromosome conformation capture assays, Nat. Methods (2021), https:/doi.org/10.1038/s41592-021-01248-7.

be achieved by exporting specific regions of cooler files to c-world dense contact matrix format using cooltools cool2cworld, and plotting using the c-world script heatmap.pl [8].

Scaling plots and chromosome folding

One of the main features of Hi-C data is that on an average, each locus interacts significantly more with proximal regions than distal regions of the genome, which is termed distance decay [4]. Differences in this distance decay between different Hi-C libraries can be detected using scaling plots, which measure contact frequencies at different genomic distances. The term "Scaling Plot" results from the characteristic scaling of certain regions of this plot, such as the power law $\sim s^{-1}$ which occurs in mitotic cells from 100 to 10 Mb [23]. As chromosomes are long polymers, scaling plots have also been used to study the principles underlying chromosome folding and demonstrate distinct folding features of mitotic chromosomes [23,24]. Different from interphase chromosomes, mitotic chromosomes show a slower contact decrease from 100kb to 10Mb and a rapid fall from 10Mb (Fig. 15.3A). This unique feature has guided simulation studies that disclosed the folding principles of mitotic chromosome, which are

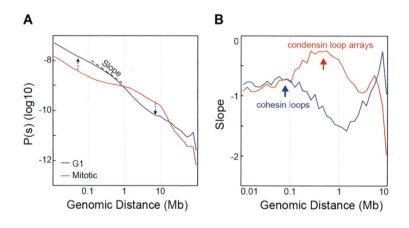

FIG. 15.3

Scaling and derivative plots reveal chromosome folding. (A) A scaling plot indicates folding differences between mitotic and G1 chromosomes of HeLa S3 cells. Red and blue lines represent mitotic and G1 chromosomes, respectively. The dashed arrows indicate scaling changes during the metaphase-to-interphase transition. The dashed line reveals that the slope at the specific genomic position (dark dot) is used to calculate the derivative plot in B. (B) A derivative plot illustrates structures of mitotic and G1 chromosomes of HeLa S3 cells. Blue and red arrows indicate cohesin loops and condensin loop arrays, signature structures of G1 and mitotic chromosomes, respectively. Both plots were generated using data from GSE133462. Since chromosome 4, 14, 17, and 21 of HeLa S3 cells have been defined as good chromosomes with less translocation signals in previous Hi-C analysis, interaction data from chromosome 4, 14, 17, and 21 of HeLa S3 cells were used here for both the plots. Mitotic and G1 HeLa S3 cells refer to the HeLa S3 cells that were arrested in prometaphase or were released from prometaphase for 8h as indicated in GSE133642.

Data from K. Abramo, A.L. Valton, S.V. Venev, H. Ozadam, A.N. Fox, J. Dekker, A chromosome folding intermediate at the condensin-to-cohesin transition during telophase, Nat. Cell Biol. 21 (11) (2019) 1393–1402, https://doi.org/10.1038/s41556-019-0406-2.

compressed chromatin loop arrays [23]. Furthermore, scaling plots of chromosomes from cells with well-controlled cycle progression demonstrated how these dense loop arrays form [24].

In interphase nuclei, the contacts at short ranges (less than 1 Mb) are associated with TADs while long-range contacts reflect compartment interactions. Therefore, changes of scaling plots at different genomic ranges reflect alterations of different chromosome features. During the metaphase-to-interphase transition (MIT), an increase of short-range interactions with a decrease in long-range interactions can reflect chromosome folding changes during the formation of interphase chromosomes [25] (Fig. 15.3A, dashed arrows).

The derivative or slope of scaling plots measures contact frequency changes at different genomic distances (Fig. 15.3A, dashed line) and can reveal cohesin loops during interphase or condensin-driven dense loop arrays in metaphase. These two different loops are reflected as peaks at different distances on scaling derivative plots (Fig. 15.3B). The size of the peak indicates the average size of cohesin loops, marked with blue arrows in Fig. 15.3B, which were first characterized by Gassler et al. [26], and were further confirmed by other studies in which the peak of cohesin loops disappears after cohesin is degraded using a degron system [27,28]. In addition, the density of cohesin loops on the chromosome can be estimated by the height of the peak [25,26]. The peak related to condensin loops has been characterized by Gibcus et al. [24].

These important structural hallmarks allow mechanistic studies of chromosome folding dynamics during the cell cycle. For example, during MIT, the cohesin loop peak at 100 kb emerges as the mitotic loop array peaks at 1 Mb disappears, revealing the formation of cohesin loops and disassembly of mitotic loop arrays [25]. Moreover, derivative plots clearly demonstrate formation of an intermediate chromosome folding state during MIT [25]. In other cases, derivative plots reveal changes in cohesin loops after the cohesin components are perturbed. For example, rapid removal of WAPL, the cohesin unloader, causes an increase in the peak location and height of the peak at 100 kb, reflecting an increase in looping interactions as a result of increased cohesin bound on chromosomes [28,29]. Taken together, signature structures can be reflected in scaling and derivative plots and these features are good indicators of chromosome folding changes during biological processes.

Compartment analysis

Compartments can be identified in Hi-C contact maps by their checkerboard pattern, where alternating blocks of chromatin have long-range interactions (Fig. 15.4A) [4]. Each block in the checkerboard is generally 1–10 Mb in length, and interactions between blocks can occur from approximately 5 Mb to the length of the chromosome, or even between chromosomes. Overall, compartments are mainly organized into two types, termed A and B [4]. High gene density and active chromatin states are found associated with A compartments whereas inactive chromatin is within B compartments [4,30]. Compartments are thus associated with cell identity, based on which genes are expressed or repressed in a particular cell type, and they are sensitive to cell state transitions [30–32].

Eigenvector deconvolution, which breaks down a matrix into a list of components that either play a significant role or can be neglected, is a simple and a fast way to identify the two compartments, that is A and B, and it has been integrated in many Hi-C analysis tools including c-world and cooltools [8,15]. In addition, K-means clustering analysis has revealed subcompartments that show stronger associations with specific histone marks [30,33,34]. Here, we focus on the basic A-B compartment analysis using cooltools.

350 Chapter 15 A practical guide for essential analyses of Hi-C data

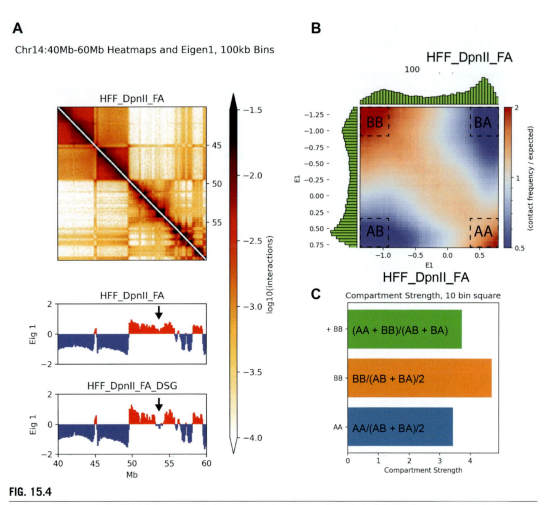

FIG. 15.4

Compartment analysis. (A) Hi-C map and eigenvector plot for a region of chr14, black arrows show bins with different compartment identities in the two samples, both in the HFFc6 cell line, which differ in cross-linking protocol used (FA only (top) or FA+DSG (bottom)), from GSE163666. (B) Saddleplot of the FA fixed sample from GSE163666, with histograms of the number of genomic bins within each eigenvector bin shown in the margins. (C) Bar plot of saddleplot strength, using a 10 bin square at each corner of the saddleplot, made with 50 eigenvector bins total. Equations to calculate compartment strength for each type of compartment are shown within the corresponding bars.

Data from B. Akgol Oksuz, L. Yang, S. Abraham, S.V. Venev, N. Krietenstein, K.M. Parsi, H. Ozadam, M.E. Oomen, A. Nand, H. Mao, R.M.J. Genga, R. Maehr, O.J. Rando, L.A. Mirny, J.H. Gibcus, J. Dekker, Systematic evaluation of chromosome conformation capture assays, Nat. Methods (2021), https:/doi.org/10.1038/s41592-021-01248-7.

Cooltools can be used to perform eigenvector deconvolution of a balanced cooler file, either in cis or in trans, resulting in three different eigenvectors by default (more can be calculated by changing the n_eigs option), each corresponding to different patterns in the data, with a score for each bin of the genome at the resolution of the original cooler file. Eigenvectors can be calculated from Hi-C libraries with low-sequence depth at larger bin sizes (up to 1 Mb), or on smaller bin sizes for deeply sequenced Hi-C libraries [15,35]. The corresponding eigenvalues of each eigenvector are then used to weigh the eigenvector amplitude and determine their order. Eigenvalues are a measure of how much of the variation in the data is captured by each eigenvector [15].

To compare across samples, the eigenvectors are phased by gene density, such that positive values will correspond to gene-dense regions, or the "A" compartment. Alternatively, eigenvectors can be ordered by correlation with gene density or another genomic feature, rather than by eigenvalue. Eigenvector 1 usually corresponds to the A and B compartments, although visual inspection is recommended, as translocations lead to very strong signal, which can complicate this result, and may lead to eigenvector 2 or 3 picking up the compartment signal. Comparing the eigenvector track at specific bins between samples can identify regions where switches in compartmentalization may occur. An example of this is shown in Fig. 15.4A, where different cross-linking protocols were used (FA only or FA+DSG), resulting in changes in compartment identity at the locus shown (black arrow) [6] (Fig. 15.4A). As discussed in detail in Akgol Oksuz et al., 2021, addition of DSG cross-linking increases detection of compartments in Hi-C with 4-bp restriction enzymes including *Dpn*II, compared to FA alone [6].

While the eigenvectors show which regions of the genome are in each compartment, they do not measure the strength of the interactions between different bins of the same type of compartment, which may be a more biologically meaningful measurement [15,25]. In order to analyze the strength of A to A (AA), B to B (BB), or A to B (AB) interactions, we can use cooltools to generate a type of plot called a saddleplot, which shows how strong the Hi-C interactions are between regions of similar or different eigenvector values [15]. In this type of plot, the eigenvectors are sorted based on their value, and Hi-C interactions are plotted based on the eigenvector value of the two sides of a pair of reads, leading to a 2D heatmap organized by eigenvector instead of by the genomic location [15]. The interactions are normalized by expected values calculated for each genomic distance, as distance decay has a large effect on interaction strength, as discussed in the Scaling Plot section (Fig. 15.4B).

These plots can also be quantified to result in a compartment strength score for AA, BB, or both AA and BB interactions, by calculating the ratio of the average interactions between the compartments of the same type versus the compartments of different types (AA + BB)/(AB + BA) for AA + BB interactions, AA/(AB + BA)/2 for AA interactions only, or BB/(AB + BA)/2 for BB interactions only (Fig. 15.4C). This score is used to indicate the overall strength of A and/or B compartments in a Hi-C sample. Comparing the compartment strength between samples can show overall changes in compartmentalization due to different biological processes.

Insulation and TAD boundaries

TADs are genomic regions on the order of hundreds of kilobases, within which contact frequencies of chromosomes are higher than outside the region. TADs appear as square blocks or triangles on the diagonal of Hi-C contact maps and TAD boundaries are enriched with CTCF and cohesin binding sites

[33,36,37]. In most studies, TADs are regarded to spatially restrict gene regulatory modules and disruption of this restriction that is defined by TAD boundaries is associated with many diseases including cancers [38–41]. TADs are self-interacting genomic regions and are insulated from each other. Based on this feature, a simple but efficient way to identify TADs from Hi-C interactions is by calculation of genome-wide insulation scores (IS) [11].

An IS is the average of contacts within a given diamond window (Fig. 17.5A, light blue box). As the diamond window slides, the IS profile of whole chromosomes can be obtained (Fig. 17.5A, light blue box and arrow, and Fig. 17.5B). As TAD boundaries flank neighboring TADs and have less contacts, IS at TAD boundaries decrease sharply and form valleys on IS profiles (Fig. 17.5A and B, red arrows). These valleys on IS profiles can be used to detect TADs [11]. However, we note that IS profile is highly dependent on the threshold for boundary detection, the resolution of the data, and the size of the sliding window used, and the integrative analysis with ChIP-seq or other data, as described below, is often useful to provide more insight. Besides IS profiles, cooltools also includes an option to calculate a metric called directionality index to detect TADs [36].

In our analysis, we usually generate an IS profile at 25–50 kb resolution and identify TAD boundaries, then perform pileup analysis of TAD boundaries to examine global TAD changes. An example using published data sets is the establishment of TADs upon the MIT in HeLa cells. In metaphase, there are no TADs in the Hi-C heatmap, but TADs appear around 2.5 h after release from prometaphase and are completely established around 8 h after release, when the majority of HeLa cells enter G1 phase [25]. The pileup of TAD boundaries indicates average TAD boundary features of HeLa cells in prometaphase and G1 phase in Fig. 15.5C, respectively [36]. The IS can also be used to measure local chromosome structure changes. For example, in hematopoietic progenitor cells, knock-out (KO) of Stag2 causes loss of cis-interactions at three Stag2-binding sites within *Ebf1* (blue arrows in Fig. 15.5D). These local chromosome changes are reflected in the IS profile with a significant reduction (difference between WT and KO plots in Fig. 15.5E) [37]. Taken together, IS and TAD boundary analysis can be used to examine short-distance interactions.

Dot calling and CTCF-CTCF loops

In the corner of TADs, strong interaction dots or stripes can be found in high-resolution Hi-C data sets [6,33]. These loci are enriched for binding motifs and ChIP-seq signal of the DNA-binding protein CTCF, and are referred to as CTCF-CTCF loops [33]. CTCF-CTCF loops are important chromosome structures that are formed by cohesin-mediated loop extrusion that is stalled at convergent CTCF-binding sites, which results in TAD formation [28,33,38–41]. It is often important to be able to measure the quantity and strength of CTCF-CTCF loops genome-wide. Besides CTCF-CTCF looping interactions, other long-distance *cis* interactions, such as enhancer and promoter interactions, can also appear as dots on the Hi-C contact map [33]. These regions of enrichment representing looping interactions can be detected using dot caller software, first developed for use on the Juicer.hic file format in HiCCUPs, and later reimplemented for the cooler file format as cooltools call-dots, and the cooltools.dotfinder module [11,33]. In both cases, dots are called as regions with higher than expected interactions at a specific distance, compared to the surrounding region.

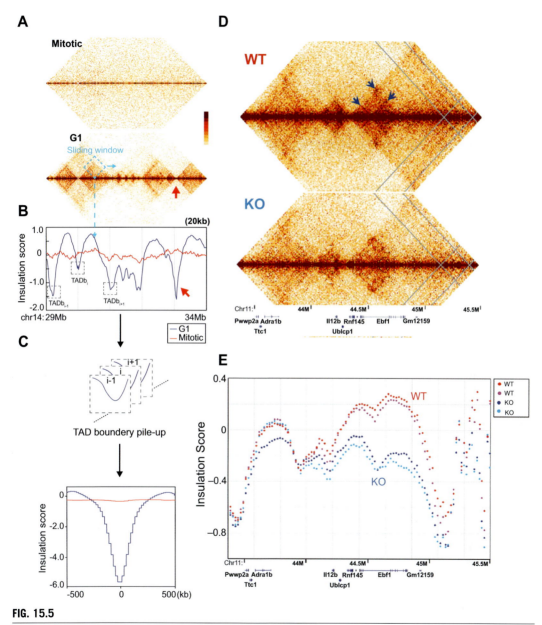

FIG. 15.5

Illustration of Hi-C contact map and insulation profiles. (A) Hi-C interaction maps for mitotic and G1 HeLa S3 cells. Data for the 29–34 Mb region of chr14 are shown at 20 kb resolution. To generate these heatmaps, cooler files were first converted to c-world contact matrix files using cool2cworld. After scaling using scalematrix.pl, the specific region was extracted by extractSubMatrices.pl and heatmaps were obtained from these matrices using heatmap.pl. Cool2cworld is a cooler file operation, while other Perl scripts are from

Figure caption continued on next page

FIG. 15.5—CONT'D

c-world. (B) Insulation profiles for the 29–34 Mb region of chr14 at 20 kb resolution. The blue and red lines represent G1 and mitotic HeLa S3 cells. The light blue diamond box indicates the sliding window, within which the mean of contacts is used to calculate the insulation score at the genomic position, and this score is shown on the insulation plot as pointed by the light blue arrow. The sliding window size here is 500 kb and the red arrows in panel A and B indicate an example of a TAD boundary. Gray dashed box indicates three TAD boundaries to illustrate TAD boundary pileup analysis in panel C. (C) Illustration of TAD boundary pile-up analysis. Left panel: insulation scores of both up- and down-stream of TAD boundaries were extracted and aggregated to generate the profile plot in right panel. Gray dashed boxes in panels B and C indicate three examples of TAD boundaries. Right panel: Profile plot of insulation score of G1 and mitotic HeLa S3 cells at TAD boundaries. 4297 TAD boundaries were identified from G1 HeLa S3 cells and used for pileup analysis. Profile plot was generated using DeepTools (Ramirez et al., 2016) with insulation scores from G1 and mitotic cells in bigwig formats. Panels A, B, and C were generated using data from GSE133462. Cell cycle states are the same as in Fig. 17.3. (D) Contact map for the region across *Ebf1*. KO represents Stag2 knock-out and the blue arrows indicate loss of cis-interaction at three loci. (E) Insulation plots across *Ebf1*, the same region as in C. The plots show two biological replicates for *Stag2* wildtype (red), and two biological replicates for *Stag2* KO (blue), as indicated. WT and KO heatmaps in D are generated from the pooled data of two replicates of each genotype, respectively. The loss of cis-interactions as pointed by blue arrows in D was reflected as a decrease of insulation scores in E. The panels are from our recent work [37].

No permission required.

Nearby pixels are then clustered together to find the center of each dot (Fig. 15.6A and B). De novo dot calling requires much more deeply sequenced libraries than compartment and TAD analysis, with published data sets of this type generally consisting of >2 billion interactions [6,33]. However, it is possible to analyze dots in shallower libraries using locations called in published deep libraries in the same cell type, or by using pairs of nearby CTCF motifs if a dot list is not available, by stacking up all of the dots or pairs of CTCF sites for an aggregate pileup analysis (Fig. 15.6A and C) [33,40]. This approach is not sufficient to determine if specific loops are gained or lost in a sample, but it will detect global changes in loop strength. Cooltools offers methods for piling up loop interactions via the snipping module, which takes a Hi-C data set in cooler format and a list of loop anchors, and aggregates observed/expected interactions at all loops (Fig. 15.6C).

The list of loop anchors can be easily modified using python packages bioframe or pandas to include specific subsets of loops based on overlap with other data sets, such as CTCF binding, gene expression, compartment type, chromatin modifications, loop size, or other characteristics of interest [11,42]. The number and strength of dots are highly dependent on the specific protocol used, with more frequent cutting and more cross-linking resulting in better detection of dots [6].

Integration of Hi-C data with ChIP-seq and RNA-seq data

We have shared our experience of essential analyses for Hi-C experiments and discussed how to characterize and interpret the results from these analyses. Our analyses cover major chromatin structures, compartments, TADs, loops and global chromosome folding. In cells, these distinct structures are not isolated and can be established or maintained by the same machinery, even though the molecular details

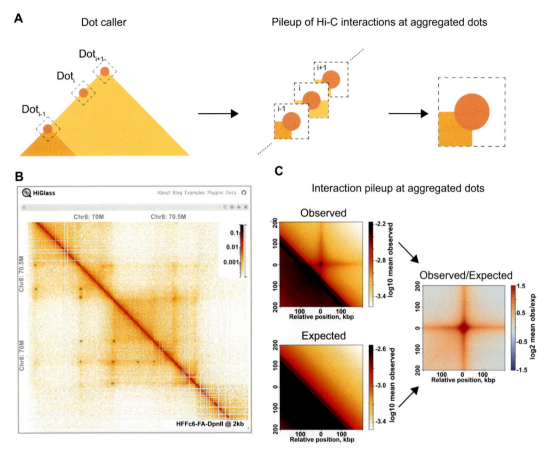

FIG. 15.6

Identifying and quantifying CTCF-CTCF loops by dot calling and aggregate pileup analysis. (A) A schematic of the dot calling method (left) and aggregate dot pileup of Hi-C interactions (middle and right). (B) Example of HiGlass visualization of a publicly available dot list from deeply sequenced HFFc6 cell line Hi-C (4DNFIWEN3L7Y) shown as rectangular domains (lower left) overlaid on published HFFc6 Hi-C (GSE163666). (C) Example of aggregate dot pileups from all-called dots genome-wide (4DNFIWEN3L7Y). Observed Hi-C pileup (top left), Expected Hi-C pileup (top left), Log2(Observed/Expected) ratio (right). Note that using the Observed/Expected ratio allows for normalization of distant dependent contact strength, and improved visualization of the dot enrichment.

Data from B. Akgol Oksuz, L. Yang, S. Abraham, S.V. Venev, N. Krietenstein, K.M. Parsi, H. Ozadam, M.E. Oomen, A. Nand, H. Mao, R.M.J. Genga, R. Maehr, O.J. Rando, L.A. Mirny, J.H. Gibcus, J. Dekker, Systematic evaluation of chromosome conformation capture assays, Nat. Methods (2021), https://doi.org/10.1038/s41592-021-01248-7.

can be different. Thus, integrative analyses of these distinct structural changes in biological contexts allow a comprehensive interpretation of chromosome dynamics.

For example, cohesion-driven loop extrusion has been proposed to mediate the formation of CTCF loops and TADs [28,38–41]. However, our recent work discovered that cohesin complexes at CTCF loops display distinct biochemical states from cohesin complexes within TADs, revealing that cohesin engages chromatin in different ways [27]. On the other hand, attraction of chromosome domains with similar states has been proposed to mediate compartmentalization [10,30,43]; however, loss of cohesin or NIPBL, a cohesin loader, can enhance compartmentalization [27,44], revealing that loop extrusion is also involved in regulating compartmentalization. Thus, examining compartment boundaries that overlap with TAD boundaries is critical to differentiate between these two distinct mechanisms in chromosome dynamics. Our recent work has also integrated cis/trans ratios to determine the extent of expansion in a region of a chromosome territory into other chromosomes due to active transcription since trans interactions increase when regions from different chromosomes mingle with each other [45]. Taken together, integrative analysis of different features from Hi-C matrices can dissect molecular mechanisms underlying these features in detail.

Hi-C provides molecular details of genome architecture, however many questions remain unaddressed. Two outstanding questions are: (1) How does chromatin state define chromosome organization? and (2) How does chromosome organization mediate transcription regulation?

ChIP-seq data provide chromatin state information including epigenetic modifications of chromatin and binding of chromatin-associated proteins, while RNA-seq measures transcription and splicing activities. Integration of ChIP-seq and RNA-seq with Hi-C data may allow the field to address these two questions.

A major challenge for integrative analyses is the resolution difference between Hi-C data and ChIP-seq or RNA-seq data. For example, ChIP-seq peaks of transcription factors are in the range of 200-500 bp while the bin size of most Hi-C data is a minimum of 20 kb, based on our experience. Therefore, historically, ChIP-seq data have been used to perform enrichment analysis in chromosome structures. At TAD boundaries, CTCF and cohesion-shared binding sites are enriched [33,46,47], and activating chromatin marks like H3K4me1 and H3K27ac were found enriched in A compartments [33]. One solution to this challenge is to increase sequencing depth of Hi-C to integrate with ChIP-seq and RNA-seq data. For example, recent work obtained around 800 million Hi-C reads to reach 2 kb resolution and generated IS using a 20 kb sliding window, allowing integrative analyses of insulation score with cohesin or CTCF ChIP-seq and RNA-seq data. This work clearly shows that three types of chromatin boundaries (CTCF sites, TSSs, and TTSs) can regulate the cohesin traffic pattern [48]. The related 3C-based method Micro-C can reach single-nucleosome resolution and is thus an alternative solution for this challenge [3,49,50]. Although other 3C-based methods, such as 5C and Tri-C, can also be considered for this challenge, they are restricted to specific genomic regions and genome-wide analyses cannot be performed [2,51].

IS measures contact frequencies of local chromatin, thus allowing monitoring of local chromosome structural changes when chromatin state and transcription activity change during biological events [37,52]. Genome-wide IS changes were shown to be close to normally distributed and the 95% confidential interval of IS changes allows identification of loci with significant local chromosome structural changes [37,52]. Integrative analyses can identify chromosome structural changes associated with chromatin state and transcription activity changes. However, as IS plots are obtained using sum or average of contacts within a given sliding window [53], contact pattern changes cannot be captured. Contact pattern changes can reflect interaction rewiring between enhancers and promoters, thus delivering important information about gene regulation. Integrating contact pattern analysis with IS plots will provide genome-wide organization information of chromosomes in greater detail.

In summary, we have described some essential and useful analysis steps for Hi-C libraries, which are a good starting point for any Hi-C project. Depending on the project, this may be sufficient to show changes between samples, and to support or refute a hypothesis, or the researchers may desire to further explore specific results or regions of interest using more sophisticated analysis approaches.

Data use

The data for Figs. 2, 4, and 6 are from HFFc6-*Dpn*II (GSE163666), which is part of 4DN data set. The data for Figs. 3 and 5 are from HeLa S3 (GSE133462). The loop list from 4DN used in Fig. 6: 4DNFIWEN3L7Y.bedpe.gz.

Computational resources

Distiller: https://github.com/open2c/distiller-nf
HiGlass: https://higlass.io/
Cooler: https://cooler.readthedocs.io/en/latest/
Cooltools: https://cooltools.readthedocs.io/en/latest/index.html
C-world: https://github.com/dekkerlab/cworld-dekker
Perl, Python 3+, Miniconda, Docker and Jupyter.

Acknowledgments

We are grateful to Dr. Job Dekker for his support and guidance for our studies researching genome architecture and functions, to members of the Dekker lab for help and discussions of analytic methods of Hi-C, and to Drs. Allana Schooley, Anne-Laure Valton, and Sergey V. Venev, and Mr. Bastiaan Dekker for their critical comments. We acknowledge support from the National Institutes of Health Common Fund 4D Nucleome Program (DK107980 and HG011536 to Dr. Job Dekker) and the National Human Genome Research Institute (HG003143 to Dr. Job Dekker). Y.L. is also supported from the National Cancer Institute (PS-OC U54 CA143869 to Dr. Job Dekker), and E.H. is supported from the National Cancer Institute (F32 CA224689).

References

[1] J. Dekker, K. Rippe, M. Dekker, N. Kleckner, Capturing chromosome conformation, Science 295 (2002) 1306–1311. https://doi.org/10.1126/science.1067799.
[2] J. Dostie, T.A. Richmond, R.A. Arnaout, R.R. Selzer, W.L. Lee, T.A. Honan, E.D. Rubio, A. Krumm, J. Lamb, C. Nusbaum, R.D. Green, J. Dekker, Chromosome conformation capture carbon copy (5C): a massively parallel solution for mapping interactions between genomic elements, Genome Res. 16 (2006) 1299–1309. https://doi.org/10.1101/gr.5571506.

[3] T.H.S. Hsieh, A. Weiner, B. Lajoie, J. Dekker, N. Friedman, O.J. Rando, Mapping nucleosome resolution chromosome folding in yeast by micro-C, Cell 162 (2015) 108–119. https://doi.org/10.1016/j.cell.2015.05.048.
[4] E. Lieberman-Aiden, N.L. Van Berkum, L. Williams, M. Imakaev, T. Ragoczy, A. Telling, I. Amit, B.R. Lajoie, P.J. Sabo, M.O. Dorschner, R. Sandstrom, B. Bernstein, M.A. Bender, M. Groudine, A. Gnirke, J. Stamatoyannopoulos, L.A. Mirny, E.S. Lander, J. Dekker, Comprehensive mapping of long-range interactions reveals folding principles of the human genome, Science 326 (2009) 289–293. https://doi.org/10.1126/science.1181369.
[5] Z. Zhao, G. Tavoosidana, M. Sjölinder, A. Göndör, P. Mariano, S. Wang, C. Kanduri, M. Lezcano, K.S. Sandhu, U. Singh, V. Pant, V. Tiwari, S. Kurukuti, R. Ohlsson, Circular chromosome conformation capture (4C) uncovers extensive networks of epigenetically regulated intra- and interchromosomal interactions, Nat. Genet. 38 (2006) 1341–1347. https://doi.org/10.1038/ng1891.
[6] B. Akgol Oksuz, L. Yang, S. Abraham, S.V. Venev, N. Krietenstein, K.M. Parsi, H. Ozadam, M.E. Oomen, A. Nand, H. Mao, R.M.J. Genga, R. Maehr, O.J. Rando, L.A. Mirny, J.H. Gibcus, J. Dekker, Systematic evaluation of chromosome conformation capture assays, Nat. Methods (2021). https://doi.org/10.1038/s41592-021-01248-7.
[7] N. Abdennur, L.A. Mirny, Cooler: scalable storage for Hi-C data and other genomically labeled arrays, Bioinformatics 36 (2020) 311–316. https://doi.org/10.1093/bioinformatics/btz540.
[8] B.R. Lajoie, J. Dekker, N. Kaplan, The Hitchhiker's guide to Hi-C analysis: practical guidelines, Methods 72 (2015) 65–75. https://doi.org/10.1016/j.ymeth.2014.10.031.
[9] J.T. Robinson, D. Turner, N.C. Durand, H. Thorvaldsdóttir, J.P. Mesirov, E.L. Aiden, Juicebox.js provides a cloud-based visualization system for Hi-C data, Cell Syst. 6 (2018). https://doi.org/10.1016/j.cels.2018.01.001. 256–258.e1.
[10] J. Nuebler, G. Fudenberg, M. Imakaev, N. Abdennur, L.A. Mirny, Chromatin organization by an interplay of loop extrusion and compartmental segregation, Proc. Natl. Acad. Sci. U. S. A. 115 (2018) E6697–E6706. https://doi.org/10.1073/pnas.1717730115.
[11] S. Venev, N. Abdennur, A. Goloborodko, I. Flyamer, G. Fudenberg, J. Nuebler, A. Galitsyna, B. Akgol, S. Abraham, P. Kerpedjiev, et al., open2c/cooltools: v0.4.1 (Zenodo), 2021.
[12] R. Dale, B. Grüning, A. Sjödin, J. Rowe, B.A. Chapman, C.H. Tomkins-Tinch, R. Valieris, B. Batut, A. Caprez, T. Cokelaer, D. Yusuf, K.A. Beauchamp, K. Brinda, T. Wollmann, G.L. Corguillé, D. Ryan, A. Bretaudeau, Y. Hoogstrate, B.S. Pedersen, S.V. Heeringen, M. Raden, S. Luna-Valero, N. Soranzo, M.D. Smet, G.V. Kuster, R. Kirchner, L. Pantano, Z. Charlop-Powers, K. Thornton, M. Martin, M.V.D. Beek, D. Maticzka, M. Miladi, S. Will, K. Gravouil, P. Unneberg, C. Brueffer, C. Blank, V.C. Piro, J. Wolff, T. Antao, S. Gladman, I. Shlyakhter, M.D. Hollander, P. Mabon, W. Shen, J. Boekel, M. Holtgrewe, D. Bouvier, J.R. de Ruiter, J. Cabral, S. Choudhary, N. Harding, R. Kleinkauf, E. Enns, F. Eggenhofer, J. Brown, P.J.A. Cock, H. Timm, C. Thomas, X.O. Zhang, M. Chambers, N. Turaga, E. Seiler, C. Brislawn, E. Pruesse, I. Fallmann, J. Kelleher, H. Nguyen, L. Parsons, Z. Fang, E.B. Stovner, N. Stoler, S. Ye, I. Wohlers, R. Farouni, M. Freeberg, J.E. Johnson, M. Bargull, P.R. Kensche, T.H. Webster, J.M. Eppley, C. Stahl, A.S. Rose, A. Reynolds, L.B. Wang, X. Garnier, S. Dirmeier, M. Knudsen, J. Taylor, A. Srivastava, V. Rai, R. Agren, A. Junge, R.V. Guimera, A. Khan, S. Schmeier, G. He, L. Pinello, E. Hägglund, A.S. Mikheyev, J. Preussner, N.R. Waters, W. Li, J. Capellades, A.T. Chande, Y. Pirola, S. Hiltemann, M.L. Bendall, S. Singh, W.A. Dunn, A. Drouin, T.D. Domenico, I.D. Bruijn, D.E. Larson, D. Chicco, E. Grassi, G. Gonnella, J.B.L. Wang, F. Giacomoni, E. Clarke, D. Blankenberg, C. Tran, R. Patro, S. Laurent, M. Gopez, B. Sennblad, J.A. Baaijens, P. Ewels, P.R. Wright, O.M. Enache, P. Roger, W. Dampier, D. Koppstein, U.K. Devisetty, T. Rausch, M. Cornwell, A.E. Salatino, J. Seiler, M. Jung, E. Kornobis, F. Cumbo, B.K. Stöcker, O. Moskalenko, D.R. Bogema, M.L. Workentine, S.J. Newhouse, F.D.V. Leprevost, K. Arvai, J. Köster, Bioconda: sustainable and comprehensive software distribution for the life sciences, Nat. Methods 15 (2018) 475–476. https://doi.org/10.1038/s41592-018-0046-7.

[13] A. Goloborodko, S. Venev, N. Abdennur, D. Tommaso, P. mirnylab/distiller-nf: v.0.3.3 (Zenodo), 2019.
[14] H. Li, R. Durbin, Fast and accurate short read alignment with Burrows-Wheeler transform, Bioinformatics 25 (2009) 1754–1760. https://doi.org/10.1093/bioinformatics/btp324.
[15] M. Imakaev, G. Fudenberg, R.P. McCord, N. Naumova, A. Goloborodko, B.R. Lajoie, J. Dekker, L.A. Mirny, Iterative correction of Hi-C data reveals hallmarks of chromosome organization, Nat. Methods 9 (2012) 999–1003. https://doi.org/10.1038/nmeth.2148.
[16] J.M. Belton, R.P. McCord, J.H. Gibcus, N. Naumova, Y. Zhan, J. Dekker, Hi-C: a comprehensive technique to capture the conformation of genomes, Methods 58 (2012) 268–276. https://doi.org/10.1016/j.ymeth.2012.05.001.
[17] H. Belaghzal, J. Dekker, J.H. Gibcus, Hi-C 2.0: an optimized Hi-C procedure for high-resolution genome-wide mapping of chromosome conformation, Methods 123 (2017) 56–65. https://doi.org/10.1016/j.ymeth.2017.04.004.
[18] D.L. Lafontaine, L. Yang, J. Dekker, J.H. Gibcus, Hi-C 3.0: improved protocol for genome-wide chromosome conformation capture, Curr. Protocols 1 (2021). https://doi.org/10.1002/cpz1.198.
[19] K.C. Akdemir, L. Chin, HiCPlotter integrates genomic data with interaction matrices, Genome Biol. 16 (2015) 198. https://doi.org/10.1186/s13059-015-0767-1.
[20] J. Wolff, L. Rabbani, R. Gilsbach, G. Richard, T. Manke, R. Backofen, B.A. Grüning, Galaxy HiCExplorer 3: a web server for reproducible Hi-C, capture Hi-C and single-cell Hi-C data analysis, quality control and visualization, Nucleic Acids Res. 48 (2020) W177–W184. https://doi.org/10.1093/NAR/GKAA220.
[21] P. Kerpedjiev, N. Abdennur, F. Lekschas, C. McCallum, K. Dinkla, H. Strobelt, J.M. Luber, S.B. Ouellette, A. Azhir, N. Kumar, J. Hwang, S. Lee, B.H. Alver, H. Pfister, L.A. Mirny, P.J. Park, N. Gehlenborg, HiGlass: web-based visual exploration and analysis of genome interaction maps, Genome Biol. 19 (2018). https://doi.org/10.1186/s13059-018-1486-1.
[22] J.D. Hunter, Matplotlib: a 2D graphics environment, Comput. Sci. Eng. 9 (2007) 90–95. https://doi.org/10.1109/MCSE.2007.55.
[23] N. Naumova, M. Imakaev, G. Fudenberg, Y. Zhan, B.R. Lajoie, L.A. Mirny, J. Dekker, Organization of the mitotic chromosome, Science 342 (2013) 948–953. https://doi.org/10.1126/science.1236083.
[24] J.H. Gibcus, K. Samejima, A. Goloborodko, I. Samejima, N. Naumova, J. Nuebler, M.T. Kanemaki, L. Xie, J.R. Paulson, W.C. Earnshaw, L.A. Mirny, J. Dekker, A pathway for mitotic chromosome formation, Science 359 (2018). https://doi.org/10.1126/science.aao6135.
[25] K. Abramo, A.L. Valton, S.V. Venev, H. Ozadam, A.N. Fox, J. Dekker, A chromosome folding intermediate at the condensin-to-cohesin transition during telophase, Nat. Cell Biol. 21 (2019) 1393–1402. https://doi.org/10.1038/s41556-019-0406-2.
[26] J. Gassler, H.B. Brandão, M. Imakaev, I.M. Flyamer, S. Ladstätter, W.A. Bickmore, J.M. Peters, L.A. Mirny, K. Tachibana, A mechanism of cohesin-dependent loop extrusion organizes zygotic genome architecture, EMBO J. 36 (2017) 3600–3618. https://doi.org/10.15252/embj.201798083.
[27] Y. Liu, J. Dekker, CTCF–CTCF loops and intra-TAD interactions show differential dependence on cohesin ring integrity, Nat. Cell Biol. 24 (2022) 1516–1527. https://doi.org/10.1038/s41556-022-00992-y.
[28] G. Wutz, C. Várnai, K. Nagasaka, D.A. Cisneros, R.R. Stocsits, W. Tang, S. Schoenfelder, G. Jessberger, M. Muhar, M.J. Hossain, N. Walther, B. Koch, M. Kueblbeck, J. Ellenberg, J. Zuber, P. Fraser, J.M. Peters, Topologically associating domains and chromatin loops depend on cohesin and are regulated by CTCF, WAPL, and PDS5 proteins, EMBO J. 36 (2017) 3573–3599. https://doi.org/10.15252/embj.201798004.
[29] J.H.I. Haarhuis, R.H. van der Weide, V.A. Blomen, J.O. Yáñez-Cuna, M. Amendola, M.S. van Ruiten, P.H.L. Krijger, H. Teunissen, R.H. Medema, B. van Steensel, T.R. Brummelkamp, E. de Wit, B.D. Rowland, The Cohesin release factor WAPL restricts chromatin loop extension, Cell 169 (2017) 693–707.e14. https://doi.org/10.1016/j.cell.2017.04.013.
[30] E.M. Hildebrand, J. Dekker, Mechanisms and functions of chromosome compartmentalization, Trends Biochem. Sci. 45 (2020) 385–396. https://doi.org/10.1016/j.tibs.2020.01.002.

[31] S. Sati, B. Bonev, Q. Szabo, D. Jost, P. Bensadoun, F. Serra, V. Loubiere, G.L. Papadopoulos, J.C. Rivera-Mulia, L. Fritsch, P. Bouret, D. Castillo, J.L. Gelpi, M. Orozco, C. Vaillant, F. Pellestor, F. Bantignies, M.A. Marti-Renom, D.M. Gilbert, J.M. Lemaitre, G. Cavalli, 4D genome rewiring during oncogene-induced and replicative senescence, Mol. Cell 78 (2020) 522–538.e9. https://doi.org/10.1016/j.molcel.2020.03.007.

[32] L. Tan, W. Ma, H. Wu, Y. Zheng, D. Xing, R. Chen, X. Li, N. Daley, K. Deisseroth, X.S. Xie, Changes in genome architecture and transcriptional dynamics progress independently of sensory experience during postnatal brain development, Cell 184 (2021) 741–758.e17. https://doi.org/10.1016/j.cell.2020.12.032.

[33] S.S.P. Rao, M.H. Huntley, N.C. Durand, E.K. Stamenova, I.D. Bochkov, J.T. Robinson, A.L. Sanborn, I. Machol, A.D. Omer, E.S. Lander, E.L. Aiden, A 3D map of the human genome at kilobase resolution reveals principles of chromatin looping, Cell 159 (2014) 1665–1680. https://doi.org/10.1016/j.cell.2014.11.021.

[34] G. Spracklin, N. Abdennur, M. Imakaev, N. Chowdhury, S. Pradhan, L. Mirny, J. Dekker, Diverse silent chromatin states modulate genome compartmentalization and loop extrusion barriers, Nat. Struct. Mol. Biol. 30 (2023) 38–51. https://doi.org/10.1038/s41594-022-00892-7.

[35] H. Belaghzal, T. Borrman, A.D. Stephens, D.L. Lafontaine, S.V. Venev, Z. Weng, J.F. Marko, J. Dekker, Liquid chromatin Hi-C characterizes compartment-dependent chromatin interaction dynamics, Nat. Genet. 53 (2021) 367–378. https://doi.org/10.1038/s41588-021-00784-4.

[36] F. Ramírez, D.P. Ryan, B. Grüning, V. Bhardwaj, F. Kilpert, A.S. Richter, S. Heyne, F. Dündar, T. Manke, deepTools2: a next generation web server for deep-sequencing data analysis, Nucleic Acids Res. 44 (2016) W160–W165. https://doi.org/10.1093/nar/gkw257.

[37] A.D. Viny, R.L. Bowman, Y. Liu, V.P. Lavallée, S.E. Eisman, W. Xiao, B.H. Durham, A. Navitski, J. Park, S. Braunstein, B. Alija, A. Karzai, I.S. Csete, M. Witkin, E. Azizi, T. Baslan, C.J. Ott, D. Pe'er, J. Dekker, R. Koche, R.L. Levine, Cohesin members Stag1 and Stag2 display distinct roles in chromatin accessibility and topological control of HSC self-renewal and differentiation, Cell Stem Cell 25 (2019) 682–696.e8. https://doi.org/10.1016/j.stem.2019.08.003.

[38] G. Fudenberg, N. Abdennur, M. Imakaev, A. Goloborodko, L.A. Mirny, Emerging evidence of chromosome folding by loop extrusion, Cold Spring Harb. Symp. Quant. Biol. 82 (2017) 45–55. https://doi.org/10.1101/sqb.2017.82.034710.

[39] G. Fudenberg, M. Imakaev, C. Lu, A. Goloborodko, N. Abdennur, L.A. Mirny, Formation of chromosomal domains by loop extrusion, Cell Rep. 15 (2016) 2038–2049. https://doi.org/10.1016/j.celrep.2016.04.085.

[40] S.S.P. Rao, S.-C. Huang, B.G.S. Hilaire, J.M. Engreitz, E.M. Perez, K.-R. Kieffer-Kwon, A.L. Sanborn, S.E. Johnstone, G.D. Bascom, I.D. Bochkov, X. Huang, M.S. Shamim, J. Shin, D. Turner, Z. Ye, A.D. Omer, J.T. Robinson, T. Schlick, B.E. Bernstein, R. Casellas, E.S. Lander, E.L. Aiden, Cohesin loss eliminates all loop domains, Cell (2017). https://doi.org/10.1016/j.cell.2017.09.026. 305–320.e24.

[41] A.L. Sanborn, S.S.P. Rao, S.-C. Huang, N.C. Durand, M.H. Huntley, A.I. Jewett, I.D. Bochkov, D. Chinnappan, A. Cutkosky, J. Li, K.P. Geeting, A. Gnirke, A. Melnikov, D. McKenna, E.K. Stamenova, E.S. Lander, E.L. Aiden, Chromatin extrusion explains key features of loop and domain formation in wild-type and engineered genomes, Proc. Natl. Acad. Sci. (2015) E6456–E6465. https://doi.org/10.1073/pnas.1518552112.

[42] Pandas-Team, pandas-dev/pandas: Pandas, 2020.

[43] M. Falk, Y. Feodorova, N. Naumova, M. Imakaev, B.R. Lajoie, H. Leonhardt, B. Joffe, J. Dekker, G. Fudenberg, I. Solovei, L.A. Mirny, Heterochromatin drives compartmentalization of inverted and conventional nuclei, Nature 570 (2019) 395–399. https://doi.org/10.1038/s41586-019-1275-3.

[44] W. Schwarzer, N. Abdennur, A. Goloborodko, A. Pekowska, G. Fudenberg, Y. Loe-Mie, N.A. Fonseca, W. Huber, C.H. Haering, L. Mirny, F. Spitz, Two independent modes of chromatin organization revealed by cohesin removal, Nature 551 (2017) 51–56. https://doi.org/10.1038/nature24281.

[45] S. Leidescher, J. Ribisel, S. Ullrich, Y. Feodorova, E. Hildebrand, S. Bultmann, S. Link, K. Thanisch, C. Mulholland, J. Dekker, et al., Spatial Organization of Transcribed Eukaryotic Genes, 2021.

[46] J.R. Dixon, S. Selvaraj, F. Yue, A. Kim, Y. Li, Y. Shen, M. Hu, J.S. Liu, B. Ren, Topological domains in mammalian genomes identified by analysis of chromatin interactions, Nature 485 (2012) 376–380. https://doi.org/10.1038/nature11082.

[47] E.P. Nora, B.R. Lajoie, E.G. Schulz, L. Giorgetti, I. Okamoto, N. Servant, T. Piolot, N.L. Van Berkum, J. Meisig, J. Sedat, J. Gribnau, E. Barillot, N. Blüthgen, J. Dekker, E. Heard, Spatial partitioning of the regulatory landscape of the X-inactivation centre, Nature 485 (2012) 381–385. https://doi.org/10.1038/nature11049.

[48] A.-L. Valton, S.V. Venev, B. Mair, E.S. Khokhar, A.H.Y. Tong, M. Usaj, K. Chan, A.A. Pai, J. Moffat, J. Dekker, A cohesin traffic pattern genetically linked to gene regulation, Nat. Struct. Mol. Biol. 29 (2022) 1239–1251. https://doi.org/10.1038/s41594-022-00890-9.

[49] T.H.S. Hsieh, C. Cattoglio, E. Slobodyanyuk, A.S. Hansen, O.J. Rando, R. Tjian, X. Darzacq, Resolving the 3D landscape of transcription-linked mammalian chromatin folding, Mol. Cell 78 (2020) 539–553.e8. https://doi.org/10.1016/j.molcel.2020.03.002.

[50] N. Krietenstein, S. Abraham, S.V. Venev, N. Abdennur, J. Gibcus, T.H.S. Hsieh, K.M. Parsi, L. Yang, R. Maehr, L.A. Mirny, J. Dekker, O.J. Rando, Ultrastructural details of mammalian chromosome architecture, Mol. Cell 78 (2020) 554–565.e7. https://doi.org/10.1016/j.molcel.2020.03.003.

[51] A.M. Oudelaar, J.O.J. Davies, L.L.P. Hanssen, J.M. Telenius, R. Schwessinger, Y. Liu, J.M. Brown, D.J. Downes, A.M. Chiariello, S. Bianco, M. Nicodemi, V.J. Buckle, J. Dekker, D.R. Higgs, J.R. Hughes, Single-allele chromatin interactions identify regulatory hubs in dynamic compartmentalized domains, Nat. Genet. 50 (2018) 1744–1751. https://doi.org/10.1038/s41588-018-0253-2.

[52] B.N. Zhang, Y. Liu, Q. Yang, P.Y. Leung, C. Wang, T.C.B. Wong, C.C. Tham, S.O. Chan, C.P. Pang, L.J. Chen, J. Dekker, H. Zhao, W.K. Chu, Rad21 is involved in corneal stroma development by regulating neural crest migration, Int. J. Mol. Sci. 21 (2020) 1–17. https://doi.org/10.3390/ijms21207807.

[53] E. Crane, Q. Bian, R.P. McCord, B.R. Lajoie, B.S. Wheeler, E.J. Ralston, S. Uzawa, J. Dekker, B.J. Meyer, Condensin-driven remodelling of X chromosome topology during dosage compensation, Nature 523 (2015) 240–244. https://doi.org/10.1038/nature14450.

CHAPTER 16

Epigenetics in the classroom

Khadijah Makky

Department of Biomedical Sciences, Marquette University, Milwaukee, WI, United States

Epigenetics in undergraduate biology curriculum: Why, when, where, and how

The field of epigenetics is evolving rapidly, with novel wet lab and in silico methods arising to expand our understanding of genetic regulation. These new techniques present a double-edged sword: our capacity to probe the DNA is greater than ever before, but critical evaluation of published literature can be challenging given bespoke methods and/or the use of complex computational models. As such, providing a foundational understanding of epigenetic modifications and methods for probing these modifications catapults undergraduate students to rigorously analyze scientific literature in the field and to demand reproducibility of themselves if conducting their own epigenetic research.

Epigenetic regulation of gene expression plays a role in cellular and molecular processes, behavioral neuroscience, developmental biology, nutrition/metabolic processes, and environmental science. Epigenetics bridges these areas of study with human health and disease. It has significant implications for our understanding of cancer, complex genetic disorders, and social adversity, and the debate over nature versus nurture. For example, many epigenetic marks, such as DNA methylation and histone modifications, currently can be considered potential markers of cancer development and progression [1].

In 2019, Kang and colleagues predicted that epigenetics research would represent approximately 20.7% of all published genetics papers by 2029 [2]. Because of the profound significance of epigenetic regulation and the rapid growth of the field, it is critical to integrate core concepts of epigenetic regulation into the biology curriculum delivered to high school students and, crucially, undergraduates. Owing to the emerging and interdisciplinary nature of this field, the topic presents complexities for learners and educators [3]. Therefore, in addition to including epigenetics in school science curricula, there is a need for professional development and training for teachers, professors, and other educators to present the subject clearly and understandably, using precise language and relevant, creative, and engaging learning activities [4].

Educators should carefully design their epigenetics curriculum, with consideration for:

(1) Situational factors of their course (students' educational level, background knowledge, length of the course, reasonable and accessible modes of assessment, and so on).
(2) Designing the unit for active and significant learning (including several activities for reinforcing a concept, for example, refer to Ref. [5]).
(3) Minimizing any students' misconceptions (touched upon further below).

Chapter 16 Epigenetics in the classroom

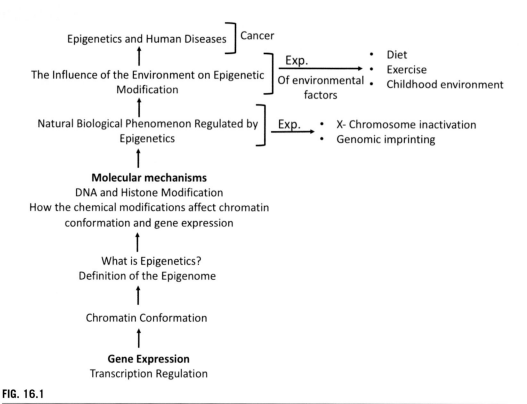

FIG. 16.1

Epigenetics unit outline. The chart shows how an instructor may organize a unit when teaching epigenetics.

How is the chapter organized?

The chapter provides an outline for instructors to follow when designing and delivering an epigenetics unit or set of epigenetics lectures. Instructors can determine the level of detail that is appropriate for the class. The chart in Fig. 16.1 shows how an epigenetics unit can be organized in chronological order. The order should be natural for students to follow logically and to understand the basic concepts of epigenetics. The topics are arranged to start with foundational knowledge and culminate with integrating the information talking about epigenetics and human diseases. The chapter provides content for each area and ideas for best practices to present it to students in a collaborative manner that allows them to engage and participate in their learning.

Where is it appropriate to introduce epigenetics in the biology curriculum?

In the last 100 years, the teaching of genetics—including the flow of information and Mendelian inheritance—has been a standard part of the biology curriculum both at the high school and undergraduate levels, while epigenetics is a relatively new topic. The Next Generation Science Standards (NGSS) provide guidance on updating science curricula. It should be noted that while epigenetics is a critical part of any complete biology or biomedical science curriculum, it is also an essential component for psychology majors, for whom epigenetics can be presented in an abbreviated way as a biological mechanism by which the environment can influence human behavior.

Elementary and high school curricula
In NGSS, Esser and colleagues suggest that, from kindergarten to grade 12 education, the different epigenetic unit components can fit within discussions of heredity (inheritance and variation of traits) lessons. The consensus is that epigenetics can be taught as part of the genetics unit or, at the high school level, as a miniunit for a more detailed look of epigenetic mechanisms. Educators seeking more detailed guidance on inclusion of epigenetics in their biology curriculum, particularly with rigorous scientific practice in mind, should refer to Esser et al., where the epigenetic education module is discussed in alignment with science practices, crosscutting concepts, and life sciences disciplinary core ideas (refer to "Framework for K–12 Science Education: Practice, Crosscutting Concepts, and Core Ideas…") [6] (See Fig. 16.2 and Box 16.1).

Undergraduate curricula
According to the 2011 report by Vision and Change on scientific curriculum, one of the five core concepts that all undergraduate biology students should master is "Information Flow, Exchange, and Storage," making genetics a foundational topic for 21st-century biology education [7]. Epigenetics can fit under this core concept, as part of the genetics unit, as a standalone unit, or as a standalone course.

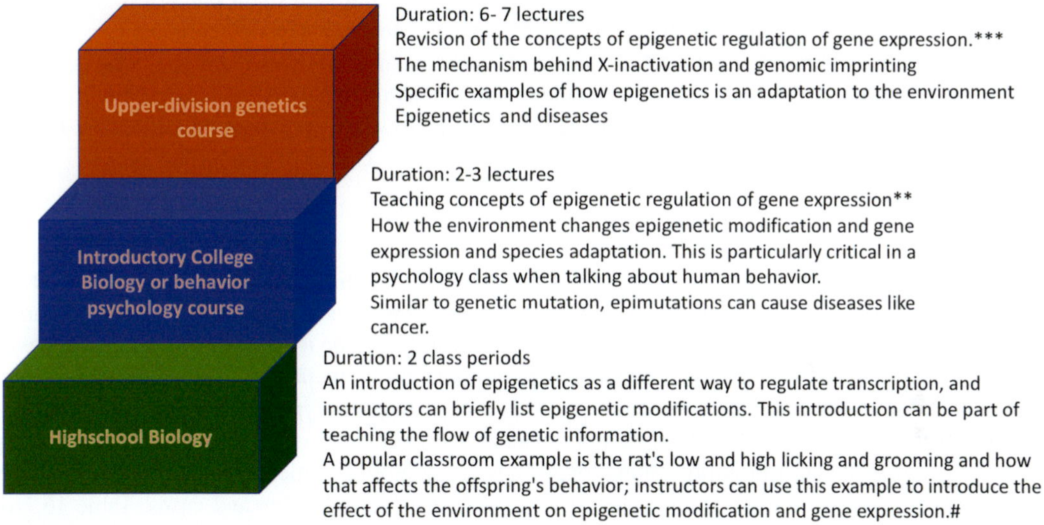

Upper-division genetics course
Duration: 6-7 lectures
Revision of the concepts of epigenetic regulation of gene expression.***
The mechanism behind X-inactivation and genomic imprinting
Specific examples of how epigenetics is an adaptation to the environment
Epigenetics and diseases

Introductory College Biology or behavior psychology course
Duration: 2-3 lectures
Teaching concepts of epigenetic regulation of gene expression**
How the environment changes epigenetic modification and gene expression and species adaptation. This is particularly critical in a psychology class when talking about human behavior.
Similar to genetic mutation, epimutations can cause diseases like cancer.

Highschool Biology
Duration: 2 class periods
An introduction of epigenetics as a different way to regulate transcription, and instructors can briefly list epigenetic modifications. This introduction can be part of teaching the flow of genetic information.
A popular classroom example is the rat's low and high licking and grooming and how that affects the offspring's behavior; instructors can use this example to introduce the effect of the environment on epigenetic modification and gene expression.#

FIG. 16.2
Sequential teaching of epigenetic concepts and applications. We suggest the following sequence of teaching epigenetics from high school to college/university. Instructors can use this structure as a guide, but they can add or drop topics based on their student population's incoming knowledge. **Instructors can integrate epigenetic regulation as part of the unit in transcription regulation. ***Instructors can expand on what we presented in this chapter and discuss different remodeling complexes that are part of epigenetic regulation. #The example is discussed at the end of the chapter; more can be found in the following link: https://learn.genetics.utah.edu/content/epigenetics/rats.

> **Box 16.1 Epigenetics teaching in a biology curriculum**
>
> Epigenetics is an essential component of a comprehensive biology curriculum. The abstract and complex nature of this topic, combined with the field's rapid pace of growth, outpaces most resources abilities to update their content regularly and completely. As such, a solid basic framework should be established by educators, scaffolded by updated information to provide a rigorous teaching unit.
>
> The interdisciplinary nature of epigenetics can prepare undergraduate students for future careers in many fields, such as medicine, environmental sciences, nutritional sciences, or psychology/psychiatry. As with many Mendelian and molecular genetic concepts, epigenetics should be introduced early in the high school curriculum, preparing students for more advanced curriculum and application learning in college/university. Teaching epigenetics should gradually align with students' developmental stages for significant learning. We gradually build on their understanding of epigenetic concepts and applications.

We suggest that in an introductory biology course for first-year students, epigenetics can be integrated into the unit, talking about the flow of information from genotype to phenotype. Epigenetics is a study of the regulation of gene expression at the transcriptional level. The instructor can give a simplified overview of the mechanisms by which epigenetic modifications of the chromatin can activate or silence transcription. Instructors can also point out the difference between genetic and epigenetic regulation of transcription and that the environment can influence that epigenetic regulation. In a standalone unit in an advanced human molecular genetics course, we cover the different chromatin conformations, the molecular mechanisms of epigenetics in more detail, epigenetic phenomena such as genomic imprinting and X-inactivation, and end the unit with applications such as presenting case studies and examples of how epigenetic mechanisms are implicated in diseases. In a standalone upper-division epigenetics course, we offer more complex chromatin remodeling epigenetic mechanisms. Students learn about the activating and silencing protein complexes that work in tandem with DNA methylation and histone modifications. The students are also introduced to different techniques used to study epigenetic modification. Finally, they read and analyze scientific articles on various topics, such as cancer research (Fig. 16.2).

How to design your epigenetics unit

A general framework for designing undergraduate epigenetics learning units is presented in Fig. 16.3, modified from Fink [8]. The learning outcomes associated with this design can be scaled to course level and tailored to the student population. The author used this guide in an upper-division, 400-level, human genetics course with a student population predominantly in the prehealth sciences. Epigenetics was included in the course where the concepts of genotype-to-phenotype and regulation of gene expression were revisited, and the teaching unit spanned ~6 course hours. In this design, the learning outcomes were structured to build foundational knowledge in epigenetics and facilitate learning by applying the information learned and integrating it with real-life examples and cases. Table 16.1 illustrates the topics covered for each learning outcome and can be used as a reference for educators designing their own curriculum.

One of the basic principles of significant learning is making the information relevant to students [9]. Epigenetics has become highly relevant to everyday life with many news articles/stories about its impacts on health and society. Active learning using real-life, relevant examples achieves two goals:

1. **Keeping students engaged**—epigenetics has an interdisciplinary nature and instructors should be mindful of their student population and use relevant examples to catch and sustain students' interest(s).

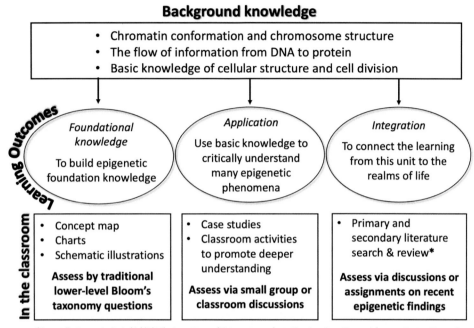

FIG. 16.3

Teaching a unit on epigenetics designed for significant learning. This chart details the key learning outcomes that should be addressed by epigenetic curricula, with suggested class-based learning tools.

Modified from L.D. Fink, Creating Significant Learning Experiences: An Integrated Approach to Designing College Courses, Jossey-Bass, 2013. 978-0787960551.

2. **Identifying and clarifying misconceptions**—active learning helps students distinguish between what they think they know and what they actually know. There has been rapid development in the field and to avoid any misconceptions, it is critical for educators to pay extra attention to the students' prior knowledge and perspective on their knowledge related to epigenetics. Introducing epigenetics in early secondary education in a simple, accurate manner provides students with a strong, accurate foundation with which to build new knowledge later in their education when epigenetics is presented in more depth.

Table 16.1 Topics covered in each learning outcome.

Foundational knowledge	Application	Integration
• What is epigenetics? • Epigenetic marks • How epigenetics regulates transcription • Methods to detect epigenetic modifications	• Epigenetic phenomenon • X-chromosome inactivation • Genomic imprinting	• How epimutations can cause disease and abnormalities, e.g., cancer • Environmental effect on epigenetic modification and how it may translate into abnormal trait

The chapter offers several human-focused and practical examples that can be used and scaled to the student populations in different courses. However, instructors can substitute examples to fit the syllabus for various courses, such as a plant biology course. **Many of the case studies in this chapter are adapted from published clinical reports; therefore, instructors can use a similar structure and use the primary literature specific to their field to select appropriate examples covering different epigenetic concepts.**

Approach to teaching epigenetics using high-impact practices

Due to epigenetics' interdisciplinary nature, students must have prior knowledge in a wide range of biological concepts for an in-depth understanding of the subject. It is valuable to provide students with a mini-review of the background information needed before starting epigenetics lectures. Instructors can start units with a form of assessment that confirms students understanding of the foundation they need. We use a class response system to assess students' background knowledge and provide some questions for self-assessment. Educators should teach epigenetics for conceptual understanding and prepare students to transfer what they learn to practical cases in their respective fields after leaving the course. This active learning modality may appear impenetrable in teaching epigenetics, but there are many published and accessible resources for instructors to achieve significant epigenetics learning. In addition to some of the examples provided in this chapter, resources are listed in Table 16.2. The educational resources in Table 16.2 for epigenetics can be used across a broad spectrum of curricular levels. This chapter presents different examples of high-impact teaching practices, such as concept maps and case studies.

Learning outcome 1: Building the epigenetics foundation knowledge

This learning outcome introduces the students to the foundations and basic concepts of epigenetics. During this first part of the unit, students can define epigenetics and connect epigenetics and regulation of transcription and gene expression. They recognize the biochemical nature of epigenetic modifications

Table 16.2 Epigenetics educational resources.

References	URL	Topics covered
Genetic Science Learning Centre	https://learn.genetics.utah.edu/content/epigenetics	Epigenetics overview Epigenetics and inheritance Genomic imprinting Epigenetics and the environment
Educational Resources for Epigenetics	https://caister.com/hsp/abstracts/epigenetics/26.html	Websites and scholarly review articles aimed at a broad audience
Teaching Epigenetic Regulation of Gene Expression is Critical in 21st-Century Science Education	https://online.ucpress.edu/abt/article-abstract/82/6/372/111547/Teaching-Epigenetic-Regulation-of-Gene-Expression?redirectedFrom=fulltext	Epigenetic mechanisms Practical classroom examples X-chromosome inactivation Activity: genomic imprinting; supplemental material

and can list them. They can identify how DNA and histone modifications can alter gene expression. It is strongly encouraged that instructors use simple models in the classroom [10,11]. To assess students' learning, instructors can use a class response system with multiple choice questions or short answers. Examples of questions are provided at the end of the section.

Regulation of gene expression and epigenetics

> With the tools and the knowledge, I could turn a developing snail's egg into an elephant. It is not so much a matter of chemicals because snails and elephants do not differ that much; **it is a matter of timing the action of genes**.
>
> Barbara McClintock

In any species, gene expression is the phenotypic manifestation of its genes and genome. It is the flow of information from the DNA code to the protein that exerts the cellular function. The group of genes expressed in any specific cell type dictates its functions and morphology. In any organism, the genome is identical in all cell types; therefore, regulation of gene expression is an essential determinant of tissue homogeneity in terms of cellular morphology and function.

What are different ways to regulate gene expression?

The regulation of gene expression can occur at the transcriptional, translational, or protein levels (Fig. 16.4).

Transcriptional regulation of gene expression

Transcription can be regulated via genetic or epigenetic regulation. Genetic regulation of transcription is mediated by the regulatory DNA sequences (cis-element) such as promoters and enhancers and the trans-elements such as noncoding RNA and transcription factors. The transcription rate depends on the interaction between the transcription factors (trans-elements) and the promotors and enhancers (cis-acting elements). Therefore, the availability of the trans-acting elements is the rate-limiting factor for transcription. In other words, the cell-specific gene expression program is determined by the expression and the availability of the trans-acting elements [12].

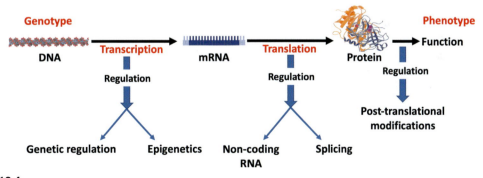

FIG. 16.4

Regulation of gene expression (central dogma). Simplified schematic representation of the flow of information from genotype to phenotype and areas where the process is regulated.

Epigenetic regulation of transcription

Epigenetics is the study of the regulation of gene expression that is dependent not only on DNA sequence but also on chromatin structure. Epigenetic regulation can affect many biological processes, such as (1) embryonic development and differentiation of various cell types in an organism, and (2) homeostasis within the human body. Epigenetic regulation can become disrupted, which leads to cancer and other diseases.

How does epigenetics work? It affects the chromatin structure by allowing either a closed or relaxed conformation.

The DNA is packaged in an ordered structure. Within the chromosome, DNA is packaged into chromatin. Chromatin is a nucleoprotein complex that packages the linear genomic DNA in a string of nucleosomes, the smallest unit of the chromatin. Each nucleosome is a 46 base pair-DNA wrapped around the histone octamers. The histones are positively charged proteins that bind to the negatively charged DNA. In the nucleosome, the N-terminal tail of each histone molecule is projecting outwards. The string of nucleosomes coil to form the chromatin fibers. Further coiling and folding of the chromatin fibers will make the chromatid. The chromatin folding and DNA packaging level depend on the DNA, and histone modifications (Fig. 16.5). The chromatin can either be tightly packed (heterochromatin) or loosely packed (euchromatin) [10].

How is the chromatin conformation connected to gene expression and epigenetics? For any particular gene, the rate of transcription will depend on its chromatin environment. Changes in chromatin conformation result in altered gene expression. A functional gene that is embedded in highly condensed chromatin may not be accessible to transcription factors. The gene is said to be silenced. A gene in an open, more relaxed conformation is more accessible to transcription factors and is actively transcribed.

Further, heterochromatin areas can be divided into constitutive and facultative heterochromatin. **Constitutive heterochromatin** is condensed regions of chromosomes that are the same in all cell types. They usually perform a structural role. Examples of constitutive heterochromatin areas are centromeres and telomeres. **Facultative heterochromatin** is an area of the genome that takes a condensed conformation in some cells but a relaxed conformation in other cells. The change in area structure regulates gene expression.

Epigenetic marks and chromatin conformation
How does the chromatin conformation change?

Chromatin conformation often changes by specific types of chemical modification of the DNA strands and histones. The pattern of DNA and histone modification that is heritable from one cell generation to the next are called **epigenetic marks**. The genetic code is like the roads that map the direction of traffic, and epigenetic marks are the traffic lights that direct when to go and when to stop. The genome encodes the genetic information necessary for human functions and traits. Epigenetic marks regulate the expression of genetic information. The epigenetic regulation of gene expression depends not only on the genetic code but also on the chromatin's conformation around the genetic code (Fig. 16.5).

Epigenetic marks
DNA methylation and histone modification

DNA methylation is a well-studied epigenetic modification. DNA methylation occurs at the CpG dinucleotides and marks the cysteine nucleotide with a methyl group. The methylation of DNA is catalyzed

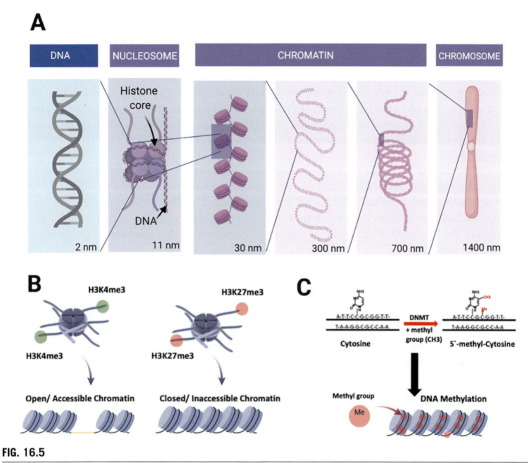

FIG. 16.5

DNA packaging and epigenetic modifications.

From I. Venkatesh, Teaching epigenetic regulation of gene expression is critical in 21st-century science education: key concepts & teaching strategies, Am. Biol. Teach. 82 (2020) 1–9.

by the enzyme De Novo Methyl Transferase 1 (DNMT 1) [13]. Adding the methyl group leads to chromatin remodeling into tight confirmation and silencing of transcription. In contrast, DNA demethylation results in a relaxed open conformation and activation of transcription (Fig. 16.5).

The effect of histone modification on gene expression is more complex than DNA methylation. Histones are a major component of the chromatin, and their post-translational modification impacts the chromatin conformation. In general, histone modifications are chemical reactions that are catalyzed by specific enzymes that act at the N-terminal tails of the proteins. Histone modification can activate or repress transcription depending on the chemical group modifier and the site of modification [14]. **Histone acetylation** by histone acetyltransferase (HAT) enzyme reduces the affinity of histone proteins for DNA and relaxes the chromatin packaging [15] whereas histone deacetylation by histone deacetylase (HDAC) enzyme condenses the chromatin. **Histone methylation** can result in both conformations depending on the histone protein and the amino acid (AA) residue that is modified, i.e., which histone is modified and which AA in the tail is methylated. For example, methylation of AA #9 on H3 is associated with heterochromatin and gene silencing, whereas methylation of AA #4 or AA #27 is associated with transcriptional activation [16]. **Histone phosphorylation** is generally associated with transcription activation. But some research shows cross-talk between histone phosphorylation and other histone modifications results in the complex regulation of chromatin conformation and gene expression [17]. Due to the rapidly evolving nature of our knowledge and understanding of histone modifications, for most courses a brief description of histone modification is sufficient. Instructors can use the Top Hat organizer described in Fig. 16.6 to self-assess their learning.

Example assessment questions
1. Some DNA sequences in our cells have a high frequency of methylated cytosines (hypermethylation); others have a low frequency of methylated cytosines (hypomethylation). Which of the two-methylation status is represented in satellite DNA, centromeric sequence?
   - Hypomethylated
   - Hypermethylated
2. Which of the following represents the chromatin conformation when genes are actively transcribed?
   - Condensed conformation
   - Relaxed/open conformation
3. Which of the following statements is true about histone tail modifications?
   - They are posttranslational modifications
   - They play a role in changing chromatin structure
   - They can activate or repress transcription epigenetically
   - All of the above

Differences	
Unique to bi-allelic genes	Unique to mono-allelic genes
Similarities	
Autosomal genes Gene expression is regulated at transcription level For both an individual will inherit paternal and maternal alleles	

FIG. 16.6

Top Hat organizer. The Top Hat organizer can help students to record similarities and differences in parallel.

Learning outcome 2: Application using basic knowledge to critically understand many epigenetic phenomena

This learning outcome is to help students bridge between epigenetics as chemical modifications and complex biological processes. Students can explain how epigenetics is the underlying mechanism for these natural phenomena and that dysregulation can have health outcomes.

Genomic imprinting—Using compare/contrast

Compare/contrast is a high-impact skill that facilitates learning [5]. When teaching genomic imprinting, this high-impact practice helps students constructively build their understanding of the phenomenon and the difference between the monoallelic and biallelic genes. Genomic imprinting is a pattern of epigenetic inheritance of autosomal genes that have unusual expression patterns. Imprinted genes are autosomal monoallelic genes that are expressed only from one of the two alleles inherited. Depending on the parent-of-origin, the allele can be expressed or silenced [18].

The concept of imprinting is challenging for students and active learning techniques will facilitate learning. One strategy is to describe first and compare it later. Start with a clear description: humans (usually) inherit two copies of each gene, one allele from each parent, each of which has the potential to be fully expressed. In contrast, a group of genes is only expressed from one allele. Depending on the gene, the expressed allele could be the paternal or the maternal one. Instructors can use the Top Hat Organizer (Fig. 16.6) to provide students with a visual organizer to help them think critically and fully understand the difference between imprinted and nonimprinted genes.

X-inactivation—Using concept maps

A well-studied example of facultative heterochromatin is the X-chromosome in females [9]. In females, each cell will receive a maternal and paternal X-chromosome. During the early stages of embryonic development, each cell will inactivate one of the X-chromosomes to equalize the X-linked gene expression between males and females. The cell will randomly inactivate the maternal or the paternal chromosome by changing the chromatin into a condensed confirmation. Therefore, the same X-chromosome will be active (relaxed confirmation) in one cell, and inactive (condensed confirmation) in another.

A concept map is a high-impact teaching practice that helps students connect multiple concepts. Here we use a concept map to connect three concepts:

(1) The differences between male and female X-chromosome numbers.
(2) A biological concept of dosage compensation to ensure that the cells of females and males have the same effective dose of genes with loci on the X-chromosome.
(3) The connection between X-inactivation and epigenetics provides an example of an epigenetic phenomenon.

For additional information on concept mapping and a tutorial, refer to the University of Guelph's McLaughlin Library page on the topic: https://guides.lib.uoguelph.ca/c.php?g=697430&p=5011748 (Fig. 16.7 and Box 16.2).

374 Chapter 16 Epigenetics in the classroom

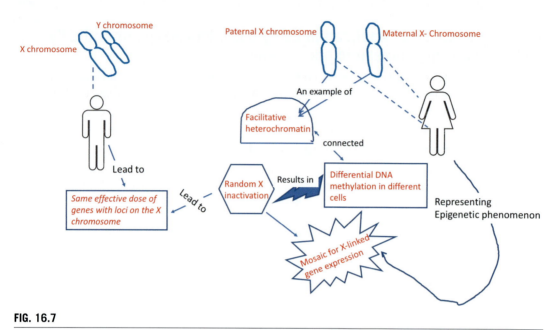

FIG. 16.7

X-inactivation concept map. High impact teaching activity to link X-inactivation to epigenetics.

Box 16.2 The following concept map illustrates X-inactivation in human females as an epigenetic phenomenon

See Table 16.3.

Table 16.3 X-inactivation concept map.

Linking words	Phrases for the nodes
An example of	Differential DNA methylation in different cells
Connected	Epigenetic phenomenon
Causes	Facultative heterochromatin
Leads to	Female/male
Representing an epigenetic phenomenon	Maternal X-chromosome
Result(s) in	Mosaic for X-linked gene expression
	Paternal X-chromosome
	Random X-inactivation
	Same effective dose of genes with loci on the Xp-chromosome
	Y chromosome

The list of links and phrases can be used to complete the following concept map.

X-chromosome inactivation phenomenon: Female mammals, including humans, inherit two X-chromosomes; one chromosome in each cell becomes inactive during embryonic development. *X inactivation is a dosage compensation process to ensure that the cells of females and males have the same effective dose of genes with loci on the X-chromosome* [12].

As many natural biological phenomena, the molecular mechanism of X-inactivation is complicated, and the details are still being unraveled by research. Although it might be challenging for instructors to effectively teach the molecular mechanism in the classroom, research has shown that teaching the molecular mechanisms of biological phenomena in the classroom can benefit students in many folds. First, it allows students to make the connection between the molecular knowledge to phenomena at the cellular levels. Many students tend to have that basic knowledge of DNA and proteins but cannot connect that to the visible phenomena. Second, it provides a framework to reason about complex systems or mechanisms to make the connections. Finally, it helps students understand how cellular phenomena can emerge from several multistep interaction cascade molecules such as RNA and proteins [1,19].

The below framework can be used in the classroom to introduce students to X-chromosome inactivation's molecular mechanism. The framework connects the foundational knowledge of epigenetics to the X-inactivation phenomenon. It shows how gene expression, protein function, and the interaction between different macromolecules can be associated with the biological phenomenon of X-inactivation. The framework is composed of several "how" questions. How does the cell know that it contains more than one X-chromosome to trigger the inactivation process? How does the cell select the X-chromosome to be inactivated? How does the cell initiate the inactivation process? Finally, how does the inactive X-chromosome remain inactive in cell progeny? Fig. 16.8 illustrates a schematic diagram to the questions that make the X-inactivation framework.

The mechanism highlights the role of Rnf12, a transcription factor, as the molecular sensor that alerts the cell to inactivate one of the X-chromosomes. The expression of Rnf12 then triggers the transcription of the RNA *Xist* from the future inactive X-chromosome, which then coats the chromosome in cis. The X-chromosome's coating recruits all the epigenetic marks and remodeling proteins to condense the X-chromosome and inactivate it. Finally, the model indicates that the process of maintaining the same X to be inactive in future cell progeny depends on the epigenetic feature of mitotic heritability.

Summarizing information—Using case studies

Students often have difficulties synthesizing information when presented with a complex scenario related to the material they memorized in fragments, making it challenging for them to make any conceptual understanding of a real-life scenario and answer questions. One teaching practice is to help students learn how to take notes and summarize the information in the case conceptually [5]. The following mini case study is used in an upper-division genetics class to help students conceptually understand the genomic imprinting phenomenon.

Practice case

Cyclin-dependent kinase inhibitor 1C (CDKN1C), located on chromosome 11, is a key negative regulator of cell growth encoded by a paternally imprinted/maternally expressed gene in humans. Loss-of-function variants in CDKN1C are associated with an overgrowth condition (Beckwith-Wiedemann Syndrome), whereas "gain-of-function" variants in CDKN1C that increase protein stability cause growth restriction as part of IMAG syndrome (intrauterine growth restriction, metaphyseal dysplasia, adrenal hypoplasia, and genital anomalies) [20].

376 Chapter 16 Epigenetics in the classroom

X inactivation occurs only if more than one x chromosome is present in a diploid cell.
How does the cell sense the presence of multiple X chromosomes?
X chromosome inactivation is Y-chromosome independent
The molecular sensor: the expression of higher **levels of Rnf12**

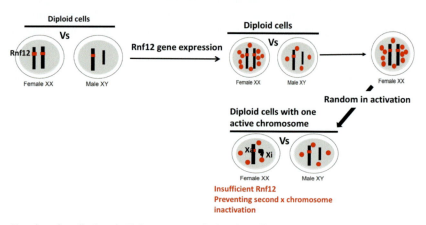

How does the cell select the X chromosome to be inactivated?
The molecular player is Xist a **17 kb Long-noncoding-RNA**
Encoded in an area of the X chromosome called XIC X-inactivation center (XIC)
- It is expressed from only one of the two X chromosomes
- It determines the chromosome that will become the inactive X

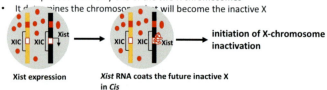

How does the cell initiate X-chromosome inactivation?

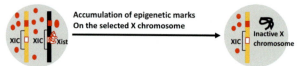

How does the inactive X chromosome remain inactive in cell progeny?
Maintenance of the inactive X, **stably through the life of the cell and its progeny**

The same X chromosome will remain inactive in next generations of cells. The process is independent of *Xist* and the XIC and depends on the depends on a combination of epigenetic marks and the mitotic heritability feature of epigenetic marks.

FIG. 16.8

The process of X-chromosome inactivation. A framework connecting the foundational knowledge of epigenetics to the process of X-chromosome inactivation.

Table 16.4 Making notes and summarizing information.

Key concepts	Supporting details
It is an imprinted or monoallelic gene	It is paternally imprinted, maternally expressed, i.e., the active allele is on the maternal chromosome
Dysregulation of gene expression leads to the disease	Over or underexpression can lead to developmental abnormalities and diseases
What are the key points we take from this clinical example about CDKN1C?	

Students were asked to read the paragraph and break down the CDKN1C information. Students should summarize their understanding in the table, but unfilled (see Table 16.4). The key concepts are information from the passage describing the characteristics of the CDKN1C gene, and the supporting details are the information supporting the concepts. The instructor can then share the table with the students and explain the contents. Finally, students are asked to answer the following question: What is one possible problem that can lead to a gain-of-function or a loss-of-function of CDKN1C in an offspring? This question prompts students to think about the outcome of inheriting an abnormal combination of CDK1C maternal and paternal alleles [21]. Instructors can use visuals such as a pedigree with chromosome 11 representation to help students answer the question and connect to disease development indicated in the paragraph.

Learning outcome 3: Integration-connecting the knowledge from this unit to the realms of life

This learning outcome is for students to connect the foundations of epigenetics learned throughout the unit and practical examples. It also illustrates how epimutation can lead to the development of a disease such as cancer. Using a case study as a team-problem-solving activity will assess students' ability to make the connection and show their ability to use evidence-based arguments to determine the cause of the problem [21].

Epigenetics and cancer

Cancer is a genetic disease and it has been studied in connection with DNA genetic mutations for decades. Cancer epigenetics has been emerging as a field since 1983, connecting DNA methylation and chromatin conformation to tumor development. Epigenetic studies have shown that the global hypomethylation or alteration in the promoters' methylation status can lead to genomic instability and alterations in gene expression, respectively. Furthermore, in the early 2000s, histone modifications have also been connected to silencing tumor suppressor loci and cancer development.

Culminating case study

Why my identical twin sister is always sick?

The following case study is based on a report published in 2012 in the journal *Epigenetics* [22]. The report describes monozygotic twins discordant for childhood cancer. The case shows how a twin pair represents an example of epigenetic somatic mosaicism due to an epimutation that most likely occurred during embryonic development. In this case, the *BRCA1* gene's bisulfite pyrosequencing shows the affected twin displayed increased methylation of the promoter compared with her sister. Subsequent bisulfite plasmid sequencing demonstrated that 13% of *BRCA1* alleles were fully methylated in the affected twin, whereas her sister displayed only single CpG errors without functional implications. We used this case study as an active-learning tool and culminating activity in the unit of epigenetics. The learning objectives of the case study:

- Integrating the foundational knowledge of epigenetics with a real-life example.
- Showing how acquiring an epimutation can affect gene expression, causing diseases such as cancer.
- Illustrating how monozygotic twins can have different epigenetic modifications during embryonic development. In some cases, one of the twins can acquire an epimutation that makes them discordant for developing diseases.

Although this case is ideal with a small class size or with small group discussions, it can be easily used in a large-class setting. In the large classes, we use a class response system for the questions to assess students learning from the case. The students were allowed to discuss questions in small groups. Some parts of the case can be done as homework and reviewed with the students in the following lecture. It could take one 50-min class period or less.

Case presentation in the classroom

The case **"Why is my identical twin sister always sick?"** can be provided to the students in advance or used as an interrupted case during a single lecture. In an interrupted case method, the different sections of the case are presented in a progressive disclosure to students working in groups. Clyde Freeman Herreid described the interrupted case method as one that promotes students' engagement in problem-solving. Therefore, we presented the case as an interrupted case study in the classroom as a problem-based learning tool to enhance students' critical thinking skills.

Case Study Part I: Introduction to Sarah

Sarah is a 40-year old female who is a monozygotic twin. Sarah and her twin sister, Norah, were born by spontaneous delivery 7 weeks before term, but they needed no intensive care or treatment. During the first 5 years of life, the development of both twins was normal. At the age of **five**, *Sarah was diagnosed with B-cell lymphoblastic leukemia. She had chemotherapy, but she had her first relapse at the age of* **seven**. *Sarah received bone marrow transplantation from her healthy twin sister, Norah. In the next 30 years, Sarah was diagnosed with thyroid carcinoma and was diagnosed with type 2 diabetes, and at the age of 34, she gave birth to a healthy daughter. Norah, on the other hand, has been healthy and has never been diagnosed with cancer. There is no family history of hereditary disease, and apart from Sarah, there were no other cancer cases in four generations. Monozygosity was confirmed by genotyping short tandem repeats and chromosome banding analysis. A karyotype test (chromosomal analysis) found both sisters to be normal. Sarah is affected by cancer and Nora is cancer-free. This fact has always been on Nora's mind and she has investigated cancer intensively.*

Part I activities

Activity # 1: Uncontrollable cell division results in cancer development. The cell can become cancerous if there is a dysregulation in the expression of genes such as tumor suppressors, oncogenes, or genes involved in apoptosis. Use the cell division concept map in Fig. 16.9 to help you think of possible causes of cancer by thinking about the expression levels of different genes that regulate cell division. In your own words, state a hypothesis as to what may have caused Sarah's cancer.

Activity #2: Cancer is a genetic disease and it can be inherited or sporadic. Use the diagram in Fig. 16.10 to show how you can test your hypothesis from Activity 1. Then, if the cancer is due to an inherited mutation, why do we see discordance between the two twin sisters?

Connecting the knowledge from this unit to the realms of life 379

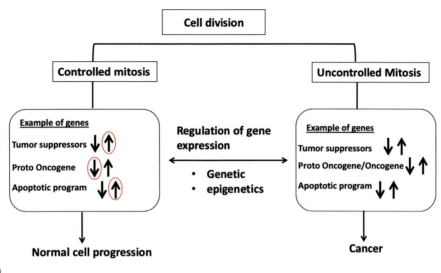

FIG. 16.9
Part I of the Case Study—Activity #1: Cancer development and gene expression. Concept Map: the *arrows* indicate if the expression of the gene is up or down.

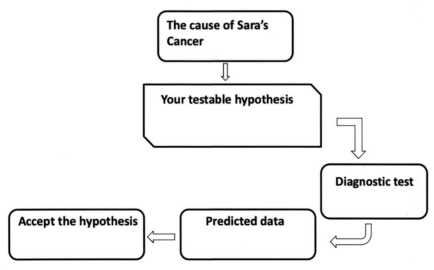

FIG. 16.10
Part I of the Case Study—Activity #2: Cancer development and gene expression. A diagram for students to state the hypothesis behind the cause of Sarah's cancer and how they can test their hypothesis.

380   Chapter 16 Epigenetics in the classroom

Part I classroom presentation and management: The instructor should give the students time to read Part I of the case (5 min). This part introduces the students to the story; the instructor can start the discussion by briefly identifying the main characters. The conversation then can focus on the following:

- Monozygosity: What it means and how the story has indicated it.
- Discordance in cancer development: The fact that one of the twins, Sarah, has had many childhood cancers whereas the other, Nora, remained healthy and cancer-free.
- Cancer review: Briefly review what cancer is and instructors can use the activity associated with Part I to help understand how cancer cells can form. It will also help students to think of a hypothesis explaining the discordance between the twin sisters.

The instructor then should refocus the students' attention back on Sarah and ask them to present a hypothesis for the cause of her cancer and how they can test the theory. The goal is for students to think that genetic mutations or epimutation can lead to cancer. The instructor should ask students to share their work and record the results on the board.

Case Study Part II: Why is my identical twin sister always sick?

Norah was worried about her sister's health. She approached Dr. Smith, an oncologist, to consult with him and look for an answer to her question: "Why am I healthy and cancer-free, but my sister is suffering from all these terrible cancers?" Dr. Smith was very approachable and helpful to Norah and started investigating the case. He noticed that the two sisters differ in height and some other minor traits, which is not unusual for monozygotic twins. But the monozygotic twins, in most cases, are concordant for genetic diseases since they share an identical genome. He thought of a few molecular tests to determine what causes Sarah's health problems, and he asked the twins to give different samples for analyses. He ordered a biochemical test to look at the levels of several tumor suppressor proteins BRCA 1 and p53 in various tissues.

Part II activity
Use the graph in Fig. 16.11 to predict the results from Dr. smith's tests. You can represent the data as a bar graph.

Part II classroom presentation: The instructor should give the students time to read Part II, about 3–5 min. Ask students to summarize their understanding of the paragraph and focus on what the physician thought about monozygotic twins and how they usually can have slightly different traits, such as height but are mostly concordant for diseases. The instructor should review the role of tumor suppressor proteins with the students and how their lower levels can lead to

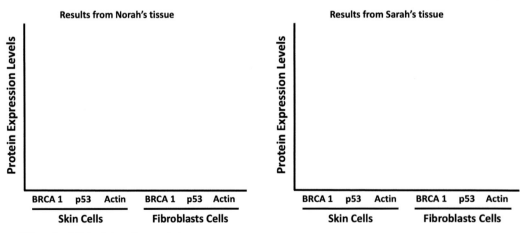

FIG. 16.11

Case Study Part II biochemical results. A diagram for students to predict the results of the biochemical test ordered by the physician.

cancer. Then, ask them to predict the results of the biochemical reactions. In a small class, the instructor can ask students to come up to the board and share their expected results and explain their reasoning. Alternatively, in a large class, the instructor can do the drawing on the board based on students' participation. Students often expect that Norah will have higher BRCA1 and p53 levels than Sarah in both cell types. Students also usually ask why the doctor is testing the levels of actin in the samples. The activity is an excellent time to explain the role controls play when analyzing protein levels and comparing samples. The next step is sharing the results with the students.

The results of the tests ordered by Dr. Smith examine if predictions of the biochemical results are correct. The result of this test is in Fig. 16.12. The graphs represent the results from two different tissues, and protein expression is presented as a ratio of levels in Sarah versus Nora.

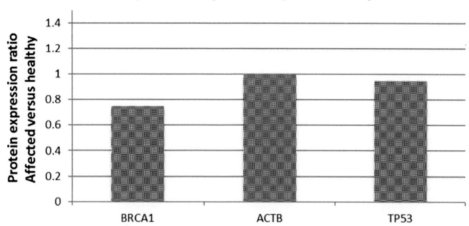

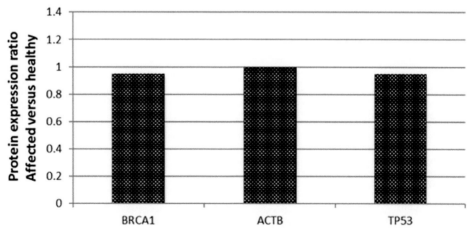

FIG. 16.12

Case Study Part III. Actual results of the biochemical assay. Constitutive protein expression of BRCA1, ACTB, and TP53 in fibroblasts and skin cells of the affected twin and her healthy sister. Protein expression is presented as a ratio of levels in Sarah versus Nora.

Chapter 16 Epigenetics in the classroom

Part II classroom presentation: The instructor should start by explaining how the protein level is presented on the Y-axis before students break into their groups and analyze them. Students usually are surprised to see the difference in the BRCA1 levels between the two different types of cells, and they wonder why. The instructor then asks students what would cause a differential regulation of BRCA1 gene expression. Can genetic mutation in the BRCA1 gene cause this differential expression of the gene?

Case Study Part III: Epigenetics and cancer

In addition to the biochemical test, Dr. Smith ordered a genetic test to look for mutations in these common tumor gene promoters. Genome sequencing from both sisters showed no genetic mutation in the BRCA1 gene or any other tumor suppressor genes. The result indicated that during embryonic development, there was no genetic variation acquired. Dr. Smith then suspected that the two sisters might differ in their epigenome. He ordered a test to look at methylation patterns for the tumor suppressor genes' promoter region using bisulfite sequencing. Bisulfite sequencing uses bisulfite treatment to determine the pattern of DNA methylation. Bisulfite treatment converts the nonmethylated cytosines to uracils. Comparing the sequence of different DNA samples will show the differences in methylation. The results of the bisulfite sequencing are shown in Fig. 16.13.

Part III activity
Analyze Fig. 16.13 and answer the following:

- Is there a clear difference in the BRCA1 promoter methylation pattern between the healthy and affected twins? Explain the difference.
- Why does Sarah have two BRCA1 methylation patterns in her cells?
- Explain why Sarah has developed childhood cancer while Nora remained utterly healthy. Is it possible for the twins' epigenomes to be different?

Part III classroom presentation. Instructors should focus on the following points:

- The possibility of acquiring genetic mutations during development. Discuss techniques that can be used to examine the BRCA 1 DNA sequence to confirm or refute the possibility of genetic mutation.
- Since the test results indicated no genetic mutations that can affect the expression of BRCA 1 in fibroblast, then the other possibility is an epimutation.
- *Epimutation is any change that can affect gene expression without changing the DNA sequence*. The instructor should start the discussion of the possible epimutation by reviewing what epimutation is and how it may have an effect on gene expression. Explain how a change in DNA methylation can alter the rate of transcription and, so too, gene expression.

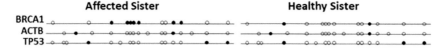

Fibroblasts

Skin Cells

FIG. 16.13

Case Study Part III: Methylation Pattern of the BRCA 1 promoter. Methylation patterns for BRCA1, ACTB, and TP53 promoter in fibroblasts and skin cells of the affected twin and healthy sister. Each *line* represents the promoter area of the corresponding gene analyzed for CpG methylation. *Filled circles* indicate methylated CpG and *open circles* unmethylated CpG sites.

When students analyze the figure and the results from the test ordered by Dr. Smith they recognize the difference in methylation easily and acknowledge the fact that the BRCA1 promoter is differentially methylated between the sisters but also between Sarah's two different cells. The instructor should discuss the difference in the expression of the BRCA1 gene between Sarah's fibroblast cells and the skin cells connecting the epigenetic alteration to the protein levels and the development of cancer.
- The next logical question to ask students is, why does Sarah have two BRCA1 methylation patterns in her cells? Using a figure such as Fig. 16.14 can help students realize that if the epimutation in a group of cells during development can lead to the differential methylation pattern of BRCA 1 promotor.

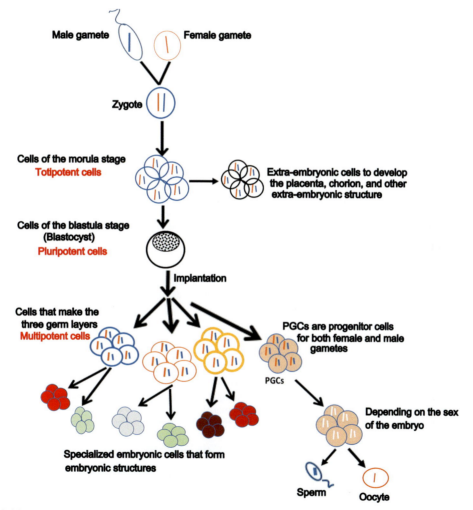

FIG. 16.14

Schematic representation of cellular development during embryonic development. This figure represents the cellular development in the embryo. If an epimutation occurred in one group of cells, it will affect only the tissues derived from these cells.

From I. Venkatesh, Teaching epigenetic regulation of gene expression is critical in 21st-century science education: key concepts & teaching strategies, Am. Biol. Teach. 82 (2020) 1–9.

Case Study Part IV: Discussion with Dr. Smith

Dr. Smith called Norah and told her that the results from all the tests came back and asked if she and Sarah could make an appointment to go to his office. Norah was happy that her questions might be answered. But she was also nervous because she was worried about what the answers might be. They had their appointment at the end of the day and Dr. Smith allowed enough time to meet with the sisters anticipating plenty of questions. "As you know, you are identical twins, which means you have a 100% identical genome, meaning that you came from a single fertilized egg, and the genetic DNA sequences in your cells are the same. However, it looks like you have differences in what we call the epigenome," Dr. Smith said.

"Epigenome?" Norah said.

Dr. Smith explained: "The genome represents the coding of the DNA carried by the chromosomes, the little sticks in each cell. The epigenome is the configuration of how these chromosomes are folded to fit in the small cellular space. You inherit your genome from your parents, but the epigenome from your parents is erased during the very early stages of development, and a new configuration is re-established. In Sarah's case, there was a mistake in her epigenome that didn't happen in yours, Norah. This mistake has put Sarah at risk of developing these multiple cancers early in life."

"Oh, so this might be good news? Sarah's kids will not inherit this problem from here, right?" Norah asked.

"Yes, it is possible that they won't have the same risk," Dr. Smith said.

Sarah smiled. She was relieved to hear that.

"Thank you so much for your time, Dr. Smith, but can I ask one more question?" Norah asked. "Can you predict anything about Sarah's future health? Is she going to have another type of cancer? Can she do anything to prevent future tumors from developing?"

Dr. Smith paused. This was a more difficult question to answer.

"We found that the mistake in the epigenome was not equally distributed in all of Sarah's tissues or cells, which is good news. At the same time, however, we cannot predict if she will be diagnosed with another cancer or if she is cancer-free. All Sarah can do is live a healthy life and avoid any environmental risks factors, like smoking or lack of exercise. That may accelerate the development of cancer." Norah and Sarah thanked Dr. Smith for his time and patience and agreed that Sarah would continue to see Dr. Smith and come for frequent check-ups.

Part IV activity

Classroom discussion questions:

1. Do you agree with Dr. Smith regarding the status of Sarah's kids? Why/why not?
2. How would you rate Dr. Smith's performance as a healthcare provider? Reflect on the things he did right and what you would change if you were providing care for Sarah.

Part IV classroom presentation: The last part of the lesson summarizes the case, and it usually generates excellent class discussions. Students can discuss the suggested questions within their small groups, followed by a general class discussion. Students are asked to reflect on their opinion of how Dr. Smith handled the case and his explanation to the twin sisters. They also reflect on his assessment of Sarah's children and if they agree with him. Here, they can use their understanding of epigenetic modifications and their fluidity during development to form their best assessment of Sarah's children.

The instructor should signify to students that there is no right or wrong answer. But there is a careful and careless answer. They should also give them time in the small group discussions to make their best argument.

Epigenetics and human behavior: Nature versus nurture

> The development of any individual is affected both by the hereditary determinants which come into the fertilized egg from two parents and also by the nature of the environment in which the development takes place.
>
> C.H. Waddingt (1957) [23,24]

In the last few decades, epigenetics has played a fundamental role in the debate of nature versus nurture and has participated in the theory that both nature and nurture control human development, physiology, behavior, and cognition. Gradually, it has become clear that epigenetics is a mechanism that can mediate the interaction between nature (genes) and nurture (the environment). It has stood the test of time that the environment can alter the DNA methylation and affect gene expression.

When talking about epigenetics, the nature versus nurture topic often is prominent at all academic levels. Nature versus nurture is particularly relevant in pedagogy related to developmental or behavioral psychology. Across academic disciplines, nature versus nurture is usually presented at a superficial level without any epigenetics review, which can be attributed to the share volume of content covered in a limited time. This can be problematic in multiple ways. First, it will hinder in-depth learning of how individuals' lifestyles can affect their long-term health and future generations' health. It is beneficial for students to preview the basics of epigenetics before discussing nature versus nurture. Second, discussing nature versus nurture without a proper understanding of epigenetics can introduce misconceptions that may be hard to correct, and instructors at all pedagogical levels should be cautious. This is particularly critical at the high school level. Instructors need to assess their understanding of epigenetics and its interaction with the environment by posing open-ended questions to students who may overestimate their knowledge if asked. It is possible to overestimate their understating by using true/false or multiple-choice questions.

Nature versus nurture and pedagogical resources

The following is a framework that we used in an epigenetic unit when teaching nature versus nurture. The section concludes the unit, and it usually takes a 75-min lecture or a 50-min lecture and a half. The class starts with the definition of nature versus nurture and how epigenetic modifications have explained the biological mechanism by which the environment can influence gene expression. Two case studies are presented, one assessing prenatal epigenetic effects, and the other assessing postnatal epigenetic effects. It is critical to disclose to students that although research-based evidence has confirmed that environmental factors can alter DNA methylation, and histone methylation patterns, influence DNA topological organization, and cause gene expression changes, the mechanisms remain unclear.

Classroom question: What is nature versus nurture?

Environmental factors can be epigenetic modifiers. Transient exposure to environmental stressors can produce persistent changes in epigenetic marks that have life-long phenotypic consequences. Environmental factors can have an epigenetic influence on pre- or postnatal development as well as influencing health during adulthood.

Prenatal case study: The Hunger Winter—Environmental influence on embryonic development

Embryonic development is a critical period for establishing epigenetic marks that regulate cellular growth, differentiation, and organogenesis. An example that demonstrates the connection between prenatal environment, epigenetic modifications, and postnatal phenotypic variations is the effect of maternal diet.

A classroom example that interests students are the study of the famine during the Dutch Hunger Winter. This is one of the rare opportunities for studying the effect of stressors on the epigenetic modifications of the human fetus.

The hunger winter case study. The Instructors should use the following mini case study to teach the students the following points:

- In humans, genetic and epigenetic analysis of health and diseases is observational and indirect rather than experimental. Scientists take advantage of natural human phenomena to explain human diseases and try to find the cause for treatment and prevention.

- Commonly these observational studies start with epidemiological studies using rigorous statistical analysis to establish a correlation between the cause and effect. These studies then lead to molecular analysis to identify specific targets to understand the cellular mechanisms.
- It is always interesting to point out to students that scientists work for years on these projects and take the effort of multiple disciplines to conclude statistically significant and accurate findings.

In addition to the suggested teaching notes discussed later, a free-to-access Dutch Hunger Winter case study with teaching resources is also available at: https://www.nsta.org/ncss-case-study/dutch-hunger-winter.

Summary of hunger winter for instructors to present. The Hunger Winter was a devastating famine that took place in a German-occupied part of the Netherlands near the end of World War II. It was an extremely cold winter that was further worsened by a German ban on food transport. Consequently, food availability dropped significantly, and caloric intake was reduced to nearly 30% of the average daily intake. By the time food supplies were restored 7 months later, about 20,000 people had died. Even though times were tough, women conceived and gave birth to babies.

From this great tragedy came remarkable scientific studies of the effects of maternal malnutrition during different periods of gestation on health in adult life.

Part I: History behind the story The instructor can definitely expand on the history behind the story [25,26]. These references go into more detail about the timing and progression of the famine. Here also the instructor can point out how scientists observe that adults who were conceived and exposed to prenatal malnutrition between 1944 and 1945 have several health issues such as diabetes and obesity [27,28].

Part II: Establishing the correlation between the cause and effect Several epidemiological studies on the Dutch famine cohort that were exposed to prenatal famine repeatedly showed correlations between maternal malnutrition and adult chronic diseases in offspring. The epidemiological data indicated the differences in the outcome between the different times of exposure during pregnancy. The data also showed time-of-exposure-dependent health outcomes in adulthood that are independent of birth weight. Many studies pointed to the exposure during the early gestation age have the strongest effect on the offspring as adults. Data from these studies indicated that if a baby was born small due to malnutrition throughout the mothers' pregnancy, they remained small for the rest of their *lives with lower obesity* rates. On the other hand, babies born an average size due to malnutrition only during the early stages of pregnancy had *higher obesity rates than average as adults* [25]. These findings conclude that health conditions found later in life can originate through adaptations by the fetus in response to malnutrition during the mother's pregnancy (Fig. 16.15).

The instructor can have a class discussion with the following question: Why would early exposure to famine cause a profound effect on adult health compared to late exposure in pregnancy? Here we want students to think critically about the sensitive periods of fetus development. Preconception and the first trimester is a sensitive period of development during which critical developmental changes occur and the main organ and structure form. In contrast, most organ and body structures have become well established during the third trimester and are less sensitive than in the early stages.

Part III: Molecular studies The instructor can conclude the case by asking students to think about the molecular changes caused by prenatal malnutrition that occurs during development and persists to adulthood, causing different health issues. They should help students bring their learning of epigenetics to the discussion and how epigenetic modifications through mitotic heritability can persist.

Connecting the knowledge from this unit to the realms of life   **387**

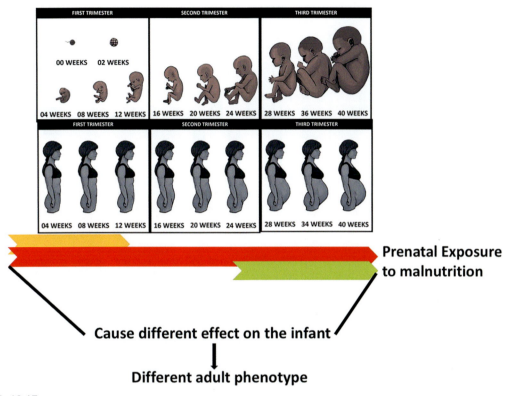

FIG. 16.15

Differential effect of prenatal exposure to the Dutch famine on adult health. The schematic diagram shows the different stages of pregnancy and prenatal development, early stage (preconception and the first trimester), *yellow arrow*, throughout the pregnancy (preconception and through the three trimesters), *red arrow*, and late-stage (the third trimester), *green arrow*. The adult health outcome depends on the prenatal timing of famine exposure.

Two points the instructor can use to lead this section:

- The instructor also can point out that animal studies can be a good start to determine if prenatal malnutrition causes molecular changes. Epigenetic studies in animal models have shown that exposure to malnutrition during prenatal development shows epigenetic changes at the level of DNA methylation in different areas of the genome, such as promoters that may explain the phenotypes observed in adult animals [29].
- In this case, investigators have a significant cohort (sample of individuals) that can be studied, the Dutch-Hunger cohort to transition from animal studies to human studies.

After 50 years and because of advanced technology, scientists can unlock the molecular cause of the health outcomes in adults exposed prenatally to malnutrition. In 2008 a study looked at the methylation of specific genes involved in metabolism. In this study, investigators determined the methylation of the DNA CpG dinucleotides by bisulfite-converted DNA—specific PCR amplification followed by mass spectrometry. The study looked at individuals who were prenatally exposed to malnutrition during the

388 Chapter 16 Epigenetics in the classroom

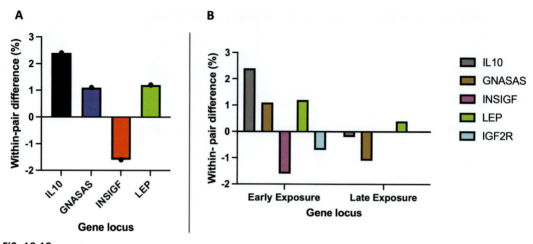

FIG. 16.16

The effect of prenatal exposure to famine on DNA methylation of CpG dinucleotide of specific metabolism-involved genes. The difference in methylation between individuals prenatally exposed to famine and their same-sex siblings. (A) The differential methylation of different loci during early exposure to famine. (B) comparing the DNA methylation of the loci in A between early- and late-prenatal exposure.

The figure was generated by summarizing data from C. Herreid, Start With a Story: The Case Study Method of Teaching College Science, illustrated ed., National Science Teachers Association, 2006. ISBN:10-1933531061.

famine and compared them to their unexposed siblings. They have two groups of individuals depending on the exposure time, early exposure group $n=60$, and late exposure group $n=62$. As controls for each exposure group, they used a matched same-sex sibling. The instructor can use Fig. 16.16 to discuss the summary of the study with the students [30]. The figure shows the difference in DNA methylation of a group of genes in siblings discordant with prenatal exposure to famine. These genes were selected based on the phenotypes observed in the epidemiological studies related to metabolism. The figure illustrates two major points: (1) the differential methylation of different genes, indicating differential impact by the famine. (2) differential effect of the famine between early prenatal and late exposure.

Instructors can discuss with students the validity of the data by asking critical questions concerning the size of the study and controls used with each group.

- The sample size is critical in accepting the study outcome. A small sample size may not be sufficient to demonstrate an accurate difference between the group with acceptable precision. We always look at the % of the targeted population in the general population.
- In human studies, finding the control sample is very critical since many factors can affect the outcome. Here they are comparing same-sex siblings, which can help scientists count for genetic, sex, and other environmental differences.

Finally, this area of research is growing, and there are more publications that instructors can use if they choose to expand this section. The following paper was published in 2014 looking at the effect of prenatal famine exposure on global methylation of the epigenome, using individuals from the Dutch famine cohort and more advanced technology [31]. Another group in 2008 looked at the impact of prenatal famine exposure on the regulation of the imprinted gene *igf2* comparing the exposed individuals from the Dutch famine cohort to their same-sex siblings [32].

Postnatal case study: Care/grooming versus neglect—Environmental influence on epigenetics in early life

Exposure to an early-life environmental modifier can influence the biology and the physiology of the brain of growing children. The commonly used example of the influence of the environment on epigenetic modification during childhood are the rodent licking and grooming studies. Many resources with lesson plans or classroom activities are available, such as the content for rodent licking, at https://learn.genetics.utah.edu/content/epigenetics/.

In advanced courses, instructors can use the original Meaney publication [33] to discuss this example. We adapted the figures from the original publication as a critical thinking activity. The instructor can start the topic by clearly stating that early life experience is a modifier that alters the DNA methylation of a molecule involved in molding adult behavior.

Before starting this section, the instructor is encouraged to remind students that brain development is a lengthy process. It occurs in multiple stages, starting during embryonic development and completing ~25 years after birth. It is hierarchically structured, meaning the later development depends on early development. Although early brain development is genetically driven, postnatal development is significantly influenced by early life experience. The experience during the first 2 years of life modulates the plasticity of the brain.

Summary of infant care/grooming/neglect behaviors for instructors to present.

- Early in life, the care received by an infant can influence the development of the neural systems regulating responses to social behavior.
- Human studies have indicated that persistent emotional and physical neglect, family conflict, and harsh environment compromise cognitive and emotional development. It also increases the risk of depression and anxiety disorders.
- Studies in rodents have confirmed the influence of caregivers on offspring's cognitive and behavioral development in adulthood. In these studies, the high level of active maternal behaviors is exemplified by high licking and grooming. In contrast, the low level of maternal behaviors is demonstrated by low licking and grooming (LG).

One very well-studied example is the influence of early life experience on the DNA methylation of the Glucocorticoid Receptor (GR) promoter in humans and rodents. Meaney's work suggested a causal relationship between the GR promoter's epigenetic modification, the GR protein expression, and the maternal effect on stress responses in the offspring. This study showed that low-caring mothers have different GR promoter methylation and stress behaviors than the offspring of high-caring mothers. The instructor can use Fig. 16.17 to walk the students through the example.

Meaney's cross-fostering experiment provides an excellent synthesis classroom activity that helps students think critically about how scientists would design experiments to test hypotheses. Students are presented with the following problem: Experimental findings suggest that the GR promoter is differentially methylated as a function of maternal care. How would you design a follow-up experiment to examine a direct relationship between maternal care and the GR promoter's methylation and confirm that the maternal care and not the biological connection is what program the methylation of the GR promoter?

Students can discuss it as a small group before the whole class discussion. The instructor can then present Fig. 16.18, which explains the cross-fostering experiment that shows the variation in maternal care directly alters the methylation of the GR promoter.

For assessing students' understanding, a table similar to what is shown in Fig. 16.19 can be used. The instructor can provide an incomplete table for students to fill in the missing information.

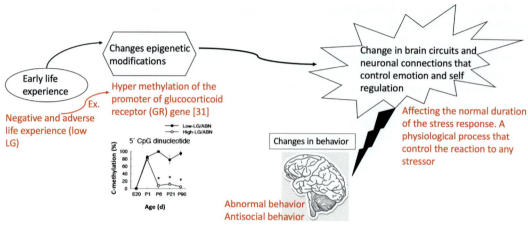

FIG. 16.17

The effect of early childhood experience on epigenetic modifications and brain development. The figure is a schematic representation of the causal relationship between early life experience, epigenetic modification, brain development, and adult behavior. The figure shows a specific example of the effect of mother's care on the methylation of the GR promoter. The methylation data is illustrated by the figure adapted from the Meaney study [33].

The methylation graph is from I.C.G. Weaver, N. Cervoni, F.A. Champagne, A.C. D'Alessio, S. Sharma, J.R. Seckl, S. Dymov, M. Szyf, M.J. Meaney, Epigenetic programming by maternal behavior, Nat. Neurosci. 7 (2004) 847–854.

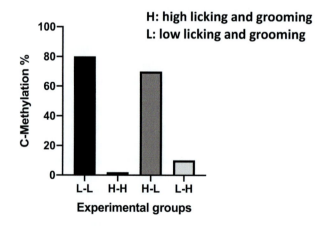

FIG. 16.18

Cross-fostering experiment and the effect of maternal care on GR methylation. The figure is adapted from the Meaney study [33] for an active learning classroom exercise.

From I.C.G. Weaver, N. Cervoni, F.A. Champagne, A.C. D'Alessio, S. Sharma, J.R. Seckl, S. Dymov, M. Szyf, M.J. Meaney, Epigenetic programming by maternal behavior, Nat. Neurosci. 7 (2004) 847–854.

Mother	DNA Methylation of the GR promoter in the offspring	Offspring phenotype
Low LG	High methylation of the 5′ CpG sites	Abnormal stress response
High LG	De-methylation of the 5′ CpG sites	Normal Stress response
Biological mother is low LG but the adoptive mother is high LG	De-methylation of the 5′ CpG sites	Normal Stress response
The biological mother is high LG, the adoptive mother is low LG	Methylation of the GR promoter	Abnormal stress response

FIG. 16.19
A representation of a table that is used to assess students' learning. Instructors can provide students with an incomplete table and ask them to fill in the missing information. We usually use the table with the *red* highlighted areas missing.

Conclusion and future considerations

The fast-growing field of epigenetics and its role as the mechanism of environmental adaptation makes it an essential part of high school and university level biology curriculum. Teaching epigenetics can challenge traditional genetic teaching and understanding of the regulation of gene expression. It is a reversible mechanism that regulates cellular processes and explains how the environment can alter gene expression and phenotype. A common goal among educators is to help students understand and appreciate the complexities of natural phenomena—this is particularly important for epigenetics as it explains many medical, environmental, and social health phenomena. We suggest that epigenetics can be taught as a unit in a biology/genetics course or as an independent course, depending on the curriculum. Instructors, particularly those teaching epigenetics for the first time, would find this chapter an excellent starting point for building lesson plans. As the field of epigenetics advances, instructors should remember to revise their lesson plans based on the updated peer-reviewed literature so as to ensure the highest level of rigor in teaching.

References

[1] Modeling Molecular Mechanisms, A framework of scientific reasoning to construct molecular-level explanations for cellular behavior, Sci. Educ. 22 (2013) 93–118, https://doi.org/10.1007/s11191-011-9379-7.
[2] J. Kang, J.R. Daines, A.N. Warren, M.L. Cowan, Epigenetics for the 21st-century biology student, J. Microbiol. Biol. Educ. 20 (2019) 1–5. https://www.ncbi.nlm.nih.gov/pmc/articles/PMC6914348/.
[3] R.J. Werner, A.D. Kelly, J.-P.J. Issa, Epigenetics and precision oncology, Cancer J. 23 (2017) 262–269, https://doi.org/10.1097/PPO.0000000000000281.
[4] M. Grace, A. Christodoulou, C. Hughes, K. Godfrey, W. Rietdijk, J. Griffiths, The case for including epigenetics in the school science curriculum and trainee teachers' views about its implications for society, Sch. Sci. Rev. 101 (2020) 27–31.
[5] J. McTighe, Teaching for Deeper Learning: Tools to Engage Students in Meaning Making, ASCD, 2020.

[6] D. Drits-Esser, M. Malone, N.C. Barber, L.A. Stark, Beyond the central dogma: bringing epigenetics into the classroom, Am. Biol. Teach. 0002-7685, 76 (2014) 365–369.
[7] C.A. Brewer, D. Smith, Vision and change in undergraduate biology education: a call to action. https://www.researchgate.net/publication/248290185_Vision_and_Change_in_Undergraduate_Biology_Education_A_Call_to_Action.
[8] L.D. Fink, Creating Significant Learning Experiences: An Integrated Approach to Designing College Courses, Jossey-Bass, 2013. 978-0787960551.
[9] S. Ambrose, M. DiPietro, M. Lovett, M. Norman, How Learning Works: Seven Research-Based Principles for Smart Teaching, Jossey-Bass, San Francisco, CA, 2010.
[10] I. Venkatesh, Teaching epigenetic regulation of gene expression is critical in 21st-century science education: key concepts & teaching strategies, Am. Biol. Teach. 82 (2020) 1–9.
[11] Center for BioMolecular Modeling-MSOE, (n.d.). https://cbm.msoe.edu/.
[12] J. Goodship, P. Chimmery, T. Strachan, Genetics and Genomics in Medicine, first ed., Garland Science, 2014.
[13] G. Hemant, D.J. Weisenberger, G. Liang, The roles of human DNA methyltransferases and their isoforms in shaping the epigenome, Genes 10 (2019) 172, https://doi.org/10.3390/genes10020172.
[14] M.S. Cosgrove, J.D. Boeke, C. Wolberger, Regulated nucleosome mobility and the histone code, Nat. Struct. Mol. Biol. 11 (2004) 1037–1043, https://doi.org/10.1038/nsmb851.
[15] X.-J. Yang, The diverse superfamily of lysine acetyltransferases and their roles in leukemia and other diseases, Nucleic Acids Res. 32 (2004) 959–976, https://doi.org/10.1093/nar/gkh252.
[16] E.L. Greer, Y. Shi, Histone methylation: a dynamic mark in health, disease and inheritance, Nat. Rev. Genetics 13 (2012) 343–357, https://doi.org/10.1038/nrg3173.
[17] B.A. Alhamwe, Histone modifications and their role in epigenetics of atopy and allergic diseases, Allergy Asthma Clin. Immunol. 14 (2018) 39, https://doi.org/10.1186/s13223-018-0259-4.
[18] M. Ishida, G.E. Moore, The role of imprinted genes in humans, Mol. Asp. Med. 34 (2013) 826–840, https://doi.org/10.1016/j.mam.2012.06.009.
[19] C.M. Trujillo, T.R. Anderson, N.J. Pelaez, An instructional design process based on expert knowledge for teaching students how mechanisms are explained, Adv. Physiol. Education 40 (2016) 265–273, https://doi.org/10.1152/advan.00077.2015.
[20] S. Berland, B.I. Haukanes, P.B. Juliusson, G. Houge, Deep exploration of a CDKN1C mutation causing a mixture of Beckwith-Wiedemann and IMAGe syndromes revealed a novel transcript associated with developmental delay, J. Med. Genet. 59 (2022) 155–164, https://doi.org/10.1136/jmedgenet-2020-107401.
[21] C. Herreid, Start With a Story: The Case Study Method of Teaching College Science, illustrated ed., National Science Teachers Association, 2006. ISBN:10-1933531061.
[22] D. Galetzka, T. Hansmann, N. El Hajj, E. Weis, B. Irmscher, M. Ludwig, B. Schneider-Rätzke, N. Kohlschmidt, V. Beyer, O. Bartsch, U. Zechner, C. Spix, T. Haaf, Monozygotic twins discordant for constitutive BRCA 1 promoter methylation, childhood cancer and secondary cancer, Epigenetics 7 (2012) 47–54.
[23] T. Erick, Epigenetics: How Nurture Shapes Our Nature. Footnotes. February 2014.
[24] C.H. Waddington, The Strategy of the Genes: A Discussion of Some Aspects of Theoretical Biology, George Allen and Unwin, London, 1957, pp. 88–104.
[25] T.J. Roseboom, J.H. van der Meulen, A.C. Ravelli, C. Osmond, D.J. Barker, O.P. Bleker, Effects of prenatal exposure to the Dutch famine on adult disease in later life: an overview, Mol. Cell. Endocrinol. 185 (2001) 93–98, https://doi.org/10.1016/s0303-7207(01)00721-3.
[26] L.S. Bleker, S.R. de Rooij, R.C. Painter, A.C.J. Ravelli, T.J. Roseboom, Cohort profile: the Dutch famine birth cohort (DFBC)—a prospective birth cohort study in the Netherlands, BMJ Open 11 (2021) 1–12, https://doi.org/10.1136/bmjopen-2020-042078.
[27] A. Vaiserman, O. Lushchak, Prenatal malnutrition-induced epigenetic dysregulation as a risk factor for type 2 diabetes, Int. J. Genomics (2019) 1–11, https://doi.org/10.1155/2019/3821409.

[28] A.C. Ravelli, J.H. van Der Meulen, C. Osmond, D.J. Barker, O.P. Bleker, Obesity at the age of 50 y in men and women exposed to famine prenatally, Am. J. Clin. Nutr. 70 (1999) 811–816, https://doi.org/10.1093/ajcn/70.5.811.

[29] P.W. Nathanielsz, Animal models that elucidate basic principles of the developmental origins of adult diseases, ILAR J. 47 (2006) 73–82, https://doi.org/10.1093/ilar.47.1.73.

[30] E.W. Tobi, L.H. Lumey, R.P. Talens, D. Kremer, H. Putter, A.D. Stein, P.E. Slagboom, B.T. Heijmans, DNA methylation differences after exposure to prenatal famine are common and timing- and sex-specific, Hum. Mol. Genet. 18 (2009) 4046–4053, https://doi.org/10.1093/hmg/ddp353.

[31] E.W. Tobi, J.J. Goeman, R. Monajemi, H. Gu, H. Putter, Y. Zhang, R.C. Slieker, A.P. Stok, P.E. Thijssen, F. Müller, E.W. van Zwet, C. Bock, A. Meissner, L.H. Lumey, P.E. Slagboom, B.T. Heijmans, DNA methylation signatures link prenatal famine exposure to growth and metabolism, Nat. Commun. 5 (2014) 1–13, https://doi.org/10.1038/ncomms6592.

[32] B.T. Heijmans, E.W. Tobi, A.D. Stein, H. Putter, G.J. Blauw, E.S. Susser, P.E. Slagboom, L.H. Lumey, Persistent epigenetic differences associated with prenatal exposure to famine in humans, Proc. Natl. Acad. Sci. USA 105 (2008) 17046–17049, https://doi.org/10.1073/pnas.0806560105.

[33] I.C.G. Weaver, N. Cervoni, F.A. Champagne, A.C. D'Alessio, S. Sharma, J.R. Seckl, S. Dymov, M. Szyf, M.J. Meaney, Epigenetic programming by maternal behavior, Nat. Neurosci. 7 (2004) 847–854.

SECTION 5

Gene editing technologies

CHAPTER 17

Genome editing technologies

Dana Vera Foss[a] and Alexis Leigh Norris[b]

[a]Wilson Lab, University of California Berkeley, Berkeley, CA, United States [b]Food and Drug Administration, Bioinformatician, Center for Veterinary Medicine, Rockville, MD, United States

Introduction

Why would we want to alter genomes? The predominant applications are in agriculture, ecology, and treating (and potentially curing) diseases. Genome editing refers to a novel suite of technologies that allows precise modification of genomes, which is drastically different from previous technologies such as mutation breeding in plants or transgene expression with uncontrolled integration sites. Genome editing can refer to insertion of new DNA into a specific location or altering the sequence at a precise location by causing nucleotide base insertions, deletions, and/or substitutions. Genome editing can be used to insert ("knock-in") a desired DNA sequence or remove ("knock-out") an undesired DNA sequence. Often in therapeutic applications, "knock-ins" are used to correct a gene mutation (such as reversing *CFTR* mutations that cause cystic fibrosis) and "knock-outs" are used to inactivate a gene that promotes the disease (such as deleting the coding sequence of *CCR5* so that certain isotypes of HIV are unable to bind CCR5 and infect cells).

Why we would want to edit genomes in a manner that reduces unintended changes is intuitive? Unintended changes may have unintended outcomes for the genome and thus unintended biological implications in cells or organisms. Editing a genome specifically allows one to make only the genetic change required. The vast majority of the impact of genome editing thus far has been in scientific research, where scientists are now able to make precise changes to the DNA of cells or organisms and understand the impact of those changes on phenotype, thereby unraveling the mysteries of the genome and increasing our understanding of biological processes. As scientists rapidly gain a firmer understanding of how to perform genome editing safely and accurately, a plethora of applications in healthcare, ecology, and food production are just around the corner.

Three main categories of genome editing technologies are in use today (Fig. 17.1): zinc finger nucleases (ZFNs; see "Zinc finger nucleases" section), transcription activator-like effector nucleases (TALENs; see "Transcriptional activator-like effector nucleases" section), and clustered regularly interspersed short palindromic repeat (CRISPR)-Cas systems (see "CRISPR-Cas systems" section). All these systems involve an endonuclease enzyme being targeted to a specific sequence of DNA where it makes a precise cut, creating a double-stranded break (DSB) in the DNA backbone. The outcome of genome editing after the cut relies upon the cellular DNA repair machinery, which can result in either a precise repair (no editing) or an error-prone repair at the cut site, in a process known as nonhomologous end joining (NHEJ) and resulting in short insertions or deletions (indels) at the site of the cut.

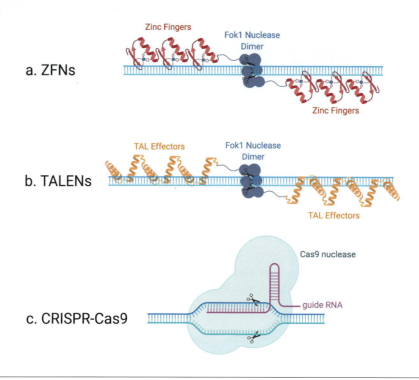

FIG. 17.1

Genome editing technologies.

Created with BioRender.com.

To insert a piece of DNA, the experimenter must provide a piece of DNA (donor DNA) with sequence complementarity (homology arms) flanking the cut site. These constructs then rely on the cellular repair machinery, by a process known as homology-directed repair (HDR), to allow the piece of DNA to be inserted at the site of the cut. Insertion of donor DNA via HDR happens at much lower frequencies than indels; thus, the outcomes of these experiments in cells or organisms result in a combination of outcomes in different cells, where some cells will have NHEJ-based indels at the cut site and some cells will have the donor DNA inserted at the cut site. The edits introduced into a cell can be inherited at a higher rate than normal by using a gene-drive approach (see "Delivery of genome editing systems" section).

Scientists track the efficiency of desired outcomes when performing genome editing in cells or in an organism by analyzing the DNA sequences and reporting percentages associated with each outcome. There are a variety of methods for analyzing the DNA sequences for editing outcomes. The most direct approach is to sequence the DNA around the intended cut site. The most accurate way to do this is by next generation sequencing (NGS) approaches and is most commonly performed as amplicon sequencing; after the genomic DNA is isolated from the cells of interest, the area around the cut site is amplified by PCR and the fragments are sequenced. One can perform NGS or Sanger sequencing for this analysis, where NGS is more expensive but more accurate. If sequencing is not available, the T7E1 assay is a method to estimate gene-editing efficiencies but is not as reliable as sequencing-based methods [1]. A critical part of genome-editing experiments involves not only checking for editing efficiencies at the intended cut site but also looking for genomic changes at off-target sites in the genome (see "Gene drive systems" section).

Zinc finger nucleases
Background
Origin
In 1985, Aaron Klug and colleagues at Cambridge discovered Cys_2His_2 zinc fingers (ZFs), which are proteins used to control gene transcription through their recognition and subsequent binding to DNA [2]. The "zinc" in ZFs refers to its cysteine (Cys) and histidine (His) amino acid binding to zinc ions. Each ZF module recognizes a specific set of the 64 possible combinations of three DNA bases. A decade later, in 1996, Srinivasan Chandrasegaran's group at Johns Hopkins reported the use of exploiting the DNA homing properties of ZFs for cutting DNA at specific target sites, by fusing a bacterial DNA nuclease (Fok1) catalytic domain to ZFs (zinc finger nuclease, ZFN) [3]. Then in 2002, Dana Carroll's lab at the University of Utah successfully used a ZFN in vitro to insert a new DNA sequence at a specific site in the genome of frog eggs [4].

Design
For genome editing, two ZFN monomers are used, where one monomer recognizes the DNA sequence upstream of the target DNA cut site and the other recognizes the DNA sequence downstream of the target DNA cut site, and they are oriented in an antisense, Fok1 catalytic domain head-to-head fashion (Fig. 17.1A). A 5–6 bp "spacer" is required to separate the adjacent DNA regions that each monomer recognizes. Each target site requires customizing two ZFs, while the Fok1 nuclease domain remains constant. For ZFNs, 3–6 finger modules are used, thus resulting in a 9–18 bp DNA sequence recognition for the ZFN. Upon ZFN monomers binding the target DNA, the Fok1 domains dimerize, resulting in cleavage of the DNA in the spacer region. The cell's machinery then repairs the DSB using homologous or nonhomologous recombination. For ZFN construction, publicly available assembly methods include context-dependent assembly (CoDA) and oligomerized pool engineering (OPEN) [5,6]. A more simplistic modular assembly, where each ZF is designed based on 3 bp and then the individual ZF modules are combined, is inferior due to contextual effects of the preceding and following ZFs in a ZFN. Therefore, designing ZFNs is complex.

Advantages
ZFNs were the first "genome editing" nucleases and have largely been eclipsed by the newer genome editing technologies of TALENs and CRISPR-Cas (see "Transcriptional activator-like effector nucleases" and "CRISPR-Cas systems" sections). However, ZFNs do have the advantages of easier delivery (due to smaller size) and likely less immunogenic potential than TALENs and CRISPR-Cas systems (due to mostly being synthetic and mammalian sequences) [7].

Limitations and important considerations
Given that the ZFs are specific to a given 9–18 bp DNA sequence, protein engineering is required for each target site. The protein engineering and assembly of ZFNs is difficult, requiring expertise that limits the use of ZFNs to specialized labs. Two important considerations when designing ZFNs are the context dependence and the triplet composition. For context dependence, one must account for the fact that specificities of individual ZFs can overlap and can depend on the context of the adjacent ZFs and DNA sequence. The triplet composition limits the available ZFs for a given sequence of interest. "GNN" triplets are ideal, as they have been more thoroughly validated, but for many target sites, one must use less stringent triplet composition requirements.

Applications

In the lab (basic and translational research)

Genome editing has revolutionized the study of biology through its ability to make precise changes to single genes, to study their function in model organisms. ZFNs were rapidly adopted to study the effects of genes and mutations in rodents, zebrafish, fruit flies, and other model organisms in the lab. Relevant resources for ZFN design and administration are available through the Zinc Finger Consortium (http://www.zincfingers.org/) (see Box 17.1). Through studying genome edited plants and animals in the lab, researchers are able to not only understand gene function but also identify biomarkers and develop therapeutic interventions for disease.

In the clinic (human disease)

In humans, clinical applications employ somatic genome editing, rather than heritable genome editing, although recently human-germline genome editing has been carried out, with significant backlash from the scientific community [10] (see Box 17.2 "Somatic vs heritable editing"). Somatic editing can be performed in vivo, where the nuclease is delivered directly into the patient, or ex vivo, where cells collected from a patient are edited in a lab and then reintroduced into the patient. The first proof-of-principle clinical application of ZFNs was in 2005, with the "fixing" of the IL2Rγ, the gene responsible for X-linked severe combined immunodeficiency (SCID), in human cells, by editing a single DNA base. Other human disease applications include engineering HIV-resistant T-cells [20] and restoring productive erythropoiesis to β-thalassemic HSC [21]. Several applications have proved successful in mouse in vivo studies [22,23]. While most applications now use newer generation genome-editing approaches (e.g., CRISPR-Cas9, see Chapter 18, sections "Generating mouse models using CRISPR-Cas9" and "Methodology"), there are still active clinical trials using ZFNs [24].

Box 17.1 The proprietary nature of ZFNs

While ZFNs have made it to the clinic, many argue that their full potential for both clinical and laboratory applications has been stymied by their proprietary nature. Sangamo Therapeutics, Inc. (https://www.sangamo.com/) owns the bulk intellectual property (IP) regarding ZFNs, after licensing patents from Johns Hopkins, Massachusetts Institute of Technology, and Scripps Research Institute, and acquiring Aaron Klug's company Gendaq. This has led to their narrow adoption by industry and lack of widespread academic research. In an attempt to remove this barrier, a group of academic researchers formed the Zinc Finger Consortium (http://www.zincfingers.org/) to provide resources and reagents for using ZFNs. This effort was cofounded by J. Keith Joung (Harvard) and Daniel Voytas (University of Minnesota). Joung stated the consortium's motivation: "*My interest is not to circumvent Sangamo's patents. I just want to make the technology available, easy to use, efficient, and robust*" [8], noting that "*if we were to stay relevant to the field we had to band together*" [9].

Sangamo, which was founded in 1995, offered its ZFNs readily available through Sigma-Aldrich's CompoZr assembly method starting in 2008, 3 years after their first ZFN-based clinical trial began. The company argues that they collaborate with dozens of academic labs and has no objections to independent efforts to develop the technology, with CEO Ed Lanphier adding "*[i]f they want to go out and work hard in this area, that's great*" [8]. However, this is not the experience of many academics. For example, Daniel Voytas (University of Minnesota) attempted to license one of Sangamo's ZFN patents in order to launch his own plant biotech firm, but the two sides failed to reach an agreement, and Sangamo complained to Scripps about the ZFN instructional website from one of their investigators, Carlos Barbas (www.scripps.edu/mb/barbas/zfdesign/zfdesignhome.php). From Barbas' perspective, Sangamo "*inhibit[s] the technology from proliferating*" [8].

> **Box 17.2 Somatic vs heritable editing**
>
> Genome editing of organisms can be performed either in the somatic cells or in the germ cells. If editing is performed only in somatic cells, the genetic changes will not be inherited in future generations. If editing is performed in sperm, eggs, or embryos, then those changes will be inherited in subsequent generations and is referred to as "germline" genome editing. Current clinical applications of genome-editing undergoing clinical trials are exclusively somatic genome editing, such as editing the cells of the retina to cure blindness or blood cells to cure diseases such as sickle cell disease [11]. These are applications that aim to cure genetic disease in the affected patient, without impacting the genetics of any future offspring from that person. Editing the human germline is much more ethically complicated than somatic genome editing, and the general consensus among scientists and clinicians is that neither the technology nor the ethical and regulatory framework surrounding it is ready for germline genome editing in humans [12,13]. While genetic engineering of human embryos for research purposes is performed by several labs all over the world [14], none have pursued implantation of the embryos toward making genetically edited babies.
>
> In November of 2018, however, Dr. He Jiankui, who was at that time a professor at the Southern University of Science and Technology in Shenzhen, China, announced to the surprise of the research community that he had performed CRISPR-based genome editing of human embryos and implanted them in human participants, resulting in the birth of two genetically edited babies [10]. Dr. He's study aimed to edit the CCR5 gene, intending to make the naturally occurring Δ32-CCR5 mutation which confers resistance to certain strains of HIV [15, p. 5]. He's clinical research has been widely condemned because there was not sufficient medical need to warrant the risks of such an untested clinical application of genome editing. Furthermore, in preimplantation genetic testing, one of the embryos exhibited mosaicism for the intended genetic edit, meaning that some of the cells contained the intended alteration in the CCR5 gene and some did not [16]. The resulting child thus does not have the intended phenotype—that is, they are not immune to certain strains of HIV. Furthermore, the resulting genetic edits were not precisely the same as any of the naturally occurring Δ32-CCR5 mutations, and thus, the biological consequences of the altered CCR5 gene remain unknown. He's research practices have been further condemned to be unethical because he did not acquire adequate informed consent from the parents enrolled in the study [17].
>
> In response to this "CRISPR babies" controversy, there has been a call from leading scientists for an international moratorium on clinically applied germline genome editing [12], although not everyone agrees. Many believe that clinical germline editing is inevitable and ethically defensible under certain conditions, and the focus should be on regulating instead of banning these developments [18]. There is much work being done to provide guidelines surrounding the ethics of genome editing of the human germline [19]. The technology presents the opportunity to cure devastating genetic diseases, but it could also be used for changes deemed as "human enhancement" or, at worst, to pursue eugenics. It remains unclear how the scientific, clinical, and regulatory communities will pursue advancement of this technology while adhering to an agreed upon set of principles and how those would be enforced globally, but a recent international effort has produced guidelines for how to move forward [13].
>
> This topic is one ripe for classroom conversation, independent research projects in high school biology, undergraduate philosophy, and debate-style projects; it is important for all citizens to be well-informed about both the promise and the bioethical implications of genome-editing technologies.

On the farm (agricultural animals)

Human intervention in evolution via selective breeding of plants and animals has spawned thousands of years of agricultural advances. Unlike human genome editing, genome editing of agricultural animals is predominantly performed on germline (rather than somatic) cells, to attain the goal of inherited traits. Genome editing offers a much faster alternative to improving agriculture than conventional breeding (artificial selection), and ZFNs were readily adopted for this purpose. Traditionally, the germline genome editing is performed on DNA in a cell line, which is then transferred into an enucleated oocyte using somatic cell nuclear transfer (SCNT; also referred to as "cloning") [25]. Newer approaches perform direct editing of the germline cells through direct delivery of the ZFN into the cells using microinjection or electroporation.

In the field (plants)
Mutation breeding, also referred to as variation breeding, began in the 1920s and continues to this day in plant breeding practices. Many of the world's current crops have been engineered by mutation breeding and are largely considered not to be a genetically modified organism (GMO). GMOs refer to crops improved through altering plant DNA with genetic engineering approaches, which has been performed for over three decades. However, genetic engineering of plants was imprecise and suffered from low homologous recombination rates as well as illegitimate homologous recombination events [26]. With ZFNs, scientists are able to precisely target genes in crops to confer resistance to disease [27] and herbicides [28].

Future directions
ZFNs were the first truly targetable genome-editing method. Most researchers and companies have shifted away from ZFNs and to newer genome-editing methods such as TALENs and CRISPR-Cas systems. The protein engineering of TALENs is much easier than ZFNs, while CRISPR-Cas is based on simpler nucleotide engineering rather than the complex protein engineering required with ZFNs.

Transcriptional activator-like effector nucleases
What they are
Origin
Like ZFNs, Transcriptional activator-like effector nucleases (TALENs) are engineered proteins consisting of a DNA-binding domain and a DNA cleavage domain, which together operate as monomers (Fig. 17.1B). In TALENs though, while the DNA cleavage domain is identical to ZFNs (Fok1 catalytic domain), the DNA binding domain is from transcription activator-like (TAL) effectors. TAL effectors were first discovered in *Xanthomonas*, a bacterial plant pathogen, by Robert Stall and colleagues in 1990 [29]. Two decades later, in 2009, two groups independently solved the DNA recognition code of TAL effectors [30,31]. A TAL effector repeat consists of 33–34 amino acids. DNA sequence recognition determined by the central amino acid residues, at positions 12 and 13 and together, is referred to as the repeat variable diresidue (RVD). One TAL effector repeat recognizes 1 bp; most naturally occurring TAL effectors contain 12–27 repeats, which in turn correlates to the length of the DNA sequence recognized.

Design
TALENs are engineered to target a specific DNA sequence through the combination of individual repeat modules. Like ZFNs, TALEN monomers must be properly oriented (Fok1 catalytic domains head-to-head, in an antisense fashion) and spaced (6–16 bp) [32]. Additionally, the number of amino acids between the DNA binding and DNA cleavage domains appears to influence efficiency. Unlike ZFNs, TALENs do not have the context effect of preceding and following TAL effectors (TALE) modules, and thus TALENs can be designed in a true modular fashion. However, the repetitive sequence of TALEs can be a challenge in TALEN engineering and assembly. A modular approach is often used for assembly; techniques include restriction enzyme and ligation (REAL), REAL-Fast, fast ligation-based automatable solid-phase high-throughput (FLASH), and Golden Gate (see Fig. 17.2) [33–35].

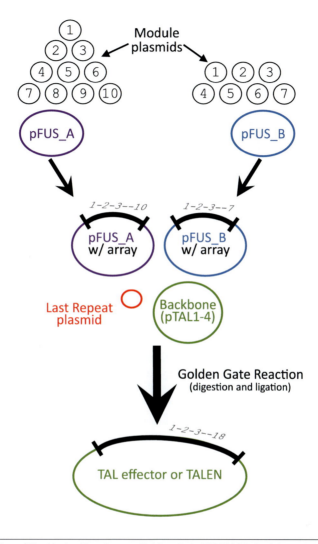

FIG. 17.2

Simplified representation of the Voytas/Bogdanove Golden Gate TALEN kit. In this example, the final array is composed of 18 TAL effector repeats. The kit can be used to make arrays from 12 to 31 repeats.

Below are a few TALEN design resources:

- GeneArt TAL Effector Search and Design Tool (ThermoFisher Scientific)
  - e-TALEN by the German Cancer Research Center (http://www.e-talen.org/E-TALEN/)
  - TAL Effector Nucleotide Targeter by Cornell University (https://tale-nt.cac.cornell.edu)

Advantages
Compared to ZFNs, TALENs are easier to engineer due to more predictable specificity; the specificity of the individual TALE modules does not appear to be influenced by neighboring repeats, in contrast to the contextual effect of ZF modules. To improve specificity, one can use "nonconventional" RVD sequences in the TALE [36]. Repeats may contain amino acid substitutions outside of the RVD, but they do not appear to affect specificity. To improve efficiency, one can use Fok1 cleavage domains that have been modified [37,38]. Like ZFNs, TALENs have the advantage of possibly being less immunogenic than CRISPR-Cas systems because they are derived mostly from bacterial species that are not human pathogens [7].

Limitations and important considerations
TALENs are larger than ZFNs, posing issues with their delivery into cells (see "Delivery of genome editing systems" section). TALENs are also sensitive to cytosine methylation, especially CpG (cytosine-guanine dinucleotide) sites. Thus, it is recommended to avoid CpGs, but otherwise one can target anywhere in the genome with TALEN-based genome editing. While TALENs demonstrate high specificity, they still require screening to confirm their specific binding and cleavage of the intended target DNA.

Applications
Genome
TALENs are used in the same applications as ZFNs: in the lab, in the clinic, on the farm, and in the fields. TALENs have been more widely adopted by researchers given their easier engineering and nonproprietary nature compared to ZFNs. In the clinic, TALENs have been used for multiple chimeric antigen receptor T-cell (CAR-T) cancer therapies [24]. On the farm, TALENs have been used to improve animal welfare, such as the hornless (polled) trait in cows [39,40]. In the field, TALENs have been used to confer disease resistance to crops, such as bacterial blight in rice [41].

Epigenome
In addition to genome editing, TALENs have also been used to "edit" the epigenome. Epigenome editing alters the regulation of a DNA sequence (e.g., gene expression) rather than altering the DNA sequence itself. In the lab, epigenome altering is used to study the function of epigenome features (e.g., enhancers and methylation) [42]. Compared to newer CRISPR-Cas genome editing, TALENs are very large and may potentially interfere with other proteins binding their epigenetic target site. While TALENs could be used in the clinic to make changes to the epigenome for therapeutic applications, most have shifted to a CRISPR-Cas-based approach for epigenome editing.

MegaTALs
TALENs have also been combined with meganucleases, an early generation genome-editing nuclease also known as homing endonucleases, to create "megaTALs." Meganucleases have high specificity owing to their recognition of long DNA stretches (12–40 bp). Meganucleases fell out of favor to ZFNs and TALENs given the limited number of recognition sequences of naturally occurring meganucleases and thus limited breadth of appropriate target sites in the genome. As well, there can be difficulty in engineering custom meganucleases to recognize a specific target site. Like ZFNs, the proprietary nature of meganucleases limits their broad application by researchers, similar to ZFNs (discussed in Box 17.1).

The chimeric MegaTAL proteins have both high efficiency and high specificity, from the meganuclease and TALE domains, respectively [43]. One disadvantage of megaTALs is their sensitivity to the epigenetic status of the target DNA sequence, including methylation and chromatin structure [44,45].

Future directions

While TALENs are still used given their high specificity, they are no longer the "easiest" method of genome editing. Thus, many applications have shifted to genome editing using CRISPR-Cas systems, which do not require protein engineering for DNA recognition, but rather much simpler RNA engineering (see "CRISPR-Cas systems" section).

CRISPR-Cas systems
Background
Origin

CRISPR-Cas systems are prokaryotic adaptive and heritable immune systems which defend against invading bacteriophages and other mobile genetic elements. The system works by detecting invading foreign DNA, cutting the DNA up into smaller pieces, and incorporating those DNA pieces at a specific location in the prokaryotic genome (termed the CRISPR locus) where it stores a library of sequences marking past infections. Keeping these sequences stored in the genome allows rapid response should the bacteria be infected with foreign DNA (such as from a bacteriophage) containing a matching sequence. The CRISPR locus is a repeating array of alternating identical repeating sequences and unique 20-nucleotide "spacer" sequences which are derived from the invading DNA pathogens [46].

This locus is transcribed into a long precursor-CRISPR-RNA (pre-crRNA) which is then processed at the repeat sequences into mature crRNAs. The mature crRNAs are then bound by the effector machinery, Cas (CRISPR-Associated) nucleases, which surveil and cut sequences of foreign DNA which are complementary to the spacer sequence, at a site termed the "protospacer." Most CRISPR-Cas systems have evolved a critical mechanism to distinguish between the sequences in the array and the invading sequences which will be targeted for cutting: the Cas nucleases only cut the complementary DNA when there is a Protospacer Adjacent Motif (PAM) sequence adjacent to the protospacer and absent from the CRISPR array. Prokaryotes have evolved a variety of CRISPR-Cas systems with variations in their mechanisms. Class 1 CRISPR-Cas systems (Type I, III, IV) consist of multicomponent Cas effector complexes, whereas Class 2 systems (Type II, V, VI) employ a single Cas nuclease to target foreign DNA [46]. The different CRISPR-Cas systems all are broadly similar in that they employ RNA sequences as guides to find their target sequence, through base pair complementarity. Cas9 is from the Type II Cas system, and it employs a two-part guide RNA sequence, where the trans-activating crRNA (tracrRNA) acts as a handle for recruiting the targeting crRNA sequence. These are covered in depth in Hille et al. [46].

Design

It was not long after the discovery of CRISPR-Cas systems in prokaryotes that the opportunity became apparent to use Cas nucleases and their targeting guide RNAs biotechnology tools, leveraging these enzymes' ability to cut at specific sequences of DNA for genome-editing applications [47]. Cas9 from *Streptococcus pyogenes* is the most well-studied and broadly applied Cas enzyme for genome editing

(Fig. 17.1C). It was the first Cas enzyme to be characterized biochemically [47] and the first to be applied for cutting genomic DNA in eukaryotic cells [48–50]. It is relatively straightforward to recombinantly express and purify this enzyme from bacteria, and it can also be expressed in mammalian cells. The guide RNA, which allows targeting the enzyme to cut specific sequences, is also easy to make, synthetically or by in vitro transcription (IVT), or can be expressed in eukaryotic cells. The guide RNA can be designed as a two-part system, analogous to the endogenous system in bacteria, with the tracrRNA remaining a consistent sequence and the crRNA being programmable depending on what sequence in the genome one would like to target. Alternatively, it is also possible to use a single piece of RNA, where the tracr and crRNAs have been engineered as a single sequence, termed a "single guide RNA" (sgRNA) [47]. When Cas9 is combined with its guide RNA it is referred to as the Cas9 ribonucleoprotein complex, or Cas9 RNP. For an in-depth discussion of experimental design considerations, see Chapter 18; for resources regarding off-target prediction, see "Unintended genomic alterations" section.

Advantages
CRISPR-Cas systems have been rapidly and broadly applied across all sectors of the life sciences, revolutionizing how scientists interrogate gene function, make model organisms, and engineer new systems. Cas enzymes are programmable nucleases, just as the previous genome editing technologies TALENs and ZFNs are, but these latter technologies have remained relatively niche compared to the widespread usage of CRISPR. This disparity is largely due to the relative simplicity of CRISPR systems and how easily programmable they are for targeting different genetic sequences. When an experimenter wants to target a specific gene, all that is required is to synthesize a new guide RNA, which is available commercially and is quite affordable for most research purposes. This is in stark contrast to TALENs and ZFNs which require re-engineering the protein itself to target different DNA sequences. Thus, it is specifically the modularity of CRISPR-Cas systems, where the nuclease remains the same but relies on the guide RNA for targeting, which is responsible for their widespread application.

Applications
In the lab
CRISPR-Cas systems have been successfully applied for genome editing in species across all kingdoms of life, but improving and increasing our understanding of these tools remains actively researched. Knocking out gene function via NHEJ remains substantially easier than "knocking in" a sequence of interest via HDR. Inserting a piece of DNA which has been supplied by the experimenter happens at lower frequencies, and thus, the outcomes of these experiments are a mixture of NHEJ outcomes and HDR insertion. Optimizing to increase the efficiency of HDR has been the focus of a vast amount of research [51], as precise control over insertion of genetic sequences would open up even more opportunities than is currently possible. Two areas of active research are manipulating the endogenous DNA repair machinery to favor HDR over NHEJ [52] as well as physically tethering the DNA templates to Cas9 for higher rates of HDR [53–58].

In the clinic—Therapeutics
CRISPR-Cas systems have been broadly applied in research settings and are just beginning to realize their potential for treating human disease in the clinic. All genome-editing technologies offer the tantalizing potential for curing a vast array of genetic diseases by fixing the underlying mutation causing disease. CRISPR-Cas9 is currently in clinical trials in North America, Europe, and China. Ongoing trials include ex vivo editing of human immune cells for making cellular therapies as cancer treatments, as well as

ex vivo editing of hematopoietic stem cells for treating blood disorders such as sickle cell anemia and β-thalassemia [11]. The first clinical trial for in vivo administration of CRISPR-Cas systems in the United States has begun and aims to treat Leber Congenital Amaurosis, which is the leading cause of inherited childhood blindness [11]. There is great excitement over these early clinical trials, which will hopefully establish CRISPR-based genome editing as a safe and effective means to cure human disease Box 17.2.

Base editing and prime editing
Base editing is an adaptation of genome editing, where instead of generating a DSB in the DNA, the target sequence can be chemically modified at a precise position to change the nucleotide identity at that site [59]. Base editors have been engineered to target both DNA (for genome editing) or RNA (for transcriptome editing), and different versions have been developed to change a C·G base pair to an A·T, or vice versa. The technology relies on a genetic fusion between a Cas9 nickase, which "nicks" one strand of the DNA target [47,60], and a cytidine deaminase or an engineered adenosine deaminase. Since the initial report of base editors from the Liu lab in 2016 [61], there have already been substantial improvements in specificity and activity levels, over several iterations [59].

Prime editing is another engineered adaptation of CRISPR-Cas9 genome editing which, similar to base editing, does not result in a DSB in the DNA, but instead allows insertion of a programmed sequence of DNA at the nicked target site. In this approach, the sgRNA is programmed with an extra flanking sequence that encodes a template, which is read by a reverse transcriptase enzyme tethered to the Cas9 nickase enzyme, to generate the desired DNA sequence which can then be incorporated into the target site; these desired edits can be single-base pair mutations or multiple-base pair insertions or deletions [62]. There is widespread excitement about base editing and prime editing technologies for both research and therapeutic applications, especially because making a precise edit while avoiding a DSB in the DNA would avoid a major drawback of genome-editing technologies, that is, the potential for large unintended genomic alterations (see "Gene drive systems" section).

Future directions
The application of CRISPR-Cas systems to genome editing for research applications has been transformative and is offering tremendous promise for therapeutic applications. Research on understanding, optimizing, and engineering CRISPR-Cas systems for increased effectiveness and diversity of application remains ongoing. Two of the most significant remaining challenges are engineering new delivery technologies to send the genome-editing components to the cells in need of correction (see "Delivery of genome editing systems" section) and understanding and controlling unintended genomic alterations (see "Gene drive systems" section). Some of the most creative application of CRISPR-Cas genome editing are gene drive applications, which is an exciting and controversial area of research (see "Gene drive systems" section).

Delivery of genome editing systems
Background
Mechanism
Performing genome editing for research purposes in cells or model organisms in a laboratory has become relatively routine and accessible for scientists all over the world. It remains, to a large extent, however, prohibitively challenging to perform genome editing in humans for clinical applications.

The biggest challenge facing clinical applications is achieving efficient and safe delivery of the genome-editing components (such as the Cas9 nuclease and its guide RNA). Unlike conventional small molecule drugs and medicines, these are large macromolecules which will not cross the cell membrane unaided. Thus, it is necessary to "deliver" the genome-editing components into the cells in need of correction.

Limitations and important considerations
For future clinical applications, in vivo administration of genome editing enzymes presents a suite of challenges [63,64]. Serum (blood) contains proteases and nucleases which degrade proteins and nucleic acids and the human adaptive immune system is very adept at detecting and reacting to foreign proteins, especially those of bacterial or viral origin. Administering CRISPR components and TALENs in vivo without generating a strong immune response remains quite challenging; most of the early clinical applications of genome editing are ex vivo therapeutics, where a patient's cells are removed from the body, edited in the laboratory, and returned to the patient. This is only possible for certain tissues, such as blood cells or hematopoietic stem and progenitor cells (HSPCs), thus ex vivo editing will only be able to treat a limited number of diseases. Furthermore, the infrastructure required to perform genome editing ex vivo is prohibitively expensive to treat a broad patient population with this approach [65]. Regardless, important therapeutic gains are being made using ex vivo editing where the first patients involved in clinical trials may be cured of sickle cell disease and β-thalassemia, thus paving the way for genome editing in the clinic.

Methods
Electroporation
In the laboratory, when performing genome editing in cells in culture, one can apply a shock of electricity to briefly open pores in the cell membrane to allow passage of the genome editing components, in a process known as electroporation. Electroporation can be performed for either plasmid DNA encoding the genome editing components or the purified, recombinant nucleases themselves as RNPs, allowing gene editing in the electroporated cells. While commonly used in to the editing of cells either in vitro or ex vivo, electroporation's invasive nature limits its use for in vivo delivery.

Viral vectors
Because of the limitations of ex vivo genome editing, there is an extensive research being performed in how to best deliver genome-editing enzymes in vivo. Many researchers package the DNA encoding genome-editing components into viral vectors, such as adeno-associated viruses (AAV) [66]. Viruses are adept at entering cells and delivering their genomes to target cells, thus engineering viruses to deliver genetic payloads to target cells is quite effective, and has been the mainstay of 30 years of gene therapy research. Because the field of genome editing is directly building on the field of gene therapy, applying viral vectors for genome editing is currently the most widespread delivery method in therapeutic development. It is critical to note, however, that viral vectors still face significant challenges in safe and effective clinical application [67]. AAVs are nonintegrating viruses, meaning they largely do not integrate their genomes into the target cell genome but instead remain episomal [66]. This is unlike their clinical predecessors, gammaretroviruses, and lentiviruses, which are integrating viruses.

Viral integration poses a risk for genotoxicity and oncogenesis in the target cells, and early gene therapy clinical trials using gammaretroviral vectors indeed have caused cancer in patients [68]. Although AAV vectors are largely nonintegrating, they do integrate into genomes at low but consistent frequencies (0.1% of total vector genomes) [69] and the risk of genotoxicity and oncogenesis using AAV vectors remains present. This risk appears to be amplified when combined with genome editing, as there are heightened rates of integration at sites of Cas9-induced DSBs [69]. Presumably, this would also be relevant for ZFNs and TALENs, which also make DSBs in the targeted DNA sequence. This is an area of active research and determining the rates of integration when combined with genome editing in therapeutic settings will be critical to understanding the safety profile of these approaches.

The size of the genetic payload that can be delivered via an AAV vector is limited by the viral packaging size, with an upper limit of approximately 4.5 kb [70]. This presents a challenge for delivering TALENs and CRISPR-Cas systems via AAV, due to their large size. Creative engineering solutions are being applied to face this hurdle, engineering both the viruses and the nucleases. ZFNs are quite a bit smaller and thus have an advantage for viral delivery. CRISPR-based base editors [59] and prime editors [62] are multicomponent systems which are too large to be encoded by AAV vectors, presenting a significant challenge in their advancement using this standard delivery method.

Another significant challenge facing viral vectors is immunogenicity. Most of the human population (~70%) has circulating neutralizing antibodies to AAV due to previous exposure to wild-type AAV, and in vivo administration of viral vectors results in an immune response to the virus [71]. The immune responses may render the treatment ineffective or cause toxicity in the patient. Although gene therapy approaches with AAV have been able to mitigate these challenges for effective therapeutics, the challenge is amplified when combined with genome-editing enzymes of bacterial origin. Long-term expression of viral proteins and nucleases of bacterial origin in transduced cells is likely to result in cell-mediated immune responses to the target cells, resulting in the immune system removing the cells which have been successfully genome-edited, as has been observed in preclinical mouse studies of AAV delivery of Cas9 to the liver [72].

ZFNs may be less immunogenic, since they are synthetic proteins derived mostly from human sequences, but still retain the bacterial Fok1 nuclease domain. Thus, immunogenicity remains an important topic of investigation for all genome-editing approaches, especially when paired with long-term expression in target cells via viral vectors. Furthermore, although large efforts are being made in engineering AAV for tissue-specific tropism [70], true specificity and selectivity for specific cell types does not yet exist. When viral vectors are administered in vivo, they enter a variety of cells and tissue types. Perhaps, the greatest challenge facing development of genome-editing therapeutics using viral vectors remains their enormous manufacturing costs, which is currently limiting their development and broad applicability as therapeutics [65].

Overall, AAV vectors are not without risks and challenges that will require further research, but still are the leading vector in development for clinical applications of genome editing. The first clinical trial for in vivo administration of CRISPR in the United States has begun, and involves administering CRISPR-Cas9 packaged in an AAV vector via subretinal injection to treat Leber Congenital Amaurosis, which is the leading cause of inherited childhood blindness (NCT03872479) [11,73]. This first in-human trial of administering CRISPR-Cas9 leverages that the eye is accessible via direct injection and is an immune-privileged organ, removing two of the largest hurdles facing in vivo genome editing and offers exciting prospects for treating eye diseases and blindness.

Nonviral delivery

Nonviral technologies being developed for in vivo administration of genome editing components are currently mostly nanoparticle-based, where the genome editing components (either encoded as mRNA or the nuclease proteins) are packaged in what is usually a lipid-based nanoparticle. Nanoparticles are efficient at entering cells, but when administered in vivo they mostly enter liver cells [74]. As such, nanoparticles are being advanced in clinical trials for delivery of CRISPR-Cas9 for treating liver diseases and show incredible promise [75]. There are many advantages of nonviral delivery for genome-editing therapeutics, perhaps the most important being that the genome-editing components will be present transiently in the target cell instead of the long-term expression characteristic of viral delivery. The transient nature of nonviral delivery of genome-editing components offers advantages in reducing rates of off-target effects and lowered immunogenic potential. See Box 17.3 for more on CRISPR delivery outcomes.

Future directions

There is a great interest in developing novel delivery platforms for genome-editing applications in cells and organisms, and this is a large area of active research and development. For ex vivo therapeutics, new methods are increasing editing rates and HDR efficiencies in difficult-to-transfect cell types such as primary T cells [78] and HSPCs [79]. Viral vectors are being engineered to have decreased immunogenicity, increased tissue tropism, and larger packaging limits [70]. Nonviral methods in development include engineering the Cas9 RNP to be able to enter target cells without a nanoparticle, which would expand the utility of nonviral methods for in vivo administration [80,81]. Engineering nanoparticles for targeting other tissues than the liver is also looking increasingly promising. Developing nonviral methods of delivery for therapeutics will be critical for base editors and prime editors, since these multicomponent systems are too large to encode in AAV vectors.

Box 17.3 Delivery context influences outcomes

A hurdle facing scientists performing genome editing in cells or organisms is that the method of delivering the genome-editing enzymes can influence the editing outcomes and other resulting phenotypes, which can make it difficult to form broad conclusions from specific editing experiments. For example, because delivery agents will differ in the amount of genome-editing enzymes delivered to, or expressed in target cells, this will result in different frequencies of on-target and off-target editing levels. As well, viral and plasmid delivery can result in high levels of sustained expression of Cas9 and its guide RNA, which may predispose to higher levels of off target editing events. Whereas when nonviral methods are used, the genome-editing machinery are delivered as either mRNA or the nucleases themselves, which are degraded quickly in target cells and thus have fewer off-target effects [76,77].

The context of delivery influencing outcomes is further challenging when performing editing experiments in cells (in vitro) in preparation for in vivo experiments in organisms. The rates of on-target and off-target editing will differ between in vitro and in vivo experiments because the effectiveness of the delivery will differ in cells versus a whole organism. It is therefore necessary to characterize the rates of editing extensively between different types of experiments. Furthermore, when performing editing experiments in vivo, the immune system becomes far more relevant; if the editing approach is sufficiently immunogenic, the cells which have been edited may be destroyed by the cell-mediated immune response [72].

Gene drive systems
What they are
Discovery
Naturally self-propagating genetic elements, or "gene drive systems," were first described in the 1920s, in mice and Drosophila flies [82,83]. Gene drives are a mechanism to propagate a specific allele/genotype throughout a population by favorably skewing/increasing the probability that the allele will be transmitted to offspring more than the natural Mendelian inheritance of 50% probability. Nearly a century after the discovery of natural gene drive systems, scientists are combining gene drive systems with genome editing.

It is important to note that gene drive systems are not being used to modify human genomes, but instead are being applied to modify nonhuman organisms in ways to control the population or capacity to spread vector-borne disease. They offer, potentially, an effective means of successfully editing the entirety (or specific population(s)) of a given species. Gene drives are inherited over successive generations, and thus in theory are inherently sustainable, where the significant investment is upfront with minimal maintenance investment thereafter. These economical aspects make the gene drive approach particularly advantageous for communities where existing tools may be costly and difficult to implement, particularly in countries with emerging economies.

Mechanism
Gene drive systems predominantly rely on newer CRISPR-Cas based genome editing. A gene drive sequence contains two elements: (1) the DNA sequence with the desired edit (introduce a mutation, inactivate or delete an existing gene, insert a new gene, etc.) and (2) the DNA sequence that will confer that the edit is inherited at higher (skewed) rate than normal, and thus "drive" or propagate the edit in the population. The second element, which confers the "drive," could include DNA encoding for the CRISPR-Cas machinery (e.g., Cas nuclease and sgRNA) and thus convert the wildtype allele(s) so that the organism is homozygous for the edit.

Safeguard strategies
Due to their self-propagating nature, upfront measures must be taken to control gene drives. Such strategies include physical confinement or molecular safeguards. Physical confinement includes preventing the escape of the genome-edited organism from a closed environment (e.g., a lab) or reducing the likelihood of spread through ecological barriers (e.g., an island). Molecular safeguards include using a synthetic target site or a split-drive approach [84]. For the synthetic target site gene drive, the gene drive's DNA sequence is not present in the genome of the organism's wild counterparts. The synthetic target site approach works well for development, but presumably not for actual field deployment. For the split drive approach, the gene drive's DNA sequence is split into two and inserted into two different regions of the genome, of which one is inherited independently and at a normal (nonskewed) rate. The discovery of AcrIIA4, an anti-CRISPR-Cas9 protein, raised the possibility of a new mechanism of inhibiting gene drives [85]; however, only proof-of-concept studies in yeast have been performed as of now [86].

Limitations
The gene drive approach relies on the free mixing of a population to spread the sequence throughout the population. For this reason, gene drives are limited to sexually reproducing organisms. Additionally,

gene drives can be eradicated through inbreeding [87]. Additional limitations include the development of resistance alleles, as well as a reliance on efficient cutting and HDR and low off-target cutting activity.

How they are used
Public health
Gene drives can be used to limit the spread of vector-borne diseases through genome editing of the pathogen's host vector(s). The host vector species is genome-edited to carry an allele that either renders the host organism an unsuitable host (reducing parasite transmission) or affects the local population dynamics of the host organism (e.g., lower survival or skewed sex ratio that reduces population size over generations). Prominent examples include the genome editing of mosquitoes and rodents for the control of malaria and tick-borne diseases, respectively.

For the mosquito-borne disease, malaria, a 2011 study by Windbichler and colleagues provided proof of principle for genome editing for gene drive systems ability for malaria disease control [88]. In 2016, Hammond et al. showed that the female sterility allele was successfully inherited at a rate of 91%–99% of progeny [89]. Because it would be inherited over successive mosquito generations, a gene drive system could potentially be a cost-effective and sustainable malaria control strategy that requires few repeat investments, particularly in communities where existing tools may be costly and difficult to implement. Given that mosquitos are hosts for many diseases beyond malaria, this approach could also be used to control other mosquito-borne diseases, such as dengue fever.

Rodents are reservoirs for many tick-borne diseases, including Lyme disease. Gene drives have been proposed to reduce the burden of tick-borne diseases through genome editing of the mice that serve as their hosts. In 2017, Prowse and colleagues estimated that through a one-time introduction of 100 genome-edited mice, they could eradicate a population of 50,000 mice within 5 years [90].

Conservation (preventing extinction)
Gene drives can be used to protect species in the native ecosystem (prevent their extinction) through the control or elimination of damaging invasive species that threaten the native flora and fauna. The threat could be predation, habitat displacement/competition, or carrying infectious diseases that put the survival of other species at risk. Prominent examples are islands, which are isolated or closed populations and which face the highest rates of extinction [91]. Growing invasive rodent populations threaten the survival of many local animals, especially birds, on islands. While rodent eradication can be performed using rodenticides, gene drives offer a potentially more targeted (e.g., rodent-specific), humane, and environmentally safe alternative. One such gene drive approach is to edit the genomes of the rodents to carry an allele that skews the sex of offspring so that reproduction is significantly impaired over subsequent generations. New Zealand is currently debating the use of gene drives in their "Predator Free 2050" initiative to protect their native biodiversity from extinction (see Box 17.4).

Future directions
While gene drives hold significant power to reduce the human toll of vector-borne disease and mitigate the effect of invasive species on biodiversity, there is much debate surrounding their use. Major concerns are unanticipated detriment to the ecological system and losing control over the gene drive

> **Box 17.4 Restoring New Zealand's birds**
>
> For millions of years, bats were the only land mammals in New Zealand. Thus, there was little evolutionary pressure on birds to build defenses to predatory mammals, and the birds were docile in comparison to birds in ecological systems with predatory mammals like rodents, cats, and dogs. But human colonization in the late 13th Century brought with mammals that have predated on the local birds to the extent that nearly a quarter is extinct. In response, conservationists rallied and in 2016, the New Zealand government announced their ambitious plan of eliminating invasive mammalian predators by 2050, as outlined in their "Predator Free 2050" initiative (http://pf2050.co.nz/).
>
> Kevin Esvelt, a leading gene drive researcher at MIT, knows the great potential of gene drives for restoring ecological systems, but he also knows the potential peril of misinformation. Stating in a 2017 article with Predator Free 2050 advisor Neil Gemmell (University of Otago): "The risk of accidental spread posed by self-propagating gene drive technologies, highlights new gene drive designs that might achieve better outcomes, and explain why we need open and international discussions concerning a technology that could have global ramifications" [92].
>
> Thus, Esvelt and colleagues have partnered with organizations to go out on the front lines and communicate the benefits and risks directly with the community members that will be impacted. In New Zealand, this engagement is especially important given the dark history between the native Māori and European colonizers that has led to mistrust. "Māori tend to have a precautionary approach because we've already had many cases of wrongdoing for the right reasons. Generally speaking, we are suspicious of any kind of genetic modification"—Aroha Te Pareake Mead, an expert of Māori perspectives on biotechnology [93].

organisms. For example, even rodents on an island will eventually leave the island, through their own efforts or human-assisted, and thus propagate their genotype, which could cause shockwaves to other ecological systems where they are an essential component.

Unintended genomic alterations
Background
Mechanism
Given that the central premise of genome editing is the cleavage of DNA by a nuclease followed by the cell machinery repairing the DSB, the process is subject to error resulting in unintended alterations. These mutations can occur in response to either or the combination of both. Nuclease specificity plays a role in ensuring that the nuclease only cuts at the target site and not at other regions of the genome. In the repair, NHEJ is much more prevalent than HDR (indeed, HDR machinery is only available in S phase of cell cycle). Given that NHEJ is error-prone, it is inherently difficult to control the edit, and often short insertions or deletions (indels) are produced [94]. This is appropriate if an inactivating indel is desired, as is the case in many genome-editing experiments. Additionally, delivery methods play a role in the type and propensity for mutations. In the case of plasmids or viral vectors used to deliver the repair template and/or nuclease, unintended integration of the plasmid or viral vector may occur.

Characteristics
On-target errors are mutations that occur at the target site, while off-target effects refer to mutations that occur in regions of the genome not at the target site. These unintended alterations vary greatly—they may delete, insert, or rearrange genetic material (Fig. 17.3). Inserted genetic material may be from the host genome or foreign DNA delivered during genome editing, such as vectors

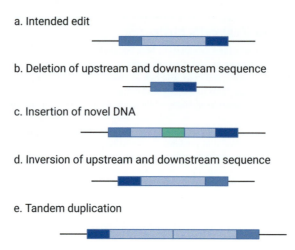

FIG. 17.3

Genomic rearrangements at/near target site.

www.biorender.com.

used for the delivery of nucleases or repair templates [69,95]. The mutations may be short or large. Short mutations include single nucleotide variants (SNVs) and short insertions and deletions (indels; 1–10 bp). Large structural rearrangements are also possible, especially at or near the target site. They have been shown to occur up to ~16 kb up- and downstream of the target site and include insertions, deletions, inversions, and translocations [94,96]. Large chromosomal truncations involving the deletion of megabases of DNA have also been shown to occur [97]. Given their size and complexity, large unintended alterations are easily missed without long-read sequencing strategies. Additionally, the orientation of DNA integrations is such that they are often difficult to detect [98] (see Box 17.5). The mutations may be present in all cells or only present in a subset ("somatic mosaic"). They may be heterozygous or homozygous, intergenic or intragenic. It is often difficult to assess or predict the effect of mutations in intergenic regions—they could be regulatory and impact gene function.

Methods

In silico prediction

Many computer algorithms are available to predict off-target effects for a given experiment (ZFN monomer pair, TALEN monomer pair, or CRISPR nuclease and guide RNA) and a reference genome for the organism to be edited. The advantages are speed, cost (often free), and ease of use. The tools are technology-specific. Free, online tools for CRISPR-Cas include: Cas-OFFinder (http://www.rgenome.net/cas-offinder/) [99]; CHOPCHOP (https://chopchop.cbu.uib.no/) [100]; and CRISPR.ML (https://crispr.ml) [101].

Limitations and important considerations include algorithm bias, the reliance on the quality of the reference genome to make accurate predictions, and these are only predictions—they can tell you where to look, but they can't tell you what will actually happen when the genome editing is performed in vitro or in vivo.

> **Box 17.5 Complexity revealed**
>
> Improvements in technology often shine a light on things previously unseen. In genome editing, a primary example of this is the extent and complexity of unintended alterations that occur. Prior to the landmark paper by Kosicki and colleagues in 2018, the focus was on off-target effects—their work shined a light on the size and complexity of unintended alterations that occur at the target site [96]. Examples of unintended genomic rearrangements at or near the target site are illustrated in Fig. 17.3. Often these rearrangements are only obvious when you are looking for them, due to limitations in previous DNA sequencing technology. With the introduction of long read DNA sequencing, researchers are no longer limited to studying DNA in ~0.5–2 kb stretches. The ability to look at 10–100+ kb stretches of DNA allows one to look far up- and downstream of the target site. These large, complex rearrangements are widely appreciated and increasingly screened for.

Biochemical and cell-based screening

During optimization of a genome-editing experiment, it is useful to test the reagents (ZFN monomer pair, TALEN monomer pair, or CRISPR nuclease and guide RNA). The cleavage (editing) may be performed on isolated DNA (biochemical) or in the cell (cell-based), and typically make use of next-generation sequencing (NGS) of the nuclease-cleaved DNA.

Biochemical and cell-based methods include:

- *B*reaks *L*abeling, *E*nrichment on *S*treptavidin, and next-generation *S*equencing (BLESS) [102]
  - *C*ircularization for *I*n vitro *R*eporting of *CL*eavage *E*ffects by Sequencing (CIRCLE-seq) [103]
    - Nuclease-*D*igested *Genome Seq*uencing (Digenome-Seq) [104]
- *Discover*y of In Situ Cas Off-targets and VE*R*ification by *Seq*uencing (DISCOVER-Seq) [105]
  - Genome-wide, *U*nbiased *I*dentification of *D*SBs *E*nabled by *Seq*uencing (GUIDE-seq) [106]
    - High-Throughput, Genome-wide, Translocation Sequencing (HTGTS) [107]
    - Integrase-Defective Lentiviral Vectors (IDLV) [108]
    - Selective enrichment and *I*dentification of Adapter-*T*agged DNA *E*nds by *Seq*uencing (SITE-Seq) [109]

An important consideration for cell-based screening is the appropriate choice of cell type, given differences across cell types that may affect nuclease activity (e.g., methylation and chromosome structure).

Screening genome-edited organisms or tissue

Screening for the editing outcomes in genome-edited organisms or tissues allows one to observe what actually happened in the organism or cell/tissue. This includes germline genome editing (such as performed in somatic cell nuclear transfer (SCNT)), and somatic genome editing, such as ex vivo, to generate genome-edited organisms or tissues. However, it only represents a single sample and thus not necessarily representative of other individuals edited using the identical method (variability of effectiveness of editing, including homoscedastic processes like delivery efficiency; DNA repair—especially in the case of NHEJ; zygosity; and variation between individuals). The choice of tissue to extract DNA from is important, especially if targeting a specific cell type (if not germline genome-edited); peripheral blood is not necessarily representative or even the best choice. It is important to have unedited controls or closely related unedited organism(s), to remove natural variation. Additionally, it is difficult to distinguish low frequency (e.g., somatic) mutations from naturally arising de novo mutations and heteroplasmy (in the case of mitochondrial editing).

A note about DNA sequencing

For identifying the mutations, NGS is often used, given its improved sensitivity over conventional Sanger sequencing and its decreasing cost. NGS can be performed on the entire genome (whole genome sequencing; WGS) or only on the sites predicted a priori (targeted sequencing). WGS benefits from no bias of site selection, while targeted sequencing allows for higher sequencing depth of coverage and lower costs. The sequencing coverage must be adequate for reliable detection of mutations; this is especially true for the detection of somatic mosaicism where the mutation occurs at a level much less than 50%. For WGS, often the minimum coverage is 30×, while targeted sequencing is often >100×—lending itself to better detection of low frequency mutations; for more about NGS coverage, see https://www.illumina.com/science/technology/next-generation-sequencing/plan-experiments/coverage.html.

With the improvements of long-read technology, both in accuracy and cost, there is a shift from short-read sequencing (e.g., Illumina) to long-read sequencing (e.g., Nanopore and PacBio), given that long reads (>10–100 kb) offer more reliable detection of large structural rearrangements. For all NGS, the bioinformatics remains unstandardized and different algorithms or parameters may yield slightly different results. Using de novo genome assembly reduces the reliance on the quality of the reference genome, which may prove especially useful for nonmodel organisms.

Future directions

We continue to understand more about unintended edits (frequency, type/landscape, factors affecting their occurrence, both in type and propensity). Efforts to mitigate with engineering nucleases for improved specificity reduce off-cutting that leads to errors introduced during DNA repair by the cell; and delivery methods of nuclease/protein/oligos to control timing/amount of nuclease activity and prevent integration of foreign DNA (e.g., plasmid) (see "Delivery of genome editing systems" section). Screening and prediction methods are often used in combination to mitigate bias of different methods during optimization. For example, using two computer programs to predict cut sites (in silico), then targeted sequencing of the predicted sites from genome editing of cell lines (in vitro), and finally screening edited organisms for the predicted sites. Base editors may offer further advantages, as these do not require a DSB in the DNA, and therefore should not result in larger genomic rearrangements [59,61].

Conclusion

Though scientists have been altering genomes through genetic engineering for decades, genome editing now offers precision, opening the doors to safer and broader clinical, agricultural, and ecological applications. Genome editing came into prominence with the introduction of ZFNs, then TALENs, and now CRISPR-Cas systems. CRISPR-Cas technology has revolutionized the field in the last decade, extending genome editing into new applications (e.g., base editing) and democratizing the use of genome editing through its technically easier approach (engineering RNA rather than engineering nuclease proteins to target a specific region of the genome). Through decades of rigorous research in understanding fundamental biology and biochemistry, the scientific community is now bringing forth these transformative technologies which offer great opportunities to treat disease and to improve human lives. The judiciousness with which these technologies should be approached around complex ethical implications must be underscored at all levels of teaching and learning.

References

[1] M. Sentmanat, S.T. Peters, C.P. Florian, J.P. Connelly, S.M. Pruett-Miller, A survey of validation strategies for CRISPR-Cas9 editing, Sci. Rep. 8 (2018). https://www.nature.com/articles/s41598-018-19441-8.

[2] J. Miller, A.D. McLachlan, A. Klug, Repetitive zinc-binding domains in the protein transcription factor IIIA from Xenopus oocytes, EMBO J. 4 (1985) 1609–1614.

[3] Y.G. Kim, J. Cha, S. Chandrasegaran, Hybrid restriction enzymes: zinc finger fusions to Fok I cleavage domain, Proc. Natl. Acad. Sci. USA 93 (1996) 1156–1160, https://doi.org/10.1073/pnas.93.3.1156.

[4] M. Bibikova, M. Golic, K.G. Golic, D. Carroll, Targeted chromosomal cleavage and mutagenesis in Drosophila using zinc-finger nucleases, Genetics 161 (2002) 1169–1175.

[5] J.E. Foley, J.-R.J. Yeh, M.L. Maeder, D. Reyon, J.D. Sander, R.T. Peterson, J.K. Joung, Rapid mutation of endogenous zebrafish genes using zinc finger nucleases made by Oligomerized Pool ENgineering (OPEN), PLoS One 4 (2009) e4348, https://doi.org/10.1371/journal.pone.0004348.

[6] J.D. Sander, E.J. Dahlborg, M.J. Goodwin, L. Cade, F. Zhang, D. Cifuentes, S.J. Curtin, J.S. Blackburn, S. Thibodeau-Beganny, Y. Qi, C.J. Pierick, E. Hoffman, M.L. Maeder, C. Khayter, D. Reyon, D. Dobbs, D.M. Langenau, R.M. Stupar, A.J. Giraldez, D.F. Voytas, R.T. Peterson, J.-R.J. Yeh, J.K. Joung, Selection-free zinc-finger-nuclease engineering by context-dependent assembly (CoDA), Nat. Methods 8 (2011) 67–69, https://doi.org/10.1038/nmeth.1542.

[7] S.H. Khan, Genome-editing technologies: concept, pros, and cons of various genome-editing techniques and bioethical concerns for clinical application, Mol. Ther. Nucleic Acids 16 (2019) 326–334, https://doi.org/10.1016/j.omtn.2019.02.027.

[8] J. Kaiser, GENE THERAPY: putting the fingers on gene repair, Science 310 (2005) 1894–1896, https://doi.org/10.1126/science.310.5756.1894.

[9] H. Pearson, Protein engineering: the fate of fingers, Nature 455 (2008) 160–164, https://doi.org/10.1038/455160a.

[10] D. Cyranoski, H. Ledford, Genome-edited baby claim provokes international outcry, Nature 563 (2018) 607–608, https://doi.org/10.1038/d41586-018-07545-0.

[11] M.P.T. Ernst, M. Broeders, P. Herrero-Hernandez, E. Oussoren, A.T. van der Ploeg, W.W.M.P. Pijnappel, Ready for repair? Gene editing enters the clinic for the treatment of human disease, Mol. Ther. Methods Clin. Dev. 18 (2020) 532–557, https://doi.org/10.1016/j.omtm.2020.06.022.

[12] E.S. Lander, F. Baylis, F. Zhang, E. Charpentier, P. Berg, C. Bourgain, B. Friedrich, J.K. Joung, J. Li, D. Liu, L. Naldini, J.-B. Nie, R. Qiu, B. Schoene-Seifert, F. Shao, S. Terry, W. Wei, E.-L. Winnacker, Adopt a moratorium on heritable genome editing, Nature 567 (2019) 165–168, https://doi.org/10.1038/d41586-019-00726-5.

[13] National Academy of Medicine, National Academy of Sciences, Heritable Human Genome Editing, The National Academies Press, Washington, DC, 2020, https://doi.org/10.17226/25665.

[14] R.A. Lea, K.K. Niakan, Human germline genome editing, Nat. Cell Biol. 21 (2019) 1479–1489, https://doi.org/10.1038/s41556-019-0424-0.

[15] E. de Silva, M.P.H. Stumpf, HIV and the CCR5-Δ32 resistance allele, FEMS Microbiol. Lett. 241 (2004) 1–12, https://doi.org/10.1016/j.femsle.2004.09.040.

[16] D. Cyranoski, First CRISPR babies: six questions that remain, Nature (2018), https://doi.org/10.1038/d41586-018-07607-3.

[17] D. Shaw, The consent form in the Chinese CRISPR Study: in search of ethical gene editing, J. Bioethical. Inq. 17 (2020) 5–10, https://doi.org/10.1007/s11673-019-09953-x.

[18] R.A. Charo, Rogues and regulation of germline editing, N. Engl. J. Med. 380 (2019) 976–980, https://doi.org/10.1056/NEJMms1817528.

[19] Committee on Human Gene Editing: Scientific, Medical, and Ethical Considerations, National Academy of Sciences, National Academy of Medicine, National Academies of Sciences, Engineering, and Medicine,

Human Genome Editing: Science, Ethics, and Governance, National Academies Press, Washington, DC, 2017, https://doi.org/10.17226/24623.

[20] E.E. Perez, J. Wang, J.C. Miller, Y. Jouvenot, K.A. Kim, O. Liu, N. Wang, G. Lee, V.V. Bartsevich, Y.-L. Lee, D.Y. Guschin, I. Rupniewski, A.J. Waite, C. Carpenito, R.G. Carroll, J.S. Orange, F.D. Urnov, E.J. Rebar, D. Ando, P.D. Gregory, J.L. Riley, M.C. Holmes, C.H. June, Establishment of HIV-1 resistance in CD4+ T cells by genome editing using zinc-finger nucleases, Nat. Biotechnol. 26 (2008) 808–816, https://doi.org/10.1038/nbt1410.

[21] J. Vierstra, A. Reik, K.-H. Chang, S. Stehling-Sun, Y. Zhou, S.J. Hinkley, D.E. Paschon, L. Zhang, N. Psatha, Y.R. Bendana, C.M. O'Neil, A.H. Song, A.K. Mich, P.-Q. Liu, G. Lee, D.E. Bauer, M.C. Holmes, S.H. Orkin, T. Papayannopoulou, G. Stamatoyannopoulos, E.J. Rebar, P.D. Gregory, F.D. Urnov, J.A. Stamatoyannopoulos, Functional footprinting of regulatory DNA, Nat. Methods 12 (2015) 927–930, https://doi.org/10.1038/nmeth.3554.

[22] K. Laoharawee, R.C. DeKelver, K.M. Podetz-Pedersen, M. Rohde, S. Sproul, H.-O. Nguyen, T. Nguyen, S.J. St Martin, L. Ou, S. Tom, R. Radeke, K.E. Meyer, M.C. Holmes, C.B. Whitley, T. Wechsler, R.S. McIvor, Dose-dependent prevention of metabolic and neurologic disease in murine MPS II by ZFN-mediated in vivo genome editing, Mol. Ther. J. Am. Soc. Gene Ther. 26 (2018) 1127–1136, https://doi.org/10.1016/j.ymthe.2018.03.002.

[23] R. Sharma, X.M. Anguela, Y. Doyon, T. Wechsler, R.C. DeKelver, S. Sproul, D.E. Paschon, J.C. Miller, R.J. Davidson, D. Shivak, S. Zhou, J. Rieders, P.D. Gregory, M.C. Holmes, E.J. Rebar, K.A. High, In vivo genome editing of the albumin locus as a platform for protein replacement therapy, Blood 126 (2015) 1777–1784, https://doi.org/10.1182/blood-2014-12-615492.

[24] A. Mullard, Gene-editing pipeline takes off, Nat. Rev. Drug Discov. 19 (2020) 367–372, https://doi.org/10.1038/d41573-020-00096-y.

[25] S. Matoba, Y. Zhang, Somatic cell nuclear transfer reprogramming: mechanisms and applications, Cell Stem Cell 23 (2018) 471–485, https://doi.org/10.1016/j.stem.2018.06.018.

[26] M. Hanin, J. Paszkowski, Plant genome modification by homologous recombination, Curr. Opin. Plant Biol. 6 (2003) 157–162, https://doi.org/10.1016/s1369-5266(03)00016-5.

[27] J.A. Townsend, D.A. Wright, R.J. Winfrey, F. Fu, M.L. Maeder, J.K. Joung, D.F. Voytas, High-frequency modification of plant genes using engineered zinc-finger nucleases, Nature 459 (2009) 442–445, https://doi.org/10.1038/nature07845.

[28] V.K. Shukla, Y. Doyon, J.C. Miller, R.C. DeKelver, E.A. Moehle, S.E. Worden, J.C. Mitchell, N.L. Arnold, S. Gopalan, X. Meng, V.M. Choi, J.M. Rock, Y.-Y. Wu, G.E. Katibah, G. Zhifang, D. McCaskill, M.A. Simpson, B. Blakeslee, S.A. Greenwalt, H.J. Butler, S.J. Hinkley, L. Zhang, E.J. Rebar, P.D. Gregory, F.D. Urnov, Precise genome modification in the crop species Zea mays using zinc-finger nucleases, Nature 459 (2009) 437–441, https://doi.org/10.1038/nature07992.

[29] G.V. Minsavage, Gene-for-gene relationships specifying disease resistance in *Xanthomonas campestris* pv. vesicatoria—pepper interactions, Mol. Plant-Microbe Interact. 3 (1990) 41, https://doi.org/10.1094/MPMI-3-041.

[30] J. Boch, H. Scholze, S. Schornack, A. Landgraf, S. Hahn, S. Kay, T. Lahaye, A. Nickstadt, U. Bonas, Breaking the code of DNA binding specificity of TAL-type III effectors, Science 326 (2009) 1509–1512, https://doi.org/10.1126/science.1178811.

[31] M.J. Moscou, A.J. Bogdanove, A simple cipher governs DNA recognition by TAL effectors, Science 326 (2009) 1501, https://doi.org/10.1126/science.1178817.

[32] C. Mussolino, R. Morbitzer, F. Lütge, N. Dannemann, T. Lahaye, T. Cathomen, A novel TALE nuclease scaffold enables high genome editing activity in combination with low toxicity, Nucleic Acids Res. 39 (2011) 9283–9293, https://doi.org/10.1093/nar/gkr597.

[33] C. Engler, R. Kandzia, S. Marillonnet, A one pot, one step, precision cloning method with high throughput capability, PLoS One 3 (2008) e3647, https://doi.org/10.1371/journal.pone.0003647.

[34] D. Reyon, S.Q. Tsai, C. Khayter, J.A. Foden, J.D. Sander, J.K. Joung, FLASH assembly of TALENs for high-throughput genome editing, Nat. Biotechnol. 30 (2012) 460–465, https://doi.org/10.1038/nbt.2170.

[35] J.D. Sander, L. Cade, C. Khayter, D. Reyon, R.T. Peterson, J.K. Joung, J.-R.J. Yeh, Targeted gene disruption in somatic zebrafish cells using engineered TALENs, Nat. Biotechnol. 29 (2011) 697–698, https://doi.org/10.1038/nbt.1934.

[36] A. Juillerat, C. Pessereau, G. Dubois, V. Guyot, A. Maréchal, J. Valton, F. Daboussi, L. Poirot, A. Duclert, P. Duchateau, Optimized tuning of TALEN specificity using non-conventional RVDs, Sci. Rep. 5 (2015) 8150, https://doi.org/10.1038/srep08150.

[37] Y. Doyon, T.D. Vo, M.C. Mendel, S.G. Greenberg, J. Wang, D.F. Xia, J.C. Miller, F.D. Urnov, P.D. Gregory, M.C. Holmes, Enhancing zinc-finger-nuclease activity with improved obligate heterodimeric architectures, Nat. Methods 8 (2011) 74–79, https://doi.org/10.1038/nmeth.1539.

[38] M. Szczepek, V. Brondani, J. Büchel, L. Serrano, D.J. Segal, T. Cathomen, Structure-based redesign of the dimerization interface reduces the toxicity of zinc-finger nucleases, Nat. Biotechnol. 25 (2007) 786–793, https://doi.org/10.1038/nbt1317.

[39] D.F. Carlson, C.A. Lancto, B. Zang, E.-S. Kim, M. Walton, D. Oldeschulte, C. Seabury, T.S. Sonstegard, S.C. Fahrenkrug, Production of hornless dairy cattle from genome-edited cell lines, Nat. Biotechnol. 34 (2016) 479–481, https://doi.org/10.1038/nbt.3560.

[40] A.E. Young, T.A. Mansour, B.R. McNabb, J.R. Owen, J.F. Trott, C.T. Brown, A.L.V. Eenennaam, Genomic and phenotypic analyses of six offspring of a genome-edited hornless bull, Nat. Biotechnol. 38 (2020) 225–232, https://doi.org/10.1038/s41587-019-0266-0.

[41] T. Li, B. Liu, M.H. Spalding, D.P. Weeks, B. Yang, High-efficiency TALEN-based gene editing produces disease-resistant rice, Nat. Biotechnol. 30 (2012) 390–392, https://doi.org/10.1038/nbt.2199.

[42] E.M. Mendenhall, K.E. Williamson, D. Reyon, J.Y. Zou, O. Ram, J.K. Joung, B.E. Bernstein, Locus-specific editing of histone modifications at endogenous enhancers, Nat. Biotechnol. 31 (2013) 1133–1136, https://doi.org/10.1038/nbt.2701.

[43] S. Boissel, J. Jarjour, A. Astrakhan, A. Adey, A. Gouble, P. Duchateau, J. Shendure, B.L. Stoddard, M.T. Certo, D. Baker, A.M. Scharenberg, megaTALs: a rare-cleaving nuclease architecture for therapeutic genome engineering, Nucleic Acids Res. 42 (2014) 2591–2601, https://doi.org/10.1093/nar/gkt1224.

[44] F. Daboussi, M. Zaslavskiy, L. Poirot, M. Loperfido, A. Gouble, V. Guyot, S. Leduc, R. Galetto, S. Grizot, D. Oficjalska, C. Perez, F. Delacôte, A. Dupuy, I. Chion-Sotinel, D. Le Clerre, C. Lebuhotel, O. Danos, F. Lemaire, K. Oussedik, F. Cédrone, J.-C. Epinat, J. Smith, R.J. Yáñez-Muñoz, G. Dickson, L. Popplewell, T. Koo, T. VandenDriessche, M.K. Chuah, A. Duclert, P. Duchateau, F. Pâques, Chromosomal context and epigenetic mechanisms control the efficacy of genome editing by rare-cutting designer endonucleases, Nucleic Acids Res. 40 (2012) 6367–6379, https://doi.org/10.1093/nar/gks268.

[45] J. Valton, F. Daboussi, S. Leduc, R. Molina, P. Redondo, R. Macmaster, G. Montoya, P. Duchateau, 5′-Cytosine-phosphoguanine (CpG) methylation impacts the activity of natural and engineered meganucleases, J. Biol. Chem. 287 (2012) 30139–30150, https://doi.org/10.1074/jbc.M112.379966.

[46] F. Hille, H. Richter, S.P. Wong, M. Bratovič, S. Ressel, E. Charpentier, The biology of CRISPR-Cas: backward and forward, Cell 172 (2018) 1239–1259, https://doi.org/10.1016/j.cell.2017.11.032.

[47] M. Jinek, K. Chylinski, I. Fonfara, M. Hauer, J.A. Doudna, E. Charpentier, A programmable dual-RNA-guided DNA endonuclease in adaptive bacterial immunity, Science 337 (2012) 816–821, https://doi.org/10.1126/science.1225829.

[48] L. Cong, F.A. Ran, D. Cox, S. Lin, R. Barretto, N. Habib, P.D. Hsu, X. Wu, W. Jiang, L.A. Marraffini, F. Zhang, Multiplex genome engineering using CRISPR/Cas systems, Science 339 (2013) 819–823, https://doi.org/10.1126/science.1231143.

[49] M. Jinek, A. East, A. Cheng, S. Lin, E. Ma, J. Doudna, RNA-programmed genome editing in human cells, eLife 2 (2013) e00471, https://doi.org/10.7554/eLife.00471.

[50] P. Mali, L. Yang, K.M. Esvelt, J. Aach, M. Guell, J.E. DiCarlo, J.E. Norville, G.M. Church, RNA-guided human genome engineering via Cas9, Science 339 (2013) 823–826, https://doi.org/10.1126/science.1232033.

[51] M. Liu, S. Rehman, X. Tang, K. Gu, Q. Fan, D. Chen, W. Ma, Methodologies for improving HDR efficiency, Front. Genet. 9 (2019), https://doi.org/10.3389/fgene.2018.00691.

[52] C.D. Yeh, C.D. Richardson, J.E. Corn, Advances in genome editing through control of DNA repair pathways, Nat. Cell Biol. 21 (2019) 1468–1478, https://doi.org/10.1038/s41556-019-0425-z.

[53] E.J. Aird, K.N. Lovendahl, A.S. Martin, R.S. Harris, W.R. Gordon, Increasing Cas9-mediated homology-directed repair efficiency through covalent tethering of DNA repair template, Commun. Biol. 1 (2018) 1–6, https://doi.org/10.1038/s42003-018-0054-2.

[54] J. Carlson-Stevermer, A.A. Abdeen, L. Kohlenberg, M. Goedland, K. Molugu, M. Lou, K. Saha, Assembly of CRISPR ribonucleoproteins with biotinylated oligonucleotides via an RNA aptamer for precise gene editing, Nat. Commun. 8 (2017) 1711, https://doi.org/10.1038/s41467-017-01875-9.

[55] X. Ling, B. Xie, X. Gao, L. Chang, W. Zheng, H. Chen, Y. Huang, L. Tan, M. Li, T. Liu, Improving the efficiency of precise genome editing with site-specific Cas9-oligonucleotide conjugates, Sci. Adv. 6 (2020) eaaz0051, https://doi.org/10.1126/sciadv.aaz0051.

[56] M. Ma, F. Zhuang, X. Hu, B. Wang, X.-Z. Wen, J.-F. Ji, J.J. Xi, Efficient generation of mice carrying homozygous double-floxp alleles using the Cas9-avidin/biotin-donor DNA system, Cell Res. 27 (2017) 578–581, https://doi.org/10.1038/cr.2017.29.

[57] S. Ma, X. Wang, Y. Hu, J. Lv, C. Liu, K. Liao, X. Guo, D. Wang, Y. Lin, Z. Rong, Enhancing site-specific DNA integration by a Cas9 nuclease fused with a DNA donor-binding domain, Nucleic Acids Res. 48 (2020) 10590–10601, https://doi.org/10.1093/nar/gkaa779.

[58] N. Savic, F.C. Ringnalda, H. Lindsay, C. Berk, K. Bargsten, Y. Li, D. Neri, M.D. Robinson, C. Ciaudo, J. Hall, M. Jinek, G. Schwank, Covalent linkage of the DNA repair template to the CRISPR-Cas9 nuclease enhances homology-directed repair, eLife 7 (2018) e33761, https://doi.org/10.7554/eLife.33761.

[59] E.M. Porto, A.C. Komor, I.M. Slaymaker, G.W. Yeo, Base editing: advances and therapeutic opportunities, Nat. Rev. Drug Discov. (2020), https://doi.org/10.1038/s41573-020-0084-6.

[60] G. Gasiunas, R. Barrangou, P. Horvath, V. Siksnys, Cas9–crRNA ribonucleoprotein complex mediates specific DNA cleavage for adaptive immunity in bacteria, Proc. Natl. Acad. Sci. USA 109 (2012) E2579–E2586, https://doi.org/10.1073/pnas.1208507109.

[61] A.C. Komor, Y.B. Kim, M.S. Packer, J.A. Zuris, D.R. Liu, Programmable editing of a target base in genomic DNA without double-stranded DNA cleavage, Nature 533 (2016) 420–424, https://doi.org/10.1038/nature17946.

[62] A.V. Anzalone, P.B. Randolph, J.R. Davis, A.A. Sousa, L.W. Koblan, J.M. Levy, P.J. Chen, C. Wilson, G.A. Newby, A. Raguram, D.R. Liu, Search-and-replace genome editing without double-strand breaks or donor DNA, Nature 576 (2019) 149–157, https://doi.org/10.1038/s41586-019-1711-4.

[63] C.A. Lino, J.C. Harper, J.P. Carney, J.A. Timlin, Delivering CRISPR: a review of the challenges and approaches, Drug Deliv. 25 (2018) 1234–1257, https://doi.org/10.1080/10717544.2018.1474964.

[64] R.C. Wilson, L.A. Gilbert, The promise and challenge of in vivo delivery for genome therapeutics, ACS Chem. Biol. 13 (2018) 376–382, https://doi.org/10.1021/acschembio.7b00680.

[65] R.C. Wilson, D. Carroll, The daunting economics of therapeutic genome editing, CRISPR J. 2 (2019) 280–284, https://doi.org/10.1089/crispr.2019.0052.

[66] D. Wang, P.W.L. Tai, G. Gao, Adeno-associated virus vector as a platform for gene therapy delivery, Nat. Rev. Drug Discov. 18 (2019) 358–378, https://doi.org/10.1038/s41573-019-0012-9.

[67] P. Colella, G. Ronzitti, F. Mingozzi, Emerging issues in AAV-mediated in vivo gene therapy, Mol. Ther. Methods Clin. Dev. 8 (2018) 87–104, https://doi.org/10.1016/j.omtm.2017.11.007.

[68] S. Hacein-Bey-Abina, A. Garrigue, G.P. Wang, J. Soulier, A. Lim, E. Morillon, E. Clappier, L. Caccavelli, E. Delabesse, K. Beldjord, V. Asnafi, E. MacIntyre, L. Dal Cortivo, I. Radford, N. Brousse, F. Sigaux,

D. Moshous, J. Hauer, A. Borkhardt, B.H. Belohradsky, U. Wintergerst, M.C. Velez, L. Leiva, R. Sorensen, N. Wulffraat, S. Blanche, F.D. Bushman, A. Fischer, M. Cavazzana-Calvo, Insertional oncogenesis in 4 patients after retrovirus-mediated gene therapy of SCID-X1, J. Clin. Invest. 118 (2008) 3132–3142, https://doi.org/10.1172/JCI35700.

[69] K.S. Hanlon, B.P. Kleinstiver, S.P. Garcia, M.P. Zaborowski, A. Volak, S.E. Spirig, A. Muller, A.A. Sousa, S.Q. Tsai, N.E. Bengtsson, C. Lööv, M. Ingelsson, J.S. Chamberlain, D.P. Corey, M.J. Aryee, J.K. Joung, X.O. Breakefield, C.A. Maguire, B. György, High levels of AAV vector integration into CRISPR-induced DNA breaks, Nat. Commun. 10 (2019) 4439, https://doi.org/10.1038/s41467-019-12449-2.

[70] C. Li, R.J. Samulski, Engineering adeno-associated virus vectors for gene therapy, Nat. Rev. Genet. 21 (2020) 255–272, https://doi.org/10.1038/s41576-019-0205-4.

[71] C. Vandamme, O. Adjali, F. Mingozzi, Unraveling the complex story of immune responses to AAV vectors trial after trial, Hum. Gene Ther. 28 (2017) 1061–1074, https://doi.org/10.1089/hum.2017.150.

[72] A. Li, M.R. Tanner, C.M. Lee, A.E. Hurley, M. De Giorgi, K.E. Jarrett, T.H. Davis, A.M. Doerfler, G. Bao, C. Beeton, W.R. Lagor, AAV-CRISPR gene editing is negated by pre-existing immunity to Cas9, Mol. Ther. 28 (2020) 1432–1441, https://doi.org/10.1016/j.ymthe.2020.04.017.

[73] H. Ledford, CRISPR treatment inserted directly into the body for first time, Nature 579 (2020) 185, https://doi.org/10.1038/d41586-020-00655-8.

[74] B.E. Givens, Y.W. Naguib, S.M. Geary, E.J. Devor, A.K. Salem, Nanoparticle based delivery of CRISPR/Cas9 genome editing therapeutics, AAPS J. 20 (2018) 108, https://doi.org/10.1208/s12248-018-0267-9.

[75] J.D. Finn, A.R. Smith, M.C. Patel, L. Shaw, M.R. Youniss, J. van Heteren, T. Dirstine, C. Ciullo, R. Lescarbeau, J. Seitzer, R.R. Shah, A. Shah, D. Ling, J. Growe, M. Pink, E. Rohde, K.M. Wood, W.E. Salomon, W.F. Harrington, C. Dombrowski, W.R. Strapps, Y. Chang, D.V. Morrissey, A single administration of CRISPR/Cas9 lipid nanoparticles achieves robust and persistent in vivo genome editing, Cell Rep. 22 (2018) 2227–2235, https://doi.org/10.1016/j.celrep.2018.02.014.

[76] S. Kim, D. Kim, S.W. Cho, J. Kim, J.-S. Kim, Highly efficient RNA-guided genome editing in human cells via delivery of purified Cas9 ribonucleoproteins, Genome Res. 24 (2014) 1012–1019, https://doi.org/10.1101/gr.171322.113.

[77] S. Ramakrishna, A.-B. Kwaku Dad, J. Beloor, R. Gopalappa, S.-K. Lee, H. Kim, Gene disruption by cell-penetrating peptide-mediated delivery of Cas9 protein and guide RNA, Genome Res. 24 (2014) 1020–1027, https://doi.org/10.1101/gr.171264.113.

[78] T.L. Roth, C. Puig-Saus, R. Yu, E. Shifrut, J. Carnevale, P.J. Li, J. Hiatt, J. Saco, P. Krystofinski, H. Li, V. Tobin, D.N. Nguyen, M.R. Lee, A.L. Putnam, A.L. Ferris, J.W. Chen, J.-N. Schickel, L. Pellerin, D. Carmody, G. Alkorta-Aranburu, D. Del Gaudio, H. Matsumoto, M. Morell, Y. Mao, M. Cho, R.M. Quadros, C.B. Gurumurthy, B. Smith, M. Haugwitz, S.H. Hughes, J.S. Weissman, K. Schumann, J.H. Esensten, A.P. May, A. Ashworth, G.M. Kupfer, S.A.W. Greeley, R. Bacchetta, E. Meffre, M.G. Roncarolo, N. Romberg, K.C. Herold, A. Ribas, M.D. Leonetti, A. Marson, Reprogramming human T cell function and specificity with non-viral genome targeting, Nature 559 (2018) 405–409, https://doi.org/10.1038/s41586-018-0326-5.

[79] Y. Wu, J. Zeng, B.P. Roscoe, P. Liu, Q. Yao, C.R. Lazzarotto, K. Clement, M.A. Cole, K. Luk, C. Baricordi, A.H. Shen, C. Ren, E.B. Esrick, J.P. Manis, D.M. Dorfman, D.A. Williams, A. Biffi, C. Brugnara, L. Biasco, C. Brendel, L. Pinello, S.Q. Tsai, S.A. Wolfe, D.E. Bauer, Highly efficient therapeutic gene editing of human hematopoietic stem cells, Nat. Med. 25 (2019) 776–783, https://doi.org/10.1038/s41591-019-0401-y.

[80] R. Rouet, L. de Oñate, J. Li, N. Murthy, R.C. Wilson, Engineering CRISPR-Cas9 RNA–protein complexes for improved function and delivery, CRISPR J. 1 (2018) 367–378, https://doi.org/10.1089/crispr.2018.0037.

[81] R. Rouet, B.A. Thuma, M.D. Roy, N.G. Lintner, D.M. Rubitski, J.E. Finley, H.M. Wisniewska, R. Mendonsa, A. Hirsh, L. de Oñate, J. Compte Barrón, T.J. McLellan, J. Bellenger, X. Feng, A. Varghese, B.A. Chrunyk, K. Borzilleri, K.D. Hesp, K. Zhou, N. Ma, M. Tu, R. Dullea, K.F. McClure, R.C. Wilson, S. Liras, V. Mascitti, J.A. Doudna, Receptor-mediated delivery of CRISPR-Cas9 endonuclease for cell-type-specific gene editing, J. Am. Chem. Soc. 140 (2018) 6596–6603, https://doi.org/10.1021/jacs.8b01551.

[82] N. Dobrovolskaia-Zavadskaia, N. Kobozieff, Sur la reproduction des souris anoures, C. R. Soc. Biol. 97 (1927) 116–119.
[83] S. Gershenson, A new sex-ratio abnormality in DROSOPHILA OBSCURA, Genetics 13 (1928) 488–507.
[84] J. Champer, J. Chung, Y.L. Lee, C. Liu, E. Yang, Z. Wen, A.G. Clark, P.W. Messer, Molecular safeguarding of CRISPR gene drive experiments, eLife 8 (2019) e41439, https://doi.org/10.7554/eLife.41439.
[85] D. Dong, M. Guo, S. Wang, Y. Zhu, S. Wang, Z. Xiong, J. Yang, Z. Xu, Z. Huang, Structural basis of CRISPR-SpyCas9 inhibition by an anti-CRISPR protein, Nature 546 (2017) 436–439, https://doi.org/10.1038/nature22377.
[86] E.M. Basgall, S.C. Goetting, M.E. Goeckel, R.M. Giersch, E. Roggenkamp, M.N. Schrock, M. Halloran, G.C. Finnigan, Gene drive inhibition by the anti-CRISPR proteins AcrIIA2 and AcrIIA4 in *Saccharomyces cerevisiae*, Microbiology 164 (2018), https://doi.org/10.1099/mic.0.000635.
[87] J.J. Bull, OUP: lethal gene drive selects inbreeding, Evol. Med. Public Health 2017 (2016) 1–16, https://doi.org/10.1093/emph/eow030.
[88] N. Windbichler, M. Menichelli, P.A. Papathanos, S.B. Thyme, H. Li, U.Y. Ulge, B.T. Hovde, D. Baker, R.J. Monnat, A. Burt, A. Crisanti, A synthetic homing endonuclease-based gene drive system in the human malaria mosquito, Nature 473 (2011) 212–215, https://doi.org/10.1038/nature09937.
[89] A. Hammond, R. Galizi, K. Kyrou, A. Simoni, C. Siniscalchi, D. Katsanos, M. Gribble, D. Baker, E. Marois, S. Russell, A. Burt, N. Windbichler, A. Crisanti, T. Nolan, A CRISPR-Cas9 gene drive system targeting female reproduction in the malaria mosquito vector Anopheles gambiae, Nat. Biotechnol. 34 (2016) 78–83, https://doi.org/10.1038/nbt.3439.
[90] T.A.A. Prowse, P. Cassey, J.V. Ross, C. Pfitzner, T.A. Wittmann, P. Thomas, Dodging silver bullets: good CRISPR gene-drive design is critical for eradicating exotic vertebrates, Proc. R. Soc. B Biol. Sci. 284 (2017), https://doi.org/10.1098/rspb.2017.0799. 20170799.
[91] B.R. Tershy, K.-W. Shen, K.M. Newton, N.D. Holmes, D.A. Croll, The importance of islands for the protection of biological and linguistic diversity, Bioscience 65 (2015) 592–597, https://doi.org/10.1093/biosci/biv031.
[92] K.M. Esvelt, N.J. Gemmell, Conservation demands safe gene drive, PLoS Biol. 15 (2017) e2003850, https://doi.org/10.1371/journal.pbio.2003850.
[93] E. Yong, New Zealand's war on rats could change the world, The Atlantic (2017).
[94] A. Rezza, C. Jacquet, A. Le Pillouer, F. Lafarguette, C. Ruptier, M. Billandon, P. Isnard Petit, S. Trouttet, K. Thiam, A. Fraichard, Y. Chérifi, Unexpected genomic rearrangements at targeted loci associated with CRISPR/Cas9-mediated knock-in, Sci. Rep. 9 (2019) 3486, https://doi.org/10.1038/s41598-019-40181-w.
[95] A.L. Norris, S.S. Lee, K.J. Greenlees, D.A. Tadesse, M.F. Miller, H.A. Lombardi, Template plasmid integration in germline genome-edited cattle, Nat. Biotechnol. 38 (2020) 163–164, https://doi.org/10.1038/s41587-019-0394-6.
[96] M. Kosicki, K. Tomberg, A. Bradley, Repair of double-strand breaks induced by CRISPR-Cas9 leads to large deletions and complex rearrangements, Nat. Biotechnol. 36 (2018) 765–771, https://doi.org/10.1038/nbt.4192.
[97] G. Cullot, J. Boutin, J. Toutain, F. Prat, P. Pennamen, C. Rooryck, M. Teichmann, E. Rousseau, I. Lamrissi-Garcia, V. Guyonnet-Duperat, A. Bibeyran, M. Lalanne, V. Prouzet-Mauléon, B. Turcq, C. Ged, J.-M. Blouin, E. Richard, S. Dabernat, F. Moreau-Gaudry, A. Bedel, CRISPR-Cas9 genome editing induces megabase-scale chromosomal truncations, Nat. Commun. 10 (2019) 1136, https://doi.org/10.1038/s41467-019-09006-2.
[98] B.V. Skryabin, D.-M. Kummerfeld, L. Gubar, B. Seeger, H. Kaiser, A. Stegemann, J. Roth, S.G. Meuth, H. Pavenstädt, J. Sherwood, T. Pap, R. Wedlich-Söldner, C. Sunderkötter, Y.B. Schwartz, J. Brosius, T.S. Rozhdestvensky, Pervasive head-to-tail insertions of DNA templates mask desired CRISPR-Cas9–mediated genome editing events, Sci. Adv. 6 (2020) eaax2941, https://doi.org/10.1126/sciadv.aax2941.

[99] S. Bae, J. Park, J.-S. Kim, Cas-OFFinder: a fast and versatile algorithm that searches for potential off-target sites of Cas9 RNA-guided endonucleases, Bioinformatics 30 (2014) 1473–1475, https://doi.org/10.1093/bioinformatics/btu048.

[100] K. Labun, T.G. Montague, M. Krause, Y.N. Torres Cleuren, H. Tjeldnes, E. Valen, CHOPCHOP v3: expanding the CRISPR web toolbox beyond genome editing, Nucleic Acids Res. 47 (2019) W171–W174, https://doi.org/10.1093/nar/gkz365.

[101] J. Listgarten, M. Weinstein, B.P. Kleinstiver, A.A. Sousa, J.K. Joung, J. Crawford, K. Gao, L. Hoang, M. Elibol, J.G. Doench, N. Fusi, Prediction of off-target activities for the end-to-end design of CRISPR guide RNAs, Nat. Biomed. Eng. 2 (2018) 38–47, https://doi.org/10.1038/s41551-017-0178-6.

[102] N. Crosetto, A. Mitra, M.J. Silva, M. Bienko, N. Dojer, Q. Wang, E. Karaca, R. Chiarle, M. Skrzypczak, K. Ginalski, P. Pasero, M. Rowicka, I. Dikic, Nucleotide-resolution DNA double-strand break mapping by next-generation sequencing, Nat. Methods 10 (2013) 361–365, https://doi.org/10.1038/nmeth.2408.

[103] S.Q. Tsai, N.T. Nguyen, J. Malagon-Lopez, V.V. Topkar, M.J. Aryee, J.K. Joung, CIRCLE-seq: a highly sensitive in vitro screen for genome-wide CRISPR-Cas9 nuclease off-targets, Nat. Methods 14 (6) (2017) 607–614, https://doi.org/10.1038/nmeth.4278.

[104] D. Kim, S. Bae, J. Park, E. Kim, S. Kim, H.R. Yu, J. Hwang, J.-I. Kim, J.-S. Kim, Digenome-seq: genome-wide profiling of CRISPR-Cas9 off-target effects in human cells, Nat. Methods 12 (2015) 237–243, https://doi.org/10.1038/nmeth.3284.

[105] B. Wienert, S.K. Wyman, C.D. Yeh, B.R. Conklin, J.E. Corn, CRISPR off-target detection with DISCOVER-seq, Nat. Protoc. 15 (2020) 1775–1799, https://doi.org/10.1038/s41596-020-0309-5.

[106] S.Q. Tsai, Z. Zheng, N.T. Nguyen, M. Liebers, V.V. Topkar, V. Thapar, N. Wyvekens, C. Khayter, A.J. Iafrate, L.P. Le, M.J. Aryee, J.K. Joung, GUIDE-seq enables genome-wide profiling of off-target cleavage by CRISPR-Cas nucleases, Nat. Biotechnol. 33 (2) (2015) 187–197, https://doi.org/10.1038/nbt.3117.

[107] R.L. Frock, J. Hu, R.M. Meyers, Y.J. Ho, E. Kii, F.W. Alt, Genome-wide detection of DNA double-stranded breaks induced by engineered nucleases, Nat. Biotechnol. 33 (2) (2015) 179–186, https://doi.org/10.1038/nbt.3101.

[108] X. Wang, Y. Wang, X. Wu, J. Wang, Y. Wang, Z. Qiu, T. Chang, H. Huang, R.-J. Lin, J.-K. Yee, Unbiased detection of off-target cleavage by CRISPR-Cas9 and TALENs using integrase-defective lentiviral vectors, Nat. Biotechnol. 33 (2015) 175–178, https://doi.org/10.1038/nbt.3127.

[109] P. Cameron, C.K. Fuller, P.D. Donohoue, B.N. Jones, M.S. Thompson, M.M. Carter, S. Gradia, B. Vidal, E. Garner, E.M. Slorach, E. Lau, L.M. Banh, A.M. Lied, L.S. Edwards, A.H. Settle, D. Capurso, V. Llaca, S. Deschamps, M. Cigan, J.K. Young, A.P. May, Mapping the genomic landscape of CRISPR–Cas9 cleavage, Nat. Methods 14 (2017) 600–606, https://doi.org/10.1038/nmeth.4284.

CHAPTER 18

Genetic modification of mice using CRISPR-Cas9: Best practices and practical concepts explained

Vishnu Hosur, Benjamin E. Low, and Michael V. Wiles
The Jackson Laboratory, Bar Harbor, ME, United States

The development of precision targetable nucleases has led to a massive acceleration in creating genetically modified mice and other species [1–6]. This began more than a decade ago with Zinc Finger Nucleases (ZFNs) [7], followed by TALENs (Transcription Activator-Like Effector Nucleases) [8–10], and then CRISPR-Cas9 (Clustered Regularly Interspaced Short Palindromic Repeats with CRISPR-associated effector protein 9) [11,12]. At present, RNA-guided CRISPR-Cas9 is the most affordable and straightforward to design, construct, and implement. This accessibility, combined with its generally high degree of targeting efficiency, has pushed CRISPR-Cas9 to the forefront of gene-editing methods. Regardless of its relative simplicity, the complexity of the resulting nuclease-derived genetic modifications, including the modified organism's phenotype, should not be underestimated [13–15].

Herein, we outline our experience using CRISPR-Cas9 to precisely and directly engineer mouse zygotes, focusing on the general methodology and screening used to characterize the resulting alleles. These screening strategies are simple, straightforward, and reproducible. While the focus of this chapter is on CRISPR-modified alleles generated in mice, these screening regimes can be applied to other organisms and to the characterization of genetic modifications resulting from ZFNs, TALENs, or any other gene-editing technology.

Genetically engineered mouse models of human disease

Mice genetically, biologically, and behaviorally resemble humans and can replicate many human genetic diseases. It is now also relatively easy to manipulate their genome, e.g., adding or removing DNA sequences with precision, to better understand the role of a human gene or protein at the organismal level and to apply the knowledge gained directly to the human condition. Additionally, for a mammal, mice have a short generation time, good fecundity, and can be maintained disease-free with relative ease. Also, an often underappreciated advantage they have over many species is that they can be maintained inbred. This last aspect means that each inbred mouse is essentially genetically identical, which eliminates the genetic noise of variation within a given background, thus enhancing experimental reproducibility across time and location. On the other hand, to investigate the influence of genetic variation on a specific phenotype, researchers can take advantage of recombinant inbred strains and highly diverse mouse strains [16,17]. In this section, we briefly outline the power of harnessing precise genetic engineering approaches in mice to accelerate discoveries in biomedical research.

International Knockout Mouse Consortium (IKMC)

In humans, there are about 21,000 protein-coding genes. The goal of the IKMC is to generate null alleles for every protein-coding gene in mice and make the new null strains widely available to biomedical researchers. Although homologous recombination in C57BL/6N (B6N) mouse embryonic stem cells resulted in the generation of many null strains prior to 2015, the IKMC has since switched to CRISPR-Cas9-mediated generation of null alleles by direct editing of B6N zygotes with higher fidelity and specificity [18].

Cancer

Substantial evidence suggests that CRISPR-Cas9 can rapidly generate technically challenging mouse models of human cancer involving multiple mutations. For instance, in Cas9 knock-in mice, simultaneous deletion of the tumor suppressor genes *p53* and *Lkb1* and installation of a missense mutation in the oncogene *Kras* using adeno-associated virus (AAV)-mediated delivery of guide RNAs (gRNAs) targeting the *p53*, *Lkb1*, and *Kras* loci, resulted in a model of lung adenocarcinoma [19]. Likewise, knockout of tumor suppressor genes *Pten* and *p53* and installation of activating point mutations in the *b-catenin* gene, following hydrodynamic injection of a plasmid co-expressing Cas9 and gRNAs, resulted in a mouse model of liver cancer [20]. Additionally, complex chromosomal rearrangements between the genes *EML4* and *ALK*, causing human non-small cell lung cancers, have been precisely modeled in mice following adenovirus-mediated delivery of Cas9 and single gRNAs (sgRNAs) targeting the *Eml4* and *Alk* genes to the lungs of adult mice [21].

More importantly, an extension of the CRISPR-Cas9 toolbox—CRISPRa (transcriptional activation of genes) and CRISPRi (transcriptional silencing of genes)—has been efficiently used to modulate gene expression in mice to test tumor progression and therapeutic resistance. Braun et al. demonstrated that tail vein injection of murine lymphoma cells (*Em-Myc p19Arf$^{-/-}$*) expressing dCas9 and gRNAs targeting the transcription start site of *p53* into B6/J mice accelerated tumor formation and significantly reduced the survival of the mice. Furthermore, transcriptional silencing of *p53* rendered lymphoma insensitive to chemotherapy in vivo [22]. The authors also showed that transcriptional activation of *Mgmt*, an enzyme whose overexpression makes tumor cells insensitive to the chemotherapy agent temozolomide (TMZ), in B-cell lymphoblastic leukemia cells (through expression of dCas9-VP16 fusion protein) enhances resistance to TMZ treatment.

Thus, CRISPR-Cas9-mediated functional modification of oncogenes or tumor suppressor genes via knock-in, knockout, or chromosomal translocations can be applied to study the development and progression of tumors rapidly. Researchers can also successfully employ CRISPRa and CRISPRi to modulate gene expression in mice to induce tumor growth and to examine therapeutic response. Especially with precision medicine efforts on the horizon, conclusive evidence from mouse models will potentially enable clinicians to either repurpose available drugs or employ novel therapeutics with improved confidence.

Alzheimer's disease (AD)

The mouse is a model organism of choice for studying AD. Although previous transgenic mouse models that overexpress proteins linked to AD have failed to recapitulate human AD pathology faithfully, direct genetic manipulation of zygotes in different inbred mouse strains by CRISPR-Cas9 has rapidly advanced the development of more accurate models of AD to analyze the function of genes associated with AD, validate disease-causing variants, and identify the underlying molecular mechanisms [23,24]. For instance, using CRISPR-Cas9 endonucleases, the *MODEL-AD* consortium aims to develop next-generation in vivo AD models based on human data and characterize and disseminate the models to the biomedical research community for preclinical therapeutic testing [25].

COVID-19

The coronavirus disease (COVID-19) caused by the severe acute respiratory syndrome coronavirus 2 (SARS-CoV-2) has resulted in hundreds of thousands of deaths, prompting an emergency response to find cures [26–28]. Mouse models that recapitulate COVID-19 are immediately needed for the preclinical evaluation of vaccines and other conceivable therapies [29,30]. Although mice can become infected with various coronaviruses, the viruses replicate poorly, and mice fail to robustly recapitulate the symptoms and pathology of the disease, including COVID-19. Michael Farzan and colleagues demonstrated that the human angiotensin-converting enzyme 2 (ACE2) is a functional receptor for SARS-CoV and that an anti-ACE2 antibody efficiently blocks viral replication [31]. Later, Stanley Perlman and co-workers generated a human *ACE2* (*hACE2*) transgenic mouse and demonstrated for the first time that SARS-CoV can induce lethal infection in *hACE2* mice, resembling the human SARS condition [32]. Using CRISPR-Cas9, we (unpublished) and others [33] have rapidly generated novel mouse models expressing *hACE2* in more than one background strain, enabling tools for studying COVID-19 pathogenesis and testing how mouse genetic background impacts COVID-19 progression.

Generating mouse models using CRISPR-Cas9

CRISPR-Cas9 nucleases from many bacterial and archaeal strains have been described. They are part of a dynamic innate defense system that protects these organisms against infection by bacterial viruses (i.e., phages) [34]. The most commonly used source of Cas9 is *Streptococcus pyogenes* (abbreviated to SpCas9 or SpyCas9), which has been co-opted and refined into a genetic modification tool for use in virtually any species. When complexed with a gRNA, the SpCas-gRNA complex, or ribonucleoprotein (RNP), induces a double-strand DNA break (DSB) at a precisely defined site. In nature, this targeting gRNA is a complex of two RNAs: crRNA (CRISPR RNA), which confers the targeting specificity of the guide, and the tracrRNA (trans-activating RNA), which provides the scaffold to allow the crRNA to complex with the Cas9 protein [35]. In 2012, Jinek et al. simplified this system for gene editing by fusing these two components into a single guide RNA, referred to as sgRNA or gRNA [12].

As a molecular tool, although either the fused (crRNA:tracrRNA) or the non-fused (sgRNA) versions work, the critical element is that 17–20 bases direct the targeting of the nuclease at the 5′ terminus of the guide, with direct homology to the genomic DNA to be targeted. Critically, for the SpCas9 RNP to function, this homologous region must also be directly proximal to a Protospacer Adjacent Motif (PAM) in the genome. The guide does not contain the PAM sequence (5′-NGG for SpCas9); instead, its presence in the genome functions to aid in the correct and rapid targeting of the specified site by the SpCas9 RNP [36]. Synthetically modified variants of SpCas9 have been designed to utilize different PAM sequences and carry additional DNA modification elements and there is a growing collection of unique Cas nucleases being isolated from other prokaryotes. The net result of these various tools means that essentially *any* desired locus can be targeted using some form of CRISPR.

The SpCas9 RNP complex binds DNA, locates its target, and then executes a DSB precisely three bases 5′ of the PAM. The cellular machinery of the targeted cells (e.g., zygote) is responsible for the repairs that follow this disruptive event. In its simplest form, the blunt-end DSB induced by SpCas9 RNP triggers the cell to repair the DNA damage via the nonhomologous end-joining (NHEJ) repair pathway. As correctly repaired DNA will continue to be targeted by the RNP, NHEJ repair will often result in the eventual insertion or deletion (indel) of a few bases, or less often, many hundreds of base pairs. This slightly serendipitous event is helpful if the goal is to disrupt a targeted gene, as such modifications usually lead to frameshift mutations or, less often, deletion of entire domains. However, the latter is more assured when two guides are used which flank the DNA segment intended to be deleted.

A more precise and versatile modification can result if the DSB is induced in the presence of donor DNA, which gets incorporated into the genome at or very near the cut site. For this to occur efficiently, the donor DNA should have sequence homology to the targeted region so it can be used as a template for repair via the homology-directed repair (HDR) pathway. CRISPR-Cas9-mediated HDR, although less efficient than NHEJ, can result in seamless incorporation of a novel sequence into the genome at or near the cut site (Fig. 18.1). This strategy dramatically enhances the utility of CRISPR-Cas9 beyond simple gene disruption, allowing precise targeted single base modifications (e.g., SNPs) or modification of tens of bases. Nevertheless, insertion of large DNA constructs (>2 kb) using CRISPR-Cas9-mediated HDR is still challenging [37].

A complete overview of CRISPR and its capabilities is outside the scope of this chapter. Our goal here is to introduce the basics of CRISPR-Cas9 for simple genetic modification, principally in mice. We outline the basic approaches and considerations in designing gRNAs, strategies for gene disruption, and cover the critical elements in building a robust screening platform to identify and characterize genetically modified mice. We hope that our chapter will serve as an entry point to the field and a foundation to facilitate the scientific investigation of fundamental biological questions.

Methodology

The majority of CRISPR-mediated genetic modifications to the mouse genome can be categorized into (i) knockout (KO), where the intent is the disruption of a gene or, less commonly, a control region of the genome by indel (insertion or deletion of a few bases of DNA) formation; or (ii) knock-in (KI), requiring the precise insertion of exogenous donor DNA into the genome. For practical reasons, it is useful to further sub-divide these categories into small versus large KOs and KIs, as there are distinct differences in the preparation, handling, and delivery of the reagents, as well as screening of the resulting alleles.

There are many variables involved in the genetic modification of mice. Here we cover the topics we believe are crucial for success. In 2016, we published a set of methods for using CRISPR-Cas9 to modify mouse strains via microinjection (MIJ) [38]. While most of those protocols remain relevant today, there have been critical advances in our methods since that publication. Most notably, there was limited availability of the requisite reagents, at least ones of consistently high quality and affordability.

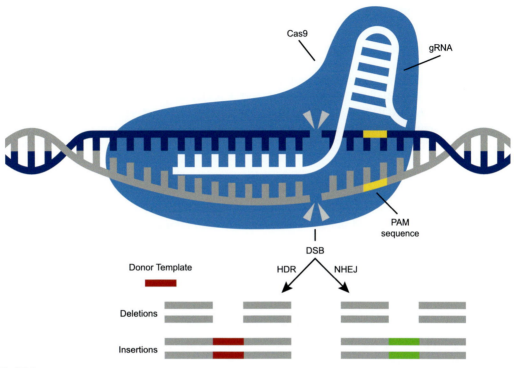

FIG. 18.1

Schematic showing CRISPR-Cas9 mediated genome editing. 17–20-nucleotide gRNA directs Cas9 to cleave its complementary DNA sequence adjacent to the PAM, generating a double-strand break. The cellular mechanism repairs the cleaved DNA by nonhomologous end-joining [NHEJ] *(right)* or homology-directed repair (HDR) in the presence of donor DNA *(left)*.

Specifically, while Cas9 protein was available from a few vendors, it often suffered lot-to-lot variation in performance. Now, Cas9 protein is widely available from many sources and is generally of consistently high quality. This advance coincided with the establishment of electroporation (EP) [39,40] as a reliable alternative to MIJ to deliver CRISPR reagents into mouse zygotes. EP relies on Cas9 protein for its utility. As a result, we have transitioned to using electroporation to generate small KO and KI modifications in mouse zygotes, a technique that was just being refined when we published our prior method set.

We previously used in vitro transcription (IVT) to synthesize our sgRNA and Cas9 mRNA. However, as quality, cost, and turnaround times have all improved, we have since progressed to using commercially available reagents for all of these reagents (guides, mRNA, Cas9 protein). The development and optimization of these resources has greatly improved embryo survivability and targeting success. The availability of these reagents combined with the establishment of EP to deliver CRISPR reagents has proven to be a reliable alternative to the more technically demanding MIJ [39]. A vital but often overlooked requirement regarding the reagents used is that they must be of the highest quality and purity. We have often seen where this has not been meticulously ensured, resulting in poor embryo survival and poor efficiency in obtaining the desired genetic modifications.

Genetic diversity in mice

The approximate timeline for establishing a new mouse line using CRISPR is outlined in Fig. 18.2. As the use of targeted nucleases has made it possible to rapidly modify the genomes of *any mouse strain* directly as zygotes, researchers now have the ability to select the most appropriate background strain for their particular experiment. We have successfully genetically modified several mouse strains, including 129S1/SvImJ, A/J, BALB/cJ, C57BL/6J, C57BL/6NJ, CAST/EiJ, DBA/2J, FVB/NJ, KK/HlJ, MRL/MpJ, NOD/ShiLtDvs, NOD/ShiLtJ, NZO/HlLtJ, PWK/PhJ SJL/J, and WSB/EiJ. Furthermore, we have performed sequential modification of previously genetically engineered mutant mouse (GEMM) strains, e.g., the immunocompromised strains NRG (NOD-$Rag1^{null}$ $IL2rg^{null}$) and NSG (NOD-*scid* $IL2rg^{null}$).

While CRISPR-Cas9 enables modification of the genomes of any mouse strain, it is important to recognize the existence of inherent strain-specific challenges. For instance, strains with low fecundity, little or no response to superovulation, or cannibalistic tendencies, may require additional resources and time to generate the desired mutant mouse line. It is also critically important to ensure that the reference sequence used when designing the guides and developing the genotyping assays accurately represents the chosen strain. Verifying that the sequence targeted in the strain of choice matches the reference (e.g., by PCR and Sanger sequencing) and contains no polymorphisms that will interfere with

Week #	Process
Week 0-3	Design guides, obtain reagents
Week 3	Perform EP
Week 6	Birth of P_0 offspring (mosaic "parental founders")
Week 9	Wean P_0's and screen
Week 12	Backcross select P_0 candidates
Week 15	Birth of N1 offspring
Week 18	Wean N1's and characterize resulting alleles
Week 21	Mate established N1 founders (second backcross or intercross)

Slower, but cleaner (preferred)

Week #	Process
Week 24	Birth of N2 offspring
Week 27	Wean N2's and screen
Week 30	Intercross N2's to homozygose (N2F1's)
Week 33	Birth of N2F1's and screen
Week 36	Wean N2F1's and screen
Week 39	HOM x HOM matings (N2F2's)

*Faster, but more chance of problems**

Week #	Process
Week 24	Birth of N1F1 offspring
Week 27	Wean N1F1's and screen
Week 30	HOM x HOM matings (N1F2's)

*e.g., fixing off-target mutations, compound heterozygosity, etc.

FIG. 18.2

Timeline for generating CRISPR-Cas9 mouse models. A slower *(left)* and a faster *(right)* approach for establishing new mouse lines.

the performance of the RNP will help prevent easily avoidable failures. Developing these assays before the CRISPR modification is attempted will also ensure the locus is correct and open to PCR amplification and Sanger sequencing, critical for characterizing the resulting alleles.

Reagent delivery to the mouse zygote

A significant rate-limiting step in CRISPR-mediated genetic modification of mice is delivering the reagents *into* the zygote. The primary barrier is the zona pellucida, a rigid extracellular matrix designed to protect the egg from many insults. As noted above, the two strategies for delivering reagents into the zygote are microinjection (MIJ) and electroporation (EP). MIJ is the older and more conventional approach and has few limitations in terms of what can be delivered to the zygote. It involves the physical insertion of a very fine glass needle into the zygote's cytoplasm, or pronucleus, delivering 1–10 picoliters of reagents. However, the approach requires expensive MIJ equipment with dedicated, well-trained staff. With EP, conditions have been established to deliver gRNAs complexed with SpCas9 protein—the RNP—into the zygote. Embryo survival and genetic modification appear higher with EP than MIJ, although more direct experimental comparisons are needed to confirm these findings. Moreover, most laboratories can perform EP with minimal investment, following published protocols and using off-the-shelf reagents. As a result, EP has become the method of choice for generating KO, dropout alleles, and small knock-ins.

In spite of multiple attempts, to date there have been no reports of successful generation of any large transgenic alleles in mice made using EP. Oligonucleotide (ssDNA) donors (<200 nt) as well as larger, long single-stranded DNA and even linearized dsDNA templates (<1 kb) have all been reported to successfully generate mutant alleles, but there have been no reports of successful EP-generated mice using a supercoiled plasmid DNA donor. Thus, if the intention is to generate a larger KI allele (>150 bp), then pronuclear MIJ is currently the most proven way to achieve this result. Regardless of the technique used, researchers must be familiar with mouse embryo generation, handling, and surgical implantation of modified zygotes into pseudopregnant females to bring mice to term. These techniques are described in great depth and detail in Manipulating the Mouse Embryo: A Laboratory Manual [41].

Guide design

Careful design of the guide is crucial to the success of any CRISPR project. There are currently various tools available to assist in the design of the guide for targeting your locus of choice [38] and many of these programs are free and accessible online. One such program, and our preferred guide design tool currently, is called Breaking-Cas [42]. Linked to ENSEMBL, this free program allows for the design of guides across many species, including different strains of mice. It returns guides ranked not only by their predicted on-target cutting efficiency but also the likelihood of off-target cutting. Further, the on- and off-target cuts predicted to result are shown in their genomic context (i.e., intergenic/intronic/exonic sequence). Thus, the researcher can make an informed decision about the consequence of the lesion if the off-target cut occurs. Breaking-Cas allows the guide to be designed for various Cas proteins, utilizing different PAMs, and even has a "custom" feature to define the PAM.

The proto-typical SpCas9 guide has 20 nucleotides of homology for targeting, adjacent to the "5′-NGG" PAM. However, we design the majority of our guides to have only an 18 base homing sequence. This truncated guide or "tru-gRNA" strategy, first described by Fu et al. [43], has been demonstrated to increase specificity and reduce the likelihood of off-target DNA damage [44]. However, one should note that while these tru-gRNAs work well with standard Cas9 protein, they may not work with some of the modified or enhanced versions of Cas9. Consequently, we use tru-gRNAs exclusively with standard Cas9 protein (PNABIO, catalog # CP01) or standard Cas9 mRNA (Trilink, catalog # L-7206).

When screening potential guides for targeting a desired region of interest, our primary consideration is to minimize potential off-target effects. In practice, we select the tru-gRNA target site unique to the locus, which cuts as close as possible to the site of our desired modification, and has the fewest number of potential off-target sites with 0, 1, or 2 mismatches. If designing a conventional guide (20 nt), then the predicted off-target events with 3 or 4 mismatches should be considered more likely to occur than what we have outlined here, which is specific for tru-gRNA guides. While Breaking-Cas will return scoring for off-targets predicted from up to 4 mismatches, the likelihood of an off-target event resulting from even a 3-base mismatch to the tru-gRNA is extremely low. The ability to quickly examine the genomic context of any such off-target site identified by Breaking-Cas is one of this program's greatest attributes. When examining the off-target events predicted for a given tru-gRNA, four primary factors warrant careful consideration:

(1) The number of events with 0 base mismatches. These can result from attempting to target a multi-copy gene (e.g., gene duplication, pseudogene) or trying to target a highly conserved/repetitive domain. Avoid having any of these, if possible, as they are highly likely to occur.
(2) The number of events with 1 or 2 base mismatches. These are less likely to occur than the on-target event but should be treated as if they will happen.
(3) Site of the predicted off-target event(s) with 0,1, or 2 base mismatches. Assuming these do occur, are they likely to disrupt a gene or otherwise impair the animal?
(4) Chromosomal context of predicted off-target events with 0,1, or 2 base mismatches. If these occur, how likely are they to segregate when the P_0 animal is backcrossed?

Ultimately, guide selection is a project-specific and subjective decision based on careful consideration of these and other critical factors. As mentioned previously, the likelihood of off-target events resulting from a well-designed tru-gRNA is low. However, one can choose to screen for any or all of the potential off-target events indicated by the software. The combination of tru-gRNAs with careful design and the use of at least one back-cross to segregate any unwanted alleles will minimize the likelihood of any off-target events.

Fig. 18.3A shows an example of an off-target analysis for a guide designed to target the *Il2rg* gene on the X chromosome. Predicted off-target events are listed in order from the highest to the lowest likelihood. In this case, the most likely off-target event to occur results from a 2-base mismatch with the guide, indicating that no off-target events from 0 or 1 base mismatches are predicted for this guide. Scrolling down on the interactive website (*not shown*) illustrates that a total of only four off-target events are expected with two base mismatches to this particular guide. Further, each of these off-target events will cut in either intronic or intergenic sequence and are unlikely to impact the mouse, even if they occur. Finally, none of these predicted off-target lesions will occur on the same chromosome as *Il2rg* and are likely to segregate when the founder animal is back-crossed. It is helpful to tabulate the results of each guide evaluation, as shown in Fig. 18.3B, and in this case, it is clear that there is little cause for concern from off-target events when using this guide.

It may be worth noting that historically, many guides were preferentially designed to begin with a 5′-G. If the desired target site did not contain one, then the guide would be artificially pre-pended with

A

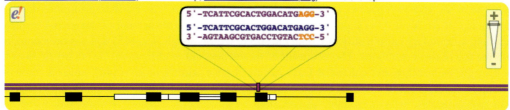

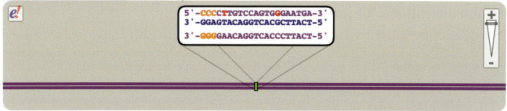

B

Off-target Evaluation	sgRNA = TCATTCGCACTGGACATG|AGG (Chr X)	Note
off-targets with 0 base mismatches:	None	Good
off-targets with 1 base mismatches:	None	Good
off-targets with 2 base mismatches:	Only 4. Scores = 9, 8.3, 5.8, 5.2 (all low)	Okay
Site of predicted off-targets:	Intronic (Ccser1); intergenic; Intergenic; Intronic (Zfp395)	Okay
Chromosomal context:	Chr 6; Chr 1; Chr 15; Chr 14	Good

FIG. 18.3

Guide evaluation using Breaking-Cas. (A) Partial screenshot from an interactive Breaking-Cas off-target analysis result. In this example, the guide designed to target Il2rg is shown with a score of 100, while the next most likely locus to be cut by this guide (with a score of 9 for comparison) contains a 2-base mismatch, indicated in *red*. If this predicted off-target event occurred, it would result in a lesion within the intronic sequence of Ccser1 on chromosome 6. As a result, this indicated off-target event is not a cause for concern as (1) it has a low chance of occurring due to the number of mismatches and the fact that this is a tru-guide, (2) if it does occur, it would hit intronic sequence and therefore be less likely to disrupt the natural function of Ccser1, (3) it is on a distinct chromosome from the target (Chr. 6 vs. Chr. X) and so any off-target events that did occur at this site would most likely segregate during the backcross(es). (B) Results of OT evaluation are summarized in table form, including three other off-target predictions not visible in the screenshot above.

one. However, this is not an actual requirement for good guide design. Instead, it is an artifact of the T7-based in vitro transcription (IVT) kits that are most commonly used to make guides in the lab. Now that there are a variety of companies from which you can economically source synthetic guides, this is no longer a necessary consideration unless you intend to make the guides yourself.

The rapid availability of chemically synthesized RNAs of lengths suitable for use as guides has been one of the critical advances facilitating the use of CRISPR-Cas9. Further, a more in-depth understanding of RNA's in vivo stability is leading to the continual refinement of the gRNA sequence, including a secondary hairpin structure onto the spacer region of sgRNAs [45] that has been combined with the use of synthetic bases. The net result of these chemical modifications is a dramatic improvement in gRNA stability and enhanced Cas9-mediated gene editing in vivo [46]. Several companies currently offer chemically modified gRNAs, and many are actively working on engineering gRNAs and Cas9 protein that are even more stable and efficient.

Key considerations for generating KO alleles

The most common strategy for gene KO by CRISPR is to target an early exon to elicit a frameshift mutation resulting in a premature termination codon, relegating the mutant transcript to be destroyed by the nonsense-mediated decay pathway. While this *can* be an effective KO strategy, it does run the risk of gene rescue by downstream alternative initiation codons or alternative splicing over the mutation using a cryptic initiation site 5′ of the lesion (Fig. 18.4). We suggest that the generation of a KO allele in mouse zygotes is best achieved through EP of Cas9 protein complexed with *two* guides—the guides should flank a domain critical to the function of the gene or be designed to remove the entire gene (i.e., dropout allele).

While it may be obvious, it is worth emphasizing that it is crucial to understand as much as possible about the gene to be targeted when designing the gene knockout project. Before you begin, consider the consequences of the gene KO. Ensure that the gene is not haploinsufficient, embryonic lethal, or essential for reproduction. Be aware of the broader context of the locus you are targeting and consider the potential for unintentional disruptions of overlapping or neighboring genes that could complicate the resulting phenotype. As mentioned previously, be sure to use the correct reference sequence for the mouse strain to be targeted when designing the guide and screening assays. If it is not available, then sequence the target locus to ensure no SNPs are present that could prevent the guides from binding. Finally, ensure that PCR can easily amplify the targeted regions so a high-quality product can be reliably obtained. For example, regions of high GC content, or low complexity repeats, should be avoided if possible.

While it is difficult to prove a negative result, best practices dictate that every effort to confirm the KO should be made when characterizing any new KO mouse model. Ideally, this is done at all three levels: DNA, RNA, and protein. Initially, a simple PCR across the cut site followed by Sanger sequencing of the amplicon can identify mosaic animals with the most favorable alleles. After germline transmission is achieved, the exact modification of each mutant allele can be characterized, and the result of the mutation can be predicted. Nevertheless, there is no guarantee that a hypomorph will not result even when an apparent frameshift mutation is identified.

Evidence for actual expression of the mutant allele can be acquired by looking for the presence of its mRNA in the appropriate tissue. Since this typically requires sacrificing an animal to collect the

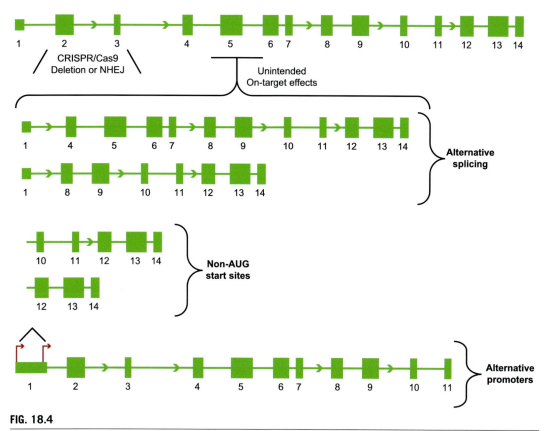

FIG. 18.4

Unintended on-target effects. CRISPR-Cas9 mediated genome editing can result in unwanted on-target effects resulting in expression of novel mRNA isoforms with functional phenotypic consequences. Alternative splicing, alternative AUG and non-AUG translation, and alterative promoters can result in truncated mRNA transcripts.

necessary tissue sample, transcript analysis is generally performed after the allele has been homozygosed and extra animals are available from the established colony. This can be achieved using RACE (Rapid Amplification of cDNA Ends) to generate cDNA templates for subsequent PCR analysis. The resulting amplicons can then be subjected to Sanger sequencing and compared to the transcript from a wild-type control animal. It is also recommended that several PCRs be performed to provide the greatest level of coverage and allow the assays the most opportunity to capture any unexpected transcripts.

At the protein level, a western blot can be used to determine if the mutant allele is translated, assuming a good, specific antibody exists for the protein generated by your gene of interest. Similar to the RNA and DNA screens, the antibody's epitope must be outside of the targeted region, ideally, 5′ of the mutation, so that it can detect any hypomorphs with alternative amino acid sequences downstream of the edited site. Alternative methods for protein detection (e.g., ELISA, flow cytometry, immunohistochemistry) can be used, though the same epitope and specificity considerations will apply.

In sum, the best practice for confirming a complete gene knockout is to select three or more promising alleles as predicted from the DNA sequence, breed to homozygosity *after* two back-crosses, then screen by RT-PCR or western to demonstrate the gene is truly knocked out. At this point, phenotypic analysis and comparison between the lines can be performed to complete the characterization of your new gene knockout mouse strain.

Verification of KO alleles

For a simple KO using just one guide, a single PCR followed by Sanger sequencing will suffice to elucidate the generated alleles. However, even in this simple situation, complications can arise. For example, if the region to be targeted is of low complexity, contains repeat elements, or has high GC content, the PCR may be challenging to perform. The resulting sequence may be difficult or impossible to interpret. Thus, it is highly recommended that the genotyping PCR be designed, performed, and optimized once the guide has been developed but before committing to having it made. Notably, the PCR product should be sequence-verified to ensure the readout will be coherent, and all design work should use the reference sequence for the *exact strain* to be targeted. Furthermore, optimizing and sequencing the PCR product using genomic DNA *directly from the intended strain* will confirm that no polymorphisms exist that could prevent either the guide or primers from binding.

As a general rule, positioning each PCR primer at least 400 bp from the cut site will capture virtually all KO alleles that tend to result from a CRISPR-induced DSB (Fig. 18.5). Moving primers further away is entirely acceptable, though as the PCR product increases in size, the product yield is likely to decrease, resulting in lower quality sequence chromatograms. High background signal will complicate the analysis of the resulting sequence data, and thus having a robust PCR product will make the screening easier to perform. As a rule of thumb, high-quality Sanger sequence data typically initiates approximately ~50 bp from the sequencing primer used and begins to decline after ~500 bp. This varies between facilities and the nature and quality of the amplicon, but it is an essential consideration when constructing the genotyping assay. To accurately deconvolute the mutant allele, high-quality sequencing peaks must be evident before the boundaries of the mutation, which will appear as overlapping peaks and can be confused with background signal (Fig. 18.6).

With a PCR product under ~1 kb, the PCR primers can be re-purposed as sequencing primers. However, as amplicon size increases, it may be necessary to design nested sequencing primers closer to (but still >200 bp away from) the targeted site. Nested primers can also be utilized if high background signal results from sequencing with the PCR primers, even on a smaller product. Using a nested primer just a few bases in from the PCR primer can often resolve low-quality sequencing data. A common cause of background signal is the inclusion of off-target amplicons in the non-homogenous PCR product. Purification of the PCR product before sequencing is also helpful in this regard, as smaller off-target amplicons will likely be removed along with primers and other contaminants. It is highly recommended that PCR products always be purified before sequencing.

Verification of dropout alleles

Including a second guide to induce a dropout will require the design of three PCRs in total, with all of the same recommendations described above (Fig. 18.7A and B). In this case, we recommend first attempting to design a long PCR that has a forward primer (F1) ~400 bp away from the *first* cut site and a reverse primer (R1) ~400 bp away from the *second*. After these primers are selected, then the second and third PCRs can be designed, ideally re-using the same primers (i.e., F1 + a new reverse primer

Key considerations for generating KO alleles 437

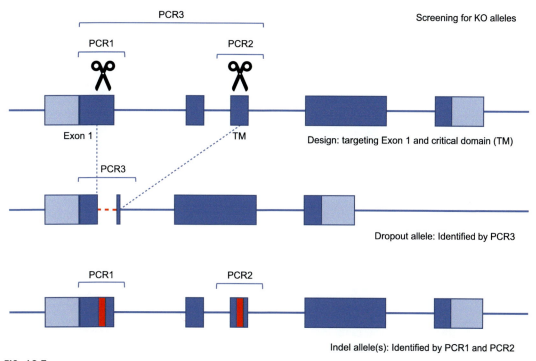

FIG. 18.5

Designing PCRs for KO alleles. There is potential for other more complicated alleles to result when targeting with two guides (e.g., duplication, inversion, insertion, etc.), which can complicate the subsequent sequence analysis of the PCR3 amplicon. These alleles can be resolved by sub-cloning (to isolate the individual allele before sequencing) or characterizing the allele after being bred to homozygosity. The same strategy can be applied for more complicated INDELs at individual cut sites if they prove too challenging to resolve by MDOC (manual deconvolution of chromatograms) analysis of the mosaic or heterozygous animals.

approximately ~400 bp from the first cut site; and R1+a forward primer ~400 bp from the second cut site). However, this is not always possible, and so three completely independent PCRs may need to be designed to build the most robust assays. All three of these PCRs will need to be performed on the putative mutants to assess the resulting alleles fully.

The dropout assay should be designed for the anticipated size of the dropout allele as the wild-type allele may be too large to produce a product. For example, if the two guides target sites 25 kb apart, the dropout PCR product from the WT allele is likely to be ~26 kb, but the expected dropout allele size will only be ~1 kb. That band should only be present if the dropout has occurred in that sample. So, optimization of this PCR before screening the actual potential mutants is not feasible. To distinguish true dropout allelic PCR products from any off-target amplicons, it is recommended that the dropout PCR is optimized using a temperature gradient so the PCR conditions can be selected to avoid false positives (i.e., screen the P_0 mice using the lowest annealing temperature that does not produce a band from WT DNA).

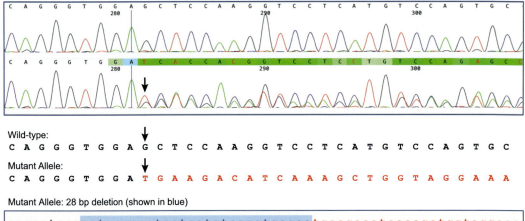

FIG. 18.6

Manual deconvolution of chromatograms. High-quality sequencing peaks from the wild-type and heterozygous alleles enable identification of overlapping peaks.

Key considerations for generating small knock-in alleles

The most proven method for generating small knock-in alleles (<150 bp) is the use of a single-stranded DNA (ssDNA) oligonucleotide donor, delivered with the RNP by EP. The donor oligo should have homology arms of at least 30 nucleotides flanking the insert, and it should be designed such that the desired allele will not be re-cut by the RNP. The potential for random off-target integration of the oligo should be considered, and the lowest effective concentration must be used to minimize this possibility.

Verification of small knock-in alleles

Small KI alleles can be assessed in the same manner as a KO allele (Fig. 18.7C). In this case, however, at least one, but preferably both, of the PCR primers must be outside of the homology arms (HA) of the donor DNA to avoid false-positive calls resulting from the detection of an off-target integration. Using an oligonucleotide donor, typically <200 nucleotides in length, is not an issue as the 400 bp recommendation more than accommodates those HAs. However, if long ssDNA is used as the donor, then the length of the HAs must be factored into the design. Ideally, both primers should be ~100 bp outside of each HA to ensure no other sequence modifications resulted from the HDR event. Verifying that the sequence is intact across the junctions of each HA and through the KI will allow for the most accurate determination of the mutant allele.

An In/Out (IO) PCR, where one primer is in the genomic region and the other is in the exogenous KI portion of the DNA donor, can be designed if the PCR across the entire donor DNA sequence is not

Key considerations for generating small knock-in alleles 439

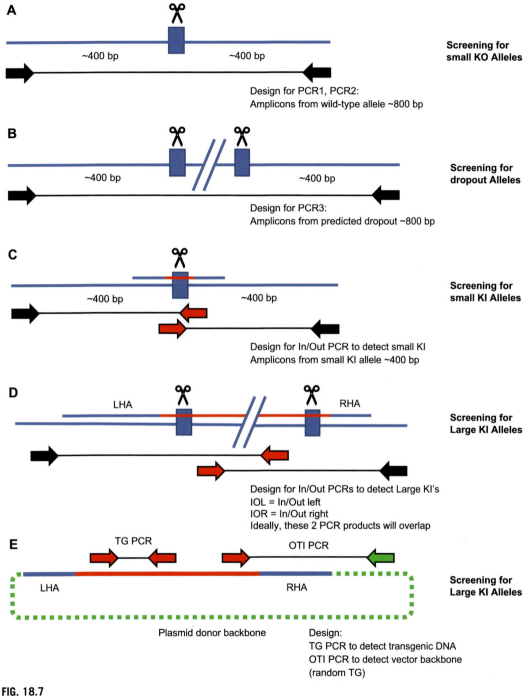

FIG. 18.7

Designing PCRs for CRISPR-modifications (A–E). Design genomic PCR primers to be at least ~400 bp distant from each targeted cut site and outside of any HA in the donor.

amenable to a robust assay (Fig. 18.7C). In this case, however, it is recommended that two overlapping PCRs—one across each HA/gDNA junction—be performed to characterize the complete allele fully. This type of assay is also beneficial for identifying P_0 candidates when the presence of the KI allele is not evident from the sequence chromatogram due to obfuscating mosaicism. Like the significant dropout PCR assay, no control will exist for this allele before it is generated in the mouse. We recommend performing the temperature gradient with WT DNA, and like the dropout PCR, using the lowest annealing temperature that did not produce any bands for screening the P_0 mice.

Key considerations for generating large knock-in alleles

It is *possible* to generate larger knock-in alleles (up to a few kb) using a dsDNA plasmid donor delivered by MIJ with the RNP into the pronucleus of fertilized oocytes; however, the efficiency is highly variable. Homology arms of at least 100 bp (often much larger, up to 5 kb) should flank the exogenous sequence, and the desired allele should not be able to be re-cut by the RNP. Optimization of the conditions for generating large KI alleles continues to be an active area of research. Indeed, the ability to use CRISPR to consistently create a large KI allele (L-KI) at *any* locus with efficiencies greater than 5% remains an elusive goal.

However, critical to the success of any MIJ project with CRISPR is the production of high purity, RNase-free prep in MIJ buffer (10 mM Tris/0.1 mM EDTA pH 7.5). In our hands, we have found the most success using plasmids generated via midiprep followed by phenol-chloroform purification to remove any traces of RNase and other contaminants. After verifying the quality of the DNA, it should be stored at $-20°C$ at a high concentration. On the day of MIJ, using a hard spin (20,000×g for 10 min at 4°C) to pellet any micro particulates, followed by a dilution using a small volume of the supernatant, will yield the highest quality MIJ prep. In turn, this will reduce the damage or toxicity to the zygote, increase the survival to liveborn numbers, and afford the highest chance of success for the large KI allele to be generated.

Verification of large knock-in alleles

As for the other types of modifications described previously, multiple PCR assays can be used to identify successful L-KI allele candidates rapidly (Fig. 18.7D). First, one transgene-specific (TG) PCR can be performed to narrow down the number of animals to screen further (Fig. 18.7E). These PCR assays can be designed for any length and against any relevant sequence of the KI, but they must be robust and unique to the transgene. However, it is helpful to design these assays for double-duty, i.e., to sequence-confirm that critical regions of the KI are as expected in the final allele (e.g., the coding sequence of the transgene). Multiple, overlapping TG PCRs can be designed for complete sequence verification later, though only one robust TG PCR needs to be performed to detect the presence of the transgene in the P_0 animals. Ideally, the TG PCR designed for this purpose should be relatively small (~500 bp), making it more robust and sensitive to detect low-efficiency KI alleles within the mosaic animals.

Plasmid DNA donors run the inherent risk of integrating randomly into the genome, and so these off-target integrations (OTI) need to be distinguished from the desired KI allele. An OTI PCR should be designed against a portion of the vector backbone to identify the presence of these random transgenics. The use of a generic PCR, for example, against the antibiotic resistance gene in the vector, is discouraged as this can lead to false-positive detection due to persistent plasmid contamination common to many molecular biology labs. Instead, an In/Out PCR approach should be used to design the OTI PCR, where one primer is unique to the project-specific portion of the DNA donor and the other primer recognizes the prokaryotic sequence of the vector backbone. This ensures the OTI PCR assay utilizes primers to generate a specific size band if this particular donor plasmid is detected in the sample.

Since random transgenics result from the shearing of the vector at any point before incorporation and often form complicated rearrangements and concatemers, there is always a chance that a single OTI PCR assay could fail to detect a random TG allele. To minimize this potential for false-negative calls, these amplicons should be small (~200–400 bp), ensuring a robust and sensitive PCR. For additional assurance, it is recommended that two OTI PCRs be designed and used, one on each side of the vector backbone sequence, i.e., OTI-L (Left) and OTI-R (Right). While the coincidence of an OTI and a KI allele will complicate the phenotype of the resulting mouse, it is not necessarily detrimental to the ultimate success of the CRISPR-KI project. There is always the potential that the OTI and KI alleles can segregate during breeding. Detecting and tracking the random TG allele by OTI PCR can be used to rescue the project and ensure that the KI mouse line generated is free of this undesired sequence.

Plasmid DNA donors used for generating L-KI alleles by homologous recombination (HR) require long HAs, typically >500 bp, and can be many kilobases in length. Thus, In/Out PCRs across these HAs will be longer than described for small KIs, and it may be difficult to generate robust overlapping In/Out Left (IOL) and Right (IOR) PCRs. If these are not feasible, then at a minimum, the IOL and IOR screening PCRs should encompass the length of the HA, and the internal primer must bind to the exogenous portion of the KI. Complete verification can be ascertained subsequently, as these PCR screens at the P_0 stage will be sufficient to identify which candidates to breed.

Because no positive control for the KI allele will be available until the actual mouse with the allele is generated, PCR optimization gradient regimes should follow the outline described for dropout, TG, and OTI assays. Thus, when screening for the allele of interest in P_0 animals, use the lowest annealing temperature that did not produce a band when the temperature gradient was performed with WT gDNA. After the mutant allele has been identified in the mouse genome, then these assays can be re-optimized using the novel sample(s), if desired.

To summarize the preparatory steps for initiating a CRISPR project, see the following *Practical methods* section. Note that all screening PCR assays should be optimized and, where applicable, sequence-verified before committing to the more expensive and time-consuming aspects of the experiment (i.e., MIJ or EP). Careful planning, design, and optimization can help avoid easily preventable project failures and ensure a smooth and coherent process for mutant allele detection, verification, and classification.

Practical methods
1. Define the project goals and decide on the region to be targeted.
2. Develop a strategy to feasibly achieve the project goals.
3. Build a reference file (e.g., in SnapGene) of the region to be targeted. It will be subsequently annotated with critical information, i.e., open reading frames, gRNA designs, PCR primer information, etc.
    a. Ideally, build the reference by importing the gene sequence information directly from a database (e.g., NCBI), which will include information on all known isoforms, allowing for a more coherent picture of the result of the intended modification using the *correct* species/genetic background;

b. Annotate the reference file with important information about the target gene, including open reading frames and critical domains. Sources of this information can vary, but MGI (Mouse Genome Informatics), Ensembl and UNIPROT are useful websites for this purpose;
c. If targeting a strain other than the source of the reference, modify the sequence file with all pertinent polymorphisms, SNPs, insertions, deletions, etc. This information is often available from MGI, as well as the Sanger Institute. If not, it is suggested that this be acquired by PCR and Sanger sequencing the target region using DNA from the source strain.

4. Design gRNA(s) using, e.g., Breaking-Cas and evaluate them for potential off-target impact.
5. Design all necessary PCRs using program of choice (e.g., IDT PrimerQuest tool) and utilize virtual PCR to assist in selecting best PCR assay [47].
6. Optimize PCRs on genomic DNA to be targeted and analyze by Sanger sequencing to verify: (a) The assay works as expected; (b) Sequence quality is high; (c) Guide recognition site is intact, as expected.
7. Order synthetic gRNA(s), and order and prepare donor DNA (oligonucleotides and plasmids).
8. Perform genetic modification by MIJ or EP into mouse zygotes.

Steps for screening mice. Screening and general timeline is outlined below and in Fig. 18.2.

1. Mice will be born ~20 days post implantation.
2. At wean, ~3 weeks of age, identify animals and collect tissue biopsy as ear notch or tail tip.
3. Extract DNA as crude lysis (purification is rarely necessary).
4. Perform PCR(s) followed by agarose gel electrophoresis using ~1/3 of the PCR product.
5. Purify remaining PCR product and use to perform Sanger sequencing.
6. Analyze chromatograms aligned to a reference:
   - For P_0 animals, simply approximate the size and position of the mutation(s);
   - At N1, thorough manual deconvolution of chromatograms can be applied to determine the precise sequence of the mutant allele. Mosaicism remains a challenge and it is important to consider that just because an allele is detected in the tissue sample from the founder animal, there is no guarantee that it will transmit through the germline. Thus, it is best to wait for N1 offspring before investing significant time and effort in characterizing the allele(s) generated.
7. Rank P_0 candidates based on the approximate allele information; identify N1 animals carrying the desired modification most clearly.
8. Back-cross P_0 candidates to wild-type to generate N1 heterozygous animals.
9. Back-cross N1 heterozygous founders to wild-type cohorts to begin establishing the mutant mouse line with the unique allele.
10. Intercross N2 animals to generate N2F1 homozygous animals.

Mosaicism

When using CRISPR-Cas9 to genetically modify mice, the active reagents are introduced into the zygote. The presumption is that the CRISPR-Cas9-mediated event occurs at this stage, leading to an organism with a single change in its genome which is reflected in all its tissues as an adult. Unfortunately, the resulting founder (P_0) animals are often a mosaic of different modifications to the target allele. This is the result of different CRISPR-mediated events (NHEJ) occurring as the mouse early embryo undergoes cleavage, leading to a collection of anywhere from 2 to ~8 or more alleles in a single animal. While this can be problematic, particularly when screening the P_0 animals, mosaicism is relatively easy to deal with in mice. By simply breeding and identifying the desired alleles in heterozygous mice from the N1 generation, mosaicism is eliminated. Although the P_0 founder animal itself is a mosaic, only *one allele* can necessarily be present in its individual (haploid) gamete (sperm or egg). As a result, it is *always* necessary to back-cross at least one generation in order to isolate and fully characterize the resulting modifications and eliminate mosaicism, and P_0 animals should *never* be intercrossed.

Unintended consequences

It is necessary to consider the potential for unintended DNA damage resulting from the process of any attempted germline editing. Particularly, in our view, certain unintended events can be minimized by simply delivering reagents at the lowest effective dose, ensuring proper design to limit off-target DSBs, selecting multiple founders, and backcrossing the founders to segregate undesired genetic modifications.

Off-targeting events can occur but are circumventable

Possible off-target effects, including large deletions [14], insertions [48], chromosomal rearrangements [49], and random integration [50,51], have been reported to occur at a low level due to CRISPR-Cas9 [52]. However, with foresight, the most blatant of these can be forestalled or screened against. This begins with the choice of region within the genome (gene) and gRNA. For most mouse backgrounds, their genomic sequence is publicly available and can be used to help reduce direct off-targets. Notably, several tools are available to detect both in vitro and in vivo off-target activity [53,54]. Also, see *Methodology* section above, which is systematically optimized to mitigate off-target effects by (1) employing optimal guide design, (2) limiting the concentration of Cas9 and gRNAs [55], (3) utilizing a gRNA with a truncated recognition sequence of 17–18 nucleotides [43], and (4) adding a secondary hairpin structure onto the spacer region of sgRNAs [45]. When necessary, our comprehensive analysis of N1 mice involves exome sequencing, 5′ RACE, western blotting, and functional rescue assays [13]. However, it still has to be conceded that collateral damage to the genome may occur and, as such, we suggest moving forward with at least two different founder lines.

Unintended on-target effects are likely but preventable

The CRISPR-Cas9-mediated generation of null alleles in mice is highly efficient. However, recent literature, using in vitro and in vivo systems, underscores the limitations of Cas9 for generating null alleles. For instance, in mouse embryonic stem cells and mouse zygotes, single sgRNAs can induce unwanted deletions of up to 600 bp, whereas simultaneous injection of two or more sgRNAs targeting larger flanking sites can produce larger deletions (up to 24.4 Mb) [14,48,56]. Interestingly, although rare, evidence also exists for insertion of nearly 1 kb DNA fragments into mouse zygotes [48]. Perhaps a more important consideration is exemplified in our recent work where we characterized a *Rhbdf1* mutant mouse strain generated by Cas9-mediated deletion of exons 2 and 3, containing the start codon (ATG). Surprisingly, we observed that the complete deletion of exons 2 and 3 did not yield null mice, but instead resulted in a novel mouse strain expressing mRNA isoforms lacking exons 2 and 3.

Furthermore, these novel transcripts utilized downstream ATG or non-ATG start codons to generate truncated proteins with phenotypic consequences [13].

Strategies to limit and detect off-target and unintended on-target effects, including guide selection, checking for incomplete coding sequences [13], design of long-range and in-out PCR, whole genome sequencing, exome sequencing, western blot, and heterologous expression in cell lines, are recommended and have been previously reviewed [51,53,54,57].

Conclusions and future perspective

CRISPR is a fast, dynamic, and innovative field. It is no longer just a precision-targeted DNA cutting tool. Numerous versions of Cas, including natural and synthetically modified variants, allow several specialized DNA modifying options. These include (a) "Dead" Cas9 (dCas9) that binds but does not cut the DNA, (b) CRISPR Nickase, which "nicks" or cuts only one strand of the targeted DNA, (c) Base editors (BE), which are designed to make precise point mutations without inducing DSBs, and (d) Prime editors (PE), which are similar to BEs but can facilitate more extensive modifications. Further, new Cas orthologs are continually being discovered and described [58]. CRISPR targeting capabilities are also being used to modulate gene expression, visualize targeted sequence (live imaging of DNA/RNA), cause epigenetic manipulation, and develop treatments [59]. CRISPR is also being modified for use as a molecular biology tool, e.g., in chromatin immunoprecipitation, sample enrichment for library preparation [60], and excitingly, diagnostic assay systems [61].

Cas9 variants

There are several new and emerging technologies for genome editing in mice. To expand the repertoire, reduce off-target effects, and improve on-target efficiency of Cas9, researchers have engineered several Cas9 variants—nickase Cas9 [62], catalytically dead Cas9 [63], dead Cas9-*Fok*I fusion [64], xCas9 [65], fiKKH saCas9fi [66], Cas9-HF1 [67], eSpCas9 [68], HypaCas9 [69], HiFi Cas9FI, Cas9-NG [70], evoCas9 [71], SlugCas9 [72]—for genome editing in diverse organisms, including vertebrates. Additionally, another alternative for Cas9 is the Cas12a or Cpf1 variant, which recognizes 5′-TTTN-3′ PAMs [73] instead of 5′-NGG-3′ recognized by Cas9. Providing more versatility to genome editing, Cpf1 induces a staggered cut generating 5′ overhangs at the cleaved sites, while Cas9 induces a straight cut generating blunt ends.

While Cas9 and Cpf1 induce lasting changes to the genome, Cas13a targets RNA, thereby allowing transient, reversible, and adaptable changes to gene expression. The RNA interference mechanisms based on short hairpin RNA (shRNA) can efficiently knock down RNA levels; however, this approach is often associated with significant off-target activity [74,75]. Abudayyeh et al. identified a novel RNA-editing tool based on the CRISPR-Cas13a system [76]. Cas13a, placed in *Leptotrichia wadei* (LwaCas13a), is a class 2 type VI RNA-guided RNA-cleaving ribonuclease that can effectively suppress transcript levels in eukaryotic cells with high efficacy and target specificity [76].

Delivery of CRISPR-Cas9 components

Highly precise tissue-specific in vivo delivery of CRISPR-Cas9 and gRNAs is essential to achieving maximum editing efficiency with limited potential off-target activity. Lentivirus, adenovirus, and adeno-associated viral (AVV) vectors are efficient vehicles for delivering CRISPR-Cas9 components;

whereas lentiviral vectors can accommodate up to ~18 kb payload, adenovirus and AAV can accommodate up to ~8 kb and ~5 kb, respectively [50]. Thus, viral vectors are a method of choice for in vivo delivery with unique advantages, including accommodation of large DNA payloads and obtaining high efficiency editing. However, viral vectors suffer from several limitations such as mutagenesis, gene rearrangements, triggering an immune response, persistent high expression of Cas9 and gRNAs resulting in significant off-target activity [77–79].

An alternative approach is to utilize clinically approved non-viral delivery vehicles based on cationic lipids and polymers. However, first-generation cationic lipids suffer from low gene editing efficiencies [80,81], limited payload capacity, rapid clearance, immune activation, and systemic toxicity [82], especially when developed for chronic therapy, which requires repeated administration due to low transfection efficiency [83]. Second-generation advanced lipid nanoparticle (LPN) delivery systems for delivery of RNA or DNA offer several unique advantages. These LPNs are non-immunogenic, non-inflammatory, easy to synthesize, carry larger payloads, enable the possibility of repeated dosing, and are highly stable in the systemic circulation [84]. Further modification of these LPNs coupled with chemically modified gRNAs has the potential for effective genome editing in vivo.

Base editing

A point mutation—the permanent substitution of one DNA base pair for another—can be deleterious to human health. Provision of a homologous donor DNA template following Cas9-induced DSBs results in efficient HDR-mediated correction of disease-causing point mutations; however, NHEJ predominates HDR following DSBs, resulting in unexpected insertions, deletions, and translocations. To circumvent the untoward effects of NHEJ, and also the need to provide a DNA repair template, Komor et al. developed a base editing approach, which does not necessitate DSBs, thereby avoiding NHEJ or HDR-mediated correction [85]. Base editors utilize a catalytically dead Cas9 (dCas9), preventing DSBs, and a gRNA complementary to the target locus. To correct point mutations, dCas9 is tethered to either a cytidine deaminase (cytosine BEs) [85] or adenosine deaminase (adenosine BEs) [86], which together can mediate transition of all four bases (C to T, A to G, T to C, and G to A), thus improving the versatility of genome editing.

Nevertheless, compared with conventional Cas9 editing, cytosine BEs induce 20-fold higher single nucleotide variants in mouse zygotes [87], necessitating additional modifications to improve their reliability. Subsequent engineering of BEs with bacteriophage Mu Gam—dsDNA end-binding protein to reduce the ability of BEs to induce unwanted indels—has greatly improved the efficacy of fourth-generation base editors (BE4) [88]. Yet compared with the high fidelity of adenosine BEs, BE4 induced significantly more single nucleotide variants and deletions in mouse zygotes [89], implying the need for further refinement of BE4. Persistent efforts to overcome the limitations of base editing and to improve genome editing potential, Anzalone et al. developed a 'search-and-replace' genome editing technology—Prime Editing [90].

Prime editing

Base editors can only induce transition point mutations—interchanges of two purines (A to G and G to A) or two pyrimidines (C to T and T to C), but not transversions—interchanges of purines for pyrimidines. Also, base editors cannot induce precise insertions or deletions. In contrast, prime editing can induce transitions, transversions, insertions, and deletions at a target locus without triggering NHEJ or

HDR, limiting off-target effects and bringing unique versatility to precise genome editing over previous approaches, including CRISPR-Cas9 [91].

Prime editors comprise a Cas9 nickase tethered to an engineered reverse transcriptase. Prime editing utilizes two gRNAs—prime-editing gRNA (pegRNA) and a conventional gRNA. Compared with the established CRISPR-Cas gene editing approaches, the pegRNA brings sophistication and versatility to prime editing. The pegRNA, but not the conventional gRNA, guides the Cas9-reverse transcriptase fusion protein to the DNA target site. Notably, the 3′ end of pegRNA has two extensions: (a) a primer binding sequence or a homology region to the target sequence, and (b) a reverse transcription template that encodes the desired mutation to be installed at the DNA target site. The conventional gRNA, even though essential to efficient prime editing, only takes part in the penultimate nicking of the non-edited DNA strand [90,91].

The first step in prime editing is nicking of the PAM-containing DNA strand (or the DNA strand to be edited) by the Cas9-reverse transcriptase-pegRNA complex. Second, the primer binding sequence on the pegRNA hybridizes with the 3′ end of the free PAM-containing DNA strand, forming a primer-template complex. Third, the reverse transcriptase transcribes the desired mutation from the pegRNA. Fourth, the newly edited 3′ end with the desired modification is incorporated at the DNA target site by a DNA-joining enzyme DNA ligase. Fifth, the penultimate step, involves correcting the heteroduplex DNA (one edited and one non-edited base) due to the newly installed modification. The conventional gRNA is designed to nick the non-edited DNA strand, which is repaired using the edited strand as a DNA template, completing the prime editing process [90,91].

Recent studies indicate that prime editing can successfully generate targeted mutations in mouse embryos with high efficiency [92], providing researchers with a novel and highly versatile toolbox for precise genome editing that does not involve NHEJ or HDR.

Acknowledgments

We acknowledge support from the National Institutes of Health under Award Number R01 CA265978 (VH), and the Director's Innovation Fund at JAX (JAX-DIF-FY17). We thank Iiro Taneli Helenius for critical reading of the manuscript. We also gratefully acknowledge assistance from Genetic Engineering Technologies at JAX. We also thank Cindy Avery, Todd Nason, Rachel Gott, Leonor Robidoux, and Mikayla Bolduc for the maintenance of mouse colonies.

References

[1] P.D. Hsu, E.S. Lander, F. Zhang, Development and applications of CRISPR-Cas9 for genome engineering, Cell 157 (2014) 1262–1278, https://doi.org/10.1016/j.cell.2014.05.010.

[2] W.Y. Hwang, Y. Fu, D. Reyon, M.L. Maeder, S.Q. Tsai, J.D. Sander, R.T. Peterson, J.R.J. Yeh, J.K. Joung, Efficient genome editing in zebrafish using a CRISPR-Cas system, Nat. Biotechnol. 31 (2013) 227–229, https://doi.org/10.1038/nbt.2501.

[3] P. Mali, L. Yang, K.M. Esvelt, J. Aach, M. Guell, J.E. DiCarlo, J.E. Norville, G.M. Church, RNA-guided human genome engineering via Cas9, Science 339 (2013) 823–826, https://doi.org/10.1126/science.1232033.

[4] J.D. Sander, J.K. Joung, CRISPR-Cas systems for editing, regulating and targeting genomes, Nat. Biotechnol. (2014) 347–355, https://doi.org/10.1038/nbt.2842.

[5] E. Zuo, Y.J. Cai, K. Li, Y. Wei, B.A. Wang, Y. Sun, Z. Liu, J. Liu, X. Hu, W. Wei, X. Huo, L. Shi, C. Tang, D. Liang, Y. Wang, Y.H. Nie, C.C. Zhang, X. Yao, X. Wang, C. Zhou, W. Ying, Q. Wang, R.C. Chen, Q. Shen, G.L. Xu, J. Li, Q. Sun, Z.Q. Xiong, H. Yang, One-step generation of complete gene knockout mice and monkeys by CRISPR/Cas9-mediated gene editing with multiple sgRNAs, Cell Res. 27 (2017) 933–945, https://doi.org/10.1038/cr.2017.81.

[6] H. Wang, H. Yang, C.S. Shivalila, M.M. Dawlaty, A.W. Cheng, F. Zhang, R. Jaenisch, One-step generation of mice carrying mutations in multiple genes by CRISPR/cas-mediated genome engineering, Cell 153 (2013) 910–918, https://doi.org/10.1016/j.cell.2013.04.025.

[7] J.A. Townsend, D.A. Wright, R.J. Winfrey, F. Fu, M.L. Maeder, J.K. Joung, D.F. Voytas, High-frequency modification of plant genes using engineered zinc-finger nucleases, Nature 459 (2009) 442–445, https://doi.org/10.1038/nature07845.

[8] J. Boch, H. Scholze, S. Schornack, A. Landgraf, S. Hahn, S. Kay, T. Lahaye, A. Nickstadt, U. Bonas, Breaking the code of DNA binding specificity of TAL-type III effectors, Science 326 (2009) 1509–1512, https://doi.org/10.1126/science.1178811.

[9] M. Christian, T. Cermak, E.L. Doyle, C. Schmidt, F. Zhang, A. Hummel, A.J. Bogdanove, D.F. Voytas, Targeting DNA double-Strand breaks with TAL effector nucleases, Genetics 186 (2010) 757–761, https://doi.org/10.1534/genetics.110.120717.

[10] R. Morbitzer, P. Römer, J. Boch, T. Lahaye, Regulation of selected genome loci using de novo-engineered transcription activator-like effector (TALE)-type transcription factors, Proc. Natl. Acad. Sci. U. S. A. 107 (2010) 21617–21622, https://doi.org/10.1073/pnas.1013133107.

[11] J.A. Doudna, E. Charpentier, Genome editing. The new frontier of genome engineering with CRISPR-Cas9, Science 346 (2014).

[12] M. Jinek, K. Chylinski, I. Fonfara, M. Hauer, J.A. Doudna, E. Charpentier, A programmable dual-RNA-guided DNA endonuclease in adaptive bacterial immunity, Science 337 (2012) 816–821, https://doi.org/10.1126/science.1225829.

[13] V. Hosur, B.E. Low, D. Li, G.A. Stafford, V. Kohar, L.D. Shultz, M.V. Wiles, Genes adapt to outsmart gene-targeting strategies in mutant mouse strains by skipping exons to reinitiate transcription and translation, Genome Biol. 21 (2020), https://doi.org/10.1186/s13059-020-02086-0.

[14] M. Kosicki, K. Tomberg, A. Bradley, Repair of double-strand breaks induced by CRISPR–Cas9 leads to large deletions and complex rearrangements, Nat. Biotechnol. 36 (2018), https://doi.org/10.1038/nbt.4192.

[15] M. Thomas, G. Burgio, D.J. Adams, V. Iyer, L. He, Collateral damage and CRISPR genome editing, PLoS Genet. 15 (2019) e1007994, https://doi.org/10.1371/journal.pgen.1007994.

[16] G.A. Churchill, D.C. Airey, H. Allayee, J.M. Angel, A.D. Attie, J. Beatty, W.D. Beavis, J.K. Belknap, B. Bennett, W. Berrettini, A. Bleich, M. Bogue, K.W. Broman, K.J. Buck, E. Buckler, M. Burmeister, E.J. Chesler, J.M. Cheverud, S. Clapcote, M.N. Cook, R.D. Cox, J.C. Crabbe, W.E. Crusio, A. Darvasi, C.F. Deschepper, R.W. Doerge, C.R. Farber, J. Forejt, D. Gaile, S.J. Garlow, H. Geiger, H. Gershenfeld, T. Gordon, J. Gu, W. Gu, G. de Haan, N.L. Hayes, C. Heller, H. Himmelbauer, R. Hitzemann, K. Hunter, H.C. Hsu, F.A. Iraqi, B. Ivandic, H.J. Jacob, R.C. Jansen, K.J. Jepsen, D.K. Johnson, T.E. Johnson, G. Kempermann, C. Kendziorski, M. Kotb, R.F. Kooy, B. Llamas, F. Lammert, J.M. Lassalle, P.R. Lowenstein, L. Lu, A. Lusis, K.F. Manly, R. Marcucio, D. Matthews, J.F. Medrano, D.R. Miller, G. Mittleman, B.A. Mock, J.S. Mogil, X. Montagutelli, G. Morahan, D.G. Morris, R. Mott, J.H. Nadeau, H. Nagase, R.S. Nowakowski, B.F. O'Hara, A.V. Osadchuk, G.P. Page, B. Paigen, K. Paigen, A.A. Palmer, H.J. Pan, L. Peltonen-Palotie, J. Peirce, D. Pomp, M. Pravenec, D.R. Prows, Z. Qi, R.H. Reeves, J. Roder, G.D. Rosen, E.E. Schadt, L.C. Schalkwyk, Z. Seltzer, K. Shimomura, S. Shou, M.J. Sillanpää, L.D. Siracusa, H.W. Snoeck, J.L. Spearow, K. Svenson, L.M. Tarantino, D. Threadgill, L.A. Toth, W. Valdar, F. Pardo-Manuel de Villena, C. Warden, S. Whatley, R.W. Williams, T. Wiltshire, N. Yi, D. Zhang, M. Zhang, F. Zou, The collaborative cross, a community resource for the genetic analysis of complex traits, Nat. Genet. 36 (2004) 1133–1137, https://doi.org/10.1038/ng1104-1133.

[17] R.A. Taft, M. Davisson, M.V. Wiles, Know thy mouse, Trends Genet. 22 (2006) 649–653, https://doi.org/10.1016/j.tig.2006.09.010.
[18] M.C. Birling, A. Yoshiki, D.J. Adams, S. Ayabe, A.L. Beaudet, J. Bottomley, A. Bradley, S.D.M. Brown, A. Bürger, W. Bushell, F. Chiani, H.J.G. Chin, S. Christou, G.F. Codner, F.J. DeMayo, M.E. Dickinson, B. Doe, L.R. Donahue, M.D. Fray, A. Gambadoro, X. Gao, M. Gertsenstein, A. Gomez-Segura, L.O. Goodwin, J.D. Heaney, Y. Hérault, M.H. de Angelis, S.T. Jiang, M.J. Justice, P. Kasparek, R.E. King, R. Kühn, H. Lee, Y.J. Lee, Z. Liu, K.C. Kent Lloyd, I. Lorenzo, A.M. Mallon, C. McKerlie, T.F. Meehan, S. Newman, L.M.J. Nutter, G.T. Oh, G. Pavlovic, R. Ramirez-Solis, B. Rosen, E.J. Ryder, L.A. Santos, J. Schick, J.R. Seavitt, R. Sedlacek, C. Seisenberger, J.K. Seong, W.C. Skarnes, T. Sorg, K.P. Steel, M. Tamura, G.P. Tocchini-Valentini, C.K.L. Wang, H. Wardle-Jones, M. Wattenhofer-Donzé, S. Wells, B.J. Willis, J.A. Wood, W. Wurst, Y. Xu, L. Teboul, S.A. Murray, A resource of targeted mutant mouse lines for 5,061 genes, bioRxiv (2019), https://doi.org/10.1101/844092.
[19] R.J. Platt, S. Chen, Y. Zhou, M.J. Yim, L. Swiech, H.R. Kempton, J.E. Dahlman, O. Parnas, T.M. Eisenhaure, M. Jovanovic, D.B. Graham, S. Jhunjhunwala, M. Heidenreich, R.J. Xavier, R. Langer, D.G. Anderson, N. Hacohen, A. Regev, G. Feng, P.A. Sharp, F. Zhang, CRISPR-Cas9 knockin mice for genome editing and cancer modeling, Cell 159 (2014) 440–455, https://doi.org/10.1016/j.cell.2014.09.014.
[20] W. Xue, S. Chen, H. Yin, T. Tammela, T. Papagiannakopoulos, N.S. Joshi, W. Cai, G. Yang, R. Bronson, D.G. Crowley, F. Zhang, D.G. Anderson, P.A. Sharp, T. Jacks, CRISPR-mediated direct mutation of cancer genes in the mouse liver, Nature 514 (2014) 380–384, https://doi.org/10.1038/nature13589.
[21] D. Maddalo, E. Manchado, C.P. Concepcion, C. Bonetti, J.A. Vidigal, Y.C. Han, P. Ogrodowski, A. Crippa, N. Rekhtman, E.D. Stanchina, S.W. Lowe, A. Ventura, In vivo engineering of oncogenic chromosomal rearrangements with the CRISPR/Cas9 system, Nature 516 (2014) 423–428, https://doi.org/10.1038/nature13902.
[22] C.J. Braun, P.M. Bruno, M.A. Horlbeck, L.A. Gilbert, J.S. Weissman, M.T. Hemann, Versatile in vivo regulation of tumor phenotypes by dCas9-mediated transcriptional perturbation, Proc. Natl. Acad. Sci. (2016) E3892–E3900, https://doi.org/10.1073/pnas.1600582113.
[23] L. Serneels, D. T'Syen, L. Perez-Benito, T. Theys, M.G. Holt, B. De Strooper, Modeling the β-secretase cleavage site and humanizing amyloid-beta precursor protein in rat and mouse to study Alzheimer's disease, Mol. Neurodegener. 15 (2020), https://doi.org/10.1186/s13024-020-00399-z.
[24] M. Takalo, R. Wittrahm, B. Wefers, S. Parhizkar, K. Jokivarsi, T. Kuulasmaa, P. Mäkinen, H. Martiskainen, W. Wurst, X. Xiang, M. Marttinen, P. Poutiainen, A. Haapasalo, M. Hiltunen, C. Haass, The Alzheimer's disease-associated protective Plcγ2-P522R variant promotes immune functions, Mol. Neurodegener. 15 (2020), https://doi.org/10.1186/s13024-020-00402-7.
[25] J.L. Silverman, J. Nithianantharajah, A. Der-Avakian, J.W. Young, S.J. Sukoff Rizzo, Lost in translation: at the crossroads of face validity and translational utility of behavioral assays in animal models for the development of therapeutics, Neurosci. Biobehav. Rev. 116 (2020) 452–453, https://doi.org/10.1016/j.neubiorev.2020.07.008.
[26] K.S. Corbett, D.K. Edwards, S.R. Leist, O.M. Abiona, S. Boyoglu-Barnum, R.A. Gillespie, S. Himansu, A. Schäfer, C.T. Ziwawo, A.T. DiPiazza, K.H. Dinnon, S.M. Elbashir, C.A. Shaw, A. Woods, E.J. Fritch, D.R. Martinez, K.W. Bock, M. Minai, B.M. Nagata, G.B. Hutchinson, K. Wu, C. Henry, K. Bahl, D. Garcia-Dominguez, L.Z. Ma, I. Renzi, W.P. Kong, S.D. Schmidt, L. Wang, Y. Zhang, E. Phung, L.A. Chang, R.J. Loomis, N.E. Altaras, E. Narayanan, M. Metkar, V. Presnyak, C. Liu, M.K. Louder, W. Shi, K. Leung, E.S. Yang, A. West, K.L. Gully, L.J. Stevens, N. Wang, D. Wrapp, N.A. Doria-Rose, G. Stewart-Jones, H. Bennett, G.S. Alvarado, M.C. Nason, T.J. Ruckwardt, J.S. McLellan, M.R. Denison, J.D. Chappell, I.N. Moore, K.M. Morabito, J.R. Mascola, R.S. Baric, A. Carfi, B.S. Graham, SARS-CoV-2 mRNA vaccine design enabled by prototype pathogen preparedness, Nature 586 (2020) 567–571, https://doi.org/10.1038/s41586-020-2622-0.
[27] A.O. Hassan, J.B. Case, E.S. Winkler, L.B. Thackray, N.M. Kafai, A.L. Bailey, B.T. McCune, J.M. Fox, R.E. Chen, W.B. Alsoussi, J.S. Turner, A.J. Schmitz, T. Lei, S. Shrihari, S.P. Keeler, D.H. Fremont, S. Greco, P.B.

McCray, S. Perlman, M.J. Holtzman, A.H. Ellebedy, M.S. Diamond, A SARS-CoV-2 infection model in mice demonstrates protection by neutralizing antibodies, Cell 182 (2020) 744–753.e4, https://doi.org/10.1016/j.cell.2020.06.011.

[28] Q. Li, X. Guan, P. Wu, X. Wang, L. Zhou, Y. Tong, R. Ren, K.S.M. Leung, E.H.Y. Lau, J.Y. Wong, X. Xing, N. Xiang, Y. Wu, C. Li, Q. Chen, D. Li, T. Liu, J. Zhao, M. Liu, W. Tu, C. Chen, L. Jin, R. Yang, Q. Wang, S. Zhou, R. Wang, H. Liu, Y. Luo, Y. Liu, G. Shao, H. Li, Z. Tao, Y. Yang, Z. Deng, B. Liu, Z. Ma, Y. Zhang, G. Shi, T.T.Y. Lam, J.T. Wu, G.F. Gao, B.J. Cowling, B. Yang, G.M. Leung, Z. Feng, Early transmission dynamics in Wuhan, China, of novel coronavirus-infected pneumonia, N. Engl. J. Med. 382 (2020) 1199–1207, https://doi.org/10.1056/NEJMoa2001316.

[29] K.H. Dinnon, S.R. Leist, A. Schäfer, C.E. Edwards, D.R. Martinez, S.A. Montgomery, A. West, B.L. Yount, Y.J. Hou, L.E. Adams, K.L. Gully, A.J. Brown, E. Huang, M.D. Bryant, I.C. Choong, J.S. Glenn, L.E. Gralinski, T.P. Sheahan, R.S. Baric, A mouse-adapted model of SARS-CoV-2 to test COVID-19 countermeasures, Nature 586 (2020) 560–566, https://doi.org/10.1038/s41586-020-2708-8.

[30] C. Lutz, L. Maher, C. Lee, W. Kang, COVID-19 preclinical models: human angiotensin-converting enzyme 2 transgenic mice, Human Genomics 14 (2020), https://doi.org/10.1186/s40246-020-00272-6.

[31] W. Li, M.J. Moore, N. Vasllieva, J. Sui, S.K. Wong, M.A. Berne, M. Somasundaran, J.L. Sullivan, K. Luzuriaga, T.C. Greeneugh, H. Choe, M. Farzan, Angiotensin-converting enzyme 2 is a functional receptor for the SARS coronavirus, Nature 426 (2003) 450–454, https://doi.org/10.1038/nature02145.

[32] P.B. McCray, L. Pewe, C. Wohlford-Lenane, M. Hickey, L. Manzel, L. Shi, J. Netland, H.P. Jia, C. Halabi, C.D. Sigmund, D.K. Meyerholz, P. Kirby, D.C. Look, S. Perlman, Lethal infection of K18-hACE2 mice infected with severe acute respiratory syndrome coronavirus, J. Virol. 81 (2007) 813–821, https://doi.org/10.1128/JVI.02012-06.

[33] S.H. Sun, Q. Chen, H.J. Gu, G. Yang, Y.X. Wang, X.Y. Huang, S.S. Liu, N.N. Zhang, X.F. Li, R. Xiong, Y. Guo, Y.Q. Deng, W.J. Huang, Q. Liu, Q.M. Liu, Y.L. Shen, Y. Zhou, X. Yang, T.Y. Zhao, C.F. Fan, Y.S. Zhou, C.F. Qin, Y.C. Wang, A mouse model of SARS-CoV-2 infection and pathogenesis, Cell Host Microbe 28 (2020) 124–133.e4, https://doi.org/10.1016/j.chom.2020.05.020.

[34] R. Barrangou, L.A. Marraffini, CRISPR-cas systems: prokaryotes upgrade to adaptive immunity, Mol. Cell 54 (2014) 234–244, https://doi.org/10.1016/j.molcel.2014.03.011.

[35] E. Deltcheva, K. Chylinski, C.M. Sharma, K. Gonzales, Y. Chao, Z.A. Pirzada, M.R. Eckert, J. Vogel, E. Charpentier, CRISPR RNA maturation by trans-encoded small RNA and host factor RNase III, Nature 471 (2011) 602–607, https://doi.org/10.1038/nature09886.

[36] S.H. Sternberg, S. Redding, M. Jinek, E.C. Greene, J.A. Doudna, DNA interrogation by the CRISPR RNA-guided endonuclease Cas9, Nature 507 (2014) 62–67, https://doi.org/10.1038/nature13011.

[37] S. Erwood, B. Gu, Embryo-based large fragment knock-in in mammals: why, how and what's next, Genes (Basel) 11 (2020) 140.

[38] B.E. Low, P.M. Kutny, M.V. Wiles, Simple, efficient CRISPR-cas9-mediated gene editing in mice: strategies and methods, Methods Mol. Biol. 1438 (2016) 19–53, https://doi.org/10.1007/978-1-4939-3661-8_2.

[39] W. Qin, S.L. Dion, P.M. Kutny, Y. Zhang, A.W. Cheng, N.L. Jillette, A. Malhotra, A.M. Geurts, Y.G. Chen, H. Wang, Efficient CRISPR/cas9-mediated genome editing in mice by zygote electroporation of nuclease, Genetics 200 (2015) 423–430, https://doi.org/10.1534/genetics.115.176594.

[40] S.E. Tröder, L.K. Ebert, L. Butt, S. Assenmacher, B. Schermer, B. Zevnik, An optimized electroporation approach for efficient CRISPR/Cas9 genome editing in murine zygotes, PLoS One 13 (2018), https://doi.org/10.1371/journal.pone.0196891.

[41] R. Behringer, M. Gertsenstein, K.V. Nagy, Manipulating the Mouse Embryo: A Laboratory Manual, Fourth Edition, Cold Spring Harbor Laboratory Press, Cold Spring Harbor, New York, 2014.

[42] J.C. Oliveros, M. Franch, D. Tabas-Madrid, D. San-León, L. Montoliu, P. Cubas, F. Pazos, Breaking-Cas-interactive design of guide RNAs for CRISPR-Cas experiments for ENSEMBL genomes, Nucleic Acids Res. 44 (2016) W267–W271, https://doi.org/10.1093/nar/gkw407.

[43] Y. Fu, J.D. Sander, D. Reyon, V.M. Cascio, J.K. Joung, Improving CRISPR-Cas nuclease specificity using truncated guide RNAs, Nat. Biotechnol. 32 (2014) 279–284, https://doi.org/10.1038/nbt.2808.
[44] S.Q. Tsai, Z. Zheng, N.T. Nguyen, M. Liebers, V.V. Topkar, V. Thapar, N. Wyvekens, C. Khayter, A.J. Iafrate, L.P. Le, M.J. Aryee, J.K. Joung, GUIDE-seq enables genome-wide profiling of off-target cleavage by CRISPR-Cas nucleases, Nat. Biotechnol. 80 (2015) 187–197, https://doi.org/10.1038/nbt.3117.
[45] D.D. Kocak, E.A. Josephs, V. Bhandarkar, S.S. Adkar, J.B. Kwon, C.A. Gersbach, Increasing the specificity of CRISPR systems with engineered RNA secondary structures, Nat. Biotechnol. 37 (2019) 657–666, https://doi.org/10.1038/s41587-019-0095-1.
[46] D.E. Ryan, D. Taussig, I. Steinfeld, S.M. Phadnis, B.D. Lunstad, M. Singh, X. Vuong, K.D. Okochi, R. McCaffrey, M. Olesiak, S. Roy, C.W. Yung, B. Curry, J.R. Sampson, L. Bruhn, D.J. Dellinger, Improving CRISPR-Cas specificity with chemical modifications in single-guide RNAs, Nucleic Acids Res. 46 (2018) 792–803, https://doi.org/10.1093/nar/gkx1199.
[47] K. Wang, H. Li, Y. Xu, Q. Shao, J. Yi, R. Wang, W. Cai, X. Hang, C. Zhang, H. Cai, W. Qu, MFEprimer-3.0: quality control for PCR primers, Nucleic Acids Res. 47 (2019) W610–W613, https://doi.org/10.1093/nar/gkz351.
[48] H.Y. Shin, C. Wang, H.K. Lee, K.H. Yoo, X. Zeng, T. Kuhns, C.M. Yang, T. Mohr, C. Liu, L. Hennighausen, CRISPR/Cas9 targeting events cause complex deletions and insertions at 17 sites in the mouse genome, Nat. Commun. 8 (2017), https://doi.org/10.1038/ncomms15464.
[49] K. Boroviak, B. Doe, R. Banerjee, F. Yang, A. Bradley, Chromosome engineering in zygotes with CRISPR/Cas9, Genesis 54 (2016) 78–85, https://doi.org/10.1002/dvg.22915.
[50] C.A. Lino, J.C. Harper, J.P. Carney, J.A. Timlin, Delivering crispr: a review of the challenges and approaches, Drug Deliv. 25 (2018) 1234–1257, https://doi.org/10.1080/10717544.2018.1474964.
[51] X.H. Zhang, L.Y. Tee, X.G. Wang, Q.S. Huang, S.H. Yang, Off-target effects in CRISPR/Cas9-mediated genome engineering, Mol. Ther. Nucl. Acids 4 (2015) e264, https://doi.org/10.1038/mtna.2015.37.
[52] V. Iyer, K. Boroviak, M. Thomas, B. Doe, L. Riva, E. Ryder, D.J. Adams, No unexpected CRISPR-Cas9 off-target activity revealed by trio sequencing of gene-edited mice, PLoS Genet. 14 (2018), https://doi.org/10.1371/journal.pgen.1007503.
[53] M. Naeem, S. Majeed, M.Z. Hoque, I. Ahmad, Latest developed strategies to minimize the off-target effects in CRISPR-Cas-mediated genome editing, Cell 9 (2020), https://doi.org/10.3390/cells9071608.
[54] J. Zischewski, R. Fischer, L. Bortesi, Detection of on-target and off-target mutations generated by CRISPR/Cas9 and other sequence-specific nucleases, Biotechnol. Adv. 35 (2017) 95–104, https://doi.org/10.1016/j.biotechadv.2016.12.003.
[55] P.D. Hsu, D.A. Scott, J.A. Weinstein, F.A. Ran, S. Konermann, V. Agarwala, Y. Li, E.J. Fine, X. Wu, O. Shalem, T.J. Cradick, L.A. Marraffini, G. Bao, F. Zhang, DNA targeting specificity of RNA-guided Cas9 nucleases, Nat. Biotechnol. 31 (2013) 827–832, https://doi.org/10.1038/nbt.2647.
[56] M.C. Birling, L. Schaeffer, P. André, L. Lindner, D. Maréchal, A. Ayadi, T. Sorg, G. Pavlovic, Y. Hérault, Efficient and rapid generation of large genomic variants in rats and mice using CRISMERE, Sci. Rep. 7 (2017), https://doi.org/10.1038/srep43331.
[57] G. Burgio, L. Teboul, Anticipating and identifying collateral damage in genome editing, Trends Genet. 36 (2020) 905–914, https://doi.org/10.1016/j.tig.2020.09.011.
[58] A. Cebrian-Serrano, B. Davies, CRISPR-Cas orthologues and variants: optimizing the repertoire, specificity and delivery of genome engineering tools, Mamm. Genome 28 (2017) 247–261, https://doi.org/10.1007/s00335-017-9697-4.
[59] L. Xu, J. Wang, Y. Liu, L. Xie, B. Su, D. Mou, L. Wang, T. Liu, X. Wang, B. Zhang, L. Zhao, L. Hu, H. Ning, Y. Zhang, K. Deng, L. Liu, X. Lu, T. Zhang, J. Xu, C. Li, H. Wu, H. Deng, H. Chen, CRISPR-edited stem cells in a patient with HIV and acute lymphocytic leukemia, N. Engl. J. Med. 381 (2019) 1240–1247, https://doi.org/10.1056/NEJMoa1817426.

[60] N.J. Hafford-Tear, Y.C. Tsai, A.N. Sadan, B. Sanchez-Pintado, C. Zarouchlioti, G.J. Maher, P. Liskova, S.J. Tuft, A.J. Hardcastle, T.A. Clark, A.E. Davidson, CRISPR/Cas9-targeted enrichment and long-read sequencing of the Fuchs endothelial corneal dystrophy–associated TCF4 triplet repeat, Genet. Med. 21 (2019) 2092–2102, https://doi.org/10.1038/s41436-019-0453-x.

[61] G. Guglielmi, First CRISPR test for the coronavirus approved in the United States, Nature (2020), https://doi.org/10.1038/d41586-020-01402-9.

[62] F.A. Ran, P.D. Hsu, C.-Y. Lin, J.S. Gootenberg, S. Konermann, A.E. Trevino, D.A. Scott, A. Inoue, S. Matoba, Y. Zhang, F. Zhang, Double nicking by RNA-guided CRISPR Cas9 for enhanced genome editing specificity, Cell (2013) 1380–1389, https://doi.org/10.1016/j.cell.2013.08.021.

[63] L.S. Qi, M.H. Larson, L.A. Gilbert, J.A. Doudna, J.S. Weissman, A.P. Arkin, W.A. Lim, Repurposing CRISPR as an RNA-γuided platform for sequence-specific control of gene expression, Cell 152 (2013) 1173–1183, https://doi.org/10.1016/j.cell.2013.02.022.

[64] J.P. Guilinger, D.B. Thompson, D.R. Liu, Fusion of catalytically inactive Cas9 to FokI nuclease improves the specificity of genome modification, Nat. Biotechnol. 32 (2014) 577–582, https://doi.org/10.1038/nbt.2909.

[65] J.H. Hu, S.M. Miller, M.H. Geurts, W. Tang, L. Chen, N. Sun, C.M. Zeina, X. Gao, H.A. Rees, Z. Lin, D.R. Liu, Evolved Cas9 variants with broad PAM compatibility and high DNA specificity, Nature 556 (2018) 57–63, https://doi.org/10.1038/nature26155.

[66] B.P. Kleinstiver, M.S. Prew, S.Q. Tsai, N.T. Nguyen, V.V. Topkar, Z. Zheng, J.K. Joung, Broadening the targeting range of *Staphylococcus aureus* CRISPR-Cas9 by modifying PAM recognition, Nat. Biotechnol. 33 (2015) 1293–1298, https://doi.org/10.1038/nbt.3404.

[67] B.P. Kleinstiver, V. Pattanayak, M.S. Prew, S.Q. Tsai, N.T. Nguyen, Z. Zheng, J.K. Joung, High-fidelity CRISPR-Cas9 nucleases with no detectable genome-wide off-target effects, Nature 529 (2016) 490–495, https://doi.org/10.1038/nature16526.

[68] I.M. Slaymaker, L. Gao, B. Zetsche, D.A. Scott, W.X. Yan, F. Zhang, Rationally engineered Cas9 nucleases with improved specificity, Science 351 (2016) 84–88, https://doi.org/10.1126/science.aad5227.

[69] J.S. Chen, Y.S. Dagdas, B.P. Kleinstiver, M.M. Welch, A.A. Sousa, L.B. Harrington, S.H. Sternberg, J.K. Joung, A. Yildiz, J.A. Doudna, Enhanced proofreading governs CRISPR-Cas9 targeting accuracy, Nature 550 (2017) 407–410, https://doi.org/10.1038/nature24268.

[70] H. Nishimasu, X. Shi, S. Ishiguro, L. Gao, S. Hirano, S. Okazaki, T. Noda, O.O. Abudayyeh, J.S. Gootenberg, H. Mori, S. Oura, B. Holmes, M. Tanaka, M. Seki, H. Hirano, H. Aburatani, R. Ishitani, M. Ikawa, N. Yachie, F. Zhang, O. Nureki, Engineered CRISPR-Cas9 nuclease with expanded targeting space, Science 361 (2018) 1259–1262, https://doi.org/10.1126/science.aas9129.

[71] A. Casini, M. Olivieri, G. Petris, C. Montagna, G. Reginato, G. Maule, F. Lorenzin, D. Prandi, A. Romanel, F. Demichelis, A. Inga, A. Cereseto, A highly specific SpCas9 variant is identified by in vivo screening in yeast, Nat. Biotechnol. 36 (2018) 265–271, https://doi.org/10.1038/nbt.4066.

[72] Z. Hu, C. Zhang, S. Wang, S. Gao, J. Wei, M. Li, L. Hou, H. Mao, Y. Wei, T. Qi, H. Liu, D. Liu, F. Lan, D. Lu, H. Wang, J. Li, Y. Wang, Discovery and engineering of small SlugCas9 with broad targeting range and high specificity and activity, Nucleic Acids Res. 49 (2021) 4008–4019, https://doi.org/10.1093/nar/gkab148.

[73] L. Gao, D.B.T. Cox, W.X. Yan, J.C. Manteiga, M.W. Schneider, T. Yamano, H. Nishimasu, O. Nureki, N. Crosetto, F. Zhang, Engineered Cpf1 variants with altered PAM specificities, Nat. Biotechnol. 35 (2017) 789–792, https://doi.org/10.1038/nbt.3900.

[74] A.L. Jackson, J. Burchard, J. Schelter, B.N. Chau, M. Cleary, L. Lim, P.S. Linsley, Widespread siRNA "off-target" transcript silencing mediated by seed region sequence complementarity, RNA 12 (2006) 1179–1187, https://doi.org/10.1261/rna.25706.

[75] T.A. Vickers, W.F. Lima, H. Wu, J.G. Nichols, P.S. Linsley, S.T. Crooke, Off-target and a portion of target-specific siRNA mediated mRNA degradation is Ago2 "slicer" independent and can be mediated by Ago1, Nucleic Acids Res. 37 (2009) 6927–6941, https://doi.org/10.1093/nar/gkp735.

[76] O.O. Abudayyeh, J.S. Gootenberg, P. Essletzbichler, S. Han, J. Joung, J.J. Belanto, V. Verdine, D.B.T. Cox, M.J. Kellner, A. Regev, E.S. Lander, D.F. Voytas, A.Y. Ting, F. Zhang, RNA targeting with CRISPR-Cas13, Nature 550 (2017) 280–284, https://doi.org/10.1038/nature24049.
[77] C.S. Manno, V.R. Arruda, G.F. Pierce, B. Glader, M. Ragni, J. Rasko, M.C. Ozelo, K. Hoots, P. Blatt, B. Konkle, M. Dake, R. Kaye, M. Razavi, A. Zajko, J. Zehnder, H. Nakai, A. Chew, D. Leonard, J.F. Wright, R.R. Lessard, J.M. Sommer, M. Tigges, D. Sabatino, A. Luk, H. Jiang, F. Mingozzi, L. Couto, H.C. Ertl, K.A. High, M.A. Kay, Successful transduction of liver in hemophilia by AAV-factor IX and limitations imposed by the host immune response, Nat. Med. 12 (2006) 342–347, https://doi.org/10.1038/nm1358.
[78] F. Mingozzi, K.A. High, Immune responses to AAV vectors: overcoming barriers to successful gene therapy, Blood 122 (2013) 23–36, https://doi.org/10.1182/blood-2013-01-306647.
[79] C.E. Thomas, A. Ehrhardt, M.A. Kay, Progress and problems with the use of viral vectors for gene therapy, Nat. Rev. Genet. 4 (2003) 346–358, https://doi.org/10.1038/nrg1066.
[80] C. Jiang, M. Mei, B. Li, X. Zhu, W. Zu, Y. Tian, Q. Wang, Y. Guo, Y. Dong, X. Tan, A non-viral CRISPR/Cas9 delivery system for therapeutically targeting HBV DNA and pcsk9 in vivo, Cell Res. 27 (2017) 440–443, https://doi.org/10.1038/cr.2017.16.
[81] J.B. Miller, S. Zhang, P. Kos, H. Xiong, K. Zhou, S.S. Perelman, H. Zhu, D.J. Siegwart, Non-viral CRISPR/Cas gene editing in vitro and in vivo enabled by synthetic nanoparticle co-delivery of Cas9 mRNA and sgRNA, Angew. Chem. Int. Ed. 56 (2017) 1059–1063, https://doi.org/10.1002/anie.201610209.
[82] J.A. Kulkarni, P.R. Cullis, R. Van Der Meel, Lipid nanoparticles enabling gene therapies: from concepts to clinical utility, Nucl. Acid Ther. 28 (2018) 146–157, https://doi.org/10.1089/nat.2018.0721.
[83] K. Romøren, B.J. Thu, N.C. Bols, Ø. Evensen, Transfection efficiency and cytotoxicity of cationic liposomes in salmonid cell lines of hepatocyte and macrophage origin, Biochim. Biophys. Acta Biomembr. 1663 (2004) 127–134, https://doi.org/10.1016/j.bbamem.2004.02.007.
[84] G. Mattheolabakis, L. Milane, A. Singh, M.M. Amiji, Hyaluronic acid targeting of CD44 for cancer therapy: from receptor biology to nanomedicine, J. Drug Target. 23 (2015) 605–618, https://doi.org/10.3109/1061186X.2015.1052072.
[85] A.C. Komor, Y.B. Kim, M.S. Packer, J.A. Zuris, D.R. Liu, Programmable editing of a target base in genomic DNA without double-stranded DNA cleavage, Nature 533 (2016) 420–424, https://doi.org/10.1038/nature17946.
[86] N.M. Gaudelli, A.C. Komor, H.A. Rees, M.S. Packer, A.H. Badran, D.I. Bryson, D.R. Liu, Programmable base editing of T to G C in genomic DNA without DNA cleavage, Nature 551 (2017) 464–471, https://doi.org/10.1038/nature24644.
[87] E. Zuo, Y. Sun, W. Wei, T. Yuan, W. Ying, H. Sun, L. Yuan, L.M. Steinmetz, Y. Li, H. Yang, Cytosine base editor generates substantial off-target single-nucleotide variants in mouse embryos, Science 364 (2019) 289–292, https://doi.org/10.1126/science.aav9973.
[88] A.C. Komor, K.T. Zhao, M.S. Packer, N.M. Gaudelli, A.L. Waterbury, L.W. Koblan, Y.B. Kim, A.H. Badran, D.R. Liu, Improved base excision repair inhibition and bacteriophage Mu Gam protein yields C:G-to-T:A base editors with higher efficiency and product purity, Sci. Adv. 3 (2017) eaao4774, https://doi.org/10.1126/sciadv.aao4774.
[89] H.K. Lee, H.E. Smith, C. Liu, M. Willi, L. Hennighausen, Cytosine base editor 4 but not adenine base editor generates off-target mutations in mouse embryos, Commun. Biol. 3 (2020), https://doi.org/10.1038/s42003-019-0745-3.
[90] A.V. Anzalone, P.B. Randolph, J.R. Davis, A.A. Sousa, L.W. Koblan, J.M. Levy, P.J. Chen, C. Wilson, G.A. Newby, A. Raguram, D.R. Liu, Search-and-replace genome editing without double-strand breaks or donor DNA, Nature 576 (2019) 149–157, https://doi.org/10.1038/s41586-019-1711-4.
[91] A.V. Anzalone, L.W. Koblan, D.R. Liu, Genome editing with CRISPR–Cas nucleases, base editors, transposases and prime editors, Nat. Biotechnol. 38 (2020) 824–844, https://doi.org/10.1038/s41587-020-0561-9.
[92] Y. Liu, X. Li, S. He, S. Huang, C. Li, Y. Chen, Z. Liu, X. Huang, X. Wang, Efficient generation of mouse models with the prime editing system, Cell Discov. 6 (2020), https://doi.org/10.1038/s41421-020-0165-z.

CHAPTER 19

CRISPR classroom activities and case studies

TyAnna L. Lovato[a] and Richard M. Cripps[b]

[a]Department of Biology, University of New Mexico, Albuquerque, NM, United States, [b]Department of Biology, San Diego State University, San Diego, CA, United States

Importance of course-based undergraduate research experiences

Significant effort has recently been placed on enhancing research experiences in undergraduate laboratory classes. Traditional laboratory classes have centered around laboratory exercises that support the lecture material and provide students with opportunities to see how known scientific facts were obtained. Over the past 20 years, there has been an increase in the number of laboratory classes for which the findings are open-ended and where the data that students generate are novel and of broader significance and useful for the scientific community. In one example, Dr. Utpal Banerjee at the University of California, Los Angeles, instructed students over several semesters in carrying out a genome-wide screen to identify genes required for *Drosophila* eye development [1]. This study uncovered new loci necessary for the development of the visual organ and exposed several hundred undergraduates to realistic research activities.

Similarly, the Howard Hughes Medical Institute (HHMI) initiated in 2008 the Science Education Alliance program, aimed at providing support to institutions wishing to provide research experience to undergraduates. This initiative, which has taken advantage of next-generation sequencing technology to support students identifying bacteriophage from soil samples, is currently in place over 100 colleges nationally and has served over 20,000 students since its inception (see https://www.hhmi.org/developing-scientists/science-education-alliance, and [2]).

Such classes are now termed course-based undergraduate research experiences (CUREs). What is the advantage of a CURE over a more traditional "canned" laboratory exercise? Much of the growth of this teaching approach has been in response to the Boyer Commission [3] that emphasized the value of research experience in enhancing the effectiveness of undergraduate teaching. Subsequent evaluations of research-based undergraduate classes have demonstrated clearly that students from the classes show significantly enhanced research skills and potential for successful graduate careers (reviewed in [4]).

Importance of CRISPR as a teaching tool

In this chapter we discuss strategies used for CUREs that involve CRISPR technology. Some approaches to teaching students about CRISPR have not involved laboratory-based activities [5]; however, given that CRISPR-Cas9 systems have been exploited for genome engineering, several instructors have developed CRISPR CUREs. We see several advantages of the laboratory-based approach, assuming that facilities and resources are available. Firstly, there is a broad consensus in the scientific community that CRISPR-based approaches are going to have a significant and sustained impact on biomedical research, and moreover CRISPR is currently being weaponized to advance human medicine. Thus, a strong appreciation of the topic is already a critical component of undergraduate biology education.

Secondly, use of CRISPR for genome editing is a multidisciplinary approach that incorporates genome-wide sequence analysis and bioinformatics, recombinant DNA technology to generate the correct reagents, and handling of cells or organisms that are being targeted. In this manner, students obtain a holistic appreciation of the research process, from design and generation of reagents through handling organisms and phenotypic analysis to interpretation and communication of their results. As such, a strong foundation is generally recommended.

Thirdly, current CRISPR-based manipulation of human DNA means that instructors can simultaneously introduce an ethical discussion of gene editing to future scientists. The ability to modify genes has an amazing potential for curing human disease; however, collaboration between science and philosophy or science policy disciplines is essential for educating future scientists to become responsible gene editors. Thus, the values of using CRISPR as a teaching tool are not only to understand molecular biology and gain research experience but to also understand the regulations, to which scientists must adhere, and the underlying reasons behind the regulations.

Approaches to teaching CRISPR

There are two main approaches to teaching CRISPR: learning about the concept and applying it experimentally. The first approach focuses on the bacteria that evolved the defense mechanism, and the second involves manipulating the mechanism and applying it to edit the genomes of different organisms. Both approaches are interdisciplinary by incorporating either microbiology or molecular biology skills and techniques. In addition to practical laboratory skills, ethical and philosophical discussions that apply to current uses of CRISPR can also be incorporated into the lesson plans. The content possibilities are vast, making CRISPR instruction an appealing scaffold for the instructor's desired learning outcomes.

One of the biggest challenges to teaching a hands-on course on CRISPR technology is time. Fortunately, there are numerous strategies to accommodate time restrictions. At the University of New Mexico, we designed the first Genome Editing course, where students applied the technology to mutate genes in *Drosophila* within a 16-week semester. The time restraint was tight. Since then, scientists and educators have developed alternative approaches in a variety of organisms, including simulations that can be carried out entirely in the classroom (https://online.ucpress.edu/abt/article/82/5/315/110288/Wet-amp-Dry-Lab-Activities-to-Introduce-Students).

Setting the stage

Prior to hands-on activities, a certain amount of background information needs to be introduced. We have found, however, that with too much information, students do not grasp all the content we are attempting to deliver. Additionally, we have discovered, from student feedback through discussion and end of the semester surveys, that a significant amount of repetition is required for a complete understanding. Thus, instead of explaining the entire CRISPR concept at once, we have been most successful in introducing concepts throughout the course of the semester. At different institutions, there are laboratory schedules of varying lengths and so the depth of information delivered is constrained by time limitation. As a consequence, it is an absolute requirement that specific and attainable learning outcomes be laid out at the beginning of the course and continuously reiterated. We have also found that there is a large variation in previous knowledge, so we begin our course by engaging students the first day in a discussion to determine where to start.

We also note that CRISPR laboratory courses can be based on two different approaches: on the one hand, classes can study the basic biological phenomenon of the CRISPR system and focus on identifying changes to the CRISPR locus as a result of acquisition of resistance to infectious particles; on the other hand, classes can apply the CRISPR technology to carry out genome editing. Despite the perceived complexity of genome editing, as described below, both approaches can be applied to classes of short or long duration.

Strategies for short time frames in bacteria

Utilizing bacteria for CRISPR experiments is appealing due to the short growth period of bacteria. To illustrate the natural CRISPR process, Hynes et al. [6] had students challenge bacteria with different lytic bacteriophages and identify strains with natural CRISPR adaptation. They then had the students carry out PCR on surviving bacterial colonies for subsequent sequencing, to identify the new repeat-spacers added to the CRISPR locus, and to compare those sequences to the bacteriophage genomes. The entire protocol only required four laboratory sessions, the longest of which was 3h.

Similarly, Militello and Lazatin [7] had students perform PCR, utilizing primers for known CRISPR loci using uncharacterized environmental *E. coli* strains. In case the sequences of the CRISPR loci in the novel strains were slightly divergent from the reference genome, which would affect PCR efficiency, the students used low primer annealing temperatures for their PCR. The students were then able to sequence their PCR products to identify differences in the incorporated sequences of the new strains, and ultimately use BLAST approaches to identify the infectious particles with which the strains had interacted. This activity was run over the course of 5 weeks.

To carry out genome editing in courses with limited duration, gene engineering kits are commercially available. A kit from Odin contains all the components necessary to introduce a genetic change in bacteria that allows it to grow in the presence of an antibiotic on which it would normally not be able to grow (http://www.the-odin.com/gene-engineering-kits/). A class using this approach can be completed over only 3–4 laboratory sessions and with minimal additional reagents or equipment required.

In 2019, Sehgal et al. [8] utilized a short time frame by preparing all of the CRISPR components ahead of time. The students then transformed the components into yeast cells which generated mutations in two genes for which there are specific, known, observable phenotypes. However, during incubation periods in the laboratory sessions, the instructors had the students design their own guide RNAs using an online tool, to ensure the students understood the process. Preparing the CRISPR components allowed for more time to incorporate theoretical and ethical assignments and discussions. Prior to the laboratory exercises, the students were given a comprehensive 50-min lecture of CRISPR background information. As with our course, some student comments noted that a large amount of information was a bit overwhelming but became more clear with reiterations at each step of the laboratory sessions.

16 week teaching strategies

Longer duration classes offer the opportunity to apply the CRISPR system to carry out genome editing approaches in more complex organisms. The duration of our Genome Editing course at the University of New Mexico is long enough for students to design their own targets, clone the target sequences into plasmids that will express the target sequence in the context of a sgRNA, inject the plasmids into *Drosophila* embryos, and screen subsequent generations of flies for heritable mutations [9]. Similar semester-long courses have also been carried out using zebrafish [10]. Our class runs 16 weeks for 2 h and 50 min on Mondays and Fridays. In order to have the students participate in every step in the process, we ask that they have some schedule flexibility such as coming in on Sunday prior to the Monday's class to inoculate bacterial cultures for the next day's activities or stopping by to collect virgin flies to mate to their generated lines. The tasks are brief but require additional commitment. Since the students work in a group, they are able to take turns for these outside tasks, and when possible, we let the class out early to compensate.

In order to efficiently use our time, lectures and primary literature reviews were interwoven into laboratory activities. The first day was the most critical because we needed to introduce basic class housekeeping, CRISPR background, and explain how to design guide RNAs for which oligonucleotide primers needed to be ordered the same day. Subsequent classes incorporated a laboratory task followed by a short lecture or paper discussion related specifically to the procedure we were carrying out. Combining detailed lecture or primary literature information related to individual laboratory procedures was the most beneficial strategy for student understanding.

General considerations for developing CRISPR in the classroom

Based on the classes described earlier that represent a subset of CRISPR classes currently ongoing, we provide the following points to consider when designing a class that uses CRISPR to effect genome editing:

1. Choose an appropriate time frame and model organism. In general, short time-frame class can benefit most from using prokaryotic or single-cell eukaryotic model organisms. Longer classes can employ multicellular eukaryotes with longer generation times. However, more complex organisms often require more specialized expertise and equipment for successful completion.
2. Choose a target gene. In many cases, CRISPR classes target genes with clearly defined or selectable mutant phenotypes, such as antibiotic resistance in bacteria, colony morphology in yeast, or visible markers in *Drosophila* (see below). The advantage to this approach is that a positive genome editing event can be clearly observed. Alternatively, classes can target novel genes, for which the mutant phenotypes are unknown. While this latter approach has the potential to generate new reagents for the biomedical research field, a drawback is that some gene knockouts lack any easily detectable phenotype.
3. Once a target gene has been selected, the target sequence must be cloned into an appropriate vector for expression in the cellular system being used. Such plasmids will differ between different model organisms, although most useful plasmids are available from resources such as Addgene (www.addgene.org). An alternative strategy is to order presynthesized sgRNAs, which are available from a number of companies. The advantage of presynthesized sgRNAs is that it saves some time; on the other hand, the generation by students of plasmids containing the correct target sequence can be a valuable lesson in cloning and sequencing that they might use later in the semester when diagnosing mutant lines.

4. Choose a Cas9 expression vector or utilize a transgenic line that expresses Cas9. Cas9 protein can be introduced into the model organism using one of several different approaches, including the introduction into cells of a plasmid encoding Cas9; injection into cells of Cas9 mRNA or protein; or injection into a transgenic line that already expresses Cas9 in germline stem cells. Cas9 vectors, cell lines, and transgenic organisms are all commercially available.
5. Determine the strategy to use for identifying mutants or mutant lines. The zebrafish approach analyses CRISPR mutants shortly after the end of embryogenesis, essentially as an F0 analysis. In other strategies, heritable genome-edited lines can be isolated, either as edited bacterial or yeast colonies on a plate, or in the context of longer-term strategies to identify and maintain mutant *Drosophila* lines.
6. Isolate DNA from potential modified lines and perform PCR/sequencing to identify introduced changes.

Classroom activities at the University of New Mexico
Designing and cloning CRISPR targets in Drosophila

To illustrate our approach in greater detail, we describe here the methodology used to design plasmids expressing the sgRNAs that will target the genes of interest. The appendix contains detailed protocols that we use for the initial cloning steps of the class.

FlyBase.org offers the complete sequence and gene structure for every gene in *Drosophila*, and students are expected to become familiar with the pages that describe their gene of interest. We ask the students to bring a laptop to class so that we can walk them through the exploration of their gene. FlyBase.org allows one to download the sequence of any *Drosophila* gene, annotated to identify exon and intron sequence, plus coding and noncoding sequences (Fig. 19.1) [11].

Students generate a Word document that contains their gene sequence, and they annotate this sequence with target sites and primers that are designed during the course of the semester (Fig. 19.2).

To generate a mutant allele of their gene, we suggest that students target a sequence early in the coding region, for example, in the first coding exon, so that they will have the greatest chance of generating null alleles for the gene. Students utilize two websites for target sequence design:

1. The first site (http://tools.flycrispr.molbio.wisc.edu/targetFinder/index.php; [12]) is designed to find CRISPR targets and whether each target has potential off-target sites in the genome (Fig. 19.3).
2. Once a target sequence is identified that lacks off-targets, it can be input into a second website (https://shigen.nig.ac.jp/fly/nigfly/cas9/cas9TargetFinder.jsp;jsessionid=6A773B233DAC4DB1E4AF3574A256B7B1; [13]) which conveniently outputs primers to order in the appropriate 5′ to 3′ orientation, eliminating the chance of ordering incorrect oligonucleotide sequences (Fig. 19.4). The oligonucleotides generated are annealed, and the oligonucleotide design incorporates 5′ overhangs for convenient ligation into the pbFv-U6.2 or pbFv-U6.3 vectors that have been cut with the restriction enzyme *Bbs*I. Multiple vectors are described at http://www.crisprflydesign.org/plasmids/ and can be purchased from the Drosophila Genomics Resource Center (DGRC; https://dgrc.bio.indiana.edu/Home).

458 Chapter 19 CRISPR in the classroom

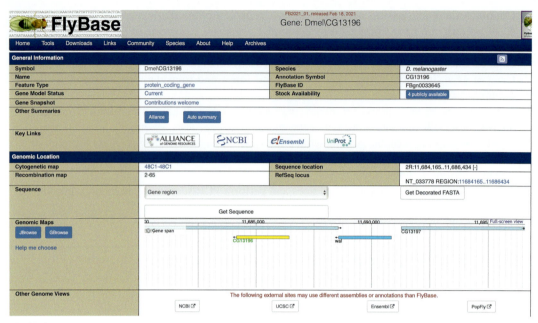

FIG. 19.1

At FlyBase.org, each gene has an individual gene page that links to known information about the gene. Specific links allow investigators to identify the chromosomal location, sequence, and predicted function of all *Drosophila* genes.

From J. Thurmond, J.L. Goodman, V.B. Strelets, H. Attrill, L.S. Gramates, S.J. Marygold, B.B. Matthews, M. Millburn, G. Antonazzo, V. Trovisco, T.C. Kaufman, B.R. Calvi, the FlyBase Consortium, FlyBase 2.0: the next generation, Nucleic Acids Res. 47 (D1) (2019) D759–D765. http://www.FlyBase.org.

The resulting clones that contain insertions with confirmed sequences can then be purified in a large-scale preparation and injected into transgenic fly embryos which express Cas9 in the germline. Teaching 15 students to inject was not practical for our timeline and instead we had each student come in for a one-hour time slot to go through the procedure and participate in the microinjections. Once G0 adult flies eclosed, they were mated to a balancer line to ensure that potential mutant alleles were not lost. Potential mutants were made homozygous in the subsequent generations, or if lethal lines were generated, then balanced lines were maintained. Details of the crosses used can be found in Adame et al. [9].

The final class activity was to identify the changes to the target genes following CRISPR. For homozygous-viable lines, this was achieved simply by PCR and by sequencing amplified DNA. For lethal lines, PCR products from heterozygotes (and therefore containing both wild-type and mutant sequences) were cloned into a TA cloning vector such as pGEM-T Easy (Promega Corp.), and multiple recombinant clones were sequenced, with the expectation that 50% of clones will have the mutant sequence.

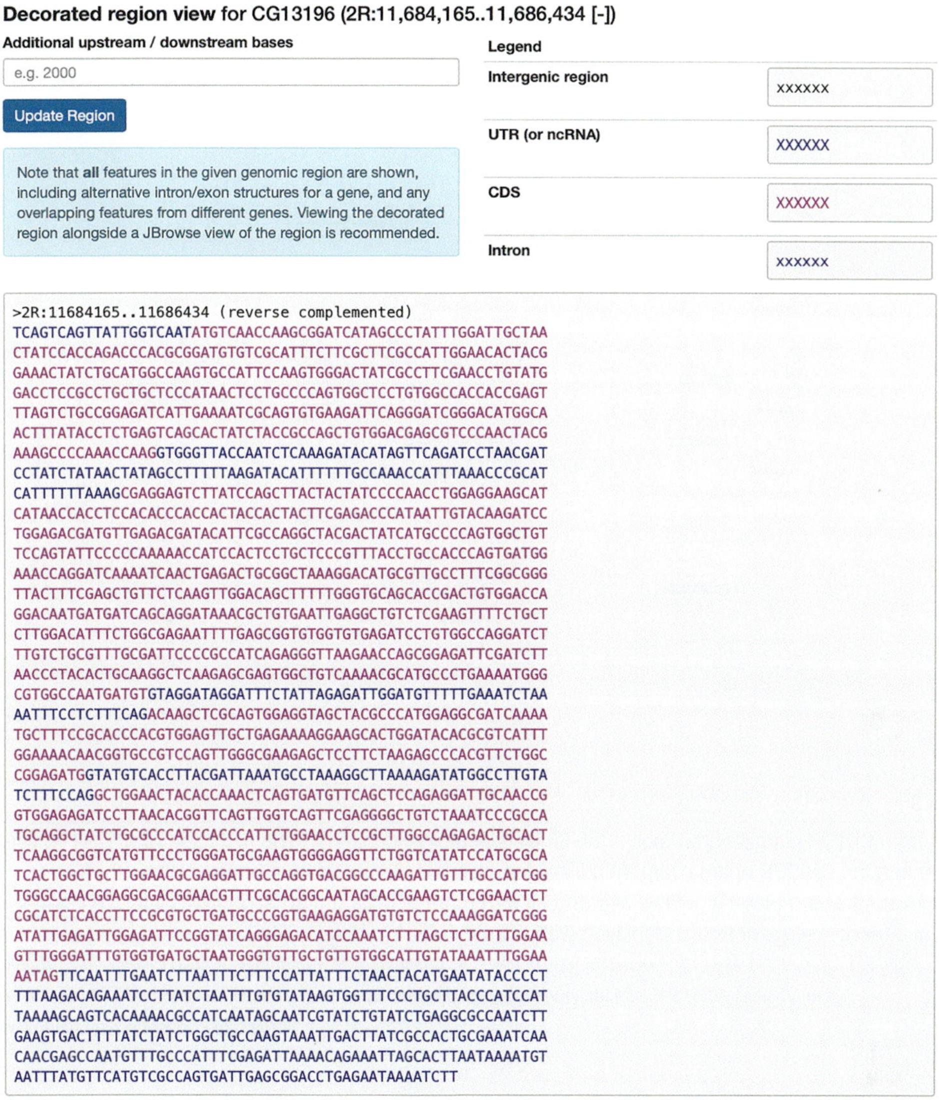

Decorated region view for CG13196 (2R:11,684,165..11,686,434 [-])

Additional upstream / downstream bases

e.g. 2000

[Update Region]

Note that **all** features in the given genomic region are shown, including alternative intron/exon structures for a gene, and any overlapping features from different genes. Viewing the decorated region alongside a JBrowse view of the region is recommended.

Legend

Intergenic region	xxxxxx
UTR (or ncRNA)	xxxxxx
CDS	xxxxxx
Intron	xxxxxx

```
>2R:11684165..11686434 (reverse complemented)
TCAGTCAGTTATTGGTCAATATGTCAACCAAGCGGATCATAGCCCTATTTGGATTGCTAA
CTATCCACCAGACCCACGCGGATGTGTCGCATTTCTTCGCTTCGCCATTGGAACACTACG
GAAACTATCTGCATGGCCAAGTGCCATTCCAAGTGGGACTATCGCCTTCGAACCTGTATG
GACCTCCGCCTGCTGCTCCCATAACTCCTGCCCCAGTGGCTCCTGTGGCCACCACCGAGT
TTAGTCTGCCGGAGATCATTGAAAATCGCAGTGTGAAGATTCAGGGATCGGGACATGGCA
ACTTTATACCTCTGAGTCAGCACTATCTACCGCCAGCTGTGGACGAGCGTCCCAACTACG
AAAGCCCCAAACCAAGGTGGGTTACCAATCTCAAAGATACATAGTTCAGATCCTAACGAT
CCTATCTATAACTATAGCCTTTTTAAGATACATTTTTTGCCAAACCATTTAAACCCGCAT
CATTTTTTAAAGCGAGGAGTCTTATCCAGCTTACTACTATCCCCAACCTGGAGGAAGCAT
CATAACCACCTCCACACCCACCACTACCACTACTTCGAGACCCATAATTGTACAAGATCC
TGGAGACGATGTTGAGACGATACATTCGCCAGGCTACGACTATCATGCCCCAGTGGCTGT
TCCAGTATTCCCCCAAAAACCATCCACTCCTGCTCCCGTTTACCTGCCACCCAGTGATGG
AAACCAGGATCAAAATCAACTGAGACTGCGGCTAAAGGACATGCGTTGCCTTTCGGCGGG
TTACTTTCGAGCTGTTCTCAAGTTGGACAGCTTTTTGGGTGCAGCACCGACTGTGGACCA
GGACAATGATGATCAGCAGGATAAACGCTGTGAATTGAGGCTGTCTCGAAGTTTTCTGCT
CTTGGACATTTCTGGCGAGAATTTTGAGCGGTGTGGTGTGAGATCCTGTGGCCAGGATCT
TTGTCTGCGTTTGCGATTCCCCGCCATCAGAGGGTTAAGAACCAGCGGAGATTCGATCTT
AACCCTACACTGCAAGGCTCAAGAGCGAGTGGCTGTCAAAACGCATGCCCTGAAAATGGG
CGTGGCCAATGATGTGTAGGATAGGATTTCTATTAGAGATTGGATGTTTTTGAAATCTAA
AATTTCCTCTTTCAGACAAGCTCGCAGTGGAGGTAGCTACGCCCATGGAGGCGATCAAAA
TGCTTTCCGCACCCACGTGGAGTTGCTGAGAAAAGGAAGCACTGGATACACGCGTCATTT
GGAAAACAACGGTGCCGTCCAGTTGGGCGAAGAGCTCCTTCTAAGAGCCCACGTTCTGGC
CGGAGATGGTATGTCACCTTACGATTAAATGCCTAAAGGCTTAAAAGATATGGCCTTGTA
TCTTTCCAGGCTGGAACTACACCAAACTCAGTGATGTTCAGCTCCAGAGGATTGCAACCG
GTGGAGAGATCCTTAACACGGTTCAGTTGGTCAGTTCGAGGGGCTGTCTAAATCCCGCCA
TGCAGGCTATCTGCGCCCATCCACCCATTCTGGAACCTCCGCTTGGCCAGAGACTGCACT
TCAAGGCGGTCATGTTCCCTGGGATGCGAAGTGGGGAGGTTCTGGTCATATCCATGCGCA
TCACTGGCTGCTTGGAACGCGAGGATTGCCAGGTGACGGCCCAAGATTGTTTGCCATCGG
TGGGCCAACGGAGGCGACGGAACGTTTCGCACGGCAATAGCACCGAAGTCTCGGAACTCT
CGCATCTCACCTTCCGCGTGCTGATGCCCGGTGAAGAGGATGTGTCTCCAAAGGATCGGG
ATATTGAGATTGGAGATTCCGGTATCAGGGAGACATCCAAATCTTTAGCTCTCTTTGGAA
GTTTGGGATTTGTGGTGATGCTAATGGGTGTTGCTGTTGTGGCATTGTATAAATTTGGAA
AATAGTTCAATTTGAATCTTAATTTCTTTCCATTATTTCTAACTACATGAATATACCCCT
TTTAAGACAGAAATCCTTATCTAATTTGTGTATAAGTGGTTTCCCTGCTTACCCATCCAT
TAAAAGCAGTCACAAAACGCCATCAATAGCAATCGTATCTGTATCTGAGGCGCCAATCTT
GAAGTGACTTTTATCTATCTGGCTGCCAAGTAAATTGACTTATCGCCACTGCGAAATCAA
CAACGAGCCAATGTTTGCCCATTTCGAGATTAAAACAGAAATTAGCACTTAATAAAATGT
AATTTATGTTCATGTCGCCAGTGATTGAGCGGACCTGAGAATAAAATCTTT
```

FIG. 19.2

The "Get Decorated Fasta" link enables investigators to download the entire sequence of the gene. This sequence is color-coded to identify coding *(Pink)* versus noncoding and intronic sequence *(Blue)*.

From J. Thurmond, J.L. Goodman, V.B. Strelets, H. Attrill, L.S. Gramates, S.J. Marygold, B.B. Matthews, M. Millburn, G. Antonazzo, V. Trovisco, T.C. Kaufman, B.R. Calvi, the FlyBase Consortium, FlyBase 2.0: the next generation, Nucleic Acids Res. 47 (D1) (2019) D759–D765. http://www.Flybase.org.

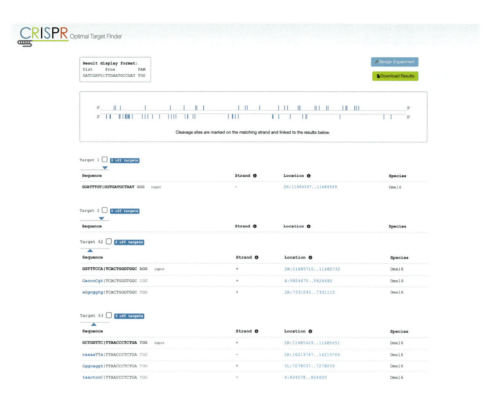

FIG. 19.3

Output from CRISPR target finder tool at the University of Wisconsin. (A) The first exon of CG13196 was inserted into the CRISPR optimal target finder algorithm. Locations can be visualized along the sequence of DNA. Hovering over a target identifies its number with more detailed information listed later. (B) Targets with potential off-target sequences are displayed with links to their locations in FlyBase.

From S.J. Gratz, F.P. Ukken, D. Rubinstein, G. Thiede, L.K. Donohue, A.M. Cummings, K.M. O'Connor-Giles, Highly specific and efficient CRISPR/Cas9-catalyzed homology directed repair in drosophila, Genetics 196 (4) (2014) 961–971. http://targetfinder.flycrispr.neuro.brown.edu/.

Recent innovations at the University of New Mexico

Since the initial description of our class [9], we have enhanced the class in a number of ways.

Introduction of video tutorials

In addition to providing written protocols, we used an iPhone to videotape procedures in real time and then edited the activities into a short movie using iMovie. This approach was designed to illustrate the subtle techniques used in several cloning activities. Students were expected to watch the videos as well as read the protocols, prior to class. The students with less experience found the videos most useful, often replaying them during the class while the more experienced students followed the written protocols.

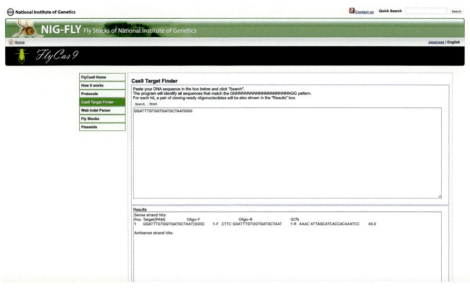

FIG. 19.4

Generation of oligonucleotide primers for cloning. When a suitable target sequence plus PAM is input into the https://shigen.nig.ac.jp/fly/nigfly/cas9/cas9TargetFinder.jsp website, the algorithm will return forward and reverse primer sequences for cloning into pBFv U6.2. Alternative 5′ overhangs would have to be added to the primer sequence if a different sgRNA vector is used.

From S. Kondo, R. Ueda, Highly improved gene targeting by germline-specific Cas9 expression in Drosophila, Genetics 195 (3) (2013) 715–721. https://shigen.nig.ac.jp/fly/nigfly/cas9/cas9TargetFinder.jsp.

Additionally, there were three instructors walking around during the lab class to offer direction. Using these approaches, we were able to cover all the learning styles and speeds.

Co-CRISPR strategy

Our laboratory is interested in genes involved in somatic and cardiac muscle development in *Drosophila*. There are a large number of *Drosophila* genes whose function is unknown, and we thought it would be beneficial to give students in our class target genes that might be of further interest in our work. In this way, students generate mutants of genes for which there are no known mutant alleles and therefore no known mutant phenotypes. These reagents should be valuable for the biomedical research community.

One drawback to this approach is that mutants for some genes might not have an obvious functional or visible phenotype. Another issue is that, despite best efforts in designing effective CRISPR targets, there is the possibility that the students will not generate a mutant allele. To address this, we have students coinject a plasmid that expresses a sgRNA to target the *ebony* gene. The lines that transmit mutant *ebony* alleles show a dramatic darkening of body color in the F1 generation when crossed to *ebony*-mutant balancer lines and, more importantly, have a higher potential for transmitting mutant alleles of the gene of interest. This co-CRISPR strategy described in Kane et al. [14] is highly efficient and enables students to visualize CRISPR success even if they are unable to detect mutations in their genes of interest. One downside to this class is the frustration experienced by students who do not obtain a mutant. They can agonize about what they did wrong during the process, and we have found that observing co-CRISPR mutant alleles being generated can mitigate these concerns.

Pairing with a graduate student for near-peer mentoring
Variation in the amount of prior knowledge and experience among undergraduate students can be a challenge for instructors. On the one hand, instructors want to ensure that all students keep up with the material, but on the other hand, it is important not to reduce the complexity of the class, which may result in more experienced students becoming disengaged. We require prerequisite classes that include a basic molecular biology background and microbiology experience. Despite this, we see a range of understanding and technical expertise. Our pilot study involved six junior or senior undergraduate students whom we recruited from the Biology Department, most of whom had prior research experience. The main obstacles in expanding the class to a regular semester laboratory were class size and knowledge base variability.

To address these issues, we designed the expanded class with ten undergraduates and five graduate students. In the first large class, undergraduates were paired to design two targets for the same gene and sat at the same laboratory table with a graduate student assigned to a different gene. This was beneficial because undergraduates had a partner who worked through tasks alongside them, and the graduate student was able to help the undergraduates if they needed assistance along the way. The undergraduates felt more comfortable asking the graduate student questions before asking instructors, and if further explanation was still necessary, it was most likely a concept or question that needed to be shared with the rest of the class. With such a layered support system we did not have any students fall behind or get lost. In fact, in all three post-pilot courses we have carried out, all students performed well as assessed by oral presentations, instructor questions, and in participation in discussions of primary literature. This type of approach to student instruction is analogous to near-peer mentoring, where students often feel more comfortable asking questions of trainees who are closer to them in the age or experience (see, for example, [15]).

Epitope tagging
In a subsequent class, the two undergraduates and a graduate student were assigned the same gene. In this new scenario, the undergraduates designed targets for mutagenesis, while the graduates designed a target for cutting the DNA along with a template for HA-tagging the gene by homology-directed repair. The design of the repair template requires a higher level of application of biological knowledge; therefore, we assigned this activity to the graduate student. This also was an effective near-peer mentoring approach, since in this case both the undergraduates and the graduate student work on the same gene and therefore can work more closely together.

Considerations of rigor and reproducibility in CRISPR classes
Central tenets of scientific research are that the studies must be rigorous in ensuring that the correct approaches and reagents have been used in order to provide robustness to the experimental design, and the results must be reproducible. As a result of the significant concerns that have been raised about rigor and reproducibility in the current scientific arena, the concepts of reliable experimental design and performance have received renewed attention [16].

Undergraduate research-based activities must therefore also address issues of rigor and reproducibility in order to provide an authentic research experience. This can be challenging in the context of an undergraduate class, where time can be short and laboratory scheduling may not lend itself to performing additional iterations of some of the activities. Moreover, since genome editing classes consist of sequential steps during the course of the semester, there may be a concern that spending time assuring rigor and reproducibility might prevent achieving the overall goals of the class, which are to generate mutant alleles for candidate genes.

We believe that many of these issues are addressed in the designs of undergraduate CRISPR classes, although not necessarily appreciated by the students. We underline some of the designs and strategies that address rigor and reproducibility, and we suggest that a goal of the instructors of CRISPR CUREs should be to underline these aspects of experimental design.

First, we stress on the students the importance of validating their reagents. The most critical of these is the sgRNA-expressing plasmid that the students generate, which they then sequence in order to verify that they have cloned into the plasmid the correct spacer sequence. Only those plasmids that are correctly validated in this manner can be used for microinjection.

We also space our laboratory activities to two classes per week. In this way, if a student or group falls behind, for example, by failing to obtain validated clones expressing the sgRNA, there is time for the students to catch up between classes.

In addition, we carry out, during the semester, a PCR of the genomic regions being targeted by the students. This approach occurs earlier in the semester, in order that we can be confident of amplifying the correct target regions of the genes to be mutated and before having mutant lines in hand. While we do not normally sequence these control PCR products, they could be sequenced to ensure that the targets' sequences present in the *Cas9*-expressing lines that we use for editing are identical to those in the reference genome (and see further comments discussed later).

A positive outcome of a CRISPR class is that we frequently generate multiple independent alleles of the same gene. The students can then also determine whether the phenotypes they observe are consistent for different alleles, which add reproducibility to their activities.

A final challenge is that it can be difficult to predict the efficacy of the CRISPR mutagenesis and crossing. This is especially germane given the complexity of the microinjection process and then subsequent crosses to generate mutant alleles. We have tried to address this issue by introducing the co-CRISPR approach (described earlier and in [14]). In this approach, students screen for mutation of the *ebony* body-color marker as a proxy for effective mutation of their gene of interest. The identification of *ebony* mutant lines confirms that the injection and crossing have been effective for one allele and, at the same time, identifies lines that have a higher probability of carrying mutations of the gene of interest.

Actions that could further improve experimental success, especially for our Drosophila class, relate largely to the *Cas9*-expressing lines that are available. These lines are not isogenic, and have not, to our knowledge, been fully sequenced. This leads to two issues: First, not knowing the genome sequence of the line that is being mutated could result in poor CRISPR efficiency if there is polymorphism between the wild-type reference genome and the genome of the *Cas9* line. In this instance, a sgRNA that is designed based on the reference genome might not work effectively if the Cas9 line has a polymorphism at the location of the protospacer sequence.

A second concern is that if the *Cas9* lines are not isogenic, second-site mutations that are present at low levels in the *Cas9* stock might become homozygosed in any mutant lines that are generated. We observed a potential example of this when we mutated the *Act79B* gene and were unable to generate homozygotes of an otherwise homozygous-viable mutation [9].

An effective solution to both of these concerns would be to generate an isogenic Cas9 line that shows strong viability, high fecundity (so that many viable embryos can be obtained for microinjection), and phenotypically normal. Efforts are ongoing in this regard. An alternative approach, which we use in our class to address the second point raised, is to cross the putative mutants to a chromosomal deficiency for the gene that is being targeted. In this manner, we are not looking for mutant phenotypes in a homozygous situation but instead in a hemizygous situation. Nevertheless, second-site mutations can still complicate further phenotypic analyses.

Conclusions

In this chapter we have described the different strategies used by a number of investigators to introduce to undergraduate students the emerging technology of CRISPR/Cas9 genome editing. We emphasize that different approaches can be used to accommodate different time constraints or levels of complexity, and we provide greater rationale and details for CRISPR classes that we have taught.

Appendix: Detailed methods used to create guide RNA plasmids for use in Drosophila

Exercise 1: Annealing oligonucleotides to generate short dsDNA inserts

The purpose of today's exercise is to take the oligonucleotides that you ordered to make the sgRNA plasmid, and combine them to make short dsDNA molecules. These molecules will be ligated into a plasmid in next week's class. For undergraduate pairs, remember you will be generating two short dsDNA molecules for your two different targeting vectors, so each student will carry out a separate reaction.

A. Re-suspend oligonucleotides
   The oligos that you designed have been lyophilized (freeze-dried) for shipment, and we need to reconstitute these before carrying out reactions.
   1. Spin your tubes of oligos briefly up to 30 s, to make sure that any powder settles to the bottom of the tube. Remember to balance the centrifuges!
   2. On the data sheet, determine how much water you should add to each tube in order to generate a 100 µM stock (micro-Molar) solutions. Hint: a column of the data sheet will tell you this value. Note: this value will be different for each tube.
   3. Add the appropriate amount of water to each tube and replace the cap.
   4. Vortex the tubes, and store on ice for short term storage, or in the freezer for long-term storage. For now, keep them on ice, because we are going to use them immediately.

B. Anneal oligos
   1. For each dsDNA to be created, label a tube and add the following components in this order:
      4 µL of your + oligo
      4 µL of your − oligo
      8 µL of water
      4 µL of 5X Phusion HF Polymerase buffer (prob. any enzyme buffer would be fine)
   2. Mix reactions by gently pipetting up and down a few times.
   3. Store on ice until all reactions are ready (wait for the rest of the class to finish).
   4. Add Grippers to tubes, and place into the 95°C heating block (Caution—hot!).

5. After 5 min, remove the block from the heating element, and leave on the bench to cool.
6. When the temperature has dropped to 30°C or below, remove the tubes.
7. Spin the tubes briefly to collect any condensation.
8. Store your annealed oligos in the freezer until next time.

Exercise 2: Phosphorylating your dsDNA, and ligating into pBFv-U6.2

We will do two procedures today: firstly, we will add phosphate groups to the 5′ ends of your dsDNA fragments, in order to improve their chances of getting ligated into the plasmid. Secondly, we will set up ligation reactions in which we will enzymatically "glue" the inserts into the plasmid. On Friday, we will transform part of the ligation mixtures into *E. coli* cells, to identify clones that have been correctly constructed.

Remember that each student will be generating a different plasmid that carries their individual targeting sequence, so each person will be doing a kinase reaction and a transformation.

A. Phosphorylation of dsDNA molecules
1. Thaw out your annealed oligos, and the T4 DNA ligase buffer, and mix each tube thoroughly. Keep on ice when not in use.
2. For each kinase reaction, add the following components in the following order:
   Annealed oligos 10 μL
   Sterile water 34 μL
   10X T4 ligase buffer 5 μL
   T4 polynucleotide kinase 1 μL
   Mix well by pipetting up and down, and stirring the mixture. Do not vortex, and avoid too many bubbles.
3. Incubate in the 37°C heating block for 30 min. After this time, store on ice.

B. Ligation
This is where we combine your DNA to be inserted with the pBFv-U6-2 vector.
1. Obtain and label a fresh tube. For each targeting construct, add the following components in the order listed:
   pBFv-U6.2 cut with *Bbs*I 1 μL
   Kinased oligos 1 μL
   Sterile water 16 μL
   10X T4 DNA ligase buffer 2 μL
   T4 DNA ligase 1 μL
   Mix well by pipetting up and down, and stirring the mixture. Do not vortex.
2. Incubate at room temperature. Later in the day, or early in the morning, we will put these into the freezer for you.

Exercise 3: Transformation of ligation products into *E. coli*

Today, we will insert your ligation products into *E. coli* cells using a method called transformation. Transformed cells will then be plated out on LB plates containing ampicillin. Since pBFv-U6.2 carries an Amp^R gene, the antibiotic will select for those cells that received an intact plasmid during the transformation protocol. The individual cells will grow to form a colony of identical cells (i.e., a clone, and hence the term cloning). On Sunday, we will select four colonies from each transformation to ultimately purify and sequence the plasmid DNA.

Each student will carry out one transformation reaction.

Before starting: fill an ice bucket with ice, and obtain one LB-amp plate per student. Remove your ligations from the freezer, thaw and mix, then store on ice.

1. Take your ice bucket to the front bench, and collect one tube of competent cells and one tube of SOC medium. Keep the cells on ice at all times.
2. Thaw the cells slowly on ice, and mix (very) gently. Label your tube of cells according to your ligation reaction, and label plates with your name, gene name, and ligation.
3. Add 2 μL of your ligation reaction to the tube of competent cells; mix by gentle pipetting and stirring.
4. Leave cells on ice for 20–60 min.
5. Heat shock cells for 90s at 42°C, then return cells to ice for 2 min.
6. Add 150 μL of SOC medium to each tube, and mix gently.
7. Incubate the tubes at 37°C for 15–60 min, to allow the cells to recover and begin expressing the ampicillin resistance gene.
8. For each ligation/transformation, add all of the cells to the appropriate plate, and spread the cells using a sterilized spreader (this will be demonstrated).
9. We will place these in the 37°C incubator right-side up until the plates are dry, then turn them over for overnight incubation.
10. Return your ligations to the freezer.
11. Colonies on the plates will be used to inoculate cultures to isolate plasmid DNA for sequencing.

Exercise 4: Picking colonies, minipreps, and sequencing

This activity involves several procedures, that we will start on Sunday and finish on the Monday afternoon.

On Sunday, we will inoculate small (3-mL) cultures with colonies from your transformation plates (four cultures per transformation), and leave these cultures to grow overnight.

On Monday, we will firstly purify plasmid DNA from each of the cultures using a technique called a miniprep, and then set up sequencing reactions of each purified plasmid. This will determine whether or not your dsDNA oligos have been inserted correctly into the plasmid.

The sequencing reactions that you set up on Monday will run in the PCR machine until later in the afternoon, and I will freeze these reactions until Friday's lab.

A. **Sunday—Picking colonies**
   1. Remove your bacterial plates from the fridge, and identify four colonies from each ligation to grow up. These colonies should be of a good size, and well-isolated from neighbors.
   2. Thaw out the 100 mg/mL Ampicillin solution, mix, and store on ice.
   3. For each culture to be grown, obtain a 15-ml snap-top tube, and label with your ligation name, followed by 1, 2, 3, or 4.
   4. To each tube, 3 mL of LB medium, and 3 μL of Amp.
   5. Using a yellow tip on a pipette, gently touch a single colony with the tip. Next, dip the tip into the appropriate culture tube. Wiggle the tip to disperse cells, and then discard tip and re-cap the tube.
   6. Repeat step 5 for your remaining three colonies.
   7. Place all four tubes containing your different clones into the shaking incubator, where they will grow overnight at 37C.

B. Monday—Minipreps
1. Before starting, label tubes and columns as follows: for each clone, label two Eppendorf (1.5-mL) tubes and one Qiagen miniprep column with your clone name (ligation followed by 1, 2, 3, or 4).
2. Remove ~1.5 mL of culture from each snap-top tube, and add to the appropriate 1.5-mL tube.
3. Spin in the microcentrifuge for 1 min to pellet cells. Pour off and discard supernatant.
4. To each tube, add 250 μL of Buffer P1, and vortex to resuspend cells.
5. To each tube, add 250 μL of Buffer P2, and mix briefly. The solution should turn blue.
6. To each tube, add 350 μL of Buffer N3, and mix briefly. The solution should turn colorless.
7. Immediately spin tubes for 5 min to pellet the cellular debris.
8. As soon as tubes have stopped spinning, remove supernatant from each tube and add to the top of the appropriately-labeled column. Discard the tubes containing the pelleted crud.
9. Spin column (plus collecting tube) for 30 s, in order that the DNA passes into the column and binds to the matrix. Discard the eluate.
10. Add 750 μL of Buffer PE to each column (plus collecting tube), and spin for 30 s. This washes the DNA while it is bound to the column. Discard the eluate.
11. Spin the column (plus collecting tube) one more time to remove trace amounts of buffer from column.
12. Discard the collecting tube, and place the column over a fresh 1.5-mL tube. Add 30 μL of Elution Buffer to each column. Do not touch the top of the binding matrix with your pipette tip, but do ensure that the water is added to the center of the binding matrix. Let the columns sit for 2 min to ensure elution of the DNA from the column.
13. Spin the column (plus 1.5-mL collecting tube) for 30 s in the centrifuge to collect the eluted DNA in the tube. Discard the column. Store the eluted DNA on ice or in the freezer.

C. Monday—Sequencing DNA
1. Collect one tiny PCR tube for each clone, and label appropriately.
2. To each tube, add the following reagents in the following order:
   Water 13 μL
   Plasmid DNA (different for each tube) 1 μL
   T3 sequencing primer (25 ng/μL) 1 μL
   5X Bigdye sequencing buffer 4 μL
   BigDye3.1 sequencing reagent 1 μL
   TOTAL 20 μL
3. Place tubes into the thermocycler, and run the sequencing program. Your reactions will be removed from the instrument once the protocol is complete, and stored in the freezer until next time.

Exercise 5: Sequencing clean-up

Today, we will return to your sequencing reactions from Monday. We will precipitate and then dry the sequencing products in order that they can be analyzed by the sequencer.

1. Remove your sequences from the freezer and thaw out. Once thawed, transfer each reaction from the tiny tubes into an appropriately-labeled regular 1.5-mL tube.
2. Add 2 μL of 3 M NaAc, pH 5.2, and mix by pipetting. This salt ensures that the DNA can be precipitated out of solution once ethanol is added.

3. Add 50 μL of 100% ethanol and mix thoroughly. Then place the tubes at −80°C for at least 10 min. These steps cause the DNA to come out of solution, and the cold enhances the efficiency of precipitation.
4. After 10 min, remove the tubes from the −80°C freezer, and warm briefly so that the plastic of the tubes is not too rigid (if you put the tubes straight into the centrifuge at this point, the centrifugal force will shatter the tubes!). Once the tubes are warmed a bit, put them into the centrifuge and spin for 15 min at full speed. IMPORTANT: orient your tubes in the centrifuge so that the hinges are pointing to the outside.
5. As soon as the centrifuge stops, remove your tubes and locate the pellet at the bottom of the tube, below the hinge. Gently pipette off the supernatant, while ensuring that your pellets do not detach and wash out of the tube.
6. Add 100 μL of 70% ethanol to your tubes, and spin again (at room temperature) for 3 min to re-stick your pellet.
7. Pipette off the ethanol, and air dry the pellet for at least 10 min.
8. Your sequencing reactions are now ready to be submitted to the Molecular Biology Facility. We will email you your sequences over the weekend.

Exercise 6: Analysis of sequences and transformation of successful clones

This exercise will use a program at https://embnet.vital-it.ch/software/LALIGN_form.html to determine if the sequences of any of your plasmids (emailed to you over the weekend) have the correct insert into the pBFv-U6.2 plasmid. You should receive up to four sequencing files corresponding to the number of clones that you sequenced.

1. Look at the sequence that was generated. If there is no sequence at all (or just a couple of Ns), then the reaction or precipitation did not work. Similarly, if there is a lot of sequence present but it is almost all Ns, then again the reaction did not work. Either way, you should move on to your next clone and hope for a good sequencing run.

A good sequencing reaction will look something like this (but not highlighted):

```
NNNNNNNNNNNNNNTCGNNNNNANNNNNATACCAACGGTGGAAAGCCNAGTGCATGCGCGGCCG-
CAAAAAANNCANNNNCTCGGTGCCACTTTTTCAAGTTGATAACGGACTANCCTTATTTTAACTTGC-
TATTTCTAGCTCTAAAACAGTCCAAGCGTGGTATCCTCGAAGTATTGAGGAAAACATACCTATATAAAT-
GATCAACATCAGGAAAGAGCAGTTGAGAATTATAAGAATTGGCAAATGGTCTTAAGAACCCTCTGCT-
TAAGATTTTCAAAATTTCCTTTAGAATCAAAGTGTCCTATTGTTCGTTTGTTGAAAACACAGTTCGATT-
TATTGACTATAATAAATTGATAGTTTTAAATATAGAGGCACGACTAAGAGAGTTTGGTTTTGTTTTGGT-
TAGGGACCAAAAAAAGTATACATAACGAATAAAAAAGGATTTAAGTGCAAATGGTAAAAAAGTGTC-
GAGTTTTTCTTGAGTTGATTGTGCTGTAATGAGACTCTGCATTCGGCATTTGACTCGGCTTTTCCTACTC-
GTGCCGTATTTCAGGCTGCAAGTCGAACCTGCAGGGAATTCGATATCAAGCTTGGATTTGTGTGCGC-
CGCACTTTCACCTCAAGTGATTGATAATTCCCAGCCTATCTGGCAGTGCCCATCGCCCAGATCAC-
CGACTGTGCAATCAGTCGGAACTGGAGCTCTCTCGCTCTGTTATCGGTTCGCTGGGGTCTCATCTC-
CGGTCCGCTGGCGGANATCAGTTCGCCAGCATCCGCCGCTCGAGGAGTCACGATCTGATCTGAGCT-
GNGNNCCATGAGCTGTCCCTATGCNGGAAACGGGTGANNNCAGCNCGTGCTGTCCAGGAAT-
GNNNNCGATCTTCANTNCTGCNANTNNNNCAANCCANNNNNAAANNATNNCNATGANNCGNNG-
GNGNCNNTNNNNCNNNNNGGGCAAANNNNNNGGNNNGNNNCTNANGCNGNNNANNGNNNNNNNCC
NNNNNNNGNNNNNNNNNNNNNNNNNNNNNNNNCNNNNNNNANNNNNNNNCNNNNNNNNNNNNNNNNNN
GNNNNNNNNNNNNNNTNNNNNNNNNNNNNNNCNTNANNGGNNNNNNANNNNNNNNNNNN
```

If none of your sequences are "good", we will come up with a contingency plan, so just check your email, or wait for us to talk about it in class.

2. Once you have identified a good sequence file, copy the largest block of readable text while avoiding the long strings of Ns at either end of your sequence (something like what I highlighted above). It does not matter if you have a few Ns in the copied sequence.
3. Go to url: http://www.ch.embnet.org/software/LALIGN_form.html And paste this sequence into the upper box. Label it with your clone name (e.g., CG1234-T1 A).
4. Copy the following **short** pBFv-U6.2 sequence, and paste it into the second box at the lalign website. Label this sequence appropriately. This sequence is also posted in Learn in the Protocols folder, and is color-coded according to the details in that file.

> GCGCGCAATTAACCCTCACTAAAGGGAACAAAAGCTGGAGCTCGACTCATCGTCCCAGAGA-
> ATCGATACCAACGGTGGAAAGCCAAGTGCATGCGCGGCCGCAAAAAAAGCACCGACTCGGTGC-
> CACTTTTTCAAGTTGATAACGGACTAGCCTTATTTTAACTTGCTATTTCTAGCTCTAAAACAG **GTCTTC**
> AACTC **GAAGAC** CCGAAGTATTGAGGAAAACATACCTATATAAATGATCAACATCAGGAAAGAGCAGTT-
> GAGAATTATAAGAATTGGCAAATGG

5. Run lalign.
6. For the sequence output, you should see something like this (but not highlighted):

\>\>pBFV-U6_2 286 bp (286 nt) Waterman-Eggert score: 940; 141.7 bits; E(1) < 4.8e-38 91.0% identity (91.0% similar) in 222 nt overlap (1-219:67-286)

> My TACCAACGGTGGAAAGCCNAGTGCATGCGCGGCCGCAAAAAANNCANNNNCTCGGTGCCA
> ::::::::::::::::: ::::::::::::::::::::::: :: :::::::::
> pBFV. TACCAACGGTGGAAAGCCAAGTGCATGCGCGGCCGCAAAAAAAGCACCGACTCGGTGCCA
> 70 80 90 100 110 120
> 70 80 90 100 110 120
> My CTTTTTCAAGTTGATAACGGACTANCCTTATTTTAACTTGCTATTTCTAGCTCTAAAACA
> ::::::::::::::::::::::::::: ::::::::::::::::::::::::::
> pBFV CTTTTTCAAGTTGATAACGGACTAGCCTTATTTTAACTTGCTATTTCTAGCTCTAAAACA
> 130 140 150 160 170 180
> 130 140 150 160 170
> My G---TCCAAGCGTGGTATCCTCGAAGTATTGAGGAAAACATACCTATATAAATGATCAAC
> : : ::: : : : : :::::::::::::::::::::::::::::::::
> pBFV GGTCTTCAAC----TCGAAGACCCGAAGTATTGAGGAAAACATACCTATATAAATGATCAAC
> 190 200 210 220 230 240
> 180 190 200 210
> My ATCAGGAAAGAGCAGTTGAGAATTATAAGAATTGGCAAATGG
> ::
> pBFV ATCAGGAAAGAGCAGTTGAGAATTATAAGAATTGGCAAATGG
> 250 260 270 280

If it doesn't look like this, let us know, and move on to your next sequencing file.

7. Focus on the region that I highlighted. In your sequence at this location should be the sequence of your **reverse** primer. Confirm that your reverse primer sequence has been inserted into the plasmid. This will confirm that you have successfully built your plasmid to knock out your gene.
8. Copy both your sequence and the alignment into the Word file that you started in the first week of classes. Make sure that it is properly labeled in the Word file, and indicate if you obtained the correct sequence.
9. Repeat this process with all of the remaining successful sequences that you obtained.
10. Decide which one of your clones will be the one with which to proceed. This should be the clone that has the cleanest correct sequence.

Acknowledgments

We acknowledge the support for our CRISPR class from the Department of Biology, University of New Mexico; from a Teaching Allocations Grant from the University of New Mexico; and from 1836718 awarded by the Genetic Mechanisms Program of the National Science Foundation to RMC. We thank Elizabeth Trujillo and Jenny Waters for the valuable comments on the manuscript.

References

[1] G.B. Call, et al., Genomewide clonal analysis of lethal mutations in the *Drosophila melanogaster* eye: comparison of the X chromosome and autosomes, Genetics 177 (2007) 689–697.
[2] S.M. Caruso, J. Sandoz, J. Kelsey, Non-STEM undergraduates become enthusiastic phage-hunters, CBE-Life Sci. Educ. 8 (2009) 278–282.
[3] Boyer Commission on Educating Undergraduates in the Research University, S. S. Kenny (Chair), Reinventing Undergraduate Education: A Blueprint for America's Research Universities, State University of New York–Stony Brook, 1998.
[4] J.T.H. Wang, Course-based undergraduate research experiences in molecular biosciences – patterns, trends, and faculty support, FEMS Microbiol. Lett. 364 (2017) fnx157.
[5] D.M. Thurtle-Schmidt, T.-W. Lo, Molecular biology at the cutting edge: a review on CRISPR/CAS9 gene editing for undergraduates, Biochem. Mol. Biol. Educ. 46 (2018) 195–205.
[6] A.P. Hynes, M.L. Lemay, L. Trudel, H. Deveau, M. Frenette, D.M. Tremblay, S. Moineau, Detecting natural adaptation of the *Streptococcus thermophilus* CRISPR-Cas systems in research and classroom settings, Nat. Protoc. 12 (3) (2017) 547–565.
[7] K.T. Militello, J.C. Lazatin, Discovery of *Escherichia coli* CRISPR sequences in an undergraduate laboratory, Biochem. Mol. Biol. Educ. (2017) 262–269.
[8] N. Sehgal, M.E. Sylves, A. Sahoo, J. Chow, S.E. Walker, P.J. Cullen, J.O. Berry, CRISPR genome editing in yeast: an experimental protocol for an upper-division undergraduate laboratory course, Biochem. Mol. Biol. Educ. 46 (2018) 592–601.
[9] V. Adame, H. Chapapas, M. Cisneros, C. Deaton, S. Deichmann, C. Gadek, T.L. Lovato, M.B. Chechenova, R.M. Cripps, An undergraduate laboratory class using CRISPR/Cas9 technology to mutate *Drosophila* genes, Biochem. Mol. Biol. Educ. 44 (2016) 263–275.

[10] J.M. Bhatt, A.K. Challa, First year course-based undergraduate research experience (CURE) using the CRISPR/Cas9 genome editing technology in zebrafish, J. Microbiol. Biol. Educ. 19 (2018), https://doi.org/10.1128/jmbe.v19i1.1245.

[11] J. Thurmond, J.L. Goodman, V.B. Strelets, H. Attrill, L.S. Gramates, S.J. Marygold, B.B. Matthews, M. Millburn, G. Antonazzo, V. Trovisco, T.C. Kaufman, B.R. Calvi, the FlyBase Consortium, FlyBase 2.0: the next generation, Nucleic Acids Res. 47 (D1) (2019) D759–D765.

[12] S.J. Gratz, F.P. Ukken, D. Rubinstein, G. Thiede, L.K. Donohue, A.M. Cummings, K.M. O'Connor-Giles, Highly specific and efficient CRISPR/Cas9-catalyzed homology directed repair in *drosophila*, Genetics 196 (4) (2014) 961–971.

[13] S. Kondo, R. Ueda, Highly improved gene targeting by germline-specific Cas9 expression in *Drosophila*, Genetics 195 (3) (2013) 715–721.

[14] N.S. Kane, M. Vora, K.J. Varre, R.W. Padgett, Efficient screening of CRISPR/Cas9-induced events in drosophila using a co-CRISPR strategy. G3: genes genomes, Genetics 7 (2017) 87–93.

[15] G. Trujillo, P.G. Aguinaldo, C. Anderson, J. Bustamante, D.R. Gelsinger, M.J. Pastor, J. Wright, L. Marquez-Magaña, B. Riggs, Near-peer STEM mentoring offers unexpected benefits for mentors from traditionally underrepresented backgrounds. PURM4.1, 2014. http://blogs.elon.edu/purm/2015/11/11/near-peer-stem-mentoring-offers-unexpected-benefits-for-mentors/.

[16] M. McNutt, Journals unite for reproducibility, Science 346 (2014) 679.

Index

Note: Page numbers followed by *f* indicate figures, *t* indicate tables, and *b* indicate boxes.

A

Academic integrity, 23
Academic misconduct, 30–32
Affymetrix, 169
Allele-specific expression (ASE) analysis, 204
Alternative splicing (AS), 205–206, 205*f*, 434
    classification of, 168
    exon skipping isoforms, 168
    in microarrays, 169
    protein diversity, 168
    proteome diversity, 168
    SpliceSeq, 168
    using RNA sequencing, 169–170
Animal research: reporting of in vivo experiments (ARRIVE) guidelines, 272
Arabidopsis thaliana, 161
ArrayExpress, 175
AS. *See* Alternative splicing (AS)
Assay of transposase accessible chromatin sequencing (ATAC-seq) assay, 334
    accessible chromatin regions, 285
    applications, 286–287
    ATAC Primer Tool, 290
    ChIPQC package, 292–293
    closed chromatin, 285, 286*f*
    computation tools, 305–308*t*
    differential peak analysis, 296–297
    experimental design, 287–288
    FastQC, 309
    filtered BAM files, 292
    functional analysis, 297
    gene regulatory network (GRN) reconstruction, 304
    Genrich, 309
    hierarchical clustering analysis (HCA), 292
    MACS2.SITE, 309
    motif mapping, 303
    MultiQC, 309
    nuclei preparation, 288–289
    nucleosome positioning, 298–299, 300*f*
    Omni-ATAC, 286, 288
    open chromatin, 285, 286*f*
    PCR amplification, 289
    PCR product size determination, 289
    peak annotation, 297
    peak calling, 293–296, 295–296*f*
    postalignment processing, 291–293, 293*f*
    preprocessing, 291
    principal component analysis (PCA), 292
    quality control, 288–289
    raw read quality control, 291
    reads alignment, 291–293
    regulatory regions, 297
    sequencing depth, 290
    sequencing saturation analyses, 290, 292
    single-cell ATAC-seq (scATAC-seq), 286–287
    step-by-step QC, 287–288
    tagmentation, 286, 289
    TF binding sites (TFBSs), 299
    transcription factors (TF)
        binding activity analysis, 303–304
        occupancy inference, 299–302
    tutorials and protocols, 305–308*t*
    visualization, 297–298
    workflow of, 286, 287*f*
Association for Biology Laboratory Education, 132
ATAC Primer Tool, 290
ATAC-seq assay. *See* Assay of transposase accessible chromatin sequencing (ATAC-seq) assay
Avian Myeloblastosis Virus (AMV), 221

B

BagFoot, 303–304
Barcode of Life Data System (BOLD), 144
Basic DNA structure lessons, 137
BGISEQ, 163
Biology (or science) education journals, 25
Biomedical and health sciences references disagreements, 6
Bio-Rad's Cloning, 133–134
Bisulfite (BS) conversion methods, 264–265, 266*f*
Bisulfite sequencing simulator/power analysis shiny app, 269
Bitter tasting ability (PTC) genotyping activity
    background, 127–129
    classroom management
        evolution of tasting ability, 131
        genotype-phenotype correlation, 130
        Hardy-Weinberg, 131
        hypothesis testing, 130–131
        online implementation, 131–132
        overview, 129
        quick guide, 129, 130*f*
        troubleshooting, 129
    learning objectives, 129
    TAS2R38 gene, 127–129, 131*f*
BlackList tool, 291

BOLD database system, 144
Breaking-Cas, 431–432, 433f
Broad Institute's Genome Aggregation Database (gnomAD), 52
Build-A-Genome (BAG), 133

C

Cap analysis of gene expression (CAGE) analysis, 164
Cell Press, 11
ChIPpeakAnno, 309, 329
ChIPseeker, 329
ChIP-seq. *See* Chromatin immunoprecipitation and sequencing (ChIP-seq)
Chi-square (χ^2) method, 54–56, 54–55t
Chromatin immunoprecipitation and sequencing (ChIP-seq), 293–294, 296–297, 301, 346–348, 352, 356
    antibody incubation, 323
    appropriate controls, 322–323
    ChIP-eluted chromatin
        broad-source factors, 325
        library preparation, 323–325, 324f
        peak calling, 326–327
        point-source factors, 325
        quality control (QC), 327–328
        quality of sequencing, 326
        raw sequencing, 326
        short-read alignment, 326
        single- or paired-end (PE) sequencing, 325
    chromatin washing, 323
    designing factors, 319–320
    differential binding analysis, 331–332
    DNA fragmentation, 321, 322f
    enzymatic digestion, 321
    formaldehyde fixation of tissue, 320
    functional enrichment analysis, 329–331
    immunoprecipitation, 320
    motif analysis, 332–333, 333f
    nuclear isolation, 321
    overview, 319
    peak annotation, 329–331
    single-cell methods, 322
    sonication, 321, 322f
    visualization of, 328–329, 330f
Chromosome conformation capture (3C) technology, 343. *See also* Hi-C analysis
Clustered Regularly Interspaced Short Palindromic Repeats (CRISPR) technologies, 9b, 12, 14, 401. *See also* CRISPR-Cas systems
ClusterProfiler, 329
Community-level disagreement, 6
Constitutive heterochromatin, 370
Contributing factors to reproducibility crisis, 9–12
Cookbook labs, 26, 32, 34b
Cooltools, 345, 349, 351–352, 354

BioConda packages, 344
    cooler file format, 343–344
Course-based undergraduate research experiences (CUREs), 32–34, 115, 142–143, 453–454
    Genomics Education Partnership (GEP)
        curriculum modules, 116
        Drosophila F element gene annotation projects, 116–117, 117–118f
        genomics education gap, 116
        insulin signaling pathway genes, 118
        members, 116
        parasitoid wasp venom genes, 119
        trained faculty, 116
    Science Education Alliance-Phage Hunters Advancing Genomics Evolutionary Science (SEA-PHAGES) program, 119
Covid-19 pandemic/SARS-CoV-2 pandemic, 8b, 14, 31, 124, 427
CRISPR-Cas9, genetic modification of mice
    base editing, 445
    Breaking-Cas, 431, 433f
    CRISPRa (transcriptional activation of genes), 426
    double-strand DNA break (DSB), 427–428
    electroporation (EP), 428–429, 431
    genetically engineered mouse models of human disease
        Alzheimer's disease (AD), 427
        cancer, 426
        coronavirus disease (COVID-19), 427
        International Knockout Mouse Consortium (IKMC), 426
    genetic diversity, 430–431, 430f
    gRNA, 427
    guide design, 431–434, 433f
    homology-directed repair (HDR) pathway, 428
    knock-in (KI), 428
    knockout (KO), 428
    KO alleles generation
        dropout alleles verification, 436–437, 439f
        protein detection, 435
        rapid amplification of cDNA ends (RACE), 434–435
        Sanger sequencing, 434–435
        transcript analysis, 434–435
        unintended on-target effects, 434, 435f
        verification of KO alleles, 436, 437–438f
    large knock-in alleles generation, 440–441
    lipid nanoparticle (LPN) delivery systems, 445
    microinjection (MIJ), 428–429, 431
    mosaicism, 442–443
    non-viral delivery vehicles, 445
    off-target effects, 443
    practical methods, 441–442
    prime editing, 445–446
    reagent delivery, 431
    schematic diagram, 428, 429f
    small knock-in alleles generation, 438–440
    SpCas9 guide, 431–432

Streptococcus pyogenes (SpCas9/SpyCas9) RNP complex, 427–428
   tissue-specific in vivo delivery, 444–445
   "tru-gRNA" strategy, 431–432
   unintended on-target effects, 443–444
   variants, 444
CRISPR/Cas9 genome editing
   background information, 454–455
   in the classroom, 456–457
   CRISPR-based manipulation of human DNA, 454
   genome editing, 455
   hands-on course, 454
   laboratory courses, 455
   rigor and reproducibility in, 462–464
   16 week teaching strategies, 456
   teaching approaches, 454
   in University of New Mexico
      Co-CRISPR strategy, 461–462
      Drosophila, designing and cloning, 457–458, 458–461*f*
      epitope tagging, 462
      near-peer mentoring, 462
      video tutorials, 460–461
CRISPR-Cas systems
   advantages, 406
   applications
      base editing, 407
      in the clinic, 406–407
      in the lab, 406
      prime editing, 407
   design, 405–406
   origin, 405
   See also CRISPR/Cas9 genome editing
Crowd-sourced undergraduate research. *See* Course-based undergraduate research experiences (CUREs)
CUREs. *See* Course-based undergraduate research experiences (CUREs)
Cybersecurity Law, 13

D

Data availability, 13
Data Security Law, 13
Demodex mite genetics, 145*b*
DESEQ2, 201–202
Developmental Origins of Health and Disease (DOHaD) hypothesis, 272
Diet-derived xenomiRs, 216
Differential expression (DE) analysis, 217–218, 225, 227, 229
Differential gene expression, 167, 194*f*, 201–202
DiffTF, 303–304
Direct-to-consumer (DTC) DNA testing, 14
DNA Barcoding, 139–140, 144
DNA Databank of Japan (DDBJ), 173
DNA hydroxymethylation, 262–263

DNA methylation, 262–264, 365–366, 370–372, 382–385, 387–389
DNA modifications
   biological features of, 262
   bisulfite (BS) conversion methods, 264–265, 266*f*
   definition, 262
   epigenome-wide association studies (EWAS) research
      bioinformatics tools, 267–268
      cell-type specificity, 265–267
      code sharing, 273
      data publication, 272
      findable, accessible, interoperable, and reusable (FAIR) data principles, 273–274
      human epidemiological studies, 269
      hypothesis testing in, 270–271
      Illumina DNA methylation arrays, 264, 268
      in vivo animal colony reporting, 272
      laser capture microdissection (LCM), 267
      mixed-effects (ME) modeling approach, 271
      power analysis, 268–269
      preregistration, 272
      reconciling paired 5-mC and 5-hmC data, 271
      reduced representation bisulfite sequencing (RRBS) data, 264, 268
      sample size planning, 268–269
      sex-related changes, 267
      temporal changes in, 267
      tissue heterogeneity, 265–267
   in health and disease, 262–263
   oxidative BS (oxBS) conversion, 265
   schematic diagram, 262, 263*f*
DNA packaging, 370, 371*f*
DNA sequencing, 416
   and analysis activities, 132–134
   at high school level
      automated sequencing machines, 138–139
      basic equipment and instrumentation, 138–139
      common features, 139–140
      content knowledge and familiarity, 138–139
      DNA Barcoding, 139–140
      essential concepts, 138–141
      field and lab work, 140
      Genetics Literacy Assessment Instrument (GLAI), 138
      laboratory exercises, 139, 140*f*
      learning goals, 138–141
      operational requirements, 140
      Teaching the Genome Generation (TtGG), 138–139
      technology considerations, 141–142
      visualizing DNA, 140–141, 141*f*
   learning opportunities, 137
   for middle school students, 138
   See also Undergraduate classroom
The *Double Helix*, 137
Drosophila F element gene annotation projects, 116–117, 117–118*f*

Drosophila melanogaster
  designing and cloning, 457–458, 458–461*f*
  guide RNA plasmids creation
    anneal oligos, 464–465
    clean-up sequencing, 467–468
    clones transformation, 468–470
    DNA sequencing, 467
    ligation, 465
    ligation products transformation into *E. coli*, 465–466
    minipreps, 467
    phosphorylation of dsDNA molecules, 465
    picking colonies, 466
    re-suspend oligonucleotides, 464
    sequences analysis, 468–470

E

Ecology and Molecular Biology classes, 144
EdgeR, 201–202
eLife, 12
Encyclopedia of DNA elements (ENCODE) consortium, 285, 290, 325, 327
Epigenetics, 272
  active learning, 366
  basic knowledge
    genomic imprinting, 373
    notes making, 375–377, 377*t*
    summarizing information, 375–377, 377*t*
    X-inactivation, 373–375, 374*b*, 374*f*, 376*f*
  curriculum design, 363
  definition, 261–262
  educational resources, 368, 368*t*
  in elementary school curricula, 365, 365*f*, 366*b*
  foundation knowledge
    chromatin conformation, 370
    epigenetic marks, 370–372, 372*f*
    epigenetic regulation of transcription, 370, 371*f*
    example assessment questions, 372
    gene expression regulation, 369, 369*f*
  in high school curricula, 365–366, 365*f*
  and human diseases, 364, 385
  interdisciplinary nature of, 366
  learning outcomes, 366, 367*f*, 367*t*
  nature *vs.* nurture
    classroom question, 385
    hunger winter case study, 385–386, 387–388*f*
    infant care/grooming/neglect behaviors, 389, 390–391*f*
  Next Generation Science Standards (NGSS), 364–365
  in school science curricula, 363
  team-problem-solving activity
    cancer epigenetics, 377, 378*b*, 382*b*
    monozygotic twins discordant for childhood cancer, 377–378, 380*b*, 384*b*
  in undergraduate curricula, 365–366
  X-inactivation concept map, 374*t*

Epigenome-wide association studies (EWAS) research
  bioinformatics tools, 267–268
  cell-type specificity, 265–267
  code sharing, 273
  data publication, 272
  findable, accessible, interoperable, and reusable (FAIR) data principles, 273–274
  human epidemiological studies, 269
  hypothesis testing in, 270–271
  Illumina DNA methylation arrays, 264, 268
  in vivo animal colony reporting, 272
  laser capture microdissection (LCM), 267
  mixed-effects (ME) modeling approach, 271
  power analysis, 268–269
  preregistration, 272
  reconciling paired 5-mC and 5-hmC data, 271
  reduced representation bisulfite sequencing (RRBS) data, 264, 268
  sample size planning, 268–269
  sex-related changes, 267
  temporal changes in, 267
  tissue heterogeneity, 265–267
European Molecular Biology Laboratory's European Bioinformatics Institute (EMBL-EBI), 175
European Nucleotide Archive (ENA) database, 175
EWAS research. *See* Epigenome-wide association studies (EWAS) research
Expression quantitative trait locus (eQTL) mapping, 53, 170–172, 202–204, 203*f*
External RNA Controls Consortium (ERCC), 200
Extracellular RNA Communication Consortium (ERCC), 212
Extracellular RNA (exRNA) sequencing, 211–213

F

Facultative heterochromatin, 370
FANTOM consortium, 164
FastQC software, 225–226, 309
Federal agencies and foundations, 142–143
Findable, accessible, interoperable, and reusable (FAIR) data principles, 273–274
Fluorescent activated nuclei sorting (FANS), 288
Frequentist *vs.* Bayesian modeling, 52–54

G

GCAT-SEEK, 133
GEDMatch, 14–15
Gender discrepancy, 60
Gene drive systems
  conservation (extinction prevention), 412
  discovery, 411
  DNA encoding, 411
  DNA sequence, 411
  limitations, 411–412
  public health, 412

safeguard strategies, 411
Gene expression
  applications, 166–173
  batch effect, 177
  expressed sequence tags (ESTs), 159
  expression quantitative trait loci assessment (eQTL), 170–172
  high-throughput transcriptomics
    microarray analysis, 159, 160f, 161–163
    sequencing-based methods, 159, 160f
  long-read sequencing, 165
  metaanalysis, 177–178
  next-generation sequencing (NGS), 163
  profiling
    definition, 166
    differential gene expression, 167
    gene and transcript quantification, 167
    lncRNA detection, 167
    PacBio, 167
    quality control (QC), 167
    reproducibility, 167–168
    splicing detection, 168
    visualization, 167
  public databases
    DNA Databank of Japan (DDBJ), 173
    European Molecular Biology Laboratory's European Bioinformatics Institute (EMBL-EBI), 175
    National Center for Biotechnology Information (NCBI), 173–175
  replicability, 176
  reproducibility, 176
  Sanger sequencing, 159
  serial analysis of gene expression (SAGE), 159
Gene Ontology (GO) enrichment analysis, 162, 230
General Data Protection Regulation (GDPR), 13
Gene set enrichment analysis (GSEA), 230
Genetic Information Nondiscrimination Act (GINA), 13
Genetic risk score (GRS). *See* Polygenic risk scores (PRS)
Genetics curriculum, 29
Genetics Literacy Assessment Instrument (GLAI), 138
Genetics Society of America (GSA), 132–133
Genetic variations, 14, 51–52, 58, 83–84, 89–90, 170, 425–426
Genome editing technologies
  categories of, 397, 398f
  CRISPR-Cas systems, 405–407
  delivery of
    electroporation, 408
    limitations, 408
    nonviral delivery, 410
    on-target and off-target editing, 410
    therapeutic gains, 408
    viral and plasmid delivery, 410
    viral vectors, 408–409
  description, 397

  gene drive systems
    conservation (extinction prevention), 412
    discovery, 411
    DNA encoding, 411
    DNA sequence, 411
    limitations, 411–412
    public health, 412
    safeguard strategies, 411
  homology-directed repair (HDR), 398
  nonhomologous end joining (NHEJ), 397, 413
  outcome of, 397
  somatic *vs.* heritable editing, 401b
  transcriptional activator-like effector nucleases (TALENs), 402–405, 403f
  unintended genomic alterations
    biochemical and cell-based screening, 415
    characteristics, 413–414, 414f
    complexity of, 415b
    delivery methods, 413
    DNA sequencing, 416
    genome-edited organisms/tissue screening, 415
    in silico prediction, 414
  zinc finger nucleases (ZFNs), 399–402
1000 Genomes Project Consortium, 7
1000 Genomes Web Browser, 84
GenomeWeb, 148–149
Genome-wide association studies (GWAS), 6, 32b, 171
  .assoc. linear file, 70
  in biomedical and social science research, 101
  for case/control association studies, 54–56, 54–55t
  Chi-square (χ^2) method, 54–56, 54–55t
  into classroom
    computational skills, 83–84
    correlation *vs.* causation, 85
    genetic contribution, 84–85
    for personality traits and preferences, 84
    phenotype, 89–90
    population diversity, 86–87
    population size, 87
    P-value, 87–88
    sample size, 88
    teaching of, 83, 84–85b, 88–89b
  DNAseI hypersensitive sites annotation, 53
  DrugY_phenotypes_for_plink.txt, 69
  expression quantitative trait loci (eQTL) mapping, 53
  fam and *bim* files, 69
  family ID (FID), 69
  frequentist *vs.* Bayesian modeling, 52–54
  genetic susceptibilities identification, 51
  genetic variants, 91–92
  genotype boxplot, 75–76, 76f
  individual ID (IID), 69
  linear regression models, 56–58, 57f
  LocusZoom plot, 73, 74–75f

Genome-wide association studies (GWAS) *(Continued)*
  logistic regression, 55–56
  Manhattan plots, 71–72, 71*f*
  METAL program, 96
  multivariate regression model, 57, 58*f*
  phenotype data, 67–69, 68*f*
  polygenic risk scores (PRS) (*see* Polygenic risk scores (PRS))
  posterior probability, 53
  prior probability, 53
  Q-Q plot, 71–73, 72*f*
  quality control (QC) procedures
    genetic QC, 59–60
    Plink primer (*see* Plink primer)
    technical QC, 59
  quantile normalization (QN), 58
  quantitative traits, 56–58, 57–58*f*
  R primer, 61, 62*f*
Genome-wide chromatin accessibility profiling, 285. *See also* Assay of transposase accessible chromatin sequencing (ATAC-seq) assay
Genome-wide polygenic score (GPS). *See* Polygenic risk scores (PRS)
Genomic Regions Enrichment of Annotations Tool (GREAT), 331
Genomics Education Partnership (GEP), 148
  curriculum modules, 116
  Drosophila F element gene annotation projects, 116–117, 117–118*f*
  genomics education gap, 116
  insulin signaling pathway genes, 118
  members, 116
  parasitoid wasp venom genes, 119
  trained faculty, 116
Genotype-phenotype associations, 56
Genotype-Tissue Expression (GTEx) Consortium, 171, 215
Genrich, 294, 296, 309
GEO-National Center for Biotechnology Information (GEO-NCBI), 173–175
GEP. *See* Genomics Education Partnership (GEP)
Glucocorticoid receptor (GR) promoter, 389
GreyListChIP package, 291
GWAS. *See* Genome-wide association studies (GWAS)

H

H3Africa Initiative, 6–7
HapMap resource, 51–52
Hardy-Weinberg equilibrium (HWE), 59
Heteroscedasticity, 271*b*
Hi-C 2.0, 346
Hi-C 3.0, 346
Hi-C analysis
  chromosome folding, 348–349, 348*f*
  cohesion-driven loop extrusion, 356
  compartment analysis, 349–351, 350*f*
  contact pattern changes, 356–357
  Cooltools, 354
    BioConda packages, 344
    cooler file format, 343–344
  CTCF-CTCF looping interactions, 352, 355*f*, 356
  data processing, 343, 345–346
  De novo dot calling, 354
  distiller, 345
  genome-wide IS changes, 356–357
  Hi-C 2.0, 346
  Hi-C 3.0, 346
  HiCCUPs, 352
  HiGlass, 346
  insulation profiles, 351–352, 353–354*f*
  integration with ChIP-seq data, 356
  integration with RNA-seq data, 356
  loop anchors, 354
  quality control (QC), 345–346
  scaling plots, 348–349, 348*f*
  schematic diagram, 343, 344*f*
  TAD boundaries, 351–352, 353–354*f*, 356
  visualization of, 346–348, 347*f*
HiCCUPs, 352
Hierarchical clustering analysis (HCA), 292
High-throughput sequencing (HTS) of nucleic acids. *See* RNA sequencing (RNA-seq)
High-throughput transcriptomics
  microarray analysis, 159, 160*f*, 161–163
  sequencing-based methods, 159, 160*f*
HiGlass, 346, 347*f*
HINT-ATAC, 301–302, 304
Histone acetylation, 371–372
Histone methylation, 371–372
Histone phosphorylation, 371–372
HMMRATAC, 294
Homology-directed repair (HDR), 398
Howard Hughes Medical Institute (HHMI) Biointeractive resources, 133
Human Genome Project (HGP), 51, 163

I

Illumina DNA methylation arrays, 264, 268
Image forensics, 4–5
Integrative Genomics Viewer (IGV), 228, 329
International Human Genome Sequencing Consortium, 51
International Knockout Mouse Consortium (IKMC), 426
International Society for Extracellular Vesicles (ISEV), 212
Ioannides' study, 23
Isolation of nuclei tagged in specific cell types (INTACT) method, 288

K

Kyoto Encyclopedia of Genes and Genomes (KEGG), 230

L

Laser capture microdissection (LCM), 267
LIMMA, 162, 201–202, 204
Lipid nanoparticle (LPN) delivery systems, 445
LocusZoom plot, 73, 74–75f
Long-read sequencing, 165

M

"MACS2.SITE" method, 293–294
Manhattan plots, 71–72, 71f
MAnorm, 332
Meaney's cross-fostering experiment, 389, 390f
Mendel's Law of Segregation, 59
Microfluidics-based electrophoresis system, 289
Minimum information about a high-throughput sequencing experiment (MINSEQE), 175
Minimum Information About a Microarray Experiment (MIAME), 175
miRBase, 227
MirGeneDB, 227
MIT DNA kits, 138
Mixed-effects (ME) modeling approach, 271
Molecular phylogenies
   classroom management
      analysis, 126–127
      basic steps of phylogeny construction, 126
      in-class activity, 125
      online implementation, 126
      preclass preparation, 125
      prelab, 126
      troubleshooting, 125–126
   learning objectives, 125
   mitochondrial gene cytochrome c oxidase subunit I (COX1), 124
   quick guide, 127, 128f
   SARS-CoV-2, origin and spread of, 124
Molecular Signatures Database (MSigDB), 230
MOODS, 303
msCentipede, 299–301
Multidimensional scaling (MDS) approach, 60
MultiQC, 309
Multivariate regression model, 57, 58f
Murine Leukemia Virus (MMLV), 221

N

National Cancer Institute's Genomic Data Commons, 13
National Center for Biotechnology Information (NCBI), 173–175
National Center for Case Study Teaching in Science (NCCSTS), 33, 132
National Institutes of Health's (NIH) *All of Us* research program, 6–7
NCBI Sequence Read Archive (SRA), 147–148
Network for Integrating Bioinformatics into Life Sciences Education (NIBLSE), 133
Next Generation Science Standards (NGSS), 364–365
Next generation sequencing (NGS) approaches, 137, 147, 163, 264, 285, 398, 416
NIEHS TaRGET II Consortium, 270
NIH Roadmap Epigenomics Program, 270
Nonhomologous end joining (NHEJ), 397, 413

O

Omni-ATAC, 286, 288
Open science, 12–15
Open Science Framework (OSF), 272

P

PacBio, 166–167
Paper-level disagreement, 6
Peer review processes, 4, 9
Personal Information Protection Law (PIPL), 13
Plink primer, 58–60, 95–96
   BED (binary PED) files, 65
   BIM files, 65
   download page, 61, 63f
   FAM files, 65
   for MacOS X, 63, 64f
   MAP files, 65
   PED files, 64–66
   test run, 66, 66–67f
   for windows, 63–64
Polygenic risk scores (PRS), 10
   in diverse ancestral populations, 100
   example script, 101–104
   missing heritability, 100
   overview of, 92–93, 92f
   predictive power, 99–100
   quality and characteristics, 100
   quality control (QC) procedures, 93
   single nucleotide polymorphisms (SNPs)
      with alleles, 92–93, 92f
      GRCh38/hg38, 96
      matching discovery, 95–96
      multiallelic single nucleotide polymorphisms (SNPs), 94–95
      palindromic *vs.* nonpalindromic single nucleotide polymorphisms (SNPs), 93–94, 94f
      pruning, 95
      sample R script, 97b, 99, 104–107
      strand flipping, 94
      target genotype files, 95–96
   software programs, 97–99, 98t
Population structure (or stratification) check, 60
Practical Analysis of Your Personal Genome, 148
"Predator Free 2050" initiative, 413

Preprint servers, 31*b*
Professional development resources, 25*b*
PRS. *See* Polygenic risk scores (PRS)
PTC genotyping activity. *See* Bitter tasting ability (PTC) genotyping activity

Q

Q-Q plot, 71–73, 72*f*
Quantification of gene expression. *See* Reverse transcription by quantitative PCR (RT-qPCR)
Quantile normalization (QN), 58
Quantitative reasoning (QR), 31
QUBES, 133

R

Reads per kilobase per million (RPKM), 199
Reduced representation bisulfite sequencing (RRBS) data, 264, 268
Registered Report, 11
Reproducibility *vs.* replicability, 5*b*
Responsible conduct of research (RCR) training, 15
  in advising, 34–35, 35*f*
  in the classroom
    academic misconduct, 31–32, 32*t*
    advanced level, 27
    authentic research experiences (AREs), 26
    case-based learning, 33*b*
    cookbook labs, 32, 34*b*
    critical reading of literature, 30*b*
    CUREs, 32
    draft revision cycles, 30
    duty to report, 28*b*
    evidence-based pedagogies, 26, 26*f*
    experiential learning (EL), 32
    explicit instruction, 29–30
    GWAS studies, 32*b*
    hypothesis testing, 29*b*
    incidental findings, 28*b*
    introductory level, 27
    nature of science, 29, 29*b*
    preprint servers, 31*b*
    professional conduct in science, 26–27
    professional development resources, 25*b*
    quantitative analyses, 31
    quantitative reasoning (QR), 31
    research integrity, 26–27
    rigorous reporting, 32, 33*b*
    science communication, 30
    scientific teaching approach, 27*b*
    sloppy science, 31–32
    stakes levels, 29
    undergraduate biology classrooms, 26
  in research lab
    accountability structures, 39–40
    communication of expectations, 37
    conflicts of interest/commitment, 40
    explicit training, 37, 38*b*
    feedback and reflection, 40–41
    habits of mind, 37
    integrity practices, 41
    interview process, 38*b*
    journal clubs, 39
    lab meetings, 39
    mentoring for equity, diversity and inclusion (EDI), 41
    mentoring networks, 35–36, 36*b*
    peer mentoring, 36
    protocols, 37
    support for mentorship, 41, 42*b*
    trainee engagement, 39, 39*b*
  responsibility for, 23, 24*f*
Retraction Watch, 6
Reverse transcription by quantitative PCR (RT-qPCR), 218
  advantages, 247
  cDNA synthesis
    of circRNA, 251–252
    of lncRNAs, 251–252
    materials, 248
    miRNA first-strand cDNA synthesis, 252
    of mRNAs, 251–252
    schematic illustration, 250, 251*f*
  flowchart, 249, 249*f*
  qPCR assay
    for circRNA, 254–255
    for lncRNA, 254–255
    materials, 248
    for microRNAs, 255
    for mRNA, 254–255
  quality, purity, and yield of RNA analysis, 250
  RNA assessment, 248
  RNA extraction, 248
  RNA-specific primers
    for circRNAs, 253–254, 253*f*
    for lncRNAs, 252–253, 253*f*
    for long RNAs, 252–253, 253*f*
    for microRNA, 254
    for mRNA, 252–253, 253*f*
  total RNA isolation procedure, 249–250
Review, promotion, and tenure (RPT) committees, 11
Rice-derived miR-168a, 216
RNA biogenesis mechanisms, 227
RNA Integrity Number (RIN), 192, 216–217
RNA sequencing (RNA-seq)
  absolute mRNA abundance quantification, 200–201
  alternative splicing events, 205–206, 205*f*
  applications, 190
  contamination, 215–216

data analysis, 164
DESEQ2, 201–202
EdgeR, 201–202
experimental confounders, 193, 194f
experimental design, 217
experimental protocol, 190–192, 191f
extracellular RNA (exRNA) sequencing, 211–213
gene and isoform expression quantification, 199–200
gene expression
    genetic determinants, 202–204, 203f
    between groups, 201–202
genomic coverage, 196
library complexity, 195
library preparation methods
    adapter attachment, 222–223
    cDNA fragmentation, 218–220
    core steps of, 218, 219f
    data analysis, 225, 225f
    database quality, 227
    data quality control, 228–229
    differential expression (DE) analysis, 217–218, 225, 227, 229
    functional analysis of, 230
    library amplification, 223
    ligation-based methods, 222
    ligation-free methods, 223
    overlapping annotations, 227
    preprocessing, 225–226
    read mapping, 226–227
    read quantification, 228
    reverse transcription, 220–222, 221f
    RNA fragmentation, 218–220
    RNA fragment distribution, 223–224
    RNA termini modification, 222–223
    unique molecular identifiers (UMIs), 223
LIMMA, 201–202
mapping and strandedness, 197–199, 198f
measurement process, 213–214, 214f
messenger RNA (mRNA)
    degradation, 189
    profiling, 189
    splicing and processing, 189
multiplexing, 195
nonsplicing-aware mapper, 197–198
principal component analysis, 201
quality control, 197, 216–217
read length and type, 196–197
replicability, 193–195
RNA quality, 192–193
sample preservation, 214–215
    cell harvest procedures, 214
    effect of extracellular ribonucleases, 214–215
    microbiological decomposition, 215

sequencing depth, 195–196, 217–218
splicing-aware mapper, 197–198
study design, 193, 194f
transcriptome analysis, 213
R package DiffBind, 331
R primer, 61, 62f
RT-qPCR. *See* Reverse transcription by quantitative PCR (RT-qPCR)

S

Sage BluePippin system, 289
Sanger sequencing, 144, 159, 398, 416, 434–436
SARS-CoV-2 pandemic/Covid-19 pandemic, 8b, 14, 31, 124, 427
School Malaise Trap Program (SMTP), 139
Science Education Alliance-Phage Hunters Advancing Genomics and Evolutionary Science (SEA-PHAGES) program, 119, 145
Scripps Coastal Reserve, 144
SEA-PHAGES program. *See* Science Education Alliance-Phage Hunters Advancing Genomics and Evolutionary Science (SEA-PHAGES) program
Sequence read archive (SRA), 175
Sequencing by synthesis, 164–165
Sequencing Explorer Series, 133–134
Serial analysis of gene expression (SAGE), 159
Single-cell ATAC-seq (scATAC-seq), 286–287
Single-cell RNA sequencing (scRNA-seq) technology, 172–173
Sloppy science, 31–32
Social media platforms, 14
SOLiD, 163
SpliceSeq, 168
Splicing-aware mapper, 197–198
16S rRNA bacterial genotyping/identification activity
    Benchling, 120–122
    BLAST search, 123
    chromatogram, 123
    classroom management, 121–122
    general biology class, 120
    introductory slides, 122
    learning outcomes, 121
    microbiology class, 120
    molecular biology class, 120
    next-generation sequencing techniques (NGS), 120
    overview demonstration, 122–123
    prelab (20 min outside of class), 122
    prelab questions, 122
    quick guide, 123, 124f
    sequence analysis activity, 120
    walkthrough sequence protocol, 122–123
Statistical significance, 72
Structured Transparent Accessible Reproducible (STAR), 11

T

Tagmentation, 286, 289
TALENs. *See* Transcriptional activator-like effector nucleases (TALENs)
Teacher training, 25
Teaching the Genome Generation (TtGG), 138–139, 141*f*
3C-based method Micro-C, 356
3-D molecular designs, 138
3D printing technologies, 12–13
TOBIAS, 299–304, 302*f*
Transcriptional activator-like effector nucleases (TALENs), 402–405, 403*f*
 advantages, 404
 applications, 404–405
 design, 402–403, 403*f*
 limitations, 404
 origin, 402
Transcriptome assembly, 165
Transcripts per million (TPM), 199
TRIzol, 215–216
"Tru-gRNA" strategy, 431–432

U

UCSC Genome Browser, 148, 329
UK Biobank, 7, 12–13, 52
Undergraduate classroom
 Association for Biology Laboratory Education, 132
 Bio-Rad's Cloning, 133–134
 bitter tasting ability (PTC) genotyping activity
  background, 127–129
  classroom management, 129–132, 131*f*
  learning objectives, 129
  TAS2R38 gene, 127–129
 course-based undergraduate research experiences (CUREs) (*see* Course-based undergraduate research experiences (CUREs))
 DNA sequencing
  and analysis activities, 132–134
  demodex mite genetics, 145*b*
  DNA sequence-based activities, 142–143, 143*f*
  engaging and exciting students, 144
  Federal agencies and foundations, 142–143
  general goals, 143
  immersive genomic sequencing, 146*b*
  promoting research careers, 145–147, 147*t*
  at Sacred Heart Academy, 142*b*
  skills and career training, 147–149
 GCAT-SEEK, 133
 Genetics Society of America (GSA), 132–133
 HHMI Biointeractive resources, 133
 molecular phylogenies
  classroom management, 125–127
  learning objectives, 125
  mitochondrial gene cytochrome c oxidase subunit I (COX1), 124
  quick guide, 127, 128*f*
  SARS-CoV-2, origin and spread of, 124
 National Center for Case Study Teaching in Science (NCCSTS), 132
 QUBES, 133
 Sequencing Explorer Series, 133–134
 16S rRNA bacterial genotyping/identification activity
  Benchling, 120–122
  BLAST search, 123
  chromatogram, 123
  classroom management, 121–122
  introductory slides, 122
  learning outcomes, 121
  next-generation sequencing techniques (NGS), 120
  overview demonstration, 122–123
  prelab (20 min outside of class), 122
  prelab questions, 122
  quick guide, 123, 124*f*
  sequence analysis activity, 120
  walkthrough sequence protocol, 122–123
Unintended genomic alterations
 biochemical and cell-based screening, 415
 characteristics, 413–414, 414*f*
 complexity of, 415*b*
 delivery methods, 413
 DNA sequencing, 416
 genome-edited organisms/tissue screening, 415
 in silico prediction, 414
US National Institutes of Health, 212

W

WashU Epigenome Browser, 329
Whole-genome bisulfite sequencing (WGBS), 264

Z

Zinc finger nucleases (ZFNs)
 advantages, 399
 applications
  in clinic (human disease), 400
  on the farm (agricultural animals), 401
  in the field (plants), 402
  in the lab (basic and translational research), 400
 context dependence, 399
 design, 399
 limitations, 399
 origin, 399
 proprietary nature of, 400*b*
 triplet composition, 399